Biotic Indicators for Biodiversity and Sustainable Agriculture

Biotic Indicators for Biodiversity and Sustainable Agriculture

Edited by

Wolfgang Büchs

Federal Biological Research Centre for Agriculture and Forestry
Institute for Plant Protection in Field Crops and Grassland
Messeweg 11/12
38104 Braunschweig, Germany

Society for Ecology
Working Group "Agroecology" at Braunschweig

2003

ELSEVIER

Amsterdam – Boston – Heidelberg – London – New York – Oxford
Paris – San Diego– San Francisco – Singapore – Sydney – Tokyo

ELSEVIER B.V.
Sara Burgerhartstraat 25
P.O. Box 211, 1000 AE Amsterdam, The Netherlands

First edition 2003

Library of Congress Cataloging in Publication Data
A catalog record from the Library of Congress has been applied for.

British Library Cataloguing in Publication Data
A catalogue record from the British Library has been applied for.

Reprinted from the Elsevier journal
***AGRICULTURE, ECOSYSTEMS AND ENVIRONMENT*, Vol. 98, Nos. 1-3 (September 2003)**

ISBN: 0-444-51551-8

∞ The paper used in this publication meets the requirements of ANSI/NISO Z39.48-1992 (Permanence of Paper).
Printed in The Netherlands.

ELSEVIER

Available online at www.sciencedirect.com

Agriculture, Ecosystems and Environment 98 (2003) v–viii

Agriculture Ecosystems & Environment

www.elsevier.com/locate/agee

Contents

Special Issue: Biotic Indicators for Biodiversity and Sustainable Agriculture

PART I
INTRODUCTION

Chapter 1 Introduction
W. Büchs

ELSEVIER

Available online at www.sciencedirect.com

Agriculture, Ecosystems and Environment 98 (2003) 1–16

Agriculture Ecosystems & Environment

www.elsevier.com/locate/agee

Biotic indicators for biodiversity and sustainable agriculture—introduction and background

Wolfgang Büchs*

Federal Biological Research Centre for Agriculture and Forestry, Institute for Plant Protection in Field Crops and Grassland, Messeweg 11/12, DE-38104 Braunschweig, Germany

Abstract

The paper is an introduction to a volume containing 45 contributions on the development of biotic indicators for sustainable land use and biodiversity. It shows the background and origins of this particular issue as well as the motivation behind it. While a detailed subject-oriented synoptic introduction is provided at the beginning of each section [Agric. Ecosyst. Environ. 98 (2003) 35; Agric. Ecosyst. Environ. 98 (2003) 79; Agric. Ecosyst. Environ. 98 (2003) 255; Agric. Ecosyst. Environ. 98 (2003) 305; Agric. Ecosyst. Environ. 98 (2003) 407; Agric. Ecosyst. Environ. 98 (2003) 371], the intention of this contribution is rather to explain the general concept, the philosophy of the structure chosen, the understanding of important terms and the usefulness of the sections with regard to their content. It therefore serves the reader with a tool to understand this special issue as a whole. The volume is divided into sections "Requirements", "Biodiversity and Habitat" (including "Soil" as a subsection), "Biodiversity and Landscape", "Experiences and Application" and "Economy". From a generic point of view a brief survey is provided on the aspects and their relative backgrounds in each section, and completed by critical comments and conclusions, particularly regarding putting the theory into practice.

Keywords: Biotic indicator; Agri-environment; Assessment; Landscape; Habitat; Application

1. Background and starting point

At international level, a current need for criteria to assess the importance of land use for the preservation of the cultural landscape exists. Regardless of the fact that the increasing land use 1000 years ago was fundamental for the development of a high level of species biodiversity in Central Europe (Piorr, 2003), current land use techniques are said to be responsible for a loss of biodiversity. During the meetings in Cardiff (June 1998) and Vienna (December 1998) the European Council underlined the importance of the development of agri-environmental indicators. Such catalogues dealing with "Criteria for Sustainability" (CSD-indicators according to the Agenda 21) or "agri-environmental indicators" are currently being discussed or have recently been developed (OECD, 2000) respectively. In spite of the great number of indicators proposed (CSD: 134; OECD: ≈250), extraordinary deficiencies exist in particular for finding indicators for biotic aspects (Mannis, 1996; Commission of the European Communities, 1998, 2000).

Starting from this recognition, an initiative for a symposium (finding of indicators for a sustainable land use in the areas landscape and biodiversity) was taken by the working group "Agroecology" of the international "Society for Ecology" in Germany, organised and co-ordinated by a group whose members mainly consisted of the collators of the sections of this special issue (Büchs, 2003a). The majority of publications within this special issue originate from contri-

* Tel.: +49-531-299-4506; fax: +49-531-299-3008.
E-mail address: w.buechs@bba.de (W. Büchs).

0167-8809/$ – see front matter © 2003 Elsevier B.V. All rights reserved.
doi:10.1016/S0167-8809(03)00068-9

butions presented at a 4-day meeting in a monastery at Freising, but were continuously adapted to the current status of discussions and completed by additional invited papers on aspects not covered deeply enough at the meeting.

It became obvious at how many Central European institutions the development of biotic indicators and questions on biodiversity in agricultural landscapes were focused on, and the partly completely different approaches made. In this regard, the need for discussions, and particularly the compilation of discussions, also became clear. However, nearly all working groups were moved by the question: How can very complex and widely differentiated scientific findings be converted into parameters that are easy to assess and which are suitable to be used as indicators for sustainable land use or biodiversity aspects? What are the options for assessing the "success" of measures that are implemented into agricultural practice in order to achieve a land use featuring maximum care of natural resources with regards to the desirable composition of species by easy, but reliable criteria?

2. Goals and open questions

The key issue of this volume is not "test organisms" as indicators for toxicological assessments mostly conducted under laboratory conditions, as they are commonly used within the registration procedure of pesticides or chemicals in general (e.g. Hassan et al., 1985, 1994; Commission of the European Communities, 1991). The goal is to find biotic indicators which enable the assessment of complex events occurring in agro-ecosystems under field conditions induced by husbandry practices and land use in general. One goal of this special issue is to contribute to the development of a strategy on how the assessment of biotic aspects in agro-ecosystems with the help of indicators can be handled and how these findings can be transferred to national as well as international panels and finally into practice. Within this scenario the role of science was defined to underlay the breakdown of goals that have been discussed and established by international committees (e.g. OECD, EU-Commission; World Trade Organisation) on a global level, to the landscape level, to individual farm level and last but not least to individual field level with scientific reliable facts. With reference to these goals the following headings were identified and forwarded to the authors to be considered within their contributions as far as possible:

- General/political requirements with respect to indicators.
- A survey of existing approaches.
- Transferability of biotic indicators (up- and down-scaling).
- Segregative or integrative approaches with respect to the application of indicators.
- Definition of hierarchic levels (priorities, ranking or "value" of indicators).
- Competing goals, risks of redundant evaluation using indicators.
- Requirements for validation, universal validity, potential of converting indicators into practice.
- Approaches of reimbursement and integration into farm economics.

With regard to the development of agri-environmental indicators, statements made by the EU-Commission to the Council of the European Parliament, by the Committee for Economics, the Committee of Social Affairs and the Committee for the Regions on 19 June 1999 pointed out that approaches which focus on single aspects will not consider the complex interrelations between agriculture and the environment. Beyond that, it was claimed for ecosystem-related approaches in order to consider positive or negative environmental effects in their entirety. In this regard the spatial indications of agri-environmental indicators should be developed from a concept of "landscape" as a partly seminatural cultural area, in which land use is practised and which is determined by the entirety of its biophysical and cultural features (Commission of the European Communities, 2000).

On the other hand the EU-Commission complains deficiencies of a spatial and thematic differentiation of agri-environmental indicators. The EU-Commission states that general and integrative indicators are available much more readily, but they provide only little evidence on the success of certain measures applied in agricultural practice. The EU-Commission judges indicators only as appropriate and effective if the conditions at a defined location can be considered, by using data which provide a clear specific and spatial differentiation (Commission of the European Communities, 2000).

Derived from the background described this volume compiles papers which cover the following topics:

- Political requirements and statements on biotic agri-environmental indicators.
- Requirements regarding agri-environmental indicators from a scientific as well as an applied point of view.
- Practical experience with the application and suitability of agri-environmental indicators (as far as they are already established).
- Scientific reviews and critical discussions of "state of the art" knowledge regarding several kinds of agri-environmental indicators for biodiversity and/or sustainable agriculture.
- Conceptions and models to show the economic effects and possibilities of practical application of such indicators.
- Original experiments on certain aspects regarding the indication of biodiversity in agro-ecosystems.

The following aspects were identified in detail for a discussion.

2.1. General

- A compilation of existing indicators which are practicable and based on the established knowledge of functions and power of agro-ecosystems.
- Identification of potential indicators free from existing demands on validation and practicability.
- Evaluation of the potential of applications, the practicability and transferability of existing indicators and new approaches of indication.
- Presentation and discussion of threshold values and reference values with regards to established indicators which are estimated as appropriate for applying in practice.
- Evaluation of the problems regarding the practical use of biotic indicators with regard to the power of scientific significance, efforts of surveying and of putting them into practice.
- Identification of the need for research, joint experiments, data survey, etc.

2.2. Requirements

- Requirements regarding indicators according to the OECD, or within the framework of CBD—"Convention of Biodiversity Development" (Agenda 21, United Nations, 1992), definition of agri-environmental indicators according to the international standard (International Standardisation Organisation, 1996) ISO 14001.
- Functional requirements and basic conditions, for instance according to the OECD, according to different professional authorities belonging to different administrative levels (e.g. federal republic, state, county or district local authority) and according to farmers. Which indicator is suitable for which purpose and for which user?
- Discussion of the problem concerning concrete evidence of a direct correlation between abiotic indicators as steering factors for determining biodiversity and the quality of the species set concerned.
- Suitability of indirect indicators as for example landscape parameters (e.g. percentage of hedgerows, field margins, etc.) as surrogates for complex biotic indicators.

2.3. Assignment of biotic indicators to different spatial units

- For which spatial level or administrative unit (e.g. EU, federal state, county, region, farm, field, natural unit, management form) can suitable approaches for a practicable (biotic) indication be defined?
- Discrepancy between the demands for the exactness of indicators at different spatial levels (global level versus ecosystem level).
- Is an upscaling, i.e. an accumulation of results derived from surveys by indicators at lower spatial levels possible?

2.4. Survey, validation and power of statements

- Possibilities and modalities of a standardisation of the survey methods.
- Validation and scientific evidence of indicators (gaps, problems, possible solutions).
- Capability of representation (e.g. transferability from one to another biological taxon).
- Problems and possibilities of simplifications to biotic indicators in respect to their power of scientific evidence.
- Problems of data aggregation: How can data losses be avoided or handled?

- The amount of time and money required for surveys in order to achieve a statistically sufficient sample size.
- Frequency in time, extent and intensity according to which a survey of indicators should be repeated.
- Capability for modelling with respect to the adaptation to field conditions and practical demands.
- Designation of competence with respect to the qualification level: Who should be allowed to survey indicators and to convert the conclusions drawn from assessment results?

2.5. *Assessment procedures*

- How do indicators become assessment tools?
- What is the basic level for assessment (e.g. baseline reference values)?
- Who defines baseline reference values, thresholds, limits? To which criteria will they be linked?
- Are these values fixed as absolute or as relative values? How are they related to regional conditions?
- What are the goals? The achievement of which ecosystem features can be checked by biotic indicators?
- Scale levels: Should only indicators of the same spatial scale level be considered in an assessment procedure?
- Overlapping and competition of different biotic indicators in their areas of competence: are redundancies and/or double assessments (carried out by different indicator approaches) senseless or do they enhance scientific evidence?
- Weighting of indicators within complex assessment systems: Establishment of priorities only within comprehensible sections and preconditioning of goals?

2.6. *Communication*

- Problem of communication between different levels dealing with indicators and assessment procedures (pure research ⇒ applied research ⇒ institutions managing conversion into practice ⇒ political decision makers ⇒ users).

3. Glossary definitions

The title and sections chosen for this volume contain terms such as "biotic indicator" or "bio-indicator", "sustainable agriculture/land use", "biodiversity", "habitat" and "landscape" which raise expectations for definitions in order to provide a clear understanding. However, the intention of this volume is not to go into detail about discussing terms; thus, only some basic and more or less well established definitions are provided, which should fulfil the intention of this volume as far as possible. A more extensive discussion of the key terms "indicator", "biodiversity" and "landscape" can be found in the synoptic introductions to the individual sections (Büchs, 2003b; Waldhardt, 2003; Osinski et al., 2003).

3.1. *Indicator*

With regard to the term biotic indicator or bio-indicator and its potential and limits, an intensive and fundamental discussion took place at the beginning of the 1980s (e.g. Arndt et al., 1987; Bick, 1982; Phillipson, 1983). However, at that time the discussion was not related to agro-ecosystems, but had its origins in the discussions on forest damage due to air pollution and the environmental impact of nuclear plants. Nevertheless, the fundamental principles can be completely transferred to biotic indicators applied in agro-ecosystems.

Similar to chemical reactions, which result in clear and unambiguous effects indicating the current status of a fluid solution or the end of a (chemical) reaction if a very small amount of an indicator substance is added, are also the requirements for a biotic indicator with respect to the characterisation of the current status (status indicators) or ongoing processes (process indicators) within ecosystems: the indicator substance is replaced by a biotic indicator (e.g. a species [community]/population/organism) that in ideal cases represents the biocoenosis to a certain extent and provides information of status and/or on changes of the ecosystem compartment considered (Büchs, 1988). However, at the time in question, a biotic indicator was mainly interpreted as an organism that reacts to harmful substances (e.g. heavy metals, air pollutants, pesticides) by changing its life functions, e.g. metabolism (reaction indicator) or by accumulating substances (accumulation indicator). The understanding of biotic indicators was focused more in the sense of test organisms (for testing toxicological effects under standardised conditions in laboratory or semi-field experiments) or as "monitor species" (e.g. lichens) which are placed outside in the environment to detect factors such as air pollution (Arndt et al., 1987).

Biotic indicators or indication systems which provide information on the current status of, or processes going on in ecosystems were introduced at species level by Hesse (1924), and Kühnelt (1943) and picked up again by Bick (1982), Dunger (1982), Kneitz (1983) and Phillipson (1983). According to Dunger (1982) and Phillipson (1983) the enlisting of single species/populations is only partially appropriate for a biotic indication because of a very limited representative value. Only if complete species assemblages or ecosystem-related taxocoenosis are included in indicator systems is the power of statements considerably increased compared to single species. Perner (2003) showed that the relation of the level of precision to required sample size (number of samples required to achieve a tolerable variance regarding data) is less favourable for a selected indicator species but much better for community parameters such as species richness, and particularly evenness, so that—within the assessment of biotic indicators—community-related indicators should be preferred to species-related parameters.

Often, species communities are very strictly correlated to certain environmental conditions so that they can be used for indication (e.g. Ellenberg et al., 1992; Büchs, 1995). "Indicator communities" can be identified (Bick, 1982), which give an indication of the intensity of certain impact factors and their combinations through presence or absence of species or the changing of dominance positions. In principal, within this context each species can be used as an indicator, but the importance of stenoecious species (species closely adapted to certain biotic/abiotic conditions) has to be particularly emphasised, since they are usually more effective indicators than euryoecious species (see Döring and Kromp, 2003; Döring et al., 2003; Nickel and Hildebrandt, 2003; Perner and Malt, 2003).

The Organisation for Economic Co-operation and Development (OECD), which has developed agri-environmental indicators in 13 different areas (e.g. biodiversity, wildlife habitats, landscape, farm management, pesticide use, nutrient use, water use, soil quality, greenhouse gases, socio-cultural issues, farm financial resources; OECD, 1999, 2000) favours the Driving Force-State-Response framework as a basic principal:

- *Driving force*: What is causing the environmental conditions to change?
- *State*: What are the effects on the environment?
- *Response*: What actions are being taken in public and private sectors to respond to changes in the state of the environment?

This three-step concept which combines the analysis of causes, effects and countermeasures to "repair" negative effects is to a certain extent reflected in the structure of this volume: particularly in section "Habitat", the development of indicators and their validation is usually based on the effects (*state*) of different husbandry practices which are the *driving forces*. Sections "Experience and Applications" and in particular "Economy" include approaches to manage changes in the environment (*response*). However, with regard to finding and developing indicators, the main focus of this volume will be the measurement and assessment of the current *state* and of developmental trends regarding environmental and natural resources in agricultural systems.

The Commission of the European Communities (2000) defines 'agri-environmental indicators' as a "generic term designating a range of indicators aiming at giving synthesised information on complex interactions between agriculture and environment. Common agri-environmental indicators are those that provide an assessment of impacts of agriculture on water quality, climate change, soil, or landscape structures". The fact that the term biodiversity is not mentioned in this definition shows how difficult this term is to tackle (see Büchs, 2003b; Büchs et al., 2003).

3.2. Sustainable agriculture

"Sustainability" (European Commission, 1998), "sustainable development" and so "sustainable agriculture" are terms that tend to be diluted and in consequence easily "abused" as it is also the case of the term "integrated" (integrated crop production, integrated pest management, etc.), due to the lack of clearly defined criteria and limitations. The World

Commission on Environment and Development defined "sustainable development" as "development which meets the needs of the present without compromising the ability of future generations to meet their own needs" (Legg, 1999). However, by the mid-1990s, there were well over 100 definitions. To avoid adding to this confusion, here a very basic definition by Stinner and House (1990) is used: sustainable agriculture (a) conserves the resources on which it depends; (b) restricts itself to a minimum input of production means, which do not have their origin in the same farming system; (c) controls pests and diseases by internal regulation processes as far as possible; (d) provides natural resources with the ability to recover from disturbances through cultivation means and harvesting by processes of natural succession. Sustainable farming requires intensive management and substantial knowledge of ecological processes.

3.3. Habitat

This is simply known as the place or type of site where an organism or population occurs naturally. The habitat of a (plant or animal) species is its "place of residence", that means the region to which it is adapted and which it is able to occupy. This does not mean so much a concrete habitat, as a habitat type including specific factors (ecological conditions) which allow the species to survive and to reproduce successfully. If the habitat quality changes (e.g. due to anthropogenic impact) or the ecological requirements of the species change, it is forced to retreat from its place of residence (Odum, 1980).

3.4. Landscape

Regarding the term landscape the European Commission (2000) admits the complexity of this term as a concept encompassing several definitions. On a higher scale level the term landscape as "an area containing a mosaic of land cover patches" (area units, covered by a single class of land cover or habitat) is adopted as a key definition. Particularly at higher scale levels and/or in planning procedures "mostly only a two-dimensional spatial configuration is considered and its influence on the landscape physiognomy, but not the vertical dimension and its effects on abiotic as well as biotic factors and visual perceptions". The term landscape is narrowly linked to the term land use which reflects the "functional dimension" of the socio-economic situation in a region (European Commission, 2000). However, especially in relation to the assessment of landscapes, the problem is that the purpose of land use (e.g. whether land use serves for agricultural, recreational or conservation purposes) is often not clearly defined and so may influence the development of classification systems, data collection, the use of (biotic) indicators and finally the assessment system in general.

4. Philosophy of structure and focus as regards contents

Aspects dealt with within this special issue on "biotic indicators for a sustainable land use" in order to provide biodiversity are considered more from an applied point of view than a scientific point of view. Reflections on a logical structure covering relevant areas led to a subdivision into the five sections "Requirements", "Biodiversity and Habitat" (including "Soil" as a subsection), "Biodiversity and Landscape", "Experiences and Applications" and "Economy", which is explained in the following.

Starting with the presentation of (political) requirements (Piorr, 2003) the structure of this volume is arranged in a kind of "bottom-up-approach", i.e. the analysis of smaller compartments in agro-ecosystems (habitat-components) is followed by the consideration of larger spatial units, the landscape level, and finally completed by existing experiences with the application of (biotic) indicators and economic impacts.

In section "Requirements" the (political) demands of committees and institutions authorised to put indicators into practice and into political frameworks at national and international levels are placed at the front of the volume. The philosophy of those committees can be demonstrated by, for example, the self-description of the OECD: "OECD work on agri-environmental indicators (AEIs) is primarily aimed at policy makers and the wider public interested in the development, trends and use of agri-environmental indicators for policy purposes. The focus of the work is in particular related to indicator definitions, methodologies and calculation of indicators". This includes the following general objectives:

- Provide information on the current state and changes in the conditions of the environment in agriculture.
- Assist policy makers in a better understanding of the links between causes and impact in agriculture and agricultural policies.
- Reform trade liberalisation and environmental measures, and help to guide responses for changing environmental conditions.
- Contribute to monitoring and evaluating the effectiveness of policies addressing agri-environmental concerns and promoting sustainable agriculture, including future-oriented perspectives of agri-environmental linkages (OECD, 2000).

Politically the following basic *requirements* are considered for the development of biotic agri-environmental indicators (L. Nellinger, Federal German Ministry for Customer Affairs, Food and Agriculture, Bonn; in litteris, 2001): (1) scientific soundness; (2) adequacy of goals; (3) political relevance; (4) comprehensibility; (5) communicability; (6) balance.

Regarding the term "*scientific soundness*", political decision makers *do not* understand a successful validation and statistical significance of messages from biotic indicators, but rather, wherever possible, the adaptation of agri-environmental indicators to environmental problems or positive environmental performances related to husbandry practices. Therefore, "scientific soundness" mainly refers to the position of an indicator within the complete set of existing indicators. This aspect also covers the avoidance of redundancies and the implementation of interfaces to other society-related information and documentation systems, as for example the environmental-economic balance sheet.

"*Adequacy of goals*" means that with indicators, the degree of fulfilment of environmental political goals should be measurable. Indicators which may be highly sophisticated from a scientific point of view, but whose message is inadequately linked to goals of environmental politics, carry a considerable risk of political mistakes.

"*Relevance for political action*" means that conditions reflected by the indicators should provide the ability to be influenced by political decisions and related adaptations by producers and consumers.

Because the great majority of decision makers (e.g. politicians, parliamentarians, journalists and ordinary citizens) are not scientifically educated, the indicators and the criteria they are based upon must be *comprehensible*, so that *communication* is possible. Even this basic criterion is often not fulfilled by scientifically excellent indicators.

To avoid mistakes in steering political processes biotic indicators should be *balanced* sectorally (i.e. regarding the different environmental problems and performances of agriculture/land use) as well as intersectorally (regarding interactions between different branches of business and the international comparison of the environmental situation in different countries). To achieve this *balance* the existence of *supraordinate evaluation scales* and *standardised report systems* are necessary.

The (political) benefit of an indicator increases if these basic demands are fulfilled, and all the more if they are accepted within international regulations. An overall precondition is a reasonable cost/benefit-relation in the case of putting the demands into practice.

These political demands elucidate that in international negotiations on indicator systems, global acting indicators with a comparatively general and almost "weak and rough" message are discussed as a priority. These indicators are able to detect (environmental) impacts or positive performances of husbandry only on a very general level in order to make efforts for international solutions and/or regulations compatible.

At international level (e.g. OECD, 2000), particularly indicators for abiotic measures have reached a high standard, and for many of them, a more or less complete series of data from nearly all Member States already exists. Indicators however, that aim to assess biodiversity and landscape matters, have reached not more than a proposal stage, and are beginning to be tested as models.

An example of the kind of indicators recently used and/or proposed at international level (e.g. by the OECD) and the way they are used are the so called "contextual indicators" (Piorr, 2003). Within the topic "farm management and the environment", information has been gathered about, e.g. "environmental whole farm management plans", "nutrient management plans", "conducting of soil tests", the "use of non-chemical and/or integrated pest control methods", "soil cover" and "land management practices". Within the topic biodiversity, information on genetic diversity is predominantly asked for, as for instance "registered number of crop varieties", "registered number of livestock breeds", "share of key crop varieties", "share of key livestock breeds" and "number of endangered crop varieties and livestock breed". With regard to species diversity, the analysis is restricted to "wildlife species on agricultural land" and "non-native species threatening agricultural production". Therefore, apart from some (vertebrate) game species and (probably mostly invertebrate) invaders, no information on the indigenous fauna and flora typical for agricultural areas (which is represented by far more than 1000 species) has to be provided to assess biodiversity!

Piorr (2003) also points out a considerable lack of reliable and consistent data, even at a supraordinate level: the result of a questionnaire among 15 European states showed that on average, 75% of the information for each category was lacking. Only statistically correct records can be provided for organic farming, even for marketing cultivars or the status of endangered livestock breeds, due to strict criteria and controls. In all other areas (with regard to conventional farming) the consistence of data is very deficient and not standardised.

Negotiations at international levels have brought another key problem to light: states that cover large areas and rely on a relatively short time of history (with regard to the recent majority of inhabitants) as for example the USA, Canada or Australia, show a completely different understanding of the term landscape. Based on large geographical units they divide strictly between landscapes for (agricultural and industrial) production and landscapes for nature conservation (H. Becker, Berlin, personal communication, 2001). This understanding which puts a lot of different ecosystems together in each category reacts upon the features of indicators which are rather coarse because a very low level of differentiation is needed. These highly globalised and generalised indicators which have recently formed the basis of international approaches for agri-environmental indications, result in unavoidable and considerable losses of information. Information derived from such indicators might be useful on the level of supraordinal landscape units, but their conversion to a single farm will be problematic.

Therefore, these kind of indicators are not transferable to rural landscapes in Europe because of completely different initial conditions: the cultural (and agriculturally used) landscapes in Europe have been developed historically over almost a 1000 years, so that particularly those where traditional structures are still present today can be judged as "cultural objects" in a very similar way to buildings (e.g. castles, towers, etc.) that create a regional identity (e.g. UNESCO Cultural Heritages of the World). In these traditionally grown cultural landscapes production, recreation and nature conservation cannot be separated but are intensively linked to each other within the same region and geographical unit. Following today's understanding in Europe, particularly the cultivation of landscapes was one of the main reasons for creating new habitats for plants and animals and so it was the most important source for an increasing biodiversity.

These basic differences derived from geographical conditions; the historical-cultural development of different countries has recently become increasingly acceptable in international negotiations on agri-environmental indicators (e.g. OECD, 2001). This means that recently, a more sensitive understanding has led to turning away from average values on a national basis and considering more seriously regional characteristics, which can be identified by special features of natural resources and regional cultural development.

Nevertheless, compared to the great number of scientifically very interesting and sound approaches regarding the development of very sensitive indicator systems, which potentially enable the exact measurement and assessment of the environmental status very exactly on several scales for single compartments of ecosystems or habitats, and for several taxa (Albrecht, 2003; Waldhardt et al., 2003; Büchs, 2003b), there is obviously a huge discrepancy between the latter and demands at international level. One solution (regarding the use of highly aggregated indicators that fulfil the demands at international level, but based on a not aggregated substructure) could perhaps to integrate them into regional agri-environmental programmes which would become more efficient through this linking.

In the context of the term *habitat* indicators and indication approaches are summarised that refer to single, clearly defined habitats within rural landscapes. In this regard the focus is clearly put on cultivated areas (e.g. fields, grasslands, orchards, vineyards, etc.). In some papers within-field structures (Brose, 2003) and uncultivated marginal structures are also considered as far as they produce interactions with the cropped areas. Furthermore, larger uncultivated areas (e.g. fallow land) which are identified as reference areas from which standards for an assessment by means of indicators can be derived (e.g. Büchs et al., 2003; Döring and Kromp, 2003; Irmler, 2003; Jeanneret et al., 2003; Perner, 2003). On the other hand, indicator concepts were also summarised in the context of the term *habitat* that refer to features of populations of one or more species (e.g. body size, body weight, fertility, etc.), for determining whether cultivated areas (fields) are the preferred habitat for feeding, for reproduction activities and so on, and whether they are within the focus of the assessment goals (Büchs et al., 2003; Döring et al., 2003; Nickel and Hildebrandt, 2003). With regard to this topic, interrelations (e.g. seasonal migrations) with other (uncultivated) habitats at various distances to the field are likely to occur (e.g. within birds), so that transitions to landscape-related indication approaches cannot be excluded (e.g. Jeanneret et al., 2003).

Within the habitat-related analysis *soil* aspects have been separated (Anderson, 2003; Ekschmitt et al., 2003; Ruf et al., 2003; Schloter et al., 2003a,b) due to the following reasons:

- The soil is a more or less self-contained medium (except its surface), unless there is no doubt that processes occurring in the soil affect other compartments of the ecosystem (e.g. vegetation, crop microclimate) significantly.
- Soil biology requires a specific methodology and terminology, which also react upon approaches to develop indicator systems.
- Processes within the soil are often slower and more sustainable (e.g. changes which are detectable rather in the long term) than in other compartments of an ecosystem; so they are to a less extent characterised by short-term invasions and emigrations. This phenomenon requires adopted indication concepts.
- As a consequence of remineralisation processes within the soil, adhesion and the release of substances, the building of aggregates, etc., the soil compartment has a stronger and more direct relation to abiotic conditions which possibly requires specific procedures of indication.

The broad range of taxa which should be considered and the requirements regarding a sufficient number of samples (which in many cases is not realistic) are the key problems for a conversion of habitat- and biodiversity-related indicator approaches into practice (Perner, 2003). An exception might be plant species that are easy to determine (Oppermann, 2003). In soil microbiology this problem is not so obvious because up to now, non-taxa-specific, but summarising parameters have been used (e.g. *C/N*-relation; Anderson, 2003). However, with increasing possibilities of the characterisation of microorganism communities (quasi-species levels and ecological types) as introduced by Schloter et al., 2003a,b, one has to be aware that soil microbiology is probably approaching a similar dilemma to the one we are currently facing concerning surveys of invertebrate communities in particular (Duelli and Obrist, 2003). Therefore, indicators based on population and/or on habitat features have more the function of providing the fundament for indicators or indication systems, which act in a far more indirect way, a fact derived from practical experience, established knowledge and/or logical deductions (e.g. application of certain measures, enhancement of landscape elements or habitat developments are estimated to increase biodiversity) (Menge, 2003; Oppermann, 2003; Roth and Schwabe, 2003). Furthermore, they are available for detailed assessments (Büchs et al., 2003; Nickel and Hildebrandt, 2003; Mueckschel and Otte, 2003; Perner and Malt, 2003; Rosenthal, 2003; Gerowitt, 2003). Heyer et al. (2003) provided the missing link to show whether the surrogate indicators are scientifically reliable and the most appropriate manner for applying them. More details are discussed in the summarising synoptic introductions by Waldhardt et al. (2003), Büchs (2003b) and Schloter et al. (2003b).

Whenever the term *landscape* is the focus for developing suitable indicators, cultivated fields are analysed in relation to the surrounding types of habitats and ecosystems. Within the section “Landscape”, several different scale levels can be the focus of analysis. The starting point for the development of indicators for *landscape* assessment can be the cultivated area itself as well as directly neighbouring habitat types (or those within a defined distance), regardless of whether they are cultivated or not, and the combination of the two which forms the "landscape matrix" (Dauber et al., 2003; Hirsch et al., 2003; Steiner and Köhler, 2003; Waldhardt and Otte, 2003). Assessment can be related to the entire area of local farms (Richter et al., 1999), to functional units (e.g. bottom land meadows, heaths; see Osinski, 2003; Jeanneret et al., 2003), to regions that are defined by natural units (e.g. Hoffmann et al., 2003; Hoffmann and Greef, 2003; Osinski et al., 2003), to administrative units of different scale levels (Osinski et al., 2003) or to surveys fixed to standardised spatial grids (Hoffmann-Kroll et al., 2003). This listing demonstrates that the approaches to biodiversity indication at landscape scale level are rather heterogeneous. The different concepts show a very different potential towards both a regionalisation and a spatial aggregation of data. Even some attempts at indicator systems is presented in the "Habitat" section—if recorded over a wide area—have the potential of being put together to form larger spatial units and can thus be used for assessments at landscape level.

In some cases the term biodiversity which originates from genetical biodiversity and species biodiversity is used for landscape complexity (or landscape patchiness defined as richness of habitat and land use types), which are secondarily linked to the species richness of those ecosystems. On the other hand indicator approaches that are strictly focused (oriented) on the population development of individual species (e.g. "mosaic indicators" by Hoffmann and Greef, 2003) may form transitions to indicators that refer specifically to the habitat quality.

Therefore, the handling of information losses as an unavoidable consequence of data aggregations can be classified as one of the key issues of "Landscape" section. Spatio-temporal processes and interactions, which are extraordinary effective at landscape level due to the extremely variable performance of cultivated habitats within seasonal changes, also play an important role.

Particularly this factor raises the question regarding the function each ecosystem (or habitat) type has in respect to its contribution to and preservation of biodiversity at landscape level in space and time. In this context it is of particular interest to see how these functions, that underlie extremely complex seasonal changes, can be assessed and how they can be integrated into indicator systems. Further details of landscape-related assessment procedures are discussed by Waldhardt (2003).

Section "Experience and Application" shows ways and modalities of how assessment procedures that are already established, or are currently being developed, handle biodiversity evaluation. Therefore, analysis was strictly concentrated on biotic indicators. Considerable deficiencies regarding applicable biotic indicators were obvious from the start.

With the exception of a few procedures (e.g. Frieben, 1998; Oppermann, 2003; Buys, 1995; Oosterveld and Guldemond, 1999) the assessment of biotic parameters (as for example biodiversity) is not carried out directly (by actual on-site recording of organisms), but is mostly derived from parameters which are assumed to be correlated to biodiversity or other ecological features which are being focussed on. Several measurements or parameters (e.g. low extent of bare fallow, low input of pesticides, high percentage of landscape structural elements, margins and structures adjacent to fields, etc.) are treated as equivalent to positive biodiversity development (Braband et al., 2003; Menge, 2003; Roth and Schwabe, 2003). The legitimacy and "validation" of this kind of indication procedure is usually derived from information selected from published references. One of the major concerns of this procedure is that most of the published experiences are based on case studies with clearly defined conditions (with regard to farming system, on-farm structures, abiotic conditions such as soil, hydrology, climate, etc.) and so, often only correspond in a very limited way to the specific requirements of the indicator chosen.

Only exceptional naturally occurring organisms as for example conspicuous plant species, butterflies or vertebrates which can also be easily recognised by amateurs are used as indicators (e.g. Oppermann, 2003; Frieben, 1998; Buys, 1995; Oosterveld and Guldemond, 1999).

Assessment procedures are very limited due to costs and practicability, so that one has to assume that, as the recent status of technical development stands, biodiversity indication will in practice be conducted mainly by using indirect parameters or surrogates derived from published basic knowledge on biodiversity matters. This patching together of results involves the risk that biodiversity indication is inexact. Koehler (1999) points out that "underestimation of taxonomy threatens the whole idea of bio-indication, because it is not sufficient to identify *invertebrates* (changed by the author) up to higher taxonomic categories (genus or family); the correct determination of species is however an absolute prerequisite. It must be emphasised that biology, ecology and consequently indicator properties are strictly related to the species". In this context it has to be mentioned that for the assessment of fresh water, an international procedure exists (saprobia-index), which was established almost 100 years ago, and has been continuously developed since (Friedrich, 1990; Marten and Reusch, 1992; Usseglio-Polatera et al., 2000). This index can be used in a modular way, which includes the on-site survey of a broad range of organisms and which allows statements on biodiversity. Each water supply office has employed at least one specialist who is exclusively responsible for water quality assessments by using biotic indicators mostly on the species level. This raises not only the question of whether (or why not) corresponding possibilities exist for establishing a similar approach for terrestrial ecosystems, at least for those under cultivation, but also the question of which priorities are established for agri-environmental procedures with reference to assessing husbandry practices. Altogether a completely new dimension arises through the fact that world-wide, labels of certification are now being used by the food industry and the retailer co-operatives of several branches which aim to guarantee an environmentally friendly and socially conscious production of food (e.g. Euro-Retailer Produce Working Group, 2002; Flower Label Program, 2002; Forest Stewardship Council, 2000; Rainforest Alliance, 2002; Pan European Forest Certificate, 2002). With reference to the Convention of Rio (United Nations, 1992), such certification labels also pick up aspects of biodiversity and its indication, but with a poor scientific background similar to the agri-environmental assessment procedures mentioned above.

Against this background it can be assumed that in the future, maintenance and conservation of biodiversity will not only be ruled and controlled by governmental institutions, but as the consequence of general tendencies of globalisation, on the basis of "private initiatives" of trade associations by establishing various labels and quality seals. These approaches will be complemented by the implementation of environmental management systems with voluntary participation (EMAS I and II; European Community, 1993, 2001). These developments demonstrate that work on biotic indicators is no longer merely a scientific end in itself, but that certain groups of society need practicable (biotic) indicators; this need has long since overtaken the current status of discussion within the scientific community.

The volume is completed with the consideration of *economical aspects* of biotic indicators. Although at first the development of biotic indicators occurred (and partially still does occur) completely separately to practical demands on a scientific level, the integration of economical aspects seems to be undeniable with respect to the key position of the economy as far as the possibilities of putting biotic indicators into practice are concerned, which finally also influences their development significantly.

Within the economic analysis, the cost-efficiency ratio of agri-environmental programmes and measures is assessed with respect to the improvement of biotic resources, by recording the success of the measures applied. Even in economics the biodiversity-related control of success (of measures applied) is not conducted directly, but indirectly for instance by specialist evidence (e.g. Marggraf, 2003), which is validated by recently developed methods of analysis as for example the Delphi-method (Richey et al., 1985).

Further aspects are strategies to integrate biodiversity-related goals into land use systems not only practically but also economically (Plachter and Werner, 1998). With reference to these aspects, parameters that are related to the production process are also considered, as for instance calculations and models of the disturbance potential (with reference to biotic resources) of different production systems, which are finally included in the entire farm-related economic calculation with the goal of developing a farm-related, economically optimised model on the basis of agreed environmental quality goals for the preservation and enhancement of biodiversity. In this regard, at the farm level the different management measures are incorporated into a field record system, from which potential conflicts with demands of nature conservation can be derived (Meyer-Aurich et al., 2003; Kantelhardt et al., 2003).

In a further step modelling techniques for the assessment of economic effects on the protection of biodiversity are tested by Herrmann et al. (2003) for different geographical scale levels. This advanced approach aims to integrate biotic indicators and control of success (of measures applied) in an optimised way into the operational course of the farm, taking into consideration economic aims. By means of "trade-off-functions" (loss of goal realisation) the level of goal achievement can be determined taking into account both economically and agriculturally based practical constraints which correspond to farm costs under the premise of agricultural production. In respect to these trade-off-functions, standard methods are tested as for instance the "Replacement Cost Method" (Pearce, 1993), the "Contingent Valuation Method" (Mitchell and Carson, 1989) and concepts such as "Save Minimum Standards" (SMS) and the "Precautionary Principle" (Meyer-Aurich et al., 2003).

Economic considerations include transaction costs, costs for information transfer regarding the control of the success of measures applied, and for the organisation and control of compensation payments. Another focus is the integration of non-tradable goods (in this respect: biotic resources) into a cost-benefit-analysis, which means money as an indicator for rational decision-making on biodiversity conservation (Bräuer, 2003). The idea is that the estimation of the (economic and social) benefits of conservation measures or ecological accomplishments by husbandry would help to promote their acceptance.

Finally, there is a demand for conceptions with which indicator-related ecological goods within entire environmental quality goals could be practically assessed economically and could be integrated into an on-farm market oriented rewarding system. It is being discussed at present whether rewarding farmers' ecological accomplishments should be integrated into agri-environmental programmes, taking into consideration market principles by trading with ecological goods as products, whose quality is described in detail by the suppliers as well as by the purchasers, and whose price can be defined as an indicator for the scarcity and need of each ecological good. In this respect the allocation of property rights to natural resources is estimated as fundamental. However, the demands and consensus within society will produce clearly defined rules for acting (Gerowitt et al., 2003).

5. Final remarks and conclusions

As can be concluded easily from the descriptions above, the intention of this volume is not a theoretical scientific discussion of terms, of biodiversity and of indicator concepts. Its goal is rather to elucidate the great range of facets and approaches with regard to questions on the development of and agreement on biotic indicator (systems), taking into special consideration the complex problems of their conversion into practice.

The term "indicator" implies in the principle the aspect of an application; i.e. something should be "indicated", from which actions are derived: a status, a process or a development in the past, at present or in the future. In this context it can be assumed that every scientist (even those conducting fundamental research) involved in the development and/or validation of an indicator will feel satisfied if "his/her" indicator or indicator system is included in a practical approach. Starting from this assumption, basically no great differences between the administrative requirements and the quality of indicators which have been developed according to scientific principles would be expected. Nevertheless, huge discrepancies between the expectations of administrative or practical institutions and scientific reality have to be admitted. Within this volume, particularly in sections "Requirements" and "Biodiversity and Habitat", these contrary expectations of political/administrative demands on the one hand and science on the other clash.

Undoubtedly, there are many highly sophisticated methods for indicating parameters (functional diversity, species diversity, habitat quality, etc.) related to several goals (environmental loads, conservation, biological control, etc.) which generally allow a gradual, very sensitive, but highly exact assessment of the actual status and/or of (future) developments in all habitat-related areas. Unfortunately, the great majority of all those advanced indicators needs to be assessed in each location specifically and requires too much labour and unaffordable skill levels, meaning that they are not practicable.

These discrepancies become particularly clear with regard to those indicators developed for habitat levels (see Büchs et al., 2003), which are connected very closely to pure scientific research due to their high and gradual resolution of environmental aspects. Beside validations, baselines and reference values, the question of practicability and conversion is of central importance in the future. To be realistic (i.e. to meet practical demands in any way), conversion into practice would require a simplification of the indication procedure to such an extent that it would affect the scientific exactness significantly, meaning that such indicator models should be capable of allowing large tolerances and error rates. The synoptic introduction to each section provides suggestions to simplify the application of indicators in the sense described above, but due to the fact that costs and labour have to be reduced to a minimum and indicators have to be applicable to large areas and several regions, a normal scientific validation will obviously not possible. Similar to technical developments, only the experience of a broad application in practice will provide information in the long term on whether an indicators function is sufficient or not. Within such practical experience specific regional/local conditions possibly lead to opposing results which exclude ways of simplification in applying indicators. Therefore, striving for a scientific standard which is too high, and perfection in developing and validating indicator systems involves the risk of paralysing progress and causing contradictory effects with regard to the protection and maintenance of biodiversity.

It will be obvious that even this extensive portrayal cannot provide final or ideal solutions for the cardinal problem of indicator development and application. However, the intention is rather to convey awareness of and understanding for the difficulties of the different uses and purposes of biotic indicators by examining several facets and areas of unsolved problems, and to initiate a horizontal (concerning different subject areas and scientific areas) and vertical (concerning pure research, applied research, practical and administrative conversion and political decision makers) transdisciplinary discussion.

Acknowledgements

In particular I would like to express gratitude to Dr. Hans Becker (Berlin) for his active accompaniment regarding the co-ordination of this volume and his continuous input into the structure and alignment of the volume. Many thanks are given to Dr. Harald Albrecht (Freising-Weihenstephan), Dr. Elisabeth Osinski, Prof. Dr. Hans-Peter Piorr (Eberswalde), Dr. Michael Schloter (Munich), Dr. Rainer Waldhardt (Gießen), and Jörg Weickel (Mainz) as chapter collators for their co-operation, passion and endurance during the 2 years of co-ordinating the project. I am indebted also to Bettina Matzdorf (Kiel) and Dr. Berthold Janßen (Marburg) for their intensive engagement at the beginning of the project.

On behalf of the chapter collators much gratitude is shown for general supports to the following collegues and institutions. The printing as spin-off book was generously supported by the STOLL-Vita-Stiftung, Waldshut-Tiengen, Germany. Improvement of the English and (in individual cases) help in translation was patiently provided by native speaker Mrs. I. Bürig (Braunschweig), Mrs. Hausdörfer (Berlin) and Miss Leicht (Mainz). The efforts of refereeing were kindly taken by:[1] Dr. habil Clemens Abs, Department of Ecology, TU Munich, Freising, Germany; Prof. Dr. H. Ahrens, Landwirtschaftliche Fakultät, Universität Halle-Wittenberg, Halle/Saale, Germany; Prof. Dr. Carl Beierkuhnlein, Fachbereich Landeskultur und Umweltschutz, Universität Rostock, Rostock, Germany; Prof. Dr. Jan Bengtsson, Department of Ecology and Crop Production, Swedish University of Agricultural Sciences, Uppsala, Sweden; Dr. Armin Bischoff, Department of Biology, University of Fribourg, Fribourg, Switzerland; Prof. Dr. S. Dabbert, Institute of Agricultural Economics, University of Hohenheim, Stuttgart, Germany; Dr. Oliver Dilly, Center of Ecology, University of Kiel, Germany; Dr. Barbara Ekbom, Department of Entomology, Swedish University of Agricultural Science, Uppsala, Sweden; Dr. Thomas van Elsen, Department of Organic Farming, University of Kassel, Witzenhausen, Germany; Dr. Bettina Frieben, Osterholz-Scharmbeck, Germany; Dr. Nikolaus Gotsch, Agronomy, ETH-Center, Zurich, Switzerland; Prof. Dr. U. Hampicke, Botanisches Institut, Universität Greifswald, Greifswald, Germany; Prof. Dr. Heikki Hokkanen, Institute of Agricultural Zoology, University of Helsinki, Helsinki, Finland; Dr. Katrin Kiehl, Department of Ecology, TU Munich, Freising, Germany; Prof. Dr. Heidrun Mühle,

[1] Colleagues being (co-)authors or chapter collators are not included in this list.

Umweltforschungszentrum Leipzig-Halle GmbH, Leipzig, Germany; Prof. Dr. Jean Charles Munch, Institut für Bodenökologie, GSF—Forschungszentrum für Umwelt und Gesundheit GmbH, Neuherberg, Germany; Prof. Juan J. Oñate, Departemento de Medio Ambiente, Universidad Europea CEES, Madrid, Spain; Kevin Paris, Agriculture Directorate, OECD, Paris, France; Prof. Begoña Peco, Departemento de Ecologia, Universidad Autonoma de Madrid, Madrid, Spain; Clive Potter, Agricultural Economic and Business Management Research Group, T.H. Huxley School of Environment, Earth Sciences and Engineering, London, UK; Prof. Dr. Janna Puumalainen, Forestry Faculty, University of Applied Science, Eberswalde, Germany; Dr. Dieter Ramseier, Geobotanical Institute, Swiss Federal Institute of Technology, Zurich, Switzerland; Prof. Dr. Winfried von Urff, Bad Zwesten, Germany; Dr. Keith F.A. Walters, Central Science Laboratory, York, UK; Dr. Dirk Wascher, Alterra b.v., Green World Research, Wageningen, The Netherlands; Dr. Kerstin Wiegand, Department of Ecology and Evolutionary Biology, Princeton University, Princeton, NJ, USA; Prof. Dr. H. Wohlmeyer, Österreichische Vereinigung für Agrarwissenschaftliche Forschung, Wien, Austria.

References

Albrecht, H., 2003. Suitability of arable weeds as indicator organisms to evaluate species conservation effects of management in agricultural ecosystems. Agric. Ecosyst. Environ. 98, 201–211.

Anderson, T.-H., 2003. Microbial eco-physiological indicators to assess soil quality. Agric. Ecosyst. Environ. 98, 285–293.

Arndt, K., Nobel, W., Schweizer, B., 1987. Bioindikatoren—Möglichkeiten, Grenzen und neue Erkenntnisse, Ulmer, Stuttgart, 388 pp.

Bick, H., 1982. Bioindikatoren und Umweltschutz. Decheniana-Hefte 26, 2–5.

Braband, D., Geier, U., Köpke, U., 2003. Bio-resource evaluation within agri-environmental assessment tools in different European countries. Agric. Ecosyst. Environ. 98, 423–434.

Bräuer, I., 2003. Money as an indicator: to make use of economic evaluation for biodiversity conservation. Agric. Ecosyst. Environ. 98, 483–491.

Brose, U., 2003. Regional diversity of temporary wetland carabid beetle communities: a matter of landscape features or cultivation intensity? Agric. Ecosyst. Environ. 98, 163–167.

Büchs, W., 1988. Stamm- und Rindenzoozönosen verschiedener Baumarten des Hartholzauenwaldes und ihr Indikatorwert für die Früherkennung von Baumschäden. Ph.D. Thesis, Math.-Nat. Faculty, University of Bonn, pp. 1–813.

Büchs, W., 1995. Tierökologische Untersuchungen als Grundlage zur Charakterisierung von Ökosystemen und Indikation von Umweltbelastungen. Habilitation Thesis. Faculty of Natural Sciences, Technical University of Braunschweig, pp. 1–312.

Büchs, W. (Ed.), 2003a. Biotic indicators for biodiversity and sustainable agriculture. Agric. Ecosyst. Environ. 98, 1–16.

Büchs, W., 2003b. Biodiversity and agri-environmental indicators—general scopes and skills with special reference to the habitat level. Agric. Ecosyst. Environ. 98, 35–78.

Büchs, W., Harenberg, A., Zimmermann, J., Weiß, B., 2003. Biodiversity, the ultimate agri-environmental indicator?—potential and limits for the application of faunistic elements as gradual indicators in agroecosystems. Agric. Ecosyst. Environ. 98, 99–123.

Buys, J.C., 1995. Naar en natuurmeetlat voor landbouwbedrijven. Centrum voor landbouw en milieu (CLM) Utrecht, No. 169, Utrecht, NL.

Commission of the European Communities, 1991. Council Directive of 15 July 1991 referring to placing plant protection products on the market (91/414/EEC). Amtsblatt der Europäischen Gemeinschaften L 230, August 19, 1991.

Commission of the European Communities, 1998. Evaluation of Agri-environment Programmes. State of Application of Regulation (EEC) No. 2078/92. DGVI Commission Working Document VI/7655/98.

Commission of the European Communities, 2000. Indicators for the Integration of Environmental Concerns into the Common Agricultural Policy. Communication from the Commission to the Council and the European Parliament COM, 2000, 20 final.

Dauber, J., Hirsch, M., Simmering, D., Waldhardt, R., Otte, A., Wolters, V., 2003. Landscape structure as an indicator of biodiversity: matrix effects on species richness. Agric. Ecosyst. Environ. 98, 321–329.

Döring, T.F., Kromp, B., 2003. Which carabid species benefit from organic agriculture?—a review of comparative studies in winter cereals from Germany and Switzerland. Agric. Ecosyst. Environ. 98, 153–161.

Döring, T.F., Hiller, A., Wehke, S., Schulte, G., Broll, G., 2003. Biotic indicators of carabid species richness on organically and conventionally managed arable fields. Agric. Ecosyst. Environ. 98, 133–139.

Duelli, P., Obrist, M.K., 2003. Diversity indicators: the choice of values and measures. Agric. Ecosyst. Environ. 98, 87–98.

Dunger, W., 1982. Die Tiere des Bodens als Leitformen für anthropogene Umweltveränderungen. Decheniana-Beihefte 26, 151–157.

Ekschmitt, K., Stierhoff, T., Dauber, J., Kreimes, K., Wolters, V., 2003. On the quality of soil biodiversity indicators: abiotic and biotic parameters as predictors of soil faunal richness at different spatial scales. Agric. Ecosyst. Environ. 98, 273–283.

Ellenberg, H., Weber, H.E., Düll, R., Wirth, V., Werner, W., Paulißen, D., 1992. Zeigerwerte der Pflanzen in Mitteleuropa. Scripta Geobot. 18, 1–258.

European Commission (Ed.), 1998. Principles and Recommendations from the European Consultative Forum on the Environment and Sustainable Development.

European Commission (Ed.), 2000. From Land Cover to Landscape Diversity in the European Union. Brussels, Belgium.

European Community (Ed.), 1993. Regulation No. 1836/93 of the European Parliament and of the Council of 29 June 1993 allowing voluntary participation by organisations in a Community eco-management and audit scheme (EMAS I). Amtsblatt der Europäischen Gemeinschaften L 168/1, July 7, 1993.

European Community (Ed.), 2001. Regulation No. 761/2001 of the European Parliament and of the Council of 19 March 2001 allowing voluntary participation by organisations in a Community eco-management and audit scheme (EMAS II). Amtsblatt der Europäischen Gemeinschaften L 114/1, April 24, 2001.

Euro-Retailer Produce Working Group (EUREP), 2002. Europäisches Handelsinstitut e.V. http://www.eurep.org or http://www.ehi.org.

Flower Label Program (FLP), 2002. Internationaler Verhaltenskodex für die sozial- und umweltverträgliche Produktion von Schnittblumen, April 25, 2002. http://www.flower-label-program.org.

Forest Stewardship Council (FSC), 2000. March 24, 2000. http://www.fsc-deutschland.de.

Frieben, B., 1998. Verfahren zur Bestandsaufnahme und Bewertung von Betrieben des Organischen Landbaus im Hinblick auf Biotop- und Artenschutz und die Stabilisierung des Agrarökosystems. Schriftenreihe Institut für Organischen Landbau 11, Verlag Dr. Köster, Berlin.

Friedrich, G., 1990. Eine Revision des Saprobiensystems. Zeitschrift für Wasser- und Abwasserforschung 23, 141–152.

Gerowitt, B., 2003. Development and control of weeds in arable farming systems. Agric. Ecosyst. Environ. 98, 247–254.

Gerowitt, B., Isselstein, J., Marggraf, R., 2003. Rewards for ecological goods—requirements and perspectives for agricultural land use. Agric. Ecosyst. Environ. 98, 541–547.

Hassan, S.A., Bigler, F., Blaisinger, P., Bogenschütz, H., Brun, J., Chiverton, P., Dickler, E., Easterbrook, M.A., Edwards, P.J., Englert, W.D., Firth, S.I., Huang, P., Inglesfield, C., Klingauf, F., Kühner, C., Ledieu, M.S., Naton, E., Oomen, P.A., Overmeer, W.P.J., Plevoets, P., Reboulet, J.N., Samsoe-Petersen, L., Shires, S.W., Stäubli, A., Stevenson, J., Tuset, J.J., van Zon, A.Q., 1985. Standard methods to test the side-effects of pesticides on natural enemies of insects and mites developed by the IOBC/WPRS Working Group "Pesticides and Beneficial Organisms". Bull. OEPP/EPPO 15, 214–255.

Hassan, S.A., Bigler, F., Bogenschütz, H., Boller, F., Brun, J., Calis, J.N.M., Coremans-Pelsener, J., Duso, C., Grove, S., Heimbach, U., Helyer, N., Hokkanen, H., Lewis, G.B., Mansour, F., van de Veire, M., Viggiani, G., Vogt, H., 1994. Results of the sixth joint pesticide testing programme of the IOBC/WPRS-Working Group "Pesticides and Beneficial Organisms". Entomophaga 39, 107–119.

Herrmann, S., Dabbert, S., Schwarz-von Raumer, H.G., 2003. Threshold values for nature protection areas as indicators for biodiversity—a regional evaluation of economic and ecological consequences. Agric. Ecosyst. Environ. 98, 493–506.

Hesse, R., 1924. Tiergeographie auf ökologischer Grundlage—Das Tier als Glied des Naturganzen. Fischer Jena, 618 pp.

Heyer, W., Hülsbergen, K.-J., Wittmann, C., Papaja, S., Christen, O., 2003. Field related organisms as possible indicators for evaluation of land use intensity. Agric. Ecosyst. Environ. 98, 453–461.

Hirsch, M., Pfaff, S., Wolters, V., 2003. The influence of matrix type on flower visitors of *Centaurea jacea* L. Agric. Ecosyst. Environ. 98, 331–337.

Hoffmann, J., Greef, J.M., 2003. Mosaic indicators—theoretical approach for the development of indicators for species diversity in agricultural landscapes. Agric. Ecosyst. Environ. 98, 387–394.

Hoffmann, J., Greef, J.M., Kiesel, J., Lutze, G., Wenkel, K.-O., 2003. Practical example of the mosaic indicators approach. Agric. Ecosyst. Environ. 98, 395–405.

Hoffmann-Kroll, R., Schäfer, D., Seibel, S., 2003. Landscape indicators from ecological area sampling in Germany. Agric. Ecosyst. Environ. 98, 363–370.

International Standardisation Organisation (ISO) 14001, 1996. Environmental management systems, Specification with guidance for use. http://www.iso.ch.

Irmler, U., 2003. The spatial and temporal pattern of carabid beetles on fields in northern Germany (Schleswig-Holstein) and their value as ecological indicators. Agric. Ecosyst. Environ. 98, 141–151.

Jeanneret, P., Schüpbach, B., Luka, H., 2003. Quantifying the impact of landscape and habitat features on biodiversity in cultivated landscapes. Agric. Ecosyst. Environ. 98, 311–320.

Kantelhardt, J., Osinski, E., Heissenhuber, A., 2003. Is there a reliable correlation between hedgerow density with agricultural site conditions? Agric. Ecosyst. Environ. 98, 517–527.

Kneitz, G., 1983. Aussagefähigkeit und Problematik eines Indikatorkonzeptes. Verhandlungen der Deutschen Zoologischen Gesellschaft, 1983, pp. 117–119.

Koehler, H.H., 1999. Predatory mites (Gamasina, Mesostigmata). In: Paoletti, M.G. (Ed.), Invertebrate Biodiversity as Bioindicators of Sustainable Landscapes. Agric. Ecosyst. Environ. 74, 395–410.

Kühnelt, W., 1943. Die Leitformenmethode in der Ökologie der Landtiere. Biologia Generalis 17, 106–146.

Legg, W., 1999. Sustainable agriculture: an economic perspective, OECD, Paris, France, 10 pp. http://www.oecd.org/agr/News/cont-8.htm.

Mannis, A., 1996. Indicators of sustainable development. European Commission DG XIII. http://cesimo.ing.ula.ve/GAIA/Reports/indics.html.

Marggraf, R., 2003. Comparative assessment of agri-environmental programmes in federal states of Germany. Agric. Ecosyst. Environ. 98, 507–516.

Marten, M., Reusch, H., 1992. Anmerkungen zur DIN "Saprobienindex" (38 410 Teil 2) und Forderung alternativer Verfahren. Natur und Landschaft 67, 544–547.

Menge, M., 2003. Experiences with the application, survey and assessment of agri-environmental indicators in agricultural practice. Agric. Ecosyst. Environ. 98, 443–451.

Meyer-Aurich, A., Zander, P., Hermann, M., 2003. Consideration of biotic nature conservation targets in agricultural land use—a case study from the Biosphere Reserve Schorfheide-Chorin. Agric. Ecosyst. Environ. 98, 529–539.

Mitchell, R.C., Carson, R.T., 1989. Using Surveys to Value Public Goods: The Contingent Valuation Method. Resources for the Future, Washington, DC, USA.

Mueckschel, C., Otte, A., 2003. Morphometric parameters: an approach for the indication of environmental conditions in calcareous grassland. Agric. Ecosyst. Environ. 98, 213–225.

Nickel, H., Hildebrandt, J., 2003. Auchenorrhyncha communities as indicators of disturbance in grasslands (Insecta, Hemiptera)—a case study from the Elbe flood plains (Northern Germany). Agric. Ecosyst. Environ. 98, 183–199.

Odum, E.P., 1980. Grundlagen der Ökologie. Georg Thieme Verlag, Stuttgart, NY, 836 pp.

OECD (Organisation for Economic Co-operation and Development) (Ed.), 1999. Environmental Indicators for Agriculture, vol. 2: Issues and Design—The York Workshop, Paris, France, 213 pp.

OECD (Organisation for Economic Co-operation and Development) (Ed.), 2000. Environmental Indicators for Agriculture, vol. 3: Methods and Results—Executive Summary, Paris, France, 53 pp.

OECD (Organisation for Economic Co-operation and Development) (Ed.), 2001. Environmental Indicators for Agriculture, vol. 3: Methods and Results, Publications Service, OECD, Paris, France.

Oosterveld, E.B., Guldemond, J.A., 1999. De natuurmeetlat gemeten. Centrum voor landbouw en milieu (CLM) Utrecht, No. 407, Utrecht, NL.

Oppermann, R., 2003. Nature balance scheme for farms—evaluation of the ecological situation. Agric. Ecosyst. Environ. 98, 463–475.

Osinski, E., 2003. Operationalisation of a landscape oriented indicator. Agric. Ecosyst. Environ. 98, 371–386.

Osinski, E., Meier, U., Büchs, W., Weickel, J., Matzdorf, B., 2003. Application of biotic indicators for evaluation of sustainable land use—current procedures and future developments. Agric. Ecosyst. Environ. 98, 407–421.

Pan European Forest Certificate (PEFC), 2002. March 17, 2002. http://www.pefc.de.

Pearce, D.W., 1993. Cost-Benefit Analysis. London, 2 ed.

Perner, J., 2003. Sample size and quality of indication—a case study using ground-dwelling arthropods in agricultural ecosystems. Agric. Ecosyst. Environ. 98, 125–132.

Perner, J., Malt, S., 2003. Assessment of changing agricultural land use: response of vegetation, ground-dwelling spiders and beetles to the conversion of arable land into grassland. Agric. Ecosyst. Environ. 98, 169–181.

Phillipson, J., 1983. Bioindicators, biological surveillance and monitoring. Verhandlungen der Deutschen Zoologischen Gesellschaft, 1983, pp. 121–123.

Piorr, H.-P., 2003. Environmental policy, agri-environmental indicators and landscape indicators. Agric. Ecosyst. Environ. 98, 17–33.

Plachter, H., Werner, A., 1998. Integrierende Methoden zu Leitbildern und Qualitätszielen für eine naturschonende Landwirtschaft. Zeitschrift für Kulturtechnik und Landentwicklung 39, 121–129.

Rainforest Alliance (RA), 2002. January 28, 2002. http://www.rainforest-alliance.org.

Richey, J.S., Man, B.W., Horth, R.R., 1985. The Delphi technique in environmental assessment. Technol. Forecast. Social Change 23, 89–94.

Richter, J., Bachinger, J., Stachow, U., 1999. Einfluss der Standortheterogenität innerhalb von Großschlägen auf die Segetalflora unter organischer und konventioneller Bewirtschaftung in Ostbrandenburg. Beiträge zur 5. Wissenschaftstagung zum Ökologischen Landbau "Vom Rand zur Mitte", pp. 416–419.

Rosenthal, G., 2003. Selecting target species to evaluate the success of wet grassland restoration. Agric. Ecosyst. Environ. 98, 227–246.

Roth, D., Schwabe, M., 2003. Method for assessing the proportion of ecologically, culturally and provincially significant areas ("OELF") in agrarian spaces used as a criterion of environmental friendly agriculture. Agric. Ecosyst. Environ. 98, 435–441.

Ruf, A., Beck, L., Dreher, P., Hund-Rinke, K., Römbke, J., Spelda, J., 2003. A biological classification concept for the assessment of soil quality: "Biological Soil Classification Scheme" (BBSK). Agric. Ecosyst. Environ. 98, 263–271.

Schloter, M., Bach, H.-J., Metz, S., Sehy, U., Munch, J.C., 2003a. Influence of precision farming on the microbial community structure and functions in nitrogen turnover. Agric. Ecosyst. Environ. 98, 295–304.

Schloter, M., Dilly, O., Munch, J.C., 2003b. Indicators for evaluating soil quality. Agric. Ecosyst. Environ. 98, 253–262.

Steiner, N.C., Köhler, W., 2003. Effects of landscape patterns on species richness—a modelling approach. Agric. Ecosyst. Environ. 98, 353–361.

Stinner, B.R., House, G.J., 1990. Arthropods and other invertebrates in conservation-tillage agriculture. Ann. Rev. Entomol. 35, 299–318.

United Nations (Ed.), 1992. Agenda 21—Conference of the United Nations on Environment and Development, Rio de Janeiro- Documents. Rio de Janeiro, Brazil, June 1992.

Usseglio-Polatera, P., Bornaud, M., Richoux, P., Tachet, H., 2000. Biomonitoring through biological traits of benthic macroinvertebrates: how to use species trait databases. In: Jungwirth, M., Muhar, S., Schmutz, S. (Eds.), Assessing the Ecological Integrity of Running Waters. Hydrobiologia 422/423, 153–162.

Waldhardt, R., 2003. Biodiversity and landscape—summary, conclusions and perspectives. Agric. Ecosyst. Environ. 98, 305–309.

Waldhardt, R., Otte, A., 2003. Indicators of plant species and community diversity in grasslands. Agric. Ecosyst. Environ. 98, 339–351.

Waldhardt, R., Simmering, D., Albrecht, H., 2003. Floristic diversity at the habitat scale in agricultural landscapes of Central Europe. Agric. Ecosyst. Environ. 98, 79–85.

PART II
REQUIREMENTS

Chapter 2 Requirements

H.-P. Piorr, H. Becker

ELSEVIER

Available online at www.sciencedirect.com

Agriculture, Ecosystems and Environment 98 (2003) 17–33

Agriculture Ecosystems & Environment

www.elsevier.com/locate/agee

Environmental policy, agri-environmental indicators and landscape indicators

Hans-Peter Piorr*

Faculty of Landscape Use and Nature Protection, Department of Agricultural Land Use, University of Applied Science, Friedrich-Ebert Str. 28, D-16225 Eberswalde, Germany

Abstract

Research for a concept of sustainable agriculture and for the sustainable use of agricultural landscapes are closely related to the development of an international acknowledged indicator framework for the analysis and valuation of the environmental situation by the OECD. Worldwide efforts are focussing on this new topic in the environmental discussion: quantifying and valuation of impacts of agricultural practice on the animated and unanimated environment to draw conclusions for agricultural policy. A key function holds the term sustainability, which is assumed to dominate future policy approaches. The growing insecurity about the environmental impacts of agricultural land use systems led to the overall goal to avoid irreversible damages by agriculture.

The development of tools for decision-makers on all decision levels are to:

1. recognise and value impacts degrading the environment;
2. assess future sources of danger;
3. develop sustainable land use systems.

Decision supporting tools are worked out since few years for diverse purposes. As a term reflecting the most relevant environmental aspects on the global level agri-environmental indicators (AEIs) are in the focus of interest. Latest since the [Environmental Indicators for Agriculture, vol. 1: Concepts and Framework. Publications Service, OECD, Paris, 1997] launched the driving force-state-response (DSR) model and a catalogue with 14 AUI areas, which resulted from international discussions started on the Rio Conference in 1992, the political significance of an environmental measuring system was recognised.

One of the focal points was the indicator issue concerning landscapes and to define indicators which are capable to describe landscapes. Actual indicator concepts for landscape of the OECD, the EU and the PAIS research project illustrate the efforts to integrate landscapes in agricultural policies.

Keywords: Agri-environmental indicators; Agricultural landscapes; Definition of landscape; Landscape indicators; Agri-environmental programmes; Good agricultural practice; OECD indicator framework; EU indicator framework; Sustainable agriculture

1. Development of environmental policy in the European Community

Agricultural policy is one of the most important and huge policy fields at European Community. In the beginning of the early stage of European Community

* Tel.: +49-3334-657307; fax: +49-3334-236316.
E-mail address: ppiorr@fh-eberswalde.de (H.-P. Piorr).

0167-8809/$ – see front matter © 2003 Elsevier Science B.V. All rights reserved.
doi:10.1016/S0167-8809(03)00069-0

agricultural policy was characterised by the so called "green revolution", including all activities to increase agricultural production. Intensification of production processes did lead to an exert pressure on natural resources and environment. Policy measures of high price level also favoured intensive agriculture and an ever increasing use of fertilisers and pesticides. Pollution of natural resources like water, soils and certain ecosystems can be considered as undesirable side effect of these policies. Further consequences are high treatment cost for environmental damage which has to be paid by public. The need to consider environmental requirements in agricultural policy firstly came into account with the "Single European Act" of 1986 as first step:

> "Increasing public awareness of the need to integrate environmental concerns into the European Community policies was given effect in the Single European Act of 1986. This required environmental protection requirements to be integrated into other policies. In 1987 the Commission produced a paper on 'Agriculture and the environment' taking up this theme". (European Commission, 1999)

Since the early 90th the new "Common Agricultural Policy" (CAP) was underlying a change of paradigm. Environmental considerations have become a major concern of the CAP. Last important editions of the European Community and therefore object to consider, related to environmental concerns in agricultural policy are:

- European Commission (1999): direction towards sustainable agriculture.
- European Commission (2000): indicators for the integration of environmental concerns into Common Agricultural Policy.
- European Commission (2001): statistical information needed for indicators to monitor the integration of environmental concerns into the Common Agricultural Policy.

2. The influence of agri-environmental programmes (AEP)

Heading for environmental friendly farming the EU has to care for policy measures which influence the decisions of farmers and production methods. In that context AEPs came into power. AEPs are a main possibility to encourage less intensive production, both to reduce market surpluses and to alleviate environmental pressure. In 1992 with the reform of CAP this kind of instruments was generated which had a specific focus on environment. Environmentally friendly production methods, as well as survive and enhancement of endangered traditional livestock breeds and cultivars are main action fields of the AEPs.

Environmentally friendly production methods cover measures like:

- Reduction or renunciation of the use of mineral fertilisers.
- Management of organic manure.
- Reduction or renunciation of the use of pesticides.
- Extension and share of grassland.
- Management of crop rotation to prevent groundwater pollution.
- Cultivation of green cover crops.
- Organic farming.
- Extensive cultivation of field margins.

Survive and enhancement of old agricultural breeds and cultivars are covered by measures like:

- Maintenance and further development of varieties of endangered animal species and rare crops.
- Preservation or improvement of the extent of ecological valuable areas.
- Preservation or improvement of high stem fruit orchards.

With Agenda 2000 it is intended, that farmers should observe a minimum level of environmental practice as part-and-parcel of the support regimes, but that any additional environmental service, beyond the basic level, should be paid for by society through the agri-environment programmes. In all EU member states AEP are in use, but extent and content of the programmes are rather different.

3. STAR-evaluation of 2078/92

First AEPs where implemented under EU-Regulation No. (EEC) 2078/92. These programmes encouraged farmers to carry out environmentally beneficial activities on their land. With the programmes is also intended to contribute to the income of farmers who provide environmental services. Examples of the type of land management activities carried out include:

- Reversion of intensively used land, such as arable or grass for silage to biologically diverse, but unprofitable extensive grassland.
- Reduction in use of nutrients (resulting in loss of yield).
- Reduction or cessation of use of pesticides (e.g. organic farming).
- Creation of nature zones taken out of production.
- Continuation of traditional environmental land management in zones liable to neglect.
- Maintenance of landscape features which are no longer viable in agricultural landscapes.

Programs under Regulation No. 2078/92 are worked out and managed by national and regional authorities. Each programme is subject to approval by the Commission. Because of the diverse site conditions throughout the EU member states, nearly each country has several AEPs running, or at least certain regional modifications of the national programme.

Germany for example has for each of the 16 federal states (Nuts 1) a certain programme, the very same situation we do have in Italy for each region (Nuts 2). To receive a statement about the influence of AEPs, the DGVI Commission Working Document (VI/765598): STATE OF APPLICATION OF REGULATION (EEC) NO. 2078/92: EVALUATION OF AGRI-ENVIRONMENT PROGRAMMES was elaborated. Tables 1 and 2 describe the implementation rate of Reg. 2078/92 in EU member states. Table 1 illustrates the proportion between the number of beneficiaries with the number of holdings which is quiet contrasting. On an average in EU-14 (excluding data from Germany), the number of farms included within programmes is 1 of 7. The new member states Austria (78.2%), Finland (77.2%) and Sweden (63.7%) have an above-average influence by their share. Luxembourg and Portugal are beyond the EU-14 level. Member states like Belgium, Greece, Spain, Italy and Netherlands, with rates of about or less than 7% are significantly below the EU average. German figures cannot be used for the comparison since the data supplied refer to contracts, and not to individual farmers.

The analysis of farmland area under AEPs generally confirms the above analysis of beneficiaries. The proportion of Austria (67.8%), Finland (86.9%) and Sweden (51.6%) lies with more than 50% above the average of EU-15 of about 20%. As well Luxembourg (75.9%), Germany (38.9%), Ireland (24.1%)

Table 1
Number of farms benefiting from Reg. 2078/92[a]

	Number of farms (10^3)	Number of beneficiaries (10^3)	Beneficiaries (%)
Belgium	71	2.0	2.8
Denmark	69	8.0	11.6
Germany	[b]	[b]	[b]
Greece	774	2.4	0.3
Spain	1278	33.9	2.7
France	735	171.0	23.3
Ireland	153	32.2	21.0
Italy	2482	176.3	7.1
Luxembourg	3	1.9	60.3
Netherlands	113	6.7	5.9
Portugal	451	137.9	30.6
UK	235	25.4	10.8
E-11	6363	597.6	9.4
Austria	222	173.4	78.2
Finland	101	77.8	77.2
Sweden	89	56.6	63.7
E-14	6774	905.4	13.4

[a] EEC (1998, modified).
[b] No data available.

Table 2
Area covered by agri-environment agreements under Reg. 2078/92[a]

	Agricultural area (10^3 ha)	Agricultural area covered by 2078 (10^3 ha)	Agricultural area covered by 2078 (%)
Belgium	1375	22.7	1.7
Denmark	2722	107.3	3.9
Germany	17335	6741.0	38.9
Greece	5741	34.8	0.6
Spain	29650	871.1	2.9
France	30170	6901.4	22.9
Ireland	4530	1089.6	24.1
Italy	16792	2291.3	13.6
Luxembourg	127	96.6	75.9
Netherlands	1848	34.5	1.9
Portugal	3960	664.2	16.8
UK	15870	2322.9	14.6
E-11	130121	21177.3	16.3
Austria	3585	2429.0	67.8
Finland	2160	1877.5	86.9
Sweden	3180	1642.2	51.6
E-14	139046	27126.0	19.5

[a] EEC (1998, modified).

and France (22.9%) exceed significantly the average. With less than 2% only, Belgium (1.7%), Greece (0.6%) and Netherlands(1.9%) are clearly below the EU average. Generally, the analysis shows a significant relation between the proportion of used agricultural area (UAA) covered by AEPs and the number of beneficiaries (as a proportion of total number of farms).

4. Good agricultural practice (GAP)

GAP is defined as reference level at least for parts of agricultural practice like pesticide use, fertiliser application and water use, in some cases also for animal husbandry. As such GAP does not cause a deterioration of natural resources under agricultural land use in general. For the purpose of agri-environmental measures, environmental benefits have to be defined as those which exceed the reference level of GAP.

If the society asks farmers to pursue environmental objectives beyond the level of GAP, which would cause income losses, the society must expect to pay for that environmental service. The Polluter-Pays-Principle is the basis of the approach. Accordingly, farmers are responsible for the compliance costs up to the reverence level of GAP, reflected in property rights. AEPs are closely connected to GAP, as already mentioned above, farmers only will be paid for environmental performance, if measures clearly surpass the level of GAP.

On the one hand the very complex relationship between agriculture and environment, marked by harmful and beneficiaries processes as well as great diversity of site conditions and production systems, had conditioned the environmental integration in the context of CAP.

On the other hand the crucial point to understand this relationship, there is the principle of "good farming practice", corresponding to the site conditions and type of farm. That means that farmers have to follow a minimum required standard (national, regional, local) without specific payments. In fact farmers should follow compulsory laws in relation to pesticide use, fertiliser application, water use and where appropriate, national or regional guidelines on good farming practice.

4.1. Code of GAP in the UK

"GAP means a practice that minimises the risk of causing pollution while protecting natural resources and allowing economic agriculture to continue" (MAFF, 1998a–c).

5. Policy need for agri-environmental indicators (AEIs)

As illustrated above a huge change and even substitution of paradigm concerning agricultural policies are perceptible. The integration of environmental concerns as well as the rural development dimension into coherent agricultural policy measures needs to be carefully designed and implemented to ensure the achievement of the goals. Policy measures, particular in this complex field should be well targeted, based on firm foundation of good data and subject to regular monitoring and evaluation.

In this context there is an increasing demand of empirical information to support this policy formulation and analysis. At national and international level the development of AEIs gains growing importance. They contribute to this process by:

- Supplying decision-makers and the general public with relevant information on the current state and trend in the environment as they affect the agricultural sector and rural development.
- Supporting decision-makers get a better grasp of the cause and effect relationships between the choices and practices of farmers and agricultural policy-makers on the one hand and the environment on the other, steering in the right direction any initiatives prompted by changes in the state of the environment.
- Assisting to monitor and assess the effectiveness of measures taken to promote sustainable agriculture.

6. Significance of indicators

Undoubtedly indicators are of growing relevance for the international co-operation. Latest since the Rio Conference in 1992 methods are in great request which enable a quantitative and qualitative environmental observation. The assumption is that an international acknowledged catalogue of indicators would receive an extreme political significance. National political strategies will have to orientate according to such measuring systems. On the international scale the comparability of environmental stress or environmental relief is heading for a real novelty in policy. As well science will have to consider e.g. the methodological development, according to these needs. Politicians will have to argue with a better informed public. The political responsibility of scientists will increase because data and models will be valuated more and more on the basis of indicator systems.

Applications of indicators are to be expected for diverse purposes (Table 3). But indicators would achieve political effectiveness only if they really contribute to policy development what implicates that indicators are powerful enough to shape political goals. Such a strong impact of indicators requires that they give information about success or failure of sustainability (Table 4). But there is a long distance to overcome before all problems occurring with indicators and indicator concepts will get past (Table 5).

Table 3
Application of indicators in the field of policies

Agri-environmental reports
International comparability of environmental concerns
National and global development plans and development strategies
National feed back on international regulations, conventions and environmental initiatives
Evaluation of progress in the achievement of environmental goals

In addition to this general attributes of indicators, the development AEIs require a "systems approach". Such an approach should cover more than one single environmental sphere or theme and should try to integrate the full range of the complex interactions between agriculture, environment and social-economic conditions. In addition a differentiated spatial approach is necessary, able to reflect regional differences. Considering this exhaustive list of requirements for an ideal indicator, it has to be admitted that this does not make their development that easy. A successive and iterative development should be envisaged.

Table 4
Demands on indicators

Agri-environmental indicators should have the following attributes
Scope of indicators
Inform about status and development of complex systems
Provide sufficient information about sustainability of land use systems
Be responsive to changes related to human activities to indicate rapidly success and failure of activities
Able to show trends over time
Work as umbrella indicators summarising different processes and/or environmental impacts
Policy relevance
Provide a representative picture of environmental, agricultural and rural conditions, pressures or society's responses
Be simple and easy to interpret for different users
Provide a basis for regional, national and international comparisons
Be either national in scope or applicable to regional issues of national significance
Assist individual decision-makers of the private sector as well as trade and industry
Analytically sound
Be theoretically well founded in technical and scientific terms
Be based on international standards and international consensus about its validity
Lend itself to being linked to economic models, forecasting and information systems
Measurability and data required
Have to be controllable
Readily available or made available at a reasonable cost/benefit ratio
Adequately documented and of known quality
Updated at regular intervals in accordance with reliable procedures
Have a threshold or reference value against which to compare it, so that users are able to assess the significance of the values associated with it

Table 5
Restrictions in application of indicators

Restrictions in application of indicators
Different availability of data and information on the national and international level
Lack of methodological standards in the international context
Different ranking of indicators on the international level and therefore different legal treatment, which leads to imbalance in the international valuation
Lack of thresholds, basis figures or reference levels, so that often no orientation is given whether trends in the environmental development are strong or weak, e.g. as for biodiversity
Spatial relation is not sufficiently considered. On the international scale the use of average values on the country level is widespreaded, so that regional environmental loads are neglected
Several indicator issues overlap so that an overestimation of single parameters is likely to be expected
Long-term monitoring programmes are necessary to realise long-term environmental changes
Indicator development is dominated by measurable parameters which distracts from those impacts dealing with non-quantifiable values like aesthetical, ethical or cultural values
Lack of models which could bridge information deficits
Application of models without realising the limits and deficits of models under others because of a lack of information
Lack of priority setting of indicators so that a ranking of strategies for the enhancement of sustainable land use of agricultural landscapes is not yet possible on the basis of agri-environmental indicators

7. The field of landscape indicators

Next to the parameters serving as indicators for agri-environmental impacts the spatial dimension revealed to be indispensable to identify the specific region where to place the measuring system and where political responses are necessary. In this context landscapes were cited as the focus of site specific environmental hot spots. Landscapes with their rather uniform biophysical properties and defined as natural spatial units should be the area which is normally completely characterised by specific environmental problems. Above that cultural features of landscapes resulting from a long history of land use were part of the landscape discussion so that even this complex term was introduced as an AEI into the international debate (OECD, 2001).

The following questions therefore have to be clarified:

- What makes the significance of agricultural landscapes?
- How to understand landscapes?
- How to select landscape indicators?

7.1. *Significance of agricultural landscapes*

An overview about landscape development exemplary for the territory of Germany reveals an impression of the interrelationship between agriculture and landscape evolution. Six hundred years after Christ (A.D.) more than 80% of the area of Germany was covered by forests. The increasing population led to cultivation of more than 75% of the land until the 14th century. Plough-land dominated on 70% of the area.

Pestilence killed more than a third of the population until the middle of 15th century. The decline of demand in foodstuff supply was followed by an increase of forests of 30%. Today at about 25% of the German territory are covered by forests. These dynamic landscape changes always based on a change of the agricultural development and were always accompanied by environmental impacts. The type of land use was mainly responsible for several of the heaviest environmental disasters. One of those in the 14th century was characterised by the worst erosion incidents humans have ever passed through in Central Europe. Between 1313 and 1350 all over Germany at about 67 billion tons of soil were eroded. An average of at about 5 cm of the top soil layer disappeared. The cause was the one-sided cultivation of cereals which even in the three-field system was not really supported by the 1 year fallow. The fallow was stressed by too high numbers of livestock and showed hardly a green cover. Poor stands of cereal as well were not able to protect sufficiently the soil. Mineralisation was increased by intensive soil cultivation leading to losses of soil carbon contents. Often landscape illustrations from that period illustrate treeless landscapes where each square meter was intensively used for agricultural production.

An extension of cropping systems by introduction of legumes and potatoes was not worth mentioning before the 18th century. The period with highest diversity of agricultural land use types obviously was established at about 1925 (Bork et al., 1998; Piorr, 2002). Concluding crucial aspects are:

- Landscape development in Germany and Middle Europe was always driven by agricultural land use and the cultivation of land.
- There is a huge heritage coming from the interaction between biophysical features of landscapes and agriculture which together formed cultural landscapes.
- Landscapes were always governed by dynamic processes.
- Landscape change was the reason for heavily environmental impacts.

7.2. Agricultural landscapes in Europe today

Significance of agricultural landscapes for a sustainable development of European landscapes follows from their high acreage (Fig. 1). Less than 5% of the area of the European Union member countries could be labelled as natural landscapes. 77% of the area are dominated by agriculture and forestry. Less than 20% of the area of European countries are exploited by industrial and urban use.

The entire territory of EU member countries covering natural, rural, urban, peri-urban areas and land, inland water, marine areas as well as landscapes with outstanding qualities or everyday characteristics or degraded landscapes is matter of the European Landscape Convention which was signed at Florence,

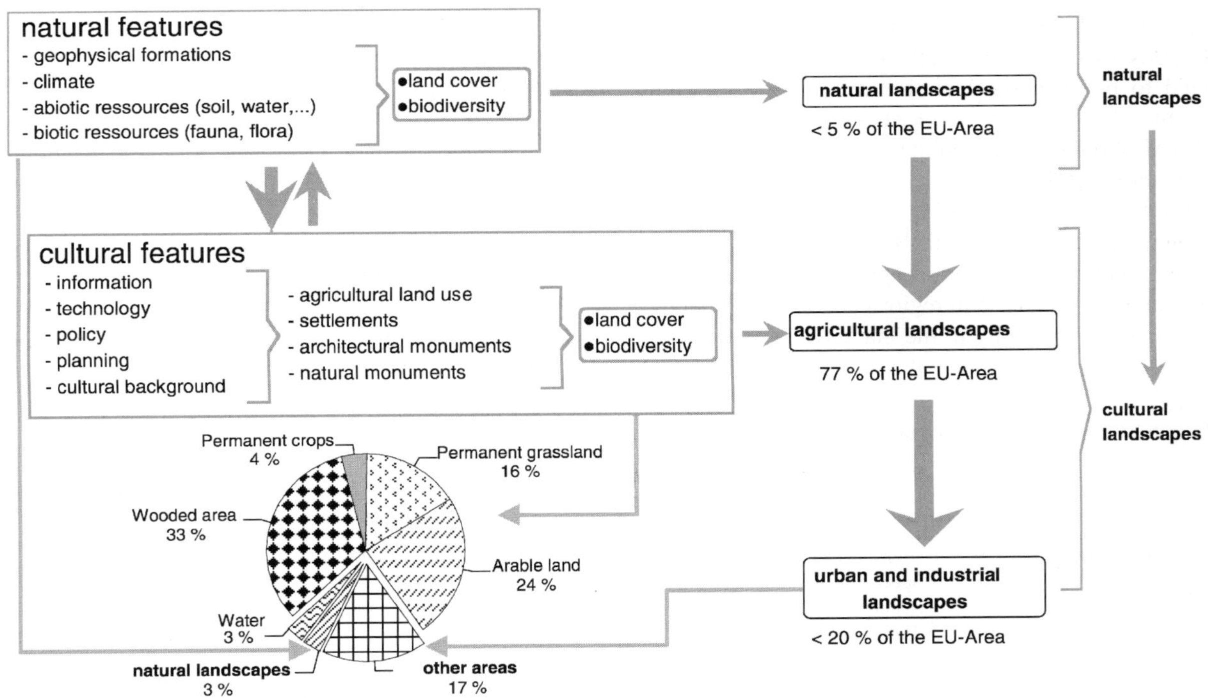

Fig. 1. Shares of natural landscapes, agricultural landscapes (agriculture and forestry) and urban/industrial landscapes in EU member states.

October 2000 (Council of Europe, 2000). The aims of this convention are to promote landscape protection, management and planning, and to organise European co-operation on landscape issues. The preamble of the convention focuses in relation to environmental concerns:

- a balanced and harmonious relationship between social needs, economic activity and the environment;
- ... the public interest role in the cultural, ecological, environmental and social fields ...;
- that landscape "... is a basic component of the European natural and cultural heritage, contributing to human well-being and consolidation of the European identity".

7.3. Dominant landscape changes today

In general the change of the CAP is leading to price cuts, a development which is continued by the Agenda 2000 resolutions. Agriculture therefore is threatened on marginal sites with low soil productivity (Table 6). Experts are talking about 20–30% of the today farmed land to be abandoned in the next decade. In some of the federal states of Germany like Brandenburg and Mecklenburg-Vorpommern or in regions with high amounts of grassland like in hilly and mountainous landscapes more than 65% of the agricultural land would become fallow land (Bork et al., 1995). On the other hand concentration of agriculture and intensification would increase in regions with high productive soils and in proximity to the market.

Landscapes are obviously affected in several respects. Independent of the direction of landscape development whether an intensification or an extensification are an outcome of future trends, environmental impacts have to be taken into consideration (Table 6). Both situations demand a specific reaction which means agricultural policy has to find tools to influence the changes on the landscape level. Therefore different works have been done contributing information about landscapes and landscape relevant topics like the linkages to biodiversity and habitat questions.

7.4. Landscape—mystic view or scientific approach

7.4.1. Background of the discussion

Landscapes integrate a multitude of properties, they are object of a huge number of organisms including those changes evolved by human activities. Hence they could be seen as individuals formed according a long history of evolution. It does not wonder that such a complex body was admired primarily by poets and painters (Eberle, 1979; Fischer, 1985; Hard, 1970, 1985; Vos, 2000). Landscapes in the perception of artists are often to conceive as individuals. This individuality derives from the unique of the coincidence of geo-morphological properties, site specific biotic fittings and the creative anthropogenic formation leading to their social and economic configuration. Thus landscape individuality is organised in a form that no pendant could be found globally.

Landscapes provide the framework for regional features, which is connected with the emotional relation

Table 6
Impacts of trends of the agricultural development in different agricultural landscapes

Impacts	Production potential of landscape	
	Landscape with high agricultural productivity	Landscape with low agricultural productivity
Production process	High intensity of production Specialisation Reduction of crop diversity Expansion of field sizes	Withdrawal of agricultural production
Landscape development	Loss of cultural features Decrease of aesthetical amenities	(a) Afforestation (b) Natural succession
Environmental impact	Loss of biodiversity	(a) Loss of biodiversity, decrease of environmental damages
	Increase of abiotic environmental damages	(b) Same like (a) but first increase of biodiversity during transition into forest

of their inhabitants. Often this relationship is accompanied by regional typical cultivation of traditions, such as dialect, architecture, regional costumes and regional song-cycles.

The term regio derives from the Latin and is equivalent in true sense to the word settlement area but was translated during the Middle Age with landscape (Hard, 1970). A relation is given to modern terms like region and regionalism which were introduced since stronger efforts are undertaken for integration of regional interests into policy. The significance of these terms is given with the relation to regional specific matter cycles, regional ecosystems and regional founded cultural, social and economic characteristics. This correspondence with landscape specific characteristics stands to reason to use the terms landscape and region as synonym in the European context.

Here we find the source for the identification of people with their landscape and home. This is the reason why landscapes and their destruction are increasingly realised. Deficits are the origin of the landscape discussion. In this context investigations have to focus on a "positive" definition of landscapes, to give advice on those characteristics of landscapes which should maintain for future generations. Especially the fusion of European nations is filled with fears because of the risk of a decrease of regional individualism and identity. This fear calls for a Europe of regions and thus a Europe of landscapes.

7.4.2. Definition of landscape

Diverse approaches were made to define landscapes which basically seem to consist of a conglomerate of physical, biological and cultural elements. Beginning with the aesthetic approach widely spreaded in the English-speaking world, Haber (1995) could be cited with the definition: "Landscape is an extensive area of sceneries viewed from a single place". A similar approach was given by Steiner (1991): "Landscape is all the natural features such as fields, hills, forests and water that distinguish one part of the earth from another part. Usually, a landscape is that portion of land or territory which the eye can comprehend in a single view, including all its natural characteristics".

The European Landscape Convention (Council of Europe, 2000) says: "Landscape means an area as perceived by people, whose character is the result of the action and interaction of natural and/or human factors". Wascher et al. (1999) widened this proposal by an extensive definition: "Landscapes are the concrete and characteristic products of the interaction between human societies and culture with the natural environment. As such, landscapes can be identified as spatial units where region-specific elements and processes reflect natural and cultural goods or history in a visible, spiritual and partly measurable way. Because the underlying human and natural processes are subject to change and evolution, landscapes are dynamic systems". Accepting these different views the short term definition reads:

> Landscapes are individual spatial units where typical soil, relief and climatic conditions provided the environment to be formed by natural processes or by both natural processes and management processes by man. As such, landscapes reflect the interaction of dynamic natural and human impacts.

Heading for a structured overview about landscapes several approaches have to be considered. In the context of a framework for agricultural landscapes natural landscapes and urban and industrial landscapes can be excluded. In between these both extreme different landscapes which are characterised by a decrease of naturalness, agricultural landscapes are positioned (Fig. 2). Agricultural landscapes are the carrier of three items including:

- *structures* of landscapes such as their natural physical, environmental land use and man-made features;
- *functions* such as their environmental buffer function, function as a place to live or to recreate or their economic function;
- *values* concerning values society places on agricultural landscape such as cultural values or the costs for farmers in their maintenance.

This mixture of properties of agricultural landscapes constitutes the basis for processes which were grouped by Blom (1999) in to three categories:

- Expansion—withdrawal, e.g. change in area of agricultural land.
- Intensification—extensification, e.g. proportion of arable land with high intensity crops.
- Concentration—marginalisation, e.g. proportion of holdings with an acreage of arable land which exceeds the national or regional average.

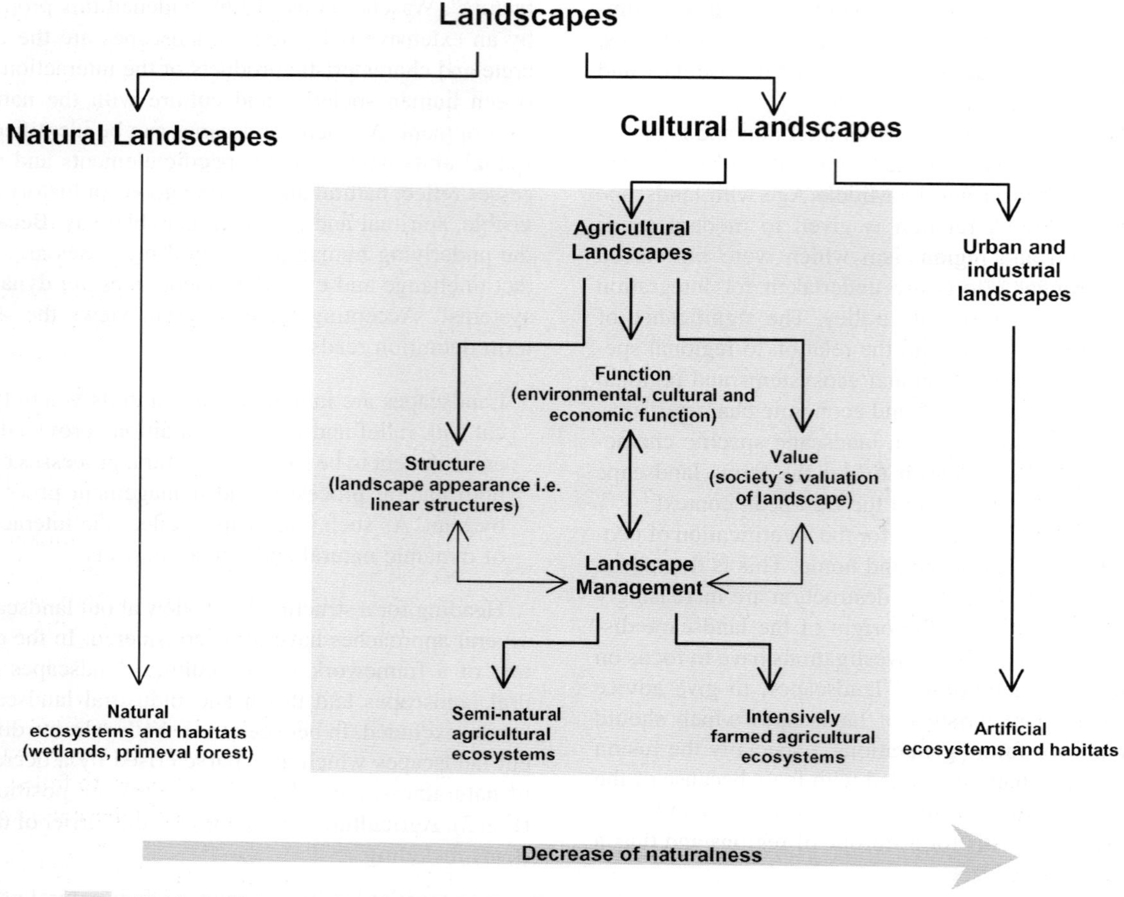

Fig. 2. Natural and cultural landscapes in agricultural context (modified according to OECD and Piorr cited in OECD, 2001).

These processes are linked together so that intensification and concentration are interrelated with marginalisation and withdrawal of agriculture in other areas. They could be applied on all scale levels and therefore be used on the field, farm, regional and national level. The impact of these processes on the change of landscape is evident and can be assessed by different indicators.

Landscape properties, landscape use and the accompanying processes lead to more or less intensively agricultural land use which are characterised by semi-natural agricultural ecosystems or intensively farmed agricultural ecosystems. Landscape characteristics and processes create a spatial matrix of dynamic land use types which is linked to the presence of habitats and biotopes. Therefore biodiversity is directly related to landscape characteristics (Piorr et al., 1998).

7.5. Landscape indicators

7.5.1. OECD landscape indicators

One of the most acknowledged indicator concepts was launched in the OECD Reports: Environmental Indicators for Agriculture, vol. 1: Concepts and Framework (OECD, 1997) and vol. 2: Issues and Design (OECD, 1999). The choice of indicators was adjusted to the driving force-state-response (DSR) framework which identifies (Table 7):

Table 7
OECD landscape indicators (OECD, 2001)

Physical appearance and structure of landscape
Physical elements, environmental features and land use patterns
Man-made objects (cultural features)
Landscape typologies
Landscape management
Landscape costs and benefits (values)

- Driving force indicators, focusing on the causes of change in environmental conditions in agriculture, such as changes in farm financial resources and pesticide use.
- State indicators, highlighting the effects of agriculture on the environment, for example, impacts on soil, water, and biodiversity.
- Response indicators covering the actions taken to respond to the changes in the state of the environment, such as changes in agri-environmental research expenditure.

Changes in the *structure of agricultural landscapes* are capable to show trends towards an increasing homogenisation of landscape structures, closely related to structural changes and intensification of agriculture. The process towards landscape homogeneity could be slowing or in reverse in some cases, as was proved in some countries. Since the late 1980s some countries have adopted measures to monitor and maintain agricultural landscapes, such as tracking trends in cultural landscape features, and providing payments for restoration of typical landscape features. Some countries are also establishing landscape typologies to provide a framework and reference base to assess landscapes, by combining information on the structure of landscapes (OECD, 2001; Blom, 2001; Blom and Ihse, 2001; Countryside Survey, 2000; Fjellstad et al., 2001; NIJOS, 2001; Slak and Lee, 2001).

Publicly funded landscape management schemes concern expenditures for enhancing biodiversity, habitat and landscape conservation. A lot of examples could be given for the support of cultural features in the local context. Development of rural areas often includes public access requirements in landscape schemes.

The *value society places on landscape* was regionally investigated by public opinion surveys. Different studies reveal that agricultural landscapes are highly valued in many cases. Significant results are that landscapes with greater heterogeneity and 'traditional' elements are given a higher value over more uniform and newer landscapes, while landscapes perceived as overcrowded have a lower value (OECD, 2001).

7.5.2. EU landscape indicators

Earlier before the OECD several European countries had started activities in setting up landscape conservation schemes and landscape indicators to follow the development of landscapes. Primarily these initiatives were dedicated for planning purposes and the delimitation of nature conservation areas. Table 8 shows an overview about landscape monitoring in EU member countries.

Based on national and regional experiences the European Commission took up the proposals of the

Table 8
Landscape monitoring in EU member countries

Country	Landscape survey	Source
UK	Countryside Survey	Countryside Survey (2000)
Norway	Tilstandsovervåkning og resultatkontroll I jordbrukets kulturlandskap	Fjellstad et al. (2001), NIJOS (2001)
Sweden	Landscape inventory and monitoring (LiM)	Swedish Environmental Protection Agency (1998), Blom and Ihse (2001)
Germany	Ecological area sampling	Hoffmann-Kroll et al. (1998), Statistisches Bundesamt and Bundesamt für Naturschutz (2000)
Finland	Monitoring of the visual agricultural landscape	MYTAS (2000)
Denmark	Danish monitoring system for small biotopes	Brandt et al. (2001)
France	TerUti	Slak and Lee (2001), Slak et al. (2001)
Austria	Kulturlandschaftsforschungprogramm	Wrbka (2000), Wrbka et al. (1997)
Italy	Regional landscape monitoring	Fais (1997)

OECD and suggested more concrete indicators in an OECD Expert Meeting, May 1999, in Paris (Morard et al., 1999). The approach included several goals:

- The existing determining links between biodiversity, natural habitats and landscapes should be considered.
- Representative landscape indicators should be selected for which data are available.
- Landscape indicators will need a certain flexibility in order to be able to be adapted to the very different agri-environmental conditions within and between the EU member countries.
- Landscape indicators will require values of reference or threshold values in order to compare the development of these indicators at a later stage.

Following these premises a first proposal from the view of the EU member states for the monitoring of landscapes was launched (Table 9). In this proposal the biophysical characteristics refer to abiotic (e.g. geology, soil) and biotic (e.g. vegetation type, habitat) properties of a landscape. The close interaction between these elements determine the specific natural character and the natural potential of land. The cultural characteristics of landscapes refer to the anthropogenetic impact on landscapes to consider the cultural heritage which is among others essential for the cultural identity of local people. The land management functions in particular designate the public and private initiatives aiming at preserving the quality of landscapes.

High emphasis in the EU concept was given to data availability. A huge amount of statistical data related to administrative units are available at European level on Nuts 2 and 3, referring e.g. in Germany to Regierungsbezirke or rural districts. The use of this information for landscape issues is in certain cases restricted because administrative areas do not necessarily coincide with landscape boundaries. Geographical data refer to georeferenced information with high spatial precision. A good example for the EU territory is the Corine Land Cover (CLC). Another common tool used for the generation of such geographical data are remote sensing images and aerial photographs.

There are certain restrictions given concerning the data availability. Commonly statistical data informing about biophysical characteristics, land cover and land use are provided by mapped data about geological formations, ecosystems and habitat types, long time series about land use and land cover, etc. However in specific analysis of landscapes, these data are of limited significance particularly when specific spatial information is required. Land cover information retrieved from satellite images or aerial photographs are not interpreted in all necessary details or linkages. As well not nearly as much as needed the European territory data are evaluated. Another problem appears with the lack of standardised treatment and analysis method of these images. No complete and exhaustive inventory on cultural features exists at European level. Even at national or local level inventories of relevant cultural features as typical landscape elements have to be conducted respecting national preferences. Data about areas under public or private commitment to maintain or enhance landscape seem to be available because of its high policy relevance, the increasing significance of already or potential areas under certain environmental commitments or protection status.

7.5.3. PAIS landscape indicators

A recent initiative of the European Community is the project proposal on agri-environmental indicators (PAISs). The objective of the PAIS project is to contribute to the efforts of the European Commission and to propose a core set of indicators within the domain of landscapes, rural development and agricultural practice, which are applicable in a meaningful way at EU level. These three themes require special attention, due to the fact that, compared to other indicator issues, these are still at a basic conceptual development stage and lack appropriate statistical data.

Five working steps led to a list of feasible landscape indicators applicable at European level, based on a comprehensive inventory of existing landscape indicators, used in national and regional surveys throughout Europe and their successive analysis (Eiden, 2001):

- Information acquisition.
- Development of a landscape indicator classification scheme.
- Creation of a landscape indicator assessment/evaluation scheme.
- Establishment of a landscape indicator data base.
- Final compilation of landscape indicator set.

A classification of the landscape indicators in this concept was gained by grouping the indicators on three

Table 9
Indicator issues and measurable parameters for landscape indicators on EU level (Morard et al., 1999)

Issue		Meaning/significance	Attributes/variables	Examples for indicators
Land characteristics	Natural biophysical features	Enables elaboration of a landscape classification system	Soil type, landform (slope, elevation) climate (temperature, rainfall), water bodies	
	Environmental appearance	Landscape pattern and structure	Georeferenced land cover	Landscape metrics (patch density, edge density, Shannon index, etc.)
		Visual and aesthetic value		To be developed on long-term basis: perception, frequency of visitors, etc.
	Land type features	Land cover/land use change	Georeferenced land cover/land use flow matrices	Expansion/withdrawal
				Intensification/extensification Concentration/marginalisation
Cultural features		Cultural identity, regional specific character, cultural assets, etc.	Inventory of cultural landscape features: architectural, historic, hedgerows, stone walls, etc.	Number and status of point features, length of linear features, surface of aerial features, share of regional specific land use patterns
Management functions		Landscape protection areas, nature conservation areas	Area estimates	Area under commitment related to total agricultural area

Table 10
Classification scheme for landscape indicators (Eiden, 2001)

Landscape dimension	Thematic indicator group	Indicator item
Landscape features	Landscape composition, i.e. LC/LU components comprising the landscape, contextual information	Stock and change of broad land cover categories Stock and flow land cover/land use matrices
	Landscape configuration, i.e. structural arrangement of landscape elements	Fragmentation Diversity Edges Shape
	Natural landscape features (state and change)	Stock and change of biotopes and habitats Hemerobie (naturalness) Habitat/biotope fragmentation Habitat/biotope diversity Habitat/biotope quality
	Historical-cultural landscape features (state and change) statistics/inventories	Point features Linear features Area features
	Present—cultural landscape features (state and change)	Point features Linear features Area features
Human perception	Visual and aesthetic value	–
Landscape management, conservation and protection schemes	Cultural landscape protection/conservation Nature conservation/protection	– –

different levels (Table 10). Level one reflects the "landscape dimension":

- Landscape features: as the elements composing a landscape and which can be described (biophysical objects).
- Human perception: indicators dealing with the perception of landscape by different stakeholders.
- Landscape management and conservation: indicators reflecting landscapes as a subject for management, planning, conservation or protection.

The second level describes thematic groups where particularly the landscape features and landscape management schemes are taken into account:

- Formal landscape features.
- Natural landscape features.
- Cultural-historical landscape features.
- Present anthropogenic landscape features.
- Cultural landscape protection/conservation.
- Nature conservation/protection.

On the third level a further differentiation is done in order to class the indicators thematically in terms of indicator items.

Thirty six landscape indicators were found out to be the most relevant in EU member countries, guaranteeing the best chances to gain in the nearer future the necessary data to inform these indicators. Focussing on the relevance of landscape indicators concerning biodiversity no solution can be given at the moment. Table 11 shows the overview of those landscape indicators extracted from this list and related more or less to biodiversity. Accordingly 26 of 36 landscape indicators provide relevant information about habitat and biotope properties of landscapes. For future working steps a ranking about the significance of landscape indicators directly related to biodiversity could give an indication where to find the most relevant landscape characteristics for nature protection purposes. Obviously the PAIS project made a big step forward evaluating selection criteria on the landscape level, the problem about data availability

Table 11
Landscape indicators related to biodiversity (Eiden, 2001, modified)

Landscape domain	Indicator theme	Indicator
Formal landscape features	Landscape composition	Stock and change of UAA Stock and change of arable land Stock and change of grassland Stock and change of forest areas Stock and change of semi-natural and natural land Stock and change of built up areas Extensification rate Intensification rate Afforestation rate
	Landscape configuration	Diversity indices Patch shape of agricultural parcels Length and distribution of different edges Fragmentation indices
	Natural landscape features	Stock and change of broad, semi-natural and natural habitats/biotopes Stock and change of valuable biotopes and habitats in agricultural landscapes (area features) managed by farmers Stock and change of linear habitats and biotopes in agricultural landscapes Stock and change of point habitats and biotopes in agricultural landscapes
	Historical-cultural landscape features	Stock and change of historical—cultural landscape area features Stock and change of historical—cultural landscape linear features Stock and change of historical—cultural landscape point features
Landscape management conservation and protection		Change in the percentage of financial expenditure of agri-environmental schemes (per hectare or per farmer involved) Share of area covered by agri-environmental schemes from total UAA Area under specific farming or management practices aiming at landscape conservation (traditional agricultural land use practices) Length of "green" linear landscape features maintained and/or restored by farmers Number of farmers participating in training programmes concerned environmental friendly management practices, landscape conservation, etc. UAA within protected sites (according to IUCN categories)

and the interrelationship between different indicator issues.

7.6. Conclusions

Evidently the change in agricultural policy in the EU as well as globally demand new efforts on the field of monitoring systems which are capable to contribute to an information framework which is necessary for the development of policies. On behalf of scientists and experts indicator systems have to be completed to provide internationally approved data which inform about trends in the environmental change and indicate fields where decision-makers have to react. Existing indicator proposals which are all to be seen in the line of a longer development chain on the international level give rise to inference that policy supporting indicator systems are available in the nearer future.

Even in the field of landscape indicators as one of the most complex and complicated topics progresses can no longer be ignored. First of all the significance of landscapes meanwhile is internationally acknowledged as a field of policy. Landscapes are the carrier of a multitude of characteristics because they are the melting pot of natural elements forming since centuries together with human activities cultural landscapes which dominate the face of Europe. Difficulties to find classification systems and indicators for a standardised European wide description and evaluation of the characteristics of landscapes are identified. New approaches to overcome these difficulties are seizable last but not least because of the rapid progress of different types of Geographical Information Systems. On the other hand an increasing challenge is growing with the demand of a better understanding of interrelations between diverse indicator areas as here discussed exemplary for landscape indicators and biodiversity indicators. Sustainable land use is dealing with a high complex system which not only affords an innovative policy framework but as well still before a higher input of scientific knowledge.

References

Blom, G., 1999. Landscape processes. Contribution to the OEDC York Conference, UK. Cited in OECD (1999): Environmental Indicators for Agriculture, vol. 3: Methods and Results. Publications Service, OECD, Paris, pp. 375–376.

Blom, G., 2001. Evaluation and monitoring environmental goals of the national CAP. Unofficial Working Document, 26 pp.

Blom, G., Ihse, M., 2001. The Swedish LiM project. In: Nordic Council of Ministers (2001): Strategic Landscape Monitoring for the Nordic Countries. Temanord Environment 523, Copenhagen, pp. 20–30.

Bork, H.-R., Dalchow, C., Kächele, H., Piorr, H.-P., Wenkel, K.-O., 1995. Agrarlandschaftswandel in Nordost-Deutschland. Ernst und Sohn Verlag, Berlin.

Bork, H.-R., Bork, H., Dalchow, C., Faust, B., Piorr, H.-P., Schatz, T., 1998. Landschaftsentwicklung in Mitteleuropa. Klett-Perthes, Stuttgart.

Brandt, J., Holmes, E., Agger, P., 2001. Integrated monitoring on a landscape scale—lessons from Denmark. In: Strategic Landscape Monitoring for the Nordic Countries. Temanord Environment 523, Copenhagen, pp. 31–41.

Council of Europe, 2000. European Landscape Convention, ETS No. 176. http://www.stichtingncm.nl/pdf/Convention.PDF.

Countryside Survey, 2000. Department of the Environment, Transport and the Regions, London, UK.

Eberle, M., 1979. Individuum und Landschaft. Zur Entstehung und Entwicklung der Landschaftsmalerei. Anabas Verlag, Gießen.

EEC, 1998. State of Application of Regulation (EEC) No. 2078/92. Evaluation of Agri-Environment Programmes, DGVI Commission Working Document, Brussels.

Eiden, G., 2001. Landscape indicators. In: Eiden, G., Bryden, J., Piorr, H.-P. (Eds.), Proposal on Agri-Environmental Indicators (PAIS). Final Report of the PAIS Project, EUROSTAT, Luxembourg, pp. 4–92.

European Commission, 1999. Directions towards sustainable agriculture. Communication to the Council, the European Parliament, the Economic and Social Committee and the Committee of the Regions, COM (1999) 22 Final, January 1, 1999. http://europa.eu.int/comm/agriculture/envir/9922/9922_en.pdf.

European Commission, 2000. Indicators for the integration of environmental concerns into the Common Agricultural Policy. Communication to the Council, the European Parliament, COM 2000 (20) Final, January 26, 2000. http://europa.eu.int/comm/agriculture/envir/com20/20_en.pdf.

European Commission, 2001. Statistical information needed for the indicators to monitor the integration of environmental concerns into the Common Agricultural Policy. Communication to the Council, the European Parliament, COM 2001 (144) Final. http://europa.eu.int/eur-lex/en/com/cnc/2001/com2001_0144en01.pdf.

Fais, A., 1997. Landscape indicators. A short review of the Italian experience. Document Presented at the EUROSTAT Meeting on Landscape Indicators. Land Use Statistical Working Group, EUROSTAT, Luxembourg.

Fischer, H., 1985. Naturwahrnehmung im Mittelalter und Neuzeit. Landschaft und Stadt 17, 97–110.

Fjellstad, W., Mathiesen, H., Stockland, J., 2001. Monitoring Norwegian agricultural landscapes—the 3Q programme. In: Nordic Council of Ministers (2001): Strategic Landscape Monitoring for the Nordic Countries. Temanord Environment 523, Copenhagen, pp. 19–28.

Haber, W., 1995. Landschaft. In: Akademie für Raumforschung und Landesplanung (Hrsg.): Handwörterbuch der Raumordnung, Hannover, pp. 597–602.

Hard, G., 1970. Die "Landschaft" der Sprache und die "Landschaft" der Geographen. Semantische und forschungslogische Studien zu einigen zentralen Denkfiguren in der deutschen geographischen Literatur. Colloquium Geographicum 11, Geographisches Institut der Universität Bonn, Dümmler Verlag Bonn.

Hard, G., 1985. Die Landschaft des Künstlers und die des Geographen. In: Hoffmann, D. (Hrsg.): Landschaftsbilder, Landschaftswahrnehmung, Landschaft. Die Rolle der Kunst in der Geschichte der Wahrnehmung unserer Landschaft. Evangelische Akademie Loccum, Loccumer Protokolle 3, pp. 122–139.

Hoffmann-Kroll, R., Schäfer, D., Seibel, S., 1998. Biodiversität und Statistik-Ergebnisse des Pilotprojektes zur Ökologischen

Flächenstichprobe. Wirtschaft und Statistik, January 1998, pp. 60–75.

MAFF, 1998a. Code of good agricultural practice for the protection of soil. UK Ministries of Agriculture, Food and Fisheries.

MAFF, 1998b. Code of good agricultural practice for the protection of water. UK Ministries of Agriculture, Food and Fisheries.

MAFF, 1998c. Code of good agricultural practice for the protection of air. UK Ministries of Agriculture, Food and Fisheries.

Morard, V., Vidal, C., Eiden, G., Lucas, S., Piorr, H.-P., Stott, A., Blom, G., Fjellstad, W., Fais, A., 1999. Landscape indicators. OECD-Room Document No. 3, OECD Expert Meeting on Biodiversity. Wildlife Habitat and Landscape, Paris, May 1999.

MYTAS, 2000. Finnish agri-environmental programme. Ministry of Agriculture, Forestry and Fishery, Helsinki.

NIJOS, 2001. Tilstandsovervåkning og resultatkontroll I jordbrukets kulturlandskap—Rapport for prosjektåtet 1999–2000—Hedland og Oppland. Norwegian Institute of Land Inventory, Ås, Norway.

OECD, 1997. Environmental Indicators for Agriculture, vol. 1: Concepts and Framework. Publications Service, OECD, Paris.

OECD, 1999. Environmental Indicators for Agriculture, vol. 2: Issues and Design—The York Workshop. Publications Service, OECD, Paris.

OECD, 2001. Environmental Indicators for Agriculture, vol. 3: Methods and Results. Publications Service, OECD, Paris.

Piorr, H.-P., 2002. Entwicklung der pflanzenbaulichen Flächennutzung in den Agrarlandschaften Deutschlands. Schriftenreihe des Bundesministeriums für Verbraucherschutz, Ernährung und Landwirtschaft Reihe A: Angewandte Wissenschaft Heft 494, Biologische Vielfalt mit der Land- und Forstwirtschaft, pp. 127–135.

Piorr, H.-P., Glemnitz, M., Kächele, H., Mirschel, W., Stachow, U., 1998. Impacts of the European union reform policy after 1996 in north-east Germany: landscape change and wildlife conservation. In: Dabbert, S., Dubgaard, A., Slangen, L., Withby, M. (Eds.), The Economics of Landscape and Wildlife Conservation. CAB International, Oxon, pp. 217–231.

Slak, M.F., Lee, A., 2001. L'agriculture s'intensifie, l'urbain s'étend... et l'occupation du territoire enregistre les transformations à l'œuvre, vol. 1. Agreste Cahiers, March 2001.

Slak, M.F., Lee, A., Michel, P., 2001. L'évolution des structures d'occupation du sol vue par TerUti, vol. 1. Agreste Cahiers, March 2001.

Statistisches Bundesamt, Bundesamt für Naturschutz, 2000. Konzepte und Methoden zur Ökologischen Flächenstichprobe. Angewandte Landschaftsökologie, Heft 33.

Steiner, F., 1991. The Living Landscape: An Ecological Approach to Landscape Planning. McGraw-Hill, New York.

Swedish Environmental Protection Agency, 1998. A Swedish Countryside Survey. Monitoring landscape features, biodiversity and cultural heritage, Stockholm.

Vos, W., 2000. A history of European landscape painting. In: Klijn, J., Vos, W. (Eds.), From Landscape Ecology to Landscape Science. Kluwer Academic Publisher, Dordrecht, pp. 81–96.

Wascher, D.M., Piorr, H.-P., Kreisel-Fonck, A., 1999. Agricultural landscape. Discussion Paper Presented at the OECD Workshop on Agri-Environmental Indicators, York, UK, September 22–25, 1998, OECD Document, COM/AGR/CA/ENV/EPOC (98) 81.

Wrbka, T., 2000. Presentation of landscape level monitoring in Austria. Workshop on Integration of Partial Coverage and Full Coverage Landscape Monitoring Information, October 23–24, 2000, Tune, Denmark.

Wrbka, T., Szerencsits, E., Reiter, K., 1997. Classification of Austrian cultural landscapes—implications for nature conservation and sustainable development. In: Miklos, L. (Ed.), Sustainable Cultural Landscapes in the Danube-Carpathian Region, Proceedings of the Second International Conference on Culture and Environment, Banska Stiavnica, Sk, pp. 31–41.

PART III
BIODIVERSITY AT DIFFERENT SCALE LEVELS

Chapter 3 Biodiversity and Habitat

W. Büchs, H. Albrecht, R. Waldhardt

Synoptic Introductions

ELSEVIER

Available online at www.sciencedirect.com

Agriculture, Ecosystems and Environment 98 (2003) 35–78

Agriculture Ecosystems & Environment

www.elsevier.com/locate/agee

Biodiversity and agri-environmental indicators—general scopes and skills with special reference to the habitat level

Wolfgang Büchs*

Federal Biological Research Centre for Agriculture and Forestry, Institute for Plant Protection in Field Crops and Grassland, Messeweg 11/12, DE-38104 Braunschweig, Germany

Abstract

This synoptic review has the intention to summarise and highlight results in respect to the area "biodiversity and habitat" considering the current "state of the art" with regard the development of biotic indicator approaches that refer to the fauna of agro-ecosystems.

Most political statements (e.g. "Convention of Biodiversity Development": Agenda 21, Convention of Rio 1992) as well as existing approaches regarding the development of biotic agri-environmental indicators focus mainly and almost exclusively on "biodiversity", the enhancement of which is the overall target for the development of agricultural landscapes towards sustainability. In this regard the understanding of the term "biodiversity" is rather different: it is largely interpreted as species richness, only occasionally as the richness of varieties, cultivars or genetical expressions (e.g. microorganisms).

A survey on the understanding of biotic indicators is presented including aspects of nomenclature, categorisation and definitions as well as preconditions and rules for their use. Requirements for different taxa (mostly invertebrates) to act as biotic indicators are summarised and several attempts to use animals on the population and/or community level as biotic indicators for biodiversity or other goals in agro-ecosystems are discussed critically as well as their replacement by surrogate indicators. Regarding the sensitivity of indicators it will be highlighted that it is necessary to evaluate and to compare cultivation intensities, that are rather similar (e.g. gradual differences in conventional managed farms and landscapes) as well as those displaying much larger differences (e.g. organic farming versus conventional farming), because conventional farms currently cover nearly 90% of the agriculturally used area in Europe. Other fundamental problems regarding the development of biotic indicators as baselines, sample size, frequency of surveys, mutual neutralisation of indicators, double assessments, selection of taxa, etc. are addressed. Finally an assessment of the practicality, the power of indication statements and the remaining work required to validate indicators will be provided for discussion, as well as suggestions for a simplification of indicator systems in order to minimise the input needed for data recording.

Keywords: Biodiversity assessment; Biotic indicator; Cultivated area; Agriculture; Animals

1. Introduction—"biodiversity": a term made for confusion?

Already in the 1970s in central Europe various societies for nature protection (e.g. World Wildlife Fund (WWF), Bund für Umwelt und Natuschutz Deutschland (BUND), Naturschutzbund (NABU)) as well as departments for environmental protection in some federal states used the slogan "species richness means quality of life" to stand for the conservation and protection of the diversity of life forms (species, varieties, genetical resources), especially in rural landscapes. The term "biodiversity", however, became

* Tel.: +49-531-299-4506; fax: +49-531-299-3008.
E-mail address: w.buechs@bba.de (W. Büchs).

0167-8809/$ – see front matter
doi:10.1016/S0167-8809(03)00070-7

more public only after the signing of the "Convention for Biodiversity" (e.g. UNEP-Conference Nairobi, Kenya, May 1992; "Earth Summit" Rio de Janeiro, Brazil, June 1992) by 168 countries.

Today "biodiversity" is a term familiar to many: hardly any research programme with an ecological intention is able to manage without using the term "biodiversity". Similar to the term "ecology" which was coined more than 30 years ago, the term "biodiversity" has also been picked up by several groups of society in central Europe with completely different goals. Examples are as follows:

- Call for a year of "International Biodiversity Observation 2001–2002", an initiative by Diversitas, an international consortium of environmental protection.
- Founding of an international journal "Biodiversity and Conservation" by the Ulmer-Verlag, Stuttgart, Germany.
- Call for Research & Development proposals by the European Commission within the fifth framework-programme "Biodiversity, Climate and Global Change" (Direction Generale VI, Brussels, Belgium).
- "Functional Importance of Biodiversity" as a focus of the international congress of the Gesellschaft für Ökologie (Society of Ecology) in Basel, Switzerland, August 2001.
- "Clearing House Mechanism—Biodiversity" by the Federal Office for Nature Conservation in Germany.
- Sounding out "Management of Biodiversity" within the Framework "Socio-Ecological Research" of the Federal Ministry of Education and Research in 2001.
- "Species Richness in Agro-Ecosystems" as a focus for a meeting of the working group of the senate of the research institutions of the Federal Ministry for Consumer Protection, Food and Agriculture.

This list could be extensively continued.

While in ecological sciences the use of the term "diversity" as a precursor of the actual term "biodiversity" has a scientific background and is fixed to clearly defined rules which have been derived from information theories (Shannon, 1948; Wiener, 1948; Shannon and Weaver, 1963), "secondary users" such as politicians and other groups of society or other branches of science that do not belong to ecological sciences treat the term "biodiversity" in a very individual and sometimes confusing manner.

Following scientific understanding, (bio-)diversity consists of two components, the diversity component and the expression of dominance structure (frequency and percentage of each element within the whole subset considered). So, the same level of biodiversity can be achieved by a considerable richness of (different) elements or by less richness but a balanced frequency of each element. An exact interpretation is possible using the term "evenness" (probability of selecting a certain element taken from a whole subset; Stugren, 1978).

However, in common use (and mainly by "secondary users") the "frequency"-component and its interpretation by the mathematically clearly defined term "evenness" is mostly neglected, so that the term "biodiversity" is actually very often used to express in an almost diffuse sense the number of different elements (mostly species) within a subset (in many cases also not clearly defined) underlayed by a "the-more-the-better"-interpretation.

More recent interpretations of the term "biodiversity" are not only restricted to "species richness", but are also related to varieties, races, life forms and genotypes as well as landscape units, habitat types, structural elements (e.g. shrubs, stonewalls, hedgerows, ponds), crop or land use diversity, etc. Finally, the term (bio-)diversity is used in areas with only a very indirect relation to the biological component of biodiversity (e.g. diversity of professions, building styles or types of cars in a defined region or community).

Hence, the generic term "biodiversity" forms an hierarchic system relevant for different scale levels (population, species, biocoenosis, habitat, landscape), compositions (sum of elements of a genome, a population, a species community, an ecosystem or a landscape) and functions (processes that run in different scale levels within the hierarchic system mentioned above) as expressed by Noss (1990).

The quality and/or quantity of a component of a higher (scale or hierarchic) level has a direct effect on the quality and/or quantity of components in lower (scale or hierarchic) levels (Waldhardt and Otte, 2000). For example, a change of the landscape pattern (*structural component*) as a result of a change in land use (*functional component*) affects the species composition (*compositional component*) and finally, processes running in ecosystems (*functional component*). However, the interrelations between scale (hierarchic) levels can also be the other way round, as so called "ecosystem engineers" (Jones and Lawton, 1995) may influence structures and processes in ecosystems (e.g. wild deer and big game by grazing and destruction of tree bark, the beaver by stemming the flow of streams and rivers or finally birds by distributing seeds and insects by pollination).

This background explains:

- that "biodiversity" is based upon many interlinked mechanisms which depend on the heterogeneity or "richness" of their elements in the same way as on the development of functional processes;
- that the knowledge of rules within and between the components and the hierarchic scale levels is a basic requirement for a sound interpretation of the data recorded, and for the development of advanced concepts on biodiversity management.

Particularly in applied research, functional aspects of biodiversity and consequently also structural components of agricultural ecosystems are increasingly important. Although there is no doubt about the ethical justifications of maintaining and recovering biodiversity (e.g. introduction of structural and not cultivated landscape elements in the marginal areas of cultivated fields), only little is known about whether a certain (higher or lower) biodiversity level or different dominance structures of species communities do influence the functionality of food webs, or whether the quality (and quantity) of the biological control (by natural enemies) of pest organisms in agro-ecosystems is affected (e.g. Scheu, 1999, 2001). The opinion even exists that an efficient regulation of pest organisms could be better achieved by promoting a high abundance of a limited number of predator or parasitoid species rather than by supporting a maximised diversity of these natural enemies accompanied by a low abundance level of each species (Wetzel, 1993).

Gaston (1996) highlighted three general points of view regarding biodiversity:

- Biodiversity as a concept (expressed as the "variety of life" it is completely abstract and extremely difficult to understand).
- Biodiversity as a measurable entity.
- Biodiversity as a social/political construct.

Biological diversity is concentrated in areas inhabited by socio-economically marginal and traditional societies, and so it is a key indicator of sustainability and buffering capacity: highly diverse ecosystems are, for instance, more efficient in capturing energy, water, nutrients and sediments than homogenous systems (Saxena et al., 1999). Thus, the high technical standard of central European societies seems to be a contradiction to achieve considerable biodiversity levels. However, within each level gradual differences are realised, so that the goal has to be the upper end of the level that is achievable under the conditions given. This is also valid for farmland.

2. Biotic indicators—approaches for definitions, nomenclature, requirements and rules

Paoletti (1999a,b) defines a bioindicator as "a species or assemblage of species that is particularly well matched to specific features of the landscape and/or reacts to impacts and changes"; the term can be described as a label for a particular situation and environmental condition.

"Biodiversity" will never be more than an umbrella for the total range of life expressions, and cannot be measured per se (Von Euler, Vancouver, Canada, personal communication, 2000*).[1] As research has shown biodiversity per se has no operational definition; therefore, an indicator to assess "biodiversity" as a whole is theoretically and practically impossible (e.g. Duelli and Obrist, 2003; Lawton et al., 1998; Watt et al., 1997). Thus, the development of 'biodiversity assessment tools', or sets of indicators, that together will allow the estimation of trends in biodiversity, is suggested (Watts, Scotland, UK, personal communication, 2000*).

Besides species diversity "biodiversity assessment can rely on the number of life forms, (plant) functional types, strategic types and other typological units as well as other qualities of biodiversity as the variability between objects (e.g. heterogeneity, similarity) or the ecological complexity (resulting from interactions

[1] Citations marked with an asterisk and "personal communication" refer to statements within an electronic conference on "Biodiversity Assessment and Indicators", November 1999–January 2000 (http://www.gencat.es/mediamb/bioind).

between the units). Concluding, it is necessary to relate biodiversity indicators within a defined space and time to specific criteria as, for example, morphological or more general spatial structures, phylogenetic similarity or functional traits" (Beierkuhnlein, Rostock, Germany, personal communication, 2000*).

Majer (1983) and Pankhurst (1994) stated that indicators: (a) should be holistic, but closely related to the assessment goals, (b) are important to the structure and function of the agro-ecosystem, (c) are a response to a range of environmental stresses, (d) can be easily measured, quantified and interpreted and (e) show an integrative potential in the long term (Lobry de Bruyn, 1997).

2.1. Preconditions to biotic indicators

Beierkuhnlein (Rostock, Germany, personal communication, 2000*) has developed four preconditions to judge a biotic indicator as appropriate:

- Occurrence and distribution: "biodiversity indicators have to be common and widespread. They should occur under the different environmental conditions that are of relevance according to the qualities of biodiversity under consideration".
  Comment: The second sentence is essential to avoid misinterpretation. Nevertheless, there is still the danger of misunderstandings, and it depends on the purpose an indicator is created for: For instance, a species that indicates a status of higher intensification might be common and widespread whereas a species typical for a status indicating the "maturity" of a habitat, which is rarely achieved (e.g. so called "primeval forest relict species" in forests) per se cannot be "widespread and common", because of its special demand on the habitat status. So, "common and widespread" can only be regarded as a theoretical value revealing geographic distribution, climate, etc. that makes the occurrence of the species potentially possible.
- Persistence: "biodiversity indicators have to be closely connected to certain areas or spatial units. It seems to be problematic to look at species which use different habitat types or communities with low spatial and temporal constancy. The life time of indicators has to cover the life time of the objects whose diversity is addressed".
  Comment: This criterion seems to be rather contradictory with regards to the demand for an indicator to be widespread and common. Secondly, if species that use different or fast changing habitats were excluded, it would be difficult to assess the biodiversity of agricultural habitats as habitats of "low spatial and temporal consistence" per se. For instance, limicoles that use a pond habitat as a food source for a short time or epigaeic predators temporarily invading a field to feed on pests create a certain diversity and so can be useful indicators for functional traits of a habitat.
- Identification: "indicators of biodiversity must be easy to identify. There should be an agreement concerning their classification and terminology. Objects with only temporal or cryptic occurrence or unclear terminology are not appropriate". Nor are objects which are costly and accessible to only a few experts.
- Sensitivity: "biodiversity indicators have to react sensitively to changes in certain forms of biodiversity. They can indicate qualitative and quantitative aspects of comparable units. For instance, the occurrence of certain species of trees (e.g. *Salix* spp.) might indicate a large variety of insects connected with this habitat, the occurrence of other species (e.g. *Fagus*) might indicate low diversity. Or, the occurrence of certain communities might indicate a certain degree of the intensity of land use and thereby a certain diversity of other communities".
  Comment: This criterion can also be contradictory to some criteria mentioned above; for instance, a "common and widespread" species which presumably should have a high tolerance to several conditions may not be able to fulfil the demands on sensitivity.

The critical comments show that it is nearly impossible to create basic rules or criteria for biodiversity indicators due to the high diversity of goals indicators have to serve for. Concluding, the definitions, limits and basic demands which are related to an indicator have to be stated case by case. It, therefore, seems to make little sense to develop general definitions or criteria for biotic indicators: definitions and/or descriptions should be stated instead as the case arises (occurrence and distribution, persistence, identification, sensitivity and many more).

The problem of using focal species, umbrella species, flagship species: keystone species. Indicators have been used to monitor economics for many years but no economist depends on a single indicator. Similarly, several ecologists have argued for sets of indicators (e.g. Stork, 1995; Stork and Eggleton, 1992; Ferris and Humphrey, 1999).

Lambeck (1997) suggested selecting an array of "focal species" which are specialised and sedentary ($\geq$K-strategists) thus showing a considerable sensitivity towards area, resources, dispersal and process limitation. If the landscape was managed for the conservation of these species most other species would also be conserved. The concept to use "focal species" has been applied mostly to nature conservation purposes up to now and has focused on (large) threatened species with a high sensitivity (e.g. top predators) of public interest. However, it is also imaginable for the assessment of beneficial potential in an agricultural habitat. According to Aarts (Nijmegen, The Netherlands, personal communication, 2000*) "focal species are just very sensitive creatures, but not intrinsically important to the functioning or the stability of the ecosystem" (which are covered by the term "keystone species"; e.g. Naeem and Li, 1997; Scheu, 1999). Thus, due to the experience of a lot of scientists, for example, in central Europe the carabid beetle *Carabus auratus* can be assumed as a species which indicates an acceptable standard of an agro-ecosystem with regards to epigaeic predator activity including some traits on functional biodiversity "ecosystem health". Due to the fact that the assessment of biodiversity per se will not be realistic in particular for routine evaluations, the restriction to a set of "focal species" which are easy to identify and to assess seems to be one solution to handle the problem.

However, when using "focal species", "flagship species" or "umbrella species" as representative units for a certain status, a serious validation of the potentials and limits (concerning all aspects) is essential. This essential precondition is usually neglected; "focal species" are used for more superficial reasons such as public acceptance, extent, and a more or less nebulous concept of their ecological potential. In most cases the second step (application as an indicator) is often carried out before the first (validation of potentials and limits as an indicator), i.e. a correlation between these species and biodiversity is claimed, but remains untested as well as the criteria for the choice of those umbrella and flagship species.

Andelman and Fagan (2000) who analysed the effectiveness (regarding number of protected species and costs) of a range of different surrogate schemes (e.g. large carnivores, charismatic species, key-species and wide-ranging species) on three different geographical scale levels (habitat ecoregion [= natural unit] and state) found a limited suitability of umbrella and flagship species as surrogates for regional biodiversity. According to Good (Cork, Ireland, personal communication, 2000*) flagship species might have a potential to indicate threatened species; they will rarely represent the complexity of habitat species diversity, in particular regarding the interpretation of management impacts. He points out that for habitat species diversity, indicators of ecological integrity are needed, interpreted as the specialist ability of ecosystems to deal with their climatic and edaphic environment, as well as specific soil/organic microenvironments with their own diversity of soil microorganisms (see Schloter et al., 2003).

The "focal species approach", and even more the "keystone species approach", depend upon knowledge of the ecology of the chosen species, which is available for taxonomically and faunistically well known groups (e.g. vertebrates, carabids, spiders, Rhopalocera) but may be sparse for the majority of invertebrate taxa (e.g. mites, Diptera, parasitic Hymentoptera and Coleoptera in general).

If focal/keystone species are introduced for management of biodiversity, regular monitoring is required. However, hardly any evidence can be found in recent literature on the intervals at which the monitoring of biodiversity or sustainable land use should be repeated, although no habitat type is free of changes. But regular monitoring intervals also imply that not only an actual status or the actual functionality of a process is recorded, but that these assessments are essentially combined with a well defined target to be achieved.

2.2. *Categorisation of indicators*

Döring et al. (2003, and in litteris) separates "goal indicators" and "status indicators". As "goal-oriented indicators", species are described which show that a defined status is achieved, a "status-oriented indicator" says something about the actual status of the field (e.g. regarding biodiversity). Stenotopic field species among carabids are suggested as "goal-oriented indicators", xerophilic, herbivorous arthropod species as "status-oriented indicators" for biodiversity, and the zonation types of the weed flora communities for management intensity. However, an indicator is not intrinsically a "goal" or a "status" indicator: the purpose for which an indicator is used is defined by the researcher through the goals or results which should be achieved and not by the indicator itself. For instance, producing a certain zonation type of field vegetation can also be a goal, on the other hand the percentage of stenotopic field carabids can provide information on the actual status of a field.

2.3. Indicator purposes

Duelli and Obrist (2003) differ between indicators *for* biodiversity (e.g. measuring the effect of heavy metal contamination on biodiversity as such) and indicators *from* biodiversity (e.g. effects of a heavy metal contamination on a certain taxonomic group). However, a good indicator for environmental pollution (e.g. heavy metals) must not indicate biodiversity (if it is not linearly correlated to biodiversity).

The complex background of using bioindicators for biodiversity can be summarised by the following:

- there is no consensus on how to use bioindicators;
- there is no indicator for biodiversity as a whole;
- each aspect of biodiversity (and also each goal for using indicators) needs its own indicator with very specific and well defined features and agreements on the mode of application.

For this purpose Duelli and Obrist (2003) developed a hierarchy of indicators and distinguished between three kinds of indicators for "*nature protection*", "*plant protection*" and "*ecological resilience*". Each of these three goals requires a different kind of indicator which will provide different values for biodiversity.

Nature protection (diversity of threatened species), tested by the number of threatened or rare species; (problems: national/international importance of threatened species, availability of Red Lists which cover only 7%; typical taxa used for conservation purposes as, e.g. butterflies, grasshoppers, birds cover only 1% of all species in agro-ecosystems so that this goal is not appropriate to evaluate agro-ecosystems).

Plant protection (diversity of beneficial organisms), tested by the abundance (short-term assessment) and species numbers (long-term assessment) of beneficials such as epigaeic and aphidophagous predators, parasitoids and key decomposers; predators are preferred to parasitoids because they are easier to assess and more is known on their ecological demands; the set up of ratios between herbivores and predators or parasitoids is conceivable following the habitat template-hypothesis of Brown and Southwood (1987).

Ecological resilience (species diversity of all organisms) tested by a set of selected taxa, e.g. Aculeate Hymenoptera, Heteroptera, and flowering plants, or a representative sample of all arthropods referred to a certain sampling method. (Problems: assessment of ecological resilience is based on the entire biodiversity as it is assumed that a higher number of different species, genes, etc. correlates with the functionality of ecosystems—for routine assessment too laborious and cost-intensive).

Finally, Duelli and Obrist (2003) recommend to pool the resulting indicators for each motivation, and the respective values to form an index which is similar to economical indices like the Dow Jones.

A similar approach, but more related to ecosystem features than anthropocentric goals has been developed by Good (Cork, Ireland, personal communication, 2000*) using soil staphylinid beetle assemblages as an indicator system for:

- *productive systems* (e.g. arable crops) considering the abundance of beneficial species (predators) only;
- *self-sustaining ecosystems* (e.g. revegetated mine waste) considering only a selected set of species associated with ecosystems not receiving external nutrient inputs and associated with litter decomposition;
- *biogeographically characteristic ecosystems* (e.g. calcareous fens) considering species which are of local occurrence and stenotopic.

Relying on McGeoch (1998), Lawton and Gaston (2001), Perner and Malt (2003) categorise indicators into *environmental indicators* (reflecting the biotic/abiotic state of an environment), *ecological indicators* (indicating impacts of environmental changes), and *biodiversity indicators* (indicating the diversity of species, taxa or entire communities in relation to habitat or area).

Döring et al. (2003) distinguish between four types of indicators:

- biotic indicators for abiotic status;
- biotic indicators to evaluate husbandry practices;
- goal parameters which can be derived from the (protection) goals agreed on beforehand;
- correlates or surrogates of goal parameters that minimise time, effort and costs with the disadvantage of information losses due to the simplification of data.

2.4. Assigning values to indicators—establishment of priorities

Assigning a "value" to different (indicator) species and other biodiversity facets and in consequence, the establishment of clear priorities seems unavoidable. In the absence of an open and formal value system, priorities will still be made, but based on hidden value systems which are beyond critical examination. (Von Euler, Vancouver, Canada, personal communication, 2000*).

Alpha-diversity (richness of species, gene alleles or other "taxonomical" units) can be interpreted in two ways according to Duelli and Obrist (2003):

(a) Each species (or other taxonomical unit) is valued equally: the species number correlates linearly to biodiversity.
(b) Each species is valued equitably (e.g. rare or threatened species for conservation purposes or predators for pest control) depending on its significance for reaching the goal. Regarding conservation purposes the ranked value depends, for instance, on the spatial level at which a species is threatened (e.g. local, national, international, global), with regards to biocontrol purposes (plant protection) on the predatory capacity, its local abundance and the gradual preference of a certain prey.

Within the latter (species as natural enemies in pest control) the term "functional biodiversity" becomes more significant. According to Duelli and Obrist (2003) "structural and functional diversity is somehow reflected in the number of species". It is assumed that more trophic levels (or functional groups, respectively) automatically include more species, and that higher structural diversity (in its widest sense; e.g. Noss, 1990) will provide more ecological niches. Apart from conservation purposes, for instance, biocontrol would be more efficient and in the long term more sustainable if a higher (species-)biodiversity is achieved. This reflects the "*niche-complement-hypothesis*". Other hypotheses (Lawton, 1994) like the "*redundancy hypothesis*" (see Perner and Malt, 2003) assume that functional effects of biodiversity show an asymptotic function, i.e. if a certain number of species is exceeded, more species will not improve ecosystem functions; the "*idiosyncratic hypothesis*" supposes that functional effects of biodiversity are not predictable and thus occur stochastically on the basis of some "keystone species" which catalyse ecosystem functions, whereas other species are redundant (see Scheu, 1999). The latter hypothesis is subliminal and is often used in applied agricultural research relying on "beneficials" (e.g. Wetzel, 1993) which are determined as "keystone elements" from an anthropocentric point of view. However, apart from some more general and model-like assumptions that a higher biodiversity would narrow the food web so that a functional replacement of an extinct species could be counterbalanced more easily (Naeem et al., 1994; Naeem and Li, 1997;[2] McGrady-Steed et al., 1997; Tilman et al., 1997), very little scientifically sound data from field experiments can be found in recent literature to support this assumption (e.g. Maraun et al., 1998).

2.5. Taxa and their suitability as indicators

Appropriate indicator species should perform:

- a low coefficient of variance for the mean number of individuals recorded per site;
- a high degree of habitat preference for the habitat considered (Perner, 2003);

[2] All these results are elaborated under very artificial circumstances (in laboratory microcosms) and on the basis of microorganisms (e.g. bacteria, algae and protists) and at least Collembola as representatives of decomposers and mesofauna (Naeem et al., 1994; Naeem and Li, 1997), on the basis of plant communities (Tilman et al., 1997) or in aquatic ecosystems (McGrady-Steed et al., 1997). However, producing an artificial "biodiversity" by adding selected species of functional groups to a microcosm according to the belief of the researcher is far away from the complex interrelations which are in reality in the field, particularly if higher taxonomic levels are considered.

- a potential high biodiversity;
- a good niche separation (Nickel and Hildebrandt, 2003).

In a 3-year survey on the conversion of arable land into grassland with six different management systems and its effect on vegetation, spiders and beetles, it could be shown by Perner and Malt (2003) that:

- Regardless of taxon, invertebrates were more suitable than vegetation for showing a difference in the effects of the conversion types.
- Spiders (Arachnida: Araneae) and beetles (Insecta: Coleoptera) showed clear reactions to changes of microclimatic conditions and soil humidity; hydrophilic spider and beetle species were identified as "quick assessment tools", referring to the high potential of spiders to recolonise rapidly by ballooning, or the ability of macropterous beetles to fly.
- The advantage of invertebrates as indicators compared to vascular plants resulted in a shorter delay period for the reaction of the taxa to changes that could be correlated mainly to microclimate and soil moisture (an indication of these changes is already visible after 3–5 years). The minor suitability of vegetation as a short-term indicator is explained by the fact that the seed bank of (intensively managed) arable fields is getting continuously poorer.
- The seed distribution (spreading of seed) is heavily reduced particularly for plant species of later succession stages.
- Weed species (e.g. *Cirsium arvense*, *Galium aparine*, *Elymus repens*) are strong competitors for power compared to other plant species (particularly on humid soils), so that a development of diverse plant communities is restricted for many years after conversion.

Thus, the reaction of the vegetation is obviously less appropriate for indicating short-term regeneration processes, but more suitable for detecting long-term changes. Therefore, Perner and Malt (2003) recommend a combination of vegetational and invertebrate monitoring particularly when land use changes have just been introduced.

The indication potential of herbivorous insects has almost been neglected up to now, in particular those which are closely related to vascular plants typical for field habitats. Whereas plant seeds are mostly able to persist for years in the soil, monophagous herbivores will become (locally) extinct after just one season without finding their host plant species. Thus, compared to vascular plants, an exponential higher sensitivity to environmental changes (e.g. by husbandry practice) is assumed. Köhler (1998) and Fritz-Köhler (1996) showed, for instance, that leaf beetles (Coleoptera: Chrysomelidae) and weevils (Coleoptera: Curculionidae) feeding on field weeds are excessively endangered compared to other taxa: 5.6% of all polyphagous, 8.5% of all oligophagous and 13.1% of all monophagous species from these beetle families have been missing for more than 50 years. Furthermore, Köhler (1998) highlighted that the number of individuals per species will increase if more extensive management (e.g. no herbicides and fertilisers) is conducted.

While plant species richness can be recovered easily and rapidly from the seed potential in the soil if conditions become more favourable, herbivore insects are influenced by several factors as, for instance, management intensity, structural performance of the landscape, the connectivity of habitats, etc. Therefore, the occurrence and abundance of mono- and oligophgous herbivores can be judged as very sensitive cross-section indicators of an overall positive development of agricultural habitats.

Nickel and Hildebrandt (2003) recommend herbivorous invertebrates (in particular Auchenorrhyncha-communities) as suitable biotic indicators for habitat disturbance, particularly in comparison to plant communities due to:

- great abundance (>1000 ind/m^2 in grassland ecosystems) and large species numbers (approximately 320 in grasslands) which allow clear and gradual assessments of management effects or environmental loads;
- a high species richness positively correlated with higher species diversity of other taxa as, e.g. Heteroptera, Saltatoria, Rhopalocera (Achtziger et al., 1999), vascular plants and their structural complexity and spatial composition (Denno, 1994; Denno and Roderick, 1991; Murdoch et al., 1972);
- a rapid reaction to management intensity by changes in dominance structure and species community

(Andrzejewska, 1976, 1991; Morris, 1973, 1981a,b, 1992; Morris and Plant, 1983; Morris and Rispin, 1987; Nickel and Achtziger, 1999; Prestidge, 1982; Sedlacek et al., 1988);

- separation into several life strategies (e.g. macropterous/brachypterous; uni-/bi-/polyvoltine; mono-/oligo-/polyphagous) which are directly correlated to environmental conditions (Novotný, 1994a,b, 1995) (see r–K-selection, Table 2);
- importance as prey items for predators (e.g. spiders, ants, birds such as partridges) and hosts for parasitoids, the amount of which can be estimated as an indicator itself (see Moreby and Aebisher, 1992, Potts, 1986);
- their focus as primary consumers on the vegetation layer and their function as vectors of plant diseases (e.g. viruses, MLOs); thus, they influence the composition of plant species and their competition power (Curry, 1994);
- the suitable and easy application of sample methods for the assessment of abundance (D-Vac suction trap samples) and species composition (sweep net samples);
- the easy estimation of reproductive success by counting the number of nymphs in the samples.

The level of taxonomical knowledge required makes the use of Auchenorrhyncha as biotic indicators more difficult and is an obstacle to a broad and easy application.

Testing several invertebrate taxa Duelli and Obrist (2003) found that *bugs* (*Heteroptera*), *wild bees* (*Apidae*), and *wasps* (*Hymenoptera aculeata*) showed the best correlation, while *spiders* and *ground beetles* showed poor correlation to entire species richness when used as indicators for ecological resilience.

In Australia soil biota have been identified as bioindicators of soil sustainability in agricultural land (Pankhurst et al., 1995; Hamblin, 1992; SCARM, 1993), but as in Europe, validation by field experiments is lacking. In this context, Lobry de Bruyn (1997) stresses the importance of *ants as soil indicators* due to their species richness (1100 species) in Australia. Radford et al. (1995) and Wang et al. (1996) showed that in agricultural soils biodiversity and abundance of ants increases when minimum tillage and stubble retention is applied. Comparing different farming systems in central Europe Büchs et al. (1999) identified ants as indicators for an extremely extensive management (no pesticides, no fertilisers, extensive rotation) or set-aside. Dauber et al. (2003) include ants in an indicator concept considering a landscape matrix. However, according to Linden et al. (1994) and Lobry de Bruyn (1997) the value of ants as indicators of soil quality remains unclear. Their functions in agro-ecosystems have to be elucidated and their role in soil processes quantified; finally, taxonomic tools are required which enable non-specialists for identification. Lobry de Bruyn (1997) estimates ant indicators as part of a broader system of ecosystem process indicators.

Diptera are mostly myco- or saprophagous and fulfil keystone functions as soil-dwelling larvae within the decomposition of plant residues. However, determination is laborious and requires great skills. Diptera could be identified as indicators for management intensity and environmental loads (e.g. fertilisers and pesticides) by Franzen et al. (1997) and Weber et al. (1997) on the community as well as on the population level. It could be demonstrated that dominance structure was affected according to the pesticide and fertiliser input and became less complex and balanced the more intensive production means were used. A clear correlation of emergence rates to management intensity as regards pesticide and fertiliser input could be observed in (mostly fungivorous) gall midges (Diptera: Cecidomyiidae) and on the population level (sometimes crop-specific) for dominant Sciaridae (fungus gnats) species (e.g. *Scatopsciara vivida* (Winnertz, 1867), *Lycoriella castanescens* (Frey, 1948), *Corynoptera dubitata* (Tuomikowski, 1960)). Prescher and Büchs (1997, 2000) showed that pest species of the genus *Delia* spp. (Diptera: Anthomyiidae) were negatively correlated with increasing extensification. Currently Sciaridae are tested as monitor organisms to indicate effects of GMO-crops on decomposition processes (Büchs et al., unpublished). Frouz (1999) highlights the potential of using morphological deformities in the context of "fluctuating asymmetry"; which has been mainly tested up to date with aquatic communities (Nematocera: Chironomidae), but also in terrestrial habitats using left-right asymmetry of wing venation (see Table 1).

In contrast to other Diptera, taxa determination of adult hover-flies (Diptera: Syrphidae) is comparatively easy. Life forms of larvae represented in a

Table 1
Selection of indicators, indication systems and approaches to indications on the basis of animal populations or species communities with special reference to agro-ecosystems[a] and direct or indirect effects on biodiversity and/or sustainable agriculture

Indication parameter/level of indication	Indication goal	Taxon	References	Comments/description
Population				
Digestive enzymes	(Heavy metal) toxicity of soils	Isopoda	Joy et al. (2000)	
Nutritional conditions	Food supply (management intensity)	Carabidae	Van Dijk (1986), Wallin (1989), Chiverton (1988), Zanger et al. (1994), Langmaack et al. (2001)	Some approaches are more indirect and more related to reproduction (Van Dijk, 1986; Wallin, 1989); some allow direct assessment by gut dissection (Chiverton, 1988; Zanger et al., 1994; Langmaack et al., 2001)
Growth rate	Habitat quality	Araneae (orb-weaving spiders)	Marc et al. (1999), Nyfeller (1982), Vollrath (1988)	Assessment of habitat quality due to (estimated) food intake and (measured) growth rates. Standard calibrated values recorded in the lab were applied to field conditions
Mean body weight restricted to populations of selected species	Management intensity; disturbance	Carabidae	Büchs et al. (2003), Zanger et al. (1994)	
Mean body size restricted to populations of single species	Management intensity; disturbance	Carabidae	Büchs et al. (1999, 2003)	Body size of adults changes depending on current life conditions (e.g. prey supply). Does not function with adult stages of univoltine holometabolous insect species, because body size is fixed with metamorphosis and depends on life conditions of larval stage which possibly developed in different conditions and locations
Development of wing muscles	Disturbance; environmental stress	Carabidae, Staphylinidae	Geipel and Kegel (1989), Assing (1992)	
Egg production	Food supply; (management intensity)	Carabidae	Van Dijk (1986)	
Abundance	Environmental stress (management intensity)	Carabidae	Büchs et al. (1997), Döring et al. (2003)	Correlation of abundance to environmental stress or management intensity obvious, but no correlation between abundance and species diversity could be detected
Web size, web structure	Environmental stress (e.g. pesticide effect; prey supply)	Araneae	Retnakaran and Smith (1980), Riechert and Harp (1987), Roush and Radabaugh (1993)	
Fluctuating asymmetry	Environmental stress	Several insect taxa	Palmer and Strohbeck (1986), Warwick (1988), Clarke (1993), Krivosheina (1993, 1995), Rahmel and Ruf (1994), Vermeulen (1995)	Up to present mostly applied in urban ecology or aquatic ecosystems; no experience in agricultural ecosystems
"Ellenberg" indicator values for vascular plants	Abiotic demands (e.g. soil humidity, moderate temperature, etc.) of invertebrates	(Epigaeic) invertebrates	Perner and Malt (2003), Stumpf (personal Communication, 1996)	Survey of the vascular plant community in the catchment area of, e.g. pitfall traps and determination of their Ellenberg values. In a second step the values are related to invertebrate species recorded in the pitfall trap samples
Hibernation behaviour	Landscape structures; management intensity	Staphylinidae	D'Hulster and Desender (1984)	
Community				
Number of morphospecies	Species richness	Several invertebrate taxa	Duelli and Obrist (2003)	Range of error varies depending on the difficulties and efforts to separate species morphologically (e.g. whether preparation of sexual organs is necessary) and the skill level of the researcher
Presence and abundance of species combination	Humus and peat bog content in soils	*Rana arvalis*, *Sphagnum* spp.	Kratz and Pfadenhauer (2001)	Restricted to originally moorland locations
Number of species, mortality, parasitism	Species richness, habitat fragmentation, disturbance	Trap-nesting bees and wasps	Tscharntke et al. (1998)	Easy method to apply: artificial units of reed internodes are placed in the habitat considered. Assessment of no. of colonised reed internodes, mortality by trap-nesting bees and wasps, species set of predators and parasitoids

Comparison of regional species pool to local species composition	Effects of management/disturbance	Auchenorrhyncha, Syrphidae, soil mites (Oribatei)	Nickel and Hildebrandt (2003), Good and Speight (1991), Speight et al. (1992), Siepel (1994, 1995, 1996)	The comparison is based on certain ecological features (e.g. mono-/oligo-/polyphagous species; macro-/brachypterous species—Auchenorrhyncha) or larval life forms (Syrphidae). For mites life strategies and life-history patterns of supraregional surveys are compared to those of the study sites
Taxonomic distance, taxonomic distinctness	Sensitive indication of environmental perturbation	Ecosystem or biome-related biocoenosis; birds	Warwick and Clarke (1998, 1999), Clarke and Warwick (1998), Von Euler (1999)	Total genetic components of a biome may remain constant but be partitioned differently among the hierarchy of taxonomic units according to the age of the successional stage of the assemblage
Zoogeographic and taxonomic structure related to different spatial scales	Environmental impacts	Carabidae	Popov and Krusteva (2000)	A two-way indicator species analysis is applied at different scale levels to correlate carabid species and assemblages to different environmental loads using diversity as well as the zoogeographic and taxonomic structure of the assemblages
Dominance structure of species communities	Environmental stress on soils	Soil mites	Hagvar (1994)	Change of dominance structure in the soil microarthropod communities is suggested as an indicator of various performances of environmental stress
Classification of species into r–K-continuum	Disturbance; environmental stress	Nematoda, Gamasina	Bongers (1990), Ruf (1998)	Index of "maturity": classification of species into the r–K-continuum depending on the type of reproduction
Habitat preferences	Management intensity; disturbance	Araneae	Büchs et al. (2003)	Veritable database available regarding ecological characteristics of central European spiders: Maurer and Hänggi (1990), Nentwig et al. (2000)
Percent stenotopic field species	Management intensity; disturbance	Carabidae; Syrphidae	Döring and Kromp (2003), Haslett (1988)	See text
Percent euryoecious species	Management intensity, disturbance	Araneae	Büchs et al. (1997)	Considerably weaker than % "pioneer species"; veritable database regarding ecological characteristics of central European spiders by Maurer and Hänggi (1990) and Nentwig et al. (2000)
Percent pioneer species	Management intensity, disturbance	Araneae, Auchenorrhyncha	Büchs et al. (1997, 2003), Nickel and Hildebrandt (2003)	Among spiders mostly Linyphiidae; veritable database regarding ecological characteristics of central European spiders by Maurer and Hänggi (1990) and Nentwig et al. (2000)
Percent Lycosidae	Management intensity; disturbance	Araneae	Büchs et al. (2003)	Indicator easy to apply, because separation of wolf spiders from samples requires only low skill level; determination of species level not obligatory
Ratio of predators and prey organisms	Management intensity; environmental stress	Several arthropod taxa	Brown and Southwood (1987), Greiler and Tscharntke (1991)	Assessment based on the "habitat template-hypothesis" of Brown and Southwood (1987)
Relation of phyto- and saprophagous to carnivorous correlated to the average body size	"Habitat maturity"; management intensity	Coleoptera	Sampels (1986)	Index of "habitat maturity" was applied in vineyards and is most suitable in permanent crops such as vineyards, orchards, china grass, etc.
Percent macropterous vs. brachypterous species/individuals	"Maturity" of a habitat; management intensity	Carabidae; Auchenorrhyncha	Den Boer (1968, 1977), Döring and Kromp (2003), Gruschwitz (1981), Nickel and Hildebrandt (2003)	Macropterous carabids/plant hoppers are said to be typical for (frequently) disturbed ecosystems; brachypterous insects for more mature ones; but risk of misinterpretation if forest species occur in fields
Ash-free dry weight; ash weight	Functional importance; environmental stress; management intensity	Enchytraeidae	Van Vliet et al. (1995)	The ash-free dry weight or ash weight is suggested as a key indicator to explain the enchytraeid community structure with regard to their functional role in (agro-)ecosystems
Mean or median of body size of species community	Management intensity; disturbance	Carabidae, Staphylinidae, Araneae	Steinborn and Heydemann (1990), Köhler and Stumpf (1992), Blake et al. (1994), Büchs et al. (1997, 1999, 2003), Döring and Kromp (2003)	Two possibilities: purely species-related or individuals of each species included; often considerable differences between sexes (e.g. spiders); reference area recommended; combinations with other parameters (e.g. threatening, rarity) possible
Percent juveniles; percent nymphs	Reproductive success—management intensity; disturbance	Araneae, Auchenorrhyncha	Büchs et al. (1997, 2003), Nickel and Hildebrandt (2003)	Juvenile wolf spiders (Araneae: Lycosidae) have to be excluded

Table 1 (*Continued*)

Indication parameter/level of indication	Indication goal	Taxon	References	Comments/description
"External" surrogates				
Soil quality index	Species diversity	Carabidae	Brose (2003)	Low soil quality is correlated with high species numbers; index easy to apply
Sand content of soil; soil type	Species diversity	Carabidae, Araneae	Irmler (2003), Perner and Malt (2003), Steinborn and Meyer (1994)	Sand content is positively correlated with species richness
Field size	Species richness	Carabidae	Frieben (1998), Irmler (2003)	Irmler (2003) discovered a correlation between field size and numbers of species; Frieben (1998) constructed a (theoretical) model of the colonisable field area in relation to field size
Edge-to-area ratio of field margins	Species diversity	Carabidae and other epigaeic predators; vascular plants	Altieri (1999), Waldhardt and Otte (2003)	This edge-to-area indicator is completed by Boatman (1994) and Frieben (1998) by an index value, e.g. for the distance carabids are able to immigrate into fields
Length (and "quality") of field margins	Biodiversity (number of species)	Complete biocoenosis (incl. plants)	Irmler (2003), Waldhardt and Otte (2003)	Length is a consistent parameter, because it can be related to field size (area); quality" needs clear definition and depends on the taxon considered
Organic farming	Species richness; stenotopic field species	Several vertebrate and invertebrate taxa	Kromp (1999), Döring and Kromp (2003), Pfiffner (1997), Irmler (2003), The Soil Association (2000)	Organic farming is introduced as a surrogate indicator for biodiversity due to various results that show an increase of species numbers with a habitat-typical performance; duration of organic farming is an important additional criterion; however, assessment restricted to organic fields, an extrapolation to the status of the whole agriculturally (but not organically) used area is not possible

[a] Neither mathematical procedures (e.g. indices on similarity, diversity, (canonical) correspondence analysis, principle component analysis, TWINSPAN analysis, etc.) nor established tools of biotic assessments are considered as such in this table.

species community are the basis for a suitable indication: Müller (1991) and Ssymank (1993) showed that Syrphidae-species with zoophagous larvae (which dominate in intensively managed fields due to a high N-fertilisers input and thus, gradations of aphids) decrease in abundance with increasing extensification of crop management in favour of an increase of species with coprophagous and filtering (saprophagous) larvae. Simultaneously the number of migrating species decreases; particularly species numbers are much higher compared to intensive management (Banks, 1959; Ssymank, 1993; Knauer, 1993).

The extraordinarily high mobility of many hover fly species and their tendency to migrations and temporal invasions, particularly the aphidophagous species, restrict the possibility of relating a species set to a certain field location, or of assessing specific impacts as, for instance, pesticide and fertiliser input, tillage, cultivars, etc. (Sommaggio, 1999). Nevertheless, since it was shown that diversity of landscape structures adjacent to fields enhance individual as well as species numbers (Banks, 1959; Knauer, 1993) and result in an earlier occurrence of aphidophagous syrphids in fields (Paoletti, 1984; Krause and Poehling, 1996), syrphid populations and species communities can be assumed as suitable assessment parameters to evaluate the success of measures implied by surrogate indicators which are based, for instance, on the percentage of "Ecological Priority Areas" (EPAs) (e.g. Roth and Schwabe, 2003; Braband et al., 2003) or other landscape-oriented measures (Waldhardt, 2003).

Alvarez et al. (2001) who characterised Collembola assemblages in conventional, integrated and organic winter wheat fields at three locations in England were able to show that community composition and species dominance are obviously influenced by farming systems, but no species could be identified as indicators for differentiating between the farming systems. Büchs (1993, 1994) and Büchs et al. (1999) showed (for various crops) that Sminthurid springtails (Collembola: Symphypleona) which feed on fungi and pollen on crop plant leaves are very reliable and sensitive indicators even at higher taxonomic levels (e.g. suborder or family) which are able to meet the demands for simplification with regards to indicator development (see further).

While *oribatids* (Acarina: Oribatei) are suitable indicators for air pollution (André, 1976; Weigmann and Jung, 1992; Büchs, 1988), their suitability for agro-ecosystems is limited, because the oribatid fauna of arable soils per se is dominated by species which indicate disturbance (e.g. Brachythoniidae, Tectocepheidae, Oppiidae), meaning that up to now, no specific keystone species of orbiatid mites could be identified for agro-ecosystems (Behan-Pelletier, 1999). Koehler (1999) states that Uropodina which prey on nematodes and insect larvae are suitable indicators of soil compaction and the quantity of organic material, whereas Gamasina indicate soil conditions, disturbance and anthropogenic impact. The hemiedaphic nematode predator *Alliphis siculus* was identified as a key species in communities which indicate sustainable agriculture.

Isopoda provide information on functional aspects of decomposition processes showing clear reactions to tillage, to the supply of decaying organic materials and to pesticide input (Paoletti, 1999a,b). Since they accumulate heavy metals (Paoletti, 1999a,b; Joy et al., 2000), they are of special interest for indicating copper residues, particularly in organic grown vineyards (Wittasek, 1987). Well developed assessment procedures exist for indicating heavy metals (Joy et al., 2000; Hopkin et al., 1993). Crop management effects can be comparatively assessed by abundance, biomass and species richness of Isopoda (Paoletti, 1999a,b; Paoletti and Hassal, 1999). Büchs et al. (1999) recorded Isopoda in fields only in considerable abundances when extensive farming or set-aside was applied.

Earthworms are estimated as suitable indicators for soil structure or compaction, tillage practice, heavy metals and pesticides (Larink et al., 1993; Paoletti, 1999a,b; Kühle, 1986; Knüsting et al., 1991). According to Paoletti (1999a,b), endogaeic earthworms are more suitable for monitoring pesticide or heavy metal residues than anecique species. Knüsting et al. (1991), however, showed that for farming systems with different pesticide input levels (beside litter supply) tillage is the main factor steering the performance of earthworm communities in agro-ecosystems. Biomass, species numbers and ecological guilds (e.g. epigaeic, anecique, endogaeic) are favoured by Paoletti (1999a,b) as key indication parameters for assessing earthworms in agro-ecosystems.

Beside the integration into an r–K-continuum, Yeates and Bongers (1999) discuss biomass of bacteria-feeding *nematodes* as well as nematode taxa with striking morphological characteristics that are easy to recognise (e.g. predacious Monochidae or plant associated genera like *Paratylenchus* or *Gracilacus*) as potential indicators. Whereas the latter Tylenchida were identified in New Zealand as indicators for the burning of field soils, the Monochidae occasionally showed a contradictory population development to earthworms, so that their use as an indicator for sustainable management is not consistent.

Fundamental works on the ecological features of central European spiders (e.g. Martin, 1972, 1973; Platen et al., 1991; Maurer and Hänggi, 1990) provide a good basis for using spiders as indicators on a species community level and can be enhanced by using mathematical procedures (e.g. CCA, TWINSPAN, PCA, DECORANA, etc.; see Marc et al., 1999; Hill, 1979; Krebs, 1999; Manly, 1992; Motulsky, 1995). As it is shown in Büchs et al. (2003) the analysis of habitat preferences enables a precise assessment of disturbances and environmental stress by crop management. In particular the ratio "spiders typical for weedy fields" versus those preferring "dry meadows" and the ratio "pioneer species" (mostly Linyphiidae) versus "wolf spiders (Araneae: Lycosidae)" are reliable indicators, meaning that they can be also used for the assessment of situations in agro-ecosystems other than management intensity, as could be demonstrated within the evaluation of several seed-mixes for rotational set-aside. Furthermore, if related to a reference area the Shannon-diversity index (H_S) and the Jaccard- or Sörensen-similarity index proved to be reliable for assessing management effects on spiders. The average body size (of male spiders) also seemed to be a suitable indicator to evaluate environmental stress (Büchs et al., 1997), but is not very sensitive and functions only if fields compared are not too closely related regarding management intensity. Methodological objections are against the use of sex ratio for indication, because in the same location pitfall traps select active males to a high percentage whereas emergence traps show a balanced sex ratio. Furthermore, indication can base on the percentage of juvenile spiders (except Lycosidae) with the advantage that no species identification is necessary, on reproduction cycles (steno-, diplo-, eurychrone), phenology, preference to abiotic factors (see above), but obviously not on species richness and abundance (Büchs et al., 1997, 1999; Kleinhenz and Büchs, 1993, 1995; Stippich and Krooß, 1997). Other approaches start from huge data bases in order to relate a site-specific spider community to supraregional species sets (e.g. Canard et al., 1999; see Table 1), partly complemented by creating values such as the community index of Ruzicka and Bohac (1994) for spiders and rove beetles (Bohac, 1999; see below), which groups species according to their occurrence in more natural or more anthropogenic influenced habitats and relates them to communities typical for undisturbed climax stages. However, it seems to be questionable whether such a procedure enables the assessment of farming systems which do not differ greatly (e.g. two types of integrated farming).

Possibilities of indication approaches basing on the population level are reviewed by Marc et al. (1999). Because prey consumption rates are said to depend on prey availability it is suggested that the quantity of potential prey is able to indicate habitat quality. Field data on food intake are related to body length, body weight, reproductive rate, metabolic rate and production of excreta. Nyfeller (1982) and Vollrath (1988) assessed different habitats by establishing a correlation between standard growth rates (of certain instars), calibrated in the laboratory, and field data from different habitats. From these correlations they drew conclusions on food consumption and the food supply under field conditions. Wise (1979) showed that an additional prey supply increases the egg production of orb-weaving spiders. However, Riechert and Tracy (1975) highlighted that a lower growth rate and finally reproduction rate are not intrinsically caused by a lack of food supply, but by (micro-)climatic factors: they demonstrated that compared to forest spiders, the food intake of open field spiders was reduced due to heat stress at noon, reducing feeding activity.

The structure and size of spider webs is also used as an indicator, particularly for pesticide effects (Retnakaran and Smith, 1980), but also indicates prey availability (Roush and Radabough, 1993; Riechert and Harp, 1987). Marc et al. (1999) collected data on the indication and accumulation of heavy metals such as Cd, Cu, Zn and Pb by spiders (e.g. Larsen et al., 1994): spiders accumulated Cd, Cu, Zn at higher levels than present in contaminated soils; epigaeic spiders to a larger extent than web spiders from the vegetation level. In agro-ecosystems, heavy metals occur in sewage sludge or are applied as pesticides (Cu), particularly in organic farming (potatoes, grapes).

Büchs et al. (1997), Zimmermann and Büchs (1996, 1999) showed that *rove beetles* (Coleoptera: Staphylinidae) are suitable indicators to assess management intensity in arable crops and particularly insecticide input on the population as well as on the community. Parameters which displayed a clear positive correlation to increasing extensification were abundance (emergence from soil), "species density" (number of species per trap and sampling period) and change of dominance positions (Büchs et al., 2003). A cluster analysis (Büchs et al., 1997) proved to be the best tool to separate species communities of different levels of extensification. Also the phenology of single rove beetle species, for instance, is suitable for indicating disturbances due to management effects (e.g. insecticide input). However, it was demonstrated by Zimmermann and Büchs (1996) that due to the life cycle of rove beetle species, the effects of, for instance, insecticide applications only become apparent after a period of up to 9 months, so that survey periods should be scheduled for an appropriate length of time. Furthermore, it could be shown by Büchs et al. (unpublished) that most rove beetle species (and particularly Aleocharinae) leave their point of origin immediately after hatching, so that a survey with pitfall trap is not sufficient, but sampling has to be conducted essentially with emergence traps.

Bohac (1999) presents a complete system to indicate environmental stress on the basis of rove beetle species communities (including threshold values (in brackets)) which was validated by a comparison of data from several authors including the consideration of habitats with different anthropogenic influences. This assessment procedure considers: (a) percentage of eurotopic species (>90%), (b) frequency of species with summer and winter activity of adult beetles (>20% summer activity), (c) proportion of winged species (no wingless specimens), (d) body size classes (more than 20% "large" species), (e) preference to abiotic conditions as temperature and humidity (more than 70% thermo- and/or hygrophilic species), (f) number of peaks in seasonal activity (less than two), (g) geographical distribution (wider than Europe), (h) number of life forms as zoo-, phyto-, myceto-, saprophagous or myrmecophilic (less than 4), (i) sex ratio (>10% from 1:1), and (j) a community index which separates rove beetle species into ecological groups according to their relation to undisturbed habitats such as climax forest stages, which per se cannot be achieved in agricultural ecosystems.

The suitability of *ground beetles* (Coleoptera: Carabidae) as indicators of the ecological status of arable fields is widely accepted and was shown, for instance, by Heydemann (1955), Topp (1989), Hingst et al. (1995), Schröter and Irmler (1999), Basedow (1990), Büchs et al. (1997) and Kromp (1999). Steinborn and Heydemann (1990) observed a tendency to a "homogenisation" of the carabid community of field habitats since 1952, a tendency which is stated by Blick (personal communication, 2002) for spiders. This homogenisation induced by crop management intensity, "hides" the influences of soil types on the field carabid community, which would naturally occur.

However, as shown by Perner (2003), habitat generalists usually have a lower coefficient of variance so that it is more favourable for them to achieve a "high level of precision" with a lower sample size. The habitat generalist *Pterostichus melanarius* (dominant in central European fields) is often said to be an indicator for intensive management, large field size and a degraded carabid species community (e.g. Müller, 1991; Raskin et al., 1992; Wallin, 1985). In contrast to these authors, Hokkanen and Holopainen (1986), Fan et al. (1993) and Büchs et al. (1999) recorded higher abundance with increasing extensification. These obviously contradictory results can be explained by the fact that *P. melanarius* larvae overwintering in the field are heavily affected by tillage, but due to the fact that a part of the population (adults) leaves the fields in autumn they rapidly recolonise the fields in spring so that larval mortality can be easily compensated for (Irmler, 2003). However, these facts show that due to this ambiguous and contradictory behaviour, the use of *P. melanarius* as an indicator is very critical and almost impossible, because no clear statements are correlated with an increase or decrease in abundance.

Döring and Kromp (2003) complain that already species numbers (of carabids) were suggested as indicators (Basedow et al., 1991; Steinborn and Heydemann, 1990; Kromp, 1990, Luka, 1996), but the mode of assessment of the indicator value and its reliability varies to a great extent. To enhance reliability Döring and Kromp (2003) pooled a lot of studies and analysed them on the basis of ecological species characteristics. Comparing organic versus conventional agriculture by an index based on the Wilcoxon matched pairs ranking (Hollander and Wolfe, 1973). Döring and Kromp (2003) created the "ecological type" as a key variable to distinguish between different environmental conditions caused by husbandry practice. The "ecological type" is defined by the abiotic and biotic demands of a species and the range which is tolerated. Referring to their index, carabid species with a high index value (as *Carabus auratus*, *Acupalpus meridianus*, *Pseodophonus rufipes*, *Amara similata*, *A. familiaris*, *A. aenea*, *Poecilus versicolor*) can be used as an indicator species for carabid biodiversity.

Also Büchs et al. (2003) identified *C. auratus* (Coleoptera: Carabidae) as one of the most suitable species indicating extensification combined with a high biodiversity, as did Steinborn and Heydemann (1990) for organic farming, whereas Irmler (2003) could not find a higher activity density in organic fields compared to conventional ones. Furthermore, Basedow (1998) and Büchs et al. (1999) showed that *C. auratus* indicates extensive cultivation by increasing activity density and body weight. Beyond that *A. similata*, *A. aenea*, *A. familiaris*, *P. rufipes* and *Harpalus affinis* benefit mostly from organic agriculture (Döring and Kromp, 2003) and so they can be assumed as the most important ground beetle species able to characterise low input agro-ecosystems.

Regarding the assessment of management intensity the following features of invertebrate species communities were generally observed (e.g. Döring and Kromp, 2003; Perner and Malt, 2003; Nickel and Hildebrandt, 2003) and seem appropriate as a basis for an indication: The more extensive management intensity was conducted:

- number and percentage of specialists increased, while generalists were reduced;
- diversity increased;
- high abundance is mostly caused by a low number of generalist species which are subject to extreme fluctuations; it/they indicate/s high frequency of disturbance and is/are thus related with intensification of management.

3. Approaches for indication—critical survey and prospects

Table 1 shows selected indicator approaches which are already established or to be tested in practice. Some of them are critically discussed in the following in order to provide an example.

3.1. Direct indicators

3.1.1. r- and K-selection as possible indicators for ecosystem assessment

Undoubtfully r- and K-selection plays a role in the characterisation of agro-ecosystems with regard to management intensity (here: crop rotation).[3] Referring to species communities, r- and K-selection seems to be one of the most important key issues for a biotic assessment of disturbance or environmental loads. Apparently r-strategist features can be related to a high, K-strategist features to a low level of management intensity, environmental loads or frequency of disturbance (Table 2).

Although r-/K-characteristics go back to a basic knowledge of animal ecology, check lists lack which classify species due to their gradual position on an imaginary scale, the r–K-continuum. This tool for the assessment of species communities and population features has only just recently been converted on the basis of reproduction biology for soil taxa as Gamasina (Ruf, 1998) and Nematoda (Bongers, 1990). Bongers (1990) introduced the "maturity index" (MI) for nematodes as a weighted mean of the coloniser–persistence (c–p) values for the non-plant feeding nematodes in a sample. If the species feeding on higher plants are to be included the use of the "plant parasitic index" (PPI) is proposed, which is negatively correlated; so that an increasing ratio of PPI/MI is identified as "ecosystem enrichment" (Yeates and Bongers, 1999).

The estimation of the gradual tendency of being an r- or K-strategist could, for example, be fitted into a scale between -1 (maximum r-strategy) and $+1$

[3] Related to microorganism communities the terms r- and K-selection are used in a different way than in animal ecology: they are restricted to the concentration and limitation of the viability of a certain substance within the media which is inhabited, and growth rate (r = high concentration; no limitations of substance, high growth rate; K = low concentration, limited availability, low growth rate).

Table 2
Simplified biological and ecological characteristics of r- and K-strategists (r- and K-selection)

Parameter	r-Strategists	K-strategists
Reproduction		
Reproduction period	Semelparity (reproduction once in lifetime)	Iteroparity (repeated reproduction periods)
Season-related life-cycle of population	Polyvoltine	Univoltine
Reproduction	High reproduction rate (mass production of eggs, larvae or juveniles)	Very low reproduction rate (only single eggs, larvae, juveniles with intensive, long period of parental care
Development		
Growth, development, maturity	Fast	Slow
Body size, body weight (in relation to taxocoenosis)	Small	Large
Tendency to mutate	High	Low
Lifetime	Short	Long
Durable stages to survive uncomfortable conditions	Yes	No
Population development/behaviour		
Abundance	High	Low
Population level	High fluctuations within short periods	Constant level over longer periods
Tolerance towards apocalyptic breakdown of essential life conditions	High	Low
Preferred concentration of habitat medium	High	Low
Behavioural ecology		
Tendency to migrate	High	Low
Food	Mono-/oligophagous	Polyphagous
Inter-/intraspecific competition	Low	High
Building of territories	No	Yes
Habitat ecology		
Habitat characteristics	Rapidly changing, subject to stochastic disturbances	Mostly stable, mature, changes within regular intervals
Habitat range	Eurytopic, ubiquitous	Stenotopic
Range of biotic and abiotic habitat conditions	Euryoecious	Stenoecious
Derived effects on species community and environmental conditions		
Species diversity	Low	High
Evenness	Low	High
Frequency of disturbance	High	Low
Management intensity	High	Low
Environmental loads	High	Low

Data compiled by Bongers (1990), Ruf (1998), May (1980), Tembrock (1982), Schubert (1984), Stern and Tigerstedt (1974), Remmert (1984) and Odum (1980).

(maximum K-strategy) and the average of all species considered calculated. If the calculated value is above zero, the location is more K-dominated, if below zero more r-dominated; the relevant conclusions and assessments can be made. Taxa with good knowledge on biological and ecological requirements of the species as, for instance, all vertebrates and ground beetles, Macrolepidoptera, spiders, Saltatoria, Auchenorrhyncha, Heteroptera, etc. should have the potential to be classified into the r–K-continuum, unless regional differences have to be considered (e.g. Kühnelt's principle of regional stenoecious behaviour; see Blower, 1955; Büchs, 1988). Unless good knowledge on ecological demands is possessed, the classification can be estimated as a permanent process and is subject to continuous changes with increasing scientific knowledge.

Sampels (1986) used the ratio of zoophagous to phytophagous and saprophagous beetles in correlation with the body size mean of the on-site species community to determine the "degree of maturity" of vineyards with different management intensity in relation to adjacent semi-natural habitats. According to the habitat-template-hypothesis of Brown and Southwood (1987) biotic interactions are as, for instance, predator–prey relationships, negatively correlated with the degree of disturbance, which leads to the use of a predator–prey ratio as an indicator for environmental stress.

From the practical point of view it is more appropriate to concentrate monitoring on one or two species than on the whole species community (Döring et al., 2003). However, there was no correlation between carabid species richness and the abundance of a single species.

3.1.2. Body size

A decreasing average of body length at community level with increasing cultivation intensity is stated for several taxa (e.g. spiders, carabids, rove beetles) by Blake et al. (1994), Lorenz (1994), Büchs et al. (1997, 1999, 2003), but not by Döring and Kromp (2003), and—with regards to the population level—not by Büchs et al. (1999, 2003) either.

3.1.3. Weight

Van Vliet et al. (1995) conducted a comparative research on enchytraeids in forests and agricultural sites (tillage versus non-tillage in North Carolina, USA). Although population densities were greater in forest soils, in the arable soils the ash-free dry weight was nearly double that of forest soil. Based on the calculation that enchytraeid field soil turnover is 2180 g/m^2 per year, but only about 400 g/m^2 per year in forest soil, it can be assumed that enchytraeids have more influence on soil structure in arable fields than in forests, in spite of lower population densities. The ash-free dry weight and ash weight per enchytraeid was suggested as a key indicator for interpreting enchytraeid community structure to explain their functional role in ecosystems. More details regarding the suitability of body weight as an indication parameter are shown by Büchs et al. (2003).

3.1.4. Taxonomic distance

Warwick and Clarke (1998) propose weighting diversity indices by taxonomic distance (including phylogenetic information) as an indicator. According to their results taxonomic distinctness appears to be a more sensitive indicator of environmental perturbation than diversity indices which often remain constant over a perturbation gradient. From this point of view it can be assumed that an index like the taxonomic distinctness come closer to a 'biodiversity' index than H_S. Furthermore, total genetic components of a biome may, within limits, remain constant but be partitioned differently among the hierarchy of taxonomic units according to the age of the succession stage of the assemblage. While biodiversity is heavily affected by the type of habitat, taxonomic distinctness depends more on trophic diversity which, however, is assumed to be strongly influenced by pollution. Therefore, Clarke and Warwick (1998) judge taxonomic distinctness as more suitable to indicate environmental stress than diversity indices (e.g. Shannon-diversity index H_S).

Von Euler (Vancouver, Canada, personal communication, 2000*) believes that the "taxonomic distinctness" index meets most "biodiversity tool" requirements better then available alternatives because it relies on the average taxonomic path length between two species randomly selected from the recorded species set (Warwick and Clarke, 1998; Clarke and Warwick, 1998). Thus, the great advantage of this index is that it does not depend on sample size or the exact order of the samples and can also be applied across different studies with species lists that lack a clear sampling protocol. Therefore, it could be a major step to simplify assessments. However, developed and tested on the basis of marine nematodes up to date, this index was not applied to agriculturally used areas.

3.1.5. Comparison of regional species pool to local species composition affected by management/disturbance

A comparison of the regional species pool is another way of using species communities for biotic indication: surveying differently managed grasslands Nickel and Hildebrandt (2003) found a third of all species to be monophagous and 42% oligophagous. The comparison of the percentage of specialist feeding habits in local study sites (e.g. with different management intensities) with the regional potential number of these species allowed a very sensitive and gradual assessment.

3.1.6. Species numbers, percentage of specialists and pioneer species

Finally, species numbers, the percentage of specialists and pioneer species were identified by Nickel and Hildebrandt (2003) as suitable indicators of biotic conditions due to management intensity (in grassland ecosystems). The results of Büchs et al. (2003) for spiders and Döring and Kromp (2003) for carabid beetles led to similar conclusions: the percentage of habitat specialists and/or pioneer species was clearly correlated with management intensity of agricultural ecosystems. Curry (1987) has also pointed out that species numbers and the percentage of specialists and threatened species are negatively affected by increasing management intensity.

3.1.7. Stenotopic field species

Döring et al. (2003) and Döring and Kromp (2003) showed that ground beetles are supported by organic agriculture the more they are stenotopic field species. The enhancement of stenotopic field species as the most typical representatives for the habitat considered can be evaluated as a key criterion to evaluate the fauna of agro-ecosystems. Thus, the percentage of species specifically adopted to field conditions can be judged as a suitable indicator for an overall positive development of field habitat conditions.

3.1.8. Relation of macropterous and brachypterous species

Ground beetle species that profit to the highest extent from organic farming are usually xerophilic and tend to be macropterous (Döring and Kromp, 2003). The percentage of brachypterous carabid species was identified as an indicator for the "degree of maturity" by Gruschwitz (1981) and Rehfeldt (1984). However, Döring and Kromp (2003) state that stenotopic forest carabid species are mostly brachypterous and are—when occurring in field ecosystems—enhanced by conventional cultivation (due to microclimatic conditions because of a higher density of the vegetation). Thus, with respect to the fact that organic cultivation promotes stenotopic field species more than conventional agriculture, brachyptery cannot be used as an indicator for a status of field habitat conditions which are regarded as sustainable.

3.1.9. Invertebrate species communities related to "Ellenberg"-indicator values

According to Perner and Malt (2003) soil humidity is most suitable as a model parameter for a biodiversity assessment on the landscape level, particularly when meadows or floodplain areas are assessed. Even the indicator value for humidity developed by Ellenberg et al. (1992) (see also Hill et al., 1999) for vascular plants has quite a similar potential as an indicator but leads to less clear statements. The indicator values of Ellenberg et al. (1992) were successfully correlated to species assemblages of beetles out of pitfall trap samples when plant communities in the catchment area of each trap were surveyed at the same time (Stumpf, personal communication, 1996).

3.1.10. Correlation of plant and animal assemblages

According to Döring et al. (2003) there is a lot of evidence of a correlation between species diversity of plants and zoological taxa (Weiss and Nentwig, 1992; Zanger et al., 1994; Denys, 1997). Among carabids, omnivorous groups as, for example, Harpalinae and Zabrinae, profit mostly from a higher species richness of plants. However, a positive correlation of weed cover to species richness of carabid beetles is restricted by the fact that a higher weed density results in a higher "spatial resistance" ("Raumwiderstand"; e.g. Heydemann, 1955) that hinders carabids moving around. Pfiffner et al. (2000) showed by multivariate data-analysis that site characteristics such as plant diversity and the type of habitat can significantly influence the ground beetle fauna. This attempt is extended to flower visiting insects and other animal taxa by Kratochwil and Schwabe (2001).

The correlation of plant species richness to carabid biodiversity includes all types of habitat preferences. This allows independence regarding methods used for ecological assessment in relation to the choice of goal parameters. From these findings it could be concluded that plant species richness can potentially be used as a surrogate measure for faunistic species diversity at least for carabids. However, Döring et al. (2003) point out that weed diversity might be suitable for a rapid assessment of faunistic biodiversity, but is alone not sufficient for a more sensitive measurement (e.g. by using highly specialised carabid species as indicators), particularly if mainly conventionally managed fields (between which the differences of management intensity are more subtle) have to be evaluated, as is the usual practice. Therefore, besides plant species richness, Döring et al. (2003) used three more parameters (activity density, activity density of single carabid species and percentage of weed cover) as indicators for carabid species richness.

3.2. Surrogate indicators

3.2.1. Soil quality index ("Bodenpunktzahlen")

Brose (2003) tested a correlation between soil quality index and carabid species diversity: the lower the soil quality index was the more ground beetle species could be recorded. Some results from other authors (e.g. Irmler, 2003; Büchs et al., 1997; Harenberg, 1997) support this hypothesis. In particular Irmler (2003) and Perner and Malt (2003) showed a correlation between sand content of soil and carabid and spider species richness. Steinborn and Meyer (1994) identified soil type as the most important influencing factor affecting the carabid species community followed by crop type and management intensity. Within this ranking, the effects of cultivation techniques were stronger on loamy soils compared to sandy soils. Also for plant communities, similar relations can be assumed (e.g. Mattheis and Otte, 1994; Wicke, 1998). However, the best parameter to explain the variances is "temperature" for spiders and "soil humidity" for beetles (Perner and Malt, 2003). Sandy soils usually achieve a low soil quality index. With the precondition that a mapping of soil quality is available this indicator, which is easy to apply, has the potential for a rapid classification of agriculturally used landscapes on higher scale levels, for instance, within planning procedures or to define areas which are designed for particular environmental programmes.

However, as a factor mainly caused by natural conditions the soil index is almost independent of management intensity, so it has no educational component and will not lead farmers to change and improve their husbandry practice towards sustainability. Finally, large areas in central Europe with high yielding soils (e.g. loess soil) will again be virtually excluded from environmental programmes and approaches as they are already today. So, beyond special environmental programmes, for any subsidisation of the ecological performance of management practices, farmland should be classified into two or three classes depending on the soil quality and should be monitored and assessed separately by applying different goal criteria and thresholds of success (of an ecologically motivated crop production) as in environmental programmes.

3.2.2. Length and quality of field margins

Due to results of Wallin (1985), Gilgenberg (1986), Basedow (1988), Dennis and Fry (1992), Kinnunen et al. (1996), Pfiffner and Luka (1996) and Irmler (2003) it can be assumed that length and quality of field margins (uncultivated habitats adjacent to fields) influence the field fauna significantly. Length and quality of field margins is proposed by Waldhardt and Otte (2003) as an indicator for biodiversity from a botanical point of view.

3.2.3. Organic farming as a surrogate indicator

Döring et al. (2003) and Döring and Kromp (2003) showed that ground beetles in organic fields produce a higher diversity of stenotopic species (most of them xerophilic and typical for (sandy) field habitats) compared to conventionally managed fields. Furthermore, herbivorous species are enhanced due to the more diverse and dense weed cover. Conventional farming, however, promoted mainly populations of euryoecious and ubiquitous species and of some hygrophilous forest species due to a more dense crop plant cover. Considering other studies on this subject the supraregional validity of these indicators could be shown by Döring and Kromp (2003).

Compared to conventional crop management 34% more ground beetle species in organic fields were recorded. Within a survey of 58 fields Irmler (2003) recorded more species only in fields which had been managed organically in the long term (at least 30 years), but not in those that had only been managed organically for a short period. However, no ground beetle species predominated especially on organic fields. The differences in ground beetle composition in organic fields compared to conventional farming (e.g. Pfiffner and Luka, 1996, 1999; Pfiffner and Niggli, 1996; Hokkanen and Holopainen, 1986; Basedow, 1990; Raskin et al., 1992) are attributed to the higher abundance of prey items and lower pesticide input in extensively managed fields.

With regard to plant communities, in conventional agriculture mainly nitrophilous and ubiquitous pest weeds are enhanced. Plant communities of organic fields were "more complete" with a higher percentage of dicotyledons, and so a prolonged period with flowering plants and a gradual transition of plant communities, zonation types and reduction species numbers was observed from the field edge to the centre of the field, while in the conventional farming systems these changes occurred abruptly (Döring et al., 2003). Positive effects of weed communities on the abundance and biodiversity of beneficials (particularly parasitoids) have been highlighted by Altieri (1999). From these results it can be concluded that the specific features in species composition in organic farming systems also indicate functional consequences.

This raises the question of whether organic agriculture itself can be used as a(n) (indirect) surrogate indicator for matters of biodiversity and/or sustainable agriculture or whether adaptations have to be considered which potentially compensate for the differences in the species composition of organic and conventional farming systems or even complement them, so that—in order to provide a maximum biodiversity—both systems (i.e. conventional and organic) have their advantages.

Arguments for the installation of organic farming as a surrogate indicator to maintain biodiversity are as follows:

- The philosophy of organic farming is fundamentally oriented around the consideration of natural processes and cycles as basic elements of a sustainable manner of farming (e.g. http://www.ifoam.org).
- Organic farming is restricted to clear and fixed regulations stated by EU Regulation 2092/91 and 473/2002 (EU-Commission, 1991, 2002) or by the growers associations (e.g. Demeter, Bioland, etc.) which can be controlled easily (a control system is already well established), in contrast to other farming systems as, for instance, "integrated farming".
- Several studies show for a lot of taxa (e.g. ground beetles, birds, mammals, butterflies) that organic farming definitely enhances species numbers and biodiversity for animals as well as for plants (The Soil Association, 2000; Van Elsen, 1996, 2000; Frieben, 1997, 1998; Kromp, 1990; Döring and Kromp, 2003; Pfiffner, 1997; Pfiffner and Luka, 1999; Pfiffner and Mäder, 1997; Pfiffner and Niggli, 1996; Rösler and Weins, 1997).

Objections to using organic farming as a surrogate indicator for sustainable production are, for example:

- the use of substances containing copper, such as fungicides in potatoes and vineyards. Copper is obviously transferred to the food-chain (Wittasek, 1987; Rhee, 1997; Hopkin et al., 1986, Paoletti et al., 1988) and may affect animal biodiversity;
- intensive tillage systems due to mechanical weed control;
- a high percentage of summer crops with soil tillage in spring which affects soil-dwelling arthropods and their biodiversity negatively as shown clearly by Büchs et al. (1999).

Furthermore, differences between conventional and organic agriculture are gradual, that means there are "overlapping areas" regarding the ecological performance of both systems. In particular, a tendency to increase management intensity is observed, even in organic farming (Döring, 2000, in litteris). Döring et al. (2003) estimate organic farming as a high contribution to nature protection, but believe that it is alone not sufficient as an indicator for biodiversity.

3.2.4. Biodiversity indicators at landscape level with reference to habitat features

Indicators related to landscape issues should have the potential to be regionalised, but at the same time be suitable for blanket coverage assessments for supraregional planning. Furthermore, rules have to be defined for bottom-up assessments (conversion from a particular spot to a region).

Approaches concerning the development of a "landscape diversity index" which are founded on the hypothesis that the basic potential of the biodiversity of any location (and so of the landscape itself) depends on the diversity of land use types (and other habitats) in a defined area around the location considered. The aim of these approaches is a comparative evaluation of land use patterns to predict the effects of hypothetical land use changes using a model. They often deal with indicator species (assemblages) whose habitat requirements are considered in relation to landscape analysis. Areas with concentrations and deficiencies with respect to these habitat requirements are identified (Hawkins and Selman, 1994). Such integrated studies of patches within their surrounding land use matrix have been developed in several variations for animals (e.g. ants, beetles, birds), as well as for plants, e.g. by Dauber et al. (2003), Waldhardt and Otte (2003), Steiner and Köhler (2003) and Hirsch et al. (2003). A precondition for such an assessment is to allocate a certain value to each land use and habitat type. However, this procedure is based unavoidably to a certain extent on subjective decisions. The suitability of a modified Shannon-diversity index on the basis of taxon-related species numbers/species compositions for each land use (and habitat) type which are identified as reference values is tested (Dauber and Gießen, personal communication, 2000).

This raises the question of specific reference values for regions and/or natural units: Which target value should a "landscape diversity index" achieve in landscapes of different types and characters? Furthermore, what are the criteria to define reference values, target values and the surveyed area (circle radius) which is related to the location that should be assessed with regard to the taxon considered? Examples from a survey on "Ecological Priority Areas" in Switzerland by Jeanneret et al. (2003) show that:

- For the presence of butterflies in cultivated areas the performance of adjacent habitats is of great importance. This means that the attainability of these undisturbed areas plays a major role in the biodiversity of butterflies.
- For the presence of spiders, however (which colonise any location with considerable numbers by ballooning), the performance of adjacent habitats is less important than the diversity and spatial structure of the species community of plants (within and adjacent to the field) and the performance of husbandry practices.

4. Development of biotic agri-environmental indicators—fundamental problems

The most outstanding fundamental problems within the development of biotic agri-environmental indicators are summarised in Table 3. A selection of key issues is discussed here. The problems listed in Table 3 are multiplied by the fact that in agricultural landscapes, a focus on one or a few impact factors is not possible, but pesticides, harvest, tillage, mineral fertilisation, crop rotation, etc. are present in varying

Table 3
Problems of development and application of biotic indicators for biodiversity in agro-ecosystems

Studies on reference values and limits are lacking or can be determined only with efforts that are not feasible
Fixing of baseline references (e.g. species richness: for central Europe 1850 as year of reference defensible?)
Availability of suitable data
Sample size required often not feasible
Efforts for data survey too high even if restricted to a minimum set of data ⇒ a blanket coverage of indicator use is not possible
A high degree of uncertainty of statements has to be tolerated
Lacking standardisation of methods for data surveys and assessments
Weighing of indicators in complex assessment procedures ⇒ establishment of priorities
"Overlapping" of indicators (with regards the area covered) ⇒ risk of double assessments
Reference to spatial scale level
Need for regionalisation of indicators
Understanding and interpretation of the term "biodiversity"
Lack of models sufficiently adopted to demands of practice
Discrepancies between global requirements on international levels and the degree of sensivity and power for detailed statements on the ecosystem level
Efforts of attendant monitoring programmes
Basic requirements (possibility of standardisation, repeatability, unambiguous interpretation) not always fulfilled
Repetition of indicator surveys (frequency/time period)
Dealing with assessment errors and limitations due to bottom-up scaling and data-aggregation
Communication between pure research, applied research, institutions responsible for conversion into practice, political decision makers and users

performances depending on the type of husbandry practice (Paoletti, 1999a,b).

4.1. Development of baseline reference values

Data surveying and determining baseline reference values is one of the major problems particularly within field assessments and most serious for animal populations due to their high mobility, migrations between different habitats and changes in population density from season to season. Due to the fact that each location has its specific, almost individual (abiotic) conditions (e.g. (micro-)climate, soil type, (ground) water supply, exposition, habitat size, surroundings, vegetation structure, etc.) it is nearly impossible to draw general conclusions: each field trial has to be judged more or less as a unique "case study". Nevertheless, the elaboration of baseline reference values is theoretically possible (e.g. for species numbers by comparison of a regional special pool to the local species composition; see above). It does, however, require a tremendous amount of work to fulfil demands on statistical significance, which is usually not feasible. A way out of this problem is to relate results to a standard (control) treatment with fixed conditions (e.g. Bartels and Kampmann, 1994; Steinmann and Gerowitt, 2000; Holland et al., 1994). However, to derive general reference values, the features of a standard used as a reference and the goal criteria to be achieved have to be defined clearly and agreed on generally (i.e. supraregionally). This is usually not the case; each approach creates its own standard, meaning in the end that the standards are not comparable.

The standardisation of survey methods is essentially linked with these aspects. For invertebrates Duelli and Obrist (2003) recommend Berlese soil extractors for endogaeic taxa, pitfall traps for epigaeic arthropods and window flight traps for insects flying above the vegetation level. They prefer plots and transects as measures for recording biodiversity related to certain taxa.

4.2. Required sample size and quality of indication

Sample size influences essentially the quality of indication, thus, it is one of the most critical aspects with regard to the chances of putting biotic indicators into practice.

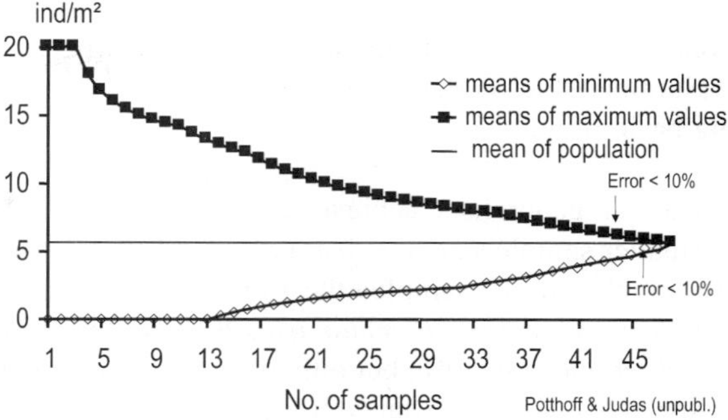

Fig. 1. Required sample size in relation to maximum possible error: means of *L. terrestris* abundance depending on the number of samples (Potthof and Judas, unpublished).

This issue which is particularly neglected in scientific attempts can often be demonstrated, for example, by an experiment by Potthoff and Judas (unpublished): on a forest soil plot of $12\,m^2$ a complete survey of the abundance of *Lumbricus terrestris* L. earthworms was conducted. The total of all 48 samples of $0.25\,m^2$ showed an abundance (area-related density) of $5.7\,ind/m^2$ on average. Fig. 1 shows the mean of *L. terrestris* abundance related to the sample size. The mean of *n* samples with the highest abundance was compared to *n* samples with the lowest abundance. It could be demonstrated that only if more than 90% of all potentially possible samples were taken (44 samples out of a total of 46 total samples) the maximum possible error was below 10%.

Financial means always limit personnel and temporal possibilities so that an appropriate balance has to be found between time-consuming field sampling, species identification, data processing and the skills expected from those who conduct the data surveys (Perner, 2003).

Perner (2003) created the "Level of Precision" (LOP) defined as the number of samples required to achieve a tolerable variance with regard to data. The definition of an LOP is quite reasonable because a precondition for the successful application of an indicator, particularly in agro-ecosystems, is that differences at species or species community level are detectable, regardless of a considerable "data noise" (mean variation) within the same farming system. This is of major importance because in practice, a comparison of extremely different farming systems is not usually made (e.g. between organic and conventional farming), but within conventional farming.

Summarising, Perner (2003) developed the following basic principles:

- Habitat generalists achieve a better LOP with a smaller sample size than habitat specialists (e.g. for a number of 10 samples the LOP for the generalist carabid species *P. melanarius* is 25%, for *Bembidion obtusum* 50%, but only 60–100% for habitat specialists as, for instance, *Calathus erratus*).
- Species communities should be preferred as indicators to indicators which rely on one species: to achieve the same LOP for a single species a higher number of samples has to be taken compared to species communities.
- Among community parameters, evenness is particularly recommended for biotic indication: Perner and Malt (2003) demonstrated for different kinds of grassland management that evenness is precise enough to distinguish between apparently rather similar management strategies.
- The required sample size varies greatly, up to a manifold of the original size: (a) from year to year (season to season), (b) from habitat to habitat, and (c) from taxon to taxon, with the consequence that in many cases differences cannot be stated as significant although they in fact occur.
- With regard to taxon, ground beetles (Coleoptera: Carabidae) require a larger sample size (pitfall traps) than spiders (Arachnida: Araneae) or Coleoptera in general to achieve the same LOP in order to assess species richness and evenness. A rather surprising result which contrasts sharply with common practice, because among field-inhabiting invertebrates, carabids are favoured as one of the easiest indicator taxons to assess.

4.3. Selection of taxa for biodiversity assessments

As mentioned above biodiversity in its entirety can never be recorded and assessed: Riecken (1992) showed that in biodiversity assessments within legally established planning procedures in Germany only a very restricted set of taxa was considered with a high risk of incorrect assessments as a consequence: overall, less than 50% of environmental assessment procedures considered zoological aspects at all, 34% of these considered only one taxon, 50% up to five and only 16% more than six and more taxonomical units (Table 4).

Table 4
Consideration of zoological taxa in environmental assessment procedures within legally established planning procedures (Riecken, 1992, 1997)

Taxon	Percentage in environmental assessments
Birds	95
Amphibians/reptiles	48
Mammals	30
Butterflies	27
Dragonflies	25
Snails	14
Grass-hoppers	11
Ground beetles	9
Other beetles	5
Bugs	5
Spiders	5
Leafhoppers	2

Regarding the practice of data surveys within repeated applications of bioindicators by McGeoch (1998) and Riecken (1992) plants and perhaps also vertebrate taxa such as birds are frequently used as indicators; however, invertebrate taxa are mostly neglected, although it is well known that invertebrate communities form the major part of biodiversity in every ecosystem and have the potential to indicate environmental stress and gradual changes (Büchs, 1995; Majer and Nichols, 1998; Wheater et al., 2000; Andersen et al., 2001).

4.4. Mutual neutralisation of indicators

A serious problem is that different indicators potentially neutralise themselves in producing opposite results particularly when used to indicate the success of certain measures applied. Different requirements of birds which nest in hedgerows and those which breed in open landscapes can be brought in as an example: while hedgerow breeders are enhanced by planting and maintaining hedgerows, the population density of open land birds is reduced and vice versa (Hoffmann and Greef, 2003). This phenomenon can be converted to all other taxons in principal (e.g. carabids, spiders, rove beetles, butterflies, etc.).

In this context, Hoffmann and Greef (2003) and Hoffmann et al. (2003) developed a kind of "mosaic indicator approach". The qualitative and quantitative assessment takes place under consideration of the historical development of the landscape. For the focus on "species" features, birds and flowering plants were selected, for the focus on "habitat" features, uncultivated areas (e.g. hedgerows, field margins, etc.) as well as areas within field locations with extreme conditions such as wetlands or arid spots (Brose, 2003; Richter et al., 1999) were considered.

4.5. Simplification of indication approaches

Studies on the basis of indicators must be technically simple and easily repeatable by different people with different levels of skill in different situations (Paoletti, 1999a,b). The main reason for assessing a facet of biodiversity with the help of an indicator, instead of measuring that target directly, is higher efficiency. Therefore, one needs to show that the loss of precision and accuracy is compensated for by gaining time and money. Because biodiversity in each facet is a moving target in time and space, indicators will need recurrent calibration. If a gain in efficiency cannot be convincingly shown because the precision and accuracy of the indicator cannot be established, then that "indicator" is not appropriate for the purpose intended. These principle thoughts of Von Euler (Vancouver, Canada, personal communication, 2000*) implicate a significant simplification of indicators, and of related assessment and survey methods. Using simplified methods (indicator systems), a considerable error probability and somewhat less clear statements have to be taken into account.

Generally, environmental monitoring programmes avoid the recording of invertebrates because of the high expenditure and costs (World Bank, 1998). Derived from Foissner's (1999) statements for Protozoa some key issues are listed which in principal can be transferred to most invertebrate in general:

- large number of species;
- costs, labour and high level of skill needed for identification;
- simple and/or computerised keys are lacking;
- time-consuming counting of abundance;
- as a contrast, for instance, to dragonflies or butterflies most invertebrate taxa are not attractive for the public and most researchers;
- unaffordable number of samples required to fulfil demands for statistical significance.

For a rational biodiversity assessment of arthropods, several procedures are suggested (Duelli and Obrist, 2003; Cranston and Hillman, 1992; Oliver and Beattie, 1996):

- To restrict sampling to a low number of taxa over a long period.
- To conduct sampling of a great range of taxa over a short period of time.
- Rapid biodiversity assessment: evaluates the number of taxa by using samples over a couple of weeks recorded with a standardised set of sampling methods, considering the whole taxonomic range but only on the basis of morphospecies.
- Sampling is not restricted to taxonomic units but to a sampling method. The more taxa a method is able to record, the higher the chance is for assessing a representative taxonomic range of the entire fauna of an ecosystem.

Therefore, Duelli and Obrist (2003) favour a reduction to an "optimised" selection of taxa for biodiversity assessments. Their concept includes a two-step approach that allows the use of all kinds of indicators with regard to their correlation to biodiversity. The best correlation was achieved by bugs and bees. It was shown that a reduction of the sampling period to five selected weeks by exceeding the number of taxa resulted in correlations which were nearly as good as those demonstrated for bugs and bees. If linear correlations of indicators to biodiversity are not focused on taxonomic units, but extended to a selected sampling method, the advantage is that the taxonomic range included in the assessment is widened. And so the chance increases of a better correlation to or representation of the entire species set of the ecosystem compartment assessed.

Simplified indication methods based on morphospecies carry the risk that a lot of species will be overlooked if taxonomically more ambitious taxa are included in the assessment (e.g. Auchenorrhyncha, Linyphiid-spiders, Diptera families like Anthomyiidae, Cecidomyiidae, Sciaridae, Coleoptera (sub)families like Aleocharinae, Cryptophagidae, Curculionidae or parasitic Hymenptera, etc.). Such a procedure might be possible for assessing ecosystems with intrinsically different features (e.g. hedgerows and fields), but in the case of comparing cultivated areas itself, whose species diversity differs only slightly (e.g. different kinds of conventionally used fields), an index based on morphospecies will be too uncertain to detect differences with the sensitivity and exactness needed and so, might fail.

In most of the many assessment procedures used currently, surrogate measures are usually used as indicators, for instance, percentages of uncultivated areas, the length of hedgerows or field margins, crop diversity, etc. (e.g. Braband et al., 2003; Menge, 2003). Demands for a simplification of methods for biodiversity assessments in land use systems are so extreme, that it can be assumed that in the end—regardless of all limitations—only surrogate indicators such as mentioned above seem to have a real chance of being applied in practice, as is already usual today (see Braband et al., 2003). Duelli and Obrist (2003) stress the fact that the use of surrogates is only possible after the conservation value has been established; they state also the lack of empirical data to test such indicators.

However, with regard to other ecosystem types of economical importance biotic indicators based on taxonomic work are not unusual at all: the evaluation of the water quality of freshwater ecosystems is highly standardised by the "Saprobien index" (whose origin can be traced back to nearly a 100 years ago). This index is applied regularly on an international scale, and is continuously developed (Friedrich, 1990; Marten and Reusch, 1992; Usseglio-Polatera et al., 2000). For freshwater habitats (or riparian habitats with contact to freshwater habitats) a status of assessments has already been achieved which is far beyond our status of the assessment of land use practices in terrestrial habitats (Innis et al., 2000):

- methods which have already shown their benefits in practice are regularly updated and cross-calibrated;
- new rapid assessment methods are in development that provide reasonable levels of accuracy for a variety of users in a variety of situations;
- assessments are currently developed for special applications with selected users;
- the degree of uncertainty is explicitly reported;
- implications of specific assessment methods on political decision making is openly addressed;
- the methods are formally tested for accuracy, cost effectiveness and practicability;
- Innis et al. (2000) show ways how to transfer models used for freshwater assessments to (semi-)terrestrial habitats.

5. Biodiversity assessment—do we know our goals and the effects of the measures we take?

The introductory phrases on the use of the term "biodiversity" shall elucidate how complex, difficult and misleading its use and, moreover, assessment procedures are, especially for agricultural ecosystems.

"Biodiversity assessment" means more than "biodiversity measurement". Biodiversity is essentially measured for a particular purpose. Biodiversity assessment is the analysis of differences between a present state and a reference one. Whatever this "reference state" may be depends on the purpose of the biodiversity assessment (Watt, Banchory, Scottland, UK, personal communication, 2000*).

It is well known (Odum, 1980) that if arable land is left to natural succession, the species diversity increases until a certain stage of succession (usually stages with shrubs and/or singles trees, that contain so to speak "a little of everything") is achieved. Only at the forest stage, as the final climax stage, does biodiversity decrease again, and is at its lowest inside of the forest ecosystem.

The year 1850 is rather often suggested as a kind of "reference year" that represents the peak of biodiversity in central Europe (Bick, 1982; Piorr, 2003). At that time the ideal of patchy and highly structured rural landscapes in central Europe occurred during conditions of subsistence and was accompanied by poverty and starvation (Konold, 1996). Derived from this knowledge, today the maximum increase of (structural) diversity and of the percentage of secondary components (e.g. landscape elements, land use types, habitat types) and of uncultivated areas is regarded as a "guarantee" for the development of a high level of biocoenotic biodiversity. In the sense described above, many models and recommendations in central Europe exist as, for instance:

- to reduce the field size (Knauer, 1993);
- to develop so called "habitat connective systems" ("Biotopverbundsysteme"), that means to increase the percentage of uncultivated areas at the field edge like field margins, tree rows, hedgerows, adjacent ditches (Jedicke, 1994; Steidl and Ringler, 1997);
- to introduce artificially sown weed strips into larger fields (Nentwig, 2000).

The endeavours to enhance the biodiversity of cultivated areas by introducing structural elements could be managed (and controlled) easily by agricultural/environmental administrations; it developed into one of the most popular measures and was integrated into several assessment procedures for evaluating the environmental impact of a farm or its husbandry practice (Braband et al., 2003). Such areas are called Ecological Priority Areas (EPAs) and are surrogate indicators of biodiversity, of which a minimum percentage of agricultural landscape is fixed (e.g. Roth and Schwabe, 2003), perhaps complemented by an index, that describes the diversity of habitat types, their dispersal and juxtaposition, and the edge density as it is proposed by the EU-Commission (Eiden et al., 2001a,b; Willems et al., 2001; Gallego et al., 2001; Steenmans and Pingborg, 2001).

This kind of action and these aims are quite understandable with regard to the intensification of crop management in central Europe, which began in the 1970s with a high increase in nitrogen and pesticide input combined with the creation of a landscape type (mostly during procedures within the reparcelling of land) whose first priority was to fulfil the technical demands for a maximum-yield-crop-production. However, there is a need for differentiation. Problems of such a "linear" understanding of biodiversity matters as well as the difficulty in selecting single species, structures or ecosystem components as representatives (indicators) for a high or low biodiversity shall be demonstrated in the following examples.

5.1. Contradiction of island-theory and mosaic-concept

According to the conclusions derived from the research on island ecology (Macarthur and Wilson, 1967) species richness increases together with an increase in the size of the area of the habitat type considered. However, following the mosaic-concept of Duelli (1997) species richness increases the more habitat types there are and the more heterogeneous habitat types are present in a defined area. According to the theoretical model, effects of the island-theory and mosaic-concept neutralise each other (see scheme in Fig. 2). In reality this model functions not as strictly as shown because of the exchange of species between different habitats (Duelli, 1992), but for a defined area it is clear: the larger each single habitat type becomes, the lower the number of different habitat types will be. Furthermore, the "quality" of species (euryoecious species versus stenoecious species) has to be considered: without any doubt an increase of species diversity will be recorded by an increasing percentage of structural elements in rural landscapes. However, most probably, habitat types which are introduced additionally, but are relatively small in size, are usually inhabited by different, but very common species in not very abundant populations, while stenoecious species in populations with a reasonable chance to survive occur only in considerable abundance if a threshold size of each habitat type is achieved (e.g. Kareiva, 1987; Kruess and Tscharntke, 1994).

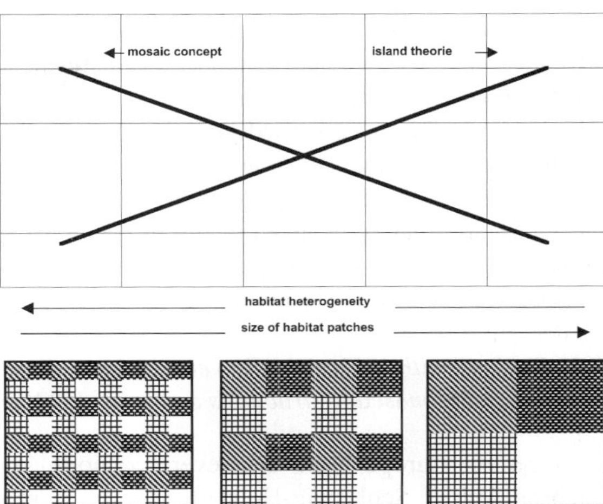

Fig. 2. Simplified model scheme to explain contradictory effects of island-theory (McArthur and Wilson, 1967) and mosaic-concept (Duelli, 1992): an increasing habitat size coincides with a decreasing habitat diversity (Duelli, 1992; see also Jedicke, 1994).

Contradictory to the mosaic-concept (Hansson et al., 1995; Duelli, 1992) Brose (2003) could not show an effect of the landscape features (e.g. heterogeneity) on the regional species biodiversity of carabids in temporary wetlands within fields, except the density itself of these wetland patches. Jeanneret et al. (2003) showed a dependence on the taxon considered: whereas in agriculturally used landscapes butterflies (Lepidoptera: Rhopalocera) are heavily influenced by the landscape structures adjacent to fields and the landscape matrix in general, the effect of structural components is less important for spiders and obviously other epigaeic (and endogaeic) arthropods.

5.2. Regions with a historically verified low percentage of landscape structures and large field size

In central Europe there are several regions (e.g. Sachsen-Anhalt, Schleswig-Holstein, Bavaria, central and north Poland, Hungarian Puszta, Austrian eastern Burgenland) that present large fields and a landscape poorly equipped with structural elements since the beginning of the 19th century. For instance, in some regions of Schleswig-Holstein, owned and managed by landlords, field plots of about 20 ha were completely normal as early as 300 years ago (Becker, 1998). Endeavours to increase biodiversity by introducing landscape elements would conflict with the historically grown type of landscape in this region (Fig. 3).

In central Europe recent field size and structural diversity (in a spatial sense) is strongly influenced by topographical conditions and the line of succession (e.g. "Realteilung", "Anerbenteilung", "Stockerbenteilung"). Such socio-economically regional differences and their impacts on biodiversity are considered, for example, within the concept of Roth and Schwabe (2003), where the goal criterion percentage of Ecological Priority Areas is also determined by historical developments: while the goal percentage in an area that is traditionally intensively used for agricultural production is fixed at a level of 10%, in a "Realteilungsgebiet" (very small fields due to equal portioning of inheritance) that is situated in a mountainous area with poor soils a goal percentage of Ecological Priority Areas of 23% was determined.

5.3. Examples of high diversity levels and habitat-related species communities on large-sized fields

After the reunion of Germany there was an intensive discussion on the faunistic importance of large-sized fields (>20 ha), which are more common in eastern Germany, versus small-sized fields (Wetzel, 1993; Poehling et al., 1994) with regard to biodiversity and field-typical expressions of zoocoenosis. In spite of all objections, recent investigations stated: on large-sized fields in eastern Germany (Wetzel, 1993; Wetzel et al., 1997; Volkmar and Wetzel, 1998; Kreuter, 2000; Hoffmann and Kretschmer, 2001; Richter et al., 1999; Stachow et al., 2001; Brose, 2003) and Hungary (Basedow et al., 1999) comparatively high species numbers and a high percentage of endangered species were recorded for several taxa. Furthermore, on large fields in agriculturally used landscapes almost completely cleared of structural elements, specialist species occurred among carabid species that were not recorded in nearby smaller fields in patchy, well structured landscapes (Table 5; Kreuter, 2000). Most of the species recorded in the centre of large fields are xerothermophilic, so that the species composition obviously shows special adaptations to the regional climatic conditions. The climate in eastern parts of central Europe (e.g. eastern Germany, Poland, Czech Republic, eastern parts of Austria and Hungary) is far more continental than in western parts (e.g. western Germany, northern parts of Switzerland, eastern parts of France and Benelux) and, therefore, the fauna contains more elements that are adopted to eastern "steppe"-conditions. These "steppe"-elements obviously meet with suitable conditions particularly within large fields in continental climates. Furthermore, the structure of these large fields develops considerably within field differences with regards, for example, to soil type, seed density, relief, exposure and soil humidity which results in a great variability of microhabitats and microclimatic conditions (Richter et al., 1999; Brose, 2003).

5.4. Examples of high levels of biodiversity in rural landscapes with a large extent of human impact

The reparcelling in the Kaiserstuhl area, a vine-growing area in south-west Germany, at the end of the 1970s, resulted in a significant reduction of the number of vineyard terraces and embankments. The resulting landscape gave the region its title of "lunar landscape", compared to the tiny structure of the old vineyards (Fig. 4a and b).

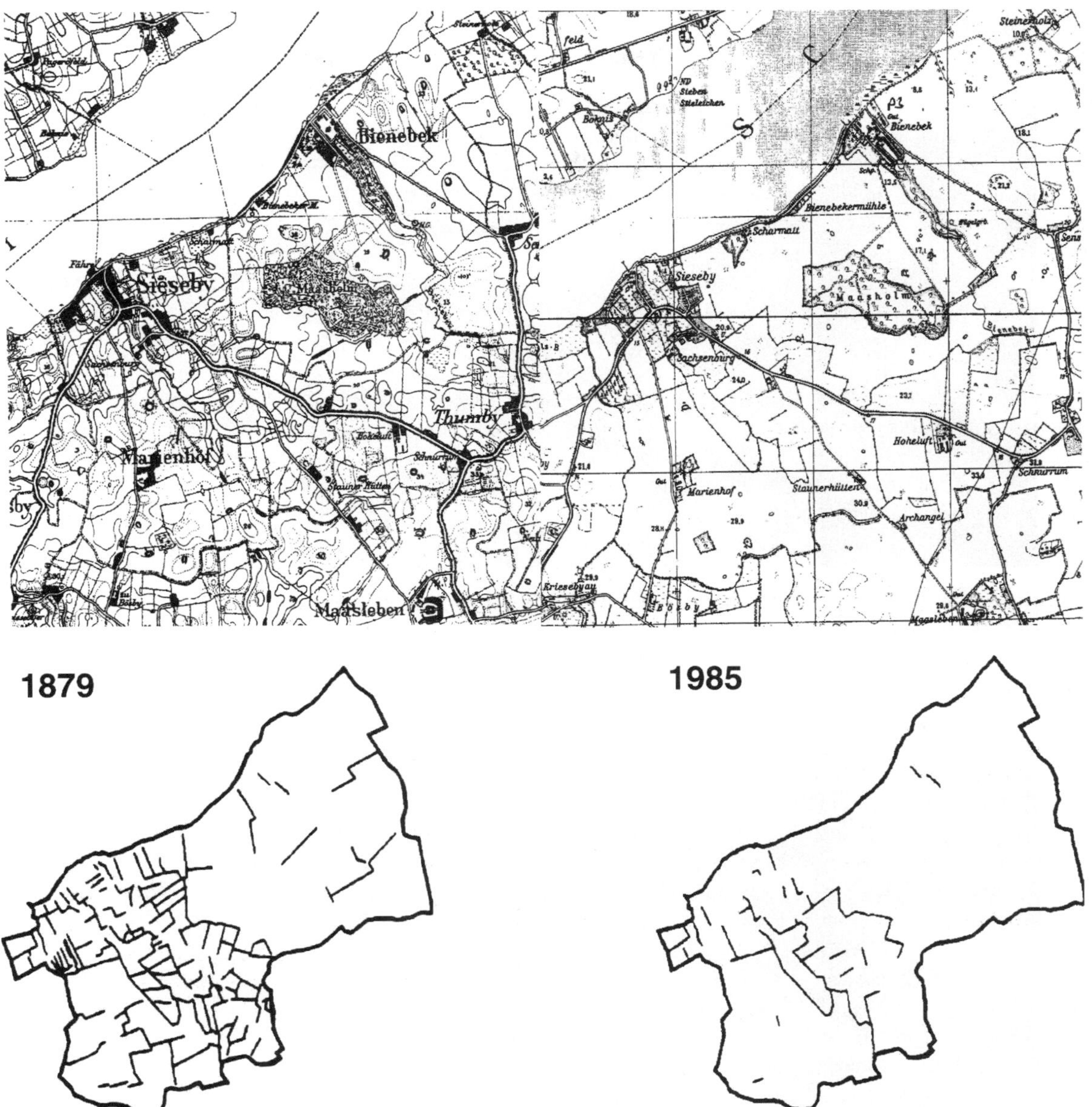

Fig. 3. Top: map (1:25,000) of the district of Bienebeck, Schleswig-Holstein, northern Germany in 1879 (left) and 1985 (right). South of the river "Schlei" the district is already 1879 divided into two parts: the south-western part is characterised by small-sized properties, the north-western part by large scale land ownership dating back to 300 years ago. Bottom: initial situations and developments of both parts between 1879 and 1985 with regard to hedgerow density (Becker, 1998).

However, long-term faunistic and floristic investigations that compared the status after reparcelling of the landscape with the semi-natural situation in a nearby nature reserve did not result in a considerable reduction of biodiversity or of the percentage of, for instance, "Red List" spider species (Table 6) and other invertebrates (Kobel-Lamparski and Lamparski, 1998; Kobel-Lamparski et al., 1999): moreover, the

Fig. 4. Vine-growing areas within the Kaiserstuhl region (south-west Germany). (a) View on large terraces and embankments in the foreground which were built approximately 20 years ago, and the dry meadows of the nature reserve "Badberg" in the background which were used as a reference (Kobel-Lamparski and Lamparski, 1998). (b) View on an original vine-growing area with small terraces and embankments near Schellingen (source: htttp://kaiserstuhl.net/cgi-kaiserstuhl/view.cgi?Titel=Schelingen&Bild=schelingen-01.jpg).

Table 5
Comparison of two (pairs of) organically managed fields in eastern Germany with different field sizes and richness of landscape structures (Kreuter, 2000, unpublished)

	Field location A	Field location B
Preconditions		
Field size	7.5/3.8 ha	42.5 ha
Landscape structure	Rich	Poor
Soil quality index "Bodenpunktzahlen"	55	30–75
Crop rotation	Winter wheat, oats, winter rye, winter wheat, alfalfa, set-aside, triticale, winter wheat	Sugar beet, yellow peas, oats, winter barley, winter wheat, summer barley, yellow peas
Ground beetle situation		
Species number		
Total	106	101
Field margin	92	89
Field centre	82	72
No. of individuals/trap × days		
Total	5.3	6.1
Field margin	3.4	5.8
Field centre	6.4	5.8

All fields belong to the same farm (Ökohof Seeben, Sachsen-Anhalt) and are situated close by (approximately 5 km).

percentage of thermophilic species increased and an intensive exchange of species between embankments and vineyards could be observed. Due to regional climatic conditions the typical fauna of the Kaiserstuhl area is xerothermophilic. By implementation of the large embankments such locations developed in great numbers that cover 30% of the area. These large new embankments are not under pressure by human activities, whereas the former small embankments were affected to a greater extent by pesticide fumigation or shading by the vine grapes. Furthermore, the new embankments are not isolated, but linked together and grid-like, spread over the whole vine-growing area, so that they can develop different exposures and structural expressions, resulting in a great variety of microclimatic conditions. Moreover, vine grapes are a permanent crop which allows a long-term development of the species community.

Table 6
Comparison of spider communities in an 18-year-old vineyard embankment (built during intensive reparcelling) and in a dry meadow of a nearby nature reservation (Kobel-Lamparski and Lamparski, 1998)

Parameter	Dry meadow (Mesobrometum) nature reservation	Artificial vineyard embankment (18 years old)
Species numbers	81	85
No. of "Red List" species	48	40
Percentage of "Red List" species of total samples	74.9	72.0
No. of non-threatened species	33	35
ind per trap	405 ± 82	201 ± 55
Shannon-diversity index (H_s)	2.92	3.09
Evenness	0.66	0.70

5.5. *Equal arthropod biodiversity at different levels of patchiness and structural complexity of rural landscapes due to the management intensity*

Comparative investigations in a poorly structured and large-sized agricultural area in Poland and a well structured small scale region at Unterfranken (Germany) by Mühlenberg and Slowik (1997) showed that in Poland the same level of insect diversity could be achieved as in Unterfranken, although the patchiness and structural complexity in Poland was remarkably less developed than at the German location (Fig. 5). Obviously, structural complexity of the landscape is not the most decisive factor that guarantees the development of a high biodiversity level. Mühlenberg and Slowik (1997) explain this observation with a

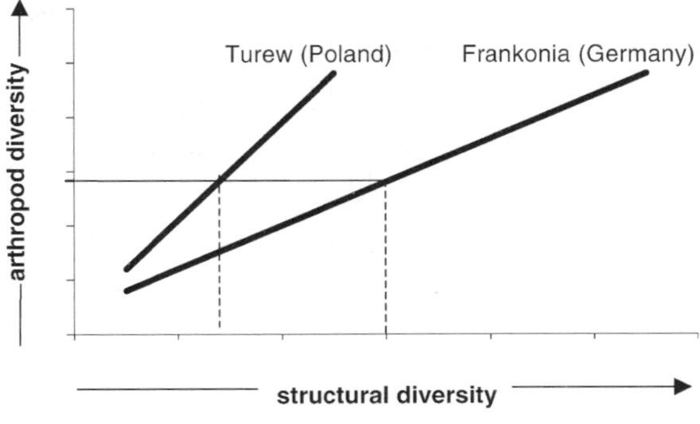

Fig. 5. Model of the development of arthropod species diversity depending on the structural diversity of the landscape according to observations made by Mühlenberg and Slowik (1997) in Poland (Turew) and Germany (Frankonia).

Table 7
Ground beetle communities (Coleoptera: Carabidae) in winter wheat fields of different size in Sachsen-Anhalt (eastern part of Germany = large-sized fields) and Lower Saxony (western part of Germany = small-sized fields) in 1991–1993 (Volkmar et al., 1994)

Study site	Barnstädt (Sachsen-Anhalt)			Peißen (Sachsen-Anhalt)			Hötzum (Lower Saxony)		
Sampling period	7 May to 6 August	9 April to 23 July	5 April to 8 July	8 May to 7 August	8 April to 21 July	7 April to 8 July	8 May to 7 August	9 April to 21 July	6 April to 6 July
Year	1991	1992	1993	1991	1992	1993	1991	1992	1993
Field size (ha)	141.0	139.0	43.0	24.0	24.0	24.0	6.0	6.0	6.0
No. of species	32	26	37	36	38	39	19	22	24
ind per trap	50.0	42.5	354.9	97.6	117.3	183.2	590.9	30.3	58.2
Shannon-diversity index (H_S)	2.59	2.60	2.14	2.48	2.74	2.85	1.09	1.97	1.18
Evenness	0.75	0.8	0.59	0.69	0.75	0.78	0.37	0.64	0.37
No. of dominant species	5	6	6	7	6	8	4	3	2
Dominant carabid species[a]	*Harpalus aeneus* (17.8%) *Pseudophonus rufipes* (16.9%) *Anchomenus dorsalis* (13.1%) *Demetrias atricapillas* (10.2%) *Amara familiaris* (9.3%) Other carabid species (32.7%)	*Calosoma auropunctatum* (14.6%) *Pseudophonus rufipes* (12.8%) *Anchomenus dorsalis* (12.0%) *Harpalus aeneus* (11.5%) *Poecilus cupreus* (11.5%) *Harpalus distinguendus* (11.0%) Other carabid species (26.6%)	*Poecilus cupreus* (35.8%) *Harpalus distinguendus* (17.1%) *Anchomenus dorsalis* (12.5%) *Microlestes minutulus* (9.0%) *Harpalus aeneus* (6.4%) *Poecilus punctulatus* (3.8%) Other carabid species (15.4%)	*Poecilus cupreus* (21.1%) *Amara familiaris* (17.3%) *Pseudophonus rufipes* (10.7%) *Bembidion lampros* (9.4%) *Harpalus aeneus* (8.9%) *Pterostichus melanarius* (8.4%) *Calathus fuscipes* (7.7%) Other carabid species (16.4%)	*Poecilus cupreus* (19.5%) *Anchomenus dorsalis* (12.0%) *Amara familiaris* (11.5%) *Harpalus aeneus* (8.7%) *Pterostichus melanarius* (6.1%) *Calosoma auropunctatum* (5.9%) Other carabid species (36.4%)	*Calathus fuscipes* (14.5%) *Harpalus aeneus* (11.5%) *Poecilus punctulatus* (9.5%) *Poecilus cupreus* (9.4%) *Anchomenus dorsalis* (6.7%) *Microlestes minutulus* (6.3%) *Pterostichus melanarius* (5.8%) *Calosoma auropunctatum* (5.5%) Other carabid species (30.7%)	*Pterostichus melanarius* (52.9%) *Trechus quadristriatus* (38.4%) *Clivina fossor* (2.5%) *Pterostichus niger* (2.4%) Other carabid species (3.8%)	*Pterostichus melanarius* (46.2%) *Loricera pilicornis* (15.5%) *Anchomenus dorsalis* (11.2%) Other carabid species (27.1%)	*Pterostichus melanarius* (74.7%) *Clivina fossor* (6.9%) Other carabid species (18.4%)

Compared to the eastern fields the western fields were situated in a more structured landscape, but managed with higher levels of N-fertilisers and pesticides.
[a] Percentage of dominance only listed if species was dominant according to Engelmann (1978).

generally far more extensive land use (crop management) in Poland compared to Germany. This conclusion is stated by Volkmar et al. (1994), who recorded significantly lower species numbers of ground beetles (Coleoptera: Carabidae) in comparatively small, but intensively managed fields (6 ha) in the West Germany "Hildesheimer Börde" in comparison to large-sized fields (24 and 141 ha) in the East Germany "Magdeburger Börde" (Table 7). Moreover, derived from the knowledge on the ecology of the ground beetles recorded, most species of the small-scaled western location could be classified as "indicators for intensification" (Büchs et al., 1999).

The intention of the examples above is to demonstrate that environmental measures which aim to maximise "biodiversity" as the nominal number of species without consideration of quality aspects (e.g. stenoecious or habitat-specific species) or other topics belonging to the hierarchic model of Noss (1990), and which are applied on a nation-wide scale without any differentiation could be misleading. It was shown that environmental goals and measures have to be adopted very specifically depending on the regional geographical unit, and that landscape-related historical and socio-economic developments have to be considered.

6. Conclusion

There are numerous ecological parameters for both the population and community level that are suitable as assessment criteria in the outlined area. The definition of which goal the assessment should be orientated towards is decisive for its methods and modalities. It has to be defined which qualities of a system are assessed as good or worthwhile and which as bad or unconditional. However, if (species-)diversity and/or a maximum of "rare" and "endangered" species are not identified as the positive goal criterion, but the development of a species community typical for the investigated ecosystem or geographical unit, respectively, then it can be assumed that all parameters which describe reproduction conditions or reproduction success of (animal) populations in any form will belong to the essential assessment criteria. For instance, only healthy populations (precondition: e.g. a sufficient food basis) which have the possibility of reproducing successfully (precondition: e.g. an area size adapted to the demands of the species and suitable habitat resources) safeguard the survival of a species. Therefore, (if applied to the whole species community of an ecosystem) healthy populations guarantee the conservation of biodiversity, but also of beneficial aspects (e.g. predation, parasitation, decomposition) on a high level. Therefore, the "physical fitness" of populations of wild animal and plant species typical in agro-ecosystems can be estimated as a decisive measure for the seriousness of the conversion of consumer protection interests into action with regard to food production.

A number of examples shows that "biodiversity" in the sense of maximising the number of species, varieties, genotypes, etc. is a misleading and, as such, not a suitable criterion for indication approaches. Moreover, the review-like analysis demonstrated that "biodiversity" is a much more complex term that consists in various facets each of them contributing to "biodiversity" in its entirety like pieces of a jigsaw puzzle. However, it could be also shown that biotic indicator approaches relying on such "puzzle pieces" (like, for instance, parameters relying on reproduction and/or physical fitness) might be thoroughly appropriate to give indications of the development of biodiversity as such.

With regard agro-ecosystems the central platform for the action of these organisms (i.e. their functional importance) is the cultivated area itself. Therefore, highest priority has to be granted to the preservation of the essential resources required by those agrobiocoenoses that are typical for the region and the relevant geographical unit, so that assessments should not be restricted to marginal or uncultivated habitats in rural landscapes, but mainly focus on the cropped area itself.

Although ecological sciences are obvious able to provide a set of indicators suitable for a detailed and predictive analysis of nearly any situation affecting the agro-ecosystems, they do not hit the demands of those institutions (mostly administrations) who need to apply such indicators in practice. Spoiled by the easy handling of abiotic indicators administrations expect that:

- (biotic) indicators are easy to assess and to understand;

- interpretations, statements and predictions are safe and easy, i.e. that they can be expressed by numerical measures and baseline reference values as well as thresholds which are clearly defined so that they are legally valid;
- best one (biotic) indicator is able to represent and characterise an ecosystem as a whole and all processes going off within;
- as far as possible no extra data surveys have to be conducted.

In a nutshell, derived from experience with abiotic indicators, also biotic indicators are expected to produce omnipotential "technical" values.

Furthermore, in order to be able to manage the continuously increasing number of (biotic and abiotic) indicators and in order to fit in the financial limitations, administration tend to aggregate data and to reduce the number of indicators that are used. This tendency, which is basically understandable, increases with the administration level and ends up in the use of highly aggregated indicators the power of statement of which is rather weak and general (e.g. number of bird species with breeding records or percentage of organic farming). These scenarios lead at least to three consequences with regard biotic indicators:

(a) The status of each biotic indicators has to be reviewed: it has to be critically discussed whether agri-environmental biotic indicators as recently still more or less neglected parts within established assessment systems a more important role has to be acknowledged as it is currently realised within the control of the quality of freshwater habitats: nearly each local office for freshwater control has someone employed who is responsible for biotic assessment of freshwater ecosystems by the "Saprobia index" which, for instance, needs a regular update of data by field surveys and a rather detailed taxonomic work. The fact that budget will not be increased in order to fulfil these new tasks is an argument frequently used against the intensification of labour in the area of routine biotic assessment procedures in terrestrial ecosystems. However, possibilities of reorganisation seem not to be fully exhausted and the example of fresh water management shows to what extent biotic assessments can be realised.

(b) The paper showed clearly that depending on the geographic level considered (e.g. if an indicator is used to assess the environmental situation of a field, of a whole farm, a region or a country) and on the goals that should be achieved an indicator has to fulfil different requirements. Therefore, the establishment of a hierarchic system of indicators (particularly biotic ones) including their linking-up considering the different levels/goals mentioned is needed as well on the scientific as on the administrative side. This should be linked with automatic control procedures (represented by committees with different composition regarding the kind and level of experts) which safeguard a regular control of quality and suitability of those indicators in use, but also check indicator approaches not in official use in order to guarantee that developments are not overlooked, and which have the power to exchange indicators that are identified as inappropriate or if the focus of society has changed, respectively. However, as far as it can be judged up to date a consistent and well accepted (hierarchical) system particularly of biotic indicators which covers and integrates all "geographic" levels and different goals of assessment is completely lacking.

(c) If the current practice will be continued, surrogate indicators will be the only choice to hit the demands of the administration mentioned above. However, the misleading potential of those indicators was illustrated in Sections 5.1–5.5. Nevertheless, under the current conditions surrogate indicators will be the only way out of the dilemma of biotic assessments. Thus, the main focus in research on biotic indicators for the forthcoming years seems to be whether statements on and predictions of biodiversity development derived from surrogate indicators (as, for instance, percentage of "ecological priority areas", length of hedgerows, field margins, weed strips, field size, pesticide and fertiliser input) are correlated to quantitative (number of species, varieties, genotypes and their abundance) as well as qualitative (e.g. fitness, percentage of stenotopic, rare or functionally important species, functional responses, etc.) features of agricultural areas (e.g. Heyer et al., 2003).

Acknowledgements

Many thanks to I. Bürig (Braunschweig) for her intensive engagement to improve the English.

References

Achtziger, R., Nickel, H., Schreiber, R., 1999. Auswirkungen von Extensivierungsmaßnahmen auf Zikaden, Wanzen, Heuschrecken und Tagfalter im Feuchtgrünland. Schriftenreihe des Bayerischen Landesamtes für Umweltschutz 150, 109–131.

Altieri, M.A., 1999. The ecological role of biodiversity in agro-ecosystems. Agric. Ecosyst. Environ. 74, 19–31.

Alvarez, T., Frampton, G.K., Goulson, D., 2001. Epigaeic Collembola in winter wheat under organic, integrated and conventional farm management regimes. Agric. Ecosyst. Environ. 83 (1–2), 95–110.

Andelman, S.J., Fagan, W.F., 2000. Umbrellas and flagships: efficient conservation surrogates or expensive mistakes? Proc. Natl. Acad. Sci. U.S.A. 97 (11), 5954–5959.

Andersen, A.N., Ludwig, J.A., Lowe, L.M., Rentz, D.C.F., 2001. Grasshopper biodiversity in Australian tropical savannahs: responses to disturbance in Kakadu National Park. Aust. Ecol. 26, 213–222.

André, H.M., 1976. Introduction a l'etude écologique des communautés de microarthropodes corticoles sonmises a la pollution atmospherique I. Les microhabitats corticoles. Bull. Ecol. 7, 431–444.

Andrzejewska, L., 1976. The influence of mineral fertilization on the meadow phytophagous fauna. Pol. Ecol. Stud. 2, 93–109.

Andrzejewska, L., 1991. Formation of Auchenorrhyncha communities in diversified structures of agricultural landscapes. Pol. Ecol. Stud. 17, 267–287.

Assing, V., 1992. Zur Bionomie von *Xantholinus rhenanus* COIFF. und anderen bodenbewohnenden Xantholinen (Col., Staphylinidae) in Nordwestdeutschland. Zoologische Jahrbücher für Systematik 115, 495–508.

Banks, C.J., 1959. Experiments with suction traps to assess the abundance of Syrphidae (Diptera), with special reference to aphidophagous species. Entomologia Experimentalis et Applicata 2, 110–124.

Bartels, G., Kampmann, T. (Eds.), 1994. Auswirkungen eines langjährigen Einsatzes von Pflanzenschutzmittel bei unterschiedlichen Intensitätsstufen und Entwicklung von Bewertungskriterien. Mitteilungen aus der Biologischen Bundeanstalt für Land- und Forstwirtschaft 295, 1405.

Basedow, T., 1988. Feldrand, Feldrain und Hecke aus der Sicht der Schädlingsregulation. Mitteilungen aus der Biologischen Bundesanstalt für Land- und Forstwirtschaft 247, 129–137.

Basedow, T., 1990. Jährliche Vermehrungsraten von Carabiden und Staphyliniden bei unterschiedlicher Intensität des Ackerbaus. Zoologische Beiträge 33, 459–477.

Basedow, T., 1998. Langfristige Bestandsveränderungen von Arthropoden in der Feldflur, ihre Ursachen und deren Bedeutung für den Naturschutz, gezeigt an Laufkäfern (Carabidae) in Schleswig-Holstein, 1971–1996. Schriftenreihe für Landschaftspflege und Naturschutz 58, 215–227.

Basedow, T., Braun, C., Lühr, A., Naumann, J., Norgall, T., Yanes, G., 1991. Biomasse und Artenzahl epigäischer Raubarthropoden auf unterschiedlich intensiv bewirtschafteten Weizen- und Rübenfeldern: Unterschiede und ihre Ursachen. Ergebnisse eines dreistufigen Vergleichs in Hessen, 1985 bis 1988. Zoologische Jahrbücher für Systematik 118, 87–116.

Basedow, T., Tóth, F., Kiss, J., 1999. The species composition and frequency of spiders in fields of winter wheat in Hungary (north-east of Budapest) and in Germany (north of Frankfurt/M.). An attempt of comparison. Mitteilungen der Deutschen Gesellschaft für allgemeine und angewandte Entomologie 12, 166–263.

Becker, W., 1998. Die Eigenart der Kulturlandschaft—Bedeutung und Strategien für die Landschaftsplanung. VWF Verlag für Wissenschaft und Forschung, Berlin, 281 pp.

Behan-Pelletier, V.M., 1999. Oribatid mite biodiversity in agro-ecosystems role for bioindication. Agric. Ecosyst. Environ. 74, 411–423.

Bick, H., 1982. Bioindikatoren und Umweltschutz. Decheniana-Hefte 26, 2–5.

Blake, S., Foster, G.N., Eyre, M.D., Luff, M.L., 1994. Effects of habitat type and grassland management practices on the body size distribution of carabid beetles. Pedobiologia 38, 502–512.

Blower, G.J., 1955. Centipedes and millipedes as soil animals. In: Kevan, D.E. (Ed.), Soil Zoology. London, pp. 138–151.

Boatman, N., 1994. Field Margins: Integrating Agriculture and Conservation. British Crop Protection Council, Surrey, UK, 404 pp.

Bohac, J., 1999. Staphylinid beetles as bioindicators. In: Paoletti, M.G. (Ed.), Invertebrate Biodiversity as Bioindicators of Sustainable Landscapes. Agric. Ecosyst. Environ. 74, 357–372.

Bongers, T., 1990. The maturity index: an ecological measure of environmental disturbance based on nematode species composition. Oecologia 83, 14–19.

Braband, D., Geier, U., Köpke, U., 2003. Bio-resource evaluation within agri-environmental assessment tools in different European countries. In: Büchs, W. (Ed.), Biotic Indicators for Biodiversity and Sustainable Agriculture. Agric. Ecosyst. Environ. 98, 423–434.

Brose, U., 2003. Regional diversity of temporary wetland carabid beetle communities: a matter of landscape features or cultivation intensity? In: Büchs, W. (Ed.), Biotic Indicators for Biodiversity and Sustainable Agriculture. Agric. Ecosyst. Environ. 98, 163–167.

Brown, V.K., Southwood, T.R.E., 1987. In: Gray, A.J., Crawley, M.J., Edwards, D.J. (Eds.), Secondary Succession: Patterns and Strategies. Blackwell Scientific Publications, Oxford, pp. 315–337.

Büchs, W., 1988. Stamm- und Rindenzoozönosen verschiedener Baumarten des Hartholzauenwaldes und ihr Indikatorwert für die Früherkennung von Baumschäden. Ph.D. thesis, University of Bonn, pp. 1–813.

Büchs, W., 1993. Auswirkungen unterschiedlicher Bewirtschaftungsintensitäten auf die Arthropodenfauna von Winterweizenfeldern. Verhandlungen der Gesellschaft für Ökologie 22, 27–34.

Büchs, W., 1994. Effects of different input of pesticides and fertilizers on the abundance of arthropods in a sugar beet crop: an example for a long-term risk assessment in the field. In: Donker, M., Eijsackers, H., Heimbach, F. (Eds.), Ecotoxicology of Soil Organisms. Lewis Publishers, Boca Raton, FL, USA, pp. 303–321.

Büchs, W., 1995. Tierökologische Untersuchungen als Grundlage zur Charakterisierung von Ökosystemen und Indikation von Umweltbelastungen. Habilitation thesis, Faculty of Natural Sciences, Technical University of Braunschweig, pp. 1–312.

Büchs, W., Harenberg, A., Zimmermann, J., 1997. The invertebrate ecology of farmland as a mirror of the intensity of the impact of man? An approach to interpreting results of field experiments carried out in different crop management intensities of a sugar beet and an oil seed rape rotation including set-aside. Biol. Agric. Horticult. 15, 83–107.

Büchs, W., Harenberg, A., Prescher, S., Weber, G., Hattwig, F., 1999. Entwicklung von Evertebratenzönosen bei verschiedenen Formen der Flächenstillegung und Extensivierung. In: Büchs, W. (Ed.), Nicht bewirtschaftete Areale in der Agrarlandschaft—ihre Funktionen und ihre Interaktionen mit landnutzungsorientierten Ökosystemen. Mitteilungen aus der Biologischen Bundesanstalt für Land- und Forstwirtschaft Berlin Dahlem 368, 9–38.

Büchs, W., Harenberg, A., Zimmermann, J., Weiß, B., 2003. Biodiversity, the ultimate agroenvironmental indicator? Potential and limits for the application of faunistic elements as gradually indicators in agro-ecosystems. In: Büchs, W. (Ed.), Biotic Indicators for Biodiversity and Sustainable Agriculture. Agric. Ecosyst. Environ. 98, 99–123.

Canard, A., Marc, P., Ysnel, F., 1999. An experimental system to test invertebrate biodiversity by means of spiders. Bull. Br. Arachnol. Soc. 1998, 319–323.

Chiverton, P.A., 1988. Searching behaviour and cereal aphid consumption by *Bembidion lampros* and *Pterostichus cupreus*, in relation to temperature and prey density. Entomologia Experimentalis et Applicata 47, 173–182.

Clarke, G.M., 1993. Fluctuating asymmetry of invertebrate population as a biological indicator of environmental quality. Environ. Pollut. 82, 207–211.

Clarke, K.R., Warwick, R.M., 1998. A taxonomic distinctness index and its statistical properties. J. Appl. Ecol. 35 (4), 523–531.

Cranston, P.S., Hillman, T., 1992. Rapid assessment of biodiversity using biological diversity technicians. Aust. Biol. 5, 144–154.

Curry, J.P., 1987. The invertebrate fauna of grassland and its influence on productivity. 1. The composition of fauna: grass and forage. Science 42, 103–120.

Curry, J.P., 1994. Grassland Invertebrates—Ecology. Influence on Soil Fertility and Effect on Plant Growth. Chapman & Hall, London.

Dauber, J., Hirsch, M., Simmering, D., Waldhardt, R., Otte, A., Wolters, V., 2003. Landscape structure as an indicator of biodiversity: matrix effects on species richness. In: Büchs, W. (Ed.), Biotic Indicators for Biodiversity and Sustainable Agriculture. Agric. Ecosyst. Environ. 98, 321–329.

Den Boer, P.J., 1968. Zooökologisch onderzoek op het Biologisch Station Wijster, 1959–1967. Miscellaneous paper. Wageningen 2, 161–181.

Den Boer, P.J., 1977. Dispersal power and survival: carabids in a cultivated countryside. Miscellaneous paper. Wageningen 14, 1–190.

Dennis, P., Fry, G.L.A., 1992. Field margins: can they enhance natural enemy population densities and general arthropod diversity on farmland? Agric. Ecosyst. Environ. 40, 95–115.

Denno, R.F., 1994. Influence of habitat structure on the abundance and diversity of planthoppers. In: Denno, R.F., Perfect, T.J. (Eds.), Planthoppers—Their Ecology and Management. Chapman & Hall, New York, pp. 140–160.

Denno, R.F., Roderick, G.F., 1991. Influence of patch size, vegetation structure and host plant architecture on the diversity, abundance and life history styles of sap-feeding herbivores. In: Bell, S.S., McCoy, E.D., Muchinsky, H.R. (Eds.), Habitat Structure: The Physical Arrangement of Objects in Space. Chapman & Hall, New York, pp. 169–196.

Denys, C., 1997. Fördern Ackerrandstreifen die Artendiversität in einer ausgeräumten Agrarlandschaft? Untersuchungen am Beispiel der Insektengemeinschaften am Beifuß (*Artemisia vulgaris* L.). Mitteilungen der Deutschen Gesellschaft für allgemeine und angewandte Entomologie 11, 69–72.

D'Hulster, M., Desender, K., 1984. Ecological and faunal studies of Coleoptera in agricultural land. IV. Hibernation of Staphylinidae in agro-ecosystems. Pedobiologia 26, 65–73.

Döring, T.F., Kromp, B., 2003. Which carabid species benefit from organic agriculture? A review of comparative studies in winter cereals from Germany and Switzerland. In: Büchs, W. (Ed.), Biotic Indicators for Biodiversity and Sustainable Agriculture. Agric. Ecosyst. Environ. 98, 153–161.

Döring, T.F., Möller, A., Wehke, S., Schulte, G., Broll, G., 2003. Biotic indicators of carabid species richness on organically and conventionally managed arable fields. In: Büchs, W. (Ed.), Biotic Indicators for Biodiversity and Sustainable Agriculture. Agric. Ecosyst. Environ. 98, 133–139.

Duelli, P., 1992. Mosaikkonzept und Inseltheorie in der Kulturlandschaft. Verhandlungen der Gesellschaft für Ökologie 21, 379–384.

Duelli, P., 1997. Biodiversity evaluation in agricultural landscapes: an approach at two different scales. Agric. Ecosyst. Environ. 62, 81–91.

Duelli, P., Obrist, M.K., 2003. Biodiversity indicators: the choice of values and measures. In: Büchs, W. (Ed.), Biotic Indicators for Biodiversity and Sustainable Agriculture. Agric. Ecosyst. Environ. 98, 87–98.

Eiden, G., Kayadkanian, M., Vidal, C., 2001a. Capturing landscape structures: tools. In: EU-Commission (Ed.), From Land Cover to Landscape Diversity in the European Union. Brussels, Belgium, 11 pp.

Eiden, G., Kayadkanian, M., Vidal, C., 2001b. Quantifying landscape structures: spatial and temporal dimensions. In: EU-Commission (Ed.), From Land Cover to Landscape Diversity in the European Union. Brussels, Belgium, 14 pp. (+2 tables and 16 maps).

Ellenberg, H., Weber, H.E., Düll, R., Wirth, V., Werner, W., Paulißen, D., 1992. Zeigerwerte der Pflanzen in Mitteleuropa. Scripta Geobotanica 18, 1–258.

Engelmann, H.D., 1978. Zur Dominanzklassifizierung von Bodenarthropoden. Pedobiologia 18, 378–380.

European Commission (Ed.), 1991. Verordnung (EWG) Nr. 2092/91 des Rates vom 24. Juni 1991 über den ökologischen Landbau und die entsprechende Kennzeichnung der landwirtschaftlichen Erzeugnisse und Lebensmittel. Amtsblatt der Europäischen Gemeinschaft No. L 198, 22 July 1991, p. 1.

European Commission (Ed.), 2002. Verordnung (EWG) Nr. 473/2002 der Kommission zur Veränderung der Verordnung (EWG) Nr. 2092/91 des Rates über den ökologischen Landbau. Amtsblatt der Europäischen Gemeinschaft No. L 75, 16 March 2002, p. 21.

Fan, Y., Liebman, M., Groden, E., Alford, A.R., 1993. Abundance of carabid beetles and other ground dwelling arthropods in conventional versus low input bean cropping systems. Agric. Ecosyst. Environ. 43, 127–139.

Ferris, R., Humphrey, J.W., 1999. A review of potential biodiversity indicators for application in British forests. Forestry 72, 313–328.

Foissner, W., 1999. Soil protozoa as bioindicators: pros and cons, methods, diversity, representative examples. In: Paoletti, M.G. (Ed.), Invertebrate Biodiversity as Bioindicators of Sustainable Landscapes. Agric. Ecosyst. Environ. 74, 95–112.

Franzen, J., Weber, G., Büchs, W., Larink, O., 1997. Langzeiteinfluß von Pflanzenschutzmitteln auf Dipteren mit bodenlebenden Entwicklungsstadien. Berichte über Landwirtschaft 75, 291–328.

Frieben, B., 1997. Arten- und Biotopschutz durch Organischen Landbau. In: Weiger, H., Willer, H. (Eds.), Naturschutz durch ökologischen Landbau. Bad Dürkheim, Holm, Germany, pp. 73–92.

Frieben, B., 1998. Verfahren zur Bestandsaufnahme und Bewertung von Betrieben des Organischen Landbaus im Hinblick auf Biotop- und Artenschutz und die Stabilisierung des Agrarökosystems. Schriftenreihe Institut für Organischen Landbau 11, Verlag Dr. Köster, Berlin.

Friedrich, G., 1990. Eine Revision des Saprobiensystems. Zeitschrift für Wasser- und Abwasserforschung 23, 141–152.

Fritz-Köhler, W., 1996. Blatt- und Rüsselkäfer an Ackerunkräutern. Ökologie und Biogeographie in Mitteleuropa und Untersuchungen an ungespritzten Ackerrandstreifen. Agrarökologie 19, 1–138.

Frouz, J., 1999. Use of soil dwelling Diptera (Insecta, Diptera) as bioindicators: a review of ecological requirements and response to disturbance. In: Paoletti, M.G. (Ed.), Invertebrate Biodiversity as Bioindicators of Sustainable Landscapes. Agric. Ecosyst. Environ. 74, 167–186.

Gallego, F.J., Escribano, P., Christensen, S., 2001. Comparability of landscape diversity indicators in the European Union. In: EU-Commission (Ed.), From Land Cover to Landscape Diversity in the European Union. Brussels, Belgium, 12 pp. (+1 map).

Gaston, K.J., 1996. Biodiversity. A Biology of Numbers and Difference. Blackwell, London, UK.

Geipel, K.-H., Kegel, B., 1989. Die Ausbildung der thoracalen Flugmuskulatur von Laufkäferpopulationen ausgewählter Straßenrandbiotope in Berlin (West). Verhandlungen der Gesellschaft für Ökologie 17, 727–732.

Gilgenberg, A., 1986. Die Verteilungsmuster der Carabiden- und Staphylinidenfauna verschieden bewirtschafteter landwirtschaftlicher Flächen sowie eines Waldes. Ph.D. thesis, University of Bonn, 261 pp.

Good, J.A., Speight, M.C.D., 1991. Sites of international and national importance for invertebrate fauna: a definition proposed for use in site surveys. Bull. Ir. Biogeogr. Soc. 14, 48–53.

Greiler, H.J., Tscharntke, T., 1991. Artenreichtum von Pflanzen und Grasinsekten auf gemähten und ungemähten Rotationsbrachen. Verhandlungen der Gesellschaft für Ökologie 20, 429–434.

Gruschwitz, M., 1981. Die Bedeutung der Populationsstruktur von Carabidenfaunen für Bioindikation und Standortdiagnose. Mitteilungen der Deutschen Gesellschaft für allgemeine und angewandte Entomologie 3, 126–129.

Hagvar, S., 1994. Log-normal distribution of dominance as an indicator of stressed soil microarthropod communities? Acta Zool. Fennica 195, 71–80.

Hamblin, A.P., 1992. Environmental Indicators of Sustainable Agriculture. BRR, Canberra, Australia.

Hansson, L., Fahrig, L., Merriam, G., 1995. Mosaic Landscapes and Ecological Processes. Chapmann & Hall, London.

Harenberg, A., 1997. Auswirkungen abgestuft intensiv geführter Anbausysteme in verschiedenen Fruchtfolgen (Raps-, Zuckerrübenfruchtfolge) und einer selbstbegrünenden Dauerbrache auf Spinnen (Arachnida: Araneae). Ph.D. thesis, Faculty of Natural Sciences, Technical University of Braunschweig, Braunschweig, Germany, 276 pp.

Haslett, J.R., 1988. Qualitätsbeurteilung alpiner Habitate: Schwebfliegen (Diptera: Syrphidae) als Bioindikatoren für Auswirkungen des intensiven Skibetriebs auf alpinen Wiesen in Österreich. Zoologischer Anzeiger 220, 179–184.

Hawkins, V., Selman, P., 1994. Landscape ecological planning and the future countryside: a research note. Landsc. Res. 19 (2), 88–94.

Heydemann, B., 1955. Carabiden der Kulturfelder als ökologische Indikatoren. Berichte Wanderversammlung Deutscher Entomologen 7, 172–185.

Heyer, W., Hülsbergen, K.-J., Wittmann, C., Papaja, S., Christen, O., 2003. Field related organisms as possible indicators for evaluation of land use intensity. Agric. Ecosyst. Environ. 98, 453–461.

Hill, M.O., 1979. Twinspan—a Fortran program for arranging multivariate data in an ordered two-way table by classification of the individuals and attrbutes. Cornell University, Ecology and Systematics, Ithaca, NY, USA.

Hill, M.O., Mountford, J.O., Roy, D.B., Bunce, R.G.H., 1999. Ellenberg's indicator values for British plants. ECOFACT vol. 2, Technical Annexure. Monks Wood, Huntingdon, UK, 46 pp.

Hingst, R., Irmler, U., Steinborn, H.-A., 1995. Die Laufkäfergemeinschaften in Wald- und Agrarökosystemen Schleswig-Holsteins. Mitteilungen der Deutschen Gesellschaft für allgemeine und angewandte Entomologie 9, 733–737.

Hirsch, M., Pfaff, S., Wolters, V., 2003. The influence of matrix type on flower visitors of *Centaurea jacea* L.). In: Büchs,

W. (Ed.), Biotic Indicators for Biodiversity and Sustainable Agriculture. Agric. Ecosyst. Environ. 98, 331–337.

Hoffmann, J., Greef, J.M., 2003. Mosaic indicators—theoretical approach for the development of indicators for species diversity in agricultural landscapes. In: Büchs, W. (Ed.), Biotic Indicators for Biodiversity and Sustainable Agriculture. Agric. Ecosyst. Environ. 98, 387–394.

Hoffmann, J., Kretschmer, H., 2001. Zum Biotop- und Artenschutzwert großer Ackerschläge in Nordostdeutschland. In: Xylander, W., Wicke, G., Büchs, W., Mühle. H. (Eds.), Großräumigkeit/Kleinräumigkeit in der Agrarlandschaft, Peckiana 1, pp. 17–31.

Hoffmann, J., Greef, J.M., Kiesel, J., Lutze, G., Wenkel, K.-O., 2003. Practical example of the mosaic indicators approach. In: Büchs, W. (Ed.), Biotic Indicators for Biodiversity and Sustainable Agriculture. Agric. Ecosyst. Environ. 98, 395–405.

Hokkanen, H., Holopainen, J.K., 1986. Carabid species diversity and activity densities in biologically and conventionally managed cabbage fields. J. Appl. Entomol. 102, 353–363.

Holland, J.M., Frampton, G.K., Cilgi, T., Wratten, S.D., 1994. Arable acronyms analysed—a review of integrated arable farming systems research in western Europe. Ann. Appl. Biol. 125, 399–438.

Hollander, M., Wolfe, D.A., 1973. Nonparametric Statistical Methods. New York, USA, 503 pp.

Hopkin, S.P., Hardisty, G., Martin, M.H., 1986. The woodlouse *Porcellio scaber* as a biological indicator of zinc, cadmium, lead and copper pollution. Environ. Pollut. 11, 271–290.

Hopkin, S.P., Jones, D.Y., Dietrich, D., 1993. The isopod *Porcellio scaber* as a monitor of the bioavailability of metals in terrestrial ecosystems: towards a global woodlouse watch scheme. Sci. Total Environ. (Suppl.), 357–365.

Innis, S.A., Naiman, R.J., Elliott, S.R., 2000. Indicators and assessment methods for measuring the ecological integrity of semi-aquatic terrestrial environments. Hydrobiologia 422–423, 111–131.

Irmler, U., 2003. The spatial and temporal pattern of ground beetles (Coleoptera: Carabidae) on fields in northern Germany (Schleswig-Holstein) and their value as ecological indicators. In: Büchs, W. (Ed.), Biotic Indicators for Biodiversity and Sustainable Agriculture. Agric. Ecosyst. Environ. 98, 141–151.

Jeanneret, P., Schüpbach, B., Luka, H., 2003. Quantifying the impact of landscape and habitat features on biodiversity in cultivated landscapes. In: Büchs, W. (Ed.), Biotic Indicators for Biodiversity and Sustainable Agriculture. Agric. Ecosyst. Environ. 98, 311–320.

Jedicke, E., 1994. Biotopverbund. Ulmer, Stuttgart, Germany, 287 pp.

Jones, C.G., Lawton, J.H. (Eds.), 1995. Linking Species and Ecosystems. Chapman & Hall, New York, USA, 387 pp.

Joy, S., Maity, S.K., Joy, V.C., 2000. Digestive enzymes in *Porcellio laevis* (Isopoda: Crustacea) as indicator of heavy metal toxicity in soil—biodiversity and environment. Proceedings of the National Seminar on Environmental Biology, pp. 79–86.

Kareiva, P., 1987. Habitat fragmentation and the stability of predator–prey interactions. Nature 326, 388–390.

Kinnunen, H., Järveläinen, K., Pakkala, T., Tiainen, J., 1996. The effects of isolation on the occurrence of farmland carabids in a fragmented landscape. Ann. Zool. Fennici 33, 165–171.

Kleinhenz, A., Büchs, W., 1993. Einfluß verschiedener landwirtschaftlicher Produktionsintensitäten auf die Spinnenfauna in der Kultur Zuckerrübe. Verhandlungen der Gesellschaft für Ökologie 22, 81–88.

Kleinhenz, A., Büchs, W., 1995. Ökologische Aspekte der Spinnenzönosen von Zuckerrübenflächen unter dem Einfluß eines unterschiedlich intensiven Einsatzes landwirtschaftlicher Produktionsmittel. Mitteilungen der Deutschen Gesellschaft für allgemeine und agewandte Entomologie 9, 481–489.

Knauer, N., 1993. Ökologie und Landwirtschaft. Ulmer, Stuttgart, 280 pp.

Knüsting, E., Bartels, G., Büchs, W., 1991. Untersuchungen zu Artenspektrum,
fruchtartspezifischer Abundanz und Abundanzdynamik von Regenwürmern bei unterschiedlich hohen landwirtschaftlichen Produktionsintensitäten. Verhandlungen der Gesellschaft für Ökologie 20, 21–27.

Kobel-Lamparski, A., Lamparski, F., 1998. Sukzessionsuntersuchungen im Rebgelände des Kaiserstuhls.

Kobel-Lamparski, A., Lamparski, F., Gack, C., Straub, F., 1999. Erhöhung der Biodiversität in Rebgebieten des Kaiserstuhls durch die Verzahnung von Rebflächen und Rebböschungen. In: Büchs, W. (Ed.), Nicht bewirtschaftete Areale in der Agrarlandschaft—ihre Funktionen und ihre Interaktionen mit landnutzungsorientierten Ökosystemen. Mitteilungen aus der Biologischen Bundesanstalt für Land- und Forstwirtschaft 368, 69–78.

Koehler, H.H., 1999. Predatory mites (Gamasina, Mesostigmata). In: Paoletti, M.G. (Ed.), Invertebrate Biodiversity as Bioindicators of Sustainable Landscapes. Agric. Ecosyst. Environ. 74, 395–410.

Köhler, F., 1998. Zur Bestandssituation an Ackerunkräutern lebender Blatt- und Rüsselkäferarten in Deutschland (Coleoptera, Chrysomelidae, Curculionidae s.l.). Schriftenreihe Landesanstalt für Pflanzenbau und Pflanzenschutz Rheinland-Pfalz 6, 243–254.

Köhler, F., Stumpf, T., 1992. Die Käfer der Wahner Heide in der Niederrheinischen Bucht bei Köln (Insecta: Coleoptera). Decheniana-Beihefte 31, 499–593.

Konold, W., 1996. Naturlandschaft Kulturlandschaft—Die Veränderung der Landschaften nach der Nutzbarmachung durch den Menschen. Ecomed, Landsberg, Germany, 322 pp.

Kratochwil, A., Schwabe, A., 2001. Ökologie der Lebensgemeinschaften. UTB 8199, Ulmer, Stuttgart, 756 pp.

Kratz, R., Pfadenhauer, J., 2001. Ökosystemmanagement für Niedermoore. Strategien und Verfahren zur Renaturierung. Ulmer, Stuttgart.

Krause, U., Poehling, H.M., 1996. Overwintering, oviposition and population dynamics of hoverflies (Diptera: Syrphidae) in northern Germany in relation to small and large scale structure. Acta Jutlandica 71, 157–169.

Krebs, C.J., 1999. Ecological Methodology. Addison-Wesley, Menlo Park.

Kreuter, T., 2000. Zur Struktur, Dynamik und bioindikatorischer Eignung von Laufkäferzönosen im Nordteil des Ökohofes

Seeben. UZU-Schriftenreihe, Neue Folge, Sonderband, pp. 135–171.

Krivosheina, M.G., 1993. Variation of the wing venation in the species of the genus *Dicranomyia stephens* (Diptera, Limoniidae) under conditions of high radioactive background level. In: Proceedings of the Fourth Meeting on Species and its Productivity in Distribution Area, UNESCO MAB program, Gidrometeoizdat, St. Petersburg, pp. 267–268.

Krivosheina, M.G., 1995. Different reactions of two populations of *Pegomya tenera* Ztt. (Diptera, Anthomyiidae) breeding on fungi polluted by ^{137}Cs and ^{90}Sr. In: Bohac, J., Triska, J., Tichy, R. (Eds.), Abstracts of the Eighth International Bioindicators Symposium, Ceske Budejovice, 58 pp.

Kromp, B., 1990. Carabid beetles (Coleoptera, Carabidae) as bioindicators in biological and conventional farming in Austrian potato fields. Biol. Fertil. Soils 9, 182–187.

Kromp, B., 1999. Carabid beetles in sustainable agriculture: a review on pest control efficacy, cultivation impacts and enhancement. In: Paoletti, M.G. (Ed.), Invertebrate Biodiversity as Bioindicators of Sustainable Landscapes. Agric. Ecosyst. Environ. 74, 187–228.

Kruess, A., Tscharntke, T., 1994. Habitat fragmentation, parasitoid species loss, and biological control. Science 264, 1581–1584.

Kühle, J.C., 1986. Modelluntersuchungen zur strukturellen und ökotoxikologischen Belastung von Regenwürmern in Weinbergen Mitteleuropas (Oligochaeta: Lumbricidae). Ph.D. thesis, Math.-Nat. Faculty, University of Bonn, Bonn, 390 pp.

Lambeck, R.J., 1997. Focal species: a multi-species umbrella for nature conservation. Conserv. Biol. 11, 849–856.

Langmaack, M., Land, S., Büchs, W., 2001. Effects of different field management systems on the ground beetle coenosis in oil seed rape with special respect to ecology and nutritional status of predacious *Poecilus cupreus* (Coleoptera: Carabidae). J. Appl. Entomol. 125, 313–320.

Larink, O., Heisler, C., Söchtig, W., Lübben, B., Wickenbrock, L., 1993. Einfluß verdichteter Ackerböden auf die Bodenfauna und ihr Beitrag zur Bodenlockerung. KTBL-Schrift 362, 142–156.

Larsen, K.J., Brewer, S.R., Taylor, D.H., 1994. Differential accumulation of heavy metals by web spiders and ground spiders in an old field. Environ. Toxicol. Chem. 13 (3), 503–508.

Lawton, J.H., Gaston, K.J., 2001. Indicator species. In: Levin, S.A. (Ed.), Encyclopedia of Biodiversity, vol. 3. Academic Press, San Diego, pp. 437–450.

Lawton, J.H., Bignell, D.E., Bolton, B., Bloemers, G.F., Eggleton, P., Hammond, P.M., Hodda, M., Holt, R.D., Larsen, T.B., Mawdsley, N.A., Stork, N.E., Srivastava, D.S., Watt, A.D., 1998. Biodiversity inventories, indicator taxa and effects of habitat modification in tropical forest. Nature 391, 72–76.

Linden, D.R., Hendrix, P.F., Coleman, D.C., van Vliet, P.C.J., 1994. Faunal indicators of soil quality. In: Doran, J.W., Coleman, D.C., Bezdicek, D.F., Stewart, B.A. (Eds.), Defining Soil Quality for a Sustainable Environment. SSSA Special Publication 35, USA, pp. 91–106.

Lobry de Bruyn, L.A., 1997. The status of soil macrofauna as indicators of soil health to monitor the sustainability of Australian agricultural soils. Ecol. Entomol. 23, 167–178.

Lorenz, E., 1994. Mechanische Unkrautbekämpfungsverfahren in Zuckerrübenkulturen und ihre Nebenwirkungen auf Laufkäfer (Coleoptera, Carabidae) und andere epigäische Arthropoden. Ph.D. thesis, Faculty of Agriculture, University of Göttingen.

Luka, H., 1996. Laufkäfer: Nützlinge und Bioindikatoren in der Landwirtschaft. Agrarforschung 3 (1), 33–36.

Macarthur, R., Wilson, E.O., 1967. The Theory of Island Biography. Princeton University Press, Princeton.

Majer, J.D., 1983. Ants: bio-indicators of minesite rehabilitation, landuse and land conservation. Environ. Manage. 7, 375–383.

Majer, J.D., Nichols, O.G., 1998. Long-term recolonization patterns of ants in Western Australian rehabilitated bauxite mines with reference to their use as indicators in restoration processes. J. Appl. Ecol. 35, 161–182.

Manly, B.F.J., 1992. The Design and Analysis of Research Studies. Cambridge University Press, Cambridge.

Maraun, M.S., Visser, S., Scheu, S., 1998. Oribatid mites enhance the recovery of the microbial community after a strong disturbance. Appl. Soil Ecol. 9, 18–175.

Marc, P., Canard, A., Ysnel, F., 1999. Spiders (Araneae) useful for pest limitation and bioindication. In: Paoletti, M.G. (Ed.), Invertebrate Biodiversity as Bioindicators of Sustainable Landscapes. Agric. Ecosyst. Environ. 74, 229–273.

Marten, M., Reusch, H., 1992. Anmerkungen zur DIN "Saprobien index" (38 410 Teil 2) und Forderung alternativer Verfahren. Natur und Landschaft 67, 544–547.

Martin, D., 1972. Die Spinnenfauna des Frohburger Raumes. II. Micryphantidae. III. Linyphiidae. IV. Theridiidae. Abhandlungen und Berichte des Naturkundlichen Museums "Mauritianum" Altenburg 7 (2–3), 239–284.

Martin, D., 1973. Die Spinnenfauna des Frohburger Raumes. V. Agelenidae, Argyronetidae, Hahniidae und Hersiliidae. VI. Ctenidae, Lycosidae und Psiauridae. VII. Drassodidae, Anyphaenidae, Clubionidae und Eusparassidae. VIII. Salticidae. IX. Thomisidae und Philodromidae. X. Atypidae, Dysderidae, Sicariidae, Pholcidae, Nesticidae, Mimetidae, Dictynidae, Amaurobiidae und Uloboridae. Abhandlungen und Berichte des Naturkundlichen Museums "Mauritianum" Altenburg 8, 27–57, 127–159.

Mattheis, A., Otte, A., 1994. Ergebnisse der Erfolgskontrollen zum "Ackerrandstreifenprogramm" im Regierungsbezirk Oberbayern 1985–1991. In: Stiftung Naturschutz Hamburg, Stiftung zum Schutze gefährdeter Pflanzen, Flora und Fauna der Äcker und Weinberge, Schriftenreihe Aus Liebe zur Natur 5, 56–71.

Maurer, R., Hänggi, A., 1990. Katalog der schweizerischen Spinnen. Doc. Faun. Helvet. 12, 1–412.

May, R.M., 1980. Theoretische Ökologie. Weinheim, 284 pp.

McGeoch, M.A., 1998. The selection, testing and application of terrestrial insects as bioindicators. Biol. Rev. 73, 181–201.

McGrady-Steed, J.P., Harris, P., Morin, P., 1997. Biodiversity regulates ecosystem predictability. Nature 390, 162–165.

Menge, M., 2003. Experiences with the application, survey and assessment of agri-environmental indicators in agricultural practice. In: Büchs, W. (Ed.), Biotic Indicators for Biodiversity and Sustainable Agriculture. Agric. Ecosyst. Environ. 98, 443–451.

Moreby, S.J., Aebisher, N.J., 1992. Invertebrate abundance on cereal fields and set-aside land: implications for wild gamebird chicks. In: Clarke, J. (Ed.), Set-Aside. BCPC-Monograph No. 50, pp. 181–186.

Morris, M.G., 1973. The effect of seasonal grazing on the Heteroptera and Auchenorrhyncha (Hemiptera) of chalk grassland. J. Appl. Ecol. 10, 761–780.

Morris, M.G., 1981a. Responses of grassland invertebrates to management by cutting. III. Adverse effects of Auchenorrhyncha. J. Appl. Ecol. 18, 107–123.

Morris, M.G., 1981b. Responses of grassland invertebrates to management by cutting. IV. Positive responses of Auchenorrhyncha. J. Appl. Ecol. 18, 763–771.

Morris, M.G., 1992. Responses of Auchenorrhyncha (Homoptera) to fertilizer and liming treatments at Park Grass, Rothamsted. Agric. Ecosyst. Environ. 41, 263–283.

Morris, M.G., Plant, R., 1983. Responses of grasland invertebrates to management by cutting. V. Changes in Hemiptera following cessation of management. J. Appl. Ecol. 20, 157–177.

Morris, M.G., Rispin, W.E., 1987. Abundance and diversity of the coleopterous fauna of a calcareous grassland under different cutting regimes. J. Appl. Ecol. 24, 451–465.

Motulsky, H., 1995. Intuitive Biostatistics. Oxford University Press, Oxford, UK.

Mühlenberg, M., Slowik, J., 1997. Kulturlandschaft als Lebensraum, UTB 1947. Quelle & Meyer, 312 pp.

Müller, L., 1991. Auswirkungen der Extensivierungsförderung auf Wirbellose. Faunistisch-Ökologische Mitteilungen Suppl. 10, 41–70.

Murdoch, W., Evans, F.C., Peterson, C.H., 1972. Diversity and pattern in plants and insects. Ecology (New York) 53, 819–829.

Naeem, S., Li, S., 1997. Biodiversity enhances ecosystem reliability. Nature 390, 507–509.

Naeem, S., Thompson, S., Lawler, S., Lawton, J., Woodfin, R., 1994. Declining biodiversity can alter the performance of ecosystems. Nature 368, 734–737.

Nentwig, W. (Ed.), 2000. Streifenförmige ökologische Ausgleichsflächen in der Kulturlandschaft: Ackerkrautstreifen, Buntbrache, Feldränder. Vaö, Bern.

Nentwig, W., Hänggi, A., Kropf, C., Blick, T., 2000. Central European Spiders—Determination Key. http://zoology.unibe.ch/araneae.

Nickel, H., Achtziger, R., 1999. Wiesen bewohnende Zikaden (Auchenorrhyncha) im Gradienten von Nutzungsintensität und Feuchte. Beiträge zur Zikadenkunde 3, 65–80.

Nickel, H., Hildebrandt, J., 2003. Auchenorrhyncha communities as indicators of disturbance in grasslands (Insecta, Hemiptera)—a case study from the Elbe flood plains (northern Germany). In: Büchs, W. (Ed.), Biotic Indicators for Biodiversity and Sustainable Agriculture. Agric. Ecosyst. Environ. 98, 183–199.

Noss, R.F., 1990. Indicators for monitoring biodiversity: a hierarchical approach. Conserv. Biol. 4, 355–364.

Novotný, V., 1994a. Relation between temporal persistence of host plants and wing length in leafhoppers (Auchenorrhyncha, Hemiptera). Ecol. Entomol. 19, 168–176.

Novotný, V., 1994b. Association of polyphagy in leafhoppers (Auchenorrhyncha, Hemiptera) with unpredictable environments. Oikos 70, 223–232.

Novotný, V., 1995. Relationships between life histories of leafhoppers (Auchenorrhyncha: Hemiptera) and their host plants (Juncaceae, Cyperaceae, Poaceae). Oikos 73, 33–42.

Nyfeller, M., 1982. Field studies on the ecological role of the spiders as insect predators in agro-ecosystems (abandoned grassland, meadows and cereal fields). Ph.D. thesis, Swiss Federal Institute of Technology, Zurich, Switzerland.

Odum, E.P., 1980. Grundlagen der Ökologie. Georg Thieme, Stuttgart, New York, 836 pp.

Oliver, I., Beattie, A.J., 1996. Invertebrate morphospecies as surrogates for species: a case study. Conserv. Biol. 10, 99–109.

Palmer, A.R., Strohbeck, C., 1986. Fluctuating asymmetry: measurement, analysis, patterns. Annu. Rev. Ecol. Syst. 1, 391–421.

Pankhurst, C.E., 1994. Biological indicators of soil health and sustainable productivity. In: Greenland, D.J., Szabolcs, I. (Eds.), Soil Resilience and Sustainable Land Use. CAB International, Wallingford, pp. 331–352.

Pankhurst, C.E., Hawke, B.G., McDonald, H.J., Kirkby, C.A., Buckerfield, J.C., Michelson, P., O'Brien, K.A., Gupta, V.S.S.R., Doube, B.M., 1995. Evaluation of soil biota properties as potential bioindicators of soil health. Aust. J. Exp. Agric. 35, 1015–1028.

Paoletti, M.G., 1984. La vegetazione spontanea dell'agroecosistema ed il controllo dei fitofagi del mais. Giornate Fitopatologiche 1984, Cleub Bologna, pp. 445–456.

Paoletti, M.G., 1999a. Using bioindicators based on biodiversity to assess landscape sustainability. In: Paoletti, M.G. (Ed.), Invertebrate Biodiversity as Bioindicators of Sustainable Landscapes. Agric. Ecosyst. Environ. 74, 1–18.

Paoletti, M.G., 1999b. The role of earthworms for assessment of sustainability and as bioindicators. In: Paoletti, M.G. (Ed.), Invertebrate Biodiversity as Bioindicators of Sustainable Landscapes. Agric. Ecosyst. Environ. 74, 137–155.

Paoletti, M.G., Hassal, M., 1999. Woodlice (Isopoda: Oniscoidea): their potential for assessing sustainability and use as bioindicators. In: Paoletti, M.G. (Ed.), Invertebrate Biodiversity as Bioindicators of Sustainable Landscapes. Agric. Ecosyst. Environ. 74, 157–165.

Paoletti, M.G., Iovane, E., Cortese, M., 1988. Pedofauna bioindicators as heavy metals in five agro-ecosystems in north-east Italy. Rev. Ecol. Biol. Sol. 25 (1), 33–58.

Perner, J., 2003. Sample size and quality of bioindication: a case study using ground-dwelling arthropods in agricultural ecosystems. In: Büchs, W. (Ed.), Biotic Indicators for Biodiversity and Sustainable Agriculture. Agric. Ecosyst. Environ. 98, 125–132.

Perner, J., Malt, S., 2003. Assessment of changing agricultural land use: response of vegetation, ground-dwelling spiders and beetles to the conservation of arable land into grassland. In: Büchs, W. (Ed.), Biotic Indicators for Biodiversity and Sustainable Agriculture. Agric. Ecosyst. Environ. 98, 169–181.

Pfiffner, L., 1997. Welchen Beitrag leistet der ökologische Landbau zur Förderung der Kleintierfauna. In: Weiger, H., Willer, H. (Eds.), Naturschutz durch ökologischen Landbau. Bad Dürkheim, Holm, Germany, pp. 93–120.

Pfiffner, L., Luka, H., 1996. Laufkäfer-Förderung durch Ausgleichsflächen. Auswirkungen neu angelegter Grünstreifen und einer Hecke im Ackerland. Naturschutz und Landschaftsplanung 28, 145–151.

Pfiffner, L., Luka, H., 1999. Faunistische Erfolgskontrolle von unterschiedlichen Anbausystemen und naturnahen Flächen im Feldbau—Bedeutung des ökologischen Landbaues. In: Büchs, W. (Ed.), Nicht bewirtschaftete Areale in der Agrarlandschaft—ihre Funktionen und ihre Interaktionen mit landnutzungsorientierten Ökosystemen. Mitteilungen aus der Biologischen Bundesanstalt für Land- und Forstwirtschaft 368, 57–67.

Pfiffner, L., Mäder, P., 1997. Effects of biodynamic, organic and conventional production systems on earthworm populations. Biol. Agric. Horticult. 15, 3–10.

Pfiffner, L., Niggli, U., 1996. Effects of biodynamic, organic and conventional farming on ground beetles (Col., Carabidae) and other epigaeic arthropods in winter wheat. Biol. Agric. Horticult. 12, 353–364.

Pfiffner, L., Luka, H., Jeanneret, P., Schüpbach, B., 2000. Effekte ökologischer Ausgleichsflächen auf die Laufkäferfauna. Agrarforschung 7 (5), 212–217.

Piorr, H.-P., 2003. Environmental policy, agri-environmental indicators and landscape indicators. In: Büchs, W. (Ed.), Biotic Indicators for Biodiversity and Sustainable Agriculture. Agric. Ecosyst. Environ. 98, 17–33.

Platen, R., Moritz, M., von Broen, B., 1991. Liste der Webspinnen- und Weberknechtarten (Arach.: Araneida, Opilionida) des Berliner Raumes und ihre Auswertung für Naturschutzzwecke (Rote Liste). In: Platen, R., Sukopp, H. (Eds.), Rote Listen der gefährdeten Pflanzen und Tiere in Berlin, Landschaftsentwicklung und Umweltforschung 6, 169–205.

Poehling, H.M., Vidal, S., Ulber, B., 1994. Genug Nützlinge auf Großflächen—Wunsch oder Wirklichkeit? Pflanzenschutz-Praxis 3, 34–38.

Popov, V.V., Krusteva, I.A., 2000. Epigeobiont animal assemblages from two landscapes of the Bulgarian black sea cost: relationship to environmental gradients, assemblage structure and biodiversity. I. Ground beetles (Coleoptera: Carabidae). Acta Zool. Bulgarica 51, 81–114.

Potts, G.R., 1986. The Partridge: Pesticides, Predation and Conservation. Collins Publisher.

Prescher, S., Büchs, W., 1997. Zum Einfluß abgestufter Extensivierungsmaßnahmen und selbstbegrünender Dauerbrache auf funktionelle Gruppen der Brachycera (Diptera). Verhandlungen der Gesellschaft für Ökologie 27, 385–391.

Prescher, S., Büchs, W., 2000. Der Einfluß abgestufter Extensivierungsmaßnahmen im Ackerbau auf die Struktur der phytophagen Brachycerazönose. Mitteilungen der Deutschen Gesellschaft für allgemeine und angewandte Entomologie 12, 347–352.

Prestidge, R.A., 1982. The influence of nitrogenous fertilizer on the grassland Auchenorrhyncha. J. Appl. Ecol. 19, 735–749.

Radford, B.J., Key, A.J., Robertson, L.N., Thomas, G.A., 1995. Conservation tillage increases soil water storage, soil animal population, grain yield and response to fertiliser in the semi-arid tropics. Aust. J. Exp. Agric. 35, 223–232.

Rahmel, U., Ruf, A., 1994. Eine Feldmethode zum Nachweis von anthropogenem Stress auf natürliche Tierpopulationen: "Fluctuating asymmetry". Natur und Landschaft 69, 104–107.

Raskin, R., Glück, E., Pflug, W., 1992. Floren- und Faunenentwicklung auf herbizidfrei gehaltenen Ackerflächen. Natur und Landschaft 67, 7–14.

Rehfeldt, G.-E., 1984. Bewertung ostniedersächsischer Flussauen durch Bioindikatorsysteme—Modell einer Landschaftsbewertung. Ph.D. thesis, Faculty of Natural Sciences, Technical University of Braunschweig, Braunschweig, Germany, 259 pp.

Remmert, H., 1984. Ökologie, third ed. Springer Verlag, Berlin, 334 pp.

Retnakaran, A., Smith, L., 1980. Web-spinning in spiders is unaffected by moult-inhibiting insect-growth regulator BAY SIR 8514. Bimonthly Res. Notes 36 (4), 19–20.

Rhee, J., 1997. Effects of soil pollution on earthworms. Pedobiologia 17, 201–208.

Richter, J., Bachinger, J., Stachow, U., 1999. Einfluss der Standortheterogenität innerhalb von Großschlägen auf die Segetalflora unter organischer und konventioneller Bewirtschaftung in Ostbrandenburg. Beiträge zur 5. Wissenschaftstagung zum Ökologischen Landbau "Vom Rand zur Mitte", pp. 416–419.

Riechert, S.E., Harp, J.M., 1987. Nutritional ecology of spiders. In: Slansky, F., Rodriguez, J.G. (Eds.), Nutritional Ecology of Insects, Mites and Spiders, pp. 645–672.

Riechert, S.E., Tracy, C.R., 1975. Thermal balance and prey availability: the basis for a model relating web site characteristics to spider reproductive success. Ecology 56 (2), 265–284.

Riecken, U., 1992. Planungsbezogene Bioindikation durch Tierarten und Tiergruppen. Grundlagen und Anwendung. Schriftenreihe für Landschaftspflege und Naturschutz 36, 1–187 (Münster: Landwirtschaftsverlag, Germany).

Riecken, U., 1997. Arthropoden als Bioindikatoren in der naturschutzrelevanten Planung—Anwendung und Perspektive. Mitteilungen der Deutschen Gesellschaft für allgemeine und angwandte Entomologie 11, 45–56.

Rösler, S., Weins, C., 1997. Situation der Vogelwelt in der Agrarlandschaft und der Einfluß des ökologischen Landbaus auf ihre Bestände. In: Weiger, H., Willer, H. (Eds.), Naturschutz durch ökologischen Landbau. Bad Dürkheim, Holm, Germany, pp. 121–152.

Roth, D., Schwabe, M., 2003. Method for assessing the proportion of ecologically, culturally and provincially significant areas ("OELF") in agrarian spaces used as a criterion of environmentally friendly agriculture. In: Büchs, W. (Ed.), Biotic Indicators for Biodiversity and Sustainable Agriculture. Agric. Ecosyst. Environ. 98, 435–441.

Roush, R.S., Radabough, D.C., 1993. Web density is related to prey abundance in cellar spiders, *Pholcus phalangoides* (Füsslin) (Araneae, Pholcidae). Bull. Br. Arachnol. Soc. 9 (5), 142–144.

Ruf, A., 1998. A maturity index for predatory soil mites (Mesostigmata: Gamasina) as an indicator of environmental impacts of pollution on forest soil. Appl. Soil Ecol. 6, 447–452.

Ruzicka, V., Bohac, J., 1994. The utilization of epigaeic invertebrate communities as bioindicators of terrestrial environmental quality. In: Salanki, J., Jeffrey, D., Hughes, G.M. (Eds.), Biological Monitoring of the Environment: A Manual of Methods. CAB International, Wallingford, pp. 79–86.

Sampels, J., 1986. Die Käfer der Weinbergsvegetationsschicht und ihre Eignung als Indikatoren der Standortbelastung. Ph.D. thesis, University of Bonn, Germany, 224 pp.

Saxena, K.G., Rao, K.S., Ramakrishnan, P.S., 1999. Ecological context of biodiversity. In: Shantharam, S., Montgomery, J.F. (Eds.), Biotechnology, Biosafety and Biodiversity: Scientific and Ethical Issues for Sustainable Developments. Science, Enfield, USA, pp. 129–156.

Scheu, S., 1999. Biologische Vielfalt und Ökosystemfunktion. In: Hummel, M.E., Simon, H.-R., Scheffran, J. (Eds.), Konfliktfeld Biodiversität: Erhalt der biologischen Vielfalt—Interdisziplinäre Problemstellungen. Working Paper IANUS 7 (1999), 3–13.

Scheu, S., 2001. Plants and generalist predators as mediators between the below-ground and the above-ground system. Basic Appl. Ecol. 2, 3–13.

Schloter, M., Bach, H.-J., Metz, S., Sehy, U., Munch, J.C., 2003. Influence of precision farming on the microbial community structure and functions in nitrogen turnover. In: Büchs, W. (Ed.), Biotic Indicators for Biodiversity and Sustainable Agriculture. Agric. Ecosyst. Environ. 98, 295–304.

Schröter, L., Irmler, U., 1999. Einfluß von Bodenart, Kulturfrucht und Feldgröße auf Carabiden-Synusien der Äcker des konventionell-intensiven und des ökologischen Landbaus. Faunistisch-Ökologische Mitteilungen Supplement 27, 1–61.

Schubert, R., 1984. Lehrbuch der Ökologie. Gustav Fischer Verlag, Jena, 595 pp.

Sedlacek, J.D., Barrett, G.W., Shaw, D.R., 1988. Effects of nutrient enrichment on the Auchenorrhyncha (Homoptera) in contrasting grassland communities. J. Appl. Ecol. 25, 537–550.

Shannon, C.E., 1948. A mathematical theory of communication. Bull. Syst. Technol. J. 27, 379–423, 623–656.

Shannon, C.E., Weaver, W., 1963. The Mathematical Theory of Communication. University of Illinois Press, Urbana, 117 pp.

Siepel, H., 1994. Life-history tactics of soil microarthropods. Biol. Fertil. Soils 18, 263–278.

Siepel, H., 1995. Applications of microarthropod life-history tactics in nature management and ecotoxicology. Biol. Fertil. Soils 19, 75–83.

Siepel, H., 1996. The importance of unpredictable and short-term environmental extremes for biodiversity in oribatid mites. Biodivers. Lett. 3, 26–34.

Sommaggio, D., 1999. Syrphidae: can they be used as environmental bioindicators? In: Paoletti, M.G. (Ed.), Invertebrate Biodiversity as Bioindicators of Sustainable Landscapes. Agric. Ecosyst. Environ. 74, 343–356.

Speight, M.C.D., McLEan, L.F.G., Goeldlin de Tiefenau, P., 1992. The recognition of sites of international importance for protection of invertebrates. Environ. Encounters Ser. 14, 39–42.

Ssymank, A., 1993. Zur Bewertung und Bedeutung naturnaher Landschaftselemente in der Agrarlandschaft. Teil I. Schwebfliegen (Diptera: Syrphidae). Verhandlungen der Gesellschaft für Ökologie 22, 255–262.

Stachow, U., Büchs, W., Schultz, A., Lutze, G., Latus, C., 2001. Assessment and evaluation of biodiversity in agricultural landscapes. In: Zotz, G., Körner, C. (Eds.), Functional Importance of Biodiversity. Verhandlungen der Gesellschaft für Ökologie, 2001. Parey-Verlag, Berlin, Germany, 31 pp.

Standing Committee on Agriculture and Resource Management, 1993. Sustainable agriculture—tracking the indicators for Australia and New Zealand. SCARM Report 51, BRS, Canberra, pp. 1–62.

Steenmans, C., Pingborg, U., 2001. Anthropogenic fragmentation potential semi-natural and natural areas. In: EU-Commission (Ed.), From Land Cover to Landscape Diversity in the European Union. Brussels, Belgium, 6 pp. (+3 maps).

Steidl, I., Ringler, A., 1997. Agrotope (Teilband)—Landschaftspflegekonzept Bayern. In: Bayerisches Staatsministerium für Landesentwicklung und Umweltfragen, Bayerische Akademie für Naturschutz und Landschaftspflege, Vol. II. München, Germany, p. 11.

Steinborn, H.-A., Heydemann, B., 1990. Indikatoren und Kriterien zur Beurteilung der ökologischen Qualität von Agrarflächen am Beispiel der Carabidae (Laufkäfer). Schriftenreihe Landschaftspflege und Naturschutz 32, 165–174.

Steinborn, H.-A., Meyer, H., 1994. Einfluß alternativer und konventioneller Landwirtschaft auf die Prädatorenfauna in Agrarökosystemen Schleswig-Holsteins (Araneida, Coleoptera: Carabidae, Diptera: Dolichopodidae, Empididae, Hybotidae, Microphoridae). Faunistisch-Ökologische Mitteilungen 6, 409–438.

Steiner, N.C., Köhler, W., 2003. Effects of landscape patterns on species richness—a modelling approach. In: Büchs, W. (Ed.), Biotic Indicators for Biodiversity and Sustainable Agriculture. Agric. Ecosyst. Environ. 98, 353–361.

Steinmann, H.-H., Gerowitt, B., 2000. Ackerbau in der Kulturlandschaft—Funktionen und Leistungen. Ergebnisse des Göttinger INTEX-Projektes. Mecke Verlag, Duderstadt, Germany, 300 pp.

Stern, K., Tigerstedt, P.M.A., 1974. Ökologische Genetik. Fischer Verlag, Stuttgart.

Stippich, G., Krooß, S., 1997. Auswirkungen von Extensivierungsmaßnahmen auf Spinnen, Laufkäfer und Kurzflügelkäfer. In: Gerowitt, B., Wildenhayn, M. (Eds.), Ökologische und ökonomische Auswirkungen von Extensiverungsmaßnahmen im Ackerbau—Ergebnisse des Göttinger INTEX-Projektes 1990–1994, pp. 221–262.

Stork, N.E., 1995. Measuring and monitoring arthropod diversity in temperate and tropical forests. In: Boyle, T.J.B., Boontawee, B. (Eds.), Measuring and Monitoring Biodiversity in Temperate and Tropical Forests. CIFOR, Bogor, pp. 257–270.

Stork, N.E., Eggleton, P., 1992. Invertebrates as determinants and indicators of soil quality. Am. J. Alternative Agric. 7, 38–47.

Stugren, B., 1978. Grundlagen der allgmeinen Ökologie. Stuttgart, New York.

Tembrock, G., 1982/1983. Spezielle Verhaltensbiologie der Tiere. Gustav Fischer Verlag, Jena, 1040 pp.

Tilman, D., Knops, J., Wedlin, D., Reich, P., Ritchie, M., Siemann, E., 1997. The influence of functional diversity and composition on ecosystem processes. Science 277, 1300–1302.

The Soil Association (Ed.), 2000. The Biodiversity Benefits of Organic Farming. Bristol, UK, 34 pp.

Topp, W., 1989. Laufkäfer als Bioindikatoren in der Kulturlandschaft. Verhandlungen 9. SIEEC Gotha 1986, pp. 78–82.

Tscharntke, T., Gathmann, A., Steffan-Dewenter, I., 1998. Bioindication using trap-nesting bees and wasps and their natural enemies: community structure and interactions. J. Appl. Ecol. 35 (5), 708–719.

Usseglio-Polatera, P., Bornaud, M., Richoux, P., Tachet, H., 2000. Biomonitoring through biological traits of benthic macroinvertebrates: how to use species trait databases? In: Jungwirth, M., Muhar, S., Schmutz, S. (Eds.), Assessing the Ecological Integrity of Running Waters. Hydrobiologia 422–423, 153–162.

Van Dijk, T., 1986. On the relationship between availability of food and fecundity in carabid beetles. How far is the number of eggs in the ovaries a measure of the quantities of food in the field. Report of the Fifth European Symposium on Carabidae, pp. 105–120.

Van Elsen, T., 1996. Wirkungen des ökologischen Landbaues auf die Segetalflora—Ein Übersichtsbeitrag. In: Deipenbrock, W., Hülsbergen, K.-J. (Eds.), Langzeiteffekte des ökologischen Landbaus auf Fauna, Flora und Boden. Halle/Saale, Germany, pp. 143–152.

Van Elsen, T., 2000. Organic Farming as a challenge for the integration of agriculture and nature development. In: Stolton, S., Geier, B., McNeely, J.A. (Eds.), The Relationship between Nature Conservation, Biodiversity and Organic Agriculture. Proceedings of the International Workshop, Vignola, Italy, 1999, pp. 76–85.

Van Vliet, P.C.J., Beare, M.H., Coleman, D.C., 1995. Population dynamics and functional role of Enchytraeidae (Oligochaeta) in hardwood forests and agricultural ecosystems. Plant Soil 170 (1), 199–207.

Vermeulen, A.C., 1995. Elaborating chironomid deformities as bioindicators of toxic sediment stress: the potential application of mixture toxicity concepts. Ann. Zool. Fennici 32, 265–285.

Von Euler, F., 1999. An objective indicator of functional integrity in avian communities. In: Gustafsson, L., Weslien, J.O. (Eds.), Biodiversity in Managed Forests—Concepts and Solutions. For. Ecol. Manage. 115, 221–229.

Volkmar, C., Wetzel, T., 1998. Zum Auftreten gefährdeter Spinnen (Arachnida: Araneae) auf Agrarflächen in Mitteldeutschland. Archiv für Phytopathologie und Pflanzenschutz 31, 561–574.

Volkmar, C., Bothe, S., Kreuter, T., Lübke-Al Hussein, M., Richter, L., Heimbach, U., Wetzel, T., 1994. Epigäische Raubarthropoden in Winterweizenbeständen Mitteldeutschlands und ihre Beziehung zu Blattläusen. Mitteilungen aus der Biologischen Bundesanstalt für Land- und Forstwirtschaft Berlin Dahlem, Heft 299, 134.

Vollrath, F., 1988. Spider growth as an indicator of habitat quality. Bull. Br. Arachnol. Soc. 7 (7), 217–219.

Waldhardt, R., 2003. Biodiversity and landscape—summary, conclusions and perspectives. In: Büchs, W. (Ed.), Biotic Indicators for Biodiversity and Sustainable Agriculture. Agric. Ecosyst. Environ. 98, 305–309.

Waldhardt, R., Otte, A., 2000. Zur Terminologie und wissenschaftlichen Anwendung des Begriffs Biodiversität. Wasser Boden 52 (1–2), 10–13.

Waldhardt, R., Otte, A., 2003. Indicators of plant species and community diversity in grasslands. In: Büchs, W. (Ed.), Biotic Indicators for Biodiversity and Sustainable Agriculture. Agric. Ecosyst. Environ. 98, 339–351.

Wallin, H., 1985. Spatial and temporal distribution of some abundant carabid beetles (Coleoptera: Carabidae) in cereal fields and adjacent habitats. Pedobiologia 28, 19–34.

Wallin, H., 1989. The influence of different age classes on the seasonal activity and reproduction of four medium-sized carabid species inhabiting cereal fields. Holarctic Ecol. 12, 201–212.

Wang, D., Lowery, B., McSweeney, K., 1996. Spatial and temporal patterns of ant burrow openings as affected by soil properties and agricultural practices. Pedobiologia 40, 201–211.

Warwick, W.F., 1988. Morphological deformities in Chironomidae (Diptera) larvae as biological indicators of toxic stress. In: Evans, M.S. (Ed.), Toxic Contaminants and Ecosystem Health: A Great Lakes Focus. Wiley, New York, pp. 280–310.

Warwick, R.M., Clarke, K.R., 1998. Taxonomic distinctness and environmental assessment. J. Appl. Ecol. 35 (4), 532–543.

Watt, A.D., Stork, N.E., Eggleton, P., Srivastava, D.S., Bolton, B., Larsen, T.B., Brendell, M.J.D., Bignell, D.E., 1997. Impact of forest loss and regeneration on insect abundance and diversity. In: Watt, A.D., Hunter, M., Stork, N.E. (Eds.), Forests and Insects. Chapman & Hall, London, pp. 273–286.

Weber, G., Franzen, J, Büchs, W., 1997. Beneficial Diptera in field crops with different input of pesticides and fertilizers. In: Kromp, B., Meindl, P. (Eds.), Entomological Research in Organic Farming. Biol. Agric. Horticult. 15, 109–122.

Weigmann, G., Jung, E., 1992. Die Hornmilben (Acari, Oribatida) au Straßenbäumen in Stadtzonen unterschiedlicher Luftbelastung in Berlin. Zool. Beitr. 34, 273–287.

Weiss, E., Nentwig, W., 1992. The importance of flowering plants on sown weed strips for beneficial insects in cereal fields. Mitteilungen der Deutschen Gesellschaft für allgemeine und angewandte Entomologie 8, 133–136.

Wetzel, T., 1993. Genug Nützlinge auch auf Großflächen. Pflanzenschutz-Praxis 4, 16–19.

Wetzel, T., Volkmar, C., Lübke-Al Hussein, M., Jany, D., Richter, L., 1997. Zahlreiche "Rote-Liste-Arten" epigäischer Raubarthropoden auf grossen Agrarflächen Mitteldeutschlands. Archiv für Phytopathologie und Pflanzenschutz 31, 165–183.

Wheater, C.P., Cullen, W.R., Bell, J.R., 2000. Spider communities as tools in monitoring reclaimed limestone quarry landforms. Landsc. Ecol. 15, 401–406.

Wicke, G., 1998. Stand der Ackerrandstreifenprogramme in Deutschland. Schriftenreihe der Landesanstalt für Pflanzenbau und Pflanzenschutz 6, 55–84.

Wiener, N., 1948. Cybernetics. New York.

Willems, E., Vandevoort, C., Willekens, A., Buffaria, B., 2001. 3. Landscape and land cover diversity index. In: EU-Commission (Ed.), From Land Cover to Landscape Diversity in the European Union. Brussels, Belgium, 17 pp. (+1 table and 23 maps).

Wise, D.H., 1979. Effects of an experimental increase in prey abundance upon reproductive rates of two orb-weaving spider species (Araneae, Araneidae). Oecologia 41, 289–300.

Wittasek, R., 1987. Untersuchung zur Kupferanreicherung in der Bodenfauna eines Weinbergökosystems. In: Welz, B. (Ed.), Colloquium Atomspektrometrische Spurenanalytik, Überlingen.

World Bank (Ed.), 1998. Guidelines for Monitoring and Evaluation for Biodiversity Projects: Environment. Department Papers, Biodiversity Series No. 065.

Yeates, G.W., Bongers, T., 1999. Nematode diversity in agro-ecosystems. In: Paoletti, M.G. (Ed.), Invertebrate Biodiversity as Bioindicators of Sustainable Landscapes. Agric. Ecosyst. Environ. 74, 113–135.

Zanger, A., Lys, J.-A., Nentwig, W., 1994. Increasing the availability of food and the reproduction of *Poecilus cupreus* in a cereal field by strip management. Entomologia Experimentalis et Applicata 71, 111–120.

Zimmermann, J., Büchs, W., 1996, Management of arable crops and its effects on rove-beetles (Coleoptera: Staphylinidae) with special reference to the effects of insecticide treatments. In: Booj, C.J.H., den Nijs, L.J.M.F. (Eds.), Arthropod Natural Enemies in Arable Land. II. Survival, Reproduction and Enhancement. Acta Jutlandica 71 (2), 183–194.

Zimmermann, J., Büchs, W., 1999. Kurzflügelkäfer (Coleoptera: Staphylinidae) in unterschiedlich intensiv bewirtschafteten Ackerflächen. Agrarökologie 32, 1–154.

ELSEVIER

Available online at www.sciencedirect.com

Agriculture, Ecosystems and Environment 98 (2003) 79–85

Agriculture Ecosystems & Environment

www.elsevier.com/locate/agee

Floristic diversity at the habitat scale in agricultural landscapes of Central Europe—summary, conclusions and perspectives

Rainer Waldhardt [a,*], Dietmar Simmering [a], Harald Albrecht [b]

[a] *Division of Landscape Ecology and Landscape Planning, Justus-Liebig-University of Giessen, IFZ, Heinrich-Buff-Ring 26-32, D-35392 Giessen, Germany*

[b] *Vegetation Ecology, Department of Vegetation Ecology, Technische Universitaet Muenchen, Am Hochanger 6, D-85350 Freising-Weihenstephan, Germany*

Abstract

Referring to the agricultural history in Central Europe determinants of floristic diversity at the habitat scale and the decline in diversity over the last few decades are described. In this context the preservation of floristic diversity is stressed to be one important goal of modern, multifunctional agricultural land use. To reach this goal indicators of diversity are useful tools to evaluate the effects management practices have on floristic diversity in agro-ecosystems. However, "key indicators" that allow an easy assessment and evaluation of diversity are still lacking. Potentially, indicators of biodiversity measures at the habitat scale can be developed from a large number of parameters. An attempt to classify parameters into three major types of indicators and requirements indicators have to meet are presented.

Based on the contributions in this special issue and further references recent approaches to indicate floristic diversity at the habitat scale in agricultural landscapes of Central Europe are summarized and discussed. It is concluded that further research should focus on the validation of existing approaches and their integration into a comprehensive set of indicators and on the development of standardized modes of collection for data.

Keywords: Biodiversity; Flora; Vegetation; Indicator; Species diversity; Structural diversity; Spatio-temporal scale

1. Determinants of floristic diversity at the habitat scale

In about 5000 years of agricultural history in Central Europe, the anthropogenic transformation of natural, old-growth woodlands into mosaic landscapes with agricultural and semi-natural habitats has had a considerable side-effect in enhancing biodiversity. This particularly concerns the species diversity of vascular plants (Sukopp, 1972), which is the organism group dealt within this paper, but is also true for the diversity of biocoenoses and ecosystems (Ellenberg, 1996). More secluded aspects of biodiversity such as the genetic diversity in populations of certain species, are likely to have profited from the evolution of our rural landscape as well (McCauley, 1995; Young, 1995).

In the middle of the twentieth century, the situation changed dramatically. Traditional and diverse management practices, which have been the main driving forces for the increase and preservation of biodiversity, were given up and were replaced by modern agriculture. Intensification of agriculture by use of high-yielding crop varieties, fertilization, irrigation,

* Corresponding author. Tel.: +49-641-9937163; fax: +49-641-9937169.
E-mail address: rainer.waldhardt@agrar.uni-giessen.de (R. Waldhardt).

0167-8809/$ – see front matter © 2003 Elsevier Science B.V. All rights reserved.
doi:10.1016/S0167-8809(03)00071-9

and pesticides has contributed substantially to tremendous increases in food production over the past 50 years (Matson et al., 1997). At the same time, many marginal regions with an unfavorable climate, topography and poor soils are threatened by abandonment. Such rural landscapes, with a traditional small-scale mosaic of grassland and arable fields, and thus a high diversity of habitats, have undergone a radical change since crop production has been already widely replaced by extensively managed grasslands or even forests in the past few decades (Baldock et al., 1996; Buhler-Natour and Herzog, 1999).

These two developments affect both, abiotic resources like water and soil (Bork, 1988; Cooper, 1993), as well as components of biodiversity. For plant species a severe decline in diversity over the last few decades is reported from arable land (e.g. Moravec, 1993; Albrecht, 1995; Sutcliffe and Kay, 2000), grasslands (e.g. Meisel and Hübschmann, 1976; Meisel, 1983; Willems, 1990; Schrautzer and Wiebe, 1993) as well as from boundary structures (e.g. Burel, 1996; Steidl and Ringler, 1997; Boutin and Jobin, 1998). Both, abandonment on the one hand, and intensification of management systems on the other, are considered as the main causes for the decrease in vascular plant species richness (Korneck et al., 1998).

It may be objected that the primary function of agricultural land use is not the preservation and support of biodiversity, but the production of food and raw materials. It has to be considered, though, that nearly 40% of the entire surface area in the European Union (EU) serves agricultural purposes and most of the remainder is consumed by intensively managed forests, settlement, and traffic (Bruyas, 2002). Under such preconditions, farmland must not only be looked at from an economical aspect as merely the basis for a profitable yield. Equally, it has to fulfil important additional functions in preserving abiotic resources, in stimulating human recreation, and in conserving biodiversity. To a growing extent, this 'multifunctionality' of agriculture is recently recognized in most Central European countries (e.g. Nagy, 1997; Beaufoy, 1998; European Commission, 1999; Vos and Meekes, 1999; Wytrzens and Mayer, 1999; Pinton, 2001; Tait, 2001; Bundesamt für Naturschutz, 2002). To counteract the documented decrease in biodiversity, which is likely to decline even more in the future (Tilman et al., 2001), the EU has initiated different measures to conserve and to increase biodiversity in agricultural landscapes (Marggraf, 1998) and is committed to do so even more in the future (European Commission, 2001).

What is still lacking, however, are "key indicators" that allow an easy assessment and evaluation of biodiversity (Flather et al., 1997; Duelli and Obrist, 1998; Simberloff, 1998; Flather and Sieg, 2000). In this context, the analysis of interactions between management practices at the habitat scale, abiotic resources, socio-economic conditions and biodiversity measures

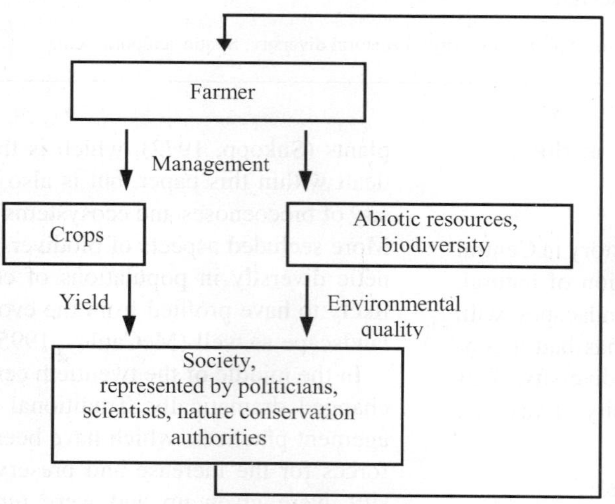

Fig. 1. By their farming practice farmers determine crop yields and environmental quality. It is for the society to evaluate and to decide on the future development of both.

(Fig. 1) is a challenge for scientific research. Management practices have a primary, direct impact on abiotic resources (e.g. the increase of nitrogen contents in soil after fertilization) and on biodiversity (e.g. the decline in the abundance of arable weeds after the application of herbicides), but may also cause many secondary, indirect effects in agro-ecosystems (e.g. the changes in zoocoenoses induced by a reduced pollen supply as a consequence of herbicide application; Bick and Brocksieper, 1979). Knowledge of such interactions can provide indicators to evaluate the effects management practices have on floristic diversity in agro-ecosystems. If such indicators would accurately reflect the impact of management practices on biodiversity, they would be useful to install and evaluate detailed funding systems 'which offer incentive payments to farmers who sign up to specific environmental commitments' (European Commission, 2001).

2. Indication of floristic diversity at the habitat scale

The quality of indicators relies on the scale, which they represent (Stein et al., 2001). Therefore, an analysis of primary and secondary interactions between land use and components of biodiversity and the identification of corresponding indicators has to consider a broad spectrum of spatial scales. However, it is of particular importance to define the spatial scale, a certain indicator is valid for. Specific relations between measures of biodiversity and their potential indicators can be detected at a 'within habitat-patch'-scale and a 'within habitat-type'-scale, as well as at the 'within habitat-mosaic' or landscape-scale, up to a regional, national or even global scale. On the different spatial scales, we are facing characteristic types of intrinsic or temporal dynamics, such as annual disturbances in specific habitats or a long-term turnover in land use at the landscape scale. Thus, interactions and relations between land use and biodiversity vary even within the respective spatio-temporal scales (Eiswerth and Haney, 2001).

Biodiversity is usually measured and expressed at three levels that are generally applicable at all spatial scales, but are especially suited to describe diversity at the habitat scale: Following Whittaker (1972), α-diversity denotes the diversity of a certain reference area (e.g. sample, habitat patch), while γ-diversity is the diversity of a larger spatial unit (e.g. habitat types, landscape) that comprises the sum of all α-diversities. While α- and γ-diversity are inventory diversities that—in case of vascular plants—can easily be measured, β-diversity, or 'differentiation diversity', characterizes the differences between the diversities (α or γ) of certain reference areas. Commonly, β-diversity is calculated using one out of a large set of available indices and statistical measures (Magurran, 1988; Waldhardt and Otte, 2003). For the differentiation of relations between aspects of biodiversity and their potential indicators, it has to be considered on which of these levels of biodiversity the relations may occur or are valid for, respectively.

2.1. Indicators of biodiversity—classification and requirements

Potentially, indicators of biodiversity measures at the habitat scale can be developed from a large number of parameters. Table 1 shows an attempt to classify a selection of parameters into three major types of indicators. We distinguish between measures of certain aspects of biodiversity that are potentially suitable for indicating overall biodiversity (Type I), measures of natural or management related environmental traits (Type II), and deductive parameters of spatially explicit data (Type III).

In order to provide comparable estimates of biodiversity measures, indicators at the habitat scale have to meet at least (some of) the following requirements:

1. Indicators should reflect correlations between management practices and certain aspects of biodiversity (e.g. species richness of vascular plants in a habitat patch) in a quantitative way.
2. Indicators may serve as surrogates for other, e.g. functional, aspects of biodiversity in a habitat patch. The suitability of a certain weed-community as a precondition for the development of a species-rich community of phytophageous insects is a conceivable example for this rather qualitative approach (Perfecto and Snelling, 1995).
3. Indicators should be valid and applicable at a defined spatial scale.
4. The indicators should be easy to assess and be receptive to repeatability with a sufficient accuracy.

Table 1
Classification of selected parameters for the indication of biodiversity at the habitat scale (with references)

Parameter	Reference
Measures for certain aspects of biodiversity	
Biotic elements (number, abundance and frequencies of)	
'Characteristic' species	Rosenthal (2003)
Functional groups	Hald (1999), Critchley (2000), Kleijn and Verbeek (2000), Waldhardt et al. (2000)
Endangered species	Lee et al. (2001), Rosenthal (2003)
Biotic structure	
Height of vegetation	Mückschel and Otte (2003)
Cover of litter in vegetation	Mückschel and Otte (2003)
Evenness indices of species distribution	Smith and Wilson (1996)
Biotic processes	
Seed production of certain populations	Mückschel and Otte (2003)
Biomass production in plant communities	Mückschel and Otte (2003)
Habitat age	Honnay et al. (1999), Dumortier et al. (2002)
Measures of natural or management related environmental traits	
Ground-water table	Rosenthal (2003)
Frequency of mowing	Gigon and Leutert (1996), Pfadenhauer and Klotzli (1996), Collins et al. (1998)
Amount of mineral fertilizer applied to fields	Alard et al. (1994), Kleijn and Verbeek (2000)
Quality and frequency of pesticide application	Boatman (1994), Schmidt et al. (1995), Snoo (1997)
Deductive parameters of spatially explicit data from	
Digital elevation models	Zimmermann and Kienast (1999)
Color-infrared (CIR) aerial photographs	Nilsen et al. (1999), Waldhardt and Otte (2003)

2.2. Recent approaches to indicate floristic diversity at the habitat scale

Within this chapter, we present new approaches for the development of indicators of floristic diversity at the habitat scale. Rosenthal (2003) and Albrecht (2003) propose to define 'characteristic' or 'vulnerable' species (i.e. biotic elements, Table 1) to assess plant species diversity in wet grassland and arable fields. In contrast, Mückschel and Otte (2003) focus on measures of structural diversity by analyzing morphological traits of certain species. Results suggest that shoot height of certain species may be a sensitive indicator for structural changes in endangered, calcareous grassland habitats.

The indicators proposed in these three papers are capable of being integrated in sophisticated instruments for the evaluation of floristic diversity at the habitat scale. A problem, the application of these indicators may cause, is their demand for a rather intensive data collection, which is consuming time and financial resources. Therefore, their future application will most likely be restricted to agricultural habitats of particular conservational interest.

To maintain or to create a satisfying amount of floristic diversity in the remaining (i.e. cultivated) area, we need sufficient management practices. A corresponding investigation was carried out by Gerowitt (2003). In a field plot experiment, she tested the effects of different management systems on the arable weed flora. A combination of reduced input and reduced tillage led to higher weed species numbers than in other systems like "good farming practice" which was characterized by a higher input of agrochemicals. The indicators presented here are yet only valid for the habitat patches that were investigated. Relations between indicators and the prevailing indicated aspects of diversity are not necessarily applicable to other habitats or habitat-types. Due to, for example, varying natural site conditions from the local up to the regional scale, a general transferability of these results cannot be assumed. Validating research on the indicators presented here has yet to be done, but this applies to most of the indicators that have been proposed in this rather new field of research. Likewise, the topic 'biodiversity at the habitat scale' is not sufficiently covered by indicators of the mere species and structural diversity presented in this chapter. Relations between land use and further aspects of biodiversity, such as genetic diversity (Young, 1995; Jelinski, 1997; Greimler and Dobes, 2000; Schubert et al., 2002), have yet been rarely studied and are still poorly understood.

3. Conclusions and perspectives

Due to the high share of land covered by agricultural utilized habitats, and the great importance of management practices for biotic elements, floristic diversity in Central Europe is only to preserve by an extensive sustainable land use. In this context, indicators of floristic diversity are helpful for the identification of valuable habitats and the evaluation of management practices for conservational purposes. But, since the relations between management practices and floristic diversity at the habitat scale are highly dependent on regional specificities and spatial as well as temporal scales, and due to the complexity of floristic diversity in terms of elements, structures and functions, a satisfying indication of floristic diversity with only one or a few indicators is not realistic or practicable. The corresponding scientific results of experimental and empirical studies in this field are diverse and partly inconsistent. Nevertheless, there are promising approaches for the development of indicators valid for specific habitat patches and habitat types on a local or even regional scale available.

In our view, further research should now focus on the validation of existing approaches and their integration into a comprehensive set or "shopping basket" (Niemela and Baur, 1998) of indicators, taking into consideration relations between floristic diversity and other components of biodiversity as discussed by Duelli and Obrist (1998), Albrecht et al. (2001) and others. The development of standardized modes of collection for data on these existing indicators seems also to be more important, than research on even more new approaches in the search for indicators. The development of a reliable set of indicators from existing approaches, which specifically considers different spatial and temporal scales and that is still practical for an evaluation of biodiversity (without simplifying the complex interdependences between land use and specific components of biodiversity), remains a challenge for scientific research.

References

Alard, D., Bance, J.F., Frileux, P.N., 1994. Grassland vegetation as in indicator of the main agro-ecological factors in a rural landscape: consequences for biodiversity and wildlife conservation in central Normandy (France). J. Environ. Manage. 42, 91–109.

Albrecht, H., 1995. Changes in the arable weed flora of Germany during the last five decades. In: Proceedings of the Ninth European Weed Research Society Symposium, Challenges for Weed Science in a Changing Europe, Budapest, pp. 41–48.

Albrecht, H., 2003. Suitability of arable weeds as indicator organisms to evaluate species conservation effects of management in agricultural ecosystems. Agric. Ecosyst. Environ. 98, 201–211.

Albrecht, H., Kühn, N., Filser, J., 2001. Site effects on plant and animal distribution at the Scheyern experimental farm. Ecol. Stud. 147, 209–227.

Baldock, D., Beaufoy, G., Brouwer, F., Godeschalk, F., 1996. Farming at the margins. Abandonment or Redeployment of Agricultural Land in Europe. Hague, London.

Beaufoy, G., 1998. The EU habitats directive in Spain: can it contribute effectively to the conservation of extensive agroecosystems? J. Appl. Ecol. 35, 974–978.

Bick, H., Brocksieper, J., 1979. Auswirkungen der Landbewirtschaftung auf die Invertebratenfauna. Landwirtsch. Angew. Wiss. 218.

Boatman, N.D., 1994. Field margins: integrating agriculture and conservation. In: Proceedings of the Symposium held at Coventry, UK, April 18–20, 1994.

Bork, H.-R.,1988. Bodenerosion und Umwelt—Verlauf, Ursachen und Folgen der mittelalterlichen und neuzeitlichen Bodenerosion. Bodenerosionsprozesse, Modelle und Simulationen. Landsch.genes. Landsch.ökol. 13.

Boutin, C., Jobin, B., 1998. Intensity of agricultural practices and effects on adjacent habitats. Ecol. Appl. 8, 544–557.

Bruyas, P., 2002. Land use-land cover: LUCAS 2001 Primary Results. Eurostat, European Community.

Buhler-Natour, C., Herzog, F., 1999. Criteria for sustainability and their application at a regional level: the case of clearing islands in the Dubener Heide nature park (Eastern Germany). Landsc. Urban Plan. 46, 51–62.

Bundesamt für Naturschutz (Ed.), 2002. Denkschrift Forschung für eine naturgerechte Landwirtschaft. Bonn.

Burel, F., 1996. Hedgerows and their role in agricultural landscapes. Crit. Rev. Plant Sci. 15, 169–190.

Collins, S.C., Knapp, A.K., Briggs, J.M., Blair, J.M., Steinhauer, E.M., 1998. Modulation of diversity by grazing and mowing in native tallgrass prairie. Science 280, 749–754.

Cooper, C.M., 1993. Biological effects of agriculturally derived surface water pollutant on aquatic systems—A review. J. Environ. Qual. 22, 402–408.

Critchley, C.N.R., 2000. Ecological assessment of plant communities by reference to species traits and habitat preferences. Biodivers. Conserv. 9, 87–105.

Duelli, P., Obrist, M.K., 1998. In search of the best correlates for local organismal biodiversity in cultivated areas. Biodiv. Conserv. 7, 297–309.

Dumortier, M., Butaye, J., Jacquemyn, H., Van Camp, N., Lust, N., Hermy, M., 2002. Predicting vascular plant species richness of fragmented forests in agricultural landscapes in central Belgium. For. Ecol. Manage. 158, 85–102.

Eiswerth, M.E., Haney, J.C., 2001. Maximizing conserved biodiversity: why ecosystem indicators and thresholds matter. Ecol. Econ. 38, 259–274.

Ellenberg, H., 1996. Vegetation Mitteleuropas mit den Alpen in ökologischer, dynamischer und historischer sicht. Ulmer, Stuttgart.

European Commission, Directorate-General of Agriculture, 1999. Contribution of the European Community on the multifunctional character of agriculture. Bruxelles.

European Commission, 2001. The Sixth Environment Action Programme of the European Community 2001–2010. http://europa.eu.int/comm/environment/newprg/index.htm.

Flather, C.H., Sieg, C.H., 2000. Applicability of montreal process criterion 1—conservation of biological diversity—to rangeland sustainability. Int. J. Sustain. Dev. World Ecol. 7, 81–96.

Flather, C.H., Wilson, K.R., Dean, D.J., McComb, W.C., 1997. Identifying gaps in conservation networks: of indicators and uncertainty in geographic-based analyses. Ecol. Appl. 7, 531–542.

Gerowitt, B., 2003. Development and control of weeds in arable farming systems. Agric. Ecosyst. Environ. 98, 247–254.

Gigon, A., Leutert, A., 1996. The dynamic keyhole-key model of coexistence to explain diversity of plants in limestone and other grasslands. J. Veg. Sci. 7, 29–40.

Greimler, J., Dobes, C., 2000. High genetic diversity and differentiation in relict lowland populations of Gentianella austriaca (A. and J. Kern.) Holub (Gentianaceae). Plant Biol. 2, 628–637.

Hald, A.B., 1999. Weed vegetation (wild flora) of long established organic versus conventional cereal fields in Denmark. Ann. Appl. Biol. 134, 307–314.

Honnay, O., Hermy, M., Coppin, P., 1999. Effects of area, age and diversity of forest patches in Belgium on plant species richness, and implications for conservation and reforestation. Biol. Conserv. 87, 73–84.

Jelinski, D.E., 1997. On genes and geography: a landscape perspective on genetic variation in natural plant populations. Landsc. Urban Plan. 39, 11–23.

Kleijn, D., Verbeek, M., 2000. Factors affecting the species composition of arable field boundary vegetation. J. Appl. Ecol. 37, 256–266.

Korneck, D., Schnittler, M., Klingenstein, F., Ludwig, G., Takla, M., Bohn, U., May, R., 1998. Warum verarmt unsere Flora? Auswertung der Roten Liste der Farn- und Blütenpflanzen Deutschlands. Schr.reihe Veg.kd. 29, 299–444.

Lee, J.T., Woddy, S.J., Thompson, S., 2001. Targeting sites for conservation: using a patch-based ranking scheme to assess conservation potential. J. Environ. Manage. 61, 367–380.

Magurran, A.E., 1988. Ecological Diversity and its Measurement. Princeton University Press, Princeton.

Marggraf, R., 1998. Die Agrarumweltprogramme der EU.—Analyse und Bewertung. Schr.reihe Landesanstalt Pflanzenbau Pflanzenschutz 6, 13–23.

Matson, P.A., Parton, W.J., Power, A.G., Swift, M.J., 1997. Agricultural intensification and ecosystem properties. Science 277, 504–509.

McCauley, D.E., 1995. Effects of population dynamics on genetics in mosaic landscapes. In: Hansson, L., Fahrig, L., Merriam, G. (Eds.), Mosaic Landscapes and Ecological Processes. Chapman & Hall, London, pp. 178–198.

Meisel, K., 1983. Zum Nachweis von Grünlandveränderungen durch Vegetationserhebungen. Tuexenia 3, 407–415.

Meisel, K., Hübschmann, A.v., 1976. Veränderungen der Acker und Grünlandvegetation im nordwestdeutschen flachland in jüngerer. Zeit. Schr.reihe Veg.kd. 10, 109–124.

Moravec, J., 1993. Biodiversity changes on an ecosystemic level—phytocoenological approach. Ekologia 12, 317–324.

Mückschel, C., Otte, A., 2003. Morphological parameters: an approach for the indication of conservation value of low productive calcareous grasslands. Agric. Ecosyst. Environ. 98, 213–225.

Nagy, G., 1997. Potential role of grasslands in sustainable land use. Acta Agron. Hung. 45, 69–83.

Niemela, J., Baur, B., 1998. Threatened species in a vanishing habitat: plants and invertebrates in calcareous grasslands in the Swiss Jura mountains. Biodivers. Conserv. 7, 1407–1416.

Nilsen, L., Brossard, T., Joly, D., 1999. Mapping plant communities in a local Arctic landscape applying a scanned infrared aerial photograph in a geographical information system. Int. J. Remote Sens. 20, 463–480.

Perfecto, I., Snelling, R., 1995. Biodiversity and the transformation of a tropical agroecosystem—ants in coffee plantations. Ecol. Appl. 5, 1084–1097.

Pfadenhauer, J., Klotzli, F., 1996. Restoration experiments in middle European wet terrestrial ecosystems: an overview. Vegetatio 126, 101–115.

Pinton, F., 2001. Conservation of biodiversity as a European directive: the challenge for France. Sociol. Rural. 41, 329–342.

Rosenthal, G., 2003. Selecting target species to evaluate the success of wet grassland restoration. Agric. Ecosyst. Environ. 98, 227–246.

Schmidt, W., Waldhardt, R., Mrotzek, R., 1995. Extensification in arable systems: Effects on flora, vegetation and soil seed bank: Results of the INTEX-project, University of Goettingen, Tuexenia, pp. 415–435.

Schrautzer, J., Wiebe, C., 1993. Geobotanische Charakterisierung und Entwicklung des Grünlandes in Schleswig-Holstein. Phytocoenologia 22, 105–144.

Schubert, P., O'Neill, R., Köhler, W., Waldhardt, R., Otte, A., 2002. Reproductive traits and genetic diversity of Arabidopsis thaliana populations originating from different agricultural regimes. Z. Pflanzenkrankh, Pflanzenschutz, Sonderh. 18, 57–66.

Simberloff, D., 1998. Flagships, umbrellas, and keystones: is single-species management passe in the landscape era? Biol. Conserv. Mar. 83, 247–257.

Smith, B., Wilson, J.B., 1996. A consumer's guide to evenness indices. Oikos 76, 70–82.

Snoo, G.R.D., 1997. Arable flora in sprayed and unsprayed crop edges. Agric. Ecosyst. Environ. 66, 223–230.

Steidl, I., Ringler, A., 1997. Agrotope (1. Teilband)—Landschaftspflegekonzept Bayern, Band II.11. Eds. Bayerisches Staatsministerium für Landesentwicklung und Umweltfragen und Bayerische Akademie für Naturschutz und Landschaftspflege, München.

Stein, A., Riley, J., Halberg, N., 2001. Issues of scale for environmental indicators. Agric. Ecosyst. Environ. 87, 215–232.

Sukopp, H., 1972. Wandel der Flora und Vegetation in Mitteleuropa unter dem Einfluß des Menschen. Ber. Landwirtsch. 50, 112–139.

Sutcliffe, O.L., Kay, Q.O.N., 2000. Changes in the arable flora of central southern England since the 1960s. Biol. Conserv. 93, 1–8.

Tait, J., 2001. Science, governance and multifunctionality of European agriculture. Outlook Agric. 30, 91–95.

Tilman, D., Fargione, J., Wolff, B., D'Antonio, C., Dobson, A., Howarth, R., Schindler, D., Schlesinger, W.H., Simberloff, D., Swackhamer, D., 2001. Forecasting agriculturally driven global environmental change. Science 292, 281–284.

Vos, W., Meekes, H., 1999. Trends in European cultural landscape development: perspectives for a sustainable future. Landsc. Urban Plan. 46, 3–14.

Waldhardt, R., Otte, A., 2003. Indicators of plant species and community diversity in grasslands. Agric. Ecosyst. Environ. 98, 339–351.

Waldhardt, R., Simmering, D., Otte, A., 2000. Standortspezifische Surrogate und Korrelate der a-Artendichten in der Grünland-Vegetation einer peripheren Kulturlandschaft Hessens. Ber. ANL 24, 79–86.

Whittaker, R.H., 1972. Evolution and measurement of species diversity. Taxon 21, 213–251.

Willems, J.H., 1990. Calcareous grasslands in continental Europe. In: Hillier, S.H., Walton, D.H.W., Wells, D.A. (Eds.), Calcareous Grasslands. Ecology Management. Bluntisham Books, Bluntisham, pp. 3–30.

Wytrzens, H.K., Mayer, C., 1999. Multiple use of alpine grassland in Austria and the implications for agricultural policy. Bodenkultur 50, 251–261.

Young, A., 1995. Landscape structure and genetic variation in plants: empirical evidence. In: Hansson, L., Fahrig, L., Merriam, G. (Eds.), IALE Studies in Landscape Ecology, vol. 2, Mosaic Landscapes and Ecological Processes.

Zimmermann, N.E., Kienast, F., 1999. Predictive mapping of alpine grasslands in Switzerland: species versus community approach. J. Veg. Sci. 10, 469–482.

PART III
BIODIVERSITY AT DIFFERENT SCALE LEVELS

Chapter 3 Biodiversity and Habitat

W. Büchs, H. Albrecht,
R. Waldhardt

Original Papers

Available online at www.sciencedirect.com

Agriculture, Ecosystems and Environment 98 (2003) 87–98

Agriculture Ecosystems & Environment

www.elsevier.com/locate/agee

Biodiversity indicators: the choice of values and measures

Peter Duelli*, Martin K. Obrist

Swiss Federal Research Institute WSL, Zürcherstrasse 111, CH-8903 Birmensdorf-Zürich, Switzerland

Abstract

Ideally, an indicator for biodiversity is a linear correlate to the entity or aspect of biodiversity under evaluation. Different motivations for assessing entities or aspects of biodiversity lead to different value systems; their indicators may not correlate at all. For biodiversity evaluation in agricultural landscapes, three indices are proposed, each consisting of a basket of concordant indicators. They represent the three value systems "conservation" (protection and enhancement of rare and threatened species), "ecology" (ecological resilience, ecosystem functioning, based on species diversity), and "biological control" (diversity of antagonists of potential pest organisms). The quality and reliability of commonly used indicators could and should be tested with a three-step approach. First, the motivations and value systems and their corresponding biodiversity aspects or entities have to be defined. In a time consuming second step, a number of habitats have to be sampled as thoroughly as possible with regard to one or several of the three value systems or motivations. The third step is to test the linear correlations of a choice of easily measurable indicators with the entities quantified in the second step. Some examples of good and bad correlations are discussed.

Keywords: Biodiversity; Indicator; Arthropods; Correlate

1. Who needs biodiversity indicators?

National and regional agencies for nature conservation, agriculture, and forestry have to monitor species diversity or other aspects of biodiversity, both before and after they spend tax money on subsidies or ecological compensation management, with the aim of enhancing biodiversity (European Community, 1997; Ovenden et al., 1998; Wascher, 2000; Kleijn et al., 2001). Similarly, international, national or regional non-governmental organisations (NGOs) may want to monitor aspects of biodiversity at different levels and scales (Reid et al., 1993; IUCN, 1994; Cohen and Burgiel, 1997). In scientific research biodiversity indicators can be used as quantifiable environmental factors. Since the biodiversity of even a small area is far too complex to be comprehensively measured and quantified, suitable indicators have to be found.

Those who are responsible for comparing and evaluating biodiversity have a strong incentive to choose a scientifically reliable and repeatable indicator, which inevitably increases costs. The financing agencies usually opt for a financially "reasonable" approach, which often results in programmes addressing only essential work. The resulting compromises make optimisation of the choice of biodiversity indicators and methods of fundamental importance.

A recent international electronic conference on biodiversity indicators (http://www.gencat.es/mediamb/bioind, 2000) has revealed widely differing views on why and what to measure and quantify.

* Corresponding author. Tel.: +41-1-739-2376; fax: +41-1-739-2215.
E-mail address: peter.duelli@wsl.ch (P. Duelli).

0167-8809/$ – see front matter © 2003 Elsevier Science B.V. All rights reserved.
doi:10.1016/S0167-8809(03)00072-0

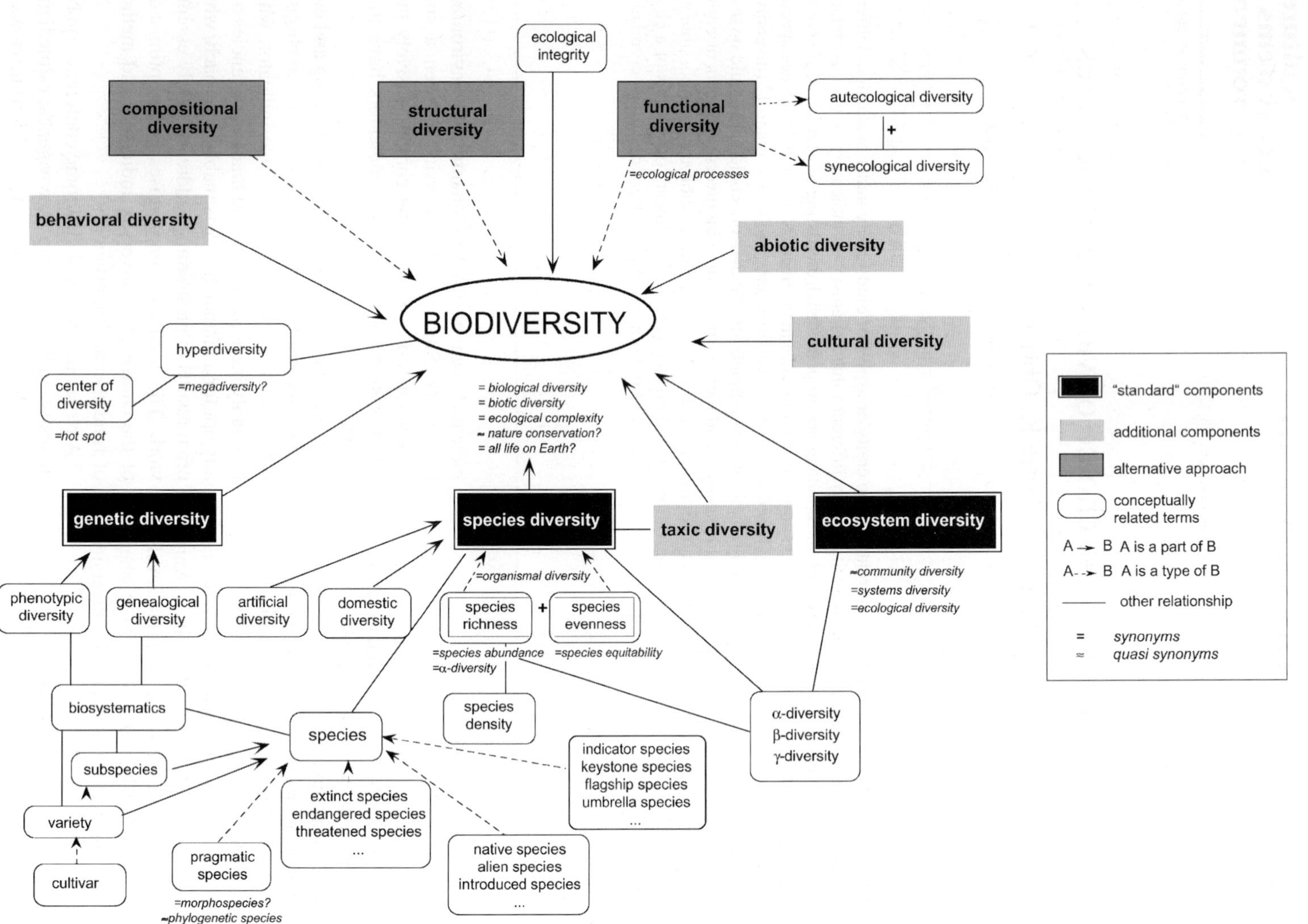

Fig. 1. Provisional domain tree of biodiversity based on the survey of 125 text documents in English (Kaennel, 1998). Concepts used by various authors to define biodiversity are in square boxes, related concepts in rounded boxes. Type and direction of conceptual relationships are indicated by arrows. Synonyms and quasi-synonyms are in italics.

2. Why is it so difficult to reach a consensus on the use of biodiversity indicators?

The complexity of all the aspects of the term biodiversity is illustrated in Fig. 1. It is obvious that no single indicator for biodiversity can be devised. Each aspect of biodiversity requires its own indicator. The difficulties for reaching a consensus on the use of biodiversity indicators are manifold. They imply differing choices for values and measures, which will be discussed here more in detail.

Terms such as biodiversity, indicator or index are not well defined and their use varies between different countries and disciplines. Dismissing research findings or scientific reports simply on the grounds of differing views on the use of particular terms (semantic discrimination) would be counterproductive, but study reports must clearly state what is meant by the terms used. A helpful review on indicator categories for bioindication is given by McGeoch (1998).

In this paper, the term indicator is used in the sense of any measurable correlate to the entity to be assessed: a particular aspect of biodiversity.

The most promising and convincing indicators of biodiversity are measurable portions of the entity that we consider to represent a target aspect of biodiversity. The term index is used here in the sense of a scaled measure for one or several concordant indicators.

3. Indicator FOR or FROM biodiversity?

A first major source of misunderstanding is, whether biodiversity itself is to be indicated, or whether certain components of biodiversity are used as indicators for something else. Until 1990, the search for bioindicators had focussed on indicators of "environmental health" or ecological processes such as disturbance, human impact, environmental or global change (Hellawell, 1986; Spellerberg, 1991; Meffe and Carroll, 1994; Dufrene and Legendre, 1997). After the world-wide launch of the term biodiversity at the Rio Convention in 1992, there was a sudden and drastic shift in the published literature towards the search for indicators of biodiversity itself (Noss, 1990; Gaston and Williams, 1993; Gaston, 1996a; Prendergast, 1997). Since then, however, the term biodiversity has sometimes been used to allude to or indicate some aspect of environmental quality.

If a species or a group of species is a good indicator for lead contamination, it may not indicate biodiversity, i.e. there may not be a linear correlate to biodiversity. It is fundamentally a contamination indicator, or an environmental indicator (McGeoch, 1998) rather than a biodiversity indicator.

However, "real" biodiversity indicators may be needed to measure the impact of e.g. lead contamination on biodiversity itself (indicator FOR biodiversity). Such an assessment is different from measuring the impact of lead on a selected taxonomic group, which had been chosen because it is especially sensitive to lead poisoning (indicator FROM biodiversity).

4. Alpha-diversity, or contribution to higher scale biodiversity?

A second major dichotomy in the value system for biodiversity indicators is the question of whether the species (or allele, or higher taxon unit) diversity of a given area is to be indicated (local, regional or national level), or if the contribution of the biodiversity of that area to a higher scale surface area (regional, national, global) is important.

In the first case (alpha-diversity, e.g. species richness of an ecological compensation area), an indicator ideally has to be a linear correlate to the biodiversity aspect or entity of the surface area in question. Each species has the same value.

In the second case, the value of the measurable units of biodiversity (alleles, species, ecosystems) depends on their rarity or uniqueness with regard to a higher level area. A nationally rare or threatened species in a local assessment has a higher conservation value than a common species, because it contributes more to regional or national biodiversity than the ubiquitous species. Thus a biodiversity indicator in the latter case not only has to count the units (alleles, species, ecosystems), but it has to value them differently and add the values.

The best known examples are red list species. For measuring alpha-diversity, they are not given a value that is greater than any other species in a plot or trap sample, but for measuring the conservation value of a plot, their higher contribution to regional, national, or even global biodiversity has to be recognised. Raised bogs are notorious for their poor species richness, but if only a few raised bogs are left within a country, the few characteristic species present in a "good bog" are of very high national importance. The problems of estimating complementarity or distinctness are addressed e.g. by Colwell and Coddington (1994) and Vane-Wright et al. (1991), endemism and spatial turnover by Harte and Kinzig (1997).

This dichotomy between "species richness" and "conservation value" is the most fervently debated issue among applied biologists concerned with biodiversity indicators, and a recurrent source of misunderstandings. It will be elaborated further in the chapter on value systems.

5. Indicator for what aspect of biodiversity?

After agreement on indicators FOR biodiversity, and a decision between "alpha-diversity" and "contribution to higher scale biodiversity", there is still potential for disagreement on "what is biodiversity?" (Gaston, 1996c). In practice, in a majority of cases, species are "the units of biodiversity" (Claridge et al., 1997). However, species diversity can be measured as simple number of species, usually of selected groups of organisms, or species richness may be combined with the evenness of the abundance distribution of the species. The best known indices are the Shannon index, the Simpson index and Fisher's alpha (Magurran, 1988). Recent observations (Duelli, unpubl.) have shown that when undergraduate biodiversity students in entomology lectures have to choose which of the two communities shown in Fig. 2 (without seeing the text below them) they consider to be more diverse, more than half of them decide for the left population, because they consider evenness to be of greater importance than species numbers. When individuals from other disciplines were asked during lectures and seminars, particularly conservationists and extension workers in agriculture and forestry, species numbers are decisive. In recent years, indices involving evenness have essentially fallen out of favour, mostly because they are difficult to interpret (Gaston, 1996c). Particularly in agriculture or forestry, single species are often collected in huge numbers with standardised methods, which results in a drastic drop of evenness and hence yields low diversity values, in spite of comparatively high species richness.

The definition of biodiversity given in the international Convention on Biological Diversity (Johnson, 1993) encompasses the genetic diversity within species, between species, and of ecosystems. Furthermore, Noss (1990) distinguished three sets of attributes: compositional, structural and functional biodiversity (see also Fig. 1). The most common approach is to measure compositional biodiversity. Presumably, both structural and functional biodiversity are either based on or lead to higher compositional

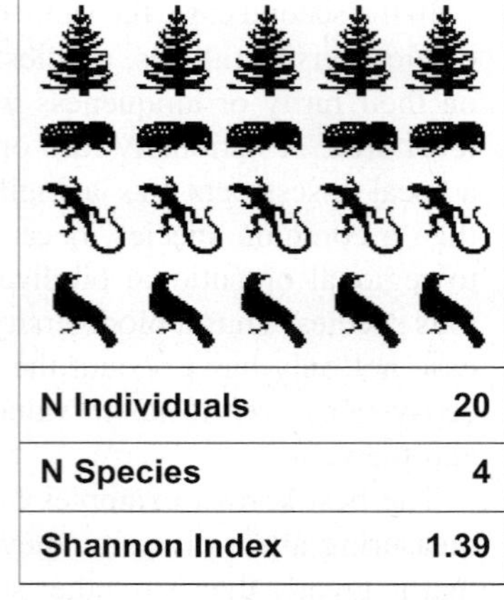

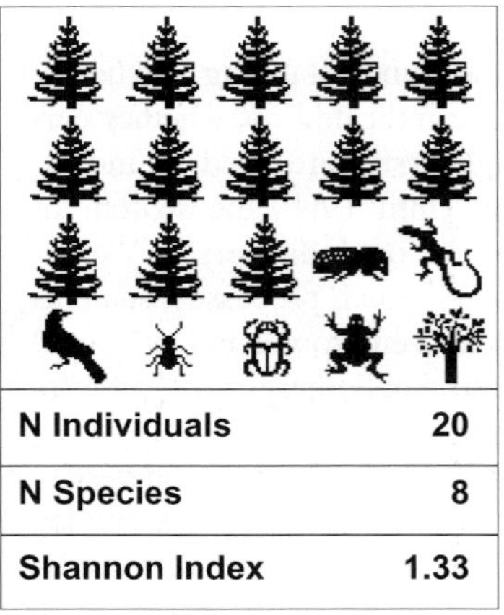

Fig. 2. "Which of the two populations do you consider to have a higher biodiversity?" A choice test for biodiversity evaluation regularly offered by the first author to students and at public lectures. For the vote, only the upper part without text is shown.

diversity. We are convinced that ecosystem diversity, as well as structural and functional diversity, is somehow reflected in the number of species present. If they are not correlated with species richness, they must be special cases and not representative as biodiversity indicators. More trophic levels will normally include more species, and a higher structural diversity will harbour more ecological niches. In fact, there is increasing evidence that at least for some taxonomic groups, species numbers are correlated with habitat heterogeneity (Moser et al., 2002), but not in others (Rykken and Capen, 1997).

For all these hierarchical separations or entities within the huge concept of biodiversity, separate comprehensible indicators can be researched and developed. In many cases, however, a rigorous scientific test may show that the conceptual entities are difficult to quantify (Prendergast, 1997; Lindenmayer, 1999; Noss, 1999), or they are basically reflected in other, better quantifiable measures of biodiversity, such as species richness (Gaston, 1996b; Claridge et al., 1997).

The aspect of intraspecific diversity is a different case. To our knowledge there is no published example of a tested correlation between inter- and intraspecific diversity.

6. Value systems

People involved in developing or using biodiversity indicators are influenced by their personal and/or professional goals. They all may want to measure or monitor biodiversity, but they address different aspects of it. Their focus depends on their motivation for dealing with biodiversity. In an agricultural context, and in an industrialised country in Europe, the three most important motivations to enhance biodiversity are

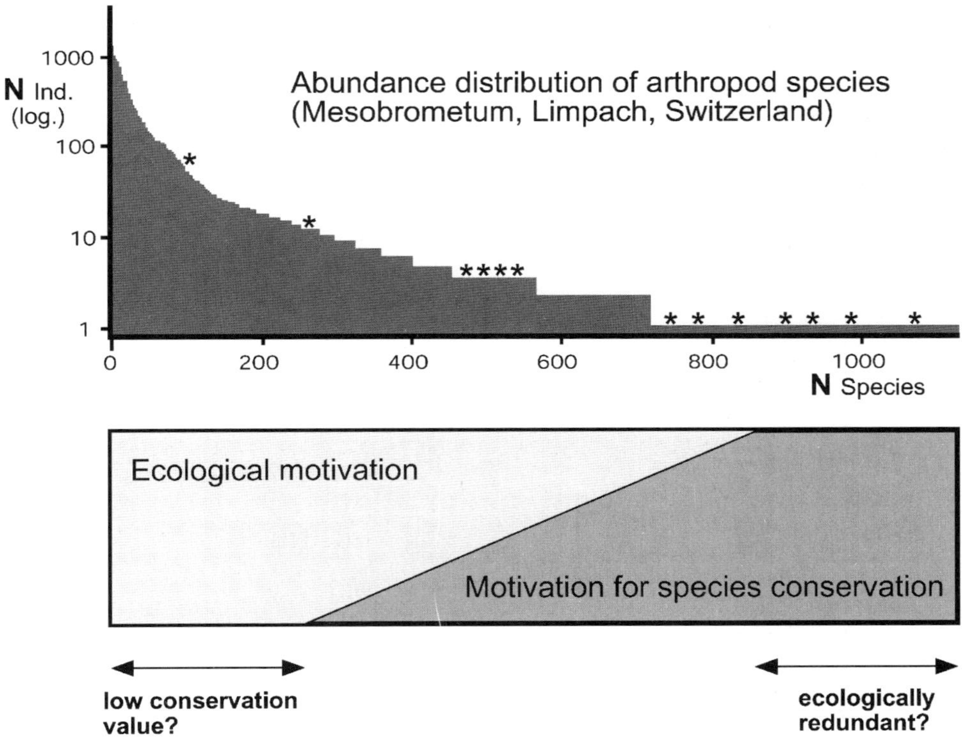

Fig. 3. Illustration of the hypothesis that abundant species usually are of higher ecological but lower conservation value, in contrast to rare and threatened species. Stars indicate red list species collected with pitfall traps, yellow water pans and window interception traps in a semidry meadow (Duelli and Obrist, 1998). Number of individuals (*N* Ind(log)) are plotted versus number of species (*N* species).

1. Species conservation (focus on rare and endangered species).
2. Ecological resilience (focus on genetic or species diversity).
3. Biological control of potential pest organisms (focus on predatory and parasitoid arthropods).

There are additional motivations, of course, but either they are closely related to the ones mentioned here, or their causal link to biodiversity is less clear (e.g. sustainability, landscape protection, cultural heritage).

Each of these three aspects of biodiversity requires its own indicators. They often do not correlate with each other or even show a negative correlation. Consequently, simply adding up different indicators may lead to misinterpretations, as long as they do not address the same aspect of biodiversity. Species conservation focusses on rare and threatened species and often regards more common species in a derogatory way as ubiquists of little interest. Ecologists, on the other hand, focus more on abundant species, because a species on the verge of extinction is likely to have less significant ecological influence. The hypothesis of an almost vicarious relationship between the motivations of "species conservation" and "ecological resilience" is illustrated in Fig. 3.

Prendergast et al. (1993) found low coincidence of species-rich areas and areas harbouring rare species for either plants, birds, butterflies or dragonflies. An investigation of carabid beetles in Scotland (Foster et al., 1997) showed that neither the number of red list species nor the number of stenotopic (faunistically interesting) species are correlated with the mean total number of carabid species in a variety of habitats such as moorland, grassland, heathland, peat, saltmarsh, bracken and swamps (Fig. 4). In an intense investigation with 51 trap stations and standardised sampling methods in field and forest habitats in Switzerland, the number of red list species of all identified arthropod groups was not significantly correlated to overall species richness per trap station (Fig. 5), while e.g. the numbers of aculeate Hymenoptera species correlated well ($R^2 = 0.88$; Fig. 6). In an assessment of the effects of ecological compensation measures in Swiss crop fields and grassland, the number of butterfly species did not show any correlation with the species numbers of spiders (Jeanneret, pers. comm.). In an effort to test the suitability of Collembola as indicators of the conservation value of Australian

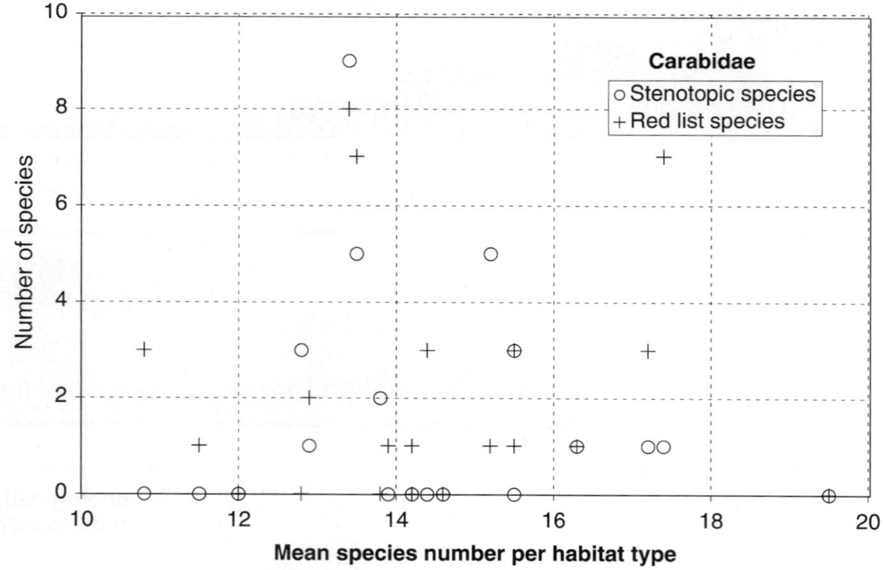

Fig. 4. Neither red list carabid species nor stenotopic carabid species are correlated significantly with the average number of carabid species collected in 18 types of habitats using pitfall traps. Data from Foster et al. (1997).

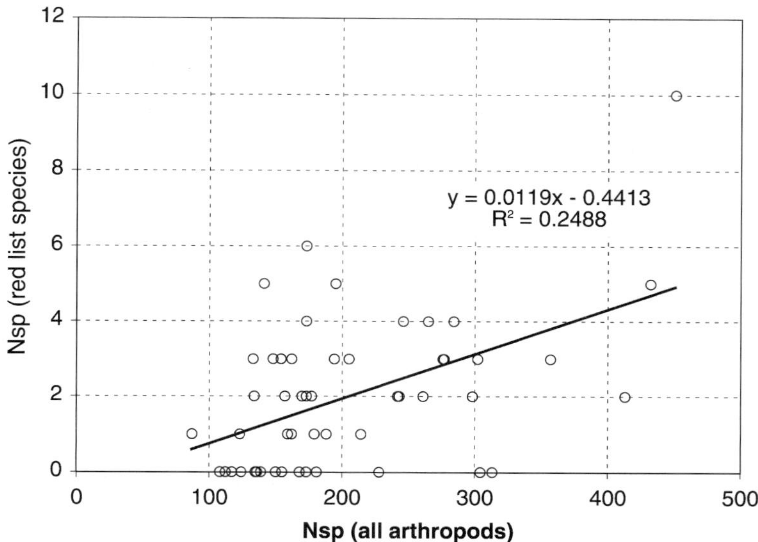

Fig. 5. No significant correlation exists between the number of red list species (from numerous arthropod taxa) and the "overall" number of arthropods collected with flight traps, pitfall traps and yellow water pans at the same 51 locations (Araneae, Coleoptera, Diplopoda, Diptera (Syrphidae only), Heteroptera, Hymenoptera (Aculeata only), Isopoda, Mecoptera, Megaloptera, Neuroptera, Raphidioptera, Thysanoptera). Data from agricultural areas (Duelli and Obrist, 1998) and forest edges (Flückiger, 1999).

grasslands, Greenslade (1997) found no correlation with species numbers of ants and carabid beetles.

The optimal approach is to select a "basket" of indicators for each motivation, similar to the Dow Jones index for the stock exchange. The measured indicators within one basket have to be fairly concordant and are pooled to form an index. The result is a set of three separate indices for the three

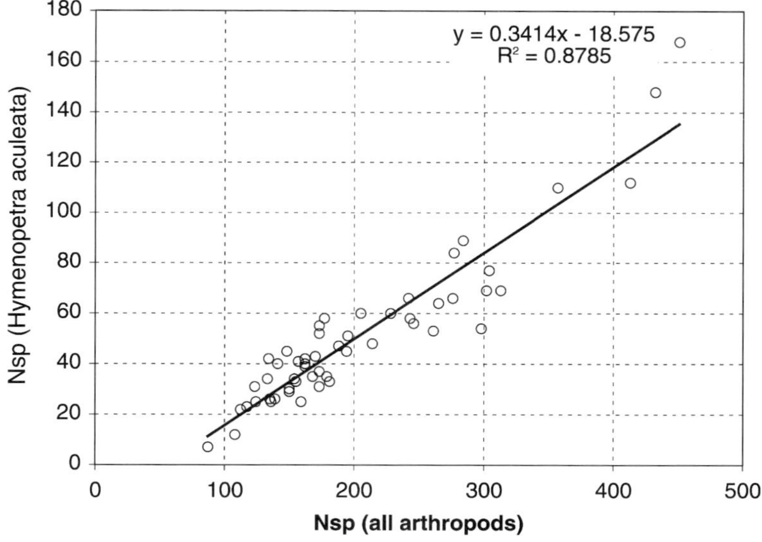

Fig. 6. Species numbers of aculeate Hymenoptera (bees, wasps and ants) show excellent correlation with the overall number of arthropod species at 51 locations (for details of data sources see Fig. 5).

basic motivations "conservation", "ecology" and "pest control".

7. How to select indicators for the three main motivations

7.1. Several steps are necessary

The most accurate indicators of biodiversity are proven linear correlates of the entity or aspect of biodiversity being evaluated. McGeoch (1998) proposed a nine-step approach for selecting bioindicators among terrestrial insects. Basically, the whole procedure can be separated into three steps. The first step is to define the aspect or entity in as quantifiable a way as possible. The second step is to actually quantify that aspect or entity in a statistically reliable number of cases. The third step is a rigorous test for linear correlation in a set of proposed indicators. The urgent need to perform a scientifically solid test has been advocated repeatedly (Balmford et al., 1996; McGeoch, 1998; Niemelä, 2000).

Starting with the first step, the three mayor motivations for protecting or enhancing biodiversity in agricultural landscapes are differentiated.

7.2. Conservation (an index based on the motivation to protect or enhance threatened species)

For assessing the value of a given habitat, e.g. an ecological compensation area, for species conservation, the entity to indicate is the accumulated conservation values (e.g. red list status) of all species present in that area. The highest values are contributed by species of national or even global importance, while the so-called ubiquists are of little value. The second step thus is a comprehensive measurement of the conservation values in a number of ecosystems or habitat types.

The third step would be to find and test the best linear correlate to that otherwise elusive entity "conservation value". The standard indicators for the conservation basket are numbers of red list species of selected taxa, weighed according to their category of threat. However, only very few of the tens of thousands of species present in a country are listed; in Switzerland they are a mere 7% of all known animal species (Duelli, 1994). Inevitably, the choice of the groups of organisms used for an inventory depends strongly on the red lists available, and on the availability of specialists to identify the listed organisms.

Lacking the information on the second step (full account of the conservation value of an area), it is not currently possible to come up with a scientifically tested indicator for that value. Nevertheless, a correlation between the cumulated conservation values of all presently available red listed species per habitat with the conservation values of single taxonomic groups, such as birds, butterflies or carabids, would at least give greater credibility to the red list species approach.

In addition to red list status (degree of threat of extinction), species values have been calculated on the bases of national or global rarity (Mossakowski and Paje, 1985) or endemism. The rationale in the context of habitat evaluation is that the presence of a nationally or globally rare species increases the biodiversity value of that habitat, because it contributes more to the conservation of national or global biodiversity than the presence of a ubiquitous species.

Only after a reliable basket of indicators for conservation value has been established, are further steps possible to test the correlative power of potential indicators such as length of hedgerows, amount of dead wood, or the surface of ecological compensation areas per unit area. Environmental diversity (ED) as a surrogate measure of the conservation value was proposed by Faith and Walker (1996), but so far there are no empirical data to test their proposal.

7.3. An index for the motivation "pest control"

For the biodiversity aspect of biological control of potential pest organisms, the first step may be to define the measurable entity as the species diversity of all predators or parasites of potential pest organisms. For short-term interests, the number of individuals of beneficial organisms may appear more important than species richness, because prey and hosts are reduced by the number of antagonistic individuals rather than by species numbers (Kromp et al., 1995; Wratten and Van Emden, 1995). However, with a longer-term perspective on maintaining a high diversity of antagonist species of potential pest organisms is certainly more important. While the species richness of predators in a small area can be assessed with reasonable accuracy and effort, the diversities of parasitoids are much harder to quantify.

The second step is therefore to test inventory methods, and selected taxa for their correlation with the above biodiversity aspect of biological control. At present species numbers of carabid and staphylinid beetles, as well as spiders, are often used as indicators because of established standardised collecting methods (Duffey, 1974; Desender and Pollet, 1988; Halsall and Wratten, 1988) and readily available keys for identification and interpretation. Specialised aphidophaga among the syrphid flies, coccinellids and Neuroptera are another option, but so far the methods are not fully standardised. Parasitoid wasps and flies are promising, but so far there is no easy way to identify them to the species level. Other possibilities for indicators to test are ratios between herbivores and predators, or parasitoids and a range of other arthropods (see e.g. Denys and Tscharntke, 2002).

7.4. An index for ecological resilience

For the basket of indicators for the motivation ecological resilience ("Balance of Nature", Pimm, 1991), the entire genetic and taxonomic spectrum of biodiversity is the entity to be indicated. The assumption is that the higher the number of alleles and species, the higher is the ecological potential of an ecosystem to react adequately to environmental change.

Here again, a first step requires quantification of a measurable proportion of local organismic diversity, which can be trusted to represent total species richness of animals and plants (alpha-diversity). Realistically, only few and small areas will ever be fully assessed. For the second and third steps, approximations with large, measurable proportions of alpha-diversity have to be used to test potential indicators.

These "ecological" indicators can be seen as indicators for ecosystem functioning (Schläpfer et al., 1999) and are representing a very basic notion of wholesale biodiversity. Most studies claiming to measure or indicate biodiversity assume that the group of organisms they investigate is somehow representative of biodiversity. However, in only very few cases has the correlation between a group or several groups of species with a more or less representative sample of all organisms been measured and published (Abensperg-Traun et al., 1996; Balmford et al., 1996; Cranston and Trueman, 1997; Duelli and Obrist, 1998).

8. Effort and costs, the limiting factors for the choice of measures

8.1. The dilemma of indicating complexity with simple measures

Large environmental monitoring programmes usually avoid using invertebrates for their indicators, although these constitute by far the largest portion of measurable biodiversity. To cut down on effort and costs, measurement of the immense richness and quantity of invertebrates has to be reduced to an optimised selection of taxa. The proposed three-step approach allows for testing all kinds of indicators for their correlation with aspects of biodiversity. The search for linear correlates of quantified entities or aspects of biodiversity is not limited to taxonomic units. Instead of choosing birds or grasshoppers as indicators, the spectrum of taxa considered can be determined by an inventory method such as Berlese soil samples or flight interception traps. The broader the taxonomic spectrum of the samples, the higher the chance of obtaining a good correlation with the entity to be assessed. Furthermore, indicators, which are not part of the organismic spectrum, can also be tested in the three-step approach: habitat diversity and heterogeneity, disturbance by traffic, neighbourhood or percentage of protected areas, etc. At present, various indicators are in use, but few of them have been tested for their correlation with aspects of biodiversity. At least in Neotropical butterflies, a positive correlation of species richness was found with composite environmental indices of heterogeneity and natural disturbance (Brown, 1997).

8.2. Plots and transects

Plots (for plants) and transects (for birds and insects such as butterflies, dragonflies and grasshoppers) are widely used relative assessment methods for the species richness of a selected group of organisms (e.g. Pollard and Yates, 1993; Wagner et al., 2000). The main advantages are that the specimens survive the inventory (important for indicating conservation value), and that large areas can be searched in a relatively short time. Scientifically, the drawback is that usually there are no voucher specimens kept for verifying the identification. Also, these popular groups (except for vascular plants) have only few species in agricultural habitats, so their species richness, even if cumulated, never reaches 1% of the local species diversity of all organisms. Their correlation power with local species diversity has never been tested. Vascular plants, on the other hand, seem to correlate reasonably well with overall organismic diversity (Duelli and Obrist, 1998). Plots and transects are low budget measures and worth testing for their correlation power in the conservation and ecology baskets of indicators.

8.3. Standardised trapping methods for arthropods

Pitfall traps for surface dwelling arthropods and various kinds of flight traps for insects are often used for biodiversity assessment in agricultural areas. Either one or a few taxonomic groups are collected over longer periods, or a larger number of taxa are sampled within a shorter collecting period. In both cases, suitable correlates have been found for the indicator basket of ecological resilience (Duelli and Obrist, 1998). Bugs (Heteroptera), and wild bees and wasps (aculeate Hymenoptera; see also Fig. 6) collected during an entire vegetation period, where highly correlated with overall species richness, while carabids and spiders in pitfall traps were not. Reducing the collecting time to five carefully selected weeks, but extending the spectrum of identified taxa (Duelli et al., 1999), yielded correlation values comparable to those of seasonal collections of bugs or bees. Tests are under way to further reduce the effort required for collecting and identifying through an adaptation of the Australian method of Rapid Biodiversity Assessment (Cranston and Hillman, 1992; Oliver and Beattie, 1996). With that method, the whole taxonomic spectrum collected within a few selected weeks in a standardised trap combination is considered, but only at the level of morphospecies, i.e. without identifying the catches to the species level (Duelli et al., unpubl.). Obviously, the resulting indicator will not be useful for the indicator baskets of conservation or pest control, where identification of the species is essential. However, it is a promising monitoring device for the indication of alpha-diversity—or the ecological resilience basket.

9. Conclusions

There is no single indicator for biodiversity. The choice of indicators depends on the aspect or entity of biodiversity to be evaluated and is guided by a value system based on personal and/or professional motivation. Each biodiversity index for a particular value system should consist of a basket of methods with one or several concordant indicators. In order to achieve greater reliability and a broader acceptance, indicators have to be tested for their linear correlation with a substantial and quantifiable portion of the entity to assess. The challenge now is to assign all the presently used or proposed indicators to a basket with a declared value system—and to test them with empirical measures.

References

Abensperg-Traun, M., Arnold, G.W., Steven, D.E., Smith, G.T., Atkins, L., Viveen, J.J., Gutter, M., 1996. Biodiversity indicators in semiarid, agricultural Western Australia. Pacific Conserv. Biol. 2, 375–389.

Balmford, A., Green, M.J.B., Murray, M.G., 1996. Using higher-taxon richness as a surrogate for species richness. I. Regional tests. Proc. R. Soc. Lond. B 263, 1267–1274.

Brown, K.S., 1997. Diversity, disturbance and sustainable use of Neotropical forests: insects as indicators for conservation monitoring. J. Insect Conserv. 1, 25–42.

Claridge, M.F., Dawah, H.A., Wilson, M.R. (Eds.), 1997. Species: The Units of Biodiversity. Chapman & Hall, London.

Cohen, S., Burgiel, S.W. (Eds.), 1997. Exploring Biodiversity Indicators and Targets under the Convention on Biological Diversity. BIONET and IUCN, Washington, DC and Gland.

Colwell, R.K., Coddington, J.A., 1994. Estimating terrestrial biodiversity through extrapolation. Phil. Trans. R. Soc. Lond. B 345, 101–118.

Cranston, P.S., Hillman, T., 1992. Rapid assessment of biodiversity using biological diversity technicians. Aust. Biol. 5, 144–154.

Cranston, P.S., Trueman, J.W.H., 1997. Indicator taxa in invertebrate biodiversity assessment. Mem. Mus. Victoria 56, 267–274.

Denys, C., Tscharntke, T., 2002. Plant–insect communities and predator–prey ratios in field margin strips, adjacent crop fields, and fallows. Oecologia 130, 315–324.

Desender, K., Pollet, M., 1988. Sampling pasture carabids with pitfalls: evaluation of species richness and precision. Med. Fac. Landbouww. Rijksuniv. Gent 53, 1109–1117.

Duelli, P., 1994. Rote Listen der gefährdeten Tierarten der Schweiz. Bundesamt für Umwelt Wald und Landschaft. BUWAL-Reihe Rote Listen. EDMZ, Bern.

Duelli, P., Obrist, M.K., 1998. In search of the best correlates for local organismal biodiversity in cultivated areas. Biodivers. Conserv. 7, 297–309.

Duelli, P., Obrist, M.K., Schmatz, D.R., 1999. Biodiversity evaluation in agricultural landscapes: above-ground insects. Agric. Ecosyst. Environ. 74, 33–64.

Duffey, E., 1974. Comparative sampling methods for grassland spiders. Bull. Br. Arach. Soc. 3, 34–37.

Dufrene, M., Legendre, P., 1997. Species assemblages and indicator species: the need for a flexible asymmetrical approach. Ecol. Monogr. 67, 345–366.

European Community, 1997. Agenda 2000, vol. L, For a Stronger and Wider EU. Office for Official Publications of the European Communities, Luxembourg.

Faith, D.P., Walker, P.A., 1996. Environmental diversity: on the best-possible use of surrogate data for assessing the relative biodiversity of sets of areas. Biodivers. Conserv. 5, 399–415.

Flückiger, P.F., 1999. Der Beitrag von Waldrandstrukturen zur regionalen Biodiversität. Doctoral Thesis. Philosophisch-Naturwissenschaftliche Fakultät, Universität Basel, Basel.

Foster, G.N., Blake, S., Downie, I.S., McCracken, D.I., Ribera, I., Eyere, M.D., Garside, A., 1997. Biodiversity in Agriculture. Beetles in Adversity? BCPC Symposium Proceedings No. 69: Biodiversity and Conservation in Agriculture, pp. 53–63.

Gaston, K.J. (Ed.), 1996a. Biodiversity: A Biology of Numbers and Difference. Blackwell Scientific Publications, London.

Gaston, K.J. (Ed.), 1996b. Species Richness: Measure and Measurement. Biodiversity: A Biology of Numbers and Difference. Blackwell Scientific Publications, London.

Gaston, K.J. (Ed.), 1996c. What is Biodiversity? Biodiversity: A Biology of Numbers and Difference. Blackwell Scientific Publications, London.

Gaston, K.J., Williams, P.H., 1993. Mapping the world's species—the higher taxon approach. Biodiv. Lett. 1.

Greenslade, P., 1997. Are Collembola useful as indicators of the conservation value of native grassland? Pedobiologia 41, 215–220.

Halsall, N.B., Wratten, S.D., 1988. The efficiency of pitfall trapping for polyphagous predatory Carabidae. Ecol. Entomol. 13, 293–299.

Harte, J., Kinzig, A., 1997. On the implications of species–area relationships for endemism, spatial turnover, and food web patterns. Oikos 80, 417–427.

Hellawell, J.M., 1986. Biological Indicators of Freshwater Pollution and Environmental Management. Elsevier, London.

IUCN, 1994. IUCN Red List Categories. Prepared by IUCN Species Survival Commission. IUCN, Gland.

Johnson, S.P., 1993. The Earth Summit: The United Nations Conference on Environment and Development (UNCED). Graham and Trotman, London.

Kaennel, M., 1998. Biodiversity: a diversity in definition. In: Bachmann, P., Köhl, M., Päivinen, R. (Eds.), Assessment of Biodiversity for Improved Forest Planning. Kluwer Academic Publishers, Dordrecht.

Kleijn, D., Berendse, F., Smit, R., Gilissen, N., 2001. Agri-environment schemes do not effectively protect biodiversity in Dutch agricultural landscapes. Nature 413, 723–725.

Kromp, B., Pflügel, G., Hradetzky, R.I.J., 1995. Estimating beneficial arthropod densities using emergence traps, pitfall traps and the flooding method in organic fields (Vienna, Austria). In: Toft, S., Riedel, W. (Eds.), Arthropod Natural Enemies in Arable Land, vol. 70. Acta Jutlandica, Aarhus University Press, Denmark, Aarhus, Denmark, pp. 87–100.

Lindenmayer, D.B., 1999. Future directions for biodiversity conservation in managed forests: indicator species, impact studies and monitoring programs. For. Ecol. Manage. 115, 277–287.

Magurran, A.E., 1988. Ecological Diversity and its Measurement. Croom Helm Limited, London.

McGeoch, M.A., 1998. The selection, testing and application of terrestrial insects as bioindicators. Biol. Rev. 73, 181–201.

Meffe, G.K., Carroll, C.R., 1994. Principles of Conservation Biology. Sinauer, Sunderland.

Moser, D., Zechmeister, H.G., Plutzar, C., Sauberer, N., Wrbka, T., Grabherr, G., 2002. Landscape patch shape complexity as an effective measure for plant species richness in rural landscapes. Landscape Ecol. 17, 657–669.

Mossakowski, D., Paje, F., 1985. Ein Bewertungsverfahren von Raumeinheiten an Hand der Carabidenbestände. Verh. Ges. Ökol. Bremen 13, 747–750.

Niemelä, J., 2000. Biodiversity monitoring for decision-making. Ann. Zool. Fenn. 37, 307–317.

Noss, R.F., 1990. Indicators for monitoring biodiversity: a hierarchical approach. Conserv. Biol. 4, 355–364.

Noss, R.F., 1999. Assessing and monitoring forest biodiversity: a suggested framework and indicators. For. Ecol. Manage. 115, 135–146.

Oliver, I., Beattie, A.J., 1996. Invertebrate morphospecies as surrogates for species: a case study. Conserv. Biol. 10, 99–109.

Ovenden, G.N., Swash, A.R.H., Smallshire, D., 1998. Agri-environment schemes and their contribution to the conservation of biodiversity in England. J. Appl. Ecol. 35, 955–960.

Pimm, S.L., 1991. The balance of nature? Ecological Issues in the Conservation of Species and Communities. University of Chicago Press, Chicago.

Pollard, E., Yates, T.J., 1993. Monitoring Butterflies for Ecology and Conservation. Chapman & Hall, London.

Prendergast, J.R., 1997. Species richness covariance in higher taxa: empirical tests of the biodiversity indicator concept. Ecography 20, 210–216.

Prendergast, J.R., Quinn, R.M., Lawton, J.H., Eversham, B.C., Gibbons, D.W., 1993. Rare species, the coincidence of diversity hotspots and conservation strategies. Nature 365, 335–337.

Reid, W.V., McNeely, J.A., Tunstall, D.B., Bryant, D.A., Winograd, M., 1993. Biodiversity Indicators for Policy-makers. WRI and IUCN, Washington, DC and Gland.

Rykken, J.J., Capen, D.E., Mahabir, S.P., 1997. Ground beetles as indicators of land type diversity in the Green Mountains of Vermont. Conserv. Biol. 11, 522–530.

Schläpfer, F., Schmid, B., Seidl, I., 1999. Expert estimates about effects of biodiversity on ecosystem processes and services. Oikos 84, 346–352.

Spellerberg, I.F., 1991. Monitor Ecological Change. Cambridge University Press, Cambridge.

Vane-Wright, R.I., Humphries, C.J., Williams, P.H., 1991. What to protect?—systematics and the agony of choice. Biol. Conserv. 55, 235–254.

Wagner, H.H., Wildi, O., Ewald, K.C., 2000. Additive partitioning of plant species diversity in an agricultural mosaic landscape. Landscape Ecol. 15, 219–227.

Wascher, D.W., 2000. Agri-environmental indicators for sustainable agriculture in Europe. ECNC Technical Report Series, Tilburg.

Wratten, S.D., Van Emden, H.F., 1995. Habitat management for enhanced activity of natural enemies of insect pests. In: Glen, D.M., Greaves, M.P., Anderson, H.M. (Eds.), Ecology and Integrated Farming Systems. Wiley, Bristol.

ELSEVIER

Available online at www.sciencedirect.com

Agriculture, Ecosystems and Environment 98 (2003) 99–123

Agriculture Ecosystems & Environment

www.elsevier.com/locate/agee

Biodiversity, the ultimate agri-environmental indicator? Potential and limits for the application of faunistic elements as gradual indicators in agroecosystems

Wolfgang Büchs*, Alexandra Harenberg, Joachim Zimmermann, Birgit Weiß

Federal Biological Research Centre for Agriculture and Forestry, Institute for Plant Protection in Field Crops and Grassland, Messeweg 11/12, DE-38104 Braunschweig, Germany

Abstract

Most political statements (e.g. convention on biodiversity), referring to existing approaches regarding the development of biotic agrienvironmental indicators focus mainly and nearly exclusively on "biodiversity", the enhancement of which is almost the overall target for the development of agricultural landscapes towards sustainability. In this regard the understanding of the term "biodiversity" is rather different: mostly it is interpreted as species richness, only occasionally as richness of varieties, cultivars or genetical expressions (e.g. micro-organisms).

In this contribution the approach of evaluating the quality of species communities in agroecosystems mainly on the basis of "biodiversity" (in the sense of species richness), will be analysed critically. General difficulties of an exact sampling or establishment of biodiversity especially of invertebrate communities in agricultural landscapes will be highlighted. Arising from this synthesis, it is demonstrated that biotic agrienvironmental indicators should not only address a maximising of species numbers in agricultural habitats, but also a sustainable and long-term stability, and the reproductive potential of populations of species typical of agroecosystems.

In this regard alternative biotic indicators for the assessment of the effects of husbandry practices are required that are based on existing knowledge of the reactions of populations and species communities to environmental stress, and which give information about the physiological fitness of populations of species that are relevant members of the agrobiocoenosis from both the qualitative (rarity, endangerment status) and functional point of view (natural enemies, parasitoids, decomposers). Regarding the sensitivity of those indicators it will be indicated that in contrast to many recent scientific experiments it is usually necessary to evaluate and compare cultivation intensities, that are quite similar (e.g. gradual differences in conventional managed farms and landscapes) as well as those displaying much larger differences (e.g. organic farming vs. conventional farming), because conventional farms currently cover nearly 90% of the area under plough in Europe. Finally an assessment of the practicality, the power of indication statements and the remaining work required to validate indicators will be provided for discussion, as well as suggestions for a simplification of indicator systems in order to minimise the input needed for data recording.

Keywords: Biodiversity assessment; Biotic indicator; Cultivated area; Agriculture; Animals

1. Introduction

Background information relating to the question posed in the title is described by Büchs (2003).

* Corresponding author. Tel.: +49-531-299-4506; fax: +49-531-299-3008.
E-mail address: w.buechs@bba.de (W. Büchs).

0167-8809/$ – see front matter © 2003 Elsevier B.V. All rights reserved.
doi:10.1016/S0167-8809(03)00073-2

Among other aspects this study showed that in contrast to the public use, in science the use of the term "(bio)diversity" follows clearly defined rules derived from procedures developed by information theory (Shannon, 1948; Shannon and Weaver, 1963). It is considered that the term "(bio)diversity" consists of two components, the diversity component (richness of species, races or genotypes) and the expression of dominance structure, which can be explained by the evenness-value that indicates the probability that one element randomly taken out of the totality of all elements belongs to a distinct group (species, race, genotype) represented within the total set. In contrast, "secondary users" (mostly located outside ecological sciences) often neglect the second (evenness-) component of biodiversity and use the term to describe the objective of maximising individual elements (number of species, varieties, genotypes) in (agro-) ecosystems.

As maximisation of, e.g. species richness in rural landscapes is undoubtedly promoted by an increasing percentage of structural elements and/or semi-natural habitats (e.g. Raskin, 1994; Frieben, 1998; Röser, 1988; Duelli, 1992) most approaches that assess the "quality" of agroecosystems start from the usage of those landscape elements or something similar (Roth and Schwabe, 2003; Frieben, 1998; Kantelhardt et al., 2003) as biodiversity indicators. Duelli (1992) has already pointed out that maximisation of habitat heterogeneity (mostly realised in small patches) in rural landscapes actually results in increased species richness because a different subset of species is introduced. Those "additional" species largely consist of common and widespread organisms with no special adaptations to field habitats. In this regard Döring et al. (2003) and Döring and Kromp (2003) demonstrated the importance of husbandry practices by showing that organically managed fields contain a higher number of stenoecious ground beetle species (species, that show specific adaptations to field habitats) and improved zonation of field specific weeds, when compared to conventionally managed crops.

This one objective of this paper is to highlight the central importance of cultivated areas in rural landscapes for the assessment of both environmental stress and biodiversity issues. This facts are emphasised because in planning activities on the community or state level so-called "Biotopverbundkonzepte" (concepts of habitat connection) are created and, within the assessment of aspects of the biotic environment of farms, cultivated areas are usually excluded and the cropped area is more or less "decorated" with structural elements, but itself not considered (e.g. Steidl and Ringler, 1997).

However, the assessment of aspects of biodiversity using biotic indicators in cultivated areas is much more difficult, when compared to uncultivated (marginal) areas such as hedgerows, field margins, fallows, etc. because effects of husbandry activities on the agroecosystems are unavoidable.

During our long-term investigations of the development and application of biotic indicators to evaluate different crop management intensities, it was shown that "biodiversity" has to be classified as a more or less uncertain and inadequate criterion for an assessment of agroecosystems if it is not enhanced with additional criteria.

2. Methods

The data presented are derived from several field experiments and interdisciplinary projects, thus only a general overview of the detailed methodology of each field experiment or project can be given here. Detailed information is available in the referenced publications.

2.1. INTEX-project

This project was based on an oilseed rape crop rotation with four different levels of extensification, and a set-aside area. The most relevant features of the different farming systems are as follows:

(I) Conventional management (according to the advice of the extension services).

(II) Integrated management (a variable extensification with a 4-year crop rotation including winter oilseed rape, winter wheat, winter rye, yellow peas or oats, respectively, an approximately 30% reduction of N-fertiliser application and 50% reduction of pesticides).

(III) Reduced management (management as in system I, but with a reduction of about 50% N-fertilisers and no use of insecticides).

(IV) Extensive management (management as in system II, but without input of mineral fertilisers or any pesticides except seed dressing).
(V) Set-aside (5-year set-aside with natural succession and no disturbances).

The results presented here were mainly recorded at a site near Eickhorst, approximately 10 km north of Braunschweig. The above farming systems were implemented in plots of 1.5–2.0 ha. More detailed information, e.g. regarding husbandry practices or basic conditions of the locations (e.g. weather, soil) are provided by Wildenhayn and Gerowitt (1997) and Teiwes (1997) and an English field guide (Research Center for Agriculture and the Environment 1992, unpublished). For the survey of the arthropod fauna four (April–October 1992), six (set-aside area) or eight (all other fields) 0.25 m^2-emergence traps (e.g. Büchs and Nuss, 2000) and pitfall traps (Barber, 1931) per field were installed. The sampling period started in April 1992 and finished in October 1995.

2.2. Set-aside tillage-project

Following re-ploughing at end of September 1994, the set-side area with natural succession (V) was included in the usual 3-year crop rotation of the surrounding fields with sugar beet, winter wheat and winter rye. In the year after the re-ploughing of the set-aside, sugar beets were grown in this area in the same way as in the surrounding fields. In order to be able to assess the effects of the re-ploughing of the set-aside area on the species composition, the adjacent sugar beet fields were included in the investigation. Thus three fields of 2.0–4.0 ha were sampled using eight emergence traps and eight pitfall traps in each field. The project was conducted from September 1994 to October 1995. For further information, see Weiß et al. (1997).

2.3. Rotational set-aside project

Plots (36 m × 50 m), with seven different seed mixes that are registered for use in set-aside areas and a treatment with natural succession, were installed in three replicates on an area of approximately 7 ha near Wendhausen, about 5 km northeast of Braunschweig. Seed mixes were selected according to the duration and intensity of the flowering period of the dicotyles. The investigation commenced during the period of rotational set-aside and was continued in the following winter wheat. Before sowing the winter wheat, no ploughing tillage was undertaken using a shallow cultivator twice and a harrow once.

Each treatment was sampled using six emergence traps and six pitfall traps (two per plot of each trap type). Sampling commenced in 1996 immediately after sowing of the seed mixes on 26 April and was completed after harvest of the winter wheat in August 1997. More details are provided by Büchs et al. (1999) and Weber and Büchs (2000).

Seed mixes (percentage of seed mix in brackets):

1. "Tübinger Mischung" (10 kg/ha): *Phacelia tanacetifolia* (40), *Sinapis alba alba* (25), *Fagopyrum esculentum* (7), *Coriandrum sativum* (6), *Calendula officinalis* (5), *Nigella* species (5), *Raphanus sativus oleiferus* (3), *Centaurea cyanus* (3), *Malva sylvestris* (2), *Borago officinalis* (2), *Anathum graveolens* (1), *Helianthus annuus* (1).
2. "Beekeepers mix" by Raiffeisen (15 kg/ha): *P. tanacetifolia* (50), *F. esculentum* (25), *H. annuus* (20), *M. sylvestris* (5).
3. 15 kg/ha: *P. tanacetifolia* (60), *Trifolium alexandrinum* (40).
4. 12 kg/ha: *Trifolium incarnatum* (83), *S.a. alba* (17).
5. 20 kg/ha: *R. sativus oleiferus* (100).
6. 100 kg/ha: *Vicia sativa* (60), *Lupinus perenne* (40).
7. 15 kg/ha: *Festuca rubra* (65), *Lolium perenne* (25), *Trifolium repens* (10).
8. Natural succession with dominating *Alopecurus myosuroides*, *Stellaria media*, *Matricaria chamomilla*, *Myosotis arvensis* a.o.

2.4. Soil conservation project

This interdisciplinary project (Ebing et al., 1991; Bartels and Kampmann, 1994) was carried out on three 12 ha fields. Each field was initially sown with a different crop of the 3-year crop rotation (sugar beet, winter wheat, winter barley) and divided into four plots with different management intensities, mainly varying in input of pesticides and fertilisers:

(I_0) Crop production without use of pesticides (except seed dressing), minimum input of mineral fertilisers.

Table 1
Species richness in farming systems with different crop management intensities. Assessment of the distribution of highest and lowest number of species in relation to management intensity (highest species number: bold underlined, background dark grey; lowest species number: bold italic, background light grey)

(a) Assessment including set-aside with natural succession

Taxon	Loc /year / crop	meth.	Farming system conv I 3	int I 2	red I 1	ext I 0	setaside	project
Spiders	Ahlum 89 SB[1]	pt	**21**	***18***	**21**	***18***		soil conservation
	Ahlum 90 WW[1]	pt	***17***	20	**26**	20		soil conservation
	Ahlum 91 WB[1]	pt	15	***11***	14	**22**		soil conservation
	Eickhorst 92 WRy[1]	pt	***37***	41	**46**	40	41	IntEx
	Eickhorst 93 WR[1]	pt	***36***	38	**40**		44	IntEx
	Eickhorst 93 YP[1]	pt		***28***		**38**	42	IntEx
	Eickhorst 94 WW[1]	pt	***35***	36	**41**		52	IntEx
	Reinshof 90 WR[2]	pt	23	***17***	**27**	26	17	IntEx
	Reinshof 91 WR[2]	pt	***21***	23	27	**31**	27	IntEx
	Reinshof 92 WR[2]	pt	30	***29***	**34**	30	33	IntEx
	Reinshof 93 WR[2]	pt	***20***	24	22	**28**	31	IntEx
	Reinshof 94 WR[2]	pt	***12***	25	24	**29**	28	IntEx
	Reinshof 90 WW[2]	pt	***22***	**26**	24		21	IntEx
	Reinshof 91 WW[2]	pt	***27***	**37**	***27***	36	27	IntEx
	Reinshof 92 WW[2]	pt	***21***	30	26	**33**	33	IntEx
	Reinshof 93 WW[2]	pt	***25***	**40**	30	34	31	IntEx
	Reinshof 94 WW[2]	pt	***13***	21	**25**	22	28	IntEx
	Marienstein 90 WW[2]	pt	31	**33**	24	***30***	30	IntEx
	Marienstein 91 WW[2]	pt	***27***	35	39	**43**	31	IntEx
	Ahlum 89 SB[1]	et	16	**24**	17	***14***		soil conservation
	Ahlum 90 WW[1]	et	***13***	16	**20**	**20**		soil conservation
	Ahlum 91 WB[1]	et	**24**	***19***	23	21		soil conservation
	Eickhorst 92 WRy[1]	et	11	***10***	11	**13**	33	IntEx
	Eickhorst 93 WR[1]	et	***12***	**21**	19		34	IntEx
	Eickhorst 93 YP[1]	et		**15**		***14***	30	IntEx
	Eickhorst 94 WW[1]	et	***15***	19	**24**		36	IntEx
	Reinshof 92 WW[2]	et	***21***	**24**	22	***21***	33	IntEx
	Reinshof 93 WW[2]	et	***19***	**36**	21	27	40	IntEx
	Reinshof 94 WW[2]	et	***18***	**28**	19	27	30	IntEx
Ground beetles	Ahlum 89 SB[3]	pt	17	17	***16***	**23**		soil conservation
	Eickhorst 93 WR[4]	pt	***38***	40	**45**			IntEx
	Reinshof 90 WR[2]	pt	***10***	12	11	**17**	12	IntEx
	Reinshof 91 WR[2]	pt	***13***	14	**16**	17	23	IntEx
	Reinshof 92 WR[2]	pt	***14***	15	15	**16**	22	IntEx
	Reinshof 93 WR[2]	pt	***15***	**22**	**22**	17	30	IntEx
	Reinshof 94 WR[2]	pt	***12***	19	18	**21**	24	IntEx
	Marienstein 90 WR[2]	pt	**20**	***15***	19	***15***	25	IntEx
	Marienstein 90 WR[2]	pt	***16***	18	20	**23**	29	IntEx
	Reinshof 90 WW[2]	pt	***9***	10	**11**		14	IntEx
	Reinshof 91 WW[2]	pt	***8***	**17**	13	14	22	IntEx
	Reinshof 92 WW[2]	pt	***7***	**19**	***7***	12	22	IntEx
	Reinshof 93 WW[2]	pt	***12***	**25**	17	18	26	IntEx
	Reinshof 94 WW[2]	pt	13	14	***12***	**18**	24	IntEx
	Marienstein 90 WW[2]	pt	**13**	11	**13**	***9***	26	IntEx
	Marenstein 91 WW[2]	pt	***15***	18	18	**20**	28	IntEx
	Ahlum 89 SB[3]	et	13	**15**	***11***	**15**		soil conservation
	Eickhorst 93 WR[4]	et	**28**	27	***26***			IntEx
Rove beetles	Ahlum 89 SB[5]	pt	***31***	38	36	**52**		soil conservation
	Ahlum 90 WW[5]	pt	51	***50***	53	**55**		soil conservation
	Ahlum 91 WB[5]	pt	45	***44***	45	**53**		soil conservation
	Ahlum 89 SB[5]	et	***46***	48	**54**	**54**		soil conservation
	Ahlum 90 WW[5]	et	**71**	***61***	67	***61***		soil conservation
	Ahlum 91 WB[5]	et	74	***70***	71	**80**		soil conservation
	Reinshof 90 WW[2]	et	**25**		***24***		29	IntEx
	Reinshof 91 WW[2]	et	***19***		**27**		18	IntEx
	Reinshof 92 WW[2]	et	22	**29**	***21***	***21***	18	IntEx
	Reinshof 93 WW[2]	et	***19***	**26**	22	20	18	IntEx
Flies	Ahlum 1989 SB[6]	et	70	***65***	76	**81**		soil conservation
	Ahlum 1992 SB[6]	et	***59***	63	**68**	60		soil conservation
	Ahlum 1990 WW[6]	et	62	***55***	**78**	70		soil conservation
	Ahlum 1991 WB[6]	et	***65***	66	**82**	74		soil conservation
	Eickhorst 1992 WRy[6]	et	26	**30**	***24***	25	58	IntEx
	Eickhorst 1993 WR[6]	et	***60***	**89**	67		65	IntEx
Bugs	Eickhorst 1992 WRy[7]	pt	14	***12***	**15**	14	23	IntEx
	Eickhorst 1993 WR[7]	pt	**13**	***8***	10		17	IntEx
	Eickhorst 1992 WRy[7]	et	6	6	***5***	**7**	28	IntEx
	Eickhorst 1993 WR[7]	et	***9***	12	**14**		28	IntEx
Leaf-hoppers	Eickhorst 1992 WRy[8]	et	**5**	2	***1***	3	13	IntEx
	Eickhorst 1992 WRy[8]	pt	**10**	**10**	9	***8***	14	IntEx

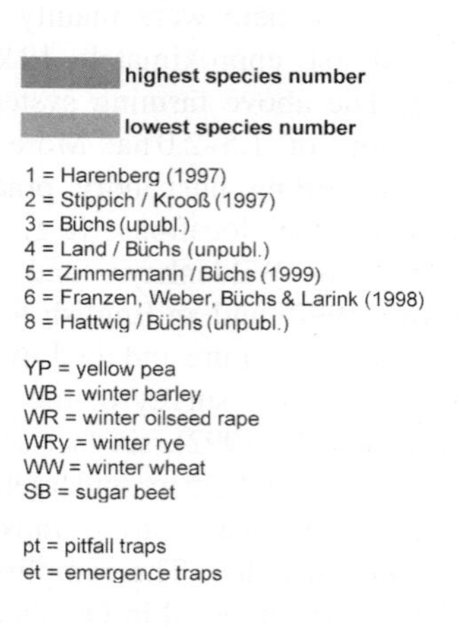

Table 1 (*Continued*)

(b) Assessment restricted to cultivated areas

taxon	loc /year / crop	meth	Farming system: conv I 3	int I 2	red I 1	ext I 0	setaside	project
Spiders	Eickhorst 92 WRy[1]	pt	37	41	46	40	41	IntEx
	Eickhorst 93 WR[1]	pt	36	38	40		44	IntEx
	Eickhorst 93 YP[1]	pt		28		38	42	IntEx
	Eickhorst 94 WW[1]	pt	35	36	41		52	IntEx
	Reinshof 90 WR[2]	pt	23	17	27	26	17	IntEx
	Reinshof 91 WR[2]	pt	21	23	27	31	27	IntEx
	Reinshof 92 WR[2]	pt	30	29	34	30	33	IntEx
	Reinshof 93 WR[2]	pt	20	24	22	28	31	IntEx
	Reinshof 94 WR[2]	pt	12	25	24	29	28	IntEx
	Reinshof 90 WW[2]	pt	22	26	24		21	IntEx
	Reinshof 91 WW[2]	pt	27	37	27	36	27	IntEx
	Reinshof 92 WW[2]	pt	21	30	26	33	33	IntEx
	Reinshof 93 WW[2]	pt	25	40	30	34	31	IntEx
	Reinshof 94 WW[2]	pt	13	21	25	22	28	IntEx
	Marienstein 90 WW[2]	pt	31	33	24	30	30	IntEx
	Marienstein 91 WW[2]	pt	27	35	39	43	31	IntEx
	Eickhorst 92 WRy[1]	et	11	10	11	13	33	IntEx
	Eickhorst 93 WR[1]	et	12	21	19		34	IntEx
	Eickhorst 93 YP[1]	et		15		14	30	IntEx
	Eickhorst 94 WW[1]	et	15	19	24		36	IntEx
	Reinshof 92 WW[2]	et	21	24	22	21	33	IntEx
	Reinshof 93 WW[2]	et	19	36	21	27	40	IntEx
	Reinshof 94 WW[2]	et	18	28	19	27	30	IntEx
Ground beetles	Reinshof 90 WR[2]	pt	10	12	11	17	12	IntEx
	Reinshof 91 WR[2]	pt	13	14	16	17	23	IntEx
	Reinshof 92 WR[2]	pt	14	15	15	16	22	IntEx
	Reinshof 93 WR[2]	pt	15	22	22	17	30	IntEx
	Reinshof 94 WR[2]	pt	12	19	18	21	24	IntEx
	Marienstein 90 WR[2]	pt	20	15	19	15	25	IntEx
	Marienstein 90 WR[2]	pt	16	18	20	23	29	IntEx
	Reinshof 90 WW[2]	pt	9	10	11		14	IntEx
	Reinshof 91 WW[2]	pt	8	17	13	14	22	IntEx
	Reinshof 92 WW[2]	pt	7	19	7	12	22	IntEx
	Reinshof 93 WW[2]	pt	12	25	17	18	26	IntEx
	Reinshof 94 WW[2]	pt	13	14	12	18	24	IntEx
	Marienstein 90 WW[2]	pt	13	11	13	9	26	IntEx
	Marenstein 91 WW[2]	pt	15	18	18	20	28	IntEx
Rove beetles	Reinshof 90 WW[2]	et	25		24		29	IntEx
	Reinshof 91 WW[2]	et	19		27		18	IntEx
	Reinshof 92 WW[2]	et	22	29	21	21	18	IntEx
	Reinshof 93 WW[2]	et	19	26	22	20	18	IntEx
Flies	Eickhorst 1992 WRy[7]	et	26	30	24	25	58	IntEx
	Eickhorst 1993 WR[7]	et	60	89	67		65	IntEx
Bugs	Eickhorst 1992 WRy[7]	et	6	6	5	7	28	IntEx
	Eickhorst 1993 WR[7]	et	9	12	14		28	IntEx
	Eickhorst 1992 WRy[7]	pt	14	12	15	14	23	IntEx
	Eickhorst 1993 WR[7]	pt	13	8	10		17	IntEx
Leaf-hoppers	Eickhorst 1992 WRy[8]	et	5	2	1	3	13	IntEx
	Eickhorst 1992 WRy[8]	pt	10	10	9	8	14	IntEx

(dark shading) highest species number

(light shading) lowest species number

1 = Harenberg (1997)
2 = Stippich / Krooß (1997)
3 = Büchs (upubl.)
4 = Land / Büchs (unpubl.)
5 = Zimmermann / Büchs (1999)
6 = Franzen, Weber, Büchs & Larink (1998)
8 = Hattwig / Büchs (unpubl.)

YP = yellow pea
WB = winter barley
WR = winter oilseed rape
WRy = winter rye
WW = winter wheat
SB = sugar beet

pt = pitfall traps
et = emergence traps

(I_1) Extensive crop production with a suboptimum input of nitrogen and low input of pesticides.

(I_2) Integrated crop production aiming for a high yield level but minimising the input of pesticides as far as possible.

(I_3) Intensive crop production using all registered agents to obtain maximum yield; prophylactic use of pesticides.

The field site was situated about 15 km south of Braunschweig on a heavy loess soil. In each management intensity five emergence traps covering 1 m^2 soil were installed. Each emergence trap was equipped withfivepitfall traps inside. In addition, six pitfall traps per plot were set in the field. The experiment was implemented between April 1989 and November 1992.

In all the above projects, pitfall and emergence traps were run for 14 day trapping periods between April and October, and monthly between November and March. Due to the development of the vegetation the location of emergence traps was changed monthly during the vegetation period (April–October). For more details, see Zimmermann and Büchs (1996, 1999), Kleinhenz and Büchs (1995) and Franzen et al. (1997).

2.5. Statistics

To describe biocoenotical effects, descriptive statistical methods were used: for classification of dominance (Engelmann, 1978; Mühlenberg, 1993), species diversity (Shannon, 1948; Shannon and Weaver, 1963), and evenness (Stugren, 1978).

Analytical statistic procedures were restricted to basic tests of significance. Depending on the result tests for normal distribution (Komolgorow–Smirnow test) a parametric t-test or a non-parametric Mann–Whitney U-test was conducted.

More detailed informations on method selection and use in the above projects are given in Büchs (1994), Büchs et al. (1997, 1999), Weiß and Büchs (1997), Weiß et al. (1997), Prescher and Büchs (1999) and Zimmermann and Büchs (1996, 1999).

3. Species richness as an uncertain measure to assess cultivated areas

Table 1a shows the number of species of spiders, ground beetles, rove beetles, flies, bugs and leafhoppers recorded using emergence traps or pitfall traps in cultivation systems with different levels of extensification. In the table, the management intensity decreases from left to right. The data were recorded as part of the project investigating "Soil Protection" of the Federal Biological Research Centre of Agriculture and Forestry, Braunschweig, and the INTEX-project coordinated by the University of Göttingen. To provide a visual overview, the cultivation system with the lowest species number has been marked in light grey with italicised numbers and the one with the highest species number in dark grey with underlined numbers.

Highest numbers of species were generally recorded in the set-aside area with natural succession, with the lowest species numbers largely found in the conventional (intensive) farming system (on the left). The table shows that habitats with extremely different life conditions (e.g. the uncultivated fallow and the cultivated areas) can be separated by species numbers with some precision.

However, if the set-aside area with natural succession is excluded from the analysis and assessments are restricted to the cultivated systems (Table 1b) an accumulation of low species numbers occur on the left side of the table, indicating that the lowest species numbers are represented in the most intensively managed farming system, especially the spiders and ground beetles. However, with regards to all other farming systems (including the very extensive management regime without input of pesticides and fertilisers, an extended crop rotation, etc.) no general trends can be observed. This indicates, that species numbers were not correlated to the farming system studied within the range integrated (II), reduced (III) or even extensively (IV) managed.

Hence, for several taxa, sampling methods and crops, it could be demonstrated that the lowest species numbers were recorded within the most intensively managed farming system, but for all other farming systems (even very extensive ones) no correlation between species richness and management intensity was found. Furthermore, in many cases the differences in species numbers between the farming systems was rather low (Table 1b). This leads to the conclusion that the species number alone is not a suitable indicator for the assessment of effects of farming systems on biocoenotical aspects of biodiversity.

4. Power and weakness of the Shannon-diversity index—dominance shifting as key problem to interpretation of biodiversity

Re-analysing the data so that biodiversity assessment is not restricted to the richness of species, races, varieties, or genotypes, etc., but using the scientific (bio)diversity index of Shannon (1948) and Shannon and Weaver (1963), which enable the dominance structure (abundance) of each species to be considered, gives a different picture.

Considering the INTEX-fields managed with farming systems with different levels of extensification, the diversity (H_s) of the spider coenosis increased ac-

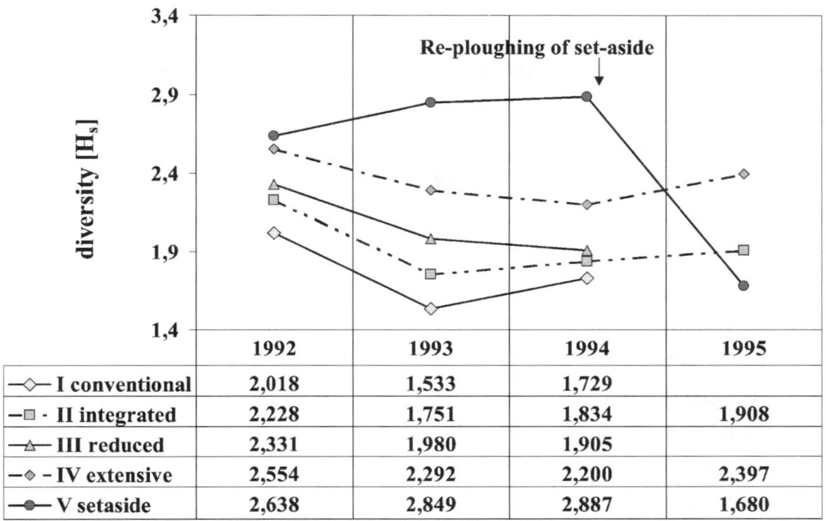

	1992	1993	1994	1995
I conventional	2,018	1,533	1,729	
II integrated	2,228	1,751	1,834	1,908
III reduced	2,331	1,980	1,905	
IV extensive	2,554	2,292	2,200	2,397
V setaside	2,638	2,849	2,887	1,680

Fig. 1. Development of the H_S index for spiders according to the management intensity (pitfall trap samples; Eickhorst, Braunschweig region, Germany, further explanations, see Section 2).

cording to the decrease of the management intensity during the 4-year study, especially in the permanent set-aside area with natural succession (V) (Fig. 1). After the re-ploughing of the set-aside area followed by cropping with sugar beet, the index of diversity (H_S) immediately decreased to the level of intensively managed fields. Thus the H_S index seems to be an exact indicator for changes of the crop management.

The sensitivity of the H_S index as an indicator is clarified, if data from the year after the re-ploughing of the set-aside area (followed by sugar beet growing) is taken into consideration: in this case a sugar beet crop rotation was grown in the fields adjacent to the former set-aside area during the whole set-aside period. After the re-ploughing of the set-aside area, all three fields were combined and managed conventionally in a uniform way. It could be shown that the diversity index (H_S) for spiders of the former set-aside area was significantly higher compared with the neighbouring areas (with a permanent intensively managed sugar beet rotation) (Fig. 2). This observation could be explained by the significantly lower numbers of the typical intensification indicator *Oedothorax apicatus* that was recorded on the former set-aside area, and the more balanced dominance structure that resulted. This example demonstrates, that even after the re-ploughing and subsequent sugar beet crops the former fallow could be identified clearly by means of the diversity index (H_S) from those sugar beet fields which had been conventionally managed for several years.

Table 2 shows an example from the dominance structure of rove beetles (Coleopterea: Staphylinidae) in winter wheat fields with extensive crop management (without pesticide and nitrogen input) vs. plots with high input of pesticides and fertilisers (for exact crop management, see Zimmermann and Büchs, 1999). The overall results demonstrate that extensively and intensively managed farming systems do not necessarily differ in abundance (ind./m^2), species numbers (more species were even recorded in the intensively managed system) and species diversity (Shannon–Wiener index). Thus, according to these criteria it would be impossible to differ between these two farming systems. Moreover, this result would lead to the conclusion that intensive farming (high input of pesticide and fertilisers) does not adversely effect populations of rove beetles or their biodiversity.

However, if the dominance position of the species is compared in addition to the dominance structure, it is clear that some species have changed their position within the structure of dominance (e.g. *Omalium*

Table 2
Example for the shifting of dominance positions of species at equal levels of diversity index (H_s): extract of the dominance structures of rove beetle communities (species >0.2% dominance) from emergence traps in an extensively and an intensively managed winter wheat field (corresponding species marked by shading and connecting lines)

Shifting of dominance at equal diversity levels
Coleoptera: Staphylinidae, emergence traps, winter wheat, Ahlum 1990
management intensity

	extensive no pesticides low input of fertilizers*		**intensive** high input of pesticides and fertilizers	
1	Aloconota gregaria	28,9**	Aloconota gregaria	32,4**
2	Atheta aegra	15.9	Atheta triangulum	14.1
3	**Coprophilus striatulus**	**15.5**	Atheta aegra	13.2
4	Oxypoda exoleta	7.4	Oxypoda exoleta	6.6
5	Atheta triangulum	6.3	Amischa analis	6.5
6	Oxypoda haemorrhoa	3.1	**Amischa decipiens**	**4.3**
7	**Omalium caesum**	**2.9**	Oxypoda haemorrhoa	4.1
8	Amischa analis	2.8	**Coprophilus striatulus**	**3.6**
9	**Lesteva longelytrata**	**2.4**	Gyrohypnus scoticus	1.8
10	Tachyporus hypnorum	1.7	**Atheta fungi**	**1.4**
11	Dinaraea angustula	1.5	Dinaraea angustula	1.3
12	Liogluta alpestris nitidula	1.2	**Omalium caesum**	**0.8**
13	**Amischa decipiens**	**1.0**	Xantholinus longiventris	0.7
14	Gyrohypnus scoticus	0.9	Omalium rivulare	0.7
15	Xantholinus longiventris	0.7	Lathrobium fulvipenne	0.7
16	Oxypoda longipes	0.7	Liogluta alpestris nitidula	0.6
17	Omalium rivulare	0.5	Amischa soror	0.6
18	Lathrobium fulvipenne	0.5	Aleochara laevigata	0.6
19	Atheta pitionii	0.5	Tachyporus hypnorum	0.5
20	Atheta elongatula	0.5	Oxypoda umbrata	0.5
21	Amischa soror	0.5	Atheta elongatula	0.5
22	Parocyusa longitarsis	0.4	Parocyusa longitarsis	0.4
23	Callicerus obscurus	0.4	Oligota pusillima	0.4
24	**Atheta fungi**	**0.4**	Atheta pitionii	0.4
25	Oligota pusillima	0.3	Aleochara bipustulata	0.4
26	Anotylus rugosus	0.3	Oxypoda longipes	0.3
27	Tachyporus nitidulus	0.2	**Lesteva longelytrata**	**0.3**
	$H_s = 2.473$ Species number = 61 Sum of individuals = 6150		$H_s = 2.443$ Species number = 70 Sum of individuals = 6482	

* No. pesticides except seed dressing; only basic fertilization of 40 kg N/ha.
** Values of dominance (%).

Spider abundance from emergence traps in sugar beet fields

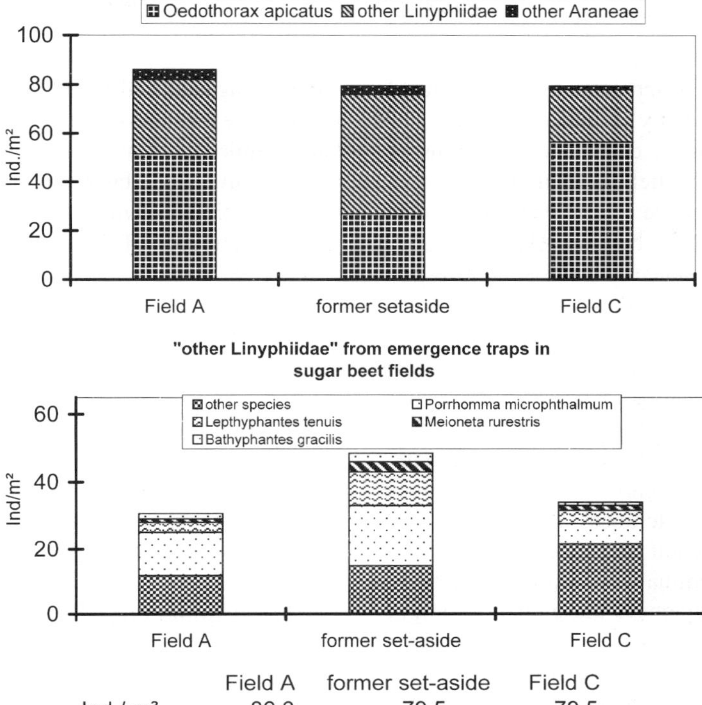

	Field A	former set-aside	Field C
Ind / m²	86.0	79.5	79.5
Species	17	17	13
Diversity (H_s)	1,52[a]	2,01[b]	1,19[c]
Evenness	0.54	0.71	0.46

Fig. 2. Diversity (H_S) and species composition of the spider communities in sugar beet fields (field A, field B) and after tillage of a 5-year-old set-side area followed by sugar beet growing (emergence trap samples, Eickhorst, Braunschweig region, Germany; H_S: different characters indicate significance, Mann–Whitney U-test; $P < 0.05$; further explanations, see Section 2).

caesum and *Amischa decipiens* or *Atheta fungi* and *Lesteva longelytrata*). Nevertheless, the diversity index (H_S) has not changed, unless a relevant shifting of species within the rove beetle community is to be considered.

The importance of such shifting of dominance for the agrobiocoenosis and its functions is shown by comparison of ground beetle species communities of cultivated fields between 1951 and 1987 (Table 3). Species that dominated in fields in the 1950s occur

Table 3
Shifting of dominance (%)[a] of ground beetle species (Coloptera: Carabidae) of arable fields with different body size between 1951 and 1987 (according to Steinborn and Heydemann, 1990)

Year	Range of body size (mm)	Dominance (%)						
		1951/1952	1977	1978	1982	1983	1984	1987
Carabus auratus	*17.0–30.0*	*48.3*	*2.1*	*2.8*	*0.8*	*1.0*	*1.0*	*1.0*
Harpalus rufipes	*9.0–16.0*	*4.8*	*3.7*	*6.3*	*0.5*	*0.6*	*0.7*	*0.5*
Bembidion tetracolum	4.9–6.1	0.5	0.5	0.5	12.9	22.2	26.5	20.0

[a] Dominance is related to entire ground beetle samples with pitfall traps; for details, see Steinborn and Heydemann (1990).

today (1987) only subrecedently (Engelmann, 1978), whereas species that occurred rather rarely in the 1950s are dominant in today's fields. These results lead to the conclusion that predominantly the loss of species (a factor that is often said to be correlated with reduction in biodiversity) is not of primary importance, but the shifting of dominance positions of individual members of the field typical species community represent the decisive changes resulting from the intensification of land-use between 1951 and 1987.

The change of dominance positions is also accompanied by a shifting of the body size: Larger species, such as *Carabus auratus*, with a body size of 17.0–30.0 mm dominated ground beetle communities in agroecosystems 50 years ago, whereas today rather small species such as *Bembidion tetracolum* with a body size of only approximately 5 mm dominate. Thus the quality of the species community changes dramatically (e.g. their tolerances and preferences with regard to environmental conditions), in line with such changes to the dominance positions and body size, and with the consequences these changes might have for functional aspects or processes taking place in agroecosystems. To illustrate such consequences, the prey species and prey size preferences the two ground beetle species mentioned will differ significantly as a result of their body size.

From this example it can be concluded that an exclusive or prior-ranking assessment of semi-natural habitats, and particularly of cultivated areas, according to recently popular assessment criteria (e.g. Roth and Schwabe, 2003; Braband et al., 2003) does not take account of major problem that we face in assessing the agricultural landscape.

For the biotic assessment of cultivated areas criteria are required that are not based only on a maximising of "biodiversity", but include more strongly structural and functional qualities of the biocoenoses according to the definition of Noss (1990).

5. Alternative community and population based parameters steering biodiversity

There are numerous ecological parameters at both the population and community level that are suitable as assessment criteria for gradual changes in agroecosystems, one of which is body size. Other examples of such parameters and assessment criteria are listed in Table 4.

The definition of the objective of the work is critical for the successful design of the assessment procedure: which qualities of a system are defined as good or worthwhile and which as poor or unimportant? If (species-)diversity and/or a maximisation of "rare" and "endangered" species are not identified as the objective, but the development of a species community typical for the ecosystem or geographical unit

Table 4
Ecological parameters steering biodiversity for potential use as assessment criteria for the indication of gradual changes in agroecosystems associated with management intensity in agricultural areas

Population level	Community level
Activity density, abundance	(Bio-)diversity
Presence, constance, frequence	Structural shaping of species communities
Period of activity	Similarity of species/dominance structure, shifting dominance, clusters, correlations, correspondences, etc.
Body size	Ratio of euryoecious/stenoecious species
Body weight	Ratio of r- and K-strategists
Feeding rates	Biotic/abiotic preferences
Growth rates	For example preferences for certain habitats/strata/millieus
Age structure	Predator/prey-relationships
Sex ratio	Rate of species-turnover
Egg production	Percentage of rare and/or endangered species
Periods/rates of reproduction	Body size (related to species communities)
Morphological features	Biomass (related to species communities)
For example makro-/brachypterism, development of wing muscles, fluctuating asymmetry	Morphological features (related to higher taxa)

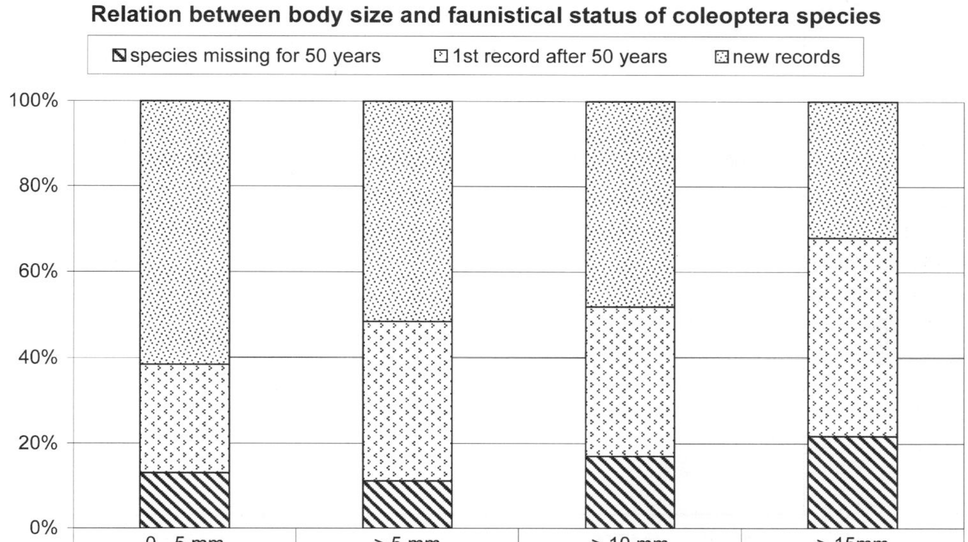

	0 - 5 mm	> 5 mm	> 10 mm	> 15mm
new records	741	235	62	25
1st record after 50 years	305	170	45	36
species missing for 50 years	158	51	22	17

Fig. 3. Body length and faunistical status of beetle species (after Köhler and Stumpf, 1992).

investigated, then it can be assumed that all parameters which describe in any form conditions favouring reproduction or reproductive success should be included in the essential assessment criteria.

Only healthy populations (precondition, e.g. a sufficient food supply) that have the potential to reproduce successfully (precondition, e.g. a sufficiently large habitat area), safeguard the survival of a species and therefore (if transferred to the whole species community of an ecosystem) guarantee the maintenance of biodiversity on a high level.

This objective fits well with the recently promoted orientation of husbandry practices towards consumer protection: at the primary level, contamination of soil, air or water, e.g. by a pesticide may be insignificant, but becomes a problem if any organism is affected with regard to its essential needs and/or the substance enters the food-chain, in which it can be transferred to man as a final consumer. Therefore, the "physical fitness" of populations of free living animal and plant species may be a decisive measure of the significance of the conversion of husbandry practices designed for consumer protection into action within food production procedures. Thus, highest priority has to be given to the preservation of the essential resources of the agrobiocoenoses that are typical for the region and the relevant geographical unit.

Two basic parameters that indicate "physical fitness" and so, might play an important role within processes that are steering biodiversity, are the body length of species and their reproductive success.

With regards body length, a change of habitat quality which can be related to an increasing environmental load, will result in the average size of all beetle species recorded decreasing (Köhler and Stumpf, 1992), i.e. the probability of species becoming more rare (first record after 50 years) or to become extinct (missing for 50 years) increases with the body length (Fig. 3). However, as in the example presented here this correlation is often very weak ($r = -0.3$, $d = 0.09$). Nevertheless, the suitability of body size as an indicator of the effects of management practices in agricultural ecosystems has frequently been highlighted. For example Blake et al. (1991) and Stippich and Krooß

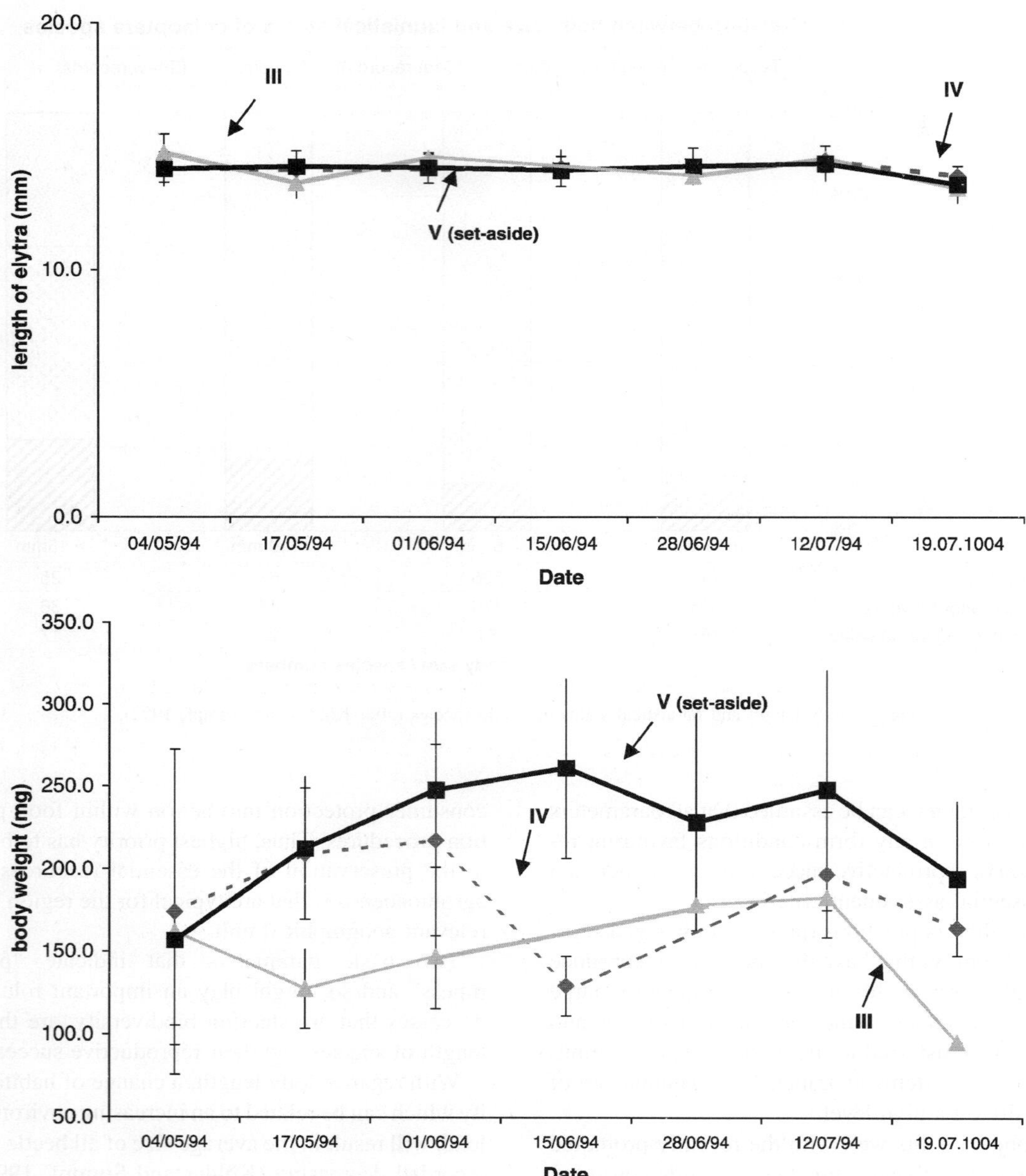

Fig. 4. Phenology of the body size (length of elytra) and body weight of *Carabus auratus* males in dependence of the management intensity. Average values of body weight (mg) and body size (mm) of *Carabus auratus* males in dependence of the management intensity (significant differences according to student's *t*-test, $P < 0.05$): dry weight (mg), 1994: farming system (see Section 2) I = 143.2 (S.D. 84.6); II = 108.3 (S.D. 23.7); III = 146.6 (S.D. 43.8); IV = 200.5 (S.D. 58.8); V = 216.2 (S.D. 66.0); significance: V ⇒ I, V ⇒ II, V ⇒ III, V ⇒ IV; IV ⇒ III, IV ⇒ II, IV ⇒ I. Dry weight, 1995: I = 112.9 (S.D. 18.6); II = 127.7 (S.D. 27.8); III = 141.6 (S.D. 37.5); IV = 147.5 (S.D. 46.0); V = 110.2 (S.D. 24.5); significance: V ⇒ IV, V ⇒ III; IV ⇒ I; III ⇒ I. Body size (length of elytra in mm), 1994: I = 14.47 (S.D. 0.96); II = 14.5 (S.D. 0.56); III = 14.24 (S.D. 0.85); IV 14.15 (S.D. 0.57); V = 14.18 (S.D. 0.66); no significant differences at all. Body size (length of elytra in mm), 1995: I = 14.67 (S.D. 0.67); II = 14.30 (S.D. 0.42); III = 14.51 (S.D. 0.70); IV 14.53 (S.D. 0.68); V = 14.03 (S.D. 0.81); significant for $P > 0.1$: V ⇒ IV, V ⇒ III, V ⇒ I.

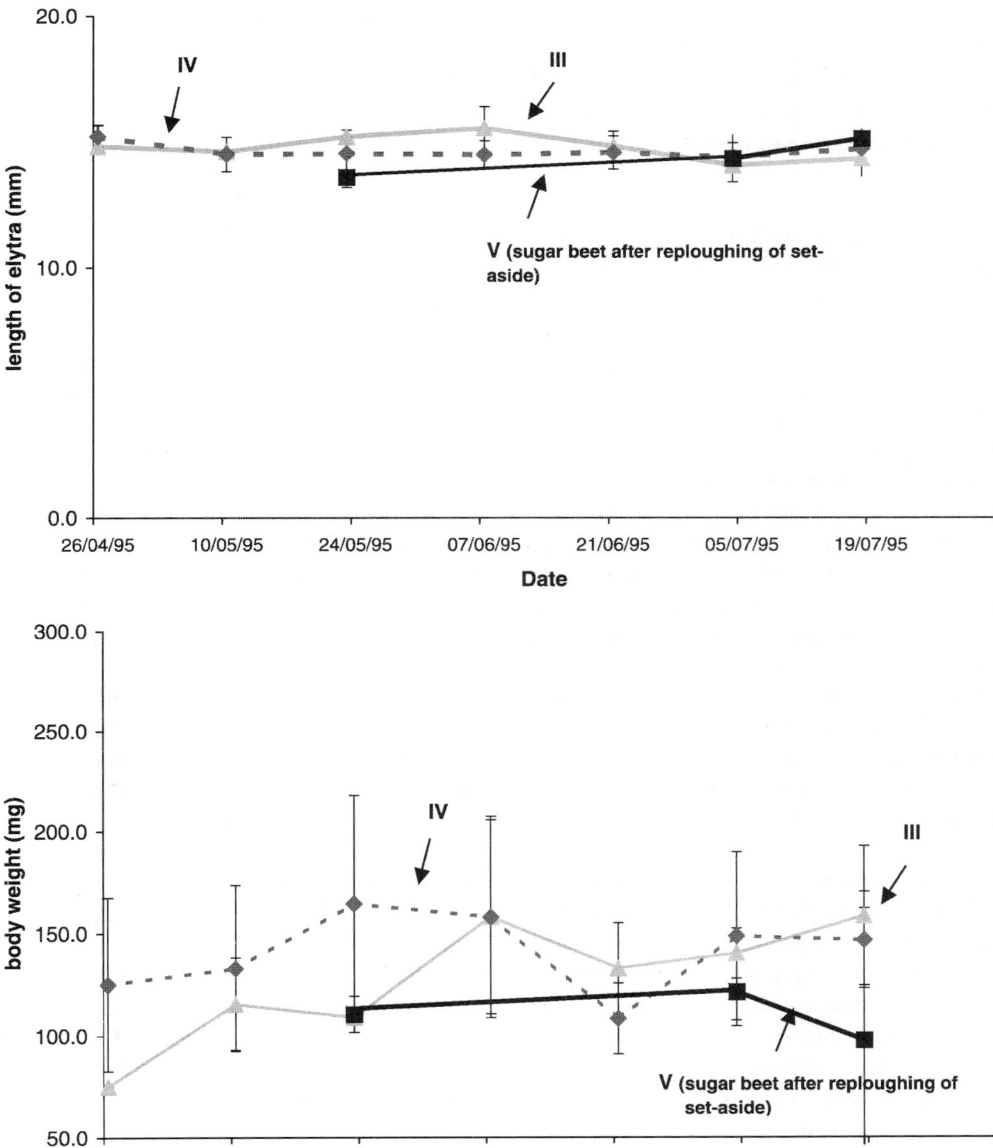

Fig. 4. (*Continued*).

(1997) showed that in fields with conventional tillage (includes ploughing), the species set of Aleocharinae (Coleoptera: Staphylinidae) and ground beetles (Coleoptera: Carabidae) was shifted towards smaller species compared to fields with reduced tillage. Blake et al. (1991) explained this phenomenon by a combination of higher mobility and shorter reproduction cycle of the smaller species. They are able to use resources that are available only briefly, more effectively than larger species that display longer and seasonally more fixed reproduction cycles. Stippich and Krooß (1997), however, suggest that larger species are more directly affected by mechanical damage caused by the plough itself. Finally, a higher exposure of larger species to pesticides (e.g. insecticides) may be assumed.

Büchs et al. (1997) demonstrated that male spider species of some species are larger with increasing extensification, but this trend was only significant if the management intensities compared differed widely, e.g. between the set-aside area with natural succession and cultivated fields. Stippich and Krooß (1997) stated that, compared to fields, in set-aside areas with natural succession the spider community shifted to families with larger species. However, this observation cannot be applied in practice when only cultivated fields are assessed or when different kinds of conventional farming systems are compared.

The question of how exact the environmental loads are indicated, if the body size is determined on the population level for a selected species that is identified as an indicator for a defined status of the ecosystem investigated, remains to be discussed. According to the results of several studies (e.g. Basedow, 1987, 1998, 2003; Steinborn and Heydemann, 1990; Büchs et al., 1997) the ground beetle species *Carabus auratus* could be identified as an indicator of an acceptable cultivated field.

Fig. 4 shows the development of the body size and the body weight of *C. auratus* males in a 5-year fallow (V), in an extensively managed farming system without pesticide and fertiliser input (IV) and in a farming system without insecticide use (III) in 2 years: one before and one after the re-ploughing of the fallow area (followed by sugar beet cultivation). The assessment was conducted on the basis of about 3000 individuals.

With regards to the phenology of the average body size (top) no significant differences can be identified between the cultivated fields and the set-aside area, and no changes in body size occur after the re-ploughing of the set-aside area followed by sugar beet cultivation, which undoubtedly represents a serious perturbation of the ecosystem.

It is known that the body size of adult beetles depends on the food supply during the larval stages and is finally fixed at metamorphosis (Freude et al., 1964). As assessments were restricted to the adult stages of the beetles which can be rather vagil, nothing is known about larval development. It is possible that the larval development had not taken place at the study site, where the adult beetles were recorded, as it is known that ground beetle species can move long distances, changing habitat or location, after exclusion (Thiele, 1977).

However, the weight of the *C. auratus* males was a very exact indicator. The weight parameter is a measure of the actual nutritional state of the adult beetle. It reflects the actual physical fitness, therefore, it is subject to short term and relatively large fluctuations of life conditions as indicated by Fig. 4.

If the insecticide free (III), the extensive (IV) farming systems, and the set-aside area with natural succession (V) are compared, the body weight of *C. auratus* males is significantly higher within the set-aside area (V) in contrast to the cultivated fields (III and IV) over the whole sampling period in 1994. The availability of food is clearly more favourable for the beetles within the set-aside area with natural succession than on cultivated fields.

The results from the subsequent year (1995), when the 5-year-old set-aside area was ploughed up and cultivated with sugar beet are also interesting: the beetles of the former set-aside area weigh only half as much as in the year before the re-ploughing of the soil. Moreover, their weight is clearly below that of the beetles from the insecticide free (III) and the extensively (IV) managed farming systems. This illustrates both the precision, and the immediate reaction of the indicator "body weight" to environmental stress as, for example, the increase of the management intensity. Furthermore, mark-release experiments with *Pterostichus melanarius* (Büchs, 1991; unpublished) in oilseed rape indicate, that in uniform structured fields during the vegetation period, the adult beetles remain in relatively small areas which they leave only in autumn when they search for hibernation sites. This indicates that adult ground beetles might be more fixed to a location than is usually assumed, improving their use as indicators. However, such features differ both between and within species, and between the range of environmental conditions which might be investigated, whereas Langmaack et al. (2001) did not find any significant differences in body weight of the ground beetle *Poecilus cupreus* in two similar farming systems which differed only in insecticide and nitrogen input but otherwise were identically managed with regard to crop

Activity periods of *Carabus auratus* (Coleoptera: Carabidae) in dependence of different management strategies

Year 1

Farming system	*Periods with C. auratus records*	*March II*	*April I*	*April II*	*May I*	*May II*	*June I*	*June II*	*July I*	*July II*	*August I*	*August II*	*Sept. I*
conventional	3			!	!					H		! T	T
integrated A	4		!							H		T	
integrated B	4	T					!					H T	
reduced	6									H		T	T
extensive	7	T										H T	
set-aside	9												

Year 2

Farming system	*Periods with C. auratus records*	*March II*	*April I*	*April II*	*May I*	*May II*	*June I*	*June II*	*July I*	*July II*	*August I*	*August II*	*Sept. I*
convention. A	1	!	!		!					H	T		T
convention. B	3						!				H T		T
integrated A	0			T			!				H T		T
integrated B	2							!			H	T	T
integrated C	2		!		!					H			T
reduced	7										H T		T
extensive	7									H	T		T
set-aside	8												

! = insecticide application T = tillage H = harvest (shaded) = sampling periods with *C. auratus* activity

pitfall traps, Eickhorst 1992 and 1993

Fig. 5. Number of sample periods with records of *C. auratus* (Coleoptera: Carabidae) in dependence of the farming system (further explanations, see figure).

rotation, seed density, cultivars, tillage, etc., Zangger et al. (1994) showed that in systems displaying greater differences such as uncultivated weed strips and integrated cereal fields, the body weight of *P. cupreus* was significantly higher in the weed strips. As a more general result, Hokkanen and Holopainen (1986) recorded a higher biomass of ground beetles in less intensively managed organic fields compared to fields managed in a conventional way.

With regard to the use of *C. auratus* as indicator of environmental stress, the temporal range and number of the sampling periods, in which this species was recorded during a growing season indicated the potential for the assessment of the intensity of crop management (Fig. 5). Under the influence of variations of pesticide input, tillage and harvesting, a clear trend of increasing duration of activity period of this predacious species, related to the degree of extensification, could be observed.

Over 2 years the lowest number of activity periods was recorded within the most intensively managed farming system, the highest number in the undisturbed area with natural succession (Fig. 5). Apparently, such fallow areas can act as a refuge in which beetles such as *C. auratus* can complete their larval development without being affected by the above-mentioned disturbances.

Body weight, duration of activity period and other parameters relying on the "fitness" of a population can be assumed to influence the success of reproduction, so that any measurements related to reproduction can be used as indicators indirectly steering biodiversity. For example, Büchs et al. (1999) were able to rank the suitability and positive effects of eight rotational set-aside types (see above) for the reproduction of ground beetles, particularly of *Carabus granulatus* by creating an "index of reproduction" which was defined as the product of activity density (ind./trap), percentage of female beetles, percentage of female beetles carrying eggs and eggs/female. This "index of reproduction" showed a wide range (e.g. the "beekeepers mix" was 85 times more suitable for reproduction compared to a mix of Phacelia and Alexandrinian clover) so that the different set-aside types could be separated easily.

Furthermore, Büchs et al. (1997) showed that the abundance of juvenile spiders could be related to the management intensity of fields in a sugar beet rotation as well as in an oilseed rape rotation. Because of the special character of the spider community of cultivated fields (which spider community is a priori dominated by so-called "pioneer species" due to the disturbance inherent in the system that periodically resets the "succession" of the agroecosystem periodically to zero) the abundance of juvenile spiders is not only comparatively easy to assess, but also a rapidly reacting parameter, that indicates actual environmental loads rather exactly. But, as was demonstrated by Büchs et al. (1997) for the winter barley, in a crop without insecticide input in the year of harvest, an overcompensation is likely to occur (due to recolonisation effects) if a certain environmental stress is discontinued.

Because of the small degrees of disturbance and environmental loads, the species composition of long-term set-aside areas with natural succession or other uncultivated habitats can be used as baseline reference points for the assessment of the degree of sustainability of cultivated fields.

Fig. 6 demonstrates that in each of the 3 years the species composition of spiders in the cultivated areas shows increasing high similarity with the species composition of the fallow area (baseline reference) when the farming system is managed extensively. This trend can be observed for all parts of the crop rotation, i.e. it is independent of current crop, but is determined in each year by the intensity of the crop management.

Thus in principle, full-coverage extensification can achieve a similar effect to set-aside of arable land. This effect is specifically the survival of (spider) species typical for agroecosystems, combined with an increase of their biodiversity (see Fig. 1). With increasing extensification, more of the species that usually only occur in natural undisturbed and/or uncultivated areas, are able to survive within the cultivated areas. So it can be assumed, that a similarity index related to a reference area has the potential to be used as an indicator of biodiversity.

An important criterion for the assessment of the sustainability of a farming system is the extent to which the species community is represented by "pioneer elements". Pioneer species usually occur in high densities and are euryoecious, i.e. their requirements for specific habitat features are rather low, and can colonise heavily disturbed or periodically newly arising areas. Disturbed areas are inherent in agroecosystems due to the operations such as harvest, pesticide applications and tillage. However, they differ in their intensity and shape. The Lycosidae are very sensitive to such interventions, particularly to re-ploughing or other tillage measures (Hammer, 1984; Stippich and Krooß, 1997).

Over 3 years a uniform trend can be recognised (Table 5): decrease of the pioneer species occurs with increasing extensification together with a simultaneous increase of the Lycosidae as indicator for environmental stress. While the pioneer species always predominate in the cultivated fields, the permanent

Table 5
Percentage of pioneer species and wolf spiders (Araneae: Lycosidae) in farming systems with different levels of extensification[a]

Year	Group	Farming system (I–IV), set-aside (V) (%)				
		I	II	III	IV	V
1	"Pioneers"	81.8	67.6	75.0	47.7	8.4
2		95.3	94.4	81.9	79.6	15.5
3		92.0	90.1	87.3	79.0	17.7
1	Lycosidae	10.4	14.7	10.8	22.7	73.8
2		2.6	2.6	4.9	8.1	53.5
3		4.4	2.3	4.6	7.5	62.0

[a] Pitfall trap samples (6–8 traps/plot), Eickhorst, Braunschweig region, Germany, 1992–1994.

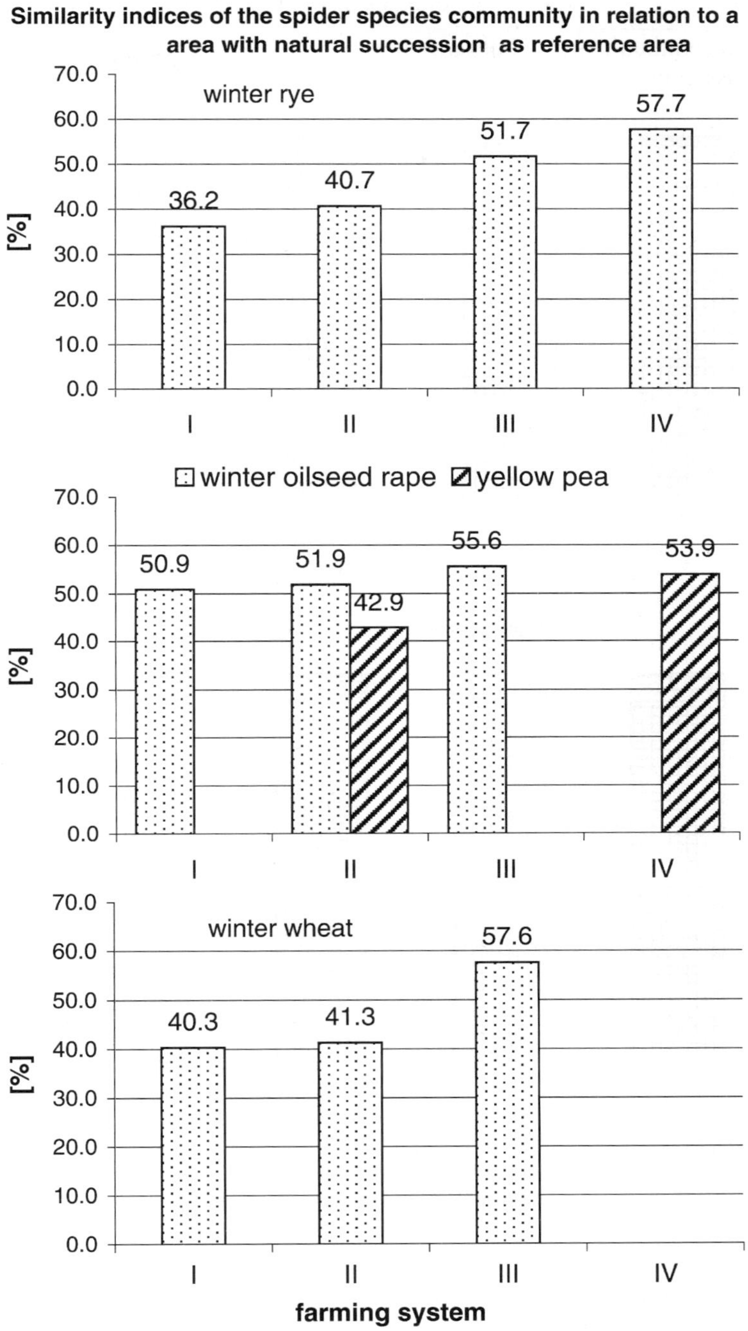

Fig. 6. Similarity of spider communities of different levels of extensification in relation to the species composition of an undisturbed set-aside area with natural succession as reference.

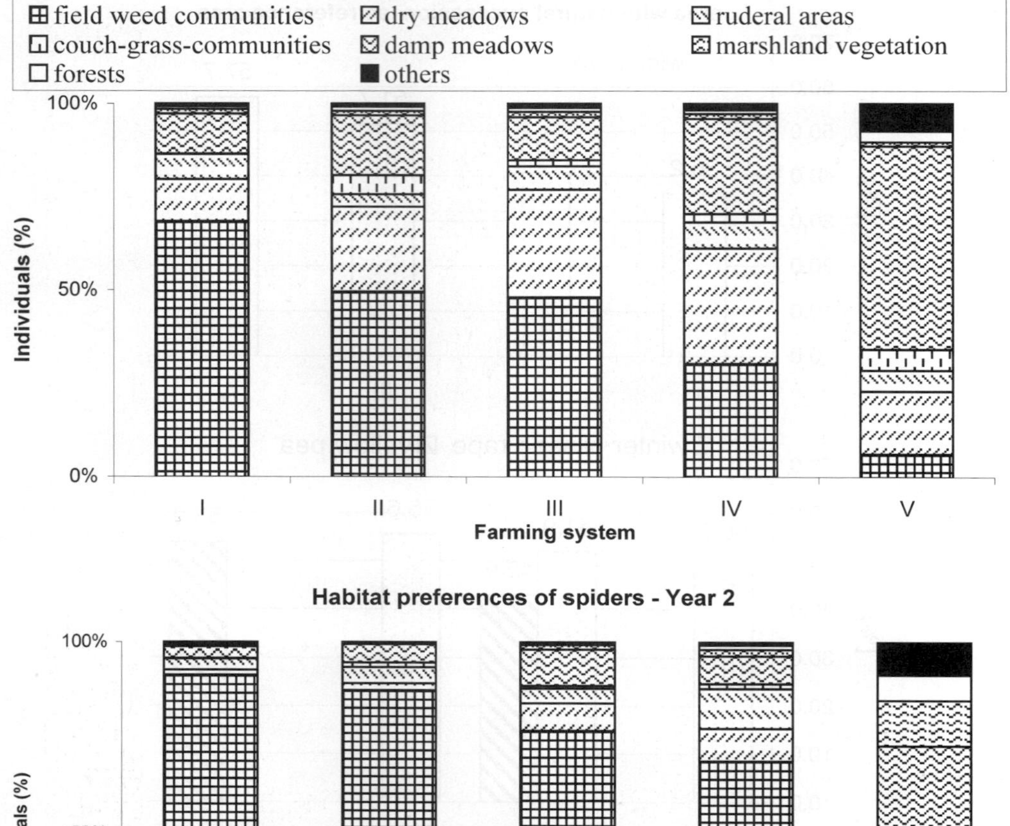

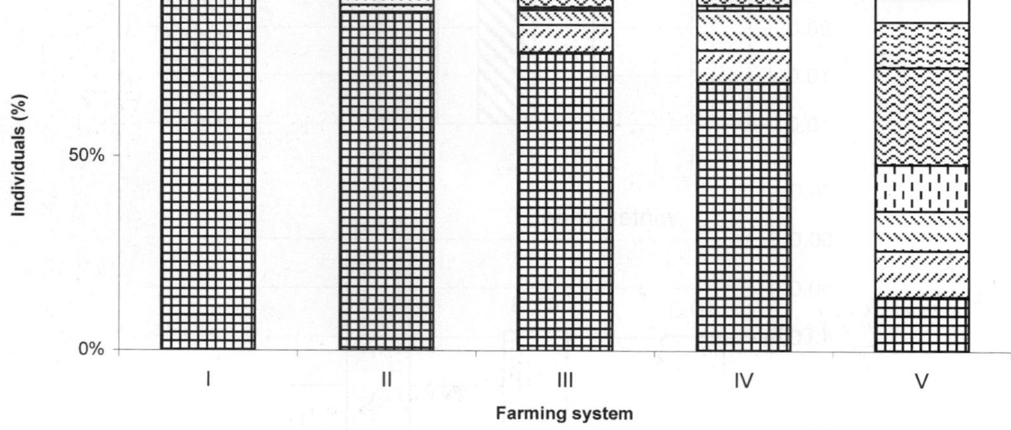

Fig. 7. Patterns of habitat preferences of spider communities in farming systems with different levels of extensification (I–IV: extensification increasing) and a set-aside area with natural succession (V). Pitfall trap samples Eickhorst (6–8 traps per plot), Braunschweig region, Germany.

set-aside area with natural succession is characterised by a larger proportion of the Lycosidae.

The discussion above dealt with habitat quality requirements of species: these ecological requirements of spiders can be differentiated at a smaller scale using details of the preferred habitats of the different species. From this base predictions of the effects of permanently intensive or a permanently extensive management on the composition and diversity of "ecological types" of spiders with regards to their habitat preference can be made. Considering different cultivation systems (Fig. 7), in all 3 years of the field experiment

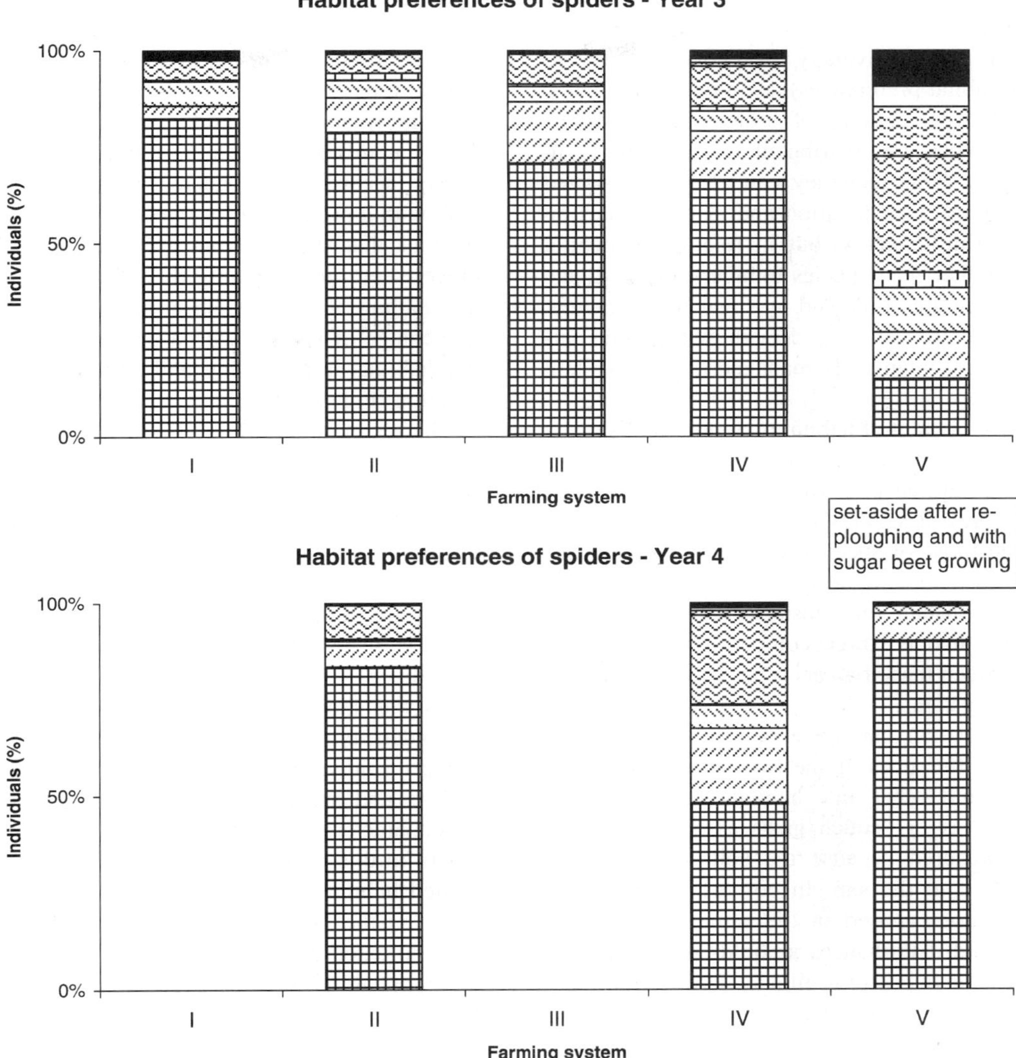

Fig. 7. (*Continued*).

the percentage of spider species typical for weed complexes in open fields (see legend of Fig. 7), decreases with increasing extensification. Simultaneously, it can be observed that higher percentages of species typical for dry meadows and species typical for damp habitats occur in fields with increasing extensification. The species typical for damp areas show particularly high percentages in the permanent set-aside with natural succession. This can be explained by the permanent vegetation cover that protects soil from drying out.

In cultivated fields, spider species that prefer damp areas occur only in rather small numbers as expected, but even their percentage is increasing the more extensively the crops are managed. This observation may be explained by the increase in weed cover with the level of extensification (Steinmann et al., 1997; Otte, 1990).

It is also striking that in the permanent set-aside area (V) with natural succession, the spider species represent a higher diversity of ecotypes (derived from their habitat preferences), as in cultivated fields. Similarly, the percentages of preferences to different habitat types are distributed more regularly to each category. This tendency can also be observed with increasing extensification and also indicates an increase of (bio)diversity, but in relation to habitat preference rather than species numbers. After tillage of the set-aside area followed by sugar beet growing, the habitat preferences of spiders show the same pattern as the most intensively managed farming system I (Fig. 7).

If the development of habitat preferences of spiders is considered over a period of 3 or 4 years (Fig. 8), it is striking that, despite changes from year to year, the percentages of species with preferences for field habitats and those of species with preference for dry habitats show a remarkable symmetry in all cultivated areas. This constant relationship over several years indicates that there is a direct competition or a dynamic balance between these two ecotypes of spiders within the farming systems.

This observation can be explained by the similarities and differences of the two types of habitats: both fields and other dry habitats (dry meadows or ruderal areas or "couch grass associations") are thermally favoured because the soil surface is directly exposed to the sun. In contrast to fields dry habitats are not exposed in the same extent to interventions and disturbances as fields resulting from management practices (e.g. tillage, fertilisation, pesticide use, harvest). Spider species that prefer field conditions are adapted to such interventions. The other dry habitats, however, build a permanent vegetation cover, that will be mowed at most once a year or used as pasture, but exist continuously. Hence, disturbances and other interventions are less extreme in these habitat types compared to fields (Fig. 8).

These interrelations demonstrate that, with regards to the structure of the spider coenosis, correlations begin to emerge which follow constant rules. These correlations or the ratios they are based upon, can be gradually shifted by changing the husbandry practice, and an active steering is possible, giving the potential to develop a model.

A knowledge of habitat preferences of spiders in field habitats from the above-mentioned experiments may be used for the assessment of other fields or habitat types in the rural landscape as for example areas with rotational set-aside.

If the different types of rotational set-aside are ranked according to a decreasing percentage of spiders with a preference for (weedy) field habitats, it is obvious that starting from a Phacelia/Alexandrinian clover seed mix including the "Tübinger Seed Mix" and ending at the raddish crop (see Section 2), the percentage of spiders with a preference for damp or dry habitats is increasing and so shows an opposite reaction (Fig. 9a).

The second figure (Fig. 9b), in which the rotational set-aside types are ordered in the same way, shows that the decrease of field spiders is also followed by a decreasing percentage of Linyphiid spiders (as typical pioneer species that tolerate disturbances) and an increase of the wolf spiders (Araneae: Lycosidae), which are usually classified as very sensitive indicators for disturbances and other environmental stress (see above).

As was shown in Figs. 6 and 7, high percentages of spiders typical of (weedy) field habitats, and of pioneer spiders, as well as low percentages of wolf spiders and spiders with preference for dry meadows characterise fields as intensively managed. Starting from this basis, the "ecological value" of a field related habitat can be assessed. Hence, the set-aside types on the left half of Fig. 9a and b, correspond more to extensively managed field habitats, those on the right half, however, are related more to intensively managed field habitats. Therefore, on the basis of our results from the long-term investigation of different management intensities an overall assessment of rotational set-aside areas regarding their suitability as habitat for a divers and balanced spider community is possible.

6. Final conclusion

The examples mentioned in the first part of this paper demonstrate some of the major problems in application of diversity indices:

- The index of diversity is an absolute value (i.e. without being related to a standard reference value) of

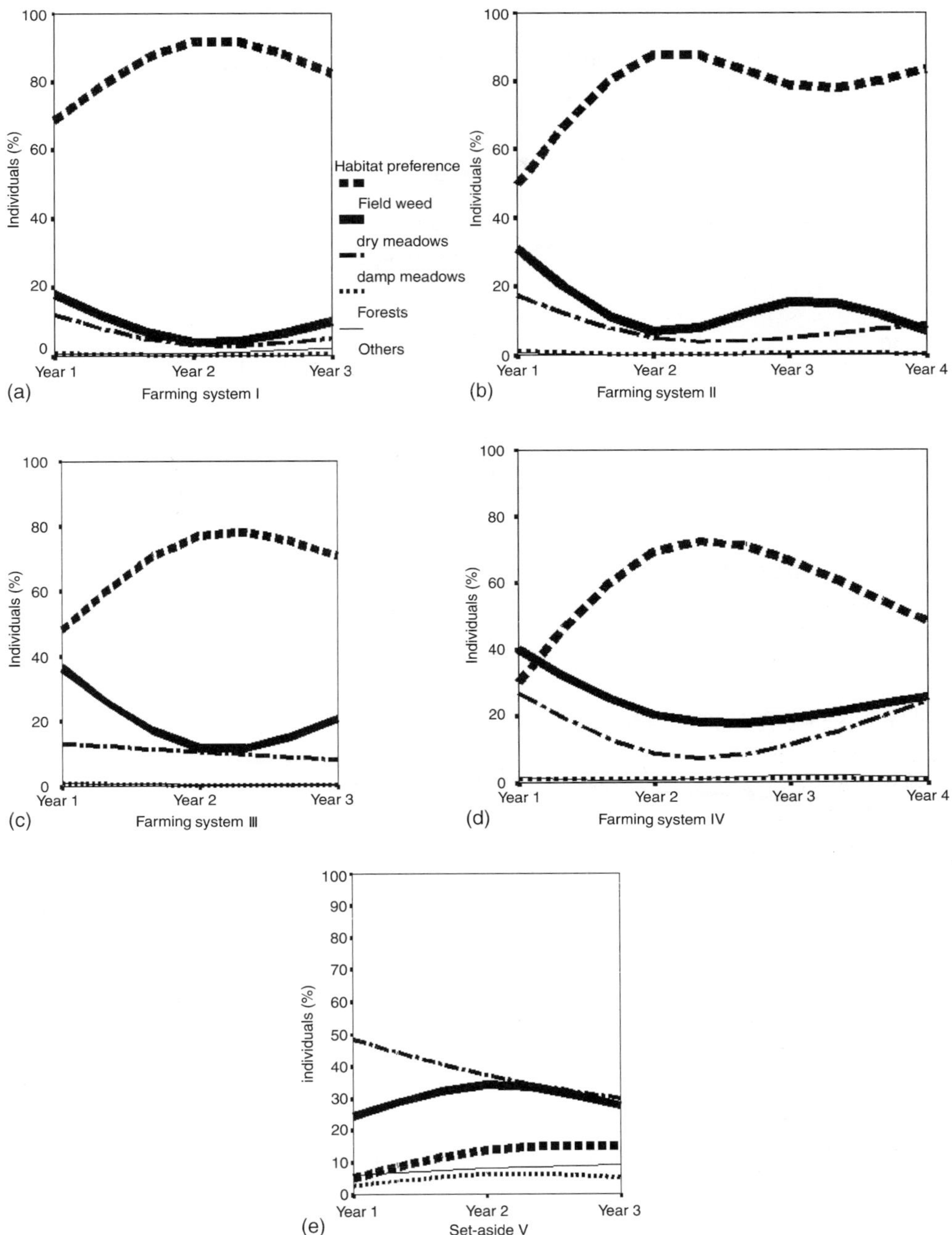

Fig. 8. (a)–(e) SPLINE-plots of the development of habitat preferences of spiders in different levels of extensification (I–IV: increasing extensification) and a set-aside area with natural succession (V).

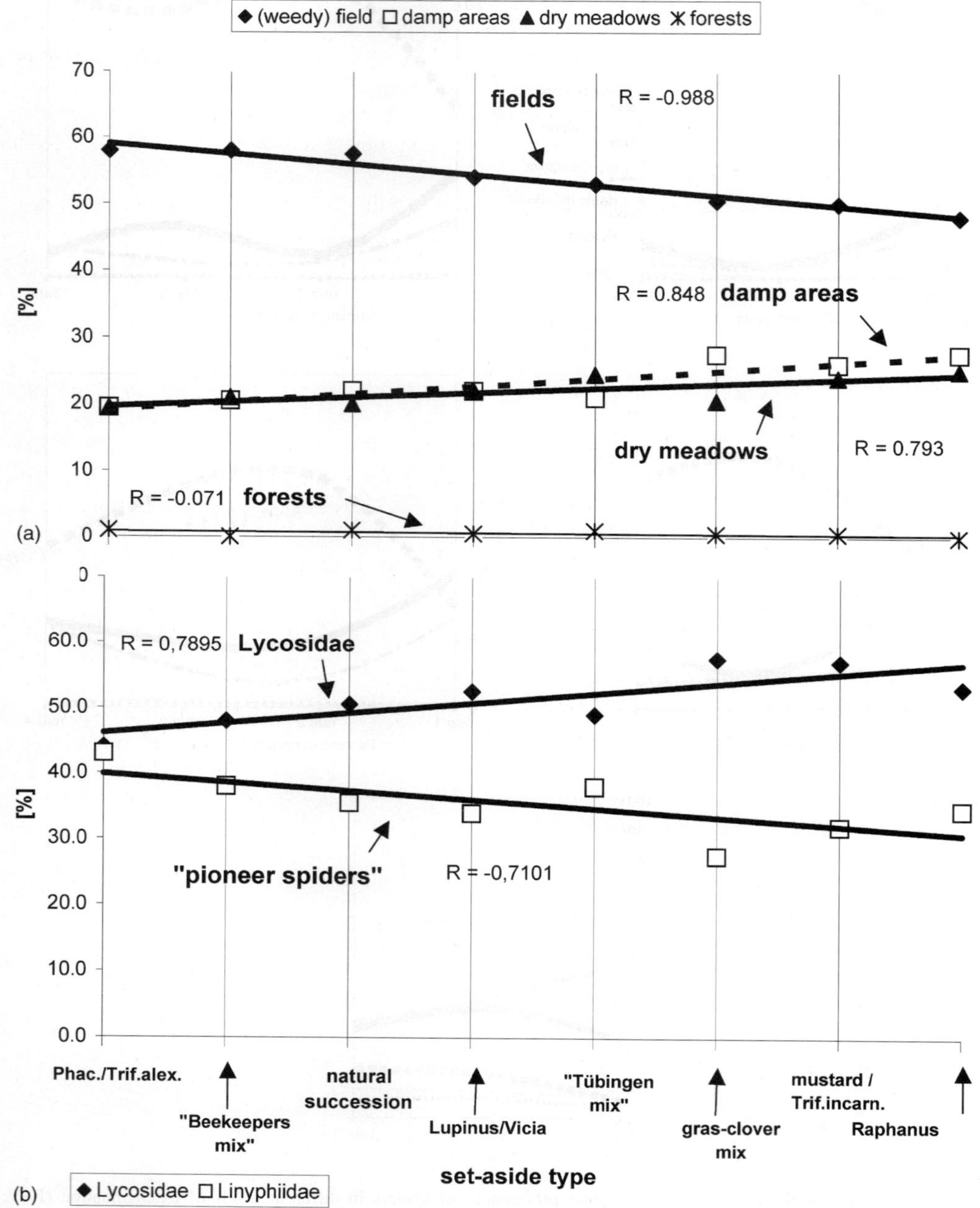

Fig. 9. Percentage of spiders with preference for (weedy) field habitats and dry or damp habitats (a) and of Linyphiidae and wolf spiders (Araneae: Lycosidae) (b) recorded in different rotational set-aside types.

low significance with regards to the power of statements.

- With regards to the species number "standard reference values" or "baselines" can be elaborated for different habitats both arithmetically (species area relation; e.g. Duelli and Obrist, 2003) and empirically (check list on the maximum number of habitat typical species that potentially can be recorded; e.g. Nickel and Hildebrandt, 2003); however, this is not possible for H_S indices, because additionally the dominance position of each species has to be considered.
- The H_S index values are not unambiguous as it could be shown exemplarity for rove beetles (Coleoptera: Staphylinidae): i.e. the H_S index is an arithmetic unit, that informs about the absolute shaping of the dominance structure, but gives alone no information, if dominance positions between species were changed.
- The input of labour to record a sufficient database to calculate the H_S index is rather high the records of individual numbers (abundances/activity density) have to be included, while a pure species survey can be done comparatively fast and exact by experienced taxonomists.
- Both species number and H_S index depend to a high extent from the experience of the researcher, the variety of methods used, the number of replicates regarding traps or samples, etc. That means, before using H_S as a value for biodiversity indication the assessment method should be standardised.

It can be concluded that an exclusive or prior-ranking assessment of biodiversity as richness of species, cultivars, genotypes, etc. by surrogates according to popular assessment criteria (e.g. "ecological priority areas"; Roth and Schwabe, 2003) does not include the major problem that we face in assessing the agricultural landscape today. Particularly for the biotic assessment of the *cultivated* areas criteria are demanded which are not based only on a pure maximising of "biodiversity", but include more structural and functional qualities of the biocoenosis according to the hierarchic components (structure, function and composition) defined by Noss (1990).

As demonstrated in Table 5 and by the examples described, there are numerous possibilities to replace the rather ill-defined and weak criterion "biodiversity" by syn- and auto-ecological parameters arising from community structure and physical fitness. These parameters enable us to implement a very sensitive gradual measurement and assessment of the biocoenotical effects of farming systems with different levels of extensification or management intensity.

Therefore, evaluation problems consist not of the limitations of the assessment of (conventionally) managed fields, but in the practical translation into action and application of the procedures within current farming practices as well as in the control of the success of the measurements that were carried out.

Acknowledgements

We are very thankful to Dr. Keith Walters (CSL, York, UK) substantial and useful comments on the manuscript and the improvement of the English. For technical assistance, we thank gratefully Ruth Polok, Elisabeth Päs, Charlotte Winkler, Johannes Siegert, Christian Kerl, Jürgen Liersch, Sylvia Urban; Mr. Riedel, Mr. Mesecke and any other persons who have been involved in some projects mentioned in the text. Many thanks also to Wolfgang Hesse, University of Göttingen, for his care for our field trials at Eickhorst and their consideration during current husbandry activities. All projects are BBA-projects except the INTEX-project, which were coordinated by the Research Centre for Agriculture and the Environment of the University of Göttingen.

References

Barber, H.S., 1931. Traps for cave inhabiting insects. J. Elish. Mitchell Sci. Soc. 46, 259–266.

Bartels, G., Kampmann, T., 1994. Auswirkungen eines langjährigen Einsatzes von Pflanzenschutzmitteln bei unterschiedlichen Intensitätsstufen und Entwicklung von Bewertungskriterien. Mitteilungen aus der Biologischen Bundesanstalt für Land- und Forstwirtschaft Berlin-Dahlem 295, 1–405.

Basedow, T., 1987. Der Einfluß gesteigerter Bewirtschaftungsintensität im Getreidebau auf die Laufkäfer (Col., Carabidae). Auswertung vierzehnjähriger Untersuchungen (1971–1984). Mitteilungen aus der Biologischen Bundesanstalt für Land- und Forstwirtschaft Berlin-Dahlem 235, 1–123.

Basedow, T., 1998. Langfristige Bestandveränderungen von Arthropoden in der Feldflur, ihre Ursachen und deren Bedeutung für den Naturschutz, gezeigt an Laufkäfern

(Carabidae) in Schleswig-Holstein, 1971–1996. Schriftenreihe Landschaftspflege und Naturschutz 58, 215–227.

Basedow, T., 2003. Befunde zur Bedeutung und Nachhaltigkeit des Ökologischen Landbaues aus zoologischer und ackerbaulicher Sicht. Peckiana 1, 1–7.

Blake, S., Foster, G.N., Eyre, M.D., Luff, M.L., 1991. Effects of habitat type and grassland management practices on the body size distribution of carabid beetles. Pedobiologia 38, 502–512.

Braband, D., Geier, U., Köpke, U., 2003. Bio-resource evaluation within agri-environmental assessment tools in different European countries. In: Büchs, W. (Ed.), Biotic Indicators for Biodiversity and Sustainable Agriculture. Agric. Ecosyst. Environ. 98, 423–434.

Büchs, W., 1991. Ergebnisse der Fang-Wiederfangmethode bei *Pterostichus* spp. mit einer Anmerkung zur erforderlichen Größe von Versuchsparzellen. D.G.a.a.E. Nachrichten 5 (2), 38.

Büchs, W., 1994. Effects of different input of pesticides and fertilizers on the abundance of arthropods in a sugar beet crop: an example for a long-term risk assessment in the field. In: Donker, M.H., Eijsackers, H., Heimbach, F. (Eds.), Ecotoxicology of Soil Organisms. Lewis Publishers, Boca Raton, FL, pp. 303–321.

Büchs, W., 2003. Biodiversity and agri-environmental indicators—general scopes and skills with special reference to the habitat level. In: Büchs, W. (Ed.), Biotic Indicators for Biodiversity and Sustainable Agriculture. Agric. Ecosyst. Environ. 98, 35–78.

Büchs, W., Nuss, H., 2000. First steps to assess the importance of epigaeic active polyphagous predators as natural enemies of oilseed rape insect pests with soil pupating larvae. IOBC/OILB wprs/srop Bull. 23 (6), 151–163.

Büchs, W., Harenberg, A., Zimmermann, J., 1997. The invertebrate ecology of farmland as a mirror of the intensity of the impact of man? An approach to interpreting results of field experiments carried out in different crop management intensities of a sugar beet and an oil seed rape rotation including set-aside. Biol. Agric. Horticult. 15 (1–4), 83–107.

Büchs, W., Harenberg, A., Prescher, S., Weber, G., Hattwig, F., 1999. Entwicklung von Evertebratenzönosen bei verschiedenen Formen der Flächenstillegung und Extensivierung. In: Büchs, W. (Ed.), Nicht bewirtschaftete Areale in der Agrarlandschaft—ihre Funktionen und Interaktionen mit landnutzungsorientierten Ökosystemen. Mitteilungen aus der Biologischen Bundesanstalt für Land-und Forstwirtschaft Berlin-Dahlem 368, 9–38.

Döring, T.F., Kromp, B., 2003. Which carabid species benefit from organic agriculture? A review of comparative studies in winter cereals from Germany and Switzerland. In: Büchs, W. (Ed.), Biotic Indicators for Biodiversity and Sustainable Agriculture. Agric. Ecosyst. Environ. 98, 153–161.

Döring, T.F., Hiller, A., Wehke, S., Schulte, G., Broll, G., 2003. Biotic indicators of carabid species richness on organically and conventionally managed arable fields. In: Büchs, W. (Ed.), Biotic Indicators for Biodiversity and Sustainable Agriculture. Agric. Ecosyst. Environ. 98, 133–139.

Duelli, P., 1992. Mosaikkonzept und Inseltheorie in der Kulturlandschaft. Verhandlungen der Gesellschaft für Ökologie 21, 379–394.

Duelli, P., Obrist, M.K., 2003. Biodiversity indicators: the choice of values and measures. In: Büchs, W. (Ed.), Biotic Indicators for Biodiversity and Sustainable Agriculture. Agric. Ecosyst. Environ. 98, 87–98.

Ebing, W., Bartels, G., Büchs, W., Eggers, T., Gottesbüren, B., Hansen, H., Heimann-Detlefsen, D., Heimbach, U., Kampmann, T., Knüsting, E., Köllner, V., Kreuzig, G., Leliweldt, B., Malkomes, H.-P., Metzler, B., Nirenberg, H.I., Oesterreicher, W., Pestemer, W., Pohl, K., Sauthoff, W., Sturhan, D., 1991. Long-term loading of field areas by pesticides during agricultural practices and impacts on soil biocoenosis. In: Proceedings of the International East–West Symposium on Contaminated Areas in Eastern Europe of the International Society of Ecotoxicology and Environmental Safety (SECOTOX), Gosen/Berlin, November 25–27, 1991, pp. 11–14.

Engelmann, H.-D., 1978. Zur Dominanzklassifizierung von Bodenarthropoden. Pedobiologia 18, 378–380.

Franzen, J., Weber, G., Büchs, W., Larink, O., 1997. Langzeiteinfluß von Pflanzenschutzmitteln auf Dipteren mit bodenlebenden Entwicklungsstadien. Berichte über Landwirtschaft 75, 291–328.

Freude, H., Harde, K.W., Lohse, G.A., 1964. Die Käfer Mitteleuropas, Vol. 1. Goecke & Evers, Krefeld.

Frieben, B., 1998. Verfahren zur Bestandsaufnahme und Bewertung von Betrieben des Organischen Landbaus im Hinblick auf Biotop- und Artenschutz und die Stabilisierung des Agrarökosystems. Schriftenreihe Institut für Organischen Landbau 11, Berlin.

Hammer, D., 1984. Synökologische Untersuchungen über die Spinnenpopulationen (Araneae) von Weinbergsflächen bei Marienthal/Ahr. Ph.D. Thesis. University of Bonn.

Harenberg, A., 1997. Auswirkungen abgestuft intensiv geführter Anbausysteme in verschiedenen Fruchtfolgen (Raps-, Zuckerrübenfruchtfolge) und einer selbstbegrünenden Dauerbrache auf Spinnen (Arachnida: Araneae). Ph.D. Thesis. Technical University of Braunschweig, 276 pp.

Hokkanen, H., Holopainen, J.K., 1986. Carabid species and activity densities in biologically and conventionally managed cabbage fields. J. Appl. Entomol. 102, 353–363.

Kantelhardt, J., Osinski, E., Heissenhuber, A., 2003. Is there a reliable correlation between hedgerow density with agricultural site conditions? In: Büchs, W. (Ed.), Biotic Indicators for Biodiversity and Sustainable Agriculture. Agric. Ecosyst. Environ. 98, 517–527.

Kleinhenz, A., Büchs, W., 1995. Ökologische Aspekte der Spinnenzönose von Zuckerrübenflächen unter dem Einfluß eines unterschiedlich intensiven Pflanzenschutz- und Düngemitteleinsatzes. Mitteilungen der Deutschen Gesellschaft für allgemeine und angewandte Entomologie 9, 481–489.

Köhler, F., Stumpf, T., 1992. Die Käfer der Wahner Heide in der Niederrheinischen Bucht bei Köln (Insecta: Coleoptera). Decheniana-Beihefte 31, 499–593.

Langmaack, M., Land, S., Büchs, W., 2001. Effects of different field management systems on the carabid coenosis in oil seed rape with special respect to ecology and nutritional status of predacious *Poecilus cupreus* L. (Col., Carabidae). J. Appl. Entomol. 125, 313–320.

Mühlenberg, M., 1993. Freilandökologie, 3rd ed. UTB, Heidelberg, 510 pp.

Nickel, H., Hildebrandt, J., 2003. Auchenorrhyncha communities as indicators of disturbance in grasslands (Insecta, Hemiptera)—a case study from the Elbe flood plains (Northern Germany). In: Büchs, W. (Ed.), Biotic Indicators for Biodiversity and Sustainable Agriculture. Agric. Ecosyst. Environ. 98, 183–199.

Noss, R.F., 1990. Indicators for monitoring biodiversity: a hierarchical approach. Conserv. Biol. 4, 355–364.

Otte, A., 1990. Die Entwicklung von Ackerwildkraut-Gesellschaften nach dem Aussetzen von Regulierungsmaßnahmen. Phytocoenologia 19, 43–90.

Prescher, S., Büchs, W., 1999. Struktur und Dynamik der Fliegenzönosen (Diptera, Brachycera) abgestuft extensiv bewirtschafteter Flächen in einer Rapsfruchtfolge. Verhandlungen der Gesellschaft für Ökologie 29, 265–269.

Raskin, R., 1994. Das Ackerrandstreifenprogramm: tierökologische und agrarökonomische Aspekte. In: Stiftung Naturschutz Hamburg, Stiftung zum Schutze gefährdeter Pflanzen (Eds.), Flora und Fauna der Äcker und Weinberge, Schriftenreihe Aus Liebe zur Natur 5, 150–157.

Röser, B., 1988. Saum-und Kleinbiotope—Ökologische Funktion, wirtschaftliche Bedeutung und Schutzwürdigkeit in Agrarlandschaften. Ecomed-Verlag, Landsberg.

Roth, D., Schwabe, M., 2003. Method for assessing the proportion of ecologically, culturally and provincially significant areas ("OELF") in agrarian spaces used as a criterion of environmentally friendly agriculture. In: Büchs, W. (Ed.), Biotic Indicators for Biodiversity and Sustainable Agriculture. Agric. Ecosyst. Environ. 98, 435–441.

Shannon, C.E., 1948. A mathematical theory of communication. Bell. Syst. Technol. J. 27, 379–423, 623–656.

Shannon, C.E., Weaver, W., 1963. The Mathematical Theory of Communication. University of Illinois Press, Urbana, IL, 117 pp.

Steidl, I., Ringler, A., 1997. Agrotope (Teilband)—Landschaftspflegekonzept Bayern. In: Bayerisches Staatsministeium für Landesentwicklung und Umweltfragen, Bayerische Akademie für Naturschutz und Landschaftspflege, Vol. II.11. München, Germany.

Steinborn, H.-A., Heydemann, B., 1990. Indikatoren und Kriterien zur Beurteilung der ökologischen Qualität von Agrarflächen am Beispiel der Carabidae (Laufkäfer). Schriftenreihe für Landschaftspflege und Naturschutz 32, 165–174.

Steinmann, H.-H., Forstreuther, C., Heitefuss, R., 1997. Auswirkungen von Extensivierungsmaßnahmen auf die Verunkrautung und deren Regulierung. In: Gerowitt, B., Wildenhayn, M. (Eds.), Ökologische und ökonomische Auswirkungen von Extensivierungsmaßnahmen im Ackerbau—Ergebnisse des Göttinger INTEX-Projektes 1990–1994. Forschungs-und Studienzentrum Landwirtschaft und Umwelt Universität Göttingen, pp. 127–154.

Stippich, G., Krooß, S., 1997. Auswirkungen von Extensivierungsmaßnahmen auf Spinnen, Laufkäfer und Kurzflügelkäfer. In: Gerowitt, B., Wildenhayn, M. (Eds.), Ökologische und ökonomische Auswirkungen von Extensivierungsmaßnahmen im Ackerbau—Ergebnisse des Göttinger INTEX-Projektes 1990–1994. Forschungs-und Studienzentrum Landwirtschaft und Umwelt Universität Göttingen, pp. 221–262.

Stugren, B., 1978. Grundlagen der Allgemeinen Ökologie. Fischer-Verlag, Stuttgart.

Teiwes, K., 1997. Standortbedingungen. In: Gerowitt, B., Wildenhayn, M. (Eds.), Ökologische und Ökonomische Auswirkungen von Extensivierungsmaßnahmen im Ackerbau. Ergebnisse des Göttinger INTEX-Projektes 1990–1994. Forschungs- und Studienzentrum Landwirtschaft und Umwelt Universität Göttingen, pp. 25–34.

Thiele, H.U., 1977. Carabid Beetles in their Environments. Springer, Berlin.

Weber, G., Büchs, W., 2000. Der Einfluß unterschiedlicher Rotationsbrachetypen auf landwirtschaftlich relevante Diptera. Mitteilungen der Deutschen Gesellschaft für allgemeine und angewandte Entomologie 12, 419–424.

Weiß, B., Büchs, W., 1997. Auswirkungen verschiedener Rotationsbrachetypen auf Struktur und Dynamik der Spinnenzönosen. Mitteilungen der Deutschen Gesellschaft für allgemeine und angewandte Entomologie 11, 147–151.

Weiß, B., Büchs, W., Harenberg, A., 1997. Entwicklung der Spinnenfauna nach Umbruch einer sechsjährigen selbstbegrünenden Dauerbrache. Verhandlungen der Gesellschaft für Ökologie 27, 379–384.

Wildenhayn, M., Gerowitt, B., 1997. Anbausysteme. In: Gerowitt, B., Wildenhyn, M. (Eds.), Ökologische und ökonomische Auswirkungen von Extensivierungsmaßnahmen im Ackerbau. Ergebnisse des Göttinger INTEX-Projektes 1990–1994. Forschungs- und Studienzentrum Landwirtschaft und Umwelt Universität Göttingen, pp. 13–23.

Zangger, A., Lys, J.-A., Nentwig, W., 1994. Increasing the availability of food and the reproduction of *Poecilus cupreus* in a cereal field by strip-management. Entomol. Exp. Apllicata 71, 111–120.

Zimmermann, J., Büchs, W., 1996. Management of arable crops and its effects on rove-beetles (Coleoptera: Staphylinidae), with special reference to the effects of insecticide treatments. Acta Jutlandica 71 (2), 183–194.

Zimmermann, J., Büchs, W., 1999. Kurzflügelkäfer (Coleoptera: Staphylinidae) in unterschiedlich intensiv bewirtschafteten Ackerflächen—Biologisch-ökologische Untersuchungen zur Interpretation ökotoxikologischer Freilanduntersuchungen. Agrarökologie 32, 1–154.

ELSEVIER

Available online at www.sciencedirect.com

Agriculture, Ecosystems and Environment 98 (2003) 125–132

Agriculture Ecosystems & Environment

www.elsevier.com/locate/agee

Sample size and quality of indication—a case study using ground-dwelling arthropods as indicators in agricultural ecosystems

Jörg Perner*

Friedrich-Schiller-University, Institute of Ecology, Dornburger Str. 159, D-07743 Jena, Germany

Abstract

A frequently cited but rarely used procedure for estimating sample size was tested on the level of both species and community parameters. The analysed field data resulted from a pitfall trap study dealing with the effects of changing agricultural land use on ground-dwelling arthropods. The relation between level of precision (LOP) and required sample size (RSS) showed significant differences between species and community parameters. For most species parameters a high precision level (5–10%) would require an unaffordable large sample size. For species that are habitat generalists realistic sample sizes around 10 allow an LOP between 25 and 50%, whereas the same sample sizes leads to a much lower LOP between 50 and 90% for selected habitat specialists. In general, the LOP–RSS relation was more favourable for the two tested community parameters (richness and evenness) than for the mean number of individuals of a species. For purposes of bioindication or comparisons of communities, the results presented herein favour the use of community parameters above species parameters. Finally, as a rule of thumb the RSSs for precision levels of 5, 10, 25 and 50% were given for both, level of species and for community parameters.

Keywords: Sample size; Level of precision; Coefficient of variation; Bioindication; Arthropods

1. Introduction

In agricultural ecology, invertebrate groups have been increasingly used as indicators for changes in agricultural land use (Duelli et al., 1999). Indicative deductions are possible on the basis of indices for (indicator-) species or for complex community parameters (e.g., measures of diversity). In all cases, successful indication requires that, e.g., differences at species- or community-level between various management practices (between group variability) are recognizable beyond the data noise within single management types (within group variability). Therefore, the precision of the sampled data has to be estimated, which is directly related to the variance of the data and thus, is a mathematical function of sample size (Tokeshi, 1993).

It is well known, that increasing sample size improves the validity and reliability of the data in question. However, personnel, temporal and financial limits apply for every study. To find a balance between these needs is particularly important for investigations of invertebrate groups that involve time-consuming field sampling, high skills and much time for species identification as well as a large amount of data processing.

A frequently cited (e.g., Manly, 1992; Motulsky, 1995; Krebs, 1999) but rarely used procedure for estimating sample size was tested on the level of both

* Tel.: +49-3641-949421; fax: +49-3641-949402.
E-mail address: bjp@uni-jena.de (J. Perner).

0167-8809/$ – see front matter © 2003 Elsevier Science B.V. All rights reserved.
doi:10.1016/S0167-8809(03)00074-4

species and community parameters in order to investigate the relationship between sample size and precision. In addition, attempts to offer suggestions for a cost-benefit balance between financial constraints and precision needs.

2. Material and methods

2.1. Statistical procedure and terminology

The precision of an arithmetical mean of a sample is usually described by the confidence interval:

$$\left(\bar{x} - t_{n-1,\alpha}\frac{s}{\sqrt{n}}, \bar{x} + t_{n-1,\alpha}\frac{s}{\sqrt{n}}\right) \quad (1)$$

where $\bar{x}$ is the arithmetical mean of the sample, s the standard deviation of the sample, $t_{n-1,\alpha}$ the student's t value for $n-1$ degrees of freedom for $1-\alpha$ level of confidence, n the sample size (for this study: the number of pitfall traps).

The half-width of this confidence interval, $d = t_{n-1,\alpha}s/\sqrt{n}$, gives an absolute measure of precision. More conveniently, the precision is expressed as a percentage proportion of the parameter in question (here: $\bar{x}$)

$$r = t_{n-1,\alpha}\frac{s}{\bar{x}\sqrt{n}}100\% \quad (2)$$

and gives a relative measure of precision. This measure is directly linked to a common measure of relative variability, the coefficient of variation $CV = s/\bar{x}$. Assuming a 95% confidence level and exploiting the fact that $t_{n-1,\alpha=5\%} \approx 2$ the formula for relative precision can be rearranged, giving

$$n \approx \left(\frac{200CV}{r}\right)^2 \quad (3)$$

as an approximate formula for the sample size required for a specified level r of relative precision. Note that *low* %-values correspond to *high* levels of relative precision. A detailed derivation of the formula for estimating necessary sample sizes is given by Manly (1992) and Krebs (1999). The required sample size (RSS) is the one needed to achieve a certain level of precision (LOP).

2.2. Study sites and analysed data

The study was carried out between 1996 and 1998 in the central Unstrut floodplain (Thuringia, Germany;

Table 1
Numbers of individuals for selected species of beetles sampled in seven different sites from 1996 to 1998 (group 1: habitat generalists of agricultural ecosystems, group 2: habitat specialists with a high preference for arable land and particularly young sown grassland sites, group 3: habitat specialists with high preference for re-wetted grassland sites)

	Site type (site abbreviations)							Total
	Arable land (A3)	Young sown grassland (A2)	Sown grassland (A1)	Permanent grassland (3-cut) (C)	Permanent grassland (1-cut) (B2)	Embankment grassland (AD)	Re-wetted grassland (fallow) (B1)	
Group 1								
P. melanarius	8802	9018	1328	4296	107	1753	1429	22437
B. obtusum	981	407	131	1860	51	101	359	2030
Group 2								
Calathus erratus	132	38	0	0	0	0	0	170
Hypera postica	26	33	1	0	0	3	0	63
Lesteva longoelytrata	54	3	0	1	0	0	0	57
Ocypus ophthalmicus	163	5	1	0	0	0	0	169
Group 3								
Bembidion lunulatum	0	0	0	0	0	3	17	20
Chlaenius nigricornis	0	0	0	0	0	7	149	156
O. helopioides	0	0	0	0	0	1	25	26
Stenus pusillus	0	0	0	0	0	1	14	15

51°08′N, 10°40′E). Seven sites of different management regimes (arable land, young sown grassland used by mowing and grazing, sown grassland with extensive cattle grazing, permanent grassland mowing 3-cuts per year, permanent grassland mowed with 1-cut per year, embankment grassland, re-wetted fallow land) were sampled by pitfall trapping. Two lines 20 m apart from each other, each consisting of five traps (4 m apart) were sampled the year around at every site. In total, these 70 traps yielded 502 species of beetles with 77,684 individuals and 120 species of spiders with 56,067 adult individuals (Malt and Perner, 2002; Perner and Malt, 2002).

Eq. (3) was used to estimate the LOP–RSS relation on species level based on numbers of individuals for selected species of beetles. Using a TWINSPAN-analysis (two-way-indicator-species-analysis; compare Hill, 1979; McCune and Mefford, 1997), three ecological groups of species were selected (Table 1). The first group includes habitat generalists, which are widespread in agricultural ecosystems and occurred in all investigated sampling sites. The second and third group was represented by habitat specialists. Species of group 2 had a high preference for arable land and partly also for young sown grassland sites (*Hypera postica*), whereas the species of group 3 were limited to re-wetted grassland and embankment grassland sites. Species of group 1 can be regarded as quantitative indicators with a fuzzy preference for certain environmental conditions (e.g., habitats). However, an

Table 2
Mean CV ±S.E. for beetles, carabids only and spiders calculated for species richness, Smith–Wilson evenness and mean numbers of individuals (see also Table 1)[a]

Taxon/parameter	Mean CV ±S.E.		
	Between sites	Between years	Total
Beetles/richness	0.143 ± 0.035	0.146 ± 0.009	0.146 ± 0.039
Beetles/Smith–Wilson evenness	0.114 ± 0.029	0.115 ± 0.022	0.115 ± 0.038
Carbids/richness	0.196 ± 0.076	0.192 ± 0.028	0.193 ± 0.083
Carabids/Smith–Wilson evenness	0.178 ± 0.071	0.183 ± 0.033	0.182 ± 0.092
Spiders/richness	0.128 ± 0.024	0.133 ± 0.026	0.132 ± 0.032
Spiders/Smith–Wilson evenness	0.105 ± 0.028	0.107 ± 0.021	0.106 ± 0.035
Beetles species/number of individuals			
Group 1			
P. melanarius	0.433 ± 0.150	0.398 ± 0.076	0.411 ± 0.172
B. obtusum	0.832 ± 0.467	0.731 ± 0.130	0.768 ± 0.478
Group 2			
Calathus erratus		1.473 ± 0.713	1.473 ± 0.713
Hypera postica[b]	0.848 ± 0.770	0.848 ± 0.199	0.848 ± 0.718
Lesteva longoelytrata		1.833 ± 1.155	1.833 ± 1.155
Ocypus ophthalmicus		1.211 ± 0.780	1.211 ± 0.780
Group 3			
Bembidion lunulatum		1.256 ± 0.307	1.256 ± 0.307
Chlaenius nigricornis		1.118 ± 0.101	1.118 ± 0.101
O. helopioides		0.943 ± 0.898	0.943 ± 0.898
Stenus pusillus		1.329 ± 0.066	1.329 ± 0.066
Mean			1.119

[a] For both community parameters and the species of group 1 the calculation of the mean CVs were performed regarding variation of the CVs between the seven sampling sites, between the 3 years and between all sites and years (total). For the species of groups 2 and 3 only the CVs of the preferred sites (A3, respectively, B1; see Table 1) were analysed. Therefore only the variation of the CVs between the 3 years within the preferred sites was calculated. Please note, that for this reason the values of those groups in columns "years" and "total" are identical.

[b] Because of a comparable number of individuals sampled in A3 and A2 (see Table 1) also a mean CV (between A3 and A2) were calculated for *H. postica*.

indication is possible if based on density changes for these species. The species of groups 2 and 3 belong to the category of qualitative indicators (indicator species in the narrower sense) and show a strong linkage to certain environmental conditions (see Müller-Motzfeld, 1989; McGeoch, 1998 for reviews).

Beside the numbers of individuals for the three selected species groups, two community parameters were analysed: the species richness and the Smith–Wilson evenness index (E_{var}). The latter index is independent of species richness and is sensitive to both rare and common species in the community (Krebs, 1999). The analysis was carried out separately for (i) all ground-dwelling beetles, (ii) for the carabids only, and (iii) the ground-dwelling spiders. The comparison of (i) and (ii) allows to validate the almost exclusive use of carabids in bioindication studies (e.g., Müller-Motzfeld, 1989) whereas (i) and (iii) compares the two most abundant taxa of ground-dwelling arthropods.

The coefficient of variation (CV) was calculated for the community parameters and the numbers of individuals recorded for the 10 traps per site and year separately. For both community parameters and the species of group 1 the calculation of the mean CVs were performed regarding variation of the CVs between the seven sampling sites (pooling CVs of different years per site), between the 3 years (pooling CVs of different sites per year) and between all sites and years (Table 2). For the species of groups 2 and 3 only the CVs of the preferred sites (A3, respectively, B1; see Table 1) were analysed. Therefore only the variation of the CVs between the 3 years within the preferred sites was calculated. Please note, that for this reason the values of those species groups in columns "years" and "total" are identical (Table 2).

In a second step, the mean CVs listed in column "total" of Table 2 were included in Eq. (3) to estimate the LOP–RSS relations for the selected species and the both community parameters.

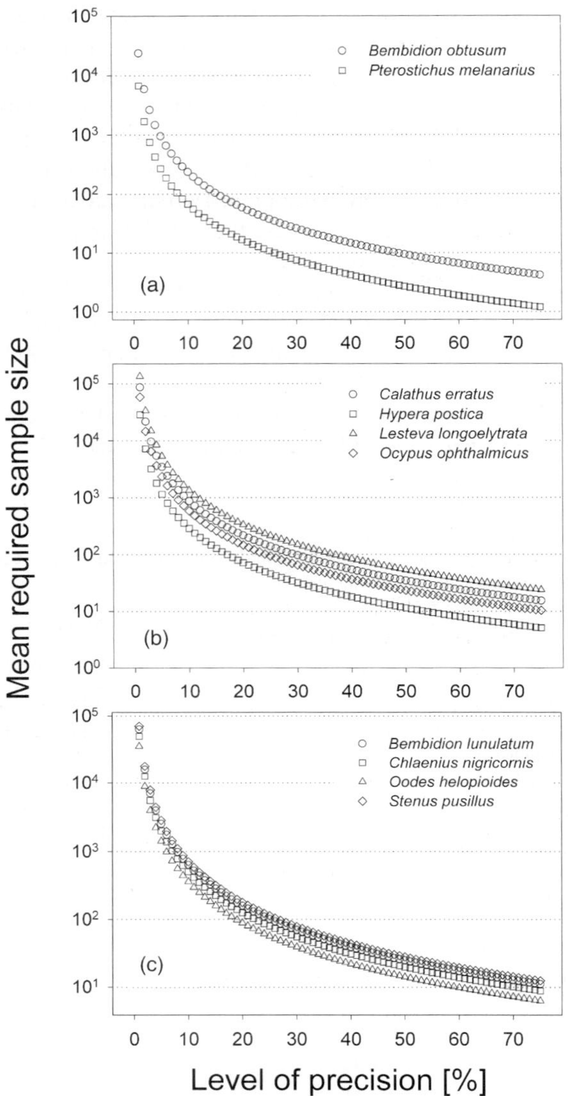

Fig. 1. Relation between LOP and RSS for ecological groups of beetles (a—group 1, habitat generalists; b—group 2, habitat specialists with a high preference for arable land; c—group 3, habitat specialists with high preference for re-wetted grassland).

3. Results

3.1. Species comparisons

With lower LOP, the RSS needed declines exponentially for all three ecological groups (Fig. 1; Table 1). There were considerable differences between the LOP–RSS relations for habitat generalists (Fig. 1a) and specialists (Fig. 1b and c). Whereas sample sizes of about 10 gave an LOP between 25 (*Pterostichus melanarius*) and 50% (*Bembidion obtusum*) for habitat generalists, the same sample size would yield only

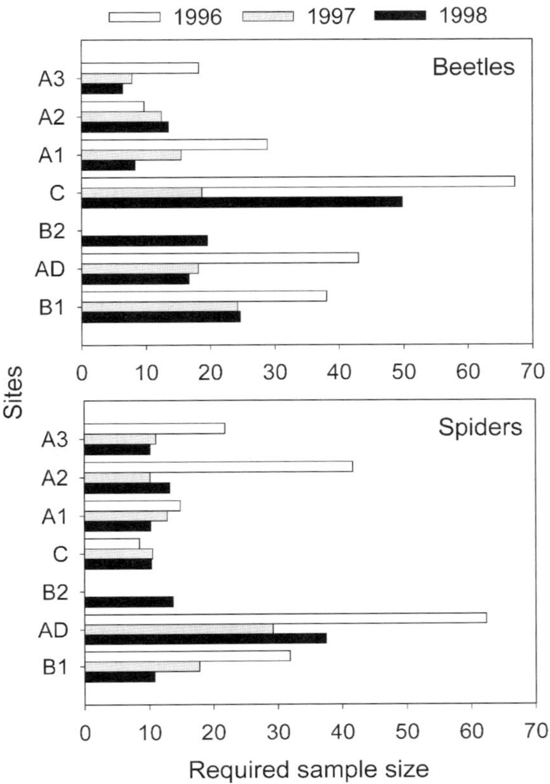

Fig. 2. Calculations for RSSs for a precision level (LOP) of 5% for beetles and spiders based on the CV caused by the Smith–Wilson evenness. Shown are the values for each of the 3 years of investigation and for each sample site.

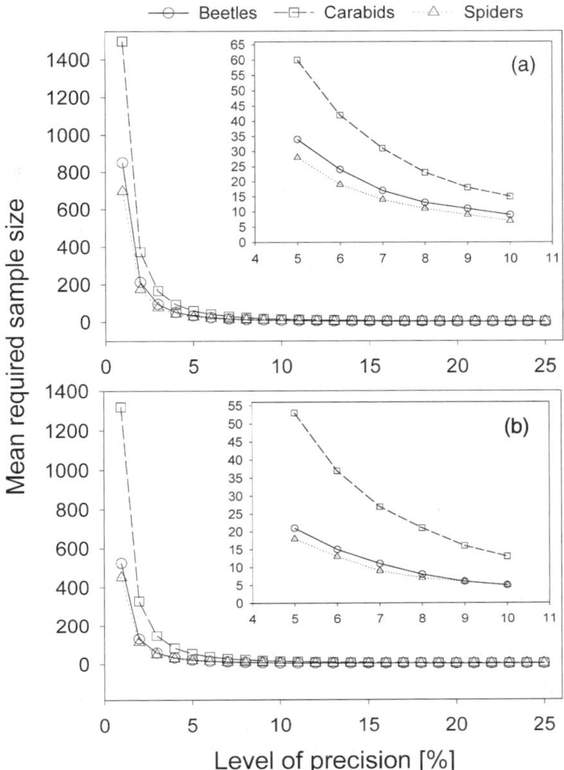

Fig. 3. Relation between LOP and mean RSS for beetles, carabids only and spiders on the basis of species richness (a) and Smith–Wilson evenness (b). Inset: enlarged plots for a range of the LOP between 5 and 10%.

an LOP between 60 and 100% for habitat specialists. On average, CV-values are very high (Table 2; mean: 1.119).

3.2. Community parameters

CV values for the two tested community parameters (richness and evenness) were distinctly smaller than those for mean numbers of individuals (Table 2). For an LOP of 5%, the RSS for both richness and evenness varies considerably within sampling sites, sampling years and taxonomic groups (Fig. 2). Despite this spatial and temporal variability, values for RSS were most often between 15 and 25 for an LOP of 5%. For all three taxonomic groups, RSS increases exponentially with LOP for both richness and evenness (Fig. 3). Whereas the plots for beetles and spider are nearly identical, carabids alone need a considerably higher RSS to reach the same LOP. In general, higher sample sizes were required for assessments on the basis of species richness than on the basis of evenness values.

4. Discussion

A procedure was tested to describe the LOP–RSS relation for abundances of selected (indicator) species and for two community parameters. The calculation used (Eq. (3)) allows to estimate the RSS necessary to obtain a specified LOP for one (!) mean of a parameter (here: numbers of individuals sampled, richness or evenness) or to assess the quality of sampled data for a given sample size. The data analysed in this study are the result of a 3-year study of sites subjected to different management regimes.

Strictly speaking, for such cases a more complicated calculation procedure should be applied to compare two or more means of a parameter (Motulsky, 1995). Beside the estimated standard deviation (s) and the minimum difference (d) between means of the parameter in question, the power index $(z_\alpha + z_\beta)^2$ has to be considered for this formula, with α as the probability of a type I error and β as the probability of a type II error (compare Krebs, 1999). However, from two reasons this procedure is hard to apply for field data. First, the procedure aims on the RSS necessary to detect a priori significant differences (with a given probability) between two means of a parameter (and to test its significance, e.g., by a Student's test). In studies like the one presented here it is not the priority to calculate the sample size that is needed to find significant differences at any cost, but to perform a cost (RSS)–benefit (LOP) analysis. To plan field studies like the one shown in this paper it is most important to have a rough idea about the quality of results (means of parameters) for a given sample size or to estimate the RSS needed for a given LOP. Second, the procedure explained above assumes that the two data sets have nearly equal standard deviations for the parameter in question, a condition hardly fulfilled for most field studies. As shown in Fig. 2, in most cases the temporal variability of the CVs within sites was higher than the spatial variability between sites. Hence it was regarded as sufficient to use Eq. (3) knowing that this procedure is not a precise calculation but a rough estimation of the LOP–RSS relation.

The results show significant differences in the LOP–RSS relations for species and community parameters. For mean numbers of individuals, affordable sample sizes between 10 and 30 cause low values of LOP, whereas for community parameters the same sample sizes would give clearly higher LOP values. For instance, if at a sample size of 10 five individuals of *Oodes helopioides* were caught at site B1, a mean LOP of 60% has to be accepted (see Fig. 1). This translates to a mean value of 5 ± 3 individuals. In contrast, the mean evenness for site B1 is 0.46, and the same sample size produces an LOP of about 7%, resulting in a much more precise evenness value of 0.46 ± 0.03.

Furthermore, the procedure presented herein can be helpful for the selection of trustworthy indicator species. Beside a high degree of habitat preference (or preference to specific conditions) a good indicator species should have a low CV for the mean number of individuals per site to allow an LOP as high as possible. The demonstrated procedure could function as a tool to select the best candidates from a pool of potential indicator species and would therefore complement the well-established TWINSPAN-analysis (Hill, 1979) or the newer and more efficient Indicator Value method after Dufrene and Legendre (1997). Occasionally, generalists of agricultural habitats (quantitative indicators; see above) were used as management intensity indicators (Büchs et al., 1999). In this study, the two generalists *P. melanarius* and *Bembidion obtusum* have indeed lower CV values than the habitat specialists and therefore a more favourable LOP–RSS relation (Fig. 1). This underlines the importance of generalists as indicator species (e.g., in the context of changes in land use) due to their better pattern of variance.

For the two analysed community parameters, three points seem to be worth of mention. First, species richness possesses a more unfavourable LOP–RSS relation than the Smith–Wilson evenness (Fig. 3). This means, e.g., that at a given sample size of 10, the LOP for mean evenness of beetles and spiders was as low as 7%, whereas the LOP for the species richness of the same groups was about 9%. This demonstrates the usefulness of this parameter for purposes of indication. As a detailed analysis of the data set demonstrated (Perner and Malt, 2003), the differences of evenness between the sites were precise enough to allow the discrimination of four sufficiently homogeneous subsets of sites. Currently the relation between productivity and evenness of a community (as one component of diversity) is under intensive (and controversial) discussion (e.g., Drobner et al., 1998; Wilsey and Potvin, 2000). In this context, a trustworthy estimation of the evenness parameter seems to be of particular importance.

Second, Fig. 3 shows approximately the same LOP–RSS relations for beetles and spiders, which makes both groups equally suitable as indicators when regarding their patterns of variance. Third, the more unfavourable LOP–RSS relation for carabids alone in comparison to beetles as a whole or spiders was unexpected (for the same LOP higher sample sizes needed!). In the light of these results, a sample size of about 15 would be necessary to obtain a mean species richness or evenness estimation with a precision of 10% if solely carabids would be used for indication. Consequently, sample sizes of about 5, as often used in practice for carabids, should be carefully evaluated.

5. Conclusions

The procedure tested herein allows a rough estimation of the LOP–RSS relation and is therefore a suitable tool to assess the quality of sampled data for a given sample size or to estimate the RSS for a given confidence level (LOP), e.g., in the context of preliminary investigations. The equation used is applicable for abundances of indicator species only as well as for community parameters. It could be demonstrated, that LOP–RSS relations are significantly better at the community level (richness, evenness) than at the species level (mean number of individuals). As a rule of thumb, RSS for LOP values of 5, 10, 25 and 50% are given for both species and community parameters in Table 3.

From the view of variance, the results recommend the utilization of community parameters to assess changes in agricultural land use. Furthermore, the procedure can be used as a tool for the selection of trustworthy indicator species from a pool of potentially usable species.

Table 3
Rules of thumb for RSSs corresponding with precision levels (LOP) of 5, 10, 25 and 50% for species and community parameters[a]

LOP (%)	RSSs			
	5	10	25	50
Species level (habitat generalists)	250–1000	50–250	10–50	3–10
Species level (habitat specialists)	1000–5000	250–1000	50–250	10–50
Community parameters	15–60	5–15	1–2	–

[a] Shown are the values for mean numbers of individuals for habitat generalists (as examples for quantitative indicators) and habitat specialists (examples for qualitative indicators) and for the two community parameters species richness and Smith–Wilson evenness.

Acknowledgements

I am very thankful to Jens Schumacher for statistical advices and to Klaus Reinhardt and Marzio Fattorini for critical comments on a previous version of the manuscript. Thanks also two anonymous referees for useful comments and Martin Schnittler and John Sloggett for checking the final version of the manuscript. I am grateful to Steffen Malt for fruitful cooperation in the project that provided the data for this paper. Fieldwork was financially supported by a grant (0339572) from the Federal Ministry of Education and Research (BMBF).

References

Büchs, W., Harenberg, A., Prescher, S., Weber, G., Hattwig, F., 1999. Entwicklung von Evertebratenzönosen bei verschiedenen Formen der Flächenstillegung und Extensivierung. In: Büchs, W. (Ed.), Nicht Bewirtschaftete Areale in der Agrarlandschaft-ihre Funktionen und ihre Interaktionen mit landnutzungsorientierten Ökosystemen. Mitteilungen aus der Biologischen Bundesanstalt für Land- und Forstwirtschaft, vol. 368, Berlin, Dahlem, pp. 9–38.

Drobner, U., Bibby, J., Smith, B., Wilson, J.B., 1998. The relation between community biomass and evenness: what does community theory predict, and can these predictions be tested? Oikos 82, 295–302.

Duelli, P., Obrist, M.K., Schmatz, D.R., 1999. Biodiversity evaluation in agricultural landscapes: above-ground insects. Agric. Ecosyst. Environ. 74, 33–64.

Dufrene, M., Legendre, P., 1997. Species assemblages and indicator species: the need for a flexible asymmetrical approach. Ecol. Monogr. 67, 345–366.

Hill, M.O., 1979. TWINSPAN—A FORTRAN Program for Arranging Multivariate Data in a Ordered Two-way Table by Classification of the Individuals and Attributes. Cornell University, Ecology and Systematics, Ithaca, NY.

Krebs, C.J., 1999. Ecological Methodology. Addison-Wesley Educational Publishers, Menlo Park.

Malt, S., Perner, J., 2002. Zur epigäischen Arthropodenfauna von landwirtschaftlichen Nutzflächen der Unstrutaue im Thüringer Becken. Teil 1: Webspinnen und Weberknechte (Arachnida: Araneae et Opiliones). Faun. Abh. Mus. Tierkde. Dresden 22, 207–228.

Manly, B.F.J., 1992. The Design and Analysis of Research Studies. Cambridge University Press, Cambridge.

McCune, B., Mefford, M.J., 1997. PC-ORD. Multivariate Analysis of Ecological Data, Version 3.0. MjM Software Design, Gleneden Beach, OR.

McGeoch, M.A., 1998. The selection, testing and application of terrestrial insects as bioindicators. Biol. Rev. 73, 181–201.

Motulsky, H., 1995. Intuitive Biostatistics. Oxford University Press, Oxford.

Müller-Motzfeld, G., 1989. Laufkäfer (Coleoptera: Carabidae) als pedobiologische Indikatoren. Pedobiologia 33, 145–153.

Perner, J., Malt, S., 2002. Zur epigäischen Arthropodenfauna von landwirtschaftlichen Nutzflächen der Unstrutaue im Thüringer Becken. Teil 2: Käfer (Coleoptera). Faun. Abh. Mus. Tierkde. Dresden 22, 261–283.

Perner, J., Malt, S., 2003. Assessment of changing agricultural land use: response of vegetation, ground-dwelling spiders and beetles to conversion of arable land into grassland. Agric. Ecosyst. Environ. 98, 169–181.

Tokeshi, M., 1993. Species abundance patterns and community structure. Adv. Ecol. Res. 24, 111–186.

Wilsey, B.J., Potvin, C., 2000. Biodiversity and ecosystem functioning: importance of species evenness in an old field. Ecology 81, 887–892.

ELSEVIER

Available online at www.sciencedirect.com

Agriculture, Ecosystems and Environment 98 (2003) 133–139

Agriculture Ecosystems & Environment

www.elsevier.com/locate/agee

Biotic indicators of carabid species richness on organically and conventionally managed arable fields

T.F. Döring [a,*], A. Hiller [b], S. Wehke [c], G. Schulte [b], G. Broll [d]

[a] *Department of Ecological Plant Protection, University of Kassel, Nordbahnhofstr. 1, D-37213 Witzenhausen, Germany*

[b] *Institute of Landscape Ecology, University of Münster, Robert-Koch-Str. 26-28, D-48149 Münster, Germany*

[c] *Institute of Geobotany, University of Trier, Universiätsring, D-54286 Trier, Germany*

[d] *Institute for Spatial Analysis and Planning in Areas of Intensive Agriculture, P.O. Box 1553, D-49364 Vechta, Germany*

Abstract

Carabids, a species rich arthropod family, potentially contribute much to biodiversity in agroecosystems, but assessing and monitoring carabid diversity is costly and time consuming. Therefore, this study aimed at finding more easily measurable parameters indicating high carabid diversity within organic and conventional management systems. Cover and number of weed species as well as activity density of single carabid species and of total carabids were investigated as potential indicators of carabid species richness. The study was carried out near Reckenfeld in Westphalia on sandy Plaggenesch soils. Three organically and four conventionally managed fields (cereals and corn) were investigated at the field margins and in the field centres from April to August 1999. Additionally, data of carabid catches and weed flora in winter cereals from an extended study in Düren (Northrhine-Westphalia) were reanalysed to validate the results. However, neither of the potential indicators showed consistently significant positive correlation with carabid diversity. This is partly attributed to the low variability of management conditions within the management systems in the studies presented.

Keywords: Carabidae; Biodiversity correlates; Bioindicators; Organic agriculture

1. Introduction

In many studies diversity of both flora and arthropod fauna has found to be higher on organically compared to conventionally managed fields (e.g., Frieben, 1997; Pfiffner, 1997). However, as biodiversity varies within each of the management systems, organic farming is not appropriate as an exclusive indicator of high biodiversity. Therefore, additional indicators are needed.

In the context of biodiversity, the term "indicator" is often used with very different definitions. These can be classified into at least four categories: (a) biotic indicators of abiotic conditions (Platen, 1995; Stumpf, 1997); (b) biotic indicators of human practices, including, e.g., pollution sensitive species (Basedow, 1990); (c) goal parameters, which are deducted from normatively set nature conservation aims and translate these into measurable features, e.g., species diversity of a certain taxon (May, 1995); (d) correlates of goal parameters, which make it possible to reduce labour and costs in assessing biodiversity and at the same time minimise loss of information. These correlates can be taxa, biotic parameters like species richness (Duelli and Obrist, 1998), or human practices influencing biodiversity (van Elsen, 1996). In this study based on data of flora and carabid fauna on arable fields in

* Corresponding author. Tel.: +49-5542-98-15-69; fax: +49-5542-98-15-64.
E-mail address: doringt@wiz.uni-kassel.de (T.F. Döring).

0167-8809/$ – see front matter © 2003 Elsevier Science B.V. All rights reserved.
doi:10.1016/S0167-8809(03)00075-6

Northrhine-Westphalia, the focus is on the category of correlates.

Carabids, a species rich arthropod family, potentially contribute much to biodiversity in agroecosystems, but assessing and monitoring carabid diversity is costly and time consuming. Therefore, this study aimed at finding more easily measurable biotic parameters indicating high carabid diversity.

2. Material and methods

2.1. Study area and field methods

The study area is located near Reckenfeld at 40–48 m above sea level, 15 km north of Münster in the region "Westfälische Bucht", with a maritime climate (average annual air temperature of 9.1 °C, annual precipitation 750 mm). Soils are mainly sandy, dominating soil types are anthrosols with plaggic horizons rich in organic matter (Plaggenesch), and gleyic podzols. Three organically and four conventionally managed arable fields were investigated (for management data see Table 1). Carabids were sampled from 19 April to 9 August 1999 with pitfall traps of 10 cm diameter (Melber, 1987). Renner's solution was used as a preservative (40% ethanol, 30% water, 20% glycerine, 10% acetic acid; Renner, 1982). Three traps were placed at the margin and in the centre of each of the seven fields (total of 42 traps). All individuals captured were identified up to species level. Vegetation was investigated according to Braun–Blanquet (Wilmanns, 1993, p. 50). Ground cover of weed vegetation was estimated visually.

Additionally, data from an extended study in Düren/Northrhine-Westphalia published by König et al. (1989) were reanalysed (30 winter cereal plots from 4 years; Winkens, 1984–1986; Wolff-Straub, 1989).

2.2. Choice of goal parameter and potential indicators

Carabid species were classified into five habitat preference groups using Thiele (1977), Koch (1989), Turin et al. (1991) and Marggi (1992): (1) stenotopic woodland species; (2) eurytopic woodland species; (3) eurytopic species; (4) eurytopic field species and (5) stenotopic field species. Species number of field species (eurytopic and stenotopic) was chosen as goal parameter for two reasons. First, field species are more typical for arable fields than woodland or eurytopic species and the promotion or protection of woodland species on arable fields is not possible. Second, edge effects could be minimised by the exclusion of woodland species from the calculations.

Four parameters were chosen as potential indicators of carabid species richness: (1) activity density of all carabids; (2) activity density of single carabid species; (3) number of plant species; (4) percentage ground cover of weed species.

2.3. Data analysis

The number of carabid individuals were $\log_{10}(x + 1)$-transformed before further statistical calculations. Statistical analysis of both studies comprised the calculation of Pearson's correlation coefficient and linear

Table 1
Management data of studied fields in Reckenfeld/Northrhine-Westphalia 1999

	Management						
	Organic			Conventional			
Crop in 1999	Faba–beans–oats	Triticale	Corn	Triticale	Winter wheat	Corn	Corn
Crop in 1998	Winter wheat	Corn	Ley	Corn	Corn	Winter barley	Corn
N-fertilisation (kg N/ha)	0	200	140	110	225	180	100
Frequency of mechanical weed control	3	2	5	0	0	0	0
Pesticides and growth regulator	–	–	–	Herbicide, growth regulator	Herbicides, insecticides, fungicides, growth regulators	Herbicides	Herbicides

regression between the goal parameter mentioned above and potential indicator parameters (Sachs, 1999; also see Sokal and Rohlf, 1995, pp. 543, 556–558). High positive correlation of a potential indicator with a goal parameter is seen as a measure of indicator reliability. However, causal relationships cannot be deducted from these calculations.

In most cases the factor "management system" (conventional vs. organic) which contributes much to the variation of diversity is already known or information is easily available. Therefore, no biotic indicators are needed to indicate the management system, but indicators are useful that predict the rest of variability of diversity within the management system. Accordingly, correlation between potential indicators and the goal parameter was calculated separately for organic and conventional fields.

3. Results

Total carabid catch was correlated significantly with the number of field species in the Düren study, but not in the Reckenfeld study (Fig. 1, Table 2). In the organic fields this correlation was higher than in the conventional ones in both studies.

Correlation between the number of individuals of single carabid species and the number of field species did not give a consistent picture for any of the species. For none of them there was significant correlation throughout both studies and both management systems (Table 3); however, the number of *Asaphidion flavipes* correlated significantly with the number of field species in the organic fields both from the Reckenfeld and the Düren study. *Pterostichus melanarius* and *Trechus quadristriatus* showed negative correlation values for the conventional fields from Reckenfeld.

The number of carabid field species correlated positively and significantly with the number of plant species only in the organic fields from Reckenfeld (Fig. 2, Table 2). For both the organic and the conventional management system the data from Düren correlation was negative, but not significant for the conventional sites.

Total cover estimates of weed vegetation could not be used to predict species numbers of carabids (data not presented). Single weed species cover estimates, especially of *Vicia hirsuta*, *Matricaria recutita* and *Stellaria media*, showed moderate correlation with species richness of field carabids (Table 4). These plant species are typical of the site specific weed community.

4. Discussion

As the total carabid catch was correlated with the goal parameter "number of carabid field species" only in the Düren study, it is not likely to be a reliable indicator of carabid diversity. Although correlation between the number of carabid species and the number

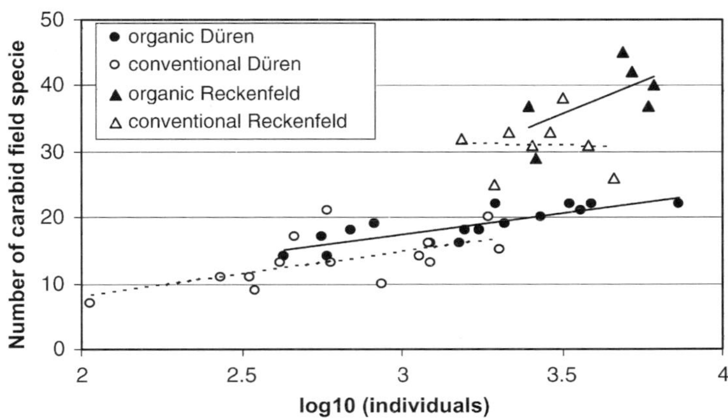

Fig. 1. Individuals of carabids ($\log(x + 1)$) caught in 44 arable fields plotted against number of carabid field species. Solid lines: organic; broken lines: conventional. For statistics see Table 2.

Table 2
Species number of carabid field species (y) as a function of two potential indicators

Indicator (x)	Site	Management	d.f.[a]	R^2	Regression equation
log (total individuals)	Reckenfeld	Organic	4	0.41, ns[b]	$y = 20.0x - 34.4$
		Conventional	6	0.00, ns	$y = -1.09x + 34.9$
	Düren	Organic	13	0.70***	$y = 6.38x - 1.7$
		Conventional	13	0.39*	$y = 6.89x - 5.7$
Plant species number	Reckenfeld	Organic	4	0.92**	$y = 0.63x + 20.7$
		Conventional	6	0.33, ns	$y = 0.30x + 27.9$
	Düren	Organic	11	0.31*	$y = -0.40x + 25.0$
		Conventional	5	0.10, ns	$y = -0.34x + 16.7$

[a] Degrees of freedom: Düren d.f. = 13, Reckenfeld organic d.f. = 4, conventional d.f. = 6.
[b] Not significant.
* Significant at $P \leq 0.05$.
** Significant at $P \leq 0.01$.
*** Significant at $P \leq 0.001$.

of individuals may be significant in other studies (e.g., Luff, 1996), residual variation remains too high for prediction purposes.

From a practical point of view it is easier to monitor only one or two carabid species than full determination of all carabid individuals. However, none of the carabid species caught in the studies from Reckenfeld and Düren can be regarded as a reliable indicator of high carabid diversity. Moreover, there is probably no (direct) causal connection between carabid species richness and the abundance of a single carabid species. Because of these restrictions, a different

Table 3
Carabid species as indicators of carabid diversity[a]

Carabid species	Reckenfeld		Düren	
	Organic	Conventional	Organic	Conventional
Agonum muelleri (Herbst 1758)	0.87*	0.23, ns[b]	0.35, ns	0.23, ns
Amara aenea (DeGeer 1774)	0.38, ns	0.72*	0.54*	0.48, ns
Amara similata (Gyll. 1810)	0.91*	0.69, ns	0.35, ns	0.70**
A. flavipes (L. 1761)	0.90*	0.11, ns	0.81***	0.36, ns
Bembidion lampros (Herbst 1784)	0.36, ns	−0.18, ns	0.56*	0.24, ns
Bembidion tetracolum Say 1823	0.47, ns	0.39, ns	0.68**	0.49, ns
Calathus fuscipes (Goeze 1777)	0.14, ns	−0.12, ns	0.53*	0.11, ns
Harpalus affinis (Schrank 1781)	0.31, ns	0.44, ns	0.25, ns	0.80***
Loricera pilicornis (F. 1775)	0.21, ns	0.24, ns	0.82***	0.69**
Nebria brevicollis (F. 1792)	−0.44, ns	0.49, ns	0.53*	0.13, ns
Poecilus cupreus (L. 1758)	0.52, ns	−0.01, ns	0.76***	0.64*
Pseudoophonus rufipes (DeGeer 1775)	−0.86, ns	0.13, ns	0.47, ns	0.68**
P. melanarius (Ill. 1789)	0.19, ns	−0.63, ns	0.56*	0.53*
Pterostichus vernalis (Panzer 1796)	−0.02, ns	0.50, ns	0.61*	0.63*
T. quadristriatus (Schrank 1781)	0.10, ns	−0.70, ns	0.63*	0.40, ns

[a] Degrees of freedom: Düren d.f. = 13, Reckenfeld organic d.f. = 4, conventional d.f. = 6. Correlation between captured individuals of carabid species ($\log_{10}(x + 1)$) and number of carabid field species. Species with correlation at least once significant.
[b] Not significant.
* Significant at $P \leq 0.05$.
** Significant at $P \leq 0.01$.
*** Significant at $P \leq 0.001$.

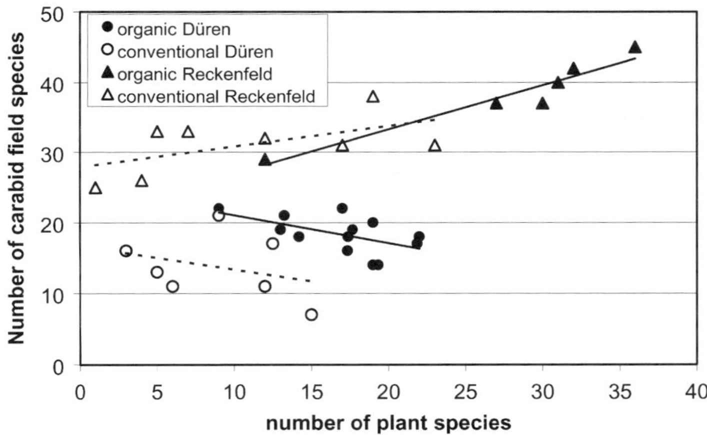

Fig. 2. Number of plant species in 44 arable fields plotted against number of carabid field species. Solid lines: organic; broken lines: conventional. For statistics see Table 2.

approach should be pursued. Ecological classification of carabid species may lead to the specification of conditions that are favourable for high species richness of carabids (Döring and Kromp, this issue).

The connection between high insect diversity and plant species richness is well known for many taxa (e.g., Weiss and Nentwig, 1992; Zanger et al., 1994; Denys, 1997). Carabid species may directly or indirectly profit from species rich vegetation. For example, the species rich subfamilies Harpalinae and Zabrinae contain many herbivorous and omnivorous species (Wachmann et al., 1995). At the same time, species numbers of both vegetation and carabids may be influenced in the same direction by practices in the agricultural management system (e.g., soil tillage; cf. Lorenz, 1994; van Elsen, 1996).

However, number of plant species number was not linked consistently to the number of carabid field species in the studies presented. Correlation was higher in the data from Reckenfeld than in the Düren study. A possible reason for this may be that the data extracted from the Düren study only comprised winter cereal plots, whereas the Reckenfeld data was based on corn as well as cereal fields, thus displaying a higher variability of management conditions. Low correlations between total weed cover estimates and number of carabid species may have resulted from the fact that higher weed cover is able to reduce activity abundance of epigaeic arthropods by reducing movement in general (Heydemann, 1956).

Table 4
Arable plant species as indicators of carabid diversity[a]

Species	Organic	Conventional	Presence
Apera spica-venti	0.60, ns[b]	0.30, ns	11
Capsella bursa-pastoris	0.53, ns	0.67, ns	7
Chenopodium album	0.00, ns	0.32, ns	10
Lamium amplexicaule	0.66, ns	–	5
Lamium purpureum	0.64, ns	–	5
M. recutita	0.71, ns	0.56, ns	9
Myosotis arvensis	0.75, ns	–, ns	5
Polygonum convolvulus	–	0.65, ns	
Polygonum lapathifolium	−0.16, ns	−0.01, ns	5
Polygonum persicaria	−0.25, ns	0.32, ns	7
S. media	0.74, ns	0.51, ns	9
Veronica arvensis	0.27, ns	0.50, ns	5
V. hirsuta	0.69, ns	0.72*	7
Viola arvensis	0.78, ns	−0.41, ns	9

[a] Degrees of freedom: organic d.f. = 4, conventional d.f. = 6. Pearson's correlation coefficient between cover estimates (%) and number of carabid field species. Plant species with presence > 4 out of 14; data from Reckenfeld.

[b] Not significant.

* Significant at $P \leq 0.05$.

5. Conclusions

If any biotic indicators for carabid diversity are to be used, they are presumably only applicable to comparisons of sites with fairly contrasting management conditions. Within more homogeneous groups of sites, such as organic winter cereal plots, there seem to be no reliable biotic indicators that are more easily measured than carabid species richness itself. For a rough biodiversity assessment of large areas or many sites, though, vegetation may presumably be sufficient (Duelli and Obrist, 1998).

Importantly, with a continuously changing agriculture, the predictive power of any indicator cannot be expected to remain unchanged; with the support of organic agriculture it may be hoped to slow down the loss of sensitive indicator taxa.

Acknowledgements

We would like to thank M. Kaiser (Münster) for patient support and helpful discussion, as well as W. Starke (Warendorf) for determination of critical taxa.

References

Basedow, T., 1990. Effects of insecticides on carabidae and the significance of these effects for agriculture and species number. In: Stork, E. (Ed.), The Role of Ground Beetles in Ecological and Environmental Studies. Andover, pp. 115–125.

Denys, C., 1997. Fördern Ackerrandstreifen die Artendiversität in einer ausgeräumten Agrarlandschaft? Untersuchungen am Beispiel der Insektengemeinschaften am Beifuß (*Artemisia vulgaris* L.). Mitt. Dtsch. Ges. Allg. Angew. Entomol. 11, 69–72.

Duelli, P., Obrist, M.K., 1998. In search of the best correlates for local organismal biodiversity in cultivated areas. Biodiv. Conserv. 7, 297–309.

Frieben, B., 1997. Arten- und Biotopschutz durch Organischen Landbau. In: Weiger, H., Willer, H. (Eds.), Naturschutz durch Ökologischen Landbau. Deukalion-Verlag, Holm, Germany, pp. 73–92.

Heydemann, B., 1956. Die Biotopstruktur als Raumwiderstand und Raumfülle für die Tierwelt. Verh. Dtsch. Zool. Ges. 1956, 332–347.

Koch, K., 1989. Die Käfer Mitteleuropas, Ökologie Bd. 1. Verlag Goecke and Evers, Krefeld, pp. 15–107.

König, W., Sunkel, R., Necker, U., Wolff-Straub, R., Ingrisch, S., Wasner, U., Glück, E., 1989. Alternativer und Konventioneller Landbau—Vergleichsuntersuchungen von Ackerflächen auf Lößstandorten im Rheinland. Schriftenr. Landesanstalt für Ökologie, Bodenordnung und Forstplanung Nordrhein-Westfalen 11, 286 pp.

Lorenz, E., 1994. Mechanische Unkrautbekämpfungsverfahren in Zuckerrübenkulturen und ihre Nebenwirkungen auf Laufkäfer (Coleoptera, Carabidae) und andere epigäische Arthropoden. Ph.D. Thesis. University of Göttingen.

Luff, M.L., 1996. Use of carabids as environmental indicators in grasslands and cereals. Ann. Zool. Fenn. 33, 185–195.

Marggi, W.A., 1992. Faunistik der Sandlaufkäfer und Laufkäfer der Schweiz (Cicindelidae & Carabidae) Coleoptera Teil 1. Doc. Faun. Helv. 13. Centre suisse de cartogr aphie de la faune, Neuchatel.

May, R.M., 1995. Conceptual aspects of the quantification of the extent of biological diversity. In: Hawksworth, D.L. (Ed.), Biodiversity—Measurement and Estimation. Chapman and Hall, London, New York, pp. 13–20.

Melber, A., 1987. Eine verbesserte Bodenfalle. Abh. Nat.wiss. Ver. Bremen 40, 331–332.

Pfiffner, L., 1997. Welchen Beitrag leistet der ökologische Landbau zur Förderung der Kleintierfauna? In: Weiger, H., Willer, H. (Eds.), Naturschutz durch Ökologischen Landbau. Deukalion-Verlag, Holm, Germany, pp. 93–120.

Platen, R., 1995. Zeigerwerte für Laufkäfer und Spinnen—eine Alternative zu herkömmlichen Bewertungssystemen. In: Riecken, U., Schröder, E. (Eds.), Biologische Daten für die Planung, Bonn-Bad Godesberg, pp. 317–328.

Renner, K., 1982. Coleopterenfänge mit Bodenfallen am Sandstrand der Ostseeküste—ein Beitrag zum Problem der Lockwirkung von Konservierungsmitteln. Faunistisch-Ökolog. Mitt. 5, 137–146.

Sachs, L., 1999. Angewandte Statistik. Springer, Berlin, 881 pp.

Sokal, R.R., Rohlf, F.J., 1995. Biometry, 3rd ed. Freeman, New York.

Stumpf, T., 1997. Neue Wege in der Bioindikation—ein ökologisches Zeigerwertsystem für Käfer. Mitt. Landesanstalt Ökologie Bodenordnung und Forstplanung Nordrhein-Westfalen 1997 (2), 53–58.

Thiele, H.U., 1977. Carabid Beetles in Their Environments. Springer/Verlag, Berlin/Heidelberg, 369 pp.

Turin, H., Alders, K., den Boer, P.J., van Essen, S., Heijerman, T., Laane, W., Penterman, E., 1991. Ecological characterization of carabid species (Coleoptera, Carabidae) in the Netherlands from thirty years of pitfall sampling. Tijdschr. Entomol. 134, 279–304.

van Elsen, T., 1996. Wirkungen des ökologischen Landbaus auf die Segetalflora—Ein Übersichtsbeitrag. In: Diepenbrock, W., Hülsbergen, K.J. (Eds.), Langzeiteffekte des ökologischen Landbaus auf Fauna, Flora und Boden, Conference Halle/Saale, pp. 143–152.

Wachmann, E., Platen, R., Barndt, D., 1995. Laufkäfer—Beobachtung, Lebensweise. Naturbuch Verlag, Augsburg, 295 pp.

Weiss, E., Nentwig, W., 1992. The importance of flowering plants on sown weed strips for beneficial insects in cereal fields. Mitt. Dtsch. Ges. Allg. Angew. Entomol. 8, 133–136.

Wilmanns, O., 1993. Ökologische Pflanzensoziologie. Quelle and Meyer Verlag, Heidelberg, 479 pp.

Winkens, H., 1984–1986. Faunistische Untersuchungen zum Alternativen Landbau—Zwischenbericht für die Jahre 1983–1986.

Untersuchte Tiergruppe: Coleoptera (partim). Report for König, et al., 1989.

Wolff-Straub, R., 1989. Vergleich der Ackerwildkraut-Vegetation alternativ und konventionell bewirtschafteter Äcker. In: König, W., Sunkel, R., Necker, U., Wolff-Straub, R., Ingrisch, S., Wasner, U., Glück, E. (Eds.), Alternativer und Konventioneller Landbau—Vergleichsuntersuchungen von Ackerflächen auf Lößstandorten im Rheinland. Schriftenr. Landesanstalt für Ökologie, Bodenordnung und Forstplanung Nordrhein-Westfalen 11, pp. 70–112.

Zanger, A., Lys, J.A., Nentwig, W., 1994. Increasing the availability of food and the reproduction of *Poecilus cupreus* in a cereal field by strip-management. Entomol. Exp. Appl. 71, 111–120.

ELSEVIER

Available online at www.sciencedirect.com

Agriculture, Ecosystems and Environment 98 (2003) 141–151

Agriculture Ecosystems & Environment

www.elsevier.com/locate/agee

The spatial and temporal pattern of carabid beetles on arable fields in northern Germany (Schleswig-Holstein) and their value as ecological indicators

Ulrich Irmler*

Ökologie-Zentrum University, Schauenburgerstr. 112, D-24118 Kiel, Germany

Abstract

Ground beetles have been often used as indicators for agricultural practices in ecological studies, but little is known about the spatial and temporal variation independent from the agricultural practices. Between 1985 and 1995 the ground beetles (Carabidae) of 53 fields in Schleswig-Holstein (northern Germany) were investigated using three replicate pitfall traps at each site. One field was studied over a period of 9 years from 1988 to 1996, and in addition two adjacent fields subjected to ecological or conventional farming methods were investigated in 1999 using 12 pitfall traps each. Overall, five assemblages of ground beetles could be differentiated, primarily by sand content of the soils and field size. The separation of the assemblages was weak (eigenvalue of 1st axis: 0.39) compared to natural ecosystems. The most common assemblage on loamy soils was dominated by *Pterostichus melanarius*. On both sandy soils and loamy soils in small fields the species *Poecilus versicolor* and *Platynus dorsalis*, *Bembidion lampros*, respectively, dominated. Eight ecologically differentiated groups of ground beetles were found. Most species correlated positively with the sand content of the soils. Only two species, *P. melanarius* and *Loricera pilicornis* showed a positive correlation with the field size. A higher species richness was observed on fields which have practised for 30 years ecological farming. Comparing the two adjacent fields with ecological and conventional farming, no difference in species richness was detected. Four species showed higher abundance on the field with ecological farming. In this analysis of a long-term dataset ground beetles did not respond to the cultivated plants, but only to the yearly climate conditions.

Keywords: Carabidae; Field community; Biodiversity; Biological farming; Long-term fluctuation

1. Introduction

Sustainable land use, including the conservation of biodiversity, is a general aim of future landscape management. Agricultural practices were thought to be responsible for the loss of species from many regions in Central Europe (e.g. Heydemann, 1986; Gall and Orians, 1992) and as a result many investigations have centred on an analysis of the agricultural impact on biodiversity (e.g. Paoletti et al., 1992; Duelli, 1997; Irmler et al., 2000). However, little is known about the relations between the species richness of agricultural fields and both the changeable agricultural practices and permanent environmental conditions of the sites.

The landscape in Schleswig-Holstein is typically dominated by arable fields (41% of total area) and grassland (32.6%). Semi-natural habitats amount to only 3.7% of the total area of the region (Heydemann, 1997). Analysing the biological composition and species richness of fields and grassland should be a primary aim of an evaluation of the biodiversity of the

* Tel.: +49-431-880-4311; fax: +49-431-880-1111.
E-mail address: irmler@ecology.uni-kiel.de (U. Irmler).

0167-8809/$ – see front matter © 2003 Elsevier Science B.V. All rights reserved.
doi:10.1016/S0167-8809(03)00076-8

region. Ground beetles (Carabidae) can play a major role as indicators of the ecological status of fields and grassland in northern Germany (Heydemann, 1955; Hingst et al., 1995; Schröter and Irmler, 1999), as they occupy a key ecological niche as natural predators of pest insects (e.g. Basedow et al., 1990; Bilde and Toft, 1999; Kromp, 1999). The ground beetles have also been proposed for the monitoring concept of the ecological area control in Germany (Hoffmann-Kroll et al., 1998).

The current study investigates the following questions: (1) How are the ground beetles assemblages separated? (2) Which environmental parameters are responsible for the separation of the carabid assemblages? (3) Which species can be evaluated as indicators for the various environmental conditions? (4) How do the environmental conditions influence the species richness of fields?

2. Material and methods

The ground beetle fauna was recorded from 52 field sites in different regions of Schleswig-Holstein (northern Germany) in the years between 1985 and 1995. Most fields investigated in 1995 were in the region near Bornhöved in central Schleswig-Holstein (Schröter and Irmler, 1999). All sites were studied during at least one vegetation period from April to October. In arable fields interruptions of the investigation were necessitated by agricultural practices, such as ploughing, harvest, etc. At each site carabids were sampled using three replicate pitfall traps filled with formaldehyde and a detergent liquid. Regular jam jars (5.6 cm diameter) were used as pitfall traps and were set in the middle of the field. Distance between pitfall traps was at least 10 m. Among these 52 sites, eight fields were managed according to the conditions of ecological farming. On these fields mineral fertiliser and pesticides were not applied, N-input was mainly provided by cultivation of legumes or internally produced manure. In addition, in 1999 two adjacent fields subjected to ecological and conventional farming methods were investigated to compare the two land uses. Winter wheat was grown on both fields, and soil type and orientation (north) were similar. The field subjected to ecological land use methods was managed by a rotation of four crops with a share of 50% legumes. In each field, 12 replicate pitfall traps were installed and sampling undertaken between 14 April and 31 July. A further field in central Schleswig-Holstein near Bornhoeved was studied over a period of 9 years from 1988 to 1996, with three pitfall traps in the centre of the site. The rotation in this field involved five different crops (maize, oat, rape, grass, rye).

The following environmental conditions were recorded: soil water content as vol.% difference between fresh weight and dry weight of a standardised volume of soil (drying at 110 °C for 24 h), pH of soil (in KCl), organic content of soil by ignition at 450 °C, sand content of soil (fraction at 63–200 μm diameter) by sieving after removing the organic content by H_2O_2 and the size of the field.

Ordination of the ground beetle data was performed using the Detrended Correspondence Analysis (DCA) (ter Braak, 1986). The following statistics were performed using the program Statistica (StatSoft, 1996): cluster analysis by unweighted pair-group average with distance method percent similarity to differentiate clusters with high similarity of species composition, differences between the environmental conditions at sites of the different assemblages by one-way ANOVA followed by a least significant difference, relations between faunal and environmental data by Spearman's correlation, correlation between abundance and climatic factors by Pearson product correlation. Diversity index was calculated as $H(s)$ using the natural logarithim Ln of the Shannon and Weaver index (Shannon and Weaver, 1963), significance of averages was analysed by Mann–Whitney U-test.

3. Results

3.1. The ground beetle assemblages

The average linkage cluster analysis of the ground beetle data resulted in five assemblages with a minimum of 40% similarity. Groups 1–3 (including 31 or 6 sites) were situated on loam or sandy loam soils (Fig. 1). The 4th and 5th group (with seven and two sites, respectively) were found on sandy soils. The correspondence analysis showed the high similarity of the ground beetle assemblages. Eigenvalues of the first and second canonical axis were 0.39 and 0.25. According to ter Braak (1986) assemblages are sufficiently

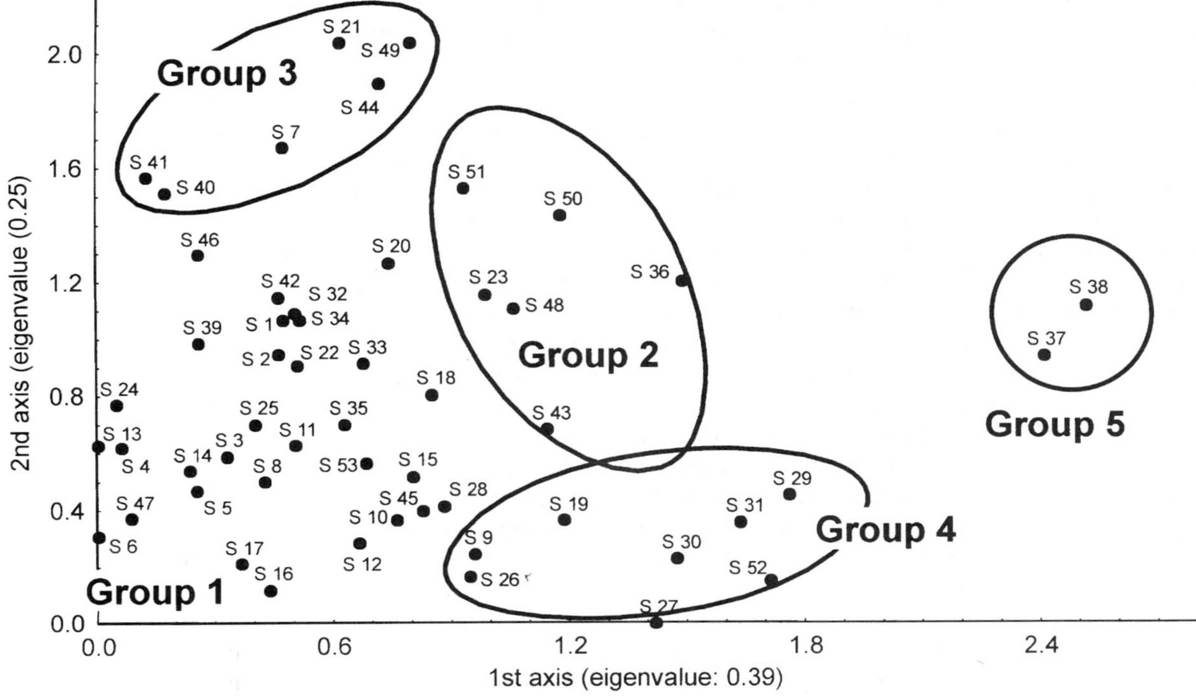

Fig. 1. Detrended canonical analysis of the 53 field sites in Schleswig-Holstein, northern Germany.

differentiated at a eigenvalue higher than 0.5. This is true for the ground beetle data only when the two axis are considered together. With the exception of the two sites of group 5, the sites are aggregated near the 0-point of the two axis. Groups 2–4 are concentrically ordered around the first group. Groups 4 and 3 in particular are differentiated by the 2nd canonical axis.

Considering the environmental factors of the five differentiated ground beetle assemblages, the sand content of the soil is the most important factor for the differentiation of the assemblages (Table 1). The soils of the sites in groups 1–3 were loam or sandy loam having a mean sand content of 50–54% and usually low variance. The soil type appears to be of lower importance only in the 2nd group, where the S.E. was greater than 50% of the mean. A differentiation of the three groups using the soil parameters was impossible. The size of the fields was the main factor

Table 1
Mean and standard error (S.E.) of the determined environmental parameters for the differentiated ground beetle assemblages[a]

	Group (number of sites)									
	1 (31)		2 (6)		3 (6)		4 (8)		5 (2)	
	Mean	S.E.	Mean	S.E.	Mean	S.E.	Mean	S.E.	Mean	S.E.
Water content (%)	20.1	4.5	23.9	12.1	22.6	3.9	17.8	3.8	14.8	0.6
Sand content (%)	**52.6**	**9.7**	**50.4**	**29.1**	**54.1**	**6.0**	**67.9**	**8.0**	**92.3**	**0.1**
Organic substance (mg/g)	4.3	4.3	5.2	2.9	3.8	0.4	4.0	1.8	4.2	0.5
Size (ha)	**10.6**	**13.0**	**3.4**	**2.9**	**21.6**	**15.6**	3.7	2.8	2.0	0.1
pH (KCl)	6.1	0.5	5.8	0.6	6.2	0.0	5.7	0.7	5.3	0.0

[a] The significant factors sand content and size of the field are in bold.

separating groups 1–3 significantly. Group 1 represented the ground beetle assemblages on loam soils with a field size of about 10 ha and a high variance. The second group was found mainly on small-sized fields with an average of 3.4 ha and the third group on large-sized fields of more than 20 ha in average. The fourth assemblage indicated a high similarity to the three groups on loam soils due to the DCA. Nevertheless, the sites were significantly different from these groups due to a high sand content (average of 68%). The fields were as small as in group two, a result of the historical land use in Schleswig-Holstein, where small farms had been established on poor soils. The last assemblage occurred on soils with extremely high sand content of more than 90%. Because agrarian use on these poor soils is scarcely possible, this group was represented only by two sites.

The species composition of these five ground beetle assemblages is shown in Table 2. Group 1 with medium-sized fields on more or less loam soils is characterised by the high dominance of *Pterostichus melanarius*, which was the only species representing more than 10% of the individuals. *P. melanarius* was also very dominant in the 3rd assemblage. The 3rd group, however, was characterised by the extremely high dominance of *Bembidion tetracolum*. The two species collectively amounted to more than 50% of all ground beetle individuals in this type of field. In the 1st group a total of four species were necessary to reach 50% of the individuals. In the 2nd assemblage on small-sized fields with loam soils the species *Platynus dorsalis* and *Bembidion lampros* dominate and the two species *P. melanarius* and *B. tetracolum* reach lower abundance than in groups 1 and 3. Diversity in group 2 is obviously higher, because eight species contribute to 50% of ground beetle individuals. This is also reflected in the diversity index, that is, 2.3 ± 0.3, compared with 1.9 ± 0.4 in group 1 and 1.8 ± 0.4 in group 3. This difference is significant at $P < 0.05$ compared to groups 1 and 3.

The high similarity between assemblages 1 and 4 is due to the high dominance of *P. melanarius* in both groups. In this group, *Poecilus versicolor* predominated, which occurred rarely on loam soils. In correspondence to assemblage 1 four species make up 50% of the individuals. The last assemblage is predominated by *Amara fulva*. This species, *P. versicolor* and *Pterostichus niger* contributed to more than 50% of the individuals. The diversity index was 2.3 ± 0.4 in group 4 and 2.5 ± 0.2 in group 5.

Table 2
Dominance of the most frequent ground beetle species in the five differentiated assemblages (S.E. in parenthesis, bold: dominate species, species richness: per site, in three replicate pitfall traps for 1 year)

	Group/number of sites/species richness				
	1/31/25 (6)	2/6/35 (8)	3/6/29 (11)	4/8/31 (3)	5/2/40 (10)
P. melanarius	**35.1**	6.2	**14.2**	**16.9**	
Platynus assimilis	+	1.8	1.2		
T. quadristriatus	4.1		6.8		
C. fossor		2.3			
Amara spreta	0.2				
Harpalus latus	0.9				
Carabus granulatus	1.8	2.5	2.1		
P. dorsalis	6.6	**16.8**	8.4	3.6	4.1
B. lampros	3.6	**14.5**	2.1	3.5	4.3
B. tetracolum	3.8	3.3	**38.5**		3.2
Blemus discus			3.2		
P. rufipes	3.1	2.4		8.2	2.5
H. affinis	3.1			7.3	
P. versicolor		3.8	1.7	**21.6**	7.0
Calathus fuscipes				3.4	3.9
Carabus cancellatus		1.6		2.5	2.6
Calathus erratus				2.6	
A. fulva					**38.7**
L. pilicornis		2.7	5.0	3.0	4.0
P. niger	4.4		3.5		5.6

3.2. Ecological preferences

The ground beetle species recorded at the 53 field sites investigated were correlated by Spearman correlation test in respect to environmental parameters, e.g. water content, sand content, content of organic substance, pH, and size of field. As a result of the correlations eight groups of ground beetles with different ecological preferences were identified (Table 3).

The first group (A) included species that prefer high soil moisture correlating positively with the water content of the soil. Additionally, they are mostly negatively correlated with the sand content of the soil, because sandy soils dry out more rapidly. Three species contributed to the second group (B) that occur on dry soils in particular. The species of the next group (C) were positively correlated with the sand content of the soil. For a few species there was also a negative

Table 3
Spearman's correlation coefficient (only significant r with $P < 0.05$) between abundance of ground beetles and environmental parameters

Species	Water	Sand	pH	Size of field	Organic substance	Ecological group
Amara familiaris	0.47					A
Trechoblemus micros	0.38	−0.47				
Patrobus atrorufus	0.34	−0.46				
Stomis pumicatus	0.38	−0.32			0.28	
C. cancellatus	−0.29					B
B. femoratum	−0.29					
Carabus coriaceus	−0.28					
C. fuscipes	−0.30	0.34				
Calathus melanocephalus	−0.44	0.46				
Synthomus truncatellus	−0.35	0.38				
Broscus cephalotes	−0.39	0.37				
Amara bifrons		0.35				
Semiophonus signaticornis		0.29			−0.28	C
Harpalus rufipalpis		0.33				
Harpalus rubripes		0.31				
P. rufipes		0.29	−0.36			
Harpalus tardus		0.44	−0.35			
Amara apricaria		0.44	−0.38			
A. fulva	−0.48	0.52	−0.38	−0.41		
C. erratus	−0.35	0.60	−0.40	−0.33		D
P. versicolor		0.32	−0.37	−0.51		
C. hortensis		0.31	−0.30	−0.50		
P. dorsalis		−0.32				
Lasiotrechus discus		−0.35				
C. auratus		−0.28				E
Acupalpus meridianus		−0.48				
T. quadristriatus		−0.29	0.28	0.45		
B. lampros			−0.33	−0.50		
A. muelleri				−0.35		
Amara aenea				−0.59		F
Demetrias atricapillus				−0.38		
Amara communis				−0.38	0.30	
P. melanarius				0.39		
L. pilicornis				0.31		G
D. globosus				−0.41	0.50	
Pterostichus strenuus					0.31	H

correlation with the water content and the pH of the soil, but the correlation with the sand content was more significant. The species of this group could be considered as being carabids that prefer sandy soils. In group (D) a positive correlation with the sand content of the soil was also found, but they were additionally negatively correlated with the size of the field and the pH of the soil. Therefore, this group contained species that were most numerous in field margins or other habitats adjacent to arable fields. This was due in particular to *Carabus hortensis*, which was caught in small abundance in arable fields with these environmental conditions, but frequently occurs in forests (Irmler, 1999). Group (E) contained five species with a negative correlation to the sand content of the soil and which could be considered as inhabitants on loam soils. The species of the remaining groups (F–H) were related to the field size. Five species were negatively correlated to the field size preferring small fields, whereas two species were positively correlated with the field size inhabiting large fields. Finally, the last ecological group (H) included two species with a positive correlation to the organic substance in the soil. These species can be considered as inhabitants of peat soils.

3.3. Impact of the farming practice

There were five arable fields in which ecological land use had been practised for more than 30 years and 27 fields subjected to conventional land use in assemblage 1. Species richness was higher on average on the ecologically managed fields than on conventionally managed fields (Fig. 2). However, the difference in the composition of the species profile was small. The low influence of farming practice was already evident from the DCA, when reflected no separation corresponding to this factor within the five assemblages. No species predominated on the ecologically managed fields at any of the sites. Even *Carabus auratus* often considered as an indicator of ecological farming (Steinborn and Heydemann, 1990), reached no higher activity density on ecologically managed fields. In assemblage 1 this species occurred on both conventionally and ecologically managed fields with a frequency of 0.6 ± 2.6 ind. per 10 trapping days and 0.7 ± 0.6 ind. per 10 trapping days ($P = 0.17$), respectively. In assemblage 3 the species was not found on conventionally used fields, whereas it occurred on ecologically used fields with 1.0 ± 1.3 ind. per 10 trapping days ($P = 0.3$).

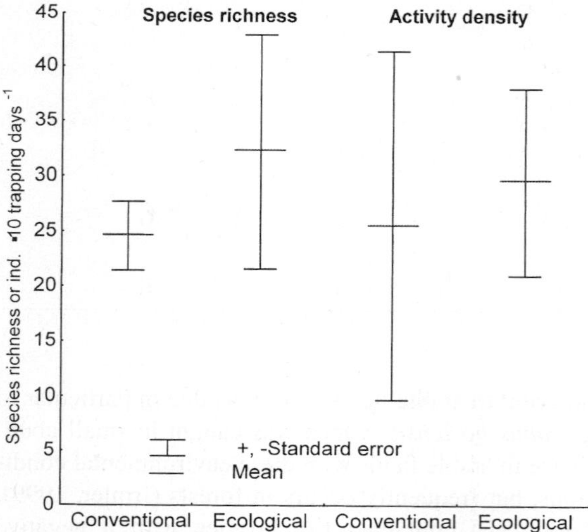

Fig. 2. Species richness per three replicate traps and abundance (ind. per 10 trapping days) in conventional and ecological farm systems of assemblage 1 (statistical analysis by t-test; species richness: $t = 3.36$, $P = 0.003$; abundance: $t = 0.61$, $P = 0.55$).

Furthermore, differences in species richness were found between the adjacent ecologically and conventionally managed fields. However, four of the total of 36 species recorded were found in significantly higher activity density on the ecologically managed field than on the conventionally managed field (Table 4). These species are common and frequently found on fields. Additionally, variance of activity density was high in these species indicating a heterogeneous distribution on the fields (Fig. 3). The activity density on the fields may also be influenced by field margins or small heterogeneous differences in environmental conditions superimposed on the influence of the farming systems.

3.4. Long-term fluctuation of ground beetles on one field

A cluster analysis was performed using ground beetle data collected over a period of 9 years on a conventionally farmed field near Bornhoeved (Fig. 4). Corresponding with the results of extensive datasets

Table 4
Activity density of ground beetles (ind. per 10 trapping days) on the two adjacent fields subjected to ecological and conventional farming practice (S.E.: standard error)

Species	Ecological		Conventional	
	Mean	S.E.	Mean	S.E.
C. granulatus	*3.5	1.5	1.2	0.7
A. muelleri	*0.8	0.6	0.1	0.1
P. melanarius	*5.4	3.0	2.7	1.3
B. tetracolum	*10.1	0.2	4.7	2.5
P. assimilis	0.3	0.4	0.4	0.4
P. dorsalis	2.3	1.1	2.0	0.9
L. pilicornis	0.3	0.2	0.6	0.2
C. fossor	0.3	0.2	0.7	0.5
P. niger	0.4	0.3	1.0	0.7
T. quadristriatus	0.2	0.2	–	–
Species richness	12.8	1.9	13.3	2.1
Mean abundance	*24.4	7.6	14.2	2.9

* Stastically significant.

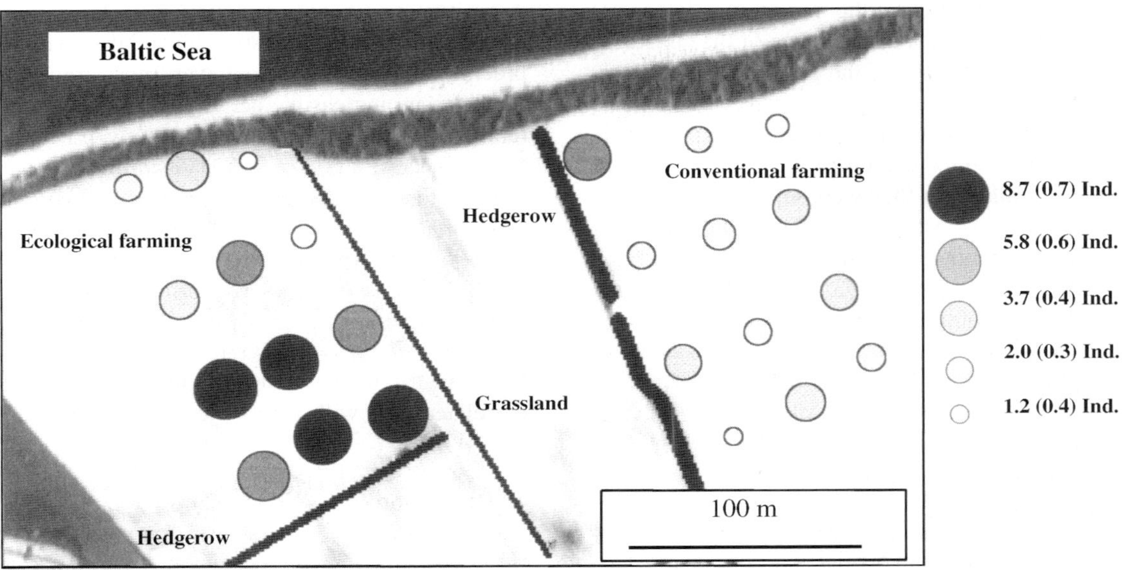

Fig. 3. Mean yearly abundance (ind. per 10 trapping days) of *P. melanarius* in the ecological and conventional farming systems. Groups of abundance (standard error in brackets) are significantly different with $P < 0.001$.

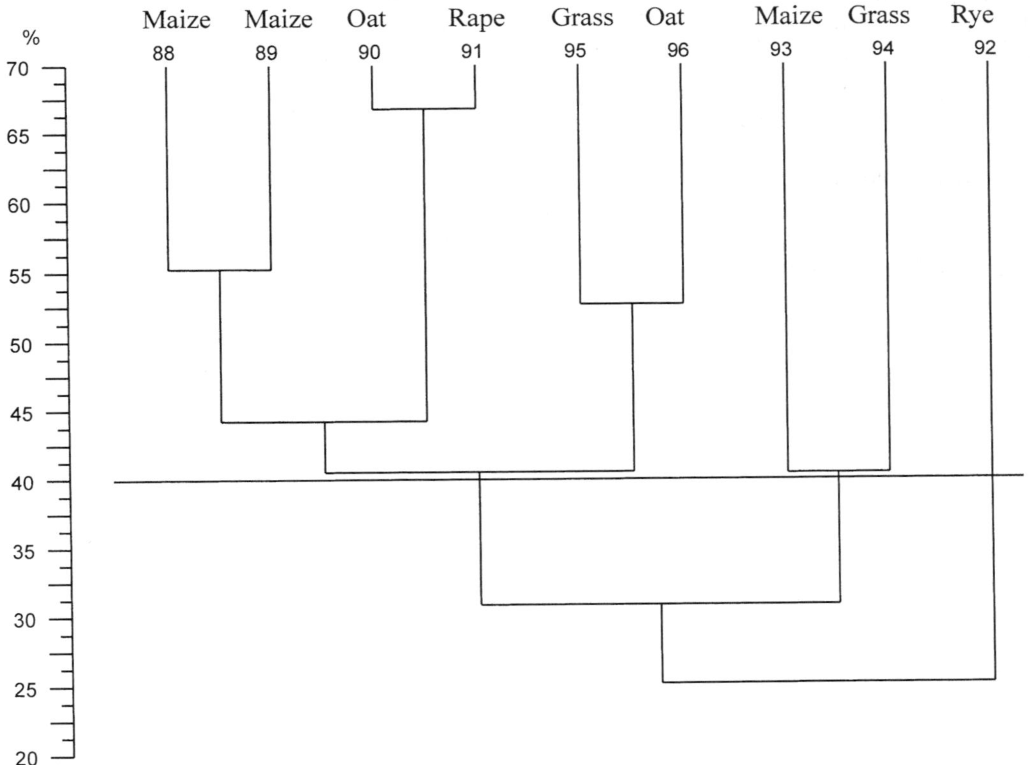

Fig. 4. Cluster analysis using the 9 years data (crops and year titled) collected at the long-term field site.

collected from 53 fields, the long-term (intensive) analysis did not indicate an influence of the cultivated crop on the composition of the ground beetle assemblages. Moreover, assemblages from successive years were marked by a high similarity. In contrast, years with the same cultivated crop, e.g. maize or oats, showed no similarity if they were separated by several years. Using a level of 40% similarity the following groups are differentiated: (1) the years 1988–1991 and 1995, 1996, (2) the years 1993 and 1994, and (3) the year 1992. In group 1, *Pseudoophonus rufipes* predominated, whereas *P. niger* and *Trechus quadristriatus* were found with highest abundance in group 2. In group 3 *P. versicolor* was the most abundant species.

No species remained on a more or less constant level of abundance or dominance during the years investigated. The most abundant species was *P. rufipes* varying by 126% of the mean between 25 ind. per 10 trapping days in 1990 and 0.1 ind. per 10 trapping days in 1993. *Agonum muelleri* was not caught between 1993 and 1995, but occurred again in 1996. The highest and lowest variance of abundance with 220 and 97% of the mean, respectively, was found for *P. versicolor* and *Loricera pilicornis*, respectively. Species richness also widely varied between 61 species in 1989 and 16 species in 1993. The 20 most abundant species were analysed to investigate the influence of the following climatic parameters: mean yearly precipitation, precipitation in May and June, mean temperature in January and June. Only half of the species were significantly correlated with one of the parameters (Fig. 5). A positive significant correlation with the mean temperature in January (five species) and the mean yearly precipitation (four species) was most frequent, indicating an increasing activity density with warm temperature in winter and high yearly precipitation. Three species, e.g. *Bembidion femoratum* (precipitation $r = 0.78$; temperature: $r = 0.89$), *Clivina fossor* (precipitation and temperature: $r = 0.70$), and *Syntomus foveatus* (precipitation: $r = 0.85$; temperature: $r = 0.77$), were significantly correlated with both climatic parameters. For *Dyschirius globosus* ($r = 0.73$) and *P. dorsalis* ($r = 0.68$) only positive relations were found to the temperature in January, while *P. niger* correlated only with yearly precipitation ($r = 0.75$).

4. Discussion

Heydemann (1955) was the first to consider soil type as a major factor influencing the composition of ground beetle communities in arable fields. He differentiated between fields on sandy and loam soils, which vary in relation to cereal or non-cereal cultivation. A similar explanation of community differentiation was adopted by later authors (e.g. Heydemann and Meyer, 1983; Brasse, 1975), although a minor difference detected between cereal and non-cereal cultivation was clearly a result of the corresponding overall reduction of species richness in fields in the late 1980s (Steinborn and Meyer, 1994). Steinborn and Meyer (1994) investigated three factors, two soil

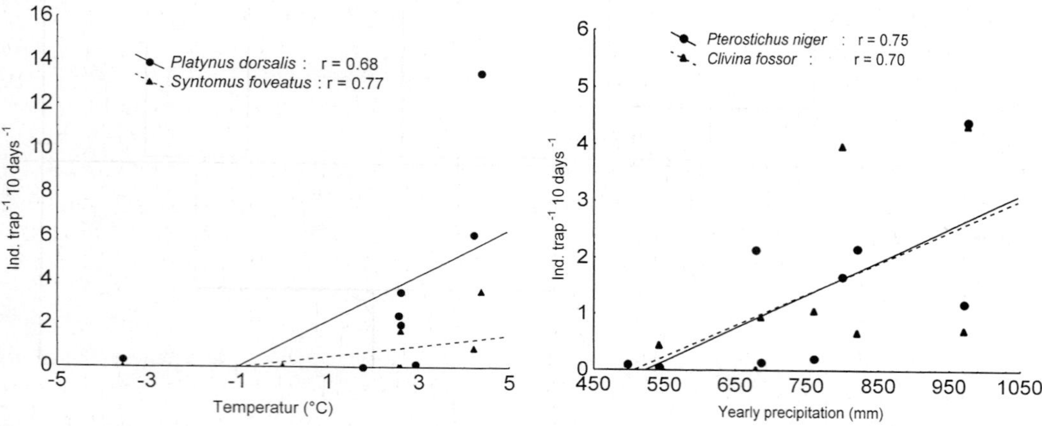

Fig. 5. Correlation between four ground beetle species and selected climatic parameters over a period of 9 years (temperature: mean temperature in January).

types, cultivated crop and management method. Soil type was the major environmental factor followed by cultivated crop and management method with decreasing influence on the composition of ground beetles and spiders. On loam soils, the influence of the agricultural practice was of higher importance than on sandy soils. In their investigations of fields in Schleswig-Holstein (northern Germany) in 1998, Schröter and Irmler (1999) found no significant influence of the cultivated crop on the composition of the ground beetle fauna. Pfiffner and Luka (1996) emphasised a small effect of different cultivated crops on ground beetle activity density, but did not comment on species composition and compared the data statistically. An analysis of a 30-year investigation of some fields in Schleswig-Holstein has indicated a homogenising trend of ground beetle assemblages in arable fields (Steinborn and Heydemann, 1990).

The small variance between the assemblages of ground beetles in fields in this study is reflected in the low eigenvalues of the canonical axis in the DCA reaching only half the value of forest assemblages (Irmler, 1999). This occurs despite the high potential of variability between fields in which different cultivated crops are present, the different management practices and the different intensity of tillage. Topp (1989) also emphasised the low variance of the ground beetle assemblages in fields compared to more natural ecosystems. Moreover, soil type as the major factor influencing the variance in the ground beetle assemblages had only a small effect. This may indicate that the constituting homogenising trend of the field fauna has also started to homogenise the primary differences between the soil types.

Even a reduction of land use intensity has almost no effect on the species composition. In North America, no differences in species composition were detected between conventional and low input cropping systems (Fan et al., 1993), with only the abundance of ground beetles differing. Differences have been found between conventional and biological cropping systems in several other investigations (Hokkanen and Holopainen, 1986; Pfiffner and Niggli, 1996) and this has been attributed to the higher abundance of ground beetle prey on fields with low land use intensity (Hokkanen and Holopainen, 1986) or to the reduction of pesticide use (Basedow, 1990; Raskin et al., 1992; Huusela-Veistola, 1996).

Pfiffner and Niggli (1996) have published data indicating a slightly higher species richness in ecological compared to organic and conventional farming. However, the presented data shows that the species richness is mainly influenced by the soil type and the size of the field. Comparison of the 53 fields indicates that ecological farming seems to effect an increase of the species richness on loam soils, if it is maintained for several decades. The field managed for 7 years under ecological standards showed no higher species richness than the adjacent conventionally managed field.

Overall, the species potential of ground beetles in arable fields is difficult to estimate, as it is essentially effected by the adjacent ecosystems. This may explain the strong impact of the field size (the second important factor beside the soil type). It is probable that the length and quality of the field margin are dominant factors effecting the composition of the ground beetle fauna. Several investigations have shown that headlands and field margins influence the field fauna (Wallin, 1985; Gilgenberg, 1986; Basedow, 1988; Dennis and Fry, 1992; Kinnunen et al., 1996; Pfiffner and Luka, 1996), with species composition varying widely between the years. Such variation is caused by the climatic conditions of the year rather than the crop grown in the field. Luff (1996) showed that different carabid species react independently from each other to annual changes on weather conditions, explaining the importance of climate. Soil type was also shown to be important for some species. In this study, *P. melanarius* and *Harpalus affinis* occurred in high abundance on soils with medium sand content (Irmler et al., 2000). Both *P. melanarius* and *H. affinis* were found in high abundance on soils with sand content between 30 and 60%, and 40 and 70%, respectively.

P. melanarius in particular has often been used as an indicator of species poor fields with high intensity land use (Raskin et al., 1992). However, Wallin (1985) reported a positive correlation between the abundance of *P. melanarius* and field size and Hokkanen and Holopainen (1986) and Fan et al. (1993) found a significantly higher abundance in fields with low land use intensity. Both results are supported by the presented study. *P. melanarius* overwinters within arable fields, but populations seem to suffer high losses due to the tillage system (Fadl and Purvis, 1998). However, the species is able to migrate rapidly into the field from the field margins, advancing between 2.6 and 5.3 m per day (Thomas et al., 1998), compensating for the losses. Thus the dominance of the species on large and usually intensively farmed fields is based not on an ability to overcome the intensive management, but on a high dispersion ability.

To estimate the species potential of ground beetles in fields more information is needed on the ecology of the species (e.g. mobility, immigration ability, dependence on climatic conditions) and the influences of the environmental conditions of the site and the adjacent field margins. In particular, evaluation of species potentials on arable fields is difficult, because variability of history and adjacent habitats is higher than in natural or semi-natural ecosystems. Many faunal investigations in arable fields compared crops or management practices on different soil types or in different years without regarding the variability between soils or the climate of different years. However, faunal differences between arable fields are very small. To give sufficient answers for the bioindication of ground beetles investigations on a very high number of fields for many years is needed. According to our investigations sand content of the soil plays the major role for the biodiversity of ground beetles on arable fields. However, the number of our sites investigated is still too low to give sufficient answers on the influence of field history, of the structure of field margins, or of adjacent ecosystems. Furthermore, future studies regarding these factors should focus on the problem how to measure the influence of field margins or adjacent habitats to generalise the results by statistical analyses.

5. Conclusions

The soil types are the major factor influencing the spatial pattern of the ground beetles in arable fields. However, concerning the results of the Detrended Correspondence Analysis even this plays only a low role in determining the composition of the ground beetle assemblages. Overall, ground beetle assemblages are very similar in arable fields independent from their soil types and agricultural practices, which reflects the trend of homogeneity in the last five decades. Area size is another important factor influencing the occurrence of ground beetle species. Several species were negatively correlated with field size. Thus, small fields contained more species than large fields and species diversity is higher. It appears, that the length of the field margin, therefore mainly determines the species richness of ground beetles in fields. Yearly variability in the composition of ground beetle assemblages is high and is independent from the crop, but correlated with different climate factors. Overall, the agricultural practices seem to play actually a minor role for the ground beetles, but this may be already a result of the biological homogeneity on fields. If ecological farming increase species diversity of ground beetles on fields, and there is some evidence in that, it will require several decades.

References

Basedow, T., 1988. Feldrand, Feldrain und Hecke aus der Sicht der Schädlingsregulation. Mitteilungen der Biologischen. Bundesanstalt für Land- und Forstwirtschaft 247, 129–137.

Basedow, T., 1990. Jährliche Vermehrungsraten von Carabiden und Staphyliniden bei unterschiedlicher Intensität des Ackerbaus. Zoologische Beiträge 33, 459–477.

Basedow, T., Liedtke, W., Rzehak, H., 1990. Die Populationsdichte der Getreideblattläuse und ihrer Antagonisten auf unterschiedlich intensiv bewirtschafteten Getreidefeldern in Schleswig-Holstein. Mitteilungen der deutschen Gesellschaft für allgemeine und angewandte Entomologie 7, 600–607.

Bilde, T., Toft, S., 1999. Prey consumption and fecundity of the carabid beetle *Calathus melanocephalus* on diets of three cereal aphids: high consumption rates of low-quality prey. Pedobiologia 43, 422–429.

Brasse, D., 1975. Die Arthropodenfauna von Getreidefeldern auf verschiedenen Böden im Braunschweiger Raum. Pedobiologia 15, 405–414.

Dennis, P., Fry, G.L.A., 1992. Field margins: can they enhance natural enemy population densities and general arthropod diversity on farmland? Agric. Ecosyst. Environ. 40, 95–115.

Duelli, P., 1997. Biodiversity evaluation in agricultural landscapes: an approach at two different scales. Agric. Ecosyst. Environ. 62, 81–91.

Fadl, A., Purvis, G., 1998. Field observations on the lifecycles and seasonal activity patterns of temperate carabid beetles (Coleoptera: Carabidae) inhabiting arable land. Pedobiologia 42, 171–183.

Fan, Y., Liebman, M., Groden, E., Alford, A.R., 1993. Abundance of carabid beetles and other ground dwelling arthropods in conventional versus low input bean cropping systems. Agric. Ecosyst. Environ. 43, 127–139.

Gall, G.A.E., Orians, G.H., 1992. Agriculture and biological conservation. Agric. Ecosyst. Environ. 42, 1–8.

Gilgenberg, A., 1986. Die Verteilungsstruktur der Carabiden- und Staphylinidenfauna verschieden bewirtschafteter landwirtschaftlicher Flächen sowie eines Waldes. Dissertation. Bonn, p. 261.

Heydemann, B., 1955. Carabiden der Kulturfelder als ökologische Indikatoren. Berichte Wanderversammlung deutsche Entomologen 7, 172–185.

Heydemann, B., 1986. Zielkonflikte zwischen Landwirtschaft, Landespflege und Naturschutz. IFOAM Zeitschrift für ökologische Landwirtschaft 56/57, 34–43.

Heydemann, B., 1997. Neuer Biologischer Atlas. Ökologie für Schleswig-Holstein und Hamburg. Wachholtz, Neumünster, p. 591.

Heydemann, B., Meyer, H., 1983. Auswirkungen der Intensivkultur auf die Fauna in den Agrarbiotopen. Schriftenreihe des Deutschen Rates für Landespflege 42, 174–191.

Hingst, R., Irmler, U., Steinborn, H.-A., 1995. Die Laufkäfergemeinschaften in Wald- und Agrarökosystemen Schleswig-Holsteins. Mitteilungen der deutschen Gesellschaft für allgemeine und angewandte Entomologie 9, 733–737.

Hoffmann-Kroll, R., Schäfer, D., Seibel, S., 1998. Die Ökologische Flächenstichprobe: ein Monitoring-Konzept des Bundes im Rahmen der Umweltökonomischen Gesamtrechnung (UGR). Ecosys. 7, 81–92.

Hokkanen, H., Holopainen, J.K., 1986. Carabid species and activity densities in biologically and conventionally managed cabbage fields. J. Appl. Entomol. 102, 353–363.

Huusela-Veistola, E., 1996. Effects of pesticide use and cultivation techniques on ground beetles (Col., Carabidae) in cereal fields. Ann. Zool. Fenn. 33, 197–205.

Irmler, U., 1999. Environmental characteristics of ground beetle assemblages in northern German forests as basis for an expert system. Zeitschrift Ökologie und Naturschutz 8, 227–237.

Irmler, U., Hanssen, U., Nötzold, R., Schröter, L., 2000. Biodiversität in der Agrarlandschaft. Bedeutung von Landschaftsstrukturen und Nutzungsänderungen. Mitteilungen der deutschen Gesellschaft für allgemeine und angewandte Entomologie 12, 311–321.

Kinnunen, H., Järveläinen, K., Pakkala, T., Tiainen, J., 1996. The effects of isolation on the occurrence of farmland carabids in a fragmented landscape. Ann. Zool. Fenn. 33, 165–171.

Kromp, B., 1999. Carabid beetles in sustainable agriculture: a review on pest control efficacy, cultivation impacts and enhancement. Agric. Ecosyst. Environ. 74, 187–228.

Luff, M.L., 1996. Use of carabids as environmental indicators in grasslands and cereals. Ann. Zool. Fenn. 33, 185–195.

Paoletti, M.G., Pimentel, D., Stinner, B.R., Stinner, D., 1992. Agroecosystem biodiversity: matching production and conservation biology. Agric. Ecosyst. Environ. 40, 3–23.

Pfiffner, L., Luka, H., 1996. Laufkäfer-Förderung durch Ausgleichsflächen. Auswirkungen neu angelegter Grünstreifen und einer Hecke im Ackerland. Naturschutz und Landschaftsplanung 28, 145–151.

Pfiffner, L., Niggli, U., 1996. Effects of bio-dynamic, organic and conventional farming on ground beetles (Col. Carabidae) and other epigaeic arthropods in winter wheat. Biol. Agric. Hortic. 12, 353–364.

Raskin, R., Glück, E., Pflug, W., 1992. Floren- und Faunenentwicklung auf herbizidfrei gehaltenen Agrarflächen. Natur und Landschaft 67, 7–14.

Schröter, L., Irmler, U., 1999. Einfluß von Bodenart, Kulturfrucht und Feldgröße auf Carabiden-Synusien der Äcker des konventionell-intensiven und des ökologischen Landbaus. Faunistisch-Ökologische Mitteilungen Supplement 27, p. 61.

Shannon, C.E., Weaver, W., 1963. The Mathematical Theory of Communication. University of Illinois Press, Urbana.

StatSoft, 1996. STATISTICA für Windows [Computer-Programm-Handbuch]. StatSoft Inc., Tulsa, OK.

Steinborn, H.-A., Heydemann, B., 1990. Indikatoren und Kriterien zur Burteilung der ökologischen Qualität von Agrarflächen am Beispiel der Carabidae (Laufkäfer). Schriftenreihe für Landschaftspflege und Naturschutz 32, 165–174.

Steinborn, H.-A., Meyer, H., 1994. Einfluß alternativer und konventioneller Landwirtschaft auf die Prädatorenfauna in Agrarökosystemen Schleswig-Holsteins (Araneida Coleoptera: Carabidae Diptera: Dolichopodidae, Empididae, Hybotidae, Microphoridae). Faunistisch-Ökologische Mitteilungen 6, 409–438.

ter Braak, C.J.F., 1986. 5 Ordination. In: Jongman, R.H.G., ter Braak, C.J.F., van Tongeren, O.F.R. (Eds.), Data Analysis in Community and Landscape Ecology. Pudoc, Wageningen.

Thomas, C.F.G., Parkinson, L., Marshall, E.J.P., 1998. Isolating the components of activity-density for the carabid beetle *Pterostichus melanarius* in farmland. Oecologia 116, 103–112.

Topp, W., 1989. Laufkäfer als Bioindikatoren in der Kulturlandschaft. Verhandlungen 9. SIEEC Gotha, 1986, pp. 78–82.

Wallin, H., 1985. Spatial and temporal distribution of some abundant carabid beetles (Coleoptera: Carabidae) in cereal fields and adjacent habitats. Pedobiologia 28, 19–34.

ELSEVIER

Available online at www.sciencedirect.com

Agriculture, Ecosystems and Environment 98 (2003) 153–161

Agriculture Ecosystems & Environment

www.elsevier.com/locate/agee

Which carabid species benefit from organic agriculture?—a review of comparative studies in winter cereals from Germany and Switzerland

Thomas F. Döring[a,*], Bernhard Kromp[b]

[a] *Department of Ecological Plant Protection, University of Kassel, Nordbahnhofstr. 1a, D-37213 Witzenhausen, Germany*

[b] *Ludwig-Boltzmann-Institute of Biological Agricultural and Applied Ecology, Rinnboeckstr. 15, A-1110 Vienna, Austria*

Abstract

Data of comparative studies about carabid beetles in organically and conventionally managed winter cereal fields of central Europe, using the pitfall trapping method, were collected from the literature and unpublished data sources and were then pooled and analysed. According to an index, which was designed to calculate how much a species benefits from organic management, *Carabus auratus* turned out to benefit most. Some *Amara* species (*A. familiaris*, *A. similata* and *A. aenea*) as well as *Pseudoophonus rufipes* and *Harpalus affinis* also showed high index values. When analysing the traits of the carabids, the habitat preference was the most important variable for the differentiation of organic and conventional management. The stronger the preference for open field, the more the species are supported by organic agriculture. For the promotion of the agricultural carabid fauna it is suggested that weedier and less densely cropped fields be tolerated.

Keywords: Carabidae; *Carabus auratus*; Winter cereals; Pitfall trap; Habitat preference

1. Introduction

Many studies have been undertaken during the last few decades to characterise the carabid fauna of central European winter cereal fields (e.g. Heydemann, 1953; Kirchner, 1960; Basedow, 1973; Lienemann, 1982; Weber, 1984; Honěk, 1988; also see Thiele, 1977). Some of them concentrate on comparing organic and conventional management practices (Letschert, 1986; Ingrisch et al., 1989; Froese, 1991; Basedow, 1987, 1998; Basedow et al., 1991; Steinborn and Meyer, 1994; Pfiffner and Niggli, 1996; Kaiser and Schulte, 1998; see also Kromp, 1999). Several carabid species were suggested as indicator species of organic or low-input management systems (Basedow et al., 1991; Steinborn and Heydemann, 1990; Kromp, 1985, 1990; Luka, 1996), but the assessment of indicator value and of reliability differs considerably. In order to give a more consistent picture, in this review data of several studies were pooled and reanalysed. Knowing which species benefit from organic husbandry of winter cereals and which do not, as well as the possible reasons for this, can help in the evaluation and design of agricultural management practices that meet the needs of these insects.

* Corresponding author. Tel.: +49-5542-98-1569; fax: +49-5542-98-1564.
E-mail address: doringt@wiz.uni-kassel.de (T.F. Döring).

0167-8809/$ – see front matter © 2003 Elsevier Science B.V. All rights reserved.
doi:10.1016/S0167-8809(03)00077-X

2. Methods

2.1. Data sources

For the comparison of the organic and the conventional cultivation system, data from pairs of fields were collected from the literature or—if the original data had not been published—directly from the investigators. One "pair" is defined as a pair of fields at one site in 1 year, except for the study of Basedow et al. (1991), where the data of a period of 4 years at one site could not be further divided. In all studies considered, pitfall traps were used to catch the carabids. Table 1 gives an overview of the study sites with information about soil type, soil texture, preserving agent and reference.

2.2. Indices for support of species by organic agriculture

For the comparison of species richness between organic and conventional management the Wilcoxon matched pairs signed rank test was calculated (Hollander and Wolfe, 1973, pp. 27–33). The same test was applied to assess differences in the activity density of each species (data not presented). However, as absolute differences within pairs tend to increase with dominance of species, comparisons between species of different dominance is difficult. Therefore, two dominance independent indices were designed and calculated for each species to determine to which degree it benefits from organic management. The first index (ε_s) counts in how many pairs the species was exclusively present in the organically managed field.

$$\varepsilon_s = \sum_{j=1}^{n} k_j \quad \text{with } k = 1 \text{ if } N_E > 0 \wedge N_C = 0;$$
$$k = 0 \text{ else}$$

where E: ecological (organic); C: conventional; N: activity density in the field; n: number of pairs; s: species.

The second index ρ_s was designed to emphasise the quantitative aspect by considering the ratio of activity density in the organic to the parallel conventional field. The geometric mean of the ratios over the pairs was transformed by $\log_{10}(x+1)$ in order to achieve clearer between species comparisons, thus ρ_s is defined as

$$\rho_s = \log \sqrt[n]{\prod_{j=1}^{n} \frac{N_{Ej}+1}{N_{Cj}+1}} = \frac{\sum_{j=1}^{n} \log(N_{Ej}+1) - \log(N_{Cj}+1)}{n}$$

The reliability of the indices increases with presence P of the species, P_s being defined as

$$P_s = \sum_{j=1}^{n} k_j \quad \text{with } k = 1 \text{ if } N_E > 0 \vee N_C = 0$$

In order to find out which ecological and biological traits of the carabid species are most important for the differentiation of organically and conventionally managed winter cereal fields, species with a certain trait (e.g. hygrophilous) were grouped and the average ρ_s-value of all the species of a group were calculated. The variables considered were humidity preference, wing dimorphism, seasonal reproduction type, habitat preference and body length. Information about the biology and ecology of the species was collected from Larsson (1939), Lindroth (1949), Thiele (1968, 1977), Freude et al. (1976), Turin et al. (1977), Kromp (1985), Luff (1987), Ingrisch et al. (1989), Koch (1989), Marggi (1992) and Wachmann et al. (1995). The diet was not analysed, because for many species of the genera *Harpalus* and *Amara*, there were no reliable data of food specialisation available, although these taxa are repeatedly called (mainly) herbivorous (e.g. Wachmann et al., 1995). Trends of ρ over ordered trait classes were tested with the non-parametric Jonckheere-test (Hollander and Wolfe, 1973, p. 120).

3. Results

3.1. Species richness

The species richness was significantly higher on the organically managed fields (Wilcoxon-test, $P < 0.001$). On average, there were 34% more species on the organic than on the conventional fields (geometric mean over all ratios of species numbers). Altogether 132 species in 121 719 captured individuals were included in the calculations.

Table 1
Comparison studies of carabid species: settings

Site(s)	Nearby city	Country and land[a]	Physiogeographic unit	Soil type (FAO)	Soil texture[b]	Preserving agent[c]	Reference	Number of pairs
Dannau/Siggen (Oldenburg)	Oldenburg/Holstein	D–SH	Ostholsteiner Hügel-land/Mittelholsteiner Geest	Luvisol	L	n.i.a.[d]	Steinborn and Meyer (1994)	1
Brodersdorf, Kreis Plön	Kiel	D–SH	Ostholsteiner Hügelland	n.i.a.	SL	F	Basedow (1987)	4
Hasenmoor/Heidmühlen (Bad Bramstedt)	Bad Bramstedt/Holstein	D–SH	Ostholsteiner Hügelland	Podzol	S	n.i.a.	Steinborn and Meyer (1994)	1
Wulksfelde/Wohldorf	Hamburg (N)	D–HH	Ostholsteiner Hügelland	n.i.a.	lS/uL	R	Olthoff et al. (1998)	1
Reckenfeld/Greven	Münster	D–NW	Sandmünsterland	Fimic Anthrosol (Esch)	(l)S	R	Döring (2000)	2
Bad Sassendorf	Soest	D–NW	Soester Boerde	Luvisol	SL	R	Brügge (1995)	2
Schwerter Ruhrtal	Dortmund	D–NW	Niedersauerland	Eutri-Cambic Fuvisol/Cambisol	L/sL	R	Flake (1996)	2
Etzweiler/Elsdorf/Düren	Düren (Köln/Aachen)	D–NW	Niederrheinische Bucht	Luvisol on Loess	LU	F	Letschert (1986), Winkens (1984–1986)	16
Massenheim/Gronau/ Dotten-feld (Bad Vilbel)	Frankfurt	D–HE	Südliche Wetterau	Luvisol	LU	B	Basedow et al. (1991) and Basedow (1995), Froese (1991)	5[e]
Therwil	Basel	CH–BA	Baselland	Luvisol on Loess		–[f]	Pfiffner and Niggli (1996)	1

[a] Germany (D); Schleswig-Holstein (SH); Nordrhine-Westfalia (NW); Hamburg (HH); Hessen (HE); Switzerland (CH); Baselland (BA).
[b] S: sand; s: sandy; L: loam; l: loamy; U: slit; u: silty.
[c] Preserving agent: R: Renner (1982): ethanol + acetic acid + glycerine + water; F: formaldehyde; B: sodium-benzoate.
[d] No information available.
[e] One pair over 4 years.
[f] Without preserving agent.

Table 2
Carabid species in winter cereals: indicator values for benefiting from ecological agriculture (ε_s, ρ_s: see Section 2.2), presence P_s and dominance (due to two-sided Wilcoxon-test)

Species (sorted by ρ_s)	ε_s	ρ_s	P_s	Dominance
All species***,[a]	0	0.37	35	100.00
Species with $\rho_s > 0.37$[b]				
C. auratus (L. 1761)***	14	1.26	25	4.85
Acupalpus meridianus (L., 1761)	6	0.91	9	0.50
A. familiaris (Duft., 1812)**	11	0.83	23	0.91
A. similata (Gyll., 1810)**	16	0.82	24	0.69
Poecilus versicolor (Sturm, 1824)	6	0.76	9	1.35
P. rufipes (DeGeer, 1774)*	8	0.75	32	2.46
A. aenea (DeGeer, 1774)**	12	0.72	26	0.66
H. affinis (Schrk., 1781)**	6	0.68	30	2.58
Bembidion lampros (Herbst, 1784)*	8	0.62	33	6.13
Bembidion properans (Steph., 1828)*	2	0.62	8	1.05
Agonum sexpunctatum (L. 1758)	12	0.59	15	0.08
Agonum muelleri (Herbst, 1784)*	4	0.56	28	2.14
A. plebeja (Gyll., 1810)	2	0.56	9	1.50
Calathus fuscipes (Goeze, 1777)	8	0.55	22	0.39
Poecilus cupreus (L., 1758)	10	0.52	28	3.94
H. distinguendus (Duft., 1812)	4	0.44	10	0.06
Nebria brevicollis (F., 1792)	10	0.39	25	1.12
Species with $0 < \rho_s < 0.37$[a]				
Anchomenus dorsalis (Pont., 1763)**	1	0.35	34	10.91
P. melanarius (Ill., 1798)*	1	0.35	34	31.33
Bembidion quadrimaculatum (L., 1761)*	12	0.35	22	1.51
Clivina fossor (L., 1758)	5	0.32	21	1.26
Stomis pumicatus (Panz., 1796)	8	0.22	18	0.41
P. vernalis (Panz., 1796)	6	0.15	20	0.17
Calathus melanocephalus (L., 1758)	6	0.13	14	0.08
Notiophilus biguttatus (F., 1779)	3	0.12	8	0.04
Dyschirius globosus (Herbst, 1784)	5	0.10	13	0.04
A. apricaria (Payk., 1790)	4	0.09	8	0.03
P. madidus (F., 1775)	2	0.08	8	0.15
C. nemoralis (Müll., 1764)	3	0.07	21	0.37
Trechus quadristriatus (Schrk., 1781)	2	0.06	33	1.60
P. strenuus (Panz., 1797)	4	0.05	9	0.03
C. granulatus (L., 1758)	4	0.04	17	0.80
Demetrias atricapillus (L., 1758)	4	0.00	23	0.15
Species with $\rho_s < 0$[a]				
Bembidion tetracolum (Say, 1823)	3	−0.05	30	10.23
H. rubripes (Duft., 1812)	3	−0.16	9	0.03
Trechoblemus micros (Herbst, 1784)	1	−0.18	9	0.06
L. pilicornis (F., 1775)*	4	−0.19	33	3.67
C. monilis (F., 1792)	0	−0.21	17	1.05
Synuchus vivalis (Ill., 1798)	4	−0.23	15	0.14
Asaphidion flavipes (L., 1761)	1	−0.26	26	2.16

[a] $\rho_s \neq 0$.
[b] $\rho_s - 0.37 \neq 0$.
* $P < 0.05$.
** $P < 0.01$.
*** $P < 0.001$.

3.2. Carabid species supported by organic agriculture

Carabid species with more than seven counts in the pairs were sorted by ρ_S-value (Table 2). *Carabus auratus*, *Harpalus affinis* and some *Amara* species (*A. aenea*, *A. similata*, *A. familiaris*) belong to the group of species which benefit more than the carabids in total from organic management (ρ_S-values > 0.37). Species that benefit less than the carabids in total from organic cultivation ($0 < \rho_S < 0.37$) are, e.g. *Demetrias atricapillus* or *Trechus quadristriatus*. A third block with $\rho_S < 0$ contains species which tend to benefit from conventional management, but there was no significant difference of the ρ_S-values to zero in these species, except for *Loricera pilicornis*. The comparison of the indices ε_S and ρ_S leads to differences in the assessment of the species, but the data seem to be consistent between the three blocks.

3.3. Traits of the carabid species

Although the standard deviation of ρ is rather high for all analysed traits, for some of them, trends can be observed (Table 3, Fig. 1). Concerning humidity preference and dispersal type, species that benefit most from organic management are xerophilous (Jonckheere-test: trend significant, $P < 0.05$), and macropterous (trend not significant). In the seasonal reproduction type there seems to be no directional pattern. The distribution of the ρ-values over the habitat preference groups revealed a clear and significant trend ($P < 0.001$, Fig. 1). The weaker the preference for woodland, the more these species are supported by organic management in winter cereals. The body length revealed no correlation with ρ (Spearman's $r_S < 0.01$). When looking at all variables together,

Table 3
Mean ρ-values (see Section 2.2) and standard deviation depending on different traits: seasonal reproduction type, wing dimorphism, and humidity preference

Trait	Mean ρ
Seasonal reproduction type	
Spring breeders	0.12 ± 0.34
Spring breeders with autumn population	0.35 ± 0.47
Autumn breeders	0.23 ± 0.36
Wing dimorphism	
Brachypterous	0.10 ± 0.58
Dimorphic	0.24 ± 0.29
Macropterous	0.28 ± 0.41
Humidity preference	
Hygrophilous	0.16 ± 0.38
Mesophilous	0.28 ± 0.46
Xerophilous	0.29 ± 0.38

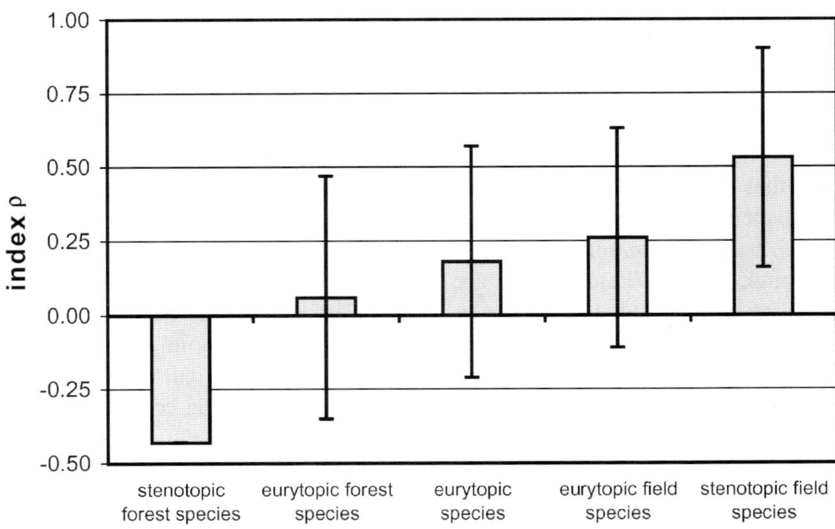

Fig. 1. Index (ρ) for support of carabid species by organic agriculture, for different habitat preference groups.

the most important variable seems to be the habitat preference, as it shows the highest range.

4. Discussion

In most cases, the species richness was higher in the organically than in the conventionally managed field. Therefore, the taxa with high ρ_S-values may carefully be used as indicator species for a higher diversity of carabid beetle communities in cereals.

C. auratus turned out to the most benefit from organic management. This carabid has already been mentioned as a sensitive indicator for extensive or organic management by Ingrisch et al. (1989), Brügge (1995), Flake (1996), Luka (1996) and Büchs et al. (1997). However, *C. auratus* is missing in the purely sandy areas due to preference for loamy soils (Gries et al., 1973). According to Basedow (1987) the species is negatively affected by mechanical weed control in spring as a usual method in organically managed fields. These results are supported by Büchs et al. (1997). They found that *C. auratus* mainly benefits from long-term set-aside (with no soil cultivation at all), due to its long larval period. A negative influence of pesticides on the populations of this species is probable according to Basedow (1987). Finally, Steinborn and Heydemann (1990) suggest that the sharp decrease of the populations of *C. auratus* between 1950s and 1980s on farmland in Holstein was caused by higher shading and reduced insolation in the cereal standing crop due to the increased fertilisation level. Until now it is not clear which factor is most important for the protection of *C. auratus* in cereals.

Among the species that benefit from organic management there are conspicuously frequent species of the genus *Amara*, here *A. aenea*, *A. familiaris* and *A. similata* (see also Kromp, 1989). These species are macropterous field species with reproduction in spring and adult hibernation. They are (at least partly) herbivorous (Thiele, 1977). Probably the food specialisation on plants is an important reason for the increased individual numbers of these species on organically managed fields. The methods of weed control in organic agriculture lead to higher cover and a greater number of weed species compared to the conventional use of herbicides (e.g. Frieben, 1997). Thus, in organically managed fields there is a higher food supply for herbivorous animals. *H. affinis*, which turned out to benefit from organic management, has already been mentioned by Basedow et al. (1991) as decreasing with higher intensity of cultivation. In its biology and ecology it resembles the *Amara* species mentioned above.

Pseudoophonus rufipes was mentioned indirectly as an indicator for lower cultivation intensity by Steinborn and Heydemann (1990). These authors found a sharp decrease after 1950. One reason for the better conditions for *P. rufipes* on the organically managed fields is possibly the omnivorous diet of this species, so that it benefits from higher weed cover. Being an autumn breeder, it is less affected by soil cultivation in spring as is typical of organic management.

According to several authors *Pterostichus melanarius* as a very euryoecous and eurytopic species seems to benefit from intense agriculture. Wallin (1985) found it more frequent on larger fields than on smaller ones. Tietze (1985) showed that populations of *P. melanarius* were higher in more intensively managed than in the more extensively treated grassland fields. By the same token, Kromp (1985) found higher numbers of the species in intensively fertilised conventional than in organically managed potato fields in Austria. Ingrisch et al. (1989, p. 121) assess *P. melanarius* as an indicator of low faunal diversity in fields. In an early review on pesticide side effects on carabids, Freitag (1979) states it is rather unsusceptible to pesticides. On the other hand Basedow et al. (1991) characterise *P. melanarius* as "decreasing with higher cultivation intensity"; Hokkanen and Holopainen (1986) found this species clearly benefiting from organic practices in a comparison study of cabbage in Germany. Considering all 35 pairs, on average the species benefits slightly from organic management. Also its dominance is higher in the organic (33.6%) than in the conventional (26.5%) fields.

The analysis of traits revealed that xerophilous species benefit more than the average of the species from organic agriculture. A possible reason for this is that the standing crop in organic agriculture might often be less dense and therefore drier than in conventional farming because of lower nitrogen fertilisation level or lower drilling density (Basedow, 1987, p. 42; Honěk, 1988; Brügge, 1995, pp. 38–40; Flake, 1996, p. 31; Pfiffner, 1997; Döring, 2000, p. 23).

The results from the analysis of the habitat preferences may possibly also be explained by the microclimatic conditions in the standing crop. Finally, this could also explain the distribution of dispersal types over organic and conventional fields, because the woodland or forest species which benefit less from organic agriculture than the field species, are often brachypterous. When considering the body length, the results differ from the experiences of Blake et al. (1994). They found that the average body length decreased clearly with increasing cultivation intensity in grassland (see also Lorenz, 1994; Büchs et al., 1997). The habitat preference seemed to be the most important variable in the differentiation of the two farming systems, because it showed the widest range of the ρ_S-values. This could be explained by the fact that the habitat preference integrates several traits. Although the type of diet was not included in the calculations, it could carefully be concluded from the results of Table 2 that the food supply for herbivorous species has a strong influence on the carabid communities on cereal fields.

The quantitative analysis of different comparison studies about carabids in organically and conventionally managed winter cereals may seem critical because the comparability of different pitfall catching methods can be doubted (Adis, 1979), especially when attracting preservatives are used. However, the calculation of the ρ_S-indices considered "relative" values (ratio of data from organic and conventional fields). With sufficient comparability within the pairs the results should therefore be preliminarily reliable. Moreover, the index refers to the activity abundance, hence it does not give completely reliable information on population size.

5. Conclusions

The more the carabid species are typical of agricultural fields (i.e. the less they are bound to woodland), the more they benefit from organic agriculture. Therefore, organic agriculture is considered to support high carabid diversity, especially referring to the typical field species. Thus, for the promotion of the typical carabid fauna of the cereal fields, organic agriculture should be supported. However, the presented analysis suggests that microclimatic conditions and food supply strongly influence the carabid fauna of cereal fields. For that reason (to a certain degree), weedier and less dense (i.e. lighter and drier) fields should be tolerated in both conventional and ecological management.

Acknowledgements

We thank S. Ingrisch (Bad Karlshafen), T. Basedow (Gießen), and A. Fonck (Aachen) for providing data for the analysis and M. Kaiser (Münster) for helpful discussion.

References

Adis, J., 1979. Problems of interpreting arthropod sampling with pitfall traps. Zool. Anz. Jena 202, 177–184.

Basedow, T., 1973. Der Einfluß epigäischer Raubarthropoden auf die Abundanz phytophager Insekten in der Agrarlandschaft. Pedobiologia 13, 410–422.

Basedow, T., 1987. Der Einfluß gesteigerter Bewirtschaftungsintensität im Getreidebau auf die Laufkäfer (Coleoptera, Carabidae)—Auswertung vierzehnjähriger Untersuchungen (1971–1984). Habilitation University, Gießen.

Basedow, T., 1995. Table of carabid catches in Dottenfeld, Gronau and Massenheim/Bad Vilbel for 1993 to 1995. University of Giessen Institute for Phytopathology and Applied Zoology. Unpublished.

Basedow, T., 1998. Langfristige Bestandsveränderungen von Arthropoden in der Feldflur, ihre Ursachen und deren Bedeutung für den Naturschutz, gezeigt an Laufkäfern (Carabidae) in Schleswig-Holstein, 1971–1996. Schriftenr. Landschaftspflege Naturschutz 58, 215–227.

Basedow, T., Braun, C., Lühr, A., Naumann, J., Norgall, T., Yanes, G., 1991. Abundanz, Biomasse und Artenzahl epigäischer Raubarthropoden auf unterschiedlich intensiv bewirtschafteten Weizen- und Rübenfeldern: Unterschiede und ihre Ursachen. Ergebnisse eines dreistufigen Vergleichs in Hessen, 1985 bis 1988. Zool. Jahrb. Syst. 118, 87–116.

Blake, S., Foster, G.N., Eyre, M.D., Luff, M.L., 1994. Effects of habitat type and grassland management practices on the body size distribution of carabid beetles. Pedobiologia 38, 502–512.

Brügge, O., 1995. Inventarisierung eines alternativ wirtschaftenden Bauernhofes bezüglich der biologischen Ausstattung in der Soester Börde. M.Sc. Thesis (Diploma) at the Institute of Landscape Ecology, University of Münster, Germany.

Büchs, W., Harenberg, A., Zimmermann, J., 1997. The invertebrate ecology of farmland as a mirror of the intensity of the impact of man?—an approach to interpreting results of field experiments carried out in different crop management intensities of a sugar beet and an oil seed rape rotation including set-aside. Biol. Agric. Hortic. 15, 83–107.

Döring, T.F., 2000. Analyse und Bewertung der Laufkäferfauna auf alternativ und konventionell bewirtschafteten Äckern im Sandmünsterland. M.Sc. Thesis (Diploma) at the Institute of Landscape Ecology, University of Münster, Germany.

Flake, A., 1996. Landschaftsökologischer Vergleich alternativer und konventioneller landwirtschaftlicher Betriebe in der Ruhraue bei Schwerte. M.Sc. Thesis (Diploma) at the Institute of Landscape Ecology, University of Münster, Germany.

Freitag, R., 1979. Carabid beetles and pollution. In: Halpern, A.L. (Ed.), Carabid Beetles—their Evolution, Natural History and Classification. Junk Publishers, Hague, pp. 507–521.

Freude, H., Harde, K.W., Lohse, G.A., 1976. Die Käfer Mitteleuropas, vol. 2, Adephaga 1, Krefeld, Germany, 302 p.

Frieben, B., 1997. Arten- und Biotopschutz durch Organischen Landbau. In: Weiger, H., Willer, H. (Eds.), Naturschutz durch ökologischen Landbau. Stiftung Ökologie und Landbau (SÖL) Deukalion-Verlag, Holm, Germany, pp. 73–92.

Froese, A., 1991. Untersuchungen über Carabiden auf unterschiedlich bewirtschafteten Ackerflächen unter Berücksichtigung des Feldrandeffekts (Coleoptera: Carabidae). Entomol. Z. 101, 213–226.

Gries, B., Mossakowski, D., Weber, F., 1973. Coleoptera Westfalica: familia Carabidae, genera *Cychrus*, *Carabus* und *Calosoma*. Abh. Westfäl. Mus. Naturkd. 35 (4), 1–80.

Heydemann, B., 1953. Agrarökologische Problematik—dargetan an Untersuchungen über die Tierwelt der Bodenoberfläche der Kulturfelder. Ph.D. Thesis. University of Kiel.

Hokkanen, H., Holopainen, J.K., 1986. Carabid species and activity in biologically and conventionally managed cabbage fields. J. Appl. Entomol. 102, 353–363.

Hollander, M., Wolfe, D.A., 1973. Nonparametric Statistical Methods. Wiley, New York, 503 p.

Honěk, A., 1988. The effect of crop density and microclimate on pitfall trap catches of Carabidae, Staphylinidae (Coleoptera) and Lycosidae (Araneae) in cereal fields. Pedobiologia 32, 233–242.

Ingrisch, S., Wasner, U., Glück, E., 1989. Vergleichende Untersuchungen der Ackerfauna auf alternativ und konventionell bewirtschafteten Flächen. In: König, W., Sunkel, R., Necker, U., Wolff-Straub, R., Ingrisch, S., Wasner, U., Glück, E. (Eds.), Alternativer und Konventioneller Landbau. Schriftenr. Landesanstalt für Ökologie, Landschaftsentwicklung und Forstplanung Nordrhein-Westfalen 11, pp. 113–271.

Kaiser, M., Schulte, G., 1998. Vergleich der Laufkäferfauna (Coleoptera, Carabidae) alternativ und konventionell bewirtschafteter Äcker in Nordrhein-Westfalen. In: Ebermann, E. (Ed.), Arthropod Biology: Contributions to Morphology, Ecology and Systematics. Biosystematics Ecol. Ser. 14, 365–384.

Kirchner, H., 1960. Untersuchungen zur Ökologie feldbewohnender Carabiden. Ph.D. Thesis. University of Cologne.

Koch, K., 1989. Die Käfer Mitteleuropas, Ökologie vol. 1. Verlag Goecke & Evers, Krefeld, pp. 15–107.

Kromp, B., 1985. Zur Laufkäferfauna (Coleoptera, Carabidae) von Äckern in drei Gegenden Österreichs unter besonderer Berücksichtigung der Bewirtschaftungsweise. Ph.D. Thesis. University of Vienna.

Kromp, B., 1989. Carabid beetle communities (Carabidae, Coleoptera) in biologically and conventionally farmed agroecosystems. Agric. Ecosyst. Environ. 27, 241–251.

Kromp, B., 1990. Carabid beetles (Coleoptera, Carabidae) as bioindicators in biological and conventional farming in Austrian potato fields. Biol. Fertil. Soils 9, 187–192.

Kromp, B., 1999. Carabid beetles in sustainable agriculture: a review on pest control efficacy, cultivation impacts and enhancement. Agric. Ecosyst. Environ. 74, 187–228.

Larsson, S.G., 1939. Entwicklungstypen und Entwicklungszeiten der dänischen Carabiden. Entomol. Medd. 20, 277–560.

Letschert, D., 1986. Untersuchungen zur Arthropoden- und Annelidenfauna von Weizen- und Zuckerrübenfeldern in einem konventionellen und einem biologisch-dynamischen Anbau. Z. Angew. Zool. 73, 93–113.

Lienemann, K., 1982. Beitrag zur Carabidenfauna landwirtschaftlich genutzter Flächen. Decheniana 135, 45–56.

Lindroth, C.H., 1949. Die fennoskandischen Carabidae - eine tiergeographische Studie. Part III. Göteborgs K. Vetensk. Vitterh. Samh. Handl. B4, Stockholm, 911 p.

Lorenz, E., 1994. Mechanische Unkrautbekämpfungsverfahren in Zuckerrübenkulturen und ihre Nebenwirkungen auf Laufkäfer (Coleoptera, Carabidae) und andere epigäische Arthropoden. Ph.D. Thesis. University of Göttingen.

Luff, M.L., 1987. Biology of polyphagous ground beetles in agriculture. Agric. Zool. Rev. 2, 237–278.

Luka, H., 1996. Laufkäfer: Nützlinge und Bioindikatoren in der Landwirtschaft. Agrarforschung 3 (1), 33–36.

Marggi, W.A., 1992. Faunistik der Sandlaufkäfer und Laufkäfer der Schweiz (Cicindelidae & Carabidae) Coleoptera Teil 1. Docum. Faunist. Helvetiae 13. Centre suisse de cartographie de la faune, Neuchatel.

Olthoff, T., Richter, J., Stieg-Lichtenberg, H., 1998. Ökologische Begleituntersuchungen auf den Gütern Wulfsdorf, Wulksfelde und Wohldorfer Hof. Report for the Environmental Agency Hamburg. Umweltbehörde Hamburg, Germany. Unpublished.

Pfiffner, L., 1997. Welchen Beitrag leistet der ökologische Landbau zur Förderung der Kleintierfauna?. In: Weiger, H., Willer, H. (Eds.), Naturschutz durch ökologischen Landbau. Ökologische Konzepte 9, Deukalion, Holm, pp. 93–120.

Pfiffner, L., Niggli, U., 1996. Effects of bio-dynamic, organic and conventional farming on ground beetles (Col. Carabidae) and other epigaeic arthropods in winter wheat. Biol. Agric. Hortic. 12, 353–364.

Renner, K., 1982. Coleopterenfänge mit Bodenfallen am Sandstrand der Ostseeküste -ein Beitrag zum Problem der Lockwirkung von Konservierungsmitteln. Faunistisch-Ökol. Mitt. (Kiel) 5, 127–146.

Steinborn, H.-A., Heydemann, B., 1990. Indikatoren und Kriterien zur Beurteilung der ökologischen Qualität von Agrarflächen am Beispiel der Carabidae (Laufkäfer). Schriftenr. Landschaftspflege Naturschutz 32, 165–174.

Steinborn, H.-A., Meyer, H., 1994. Einfluß alternativer und konventioneller Landwirtschaft auf die Prädatorenfauna in Agrarökosystemen Schleswig-Holsteins (Araneida, Coleoptera: Carabidae, Diptera: Dolichopodidae, Empididae, Hybotidae, Microphoridae). Faunistisch-Ökol. Mitt. 6, 409–438.

Thiele, H.U., 1968. Was bindet Laufkäfer an ihre Lebensräume? Naturwiss. Rundsch. 21 (2), 57–65.

Thiele, H.U., 1977. Carabid Beetles in Their Environments. Springer, Berlin, 369 p.

Tietze, F., 1985. Veränderungen der Arten- und Dominanzstruktur in Laufkäfer-taxozönosen (Coleoptera, Carabidae) bewirtschafteter Grasland-Ökosysteme durch Intensivierungsfaktoren. Zool. Jahrb. Syst. 112, 337–382.

Turin, H., Haeck, J., Hengeveld, R., 1977. Atlas of the Carabid Beetles of The Netherlands. North-Holland, Amsterdam, 229 p.

Wachmann, E., Platen, R., Barndt, D., 1995. Laufkäfer - Beobachtung, Lebensweise. Naturbuch Verlag, Augsburg, 295 p.

Wallin, H., 1985. Spatial and temporal distribution of some abundant carabid beetles (Coleoptra, Carabidae) in cereal fields in adjacent habitats. Pedobiologia 28, 19–34.

Weber, G., 1984. Die Carabidenfauna landwirtschaftlicher Nutzfläche bei Bonn (Coleoptera, Carabidae). Decheniana 137, 112–124.

Winkens, H., 1984–1986. Faunistische Untersuchungen zum Alternativen Landbau - Zwischenbericht für die Jahre 1983 bis 1986. Untersuchte Tiergruppe: Coleoptera (partim). Unpublished report.

ELSEVIER

Available online at www.sciencedirect.com

Agriculture, Ecosystems and Environment 98 (2003) 163–167

Agriculture Ecosystems & Environment

www.elsevier.com/locate/agee

Regional diversity of temporary wetland carabid beetle communities: a matter of landscape features or cultivation intensity?

Ulrich Brose*

Centre for Agricultural Landscape and Land-Use Research, Institute of Land-Use Systems and Landscape Ecology, Eberswalder Straße 84, 15374 Müncheberg, Germany

Abstract

The challenge of finding applicable indicators for sustainable agriculture requires evaluations at regional scales to lead to policy-relevant results. In this study, the regional diversity of temporary wetland carabid beetles was analysed for six landscapes of 10 km^2 each. The relative importance of landscape features and cultivation intensity for the regional diversity was compared. Total species richness was correlated with the mean soil-indices that were used as indicators of cultivation intensity. This is consistent with studies on local scales, which emphasise the importance of cultivation intensity for arthropod communities. The diversity of wetland and habitat-specific species correlated with the temporary wetlands mean duration of flooding and the density of temporary wetlands, but apart from this, there was no impact of landscape features on diversity. These results do not corroborate concepts of using indices of landscape structure as biodiversity indicators, but the importance of cultivation intensity cannot be too strongly emphasised.

Keywords: Biodiversity; Species richness; Macro-ecology; Landscape ecology; Habitat heterogeneity

1. Introduction

As a consequence of the Rio convention, biodiversity has been used as a means to assess landscape sustainability (Paoletti, 1999). On regional scales, interest was directed towards the mosaic structure of landscapes and its influence on biodiversity (Hansson et al., 1995). This has led to the mosaic concept, which predicts that species richness increases with habitat variability, i.e. the number of habitat types, and habitat heterogeneity, i.e. the number and proportional distribution of habitat patches with constant habitat variability (Duelli, 1997).

On local scales, the importance of cultivation intensity for arable land communities has been emphasised (review in Kromp, 1999). While our knowledge of animal and plant communities of arable fields has been continuously growing, we know little about the communities of the accompanying small within-field habitats, such as temporary wetlands. However, they contribute a substantial proportion to the biodiversity in agricultural landscapes. Hence, in this study the combined effects of landscape features and cultivation intensity on the regional diversity of temporary wetland carabid beetles are analysed. Due to the fact that total diversity is sometimes dubious in evaluating

* Present address: Department of Biology, Romberg Tiburon Center, San Francisco State University, 3152 Paradise Drive, Tiburon, CA 94920, USA. Tel.: +1-415-338-3742; fax: +1-415-435-7120.
E-mail address: brose@sfsu.edu (U. Brose).

0167-8809/$ – see front matter © 2003 Elsevier Science B.V. All rights reserved.
doi:10.1016/S0167-8809(03)00078-1

land-use, the diversity of obligate wetland species and habitat-specific species is also taken into account.

2. Materials and methods

The study was carried out in the agricultural landscape of north-eastern Germany (between 53°22′N, 13°34′E and 52°22′N, 14°15′E; Fig. 1). This younger pleistocene landscape is characterised by numerous types of potholes and temporary wetlands. Temporary wetlands are cultivated and typically produce crops in the driest years. During years of average or above-average precipitation, temporary wetlands retain water until spring or summer. In 1998, six landscapes of 10 km^2 each were chosen, and the fieldwork was carried out until 2000.

2.1. Landscape mosaic structure

The habitat types of the landscapes were classified into seven groups: small habitats (temporary wetlands, potholes) were recorded by their numbers and the main land-use types (meadows, arable land, forests, fallow land and marshland) by their surface area. All entities refer to the landscape area of 10 km^2. Landscape diversity was calculated as the Shannon–Wiener diversity of the main land-use types:

$$H_S = -\sum_{j=1}^{S} p_j \log p_j,$$

where S represents the number of habitat types and p_j the proportion of habitat type j (Magurran, 1988). Landscape diversity and the densities of potholes and temporary wetlands represent habitat heterogeneity. Habitat variability (number of habitat types) was equal in all landscapes.

2.2. Cultivation intensity and mean duration of flooding

The cultivation intensity of the arable fields was indicated by the soil-indices (German: mittlere Bodenzahlen), which provide information about the productivity of the soils and indicate the potential yield. Furthermore, the soil productivity determines which crops can be grown on the fields. As the application of fertilisers and pesticides depend on crop type and potential yield, both applications correlate with the soil index. The soil-indices were determined for each arable field in the study areas and, subsequently, the

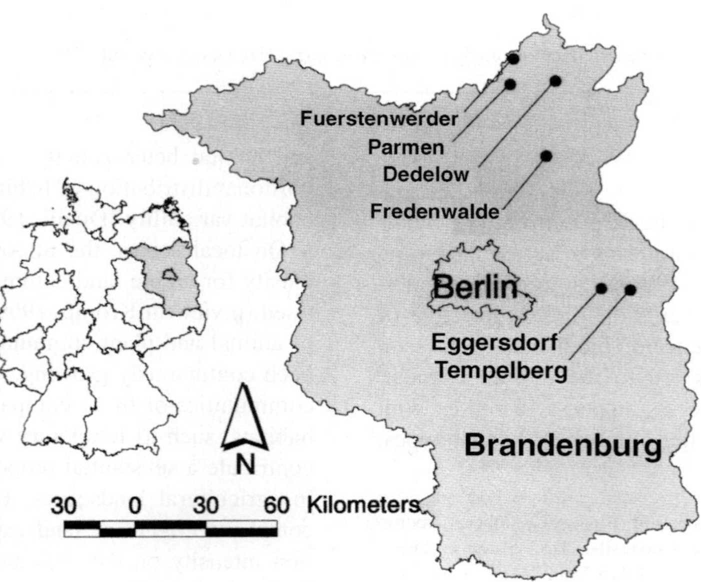

Fig. 1. Location of the landscapes studied in Brandenburg, Germany.

mean of the soil-indices was recorded for each of the landscapes studied.

From March to July 1998, the water-levels of the temporary wetlands were recorded. In the middle of each month, it was recorded for each temporary wetland in the landscapes studied whether it still contained flooded parts (Brose, 2001). The hydroperiods of the temporary wetlands ranged between 1 (drying up until mid-April) and 4 (drying up until mid-July). The mean of these hydroperiods was recorded as the mean duration of flooding in the landscapes.

2.3. Sampling design for carabid beetle diversity

In each landscape, six temporary wetlands were sampled with five pitfall traps on each site. The surface area of the temporary wetlands studied ranged between 50 and 2400 m^2 (mean = 926 + 568). Pitfall traps with a minimum separation distance of 4 m were installed from mid-April until mid-July, when harvesting and soil tillage began. To assess regional diversity, the catches were totalled within each landscape. Three variables of diversity were recorded: total species richness (all species), number of wetland species (Scheffler et al., 1999) and number of habitat-specific species (own classification based on comparisons with other agricultural habitats).

2.4. Data analysis

The data set met the assumptions of normality and linearity, and the independent variables were not correlated (Pearson's product moment correlation). Stepwise multiple regression analyses were performed with a significance level P of 0.05 for entering a variable into the model and a significance P of 0.1 to retain the variable.

3. Results

Total species richness of the landscapes studied ranged between 74 and 94 (see Table 1 for an overview of the results). There was a strong negative correlation between the mean soil-indices and total species richness (Fig. 2). However, there was no correlation between the mean soil-indices and the number of wetland and habitat-specific species, which were both positively correlated with the mean duration of flooding (Fig. 3). The results of the stepwise regression analyses are summarised in Table 2. The number of habitat-specific species also depended on the density of temporary wetlands. Apart from the density of temporary wetlands, no other landscape feature was correlated with the dependent variables.

Table 1
The study landscapes: mean soil-indices, mean duration of flooding (number of month after mid-March), landscape features and carabid beetle diversity[a]

	Egg	Temp	Dedel	Parm	Fuerst	Fred
Mean soil-indices	32.33	39.29	52.65	45.99	46.90	45.02
Mean duration of flooding	2.64	2.15	3.11	2.71	2.30	3.00
Landscape diversity	0.16	0.14	0.24	0.22	0.55	0.44
Landscape features						
Density of potholes	45	42	23	17	61	27
Density of temporary wetlands	13	17	8	19	27	15
Marshland (ha)	3.44	4.56	64.04	12.35	51.98	80.67
Meadowland (ha)	19.87	11.39	33.58	110.38	133.11	49.07
Forest (ha)	54.77	28.58	38.63	14.76	160.46	112.76
Arable land (ha)	865.35	883.06	833.47	803.48	449.80	652.66
Fallow land (ha)	3.66	20.03	0.00	0.00	41.60	38.71
Regional diversity						
Total species richness	94	89	74	78	75	84
Wetland species	42	38	43	38	38	43
Habitat-specific species	11	9	12	11	10	12

[a] The small habitat types were recorded by their numbers, the main land-use types by their surface area. Landscape diversity = Shannon–Wiener H_S of main land-use types. Study landscapes: Egg: Eggersdorf, Temp: Tempelberg, Dedel: Dedelow, Parm: Parmen, Fuerst: Fuerstenwerder, Fred: Fredenwalde.

Table 2
Results of the stepwise regression analyses

Dependent	Independent	r^2	Coefficient + S.E.	P	n
All species	Mean soil index	0.90	−1.10 + 0.19	**	6
Wetland species	Mean duration of flooding	0.681	5.82 + 2.0	*	6
Habitat-specific species	Mean duration of flooding	0.958	3.75 + 0.14	***	6
	Density of temporary wetlands	0.997	0.05 + 0.008		

* $P < 0.05$.
** $P < 0.01$.
*** $P < 0.001$.

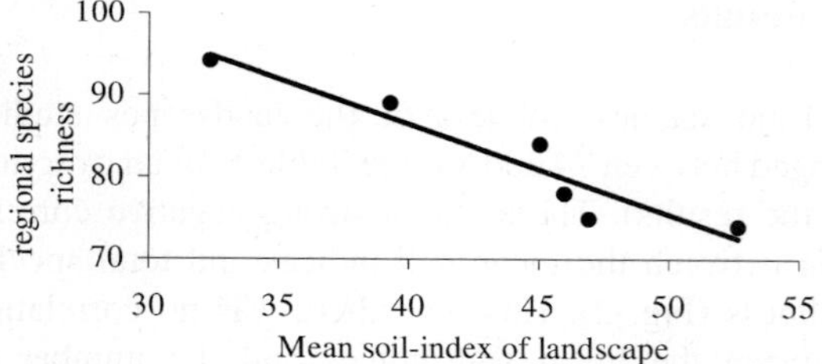

Fig. 2. The relationship between the mean soil-indices of the landscapes (indicators of land-use intensity) and the regional species richness of temporary wetlands' carabid beetle communities ($r^2 = 0.9$; $P < 0.01$, $n = 6$).

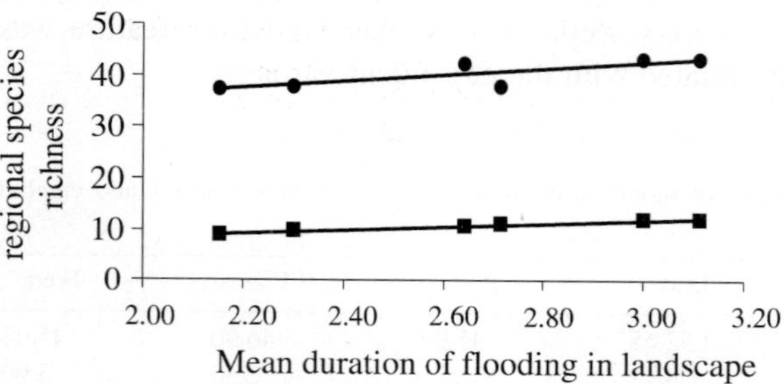

Fig. 3. The relationship between the mean duration of flooding (number of months after mid-March) in the landscapes and the regional species richness of temporary wetlands' wetland species (circles, $r^2 = 0.681$, $P < 0.05$, $n = 6$) or habitat-specific species (squares, $r^2 = 0.958$, $P < 0.001$, $n = 6$).

4. Discussion

The relationship between landscape features and regional diversity of carabid beetles was analysed. It has been hypothesised that heterogeneous landscapes have a higher regional diversity, because meta-community-dynamics lead to a faster recolonisation of vacant niches (Duelli, 1997). Apart from the density of temporary wetlands, the studied landscape features did not have an impact on regional diversity, which contradicts the mosaic concept. However, communities of arable land are distinct from those of other habitats, primarily because the sites are ploughed. Therefore, recolonisation of vacant niches is unlikely for species that belong to more natural and unploughed habitats. Empirical studies on the consequences of different landscape structures have been largely restricted to the community responses at the spatial levels of habitats or plots of few square meters (Kareiva, 1987; Wiens et al., 1993; Hansson et al., 1995; Collinge and Forman, 1998; Thies and Tscharntke, 1999). In these studies, correspondence to landscape features has mainly been reported for flying arthropods. In a study that included flying as well as ground-inhabiting arthropods, Jeanneret et al. (2000) have documented varying effects of landscape features; while butterflies were affected, spider communities were not. As in the present study the regional diversity of carabid beetles did not respond to landscape heterogeneity, this pattern might be supported at a regional level. In conclusion, there are two explanations for this lack of relationship: (i) vacant niches at temporary wetlands are not recolonised by species of more natural habitats; (ii) landscape features are generally less important for ground-inhabiting organisms.

There was a strong impact of cultivation intensity—indicated by the mean soil-indices—on total species richness. This is in accordance with studies on local scales, where similar effects have been explained by application of pesticides and differences in the land-use systems (Büchs et al., 1997; Kromp, 1999). Surprisingly, cultivation intensity has not yet been included in studies on the landscape scale, a fact that might be due to problems of finding indicators on this topic scale. The mean soil index provided a relative indicator, which is due to changes in production systems limited to concurrent comparisons of landscapes in the same region. Future research should lead to absolute indicators, which are transferable in space and time. However, the present study emphasises the importance of cultivation intensity as a key factor for regional diversity.

The diversity of wetland and habitat-specific species was strongly dependent on the mean duration of flooding. There might be two reasons: (i) a high attractiveness of landscapes with a high mean duration of flooding for potential immigrants (Duelli, 1997) and (ii) a generally high number of available niches for hygrophilous species in these landscapes.

5. Conclusion

The present study could not confirm the universal importance of landscape heterogeneity for species diversity, as predicted by the mosaic concept. Accordingly, the application of landscape heterogeneity as an indicator for sustainable land-use might be restricted to specific habitats or certain taxa. The regional diversity of carabid beetles was dependent on the mean soil-indices. However, future research is needed to create absolute parameters of cultivation intensity on the landscape scale, which may result in potential indicators for sustainable land-use. For the diversity of wetland and habitat-specific species, the mean duration of flooding was a strong predictor. This variable, which is easily accessible by aerial pictures, might be an indicator for sustainable land-use with respect to the carabid beetle communities of temporary wetlands.

Acknowledgements

I thank S. Samu, D. Wrase, S. Ehlert, S. Ellgen, I. Wolf and M. Glemnitz for their help. This work was funded through a fellowship from the state of Brandenburg.

References

Brose, U., 2001. Relative importance of isolation, area and habitat heterogeneity for vascular plant species richness of temporary wetlands in east-German farmland. Ecography 24, 722–730.

Büchs, W., Harenberg, A., Zimmermann, J., 1997. The invertebrate ecology of farmland as a mirror of the intensity of the impact of man?—An approach to interpreting results of field experiments carried out in different crop management intensities of a sugar beet and an oil seed rape rotation including set-aside. In: Kromp, B., Meindl, P. (Eds.), Entomological Research in Organic Agriculture. Biol. Agric. Hortic. 15, 83–107.

Collinge, S.K., Forman, R.T.T., 1998. A conceptual model of land conversion processes: predictions and evidence from a microlandscape experiment with grassland insects. Oikos 82, 66–84.

Duelli, P., 1997. Biodiversity evaluation in agricultural landscapes: an approach at two different scales. Agric. Ecosyst. Environ. 62, 81–91.

Hansson, L., Fahrig, L., Merriam, G., 1995. Mosaic Landscapes and Ecological Processes. Chapman & Hall, London.

Jeanneret, P., Schüpach, B., Dreier, S., Pfiffner, L., Pozzi, S., Walter, T., Bigler, F., Herzog, F., 2000. Biodiversity in cultivated landscapes: are landscape features important? In: Proceedings of the 13th IFOAM Scientific Conference.

Kareiva, P., 1987. Habitat fragmentation and the stability of predator–prey interactions. Nature 326, 388–390.

Kromp, B., 1999. Carabid beetles in sustainable agriculture: a review on pest control efficacy, cultivation impacts and enhancement. Agric. Ecosyst. Environ. 74, 187–228.

Magurran, A.E., 1988. Ecological Diversity and Its Measurement. Princeton University Press, Princeton, NJ.

Paoletti, M.G., 1999. Using bioindicators based on biodiversity to assess landscape sustainability. Agric. Ecosyst. Environ. 74, 1–18.

Scheffler, I., Kielhorn, K.H., Wrase, D.W., Korge, H., Braasch, D., 1999. Rote Liste und Artenliste der Laufkäfer des Landes Brandenburg (Coleoptera: Carabidae). Naturschutz und Landschaftspflege in Brandenburg 8, 3–27.

Thies, C., Tscharntke, T., 1999. Landscape structure and biological control in agroecosystems. Science 285, 893–895.

Wiens, J.A., Stenseth, N.C., Van Horne, B., Ims, R.A., 1993. Ecological mechanisms and landscape ecology. Oikos 66, 369–380.

ELSEVIER

Available online at www.sciencedirect.com

Agriculture, Ecosystems and Environment 98 (2003) 169–181

Agriculture Ecosystems & Environment

www.elsevier.com/locate/agee

Assessment of changing agricultural land use: response of vegetation, ground-dwelling spiders and beetles to the conversion of arable land into grassland

Jörg Perner [a,*], Steffen Malt [b,1]

[a] *Institute of Ecology, Friedrich-Schiller-University, Dornburger Str. 159, D-07743 Jena, Germany*

[b] *Centre for Agricultural Landscape and Land Use Research (ZALF e.V.), Institute of Land Use Systems and Landscape Ecology, Eberswalder Str. 84, D-15374 Müncheberg, Germany*

Abstract

An ecological indicator approach was used to examine the effects of changing agricultural land use on vegetation and ground-dwelling spider and beetle assemblages. An arable land site (control) and six differently managed grassland sites (with different time since conversion from arable land or time of current management) were comparatively investigated in a 3-year study. Whereas species richness increased with decreasing management impact for plants and spiders, the Camargo's evenness (E') index increased for all three examined assemblages. This suggests a very trustworthy community parameter for ecological indication studies. Differences did occur between the vegetation and the invertebrate groups in the assessment of the grassland sites. Ordination analysis indicated a much better separation of the different sites based on the invertebrate data than is possible on the basis of the vegetation data. These considerable differences were attributable to the short reaction time of these groups to changes in land use: ground-dwelling spiders and beetles mainly respond to changes in the microclimate and the soil-moisture. Efficient indication of restoration management is therefore possible after 3–5 years. Vegetation assemblages appeared as less powerful indicators of short-term restoration processes. We suggest that vegetation monitoring be used as a more powerful long-term approach but that it should be coupled with (short-term and sensitive) invertebrate monitoring (e.g. ground-dwelling spiders and beetles) especially at the beginning of the agricultural restoration processes.

Keywords: Araneae; Biodiversity; Bioindication; Coleoptera; Evenness; Agricultural land use changes; Restoration ecology; Species richness; Vegetation

1. Introduction

Changes in land use, habitat fragmentation, nutrient enrichment, and environmental stress often affect species diversity in ecosystems (Chapin et al., 1997; Vitousek et al., 1997). van der Putten et al. (2000) pointed out that, at both local and regional scales, land use changes are among the most immediate drivers of species diversity. Furthermore, they explained that one of the possible ways of counteracting the loss of biodiversity resulting from agricultural intensification during the last decades might be to reduce land use intensity. Duelli (1997) also mentioned that intensive agricultural management has led to an alarming level of 'ecological degradation' and that a less intensive land use could have two different effects. On the

* Corresponding author. Tel.: +49-3641-949421; fax: +49-3641-949402.

E-mail addresses: bjp@rz.uni-jena.de (J. Perner), smalt@zalf.de (S. Malt).

[1] Tel.: +49-33432-82146/82245; fax: +49-33432-82387.

0167-8809/$ – see front matter © 2003 Elsevier Science B.V. All rights reserved.
doi:10.1016/S0167-8809(03)00079-3

one hand, less intensive managed agricultural areas enrich regional species diversity while, at the same time those same areas could have a positive effect on beneficial arthropods by providing habitats after crop harvest or for hibernation. Goedmakers (1989) refers to the need to reduce overproduction and also to diminish agricultural pollution within the European Union from an economic perspective. Furthermore, Goedmakers points out that changing agricultural land into conservation areas increase substantially the ecological potential of the rural environment, as measured by species diversity or the presence of indicator species.

Bioindication has tended to be used somewhat "... loosely and has been adopted in a broad range of contexts, including the indication of habitat alteration, destruction, contamination and rehabilitation, vegetation succession, climate change and species diversity..." (McGeoch, 1998). More recently, the term and its many synonyms has been reclassified into three categories: (i) *environmental indicators* that reflect the biotic or abiotic state of an environment; (ii) *ecological indicators* that reveal evidence for, or the impacts of, environmental change; (iii) *biodiversity indicators* that specifically indicate the diversity of species, taxa, or entire communities within an area (McGeoch, 1998; Lawton and Gaston, 2001). Following this classification any attempt to identify the effects of changing agricultural land use on biotic systems should focus on *ecological indicators* defined as characteristic taxa or assemblages "... sensitive to identified environmental stress factors on biota, and whose response is representative of the response of at least a subset of other taxa present in the habitat..." (McGeoch, 1998).

Plants and plant communities are frequently used groups in ecological bioindication and monitoring (= repeated application of bioindicators; see McGeoch, 1998) studies, however inventories of animal taxa are often missing or at least reduced to vertebrate groups (e.g. birds), because of the high effort involved in investigation. There is, however, increasing evidence that invertebrates or invertebrate assemblages provide a good indication of changing environment (e.g. Majer and Nichols, 1998; Wheater et al., 2000; Andersen et al., 2001).

Using McGeoch's (1998) concept of ecological indicators we investigated agricultural land that had experienced a range of use and different times since conversion from arable land or time of current management. Studies were made from 1996 to 1999 in the Unstrut river floodplain region of Thuringia, Germany. To assess the effects of changing land use on different trophic levels as well as carrying out vegetation analysis we also investigated the spatial and temporal patterns of spiders (carnivorous) and beetles (carnivorous, detritivorous and herbivorous) in the community. Furthermore, the groundwater dynamics, nutrient balances, soil characteristics and microclimates of different sites were analysed. This paper aims to answer the following questions: (i) Do the two components of species diversity (richness and evenness) increase with decreasing management intensity within all three tested indicator groups? (ii) Are there differences in reaction time and sensitivity to changing agricultural land use between the investigated assemblages?

2. Material and methods

2.1. Study sites

The research area was located in the middle Unstrut floodplain, situated in the Thuringian basin (Germany) with annual average temperature of 9.6 °C and precipitation of <500 mm.

We investigated an arable land site (as control) and six different managed grassland sites (with different time since conversion from arable land or time of current management) in a 3-year study (see Table 1). All sites were similar in soil characteristics (Fluvisols, Clay-Vega or Vega-Clay). The groundwater level dynamics were more or less uncoupled from river table dynamics for the majority of the sites. Within the investigated sites there is a gradient in soil-moisture status ranging from 'wet' with surface near groundwater level (B1) through 'mesic humid' groundwater influenced sites with seasonal partial drought periods (A1, A2, A3, C) and sites with a groundwater level some depth from the soil surface and experiencing mild drought conditions especially in summer (B2, partly AD) to river-bank influenced sites (partly AD and C). For further details see Malt and Perner (2002) as well as Perner and Malt (2002).

Table 1
Description of sampling sites[a]

Site name	Abbreviation	Scores ordination axis 1	Location; coordinates[b]	Management/land use activities	Current management since	Flooding frequency	Groundwater level in dm min./max./mean
Arable land	A3	0	Near Altengottern; 51°08′N, 10°38′E	Conventionally managed, 1996/1997 winter wheat/1998 broad bean, 130 kg N/ha, pesticide application, harvest in August, ploughing in November	About 40 years	No inundation	7/17/14
Young sown grassland	A2	3.04	Near Altengottern; 51°08′N, 10°38′E	Mowing (one-cut in July) and low level cattle grazing (1.0 livestock unit/ha) September–April	1996, formerly arable land	No inundation	7/18/15
Permanent grassland	C	3.17	Near Sömmerda; 51°11′N, 11°05′E	Mowing, three-cuts per year (May, July, and September), liquid manure (about 50–80 kg N/ha)	1967, formerly arable land	Every 5–7 years	4/16/12
Sown grassland	A1	3.41	Near Altengottern; 51°08′N, 10°38′E	All year low level cattle grazing (1.0 livestock unit/ha) and mowing (one-cut in July)	1992, formerly arable land	No inundation	5/12/10
Re-wetted grassland (fallow)	B1	3.43	Near Thamsbrück; 51°07′N, 10°41′E	Yearly mulching in late summer/autumn	1994, formerly arable land	Since 1995 yearly inundation in spring	0/13/5
Embankment grassland	AD	4.14	Near Altengottern; 51°08′N, 10°38′E	Mowing (July) and/or sheep grazing (April and June)	About 30 years	Inundation about three times per century	14/22/19
Permanent grassland	B2	4.58	Near Thamsbrück; 51°07′N, 10°41′E	Mowing one-cut per year (July)	About 30 years, formerly arable land	Inundation about three times per century	15/21/19

Decreasing management impact →

[a] Sites arranged along decreasing dimension of management impact. The site-ranking is derived from the site scores of axis 1 using a Bray–Curtis ordination, which explained the predominant part (70%) of the variance included in the management matrix (see Section 2).

[b] Altitude of sites: 140–180 m about sea level.

2.2. Sampling methods

To compare the floristic species composition at each site, phytosociological plant samples according to Braun–Blanquet with species: area ratio after Wilmanns (1989) were carried out in ten 2 m × 2 m squares per site (except for site B2 where five 5 m × 5 m squares were taken). Plant samples were taken in spring and summer of both 1997 and 1998. Plant assemblages were classified according to the plant indicator values by Ellenberg et al. (1992).

Ground-dwelling arthropods were sampled with 10 plastic pitfall traps (∅45 mm) per site, filled to one third height with 3% formaldehyde with some drops of detergent (in winter drops of glycerol were also added). The traps were positioned in two groups along a transect with five traps (trap–trap distance = 4 m, group distance = 20 m) per group. At site B1 there was a group distance of about 150 m dictated by a site-specific humidity gradient. Pitfall trapping was carried out from May 1996 to October 1998 with trap exchanges taking place every 14 days. Spiders and beetles were sorted and determined to species level in the laboratory as well as classified to ecotypes following Maurer and Hänggi (1990) and Platen et al. (1991) for spiders, and Koch (1989 ff.), Barndt et al. (1991), Korge (1991) and Winkelmann (1991) for beetles.

The surface microclimate around the trap-transects was surveyed using ONSET data loggers (temperature, light intensity, relative humidity). The physical and chemical properties of the soil were also investigated as were surface soil-moisture (Hydra Soil Moisture Probe from VITEL Inc.) and groundwater level. These were measured twice a month. These abiotic data sets were used for direct gradient analysis (RDA, see below) to identify key determinants of the species patterns.

2.3. Statistical analysis

Management activities were coded as follows: fertilisation (0: without, 1: <50 kg N/ha, 2: 50–100 kg N/ha, 3: 100–150 kg N/ha); pesticide application (0: without, 1: with); grazing (0: without, 1: temporary—low level, 2: whole year—low level); mowing (0: without, 1: one-cut per year, 2: two-cuts per year, 3: three-cuts per year); ploughing (0: without, 1: with); mulching (0: without, 1: with); age as time since conversion from arable land or time of current management (0: arable land, 1: 1–2 years, 2: 3–10 years, 3: 11–30years). This matrix were analysed with Bray–Curtis ordination techniques (Beals, 1984; McCune and Mefford, 1997) and site scores of the ordination axis 1 were used as an integrated measure of management impact (see Section 3 and Table 1).

To compare the species patterns of the three groups in different managed field sites, in addition to species richness Camargo's indices of evenness (E'; see Krebs, 1999) were also calculated. Camargo's index is unaffected by species richness and only slightly affected by the presence of rare species in a sample. Linear regression was used to determine the relation between site management intensity and the species richness and evenness of plants, spiders and beetles.

Similarity patterns in species composition and community structure between the sampling sites were explored using non-metric multidimensional scaling (NMDS) ordination for the plant, spider and beetle data separately (McCune and Mefford, 1997). NMDS is an iterative search for a ranking and placement of n entities (samples) on k dimensions (ordination axes) that minimises the stress of the k-dimensional configuration. The "stress" value is a measure of departure from monotonicity in the relationship between the dissimilarity (distance) in the original p-dimensional space and distance in the reduced k-dimensional ordination space (Clarke, 1993). As distance measure we used the Euclidean distance and for all three analysed data sets (plants, spiders, beetles) three ordination axes (dimensions) were sufficient to achieve low stress values. To evaluate how well the distances in the ordination space represent the distances in the original (unreduced) space the coefficient of determination (r^2) were calculated. This value is a measure of the quality of data reduction, along with an assessment of how the variance explained is distributed among the primary axes.

A code replacement of the Braun–Blanquet-scale (van der Maarel, 1979) was necessary for the plant data and the resulting ordinal scale (from 1 to 9) was square-root transformed. In preparation for the multivariate statistical analyses the spider and beetle data were ln-transformed and rare species (all species < 0.01% of catch in total) were omitted to reduce data noise.

To analyse the environmental factors having a potential influence on the studied arthropod assemblages we used the redundancy analysis (RDA) ordination technique (Jongman et al., 1995). The inclusive forward selection procedure was employed to sorting out the factors explaining the most variance in the species data and finally, Monte Carlo permutation procedures were carried out for significance testing of the selected environmental factors (Jongman et al., 1995; Ter Braak and Šmilauer, 1998). To remove weather effects between the different sampling years from the data we used a partial redundancy analysis (pRDA) by entering the total yearly precipitation from May to August into the RDA as covariable.

3. Results

Using a Bray–Curtis ordination the management impact of each of the sampling sites was estimated. As 70% of the variance included in the management matrix (see Section 2) was extracted along ordination axis 1 the site scores of this axis 1 were used to rank the investigated sampling sites (Table 1). A decreasing management impact could be detected from the arable land site (A3), through the young sown grassland (A2), the permanent grassland (C), the sown grassland (A1), the re-wetted grassland (B1), and the embankment grassland (AD) to the permanent grassland site (B2).

Overall 105 plant species, 120 species of adult spiders (56,067 individuals) and 502 species of beetles (77,684 individuals) were recorded from 1996 to 1998 (Table 2). The highest total number of plant species was observed in the re-wetted grassland (B1) while numbers of spider and beetle species were highest in the embankment grassland (AD).

Whereas plant and spider species richness increases with decreasing management impact (relation for spiders not significant, but $r^2 = 0.51$, $P = 0.07$), the species richness of the beetle assemblages is unrelated to the site's management impact (Fig. 1a–c). Detailed analysis of the different trophic guilds within the beetle assemblages show that while carnivore and detritivore species richness is unrelated herbivore beetle species richness does tend to increase with decreasing management impact ($r^2 = 0.34$, $P = 0.16$). Evenness increases with decreasing management impact in all three studied model groups, although for spiders this relationship is not significant ($r^2 = 0.50$, $P = 0.07$; Fig. 1d–f).

Fig. 2 shows the mean indicator values (after Ellenberg) of plants and the total number of hygrophilous species and individuals for spiders and beetles per site. A slight increase of plant moisture indicator value is detectable from the sown grassland (A2) to the re-wetted grassland (B1) and the embankment grassland (AD), whereas the other indicator values show no such tendencies (Fig. 2a). Spiders, and especially beetles, show a clear increase of hygrophilous species and a very strong increase of hygrophilous individuals from the sown (A1, A2) to the embankment (AD) and re-wetted (B1) grasslands (Fig. 2b and c).

To analyse and visualise the differences between the sampling sites in assemblage structure of plants, spiders and beetles we used NMDS ordinations (Fig. 3a–f). The ordination pattern for the plant data

Table 2
Number of species (and individuals) for plants, spiders and beetles caught at each site from 1996 to 1998 (for detailed species lists see Malt and Perner (2002) as well as Perner and Malt (2002))

	Plant species	Spiders		Beetles	
		Species	Individuals	Species	Individuals
Arable land (A3)	20	74	6850	205	18863
Young sown grassland (A2)	23	63	8845	249	21764
Permanent grassland/three-cuts (C)	27	72	14969	244	12972
Sown grassland (A1)	32	59	7119	196	4989
Re-wetted grassland (B1)	46	67	10156	188	4967
Embankment grassland (AD)	45	81	5642	312	11720
Permanent grassland/one-cut (B2)	30	69	2486	195	2409
Total	105	120	56067	502	77684

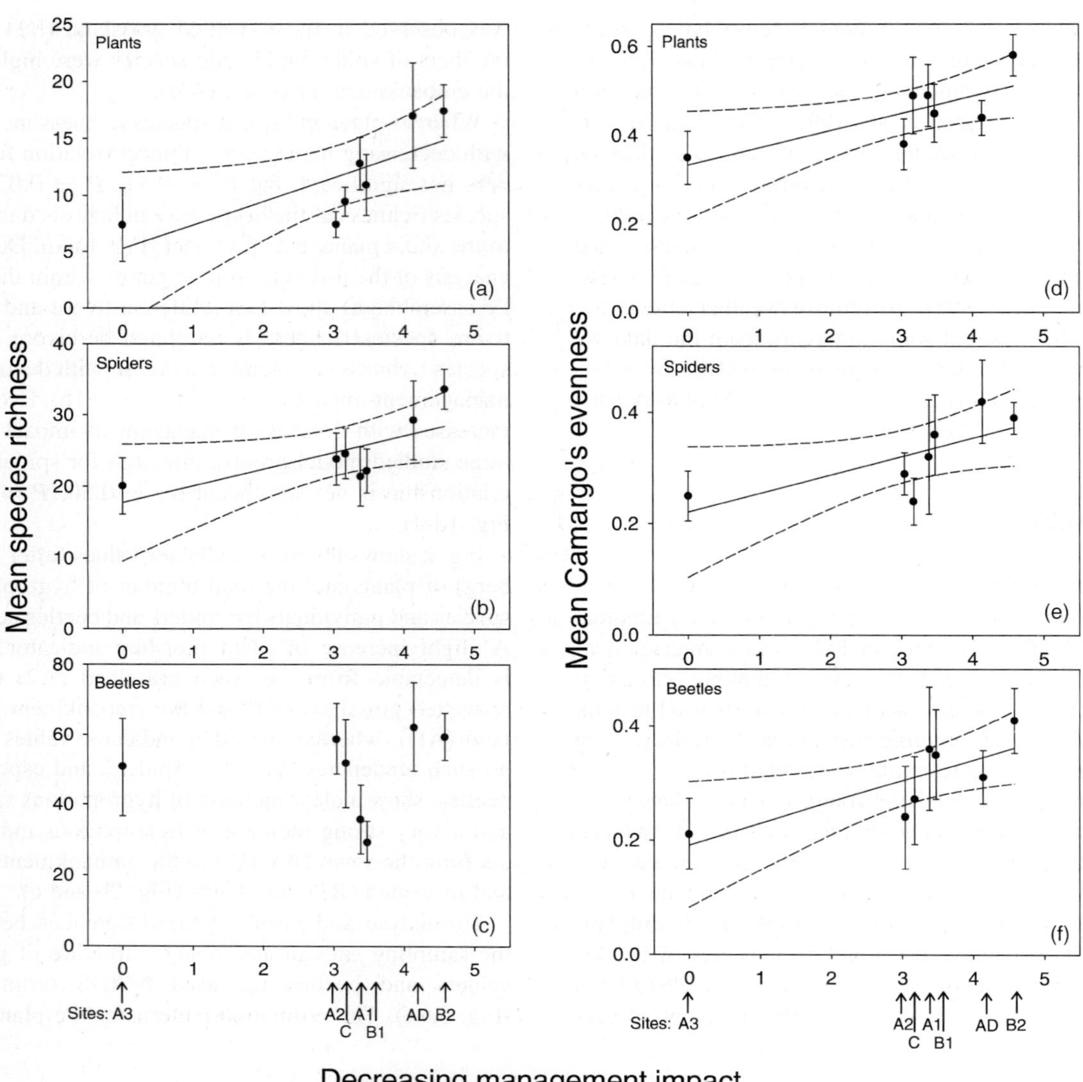

Fig. 1. Changes in mean species richness (a–c) and mean Camargo's evenness (d–f) of plants, spiders and beetles with decreasing management impact. Standard deviations (S.D.) and 95% confidence intervals of the linear models are displayed. Estimation of decreasing management impact derived from the site scores of axis 1 using a Bray–Curtis ordination, which explained 70% of the variance included in the management matrix (see Section 2). Decreasing management impact vs. mean species richness of (a) plants ($r^2 = 0.55$, $P < 0.05$), (b) spiders ($r^2 = 0.51$, $P = 0.07$), (c) beetles ($r^2 = 0.8$, $P < 0.01$) and vs. mean Camargo's evenness of (d) plants ($r^2 = 0.57$, $P < 0.05$), (e) spiders ($r^2 = 0.50$, $P = 0.07$) and (f) beetles ($r^2 = 0.60$, $P < 0.05$).

(cumulative r^2 for axis 1, 2 and 3 were 17.6, 62.2 and 77.3, respectively) shows that all sites with exception of both sites AD and B2 clumped together (Fig. 3a). Also the projection of axis 1 vs. axis 3 improves the separation of those sites only slightly (Fig. 3b). Ordination of spider (cumulative r^2 for axis 1, 2 and 3 were 36.8, 61.4 and 91.8, respectively) and beetle (cumulative r^2 for axis 1, 2 and 3 were 25.6, 62.1 and 89.0)

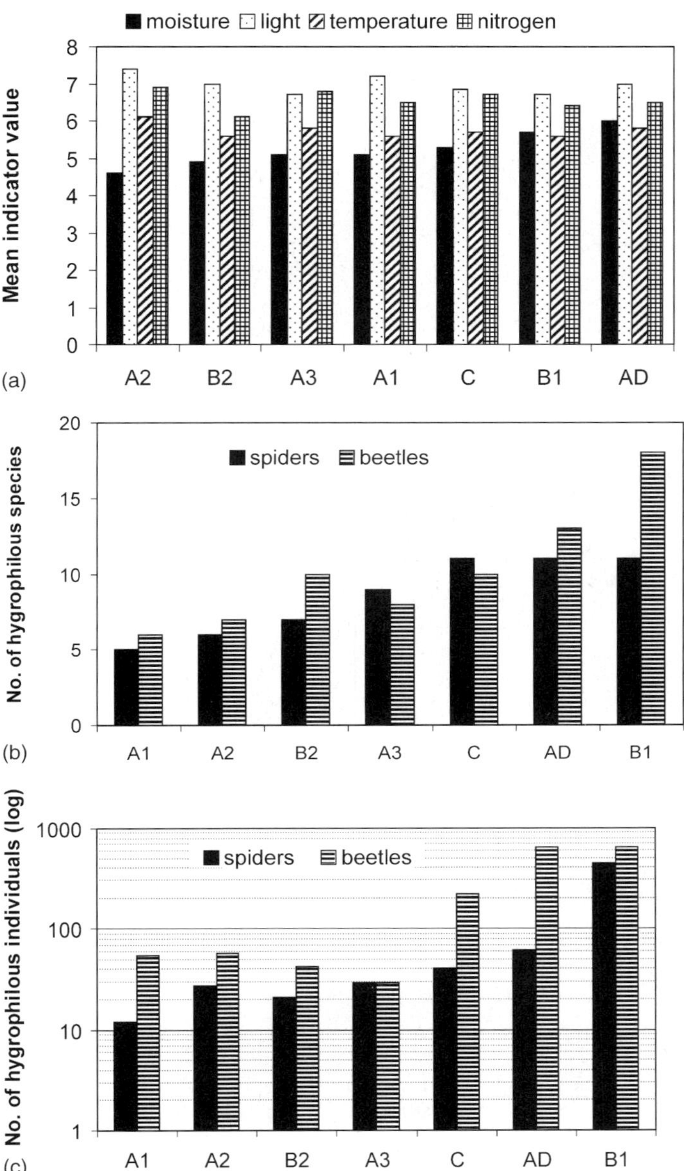

Fig. 2. Mean indicator values (after Ellenberg) of plants (a) for moisture, light, temperature and nitrogen, (b) numbers of hygrophilous species, and (c) individuals for spiders and beetles caught at each site. Sites arranged from lowest to highest moisture indicator values respectively amount of hygrophilous species.

assemblage provide an outstanding separation of data (Fig. 3c–f). Here, the examined set of agricultural land use types evokes distinct patterns of community structure. In both cases (spiders—Fig. 3c and d and beetles—Fig. 3e and f) arable (A3) and young sown grassland (A2) sites are close together. The good site separation based on spider and beetle data suggests that using procedures of direct gradient analysis to screen important (good describing) site parameters may be a promising avenue to follow. RDA (Fig. 4)

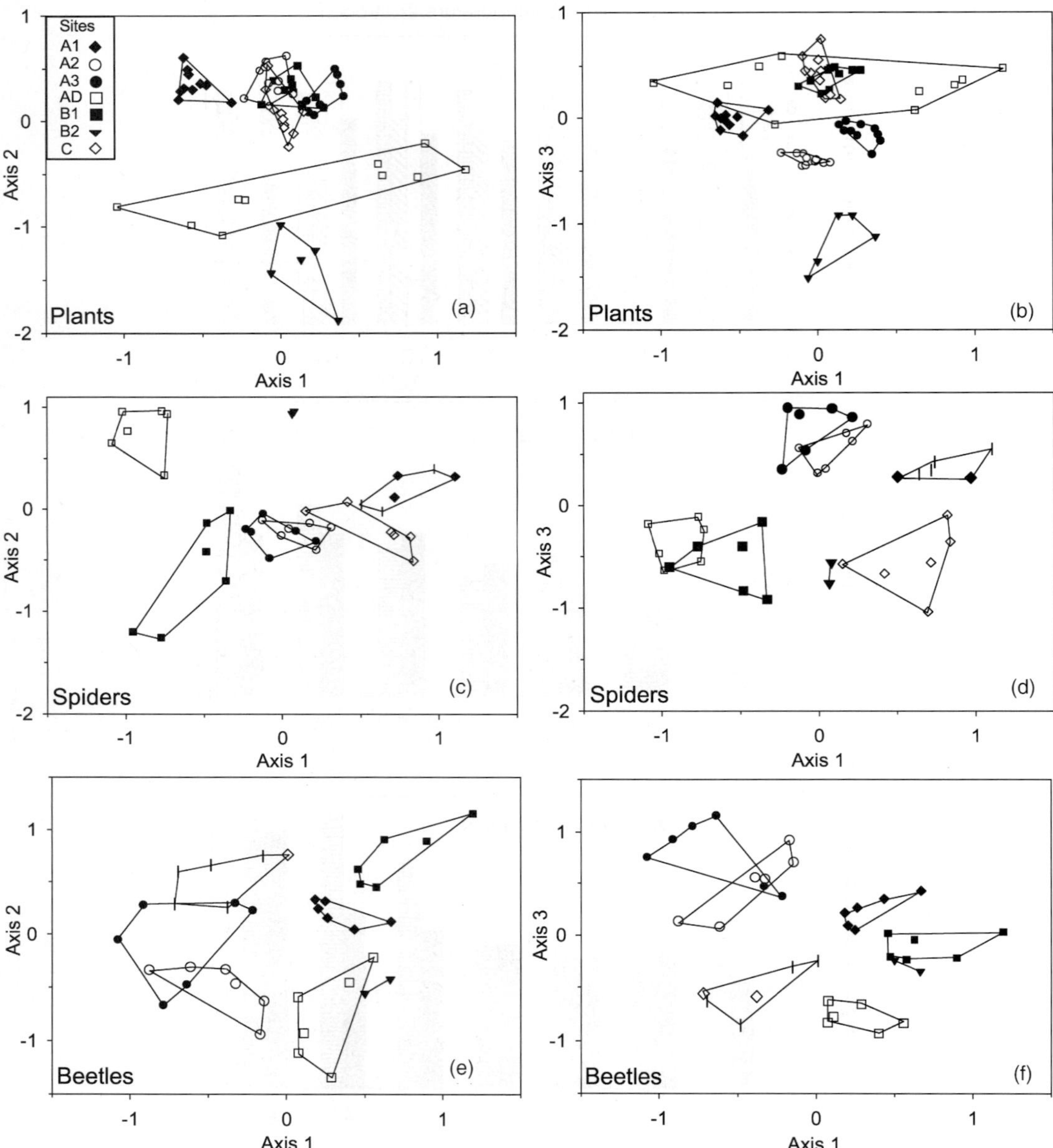

Fig. 3. NMDS ordinations of plants (a and b), spiders (c and d) and beetles (e and f) assemblages. For the plant data 10 samples per site (except site B2 only five samples) and the arthropod data six samples per site (two partial transects each with five pooled pitfall traps and 3 years, except site B2 only 1 year studied) have been plotted (see Section 2). See Table 1 for site abbreviations. Minimum stress values for (a and b) plants: axis 1 = 32.2, axis 2 = 20.6, axis 3 = 14.9, (c and d) spiders: axis 1 = 34.6, axis 2 = 16.0, axis 3 = 9.2 and (e and f) beetles: axis 1 = 36.4, axis 2 = 16.5, axis 3 = 10.0.

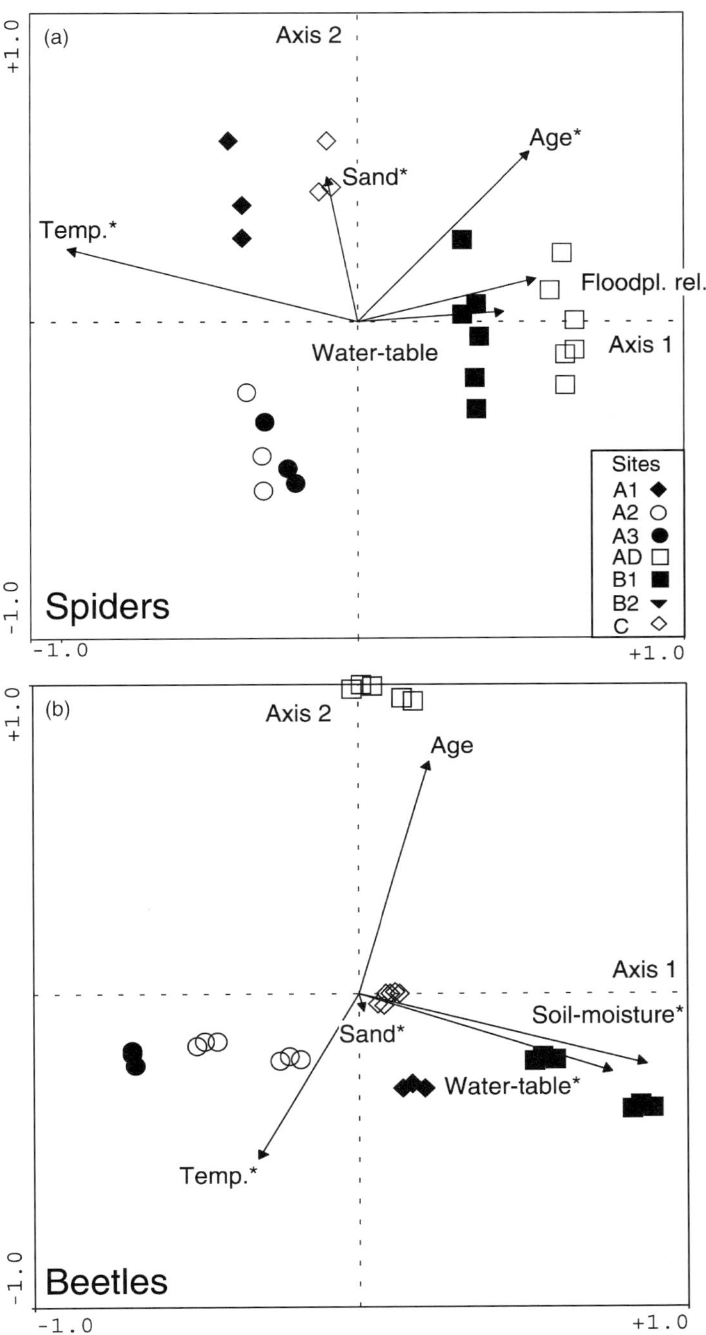

Fig. 4. pRDA of spider (a) and beetle (b) assemblages; database like Fig. 3c–f; used covariable: total of precipitation from May to August. See Table 1 for site abbreviations. Abbreviations of abiotic parameters: Age—time since conversion from arable land or time under current management, Floodpl. rel.—amount of floodplain relict habitats (e.g. backwaters, banks) within 100 m, Sand—mean amount of sand in the soil, Soil-moisture—mean amount of time per year with water saturation of the soil surface, Temp.—mean temperature differences on the soil surface during the vegetation period, Water-table—mean amount of time per year with ground-water level <100 cm; $^*P < 0.05$.

Table 3
Partial inter-set correlation coefficients of the selected environmental variables for the spider and beetles data sets[a]

Parameter	Lambda-*A*	Axis 1	Axis 2	Axis 3	Axis 4
SPIDERS—eigenvalues		0.2612	0.1411	0.1367	0.0623
Temp.*	0.227	−0.8545	0.2179	−0.2233	0.2302
Age*	0.125	0.5151	0.5398	−0.3348	−0.3941
Water-table	0.125	0.4347	0.0357	0.8422	0.1588
Sand*	0.096	−0.1082	0.4952	0.2299	−0.6868
Floodpl. rel.*	0.055	0.5279	0.1346	0.0935	−0.2183
% var. expl.	62.7				
BEETLES—eigenvalues		0.193	0.156	0.090	0.0580
Soil-moisture*	0.163	0.8312	−0.2218	0.1581	0.0066
Age	0.132	0.2064	0.7599	0.5534	−0.1282
Sand*	0.100	0.0093	−0.0598	0.7847	−0.5298
Temp.*	0.074	−0.2960	−0.5188	0.3385	0.6509
Water-table*	0.052	0.7306	−0.2432	−0.3164	−0.4457
% var. expl.	52.2				

[a] Weather effects were partialled out of the analysis by entering the total precipitation totals from May to August into the RDA as covariables. Environmental variables are given in the order of their inclusion in the model together with the additional variance each variable explains at the time it was included (lambda-*A*-value) and the total variance explained (% var. expl.) by the selected environmental variables.

illustrates that ground-water level (water-table), temperature on the soil surface during the vegetation period (Temp.), time since conversion from arable land (Age) and amount of sand in the soil (Sand) explain the majority of the variance in both spider and beetle data set. Floodplain relict habitat areas (Floodpl. rel.) and near-surface soil-moisture (Soil-moisture) identified as additional environmental factors for the spider (Fig. 4a) and beetle data set (Fig. 4b), respectively. With the exception of 'water-level' (spiders) and 'age' (beetles) all selected environmental parameters show significant relationships to patterns of species assemblages (see also Table 3).

4. Discussion

In this study we assessed the effects of changing agricultural land use on vegetation, spider and beetle assemblages. Overall, we found comparable relationships between the management impact and species richness (plants, spiders) and evenness (plants, spiders, beetles) for the different trophic levels (Fig. 1). When considered in detail specific differences were found to exist between the vegetation and the invertebrate groups (Figs. 1 and 3) in the different grassland sites (A1, A2, AD, B1, B2, C). Ecological effects at the vegetation level strongly depend on both the intensity of former arable management (e.g. nutrient balance, seed bank) and the initial conditions of grassland succession (e.g. seed mixture; van der Putten et al., 2000). Bakker and Berendse (1999) have shown that the development of species-rich vegetation on abandoned arable land is often constrained, even when natural abiotic conditions have been restored. Reasons for this constraint include a depleted seed bank, the poor seed dispersal of late-succession species as well as the fact that the first established competitive weedy species, which are already present in the seed bank of agricultural sites, prevent vegetation development for many years (Hansson and Fogelfors, 1998). Although the poor vegetation richness in our study shows considerable differences between the sites, the more quantitative measure of evenness, and especially the ordination patterns, demonstrate a weak separation of the most grassland sites (Fig. 3a and b). This is likely to have been caused by the dominance of competitive (persistent) weedy species (e.g. *Cirsium arvense*, *Elymus repens*, *Galium aparine*) and a general vegetation pattern existing among these sites at that time. While plant composition pattern show little differences between sites both invertebrate groups show clear separation between examined sites (Fig. 3c–f). It is well known that ground-dwelling spider and beetle (especially ground and rove beetles) assemblages are sensitive to habitat structure, microclimate and management (Luff and Rushton, 1989; McFerran et al., 1994; Ekschmitt et al., 1997; Topping and Lovei, 1997; Dennis et al., 1997; Wardle et al., 1999). In our study main source of variation within both invertebrate groups could be explained by environmental parameters such as 'temperature', 'age', 'water-level', 'sand', 'floodplain relict habitats' and 'soil-moisture' (Fig. 4a and b). Among these parameters temperature explained the largest proportion of variance in the spider data set and soil-moisture in the beetle data set (Table 3), which was many times mentioned as key factor for distribution patterns of ground beetle assemblages (e.g. Rykken et al., 1997). It is also well-documented that vegetation structure (height and density) is important for spiders (Greenstone, 1984; White and Hassall, 1994), and our results are not inconsistent with these findings. Vegetative structure especially frequency and time of disturbances (e.g. harvesting or mowing), would have resulted in significant microclimatic and particularly also soil surface temperature differences between the sites.

Often, if not in most cases, it is difficult (and in this context not absolutely necessary) to disentangle the causal mechanisms of response. These extracted 'factors' should therefore be seen as (statistically significant) model descriptors for system patterns. In the context of GIS-technology such descriptors should enable us to make spatial extrapolations of and provide us with predictions for success of restoration processes.

The hygrophilous habitat specialists within the spider and beetle assemblages are an useful "quick assessment tool" for monitoring the efficiency of restoration projects in agricultural floodplain landscapes (Fig. 2b and c). The high colonisation potential of spider (air ballooning) and many beetle (flight, air plankton) groups suggests that the increased number of species, and especially individuals of hygrophilous habitat specialists, may indicate the success of the changing agricultural land use. On the other hand, it is also a measure for the proportion of floodplain relict habitats in the agricultural landscape (e.g. backwaters and banks, and also embankment habitats) and their connectivity of the landscape elements, because a high amount of relict habitats should reduce the time of conversion considerably (Collinge and Forman, 1998).

Despite the recognised problems of using Ellenberg's indicator values (Ter Braak and Gremmen, 1987) the moisture value of plants produce approximately the same ranking of sites as both invertebrate groups using the numbers of hygrophilous species or individuals (Fig. 2). But in general, it seems that assessment based on vegetation tends to over-emphasise the similarities between the sites. We therefore suggest vegetation monitoring should be seen as a more powerful long-term approach complemented by a time-interrupted (e.g. 3 year intervals) short-term sensitive invertebrate monitoring (e.g. ground-dwelling spiders and beetles). This is especially true at the beginning of the restoration process, when information needed about the success of selected habitat management. Wheater et al. (2000) came to the same conclusions in the context of monitoring reclaimed limestone quarry landforms and Mattoni et al. (2000) argued that because of their high turnover and growth rates arthropods serve as probes that quickly respond to environmental change and are therefore most likely to provide the most convincing monitoring for estimating the success or failure of any given habitat restoration project.

The link between species richness and ecosystem function has been intensively discussed over the past 10 years (Johnson et al., 1996; Peterson et al., 1998). Despite a growing body of literature supporting this linkage and its use in defence of the conservation of biodiversity its importance is still debated (Chapin et al., 1997; Schwartz et al., 2000). At the heart of this debate is the notion that loss of plant and/or animal species: (1) will bring an ecosystem closer to collapse (Schwartz et al., 2000); (2) increase ecosystem vulnerability to invasions by alien species (Knops et al., 1999). In our study the species richness of plants and spiders increases with the degree of low intensity management (Fig. 1a and b). This suggests a higher level of resilience at low intensity managed sites against environmental variability. However, locally detected species richness is not only determined by site-specific attributes, but also by surrounding habitat variability (Duelli, 1997). This is of particular importance for very mobile assemblages (e.g. beetles) and can reduce the value of species richness as a trustworthy community parameter for ecological indication.

When Wagner et al. (2000) studied habitat variability and heterogeneity based on the partitioning of landscape species diversity into additive components and then linked them to patch-specific diversity measurements, they concluded, on the one hand, that landscape composition is apparently a key factor for explaining landscape species richness while, on the other hand landscape composition hardly affects evenness. As second aspect of species diversity (Smith and Wilson, 1996) the evenness provides an useful measure of the degree of resource balancing and partitioning within the assemblages. Wagner et al.'s (2000) findings support the assumption that evenness appears less 'biased' by habitat environment and may therefore be a useful and trustworthy parameter to assess the effects of site-specific changes on assemblages. In our study the Camargo's evenness (E') measure increased with decreasing management impact in all examined groups and highlighted the usefulness of this complex community parameter for ecological indication of land use change in agricultural ecosystems.

5. Conclusions

Overall, plants, spiders and beetles are all suitable as ecological indicators of restoration effects in agricultural landscapes. However, the assemblages show considerable differences in reaction times to changes in land use. Ground-dwelling spiders and beetles mainly respond to changes in the microclimate and the soil-moisture and therefore allow an efficient indication of restoration management after 3–5 years. Vegetation monitoring, on the other hand, appears as a more powerful long-term assessment. The most productive monitoring system would complement the frequently used vegetation monitoring with a sensitive short-term invertebrate monitoring especially at the beginning of agricultural restoration projects.

Acknowledgements

This study was supported by the Federal Ministry of Education and Research (BMBF-grant 0339572). Especially we would like to thank Katja Reichenbecher for doing the vegetation surveys and T. Hefin Jones for helpful comments on the manuscript and for checking the English. We also thank Jens Schumacher, Andrew Davis, John Sloggett and two anonymous referees for constructive comments on earlier drafts of this manuscript.

References

Andersen, A.N., Ludwig, J.A., Lowe, L.M., Rentz, D.C.F., 2001. Grasshopper biodiversity and bioindicators in Australian tropical savannas: responses to disturbance in Kakadu National Park. Aust. Ecol. 26, 213–222.

Bakker, J.P., Berendse, F., 1999. Constraints in the restoration of ecological diversity in grassland and heathland communities. Trends Ecol. Evol. 14, 63–68.

Barndt, D., Brase, S., Glauche, M., Gruttke, H., Kegel, B., Platen, R., Winkelmann, H., 1991. Die Laufkäfer von Berlin (West)—mit Kennzeichnung und Auswertung der verschollenen und gefährdeten Arten (Rote Liste, 3.Fassung). In: Auhagen, A., Platen, R., Sukopp, H. (Eds.), Rote Listen der gefährdeten Pflanzen und Tiere in Berlin. Landschaftsentwicklung und Umweltforschung, TU Berlin, pp. 243–276.

Beals, E.W., 1984. Bray–Curtis ordination: an effective strategy for analysis of multivariate ecological data. Adv. Ecol. Res. 14, 1–55.

Chapin, F.S., Walker, B.H., Hobbs, R.J., Hooper, D.U., Lawton, J.H., Sala, O.E., Tilman, D., 1997. Biotic control over the functioning of ecosystems. Science 277, 500–504.

Clarke, K.P., 1993. Non-parametric multivariate analyses of changes in community structure. Aust. J. Ecol. 18, 117–143.

Collinge, S.K., Forman, R.T.T., 1998. A conceptual model of land conversion processes: predictions and evidence from a microlandscape experiment with grassland insects. Oikos 82, 66–84.

Dennis, P., Young, M.R., Howard, C.L., Gordon, I.J., 1997. The response of epigeal beetles (Col, Carabidae, Staphylinidae) to varied grazing regimes on upland *Nardus stricta* grasslands. J. Appl. Ecol. 34, 433–443.

Duelli, P., 1997. Biodiversity evaluation in agricultural landscapes: an approach at two different scales. Agric. Ecosyst. Environ. 62, 81–91.

Ekschmitt, K., Wolters, V., Weber, M., 1997. Spiders, carabids, and staphylinids: the ecological potential of predatory macroarthropods. In: Benckiser, G. (Ed.), Fauna in Soil Ecosystems. Marcel Dekker, New York, pp. 307–362.

Ellenberg, H., Weber, H.E., Düll, R., Wirth, V., Werner, W., Paulißen, D., 1992. Zeigerwerte der Pflanzen in Mitteleuropa. Scripta Geobot. 18, 1–258.

Goedmakers, A., 1989. Ecological perspectives of changing agricultural land use in the European Community. Agric. Ecosyst. Environ. 27, 99–106.

Greenstone, M.H., 1984. Determinants of web spider species diversity: vegetation structural diversity vs. prey availability. Oecologia 62, 299–304.

Hansson, M., Fogelfors, H., 1998. Management of permanent set-aside on arable land in Sweden. J. Appl. Ecol. 35, 758–771.

Johnson, K.H., Vogt, K.A., Clark, H.J., Schmitz, O.J., Vogt, D.J., 1996. Biodiversity and the productivity and stability of ecosystems. Trends Ecol. Evol. 11, 372–377.

Jongman, R.H.G., Ter Braak, C.J.F., van Tongeren, O.F.R., 1995. Data Analysis in Community and Landscape Ecology. Cambridge University Press, Cambridge.

Knops, J.M.H., Tilman, D., Haddad, N.M., Naeem, S., Mitchell, C.E., Haarstad, J., Ritchie, M.E., Howe, K.M., Reich, P.B., Siemann, E., Groth, J., 1999. Effects of plant species richness on invasion dynamics, disease outbreaks, insect abundances and diversity. Ecol. Lett. 2, 286–293.

Koch, K., 1989 ff. Die Käfer Mitteleuropas. Ökologie Bd.1-6. Goecke & Evers, Krefeld.

Korge, H., 1991. Liste der Kurzflügelkäfer (Coleoptera, Staphylinidae) von Berlin (West) mit Kennzeichnung der verschollenen und gefährdeten Arten (Rote Liste). In: Auhagen, A., Platen, R., Sukopp, H. (Eds.), Rote Listen der gefährdeten Pflanzen und Tiere in Berlin. Landschaftsentwicklung und Umweltforschung, TU Berlin, pp. 277–318.

Krebs, C.J., 1999. Ecological Methodology. Addison-Wesley, Menlo Park.

Lawton, J.H., Gaston, K.J., 2001. Indicator species. In: Levin, S.A. (Ed.), Encyclopedia of Biodiversity, vol. 3. Academic Press, San Diego, pp. 437–450.

Luff, M.L., Rushton, S.P., 1989. The ground beetle and spider fauna of managed and unimproved upland pasture. Agric. Ecosyst. Environ. 25, 195–206.

Majer, J.D., Nichols, O.G., 1998. Long-term recolonization patterns of ants in Western Australian rehabilitated bauxite mines with reference to their use as indicators of restoration success. J. Appl. Ecol. 35, 161–182.

Malt, S., Perner, J., 2002. Zur epigäischen Arthropodenfauna von landwirtschaftlichen Nutzflächen der Unstrutaue im Thüringer Becken. Teil 1. Webspinnen und Weberknechte (Arachnida: Araneae et Opiliones). Faun. Abh. Mus. Tierkde Dresden 22, 207–228.

Mattoni, R., Longcore, T., Novotny, V., 2000. Arthropod monitoring for fine-scale habitat analysis: a case study of the El Segundo sand dunes. Environ. Manage. 25, 445–452.

Maurer, R., Hänggi, A., 1990. Katalog der schweizerischen Spinnen. Doc. Faunist. Helvet. 12.

McCune, B., Mefford, M.J., 1997. PC-ORD. Multivariate Analysis of Ecological Data, Version 3.0. MjM Software Design, Gleneden Beach, OR.

McFerran, D.M., Montgomery, W.I., McAdam, J.H., 1994. The impact of grazing on communities of ground-dwelling spiders (Araneae) in upland vegetation types. Proc. R. Irish Acad. 94b, 119–126.

McGeoch, M.A., 1998. The selection, testing and application of terrestrial insects as bioindicators. Biol. Rev. 73, 181–201.

Perner, J., Malt, S., 2002. Zur epigäischen Arthropodenfauna von landwirtschaftlichen Nutzflächen der Unstrutaue im Thüringer Becken. Teil 2. Käfer (Coleoptera). Faun. Abh. Mus. Tierkde Dresden 22, 261–283.

Peterson, G., Allen, C.R., Holling, C.S., 1998. Ecological resilience, biodiversity, and scale. Ecosystems 1, 6–18.

Platen, R., Moritz, B., Broen, B., 1991. Liste der Webspinnen- und Weberknechtarten (Arach.: Araneida, Opilionida) des Berliner Raumes und ihre Auswertung für Naturschutzzwecke (Rote Liste). In: Auhagen, A., Platen, R., Sukopp, H. (Eds.), Rote Listen der gefährdeten Pflanzen und Tiere in Berlin. Landschaftsentwicklung und Umweltforschung, TU Berlin, pp. 169–206.

Rykken, J.J., Capen, D.E., Mahabir, S.P., 1997. Ground beetles as indicators of land type diversity in the Green Mountains of Vermont. Cons. Biol. 11, 522–530.

Schwartz, M.W., Brigham, C.A., Hoeksema, J.D., Lyons, K.G., Mills, M.H., van Mantgem, P.J., 2000. Linking biodiversity to ecosystem function: implications for conservation ecology. Oecologia 122, 297–305.

Smith, B., Wilson, J.B., 1996. A consumer's guide to evenness indices. Oikos 76, 70–82.

Ter Braak, C.J.F., Gremmen, N.J.M., 1987. Ecological amplitudes of plant species and the internal consistency of Ellenberg's indicator values for moisture. Vegetatio 69, 79–87.

Ter Braak, C.J.F., Šmilauer, P., 1998. Canoco 4—Reference Manual and User's Guide to Canoco for Windows: Software for Canonical Community Ordination, Version 4. Microcomputer Power, Ithaca, NY.

Topping, C.J., Lovei, G.L., 1997. Spider density and diversity in relation to disturbance in agroecosystems in New Zealand, with a comparison to England. New Zeal. J. Ecol. 21, 121–128.

van der Maarel, E., 1979. Transformation of cover-abundance values in phytosociology and its effects on community similarity. Vegetatio 39, 97–114.

van der Putten, W.H., Mortimer, S.R., Hedlund, K., Van Dijk, C., Brown, V.K., Leps, J., Rodriguez-Barrueco, C., Roy, J., Len, T.A.D., Gormsen, D., Korthals, G.W., Lavorel, S., Regina, I.S., Šmilauer, P., 2000. Plant species diversity as a driver of early succession in abandoned fields: a multi-site approach. Oecologia 124, 91–99.

Vitousek, P.M., Mooney, H.A., Melillo, J.M., 1997. Human domination of earth's ecosystems. Science 277, 494–499.

Wagner, H.H., Wildi, O., Ewald, K.C., 2000. Additive partitioning of plant species diversity in an agricultural mosaic landscape. Landscape Ecol. 15, 219–227.

Wardle, D.A., Nicholson, K.S., Bonner, K.I., Yeates, G.W., 1999. Effects of agricultural intensification on soil-associated arthropod population dynamics, community structure, diversity and temporal variability over a seven-year period. Soil. Biol. Biochem. 31, 1691–1706.

Wheater, C.P., Cullen, W.R., Bell, J.R., 2000. Spider communities as tools in monitoring reclaimed limestone quarry landforms. Landscape Ecol. 15, 401–406.

White, P.C.L., Hassall, M., 1994. Effects of management on spider communities of headlands in cereal fields. Pedobiologia 38, 169–184.

Wilmanns, O., 1989. Ökologische Pflanzensoziologie. Quelle & Meyer, Heidelberg.

Winkelmann, H., 1991. Liste der Rüsselkäfer (Col.: Curculionidae) von Berlin mit Angaben zur Gefährdungssituation (Rote Liste). In: Auhagen, A., Platen, R., Sukopp, H. (Eds.), Rote Listen der gefährdeten Pflanzen und Tiere in Berlin. Landschaftsentwicklung und Umweltforschung, TU Berlin, pp. 319–358.

Available online at www.sciencedirect.com

Agriculture, Ecosystems and Environment 98 (2003) 183–199

Agriculture Ecosystems & Environment

www.elsevier.com/locate/agee

Auchenorrhyncha communities as indicators of disturbance in grasslands (Insecta, Hemiptera)—a case study from the Elbe flood plains (northern Germany)

Herbert Nickel [a], Jörn Hildebrandt [b,*]

[a] *Institute of Zoology and Anthropology, Ecology Group, Berliner Str. 28, 37073 Göttingen, Germany*
[b] *Section of Biology, Institute of Ecology and Evolutionary Biology, University of Bremen, P.O. Box 330 440, 28 334 Bremen, Germany*

Abstract

Diversity of insect communities in grasslands is often negatively correlated with management intensity. Targets of sustainable husbandry practices in agricultural systems should include the conservation of insects highly specific to certain types of grassland. In general, Auchenorrhyncha can be regarded as suitable indicators of biotic conditions in grasslands because (i) their response to management, like cutting, grazing and fertilizing, is strong and immediate, (ii) quantitative and semiquantitative sampling is relatively easy, (iii) their abundance and species numbers in grasslands are usually high, and (iv) they show different life strategies ranging from polyphagous pioneer species to strictly monophagous specialists.

This study was conducted in the floodplains of the middle course of the Elbe river (northern Germany). Its purpose was (i) a survey of local Auchenorrhyncha communities and their responses to different grassland management regimes, and (ii) a study of their suitability as indicators of habitat disturbance, particularly in comparison to plant communities. Samples were taken with the sweepnet in 1999, and with a suction apparatus in 2000. The plots included 25 sites in meadows, pastures and fallows. Altogether, 88 species were recorded. Regarding the distribution of generalist species, differences between plots being subject to high-intensity management and those being subject to low-intensity management were little pronounced. However, most plots of low-intensity pastures and fallows showed higher numbers and higher proportions of specialists. In late spring, suction sampling produced higher individual numbers in low-intensity sites. Moreover, diversity in high-intensity sites was reduced, with generalists dominating. We discuss different responses of plant and animal communities on grassland management and propose species numbers and proportion of both pioneer species and specialists as robust indicators of biotic conditions in grasslands. Furthermore, we make proposals for a future land use management of flood plain grasslands.

Keywords: Auchenorrhyncha; Grasslands; Disturbance; Management; Indicator

1. Introduction

Grassland invertebrate communities are rather complex and diverse. Species and individual numbers are particularly high among Nematoda, Enchytraeidae, Lumbricidae, Acari and Insecta, with most above-ground studies focusing on the latter (Boness, 1953; Curry, 1994; Tscharntke and Greiler, 1995). Species numbers as well as proportions of specialists and threatened species are negatively affected by the intensity of management (Curry, 1987). Sustainable use of agricultural systems should include the conservation of communities adapted to the specific

* Corresponding author. Tel.: +49-421-2184502; fax: +49-421-2184504.
E-mail addresses: hnickel@gwdg.de (H. Nickel), hildebra@uni-bremen.de (J. Hildebrandt).

0167-8809/$ – see front matter © 2003 Elsevier B.V. All rights reserved.
doi:10.1016/S0167-8809(03)00080-X

conditions in grasslands. However, arthropod responses to management can be group-specific, and the question arises which invertebrate taxa are most suitable for an indication of management effects. Morris and Rispin (1987) demonstrated that beetles in British calcareous grasslands show less clear responses to cutting than Auchenorrhyncha and Heteroptera, probably due to their less distinct niche separation in different layers of the vegetation. Important prerequisites of arthropod taxa used for monitoring biotic conditions in grasslands include both intrinsic characters, like diversity and niche separation, and extrinsic factors such as sampling and taxonomic practicability.

We chose the Auchenorrhyncha for the following reasons:

(i) They occur in high individual and species numbers. Abundances in grasslands are often high and may exceed 1000 ind./m^2 (Waloff, 1980; Curry, 1994). The total species number in German grassland ecosystems is almost 320 (see Nickel, 2003). Mown grasslands, i.e. meadows sensu stricto, altogether harbour only 120 species, with up to 40 occurring on one plot (Nickel and Achtziger, 1999).

(ii) Due to large population sizes and species numbers, Auchenorrhyncha form an important component of the grassland fauna, although functional aspects of their ecology are little studied. Most species live in the herbaceous layer. Nymphs and adults usually feed externally on the above-ground parts of plants, ingesting cell contents, phloem or xylem sap. As primary consumers, they assimilate plant biomass, damage plant tissue by oviposition and transmit plant diseases. Thus, they affect competitive relationships and species composition of plants and the direction of succession in agricultural systems (e.g. Brown, 1985; Jung et al., 2000). Moreover, they are an important prey for predators, particularly spiders (Araneae), ants (Hymenoptera, Formicidae) and songbirds (Aves, Passeres), and they are essential host organisms for parasitoids, notably the Dryinidae, Myrmaridae (both Hymenoptera), Pipunculidae (Diptera) and Strepsiptera (Waloff and Jervis, 1987).

(iii) They show specific life strategies and occupy specific spatial and temporal niches. Life strategies range from pioneers, which are usually polyphagous, macropterous and at least bivoltine, to stenotopic species, which are monophagous, usually brachypterous and monovoltine (Novotný, 1994a,b, 1995; Table 3). Depending on overwintering stage and generation numbers, maturity peaks may be reached in mid-spring, early or late summer. The species number of Auchenorrhyncha is positively correlated with species numbers of other important groups living in the herbaceous layer, like Heteroptera, Saltatoria and Rhopalocera (Achtziger et al., 1999) as well as of plants and their structural complexity (Denno, 1994; Denno and Roderick, 1991; Murdoch et al., 1972). Resource utilization is rather diverse regarding stratification and plant architecture (Andrzejewska, 1965) and, as a result, Auchenorrhyncha communities are richer in taller grasslands (Morris, 1971, 1973). Generally, host specialists are dominating, with many species being associated with grasses and sedges. Sedges (*Carex* spp.), fescue (*Festuca* spp.) and small-reed (*Calamagrostis* spp.) are among the most-favoured host plants of monophagous Auchenorrhyncha species in central Europe (Nickel, 2003). These include 1st degree monophages, being specific to a single plant species, and 2nd degree monophages, being specific to a single plant genus.

(iv) Their responses to management are immediate, with marked changes in dominance and community structure (Andrzejewska, 1991; Morris, 1981a,b; Morris and Plant, 1983). Eventually, communities become subject to selection of species tolerating the management regime (Nickel and Achtziger, 1999).

(v) Sampling of the whole species range can be done quickly on two or three dates a year. Suction samples, in particular, produce reliable estimates of abundance and dominance. Sweep-net samples can usually be compared within one study, although most epigeic species are under-represented. The effort for determination is reasonable. The knowledge of taxonomy and ecology of this group in grassland ecosystems is advanced (e.g. Waloff, 1980; Hildebrandt, 1995; Nickel and Achtziger, 1999; Nickel, 2003; see also Achtziger, 1999).

(vi) Suction samples of Auchenorrhyncha can produce a high spatial resolution due to high proportions of monophagous and oligophagous species closely associated with their hosts. Usually, reproduction on the plot can be inferred from the presence of nymphs and freshly emerged adults. In contrast, spatial resolution of pitfall and light trap catches is limited.

In this paper, we present a case study on flood plain grasslands in the middle Elbe valley (Lower Saxony, Germany). Unlike most studies focusing on direct and short-term effects of treatments (e.g. Morris, 1981a,b; Southwood and van Emden, 1967), we concentrate on long-term changes of the species composition under different management regimes in meadows, pastures and fallows. For demonstrating differences between treatments, we apply a classification scheme based upon Auchenorrhyncha life strategies, which allows an indication of habitat disturbance.

2. Sites, methods and material

The study sites are located in the lower part of the middle Elbe valley (northern Germany) (Fig. 1). In this section of the river, anthropogenic influences on river morphology and dynamics have been less severe than in most other central European river valleys, with vast areas of sandy and muddy river banks, willow scrub and grassland being subject to periodical flooding. Due to these floodings, much of the outer dike area is used as meadows and pastures, whereas most inland areas are used as fields. We studied 25 sites covering the most widespread management regimes, moisture conditions and vegetation types (Table 1):

1. Five sites of high-intensity grassland (HIG) being subject to at least two cuts per year, occasional cattle grazing and frequent mineral fertilizing. Dominating plant communities include the Lolio-Cynosuretum and the *Alopecurus pratensis* community. Moisture conditions during summer

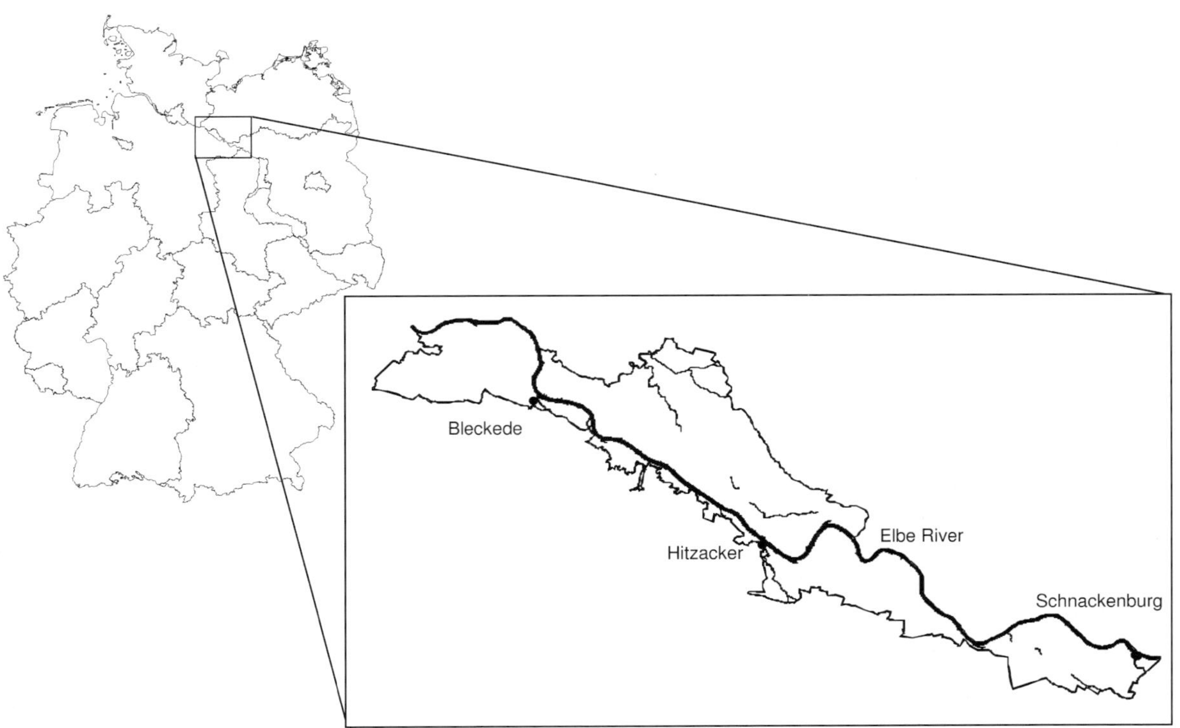

Fig. 1. Location of the study area.

Table 1
Overview of the study plots

Plot no.	Plant community/dominating plants	Moisture condition	Management regime[a]
1	Lolio-Cynosuretum	Damp	HIG
2	*Alopecurus pratensis* community	Damp	HIG
3	*Festuca rubra, Holcus lanatus, Deschampsia cespitosa*	Damp	F
4	Chrysanthemo-Rumicetum thyrsiflori	Damp	LIM
5	Cnidio-Deschampsietum (depression)	Moist	LIM
6	Cnidio-Deschampsietum	Moist	LIM
7	*Phalaris arundinacea, Alopecurus pratensis, Symphytum officinale*	Moist	F
8	*Silaum silaus* community	Moderately moist	LIM
9	Lolio-Cynosuretum	Damp	HIG
10	Lolio-Cynosuretum/Chrysanthemo-Rumicetum thyrsiflori	Damp	HIG
11	Ranunculo-Alopecuretum geniculati	Moderately wet	LIP
12	Diantho-Armerietum	Moderately dry	LIM
13	*Carex arenaria, Festuca ovina, Artemisia campestris*	Moderately dry	LIP
14	*Festuca ovina, Anthoxanthum odoratum, Agrostis capillaris*	Moderately dry	LIP
15	*Elymus repens, Carex hirta*	Moist	LIP
16	*Corynephorus canescens, Holcus lanatus, Agrostis capillaris*	Moderately dry	F
17	*Calamagrostis epigejos, Elymus repens*	Moderately dry	LIP
18	Caricetum vulpinae/Ranunculo-Alopecuretum geniculati	Moist	LIP
19	*Corynephorus canescens, Carex arenaria, Festuca ovina*	Moderately dry	LIP
20	*Alopecurus pratensis* community	Damp	LIP
21	Lolio-Cynosuretum	Damp	HIG
22	*Carex gracilis* community	Wet	F
23	*Lolium perenne, Elymus repens, Apera spica-venti*	Damp	F
24	Cnidio-Deschampsietum	Moist	LIM
25	Cnidio-Deschampsietum (depression)	Moderately wet	LIM

[a] HIG: high-intensity grassland, LIM: low-intensity meadow, LIP: low-intensity pasture, F: fallow. Plant communities after Redecker (unpublished).

range from moist to damp, occasionally moderately dry. In most years, during winter and early spring these sites become flooded, except sites no. 1 and 2.

2. Seven sites of low-intensity meadows (LIMs), harbouring endangered plant communities specific to lowland river basins ('Cnidion meadows'). This type of grassland only thrives in sites inundated for 40–100 days a year, being of major concern for conservation due to the occurrence of *Cnidium dubium* (Schkuhr) Thell., *Silaum silaus* (L.) Schinz and Thell., *Lathyrus palustris* L. and other endangered plants (Redecker, unpublished). The management regime includes two cuts a year after mid-June without use of fertilizers.
3. Eight sites of low-intensity pastures (LIPs), ranging from wet and regularly flooded to moderately dry sites, grazed by cattle over large areas. Cattle density is low and ranges from 0.9 to 1.2 livestock units per hectare. Floodings occur more or less frequently during winter and early spring.
4. Five fallow sites (F), including fallow fields, abandoned meadows and a tall sedge swamp formerly used for straw production. These sites may be mown after several years to prevent growing of shrubs. In summer, moisture conditions range from moderately wet to moderately dry. Sites no. 7 and 22, but not the remaining sites, are frequently flooded in winter and spring.

Sweepnet samples, taken on 9, 10 and 11 June, and 29 and 30 August 1999, comprised 50 sweeps per plot, respectively, done by a 32 cm × 26 cm linen net. Suction samples, taken on 30 and 31 May, and 6 and 7 August 2000, were done by an 'Eco-Vac' (made by Eco-Tech, Bonn, Germany), which is essentially a converted leaf blow apparatus. The nozzle area measured $0.015\,m^2$, the maximum air velocity was $76\,ms^{-1}$. Each sample involved placing the nozzle onto the surface for ca. 5 s, which was repeated 10 times per plot. The sample was then removed by inverting the collection bag into a linen funnel, the outlet meeting into a screw-cap glass filled with alcohol.

In order to standardize the different numbers of plots per treatment type we calculated an index value, dividing the number of threatened species by the number of replicates.

A statistical analysis of sweepnet samples in 1999 was performed using detrended canonical analysis (DCA) (Kovach, 1999). We applied a $\log_e$-transformation to the data and downweighted rare species (Jongman et al., 1995). In 1999, sites no. 24 and 25 could be sampled only on the 30 August 1999. Thus, these data were excluded from the DCA. Likewise, sites no. 3 and 19 could be sampled only in 1999.

3. Results

3.1. Ecological characteristics of the Auchenorrhyncha communities

In both years, altogether 16.613 individuals belonging to 88 species were sampled. The leafhoppers *Psammotettix confinis* (Dhlb.), *Errastunus ocellaris* (Fall.), *Deltocephalus pulicaris* (Fall.), *Arthaldeus pascuellus* (Fall.), *Notus flavipennis* (Zett.) and *Cicadula quadrinotata* (F.) were most abundant, the last two with clear preference of wet sites. Eight species occurred as single individuals only and were treated as non-resident. Eighteen species are listed in the Red Data Book of Germany, 14 of which are considered as threatened, mainly comprising specialists restricted to dry or wet habitats (Remane et al., 1998; Nickel et al., 1999).

The degree of food plant specialization appears to be high. Host specialists associated with a single plant species or genus (i.e. 1st or 2nd degree monophagous—see Table 2 for explanation) comprise more than one-third of the species total. The majority of species is 1st degree oligophagous, comprising 42%, while the remaining species are 2nd degree oligophagous or polyphagous. We compiled the information on the regional species pool, defined as the sum of species living in herbaceous vegetation known from Lower Saxony and the adjacent federal states of Schleswig–Holstein and Mecklenburg–Vorpommern (after Schiemenz, 1987, 1988, 1990; Schiemenz et al., 1996; Wagner, 1935; Nickel, 2003). A comparison reveals that the proportion of host specialists is considerably lower in the study sites than in the regional pool (Fig. 2).

Suction samples in 2000 revealed highest population densities in late spring on plot no. 7, in a fallow dominated by *Phalaris arundinacea* L., and on plot no. 15, in an LIP with dominating *Carex hirta* L. (Fig. 3). All high-intensity sites showed relatively low abundances on this date, which was reflected, in particular, by densities of eurytopic and epigeic species, e.g. *E. ocellaris* (Fall.) and *Anoscopus flavostriatus* (Don.). In August, differences to low-intensity sites were less pronounced.

Seven species were found exclusively in the suction samples, namely *Paraliburnia adela* (Fl.), *Criomorphus albomarginatus* Curt., *Agallia brachyptera* (Boh.), *Dikraneura variata* Hardy, *Rhytistylus proceps* (Kbm.), *Cicadula persimilis* (Edw.) and *Cosmotettix costalis* (Fall.). Most of these are more or less epigeic and more specific regarding habitat requirements. *D. variata* Hardy is probably an immigrant from neighbouring pine forests. Thus, in sweepnet samples, generalist species tend to be over-represented, particularly in stands of taller vegetation.

3.2. Differences between treatments

Differences in species composition are only slight between HIGs and LIMs, both treatments showing a dominance of eurytopic grass-dwelling species and only low individual numbers of a few specialists, like *Euconomelus lepidus* (Boh.) and *Mocuellus metrius* (Fl.). The two latter are mainly restricted to small patches of depressions with dominating *Eleocharis* spp. or *P. arundinacea* L., which may, for technical reasons, be difficult to reach for mowing machines. Moreover, some of these species were not recorded as nymphs, and hence, there was no breeding evidence.

Species numbers and proportion of specialists are much higher in LIPs. This increase in diversity is often caused by the patchy occurrence of additional host species, like *Festuca ovina* L., *Calamagrostis epigejos* (L.) Roth, *Hypericum perforatum* L. and others, which are absent from most mown sites. Communities

Table 2
Auchenorrhyncha catches in the study sites[a]

Treatment site no.	HIGs					LIMs							LIPs								F					Total	Host specificity	Voltinism	Status
	1	2	9	10	21	4	5	6	8	12	24	25	11	13	14	15	17	18	19	20	16	3	23	7	22				
Pioneer species																													
Empoasca pteridis (Dhlb.)	**3**	**1**		X		1	**4**	**2**	X	**1**	**1**		**1**								**1**		2			17	po	2?	
Javesella pellucida (F.)	**5**	**9**	X	**12**	**26**	**26**	**7**	**4**	**15**	**16**	**3**	X	**1**	**2**	1	**2**	X	X		**11**	X	**10**	X	**6**		156	po	2	
Laodelphax striatella (Fall.)		X	1																			1				2	o1	2	
Macrosteles cristatus (Rib.)					1																		2			3	po	2	
Macrosteles laevis (Rib.)	19	20		15	8		3		4	7	8	8	7							**4**					1	104	po	2	
Macrosteles quadripunctulatus (Kbm.)																	1									1	po?	2	3
Macrosteles sexnotatus (Fall.)	22	23		1	**54**	1	22	13	3	7	7	22	1				10			2					15	203	po	2	
Macrosteles viridigriseus (Edw.)		4					22	X	X			1	**168**	2										2		197	o1?	2	
Psammotettix alienus (Dhlb.)	2	8		2	3	X	5				3	9		1	X		X	1			3	2	1			40	o1	2	
Psammotettix confinis (Dhlb.)	5	50	**32**	**44**	**14**	**28**		**4**	58	**40**	**4**		**24**	**282**	**143**	2	32	**67**	**70**	**136**	**839**	12	**5**			1591	o1	2	
Eurytopic species																													
Anoscopus serratulae (F.)	X	X	X	X	X	**1**	X	X	X	X				X	2	2	X						X	X		5	o1	1	
Arthaldeus pascuellus (Fall.)	52	10	**29**	**28**	**17**	**3**	**3**	**11**	21	**4**	**21**	**5**	**66**	**20**	**39**	**34**	**2**	**234**	**11**	**5**	**1**	**18**	**5**	**6**	X	645	o1	2	
Deltocephalus pulicaris (Fall.)	4	11	**31**	**16**	**11**	**26**	**3**	**2**	192	**38**	2	X	**30**	6	**15**	X	X	**334**	**33**	143	**19**			**2**		918	o1	2	
Dicranotropis hamata (Boh.)																1										1	o1	2	
Errastunus ocellaris (Fall.)	**47**	1	**20**	**19**	**48**	**87**	**17**	**18**	85	**8**	**34**	**3**	**23**	**148**	**178**	**11**	**90**	**83**	**212**	**90**	**21**	**8**	**61**	**50**	X	1362	o1	2	
Euscelis incisus (Kbm.)	**4**	7	1	1	X			X	2	12	3			18	20				40	1	1		**48**	3		161	o2	2	
Javesella dubia (Kbm.)												2					1									3	o1	2	
Philaenus spumarius (L.)									1	1			1										3	**18**		24	po	1	
Streptanus aemulans (Kbm.)	X	X		X	1		X				X	X		X		1	**1**	2		2	2	**1**	**4**	**3**		17	o1	2?	
Oligotopic species																													
Agallia brachyptera (Boh.)						X	X	X																X		X	o2?	1	
Aphrodes bicincta (Schrk.)														**3**	**3**						2					8	o1?	1	
Aphrodes makarovi Zachv.				1									X			**3**							X	1		5	po	1	
Anaceratagallia ribauti (Oss.)		X		X			1			X										X			**4**			5	o2?	1	
Anoscopus albifrons (L.)																	1				6					7	o1	1	
Anoscopus flavostriatus (Don.)					X	X	X	2	X	X		X	X			**1**		X			X	**1**		**4**	X	8	o1	1	
Artianus interstitialis (Germ.)	**1**		X	1		**2**		1	1					**6**	**2**		2	1	**4**		16	1	**1**			39	o1	1	
Athysanus argentarius Metc.														1		X	1				1	1				4	o1	1	
Balclutha punctata (F.)				1		2															4	2				9	o1	1	
Chlorita paolii (Oss.)					3									**46**	1				36		1		1			88	o1	2	
Cicadella viridis (L.)								1																		1	po	1	
Cicadula quadrinotata (F.)					1		5	X			1	X		2	7	**429**	**20**	29	2		1			**9**	4	510	o1?	2	
Criomorphus albomarginatus Curt.															X											X	o2?	1	
Dikraneura variata Hardy															X											X	o1	2?	
Doratura homophyla (Fl.)		**1**			1					**3**				**52**	**20**	X	**26**		**111**	**65**	87					366	o1	2	
Doratura stylata (Boh.)														7	**9**			2	**48**			1				67	o1?	1	
Elymana sulphurella (Zett.)																							**2**			2	o1	1	
Eupteryx atropunctata (Goeze)	1	1											8										1			11	po	2	
Eupteryx notata Curt.											1			**3**	X				1							5	o2	2	
Eupteryx vittata (L.)							1	X			X	X												**4**	**1**	6	o2	2	
Eurybregma nigrolineata Scott														X		1							X			1	o1	1	
Graphocraerus ventralis (Fall.)														X	1	**3**					5		**4**			13	o1	1	
Jassargus pseudocellaris (Fl.)								1		2			1	**6**	**59**	**3**			1				**10**			83	o1	2	
Javesella obscurella (Boh.)		X					1					X	13					**9**								23	o1?	2	
Limotettix striola (Fall.)							1	1					**1**					1								4	o1	2	
Mocuellus collinus (Boh.)		4															4	1			95		9			113	o1	2	
Neophilaenus lineatus (L.)									1							4	3									8	po	1	
Neophilaenus minor (Kbm.)					1														4		9					14	o1	1	V
Notus flavipennis (Zett.)					1	1	11					**52**					7						1	7	**523**	603	o1?	2	
Psammotettix helvolus (Kbm.)										10				2					5							17	o1	2	
Psammotettix kolosvarensis (Mats.)	1	**4**	**25**	**9**		6		X	1	1				**6**	**1**	X	**69**	**45**		139	9	**1**		1		318	o1	2	3

Table 2 (*Continued*)

Treatment site no.	HIGs					LIMs							LIPs								F					Total	Host specificity	Voltinism	Status
	1	2	9	10	21	4	5	6	8	12	24	25	11	13	14	15	17	18	19	20	16	3	23	7	22				
Psammotettix nodosus (Rib.)										3																3	o1	2	
Psammotettix sabulicola (Curt.)																			8							8	o1?	2	2
Streptanus marginatus (Kbm.)																			1							1	o1	1	
Streptanus sordidus (Zett.)			1	X		X	1	X	X		4	X	**5**					**11**								22	o1	2	
Turrutus socialis (Fl.)					1					3			1						71				1			77	o1	2	
Stenotopic species																													
Acanthodelphax denticauda (Boh.)																								1		1	m1	2	3
Acanthodelphax spinosa (Fieb.)															8											8	m2	2	
Arocephalus punctum (Fl.)															X				26							26	m2?	2	
Balclutha rhenana W.Wg.												X					2							X		2	m1	1	
Cicadula flori (J. Shlb.)																									**65**	65	m2	2	V
Cicadula persimilis (Edw.)																										X	m1	2	
Cosmotettix costalis (Fall.)																							X			X	m2	1	2
Delphacinus mesomelas (Boh.)														**3**	**5**	1										9	m2	1?	
Euconomelus lepidus (Boh.)							X			1								1								2	m2?	1	3
Eupelix cuspidata (F.)														1	X						X					1	m1?	0.5	
Eupteryx calcarata Oss.	7																									7	m1?	2	
Kelisia sabulicola W.Wg.														**17**	**1**				**41**		8					67	m1	1	3
Kosswigianella exigua (Boh.)														40	**24**				**77**							141	m1	2	
Megamelus notula (Germ.)	1																								**14**	15	m2	1?	
Metalimnus formosus (Boh.)																									**26**	26	m2	1	2
Mocuellus metrius (Fl.)			X	1			2	2	X		8	1	**2**				**9**	11	1					**5**	**1**	43	m1	2?	
Mocydiopsis parvicauda Rib.														3	**9**											12	m1	1	
Muellerianella brevipennis (Boh.)	1		1	1					2														1	**3**		9	m1	1	
Muellerianella fairmairei (Perr.)																						7				7	m2	2?	D
Paluda flaveola (Boh.)																2	1							**4**		7	m2?	1	
Paraliburnia adela (Fl.)																		X								X	m1	1	3
Psammotettix excisus (Mats.)					1														17		18					36	m1?	2	3
Rhopalopyx preyssleri (H.-S.)						X								X	X	**15**										15	m1	1	
Rhopalopyx vitripennis (Fl.)					9					**8**				8	3				44		1	6	7			86	m1	2	3
Rhytistylus proceps (Kbm.)																					X					X	m1?	1	3
Ribautodelphax albostriata (Fieb.)					1	X									2					1						4	m1	2	
Ribautodelphax angulosa (Rib.)															4				8							12	m1	2	1
Ribautodelphax collina (Boh.)										1				3	**11**			2			26	11				54	m1	2	
Stenocranus major (Kbm.)											1						2	4	1							8	m1	1	
Stenocranus minutus (F.)		1																								1	m2	1	
Stroggylocephalus agrestis (Fall.)							2					1													**4**	7	m2?	1	V
Xanthodelphax straminea (Stal)									1	1				**6**	**9**			3								20	m2	1	3
Zygina hyperici (H.-S.)															3						2		3			8	m1	2	
Not determinated to species level																													
Deltocephalinae—juveniles	1		2		6	5	50		2	6	17	34	3	2	7		28	7	113	65		1	6			355			
Delphacidae—juveniles							3	1					12	1		1	1	1	8							28			
Ribautodelphax—females			1												7											8			
Ribautodelphax—juveniles															20											20			
Macrosteles—females			3	19	36	4			10	8	16	15	94					3	1		1				6	216			
Macrosteles—juveniles	7	4		13	13		1		6	7	6	2	50					2								111			
Psammotettix—females										38									29							67			
Psammotettix—juveniles	6	7	6							8									37							64			
Number of individuals	24	14	22	44	87	13	63	9	29	90	72	78	172	97	127	34	60	52	422	86	72	28	40	20	138	1893			
Number of recorded species	18	21	13	20	22	18	23	21	20	23	17	19	19	32	34	21	24	22	25	13	29	16	27	21	13	88			
Number of resident species	13	16	9	13	13	14	18	17	15	17	13	16	12	29	29	16	18	17	20	11	22	9	20	18	10	80			

[a] Numbers are given for sweep net samples in 1999 only. X: recorded in suction samples in 2000 only, bold: recorded both in sweepnet and suction samples. Host plant specificity: m1: 1st degree monophagous (on one plant species), m2: 2nd degree monophagous (on one plant genus), o1: 1st degree oligophagous (on one plant family), o2: 2nd degree oligophagous (on two plant families or up to four plant genera belonging to up to four families), po: polyphagous. Red Data Book of Germany (after Remane et al., 1998)—1: critical, 2: endangered, 3: vulnerable, V: near threatened, D: data deficient (categories adapted to the IUCN criteria). HIGs: high-intensity grassland; LIMs: low-intensity meadows; LIPs: low-intensity pastures; F: fallows; status: conservation status.

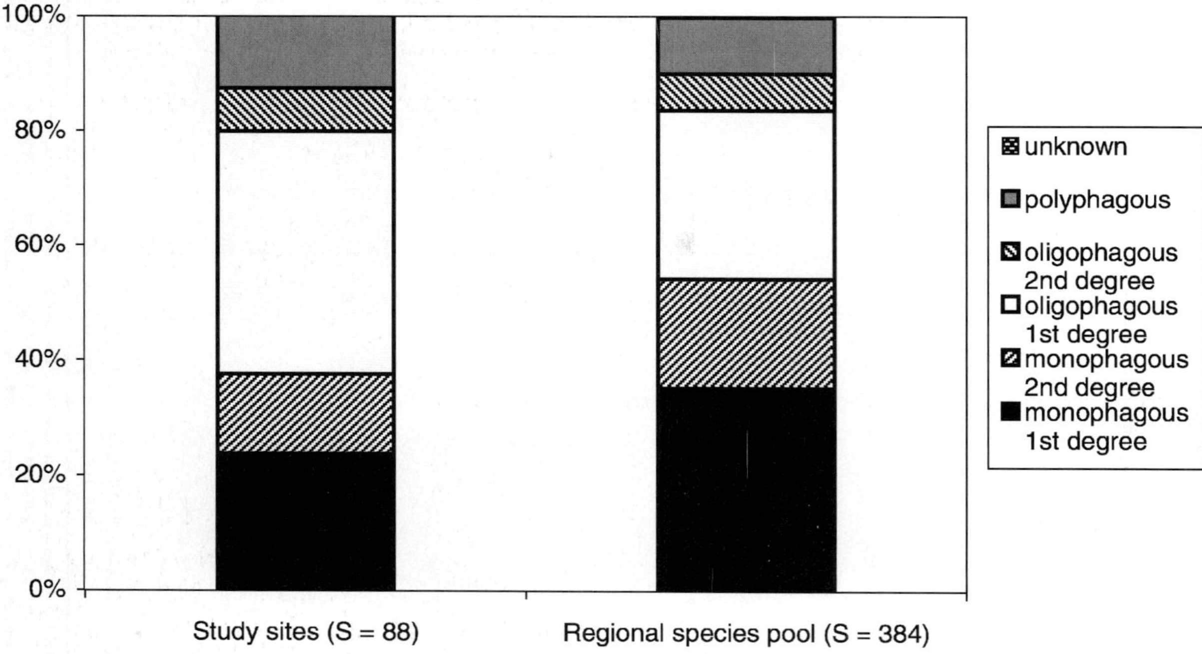

Fig. 2. Diet width of the Auchenorrhyncha species of the study sites and the regional pool.

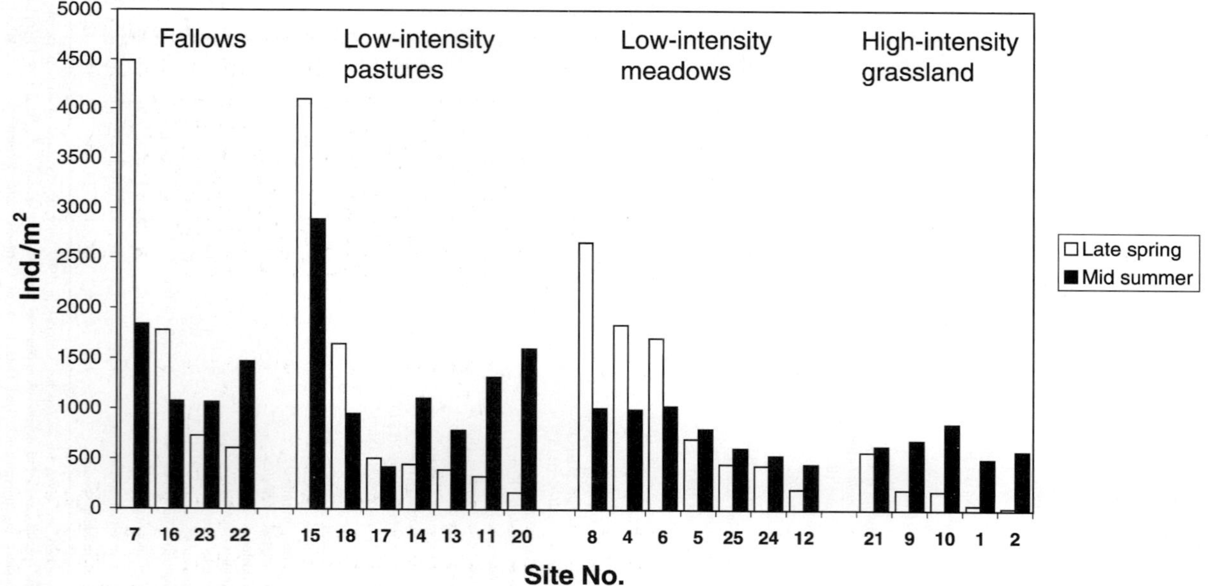

Fig. 3. Total suction sample catches of Auchenorrhyncha in study sites of the Elbe flood plains in 2000.

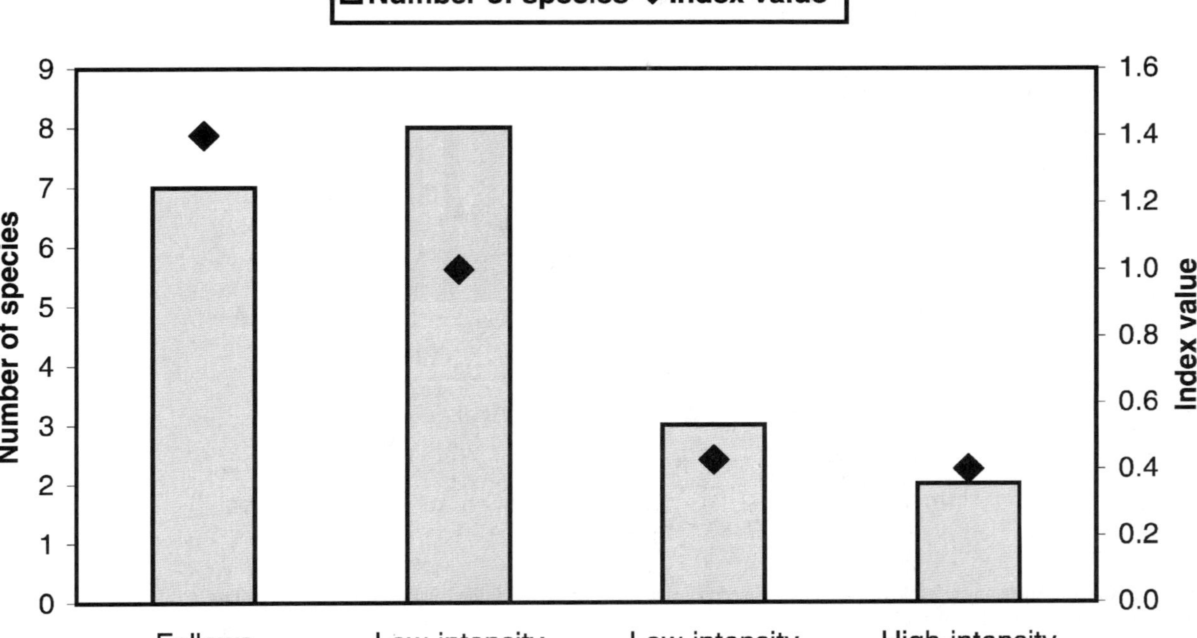

Fig. 4. Distribution of threatened species (categories 1, 2, 3, R, after Remane et al., 1998). Index value = number of species divided by the number of sites.

in fallow sites (F) show marked differences in moisture requirements. Moderately dry sites are dominated by xerophilous species, e.g. *Ribautodelphax collina* (Boh.), whereas sedge-dwelling species, like *N. flavipennis* (Zett.), *Cicadula* spp. and *Metalimnus formosus* (Boh.) are common on wet sites.

A comparison of the distribution of threatened species shows marked differences between the treatments (Fig. 4). Highest index values were found in F (=1.4) and LIP sites (=1.0); index values were lower in LIM and HIG sites (=0.4, respectively). Highest species numbers were found in LIP (eight species) and F sites (seven species). LIM and HIG sites harboured only three and two species, respectively.

The DCA (Fig. 5) reveals a dense cluster of both HIGs and LIMs. This indicates a strong and equalizing effect of a mowing regime comprising two cuts or more a year. However, there are a few exceptions: sites no. 11 and 20 are both LIP. The former is a deep depression close to the river bank being subject to most inundations among all sites, the latter was subject to heavy trampling by cattle. Site no. 3, situated in close vicinity to HIG and LIM sites, is a young fallow on a former field on highly fertile soil. Site no. 5, which is the only LIM outside the cluster, is situated in a meadow depression, where mowing depth was reduced for technical reasons. All other sites are widely scattered, suggesting that low-intensity grazing and fallowing offers a high potential for sustaining local biodiversity.

3.3. Life strategies

Achtziger and Nickel (1997) proposed a classification scheme of grassland Auchenorrhyncha species based upon their life strategies (Table 3). Accordingly, there are four groups showing distinct combinations of ecological traits and differential response to treatments.

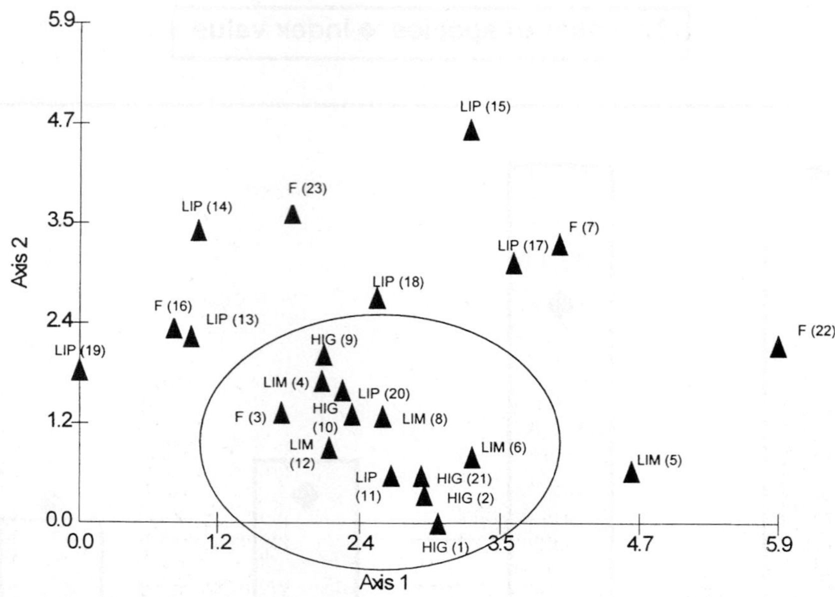

Fig. 5. DCA of Auchenorrhyncha communities (sweepnet samples 1999) (explained variance of the ordination = 38.2; axis 1 = 18.3, eigenvalue = 0.67; axis 2 = 9.1, eigenvalue = 0.3).

(i) Pioneer species are macropterous and highly mobile, often dominating in aerial catches (see della Giustina and Balasse, 1999; Waloff, 1973) although populations of *Laodelphax striatella* (Fall.) and *Javesella pellucida* (F.) in more stable habitats may occasionally comprise some brachypterous individuals. Early successional stages are usually colonized within one season, but influx of individuals is noted in all terrestrial habitats, although breeding may not be successful. All species are bivoltine in central Europe, perhaps locally even trivoltine. Most species are polyphagous; those which are oligophagus usually feed on a broad range of grasses.

(ii) Eurytopic species are widespread and common in various types of grass-dominated habitats. They occupy a wide range of moisture conditions and tolerate disturbance by mowing and grazing. Most species are bivoltine and oligophagous, feeding on grasses. Flight ability tends to be reduced, but new habitats are colonized within some years due to large population sizes along

Table 3
Auchenorrhyncha life strategies in central European grasslands

Life strategy trait	Generalists		Specialists	
	Pioneer species	Eurytopic species	Oligotopic species	Stenotopic species
Habitat preference	Mainly in early successional stages	Eurytopic in various types of grasslands	Associated with specific abiotic conditions	Associated with specific abiotic conditions
Wing length and dispersal	Always long-winged; permanent influx into most terrestrial habitats	Long- or short-winged, flight activity moderate	Long- or short-winged, flight activity moderate	Mostly short-winged; low flight activity
Host plant specialization	Usually polyphagous	Usually oligophagous on Poaceae	Usually oligophagous	1st or 2nd degree monophagous
Voltinism	At least bivoltine	Mostly bivoltine	Uni- or bivoltine	Uni- or bivoltine

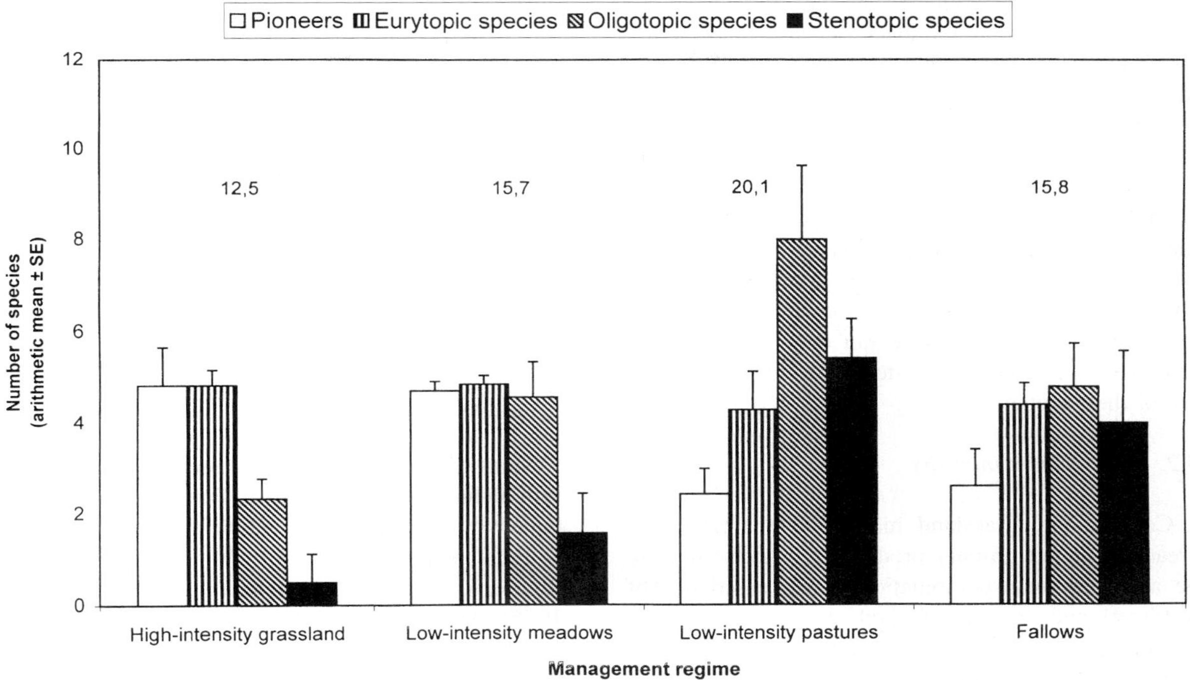

Fig. 6. Numbers of Auchenorrhyncha species under different management regimes. Numbers above grouped columns indicate the total sums of columns.

roadsides, field margins and ditches all over central European lowlands.

(iii) Oligotopic species are also widespread and show similar traits of life history, but they are restricted to a narrower range of abiotic conditions, like moisture, nutrient contents or vegetation height. Their tolerance towards disturbance by mowing, grazing and fertilizing is reduced. Flight ability is also reduced, but new habitats may be colonized within some years, depending on location and densities of the nearest resident populations. Univoltine species prevail.

(iv) Stenotopic species also occupy a narrow range of abiotic conditions, but in addition, they show a close association with certain host species or genera, i.e. they are 1st or 2nd degree monophagous. Due to specific requirements of their host plants, their distribution may be rather localized and confined to certain geological, hydrological or microclimatic features, e.g. limestone regions, valley bottoms or forest margins.

The two former groups are referred to here as generalists, the two latter as specialists. We used this scheme for characterizing the communities found in the Elbe valley. Considering the number of generalist species, differences between the four treatments are slight. However, specialists clearly favoured LIPs and fallows, with only small numbers in high-intensity sites and LIMs (Fig. 6). Interestingly, a further reduction of disturbance on the fallow sites, which are mown at most after several years, did not lead to a further increase of numbers of species and specialists. Instead, numbers declined compared to LIPs.

4. Discussion

4.1. Time scale

There is a general need for distinction between short-term and long-term effects of grassland management. Removal of plant biomass by mowing and grazing is usually followed by an immediate decrease of individual numbers of most taxa living above-ground (Curry, 1994). On the other hand, the management regime has long-termed effects on the structure and composition of the vegetation by excluding some plant species and promoting others, thus strongly affecting host availability and habitat conditions of phytophagous insects. Furthermore, management prevents immigration of shrubs and trees which would eventually lead to succession and tree cover. It should also be noted that most studies on grassland management were run over a period of a few years only. Thus, long-term developments were out of their scope.

4.2. *Management intensity*

Conventional grassland management aims at increasing the net primary production, which can only be achieved by a combination of increased use of fertilizers and removal of plant biomass by mowing or grazing. In turn, this is usually correlated with a decrease of diversity of plant species and architecture, and consequently, with a reduction of faunal diversity (see Curry, 1994). Responses of Auchenorrhyncha species numbers to an increased management intensity are also negative, although the effects of fertilizing and removal of plant material could not be disentangled (e.g. Andrzejewska, 1991; Remane, 1958). Considering species groups favouring different life strategies in Bavarian grasslands, Achtziger and Nickel (1997) and Achtziger et al. (1999) found that effects on generalist species numbers were insignificant. Specialists, however, were almost absent from high-intensity sites, but common and diverse in unfertilized meadows mown only once a year. Pioneer species, like *J. pellucida* (F.) and *Macrosteles laevis* (Rib.), have been repeatedly reported to show a strong increase of population densities in HIGs and other disturbed habitats (Andrzejewska, 1979; Novotný, 1994b).

Our data show differences in species numbers and proportions of specialists between high-intensity sites (HIG) on the one hand and low-intensity pastures (LIPs) and fallows (F) on the other hand. However, these differences were less pronounced between high-intensity treatments (HIGs) and low-intensity meadows (LIMs). These results indicate a rather severe impact of the mowing regime, even in unfertilized sites, which may lead to a long-term exclusion of a number of stenotopic Auchenorrhyncha species. Moreover, such grasslands may constitute sinks for large numbers of ovipositing individuals on dispersal flight, the offspring of which will die after the next cut. It is also noteworthy, that the proportion of stenotopic species in the study sites is lower than among the total regional fauna of grassland Auchenorrhyncha (see Fig. 2), providing further evidence for the selection of generalists by disturbance in managed flood plain grassland.

4.3. *Effects of fertilizing*

Short-term effects of mineral fertilizers on Auchenorrhyncha were studied in a Polish meadow dominated by *Arrhenatherum elatius* (L.) P. Beauv. ex J. Presl and C. Presl, *Dactylis glomerata* L., *Poa pratensis* L., and *Festuca rubra* L. (Andrzejewska, 1976). After 3 years, differences between treated and control plots were only slight in Auchenorrhyncha biomass, but marked in community structure. In the fertilized plots, generalists, like *J. pellucida* (F.), *M. laevis* (Rib.), and *Streptanus aemulans* (Kbm.), had increased. Most specialists, however, e.g. *Acanthodelphax spinosa* (Fieb.), *Ribautodelphax albostriata* (Fieb.), *Rhopalopyx preyssleri* (H.-S.), and *Athysanus argentarius* Metc. remained confined to the untreated plots or had a much higher emergence rate there.

Other short-term field experiments were made by Sedlacek et al. (1988) and Prestidge (1982), both revealing higher population densities after a few years of fertilizing, but without discussing the life strategies of species and the composition of communities. From a study of Auchenorrhyncha communities on the grass *Holcus lanatus* L., the latter author concluded that both increased food quality and plant architecture were the main factors. However, all species found by him (see also Prestidge and McNeill, 1983) were generalists. *Ribautodelphax angulosa* (Rib.) was the only exception and remains enigmatic, since it was later shown to be strictly monophagous on *Anthoxanthum odoratum* L. (den Bieman, 1987). We did not find sufficient information to classify the species found by Sedlacek et al. (1988) in Ohio, although all those which are common to Europe and North America were eurytopic or oligotopic, but none was stenotopic.

Long-term effects of nitrogen fertilizing and liming were studied by Morris (1992) at Park Grass, Rothamsted (England). Neglecting species found in less than four individuals, the communities on treated plots comprised only pioneer, eurytopic and oligotopic species as well as *C. persimilis* (Edw.), which was the only stenotopic species. Species richness and total abundance were lower on plots receiving nitrogen.

Among our study plots, only the HIG sites were conventionally treated with mineral fertilizers and liquid manure. All remaining treatments received nutrients exclusively by flooding. Our design allows only a comparison of HIG and LIM sites, both of which are mown twice a year. The LIM sites show almost identical groups of generalists, but higher numbers of specialists. However, these differences are not significant, and variation is high. Achtziger and Nickel (1997) found similar results on more homogeneous study sites, with significant differences of numbers of oligotopic and stenotopic species between fertilized and unfertilized sites. Thus, there is increasing evidence for negative long-term effects of fertilizing on Auchenorrhyncha diversity, and for positive effects on the abundance of generalists.

Curry (1994) concluded that the effects of fertilizing on grassland invertebrates are mainly a consequence of vegetational changes, i.e. increased net primary production, but reduced species number, notably in dicotyledons, combined with stronger dominance of some grasses. Reduction of Auchenorrhyncha diversity can be explained by the decrease of plant species numbers and the simplification of the vegetation structure.

4.4. *Effects of grazing and mowing*

Like mowing, intensive grazing reduces the complexity of the vegetation structure, which in turn affects arthropod habitat conditions. For instance, spider communities in British grasslands on peat soils were twice as diverse when sheep grazing was excluded (Cherrett, 1964). On the other hand, selective feeding, trampling, and dung production have additional effects by creating patches of different microhabitats. In contrast, mowing is less selective than grazing and leads to a lower heterogeneity of structures. Negative effects are mainly due to the sudden and almost complete removal of above-ground plant biomass, which is combined with a reduction of food plants, eggs, substrates for oviposition and a dramatic change of microclimatic conditions. Insects are particularly affected when mowing takes place during sensitive phases of development (Curry, 1994).

In our study sites, Auchenorrhyncha diversity was low both in HIG sites and in LIMs, indicating a severe negative influence of mowing, particularly on specialists. Adverse effects on Auchenorrhyncha as well as on Heteroptera were also found by Morris (1981a) and Morris and Lakhani (1979), particularly if mowing had been performed twice a year or in mid-summer only. Auchenorrhyncha communities of sites no. 11 and 20 showed a close similarity to HIG and LIM sites (Fig. 5), which can be explained by unusual high disturbance due to flooding and trampling by cattle in these sites. This indicates that strong disturbance effects of mowing, flooding, grazing and trampling may be rather similar in selecting a few generalist species, and that some meadow-dwelling Auchenorrhyncha may originate from natural flood plain grasslands.

We found a number of species being confined to fallows and LIPs, e.g. *Eurybregma nigrolineata* Scott, *R. collina* (Boh.), *Graphocraerus ventralis* (Fall.), *A. argentarius* Metc., and *Jassargus pseudocellaris* (Fl.). In some cases, the lack of Auchenorrhyncha species on mown sites can be easily explained by the lack of their host plants. For instance, *Zygina hyperici* (H.-S.), a leafhopper obligatorily associated with *H. perforatum* L., is absent from meadows all over central Europe, but commonly breeds in fallows and pastures, where its host plant is usually avoided by grazing animals. Thus, regular mowing causes a long-term exclusion of some species by eliminating favoured hosts. Decrease of less specific feeders can be explained by the simplification of plant architecture combined with frequent disturbance. This is probably also true for a number of *Carex*-dwelling species, like *Kelisia* spp., *Cicadula flori* (J. Shlb.), *M. formosus* (Boh.), and *C. costalis* (Fall.). However, some of these species have been recorded in unfertilized meadows of the foothills of the Alps, which are subject to a single autumn cut (Nickel, unpublished data).

In fertilized meadows in Poland with three cuts a year, individual numbers of Auchenorrhyncha declined dramatically after each cut, but recovered immediately due to survival, immigration, and emergence of the next generation. In these intensively managed sites, pioneer and eurytopic species, like *M. laevis* (Rib.), *J. pellucida* (F.), und *S. aemulans* (Kbm.), strongly dominated (Andrzejewska, 1979). We cannot demonstrate these numerical responses in our study, because of only two sampling intervals per year. But high densities of pioneer and eurytopic species in some of our sites suggest a dependance of populations on influx from surrounding habitats.

In a study on the effects of sheep grazing, Morris (1973) found significantly higher abundances of six Auchenorrhyncha species in intensively grazed sites, notably of *Anaceratagallia venosa* (Geoffr.), *Aphrodes bicincta* (Schrk.), *Arocephalus punctum* (Fl.), *R. proceps* (Kbm.), and *Eupteryx notata* Curt., whereas *Arboridia parvula* (Boh.) showed a negative response. *Stenocranus minutus* (F.), *A. flavostriatus* (Don.), and *S. aemulans* (Kbm.) were almost completely absent from grazed plots. Further species, especially those preferring tall grassland, were less abundant, but negative effects of grazing were more pronounced, when treatment was performed in autumn or winter compared to spring or summer. Effects on species of Heteroptera were not significant, although four species of grass bugs (Miridae: Stenodemini) occurred exclusively in ungrazed sites, and maximum species numbers were found on the ungrazed plots.

In our study sites, LIPs showed considerably higher species numbers of Auchenorrhyncha compared to LIMs, indicating that effects of moderate grazing (cattle densities about one livestock unit per hectare) are less severe than effects of two cuts a year. This response is even more pronounced than cessation of management, which leads to fallows often dominated by monospecific stands of tall grasses or sedges, e.g. *P. arundinacea* L. or *Carex acuta* L. Lower species numbers in fallows were also found in Bavaria in a comparison to LIMs (Achtziger et al., 1999). In contrast to our study sites, however, these meadows were mown only once a year in autumn. Maximum species diversity at intermediate disturbance levels have been observed in many ecological systems (e.g. Connell, 1978; see also Huston, 1994; Wilkinson, 1999).

We conclude that the impact of two cuts in late spring and summer on the Auchenorrhyncha fauna is rather severe, because numbers of species and, in particular, of specialists are reduced. On the other hand, grazing at low intensity offers considerable potential for higher species numbers, since cattle or sheep avoid patches harbouring less palatable plants, which may enhance diversity of plant species and architecture.

4.5. Flooding and moisture preferences

Inundations are significant constraints for the management regime in river floodplains. Mowing and grazing can take place only relatively late in the year, and cattle densities have to be kept low due to the vulnerability of the turf to trampling. Furthermore, flooding probably has a strong effect on insect communities and, hence, may blur differences between treatments.

At least 18 central European Auchenorrhyncha species are associated with habitats being subject to strong fluctuations of the water table, which may imply an ability to survive inundations (Nickel and Achtziger, 1999). In our study sites, we found evidence for further species, which showed high nymphal densities in May on plots flooded during winter and spring, e.g. *A. brachyptera* (Boh.), *A. flavostriatus* (Don.), *D. pulicaris* (Fall.), *Streptanus sordidus* (Zett.), *E. ocellaris* (Fall.), and *A. pascuellus* (Fall.). Well-designed field and laboratory experiments are needed to clarify the tolerance and requirements towards moisture and flooding in these species.

From alpine river banks, which are subject to dramatic and frequent flooding incidents, it is known that Auchenorrhyncha communities are highly adapted to river dynamics, being characterized by a high proportion of host specialists, monovoltine species and European or even Alpine endemics (Nickel, 1999). In these species, heavy population losses, which must be compensated, frequently occur due to translocation of gravel banks.

4.6. Incongruence of botanical and zoological conservation priorities

Although floristic as well as faunistic diversity is high in many nature reserves, priority sites for conservation are not necessarily identical. Most Auchenorrhyncha, including host specialists and threatened species, live on common and widespread plants rich in biomass (Nickel, 2003). On a regional scale, faunal diversity has been found to be mainly correlated with habitat diversity and small-scale disturbance (see Rosenzweig, 1995).

In our study region, the priority sites for threatened plants, i.e. flood plain meadows with *C. dubium* (Schkuhr) Thell., *L. palustris* L., and others, were clearly different from those for threatened Auchenorrhyncha. These were mainly grazed and fallow sites with only low relevance for plant conservation. Extended grazing would promote vegetation complexes dominated by *Elymus repens* (L.) Gould and other tall grasses as well as a decline of established meadow plant communities, notably the 'Cnidion meadows'. Possible implications of such an extension of grazing on animal communities include (i) an increase of specific dwellers of tall grasses and *Elymus* specialists among Auchenorrhyncha, and (ii) new or enlarged habitats for grassland songbirds, e.g. *Alauda arvensis* (L.) (Skylark) and *Saxicola rubetra* (L.) (Whinchat). Therefore, maintaining maximum local biodiversity in grasslands requires different management regimes, including grazing at low intensity and fallowing, in order to balance issues of animal and plant conservation.

5. Conclusions

(i) The study of grassland Auchenorrhyncha can provide a useful tool for demonstrating biotic conditions of the habitat. Suction sampling produces exact figures of species numbers and abundance, whereas sweepnet samples are biased towards species of higher vegetation layers. Resident species can be recognised if nymphs are present.

(ii) Short-term effects of most grassland treatments, like mowing, grazing, fertilizing and trampling, result in a decrease of individual numbers of most Auchenorrhyncha species. Populations of a few pioneer and generalist species are capable of a rapid recovery. Long-term effects include the maintenance of plant species composition and prevention of tree growth, but also a permanent exclusion of specialists, if disturbance occurs too frequently.

(iii) Although mineral fertilizing may lead to an increase in individual numbers of generalists, its main effect is a decrease in species numbers of specialists. The underlying mechanisms are not fully understood.

(iv) Mowing and grazing at low intensity may maintain maximum diversity of Auchenorrhyncha species. However, there is increasing evidence for severe negative long-term effects of two or more cuts a year and cutting in early or mid-summer. Grazing at low intensity offers a high potential for conservation, because selective feeding, trampling, dung production, and avoidance of less palatable plants creates a diverse patchwork of microhabitats. In these patches, insect species favouring taller vegetation can permanently survive. At high-intensity, grazing (like mowing) seriously reduces insect diversity and, in particular, species numbers of specialists.

(v) Fallows may act as refuges for grassland species, which do not tolerate conventional management regimes. This is particularly true for intensively utilized regions.

(vi) We suggest that total species numbers as well as the proportion and abundance of specialists, which are all negatively correlated with management intensity, may be used as robust parameters for the indication of favourable habitat conditions in grasslands. Conversely, communities showing a high proportion and abundance of pioneer species indicate severe disturbance. Their stability is often reduced as they are subject to strong fluctuations.

(vii) Management regimes aiming at zoological and botanical conservation priorities are not necessarily identical. On a regional scale, different regimes should be employed for maintaining maximum biodiversity.

(viii) The study of Auchenorrhyncha communities is complicated by the relatively high effort required for determination, e.g. compared to grasshoppers and birds. Moreover, strong overlaying effects on communities may occur in river flood plains, because their responses to mowing, fertilizing, and flooding are principally similar and are difficult to disentangle in field experiments.

(ix) Community dynamics in river flood plains may include significant population shifts between flooded and non-flooded areas. A number of species, however, can tolerate flooding events in situ.

(x) Quantitative indicative values, like in some freshwater groups, are difficult to calculate for Auchenorrhyncha, because the overall data base is too weak. Thus, a macroecological approach is needed to study whether local patterns are robust.

(xi) In general, causal relations between management measures and community patterns of invertebrates are not yet fully understood. Although much research has been done on responses of communities and single species to treatment, little is known of the implications on meso-scale population dynamics in cultivated landscapes. More information is also needed on the effects of flooding.

Acknowledgements

This study was part of the Research Project "Leitbilder des Naturschutzes und deren Umsetzung mit der Landwirtschaft", FKZ 0339581, supported by the BMBF, Projektgruppe Elbe-Ökologie, Berlin. We thank M. Schaefer, M. Sayer and J. Rothenbücher for helpful comments on the manuscript.

References

Achtziger, R., 1999. Möglichkeiten und Ansätze des Einsatzes von Zikaden in der Naturschutzforschung (Hemiptera: Auchenorrhyncha). Reichenbachia 33, 171–190.

Achtziger, R., Nickel, H., 1997. Zikaden als Bioindikatoren für naturschutzfachliche Erfolgskontrollen im Feuchtgrünland. Beitr. Zikadenkde. 1, 3–16.

Achtziger, R., Nickel, H., Schreiber, R., 1999. Auswirkungen von Extensivierungsmaßnahmen auf Zikaden, Wanzen, Heuschrecken und Tagfalter im Feuchtgrünland. Schriftenr. Bayer. Landesamt Umweltsch. 150, 109–131.

Andrzejewska, L., 1965. Stratification and its dynamics in meadow communities of Auchenorrhyncha (Homoptera). Ecologia Polska A 13, 685–715.

Andrzejewska, L., 1976. The influence of mineral fertilization on the meadow phytophagous fauna. Pol. Ecol. Stud. 2, 93–109.

Andrzejewska, L., 1979. Herbivorous fauna and its role in the economy of grassland ecosystems. I. Herbivores in natural and managed meadows. Pol. Ecol. Stud. 5, 5–44.

Andrzejewska, L., 1991. Formation of Auchenorrhyncha communities in diversified structures of agricultural landscapes. Pol. Ecol. Stud. 17, 267–287.

Boness, M., 1953. Die Fauna der Wiesen unter besonderer Berücksichtigung der Mahd. Z. Morphol. Oekol. Tiere 42, 225–277.

Brown, V.K., 1985. Insect herbivores and plant succession. Oikos 44, 17–22.

Cherrett, J.M., 1964. The distribution of spiders on the Moor House National Nature Reserve, Westmoorland. J. Anim. Ecol. 33, 27–48.

Connell, J.H., 1978. Diversity in tropical rain forests and coral reefs. Science (Washington, DC) 199, 1302–1310.

Curry, J.P., 1987. The invertebrate fauna of grassland and its influence on productivity. 1. The composition of the fauna. Grass and forage. Science (Washington, DC) 42, 103–120.

Curry, J.P., 1994. Grassland Invertebrates—Ecology, Influence on Soil Fertility and Effect on Plant Growth. Chapman & Hall, London.

della Giustina, W., Balasse, H., 1999. Gone with the wind: Homoptera Auchenorrhyncha collected by the French network of suction traps in 1994. Marburger Entomol. Publ. 3 (1), 7–42.

den Bieman, C.F.M., 1987. Host plant relations in the planthopper genus Ribautodelphax (Homoptera, Delphacidae). Ecol. Entomol. 12, 163–172.

Denno, R.F., 1994. Influence of habitat structure on the abundance and diversity of planthoppers. In: Denno, R.F., Perfect, T.J. (Eds.), Planthoppers—Their ecology and Management. Chapman & Hall, New York, pp. 140–160.

Denno, R.F., Roderick, G.F., 1991. Influence of patch size, vegetation structure and host plant architecture on the diversity, abundance and life history styles of sap-feeding herbivores. In: Bell, S.S., McCoy, E.D., Muchinsky, H.R. (Eds.), Habitat Structure: The Physical Arrangement of Objects in Space. Chapman & Hall, New York, pp. 169–196.

Hildebrandt, J., 1995. Zur Zikadenfauna im Feuchtgrünland—Kenntnisstand und Schutzaspekte. Mitt. 1. Auchenorrhyncha-Tagung, 23.9. bis 25.9.1994, Halle/Saale, pp. 5–22.

Huston, M.A., 1994. Biological Diversity—The Coexistence of Species on Changing Landscapes. Cambridge University Press, Cambridge.

Jongman, R.H.G., Ter Braak, C.J.F., van Tongeren, O.F.R., 1995. Data Analysis in Community and Landscape Ecology. Cambridge University Press, Cambridge.

Jung, G., Schädler, M., Auge, H., Brandl, R., 2000. Effects of herbivorous insects on secondary plant succession. Mitt. Dtsch. Ges. Allg. Ang. Entomol. 12, 169–174.

Kovach, W.L., 1999. MVSP—A MultiVariate Statistical Package for Windows, Ver. 3.1. Kovach Computing Services, Pentraeth, Wales.

Morris, M.G., 1971. Differences in the invertebrate fauna of grazed and ungrazed chalk grasslands. IV. Abundance and diversity of Homoptera—Auchenorrhyncha. J. Appl. Ecol. 8, 37–52.

Morris, M.G., 1973. The effect of seasonal grazing on the Heteroptera and Auchenorrhyncha (Hemiptera) of chalk grassland. J. Appl. Ecol. 10, 761–780.

Morris, M.G., 1981a. Responses of grassland invertebrates to management by cutting. III. Adverse effects of Auchenorrhyncha. J. Appl. Ecol. 18, 107–123.

Morris, M.G., 1981b. Responses of grassland invertebrates to management by cutting. IV. Positive responses of Auchenorrhyncha. J. Appl. Ecol. 18, 763–771.

Morris, M.G., 1992. Responses of Auchenorrhyncha (Homoptera) to fertiliser and liming treatments at Park Grass, Rothamsted. Agric. Ecosyst. Environ. 41, 263–283.

Morris, M.G., Lakhani, K.H., 1979. Responses of grassland invertebrates to management by cutting. II. Heteroptera. J. Appl. Ecol. 16, 77–98.

Morris, M.G., Plant, R., 1983. Responses of grassland invertebrates to management by cutting. V. Changes in Hemiptera following cessation of management. J. Appl. Ecol. 20, 157–177.

Morris, M.G., Rispin, W.E., 1987. Abundance and diversity of the coleopterous fauna of a calcareous grassland under different cutting regimes. J. Appl. Ecol. 24, 451–465.

Murdoch, W., Evans, F.C., Peterson, C.H., 1972. Diversity and pattern in plants and insects. Ecology (NY) 53, 819–829.

Nickel, H., 1999. Life strategies of Auchenorrhyncha species on river floodplains in the northern Alps, with description of a new species: *Macropsis remanei* sp. n. (Hemiptera). Reichenbachia 33, 157–169.

Nickel, H., 2003. The leafhoppers and planthoppers of Germany (Hemiptera, Auchenorrhyncha): patterns and strategies in a highly diverse group of phytophagous insects. Pensoft, Sofia and Moscow.

Nickel, H., Achtziger, R., 1999. Wiesen bewohnende Zikaden (Auchenorrhyncha) im Gradienten von Nutzungsintensität und Feuchte. Beitr. Zikadenkde. 3, 65–80.

Nickel, H., Witsack, W., Remane, R., 1999. Rote Liste der Zikaden Deutschlands (Hemiptera, Auchenorrhyncha)—Habitate, Gefährdungsfaktoren und Anmerkungen zum Areal. Beitr. Zikadenkde. 3, 13–32.

Novotný, V., 1994a. Relation between temporal persistence of host plants and wing length in leafhoppers (Auchenorrhyncha, Hemiptera). Ecol. Entomol. 19, 168–176.

Novotný, V., 1994b. Association of polyphagy in leafhoppers (Auchenorrhyncha, Hemiptera) with unpredictable environments. Oikos 70, 223–232.

Novotný, V., 1995. Relationships between life histories of leafhoppers (Auchenorrhyncha—Hemiptera) and their host plants (Juncaceae, Cyperaceae, Poaceae). Oikos 73, 33–42.

Prestidge, R.A., 1982. The influence of nitrogenous fertilizer on the grassland Auchenorrhyncha. J. Appl. Ecol. 19, 735–749.

Prestidge, R.A., McNeill, S., 1983. Auchenorrhyncha—host plant interactions: leafhoppers and grasses. Ecol. Entomol. 8, 331–339.

Remane, R., 1958. Die Besiedlung von Grünlandflächen verschiedener Herkunft durch Wanzen und Zikaden im Weser-Ems-Gebiet. Z. Angew. Entomol. 42, 353–400.

Remane, R., Achtziger, R., Fröhlich, W., Nickel, H., Witsack, W., 1998. Rote Liste der Zikaden (Homoptera, Auchenorrhyncha). In: Binot, M., Bless, R., Boye, P., Gruttke, H., Pretscher, P. (Eds.), Rote Liste gefährdeter Tiere Deutschlands. Schr.-R. Landschaftspfl. Natursch. 55, pp. 243–245.

Rosenzweig, M.L., 1995. Species Diversity in Space and Time. Cambridge University Press, Cambridge.

Schiemenz, H., 1987. Beiträge zur Insektenfauna der DDR: Homoptera—Auchenorrhyncha (Cicadina, Insecta). Teil I: Allgemeines, Artenliste; Überfamilie Fulgoroidea. Faun. Abh. (Dres.) 15, 41–108.

Schiemenz, H., 1988. Beiträge zur Insektenfauna der DDR: Homoptera—Auchenorrhyncha (Cicadina, Insecta). Teil II: Überfamilie Cicadoidea excl. Typhlocybinae et Deltocephalinae. Faun. Abh. (Dres.) 16, 37–93.

Schiemenz, H., 1990. Beiträge zur Insektenfauna der DDR: Homoptera—Auchenorrhyncha (Cicadina Insecta). Teil III: Unterfamilie Typhlocybinae. Faun. Abh. (Dres.) 17, 141–188.

Schiemenz, H., Emmrich, R., Witsack, W., 1996. Beiträge zur Insektenfauna Ostdeutschlands: Homoptera—Auchenorrhyncha (Cicadina, Insecta). Teil IV: Unterfamilie Deltocephalinae. Faun. Abh. (Dres.) 20, 153–258.

Sedlacek, J.D., Barrett, G.W., Shaw, D.R., 1988. Effects of nutrient enrichment on the Auchenorrhyncha (Homoptera) in contrasting grassland communities. J. Appl. Ecol. 25, 537–550.

Southwood, T.R.E., van Emden, H.F., 1967. A comparison of the fauna of cut and uncut grasslands. Z. Angew. Entomol. 60, 188–198.

Tscharntke, T., Greiler, H.-J., 1995. Insect communities, grasses, and grasslands. Annu. Rev. Entomol. 40, 535–558.

Wagner, W., 1935. Die Zikaden der Nordmark und Nordwest-Deutschlands. Verh. Naturwiss. Ver. Hambg. 24, 1–44.

Waloff, N., 1973. Dispersal by flight of leafhoppers. J. Appl. Ecol. 10, 705–730.

Waloff, N., 1980. Studies on grassland leafhoppers (Auchenorrhyncha, Homoptera) and their natural enemies. Adv. Ecol. Res. 11, 81–215.

Waloff, N., Jervis, M.A., 1987. Communities of parasitoids associated with leafhoppers and planthoppers in Europe. Adv. Ecol. Res. 17, 281–402.

Wilkinson, D.M., 1999. The disturbing history of intermediate disturbance. Oikos 84, 145–147.

ELSEVIER

Available online at www.sciencedirect.com

Agriculture, Ecosystems and Environment 98 (2003) 201–211

Agriculture Ecosystems & Environment

www.elsevier.com/locate/agee

Suitability of arable weeds as indicator organisms to evaluate species conservation effects of management in agricultural ecosystems

Harald Albrecht*

Vegetation Ecology, Department of Ecology, TU Muenchen-Weihenstephan, 85350 Freising, Germany

Abstract

The overall objective of this study is to examine the application of arable weeds as indicator organisms of biodiversity in agro-ecosystems to evaluate species conservation effects of management practices.

Both investigations of interactions between weeds with heterotrophic consumers and strong overall correlations between the number of weed species and the total species diversity indicate that arable weeds are "key species", the loss of which leads to serious changes in the remaining biocoenosis via habitat and food chain relations.

The assessment of the value of management measures for species conservation presupposes a strong relation of target organisms to land use practice. In arable fields, the high percentage of dormant seeds reduces the relevance of single cultivation measures against weed populations and emphasizes the significance of long lasting management including cultivation systems, crop rotations, or field edge effects.

A comparison of the number of weed species found at different times and frequencies of sampling indicates—especially in herbicide treated fields—the importance of two recording times—one before the application and one before harvest. This ensures a better estimation of the total species spectrum.

To evaluate plant species diversity in fields the number of characteristic arable weeds is proposed. In contrast to the total number of species, the typical arable weeds do not include species frequently occurring outside fields. Thus, several highly noxious species like *Cirsium arvense*, *Elymus repens* and *Galium aparine* are not positively valued. Differences in rarity and usefulness could lead to a more sophisticated evaluation of single weed species and the species spectrum. The use of different number of species as threshold values for different soil types cannot be recommended, however.

Programs which fund the results of management necessitate control measures. In Germany, which has an arable area of 11.8 million ha and an estimated average field size of 4 ha, 300 000 sites must be controlled per year when 10% of the farmers were to participate in a corresponding program. Calculating costs of € 50 per site, which includes two vegetation relevés, € 15 million have to be spent each year.

Another possibility to increase agro-biodiversity are programs which pay for the application of specific management practices (reduced fertilization, tillage, and weed control, measures of crop selection and rotation) or management systems like organic farming. As the corresponding control is less time consuming such programs are less expansive. Their positive effects on biodiversity are less specific and less reliable, however.

Keywords: Weeds; Indicator; Species conservation; Biodiversity

* Tel.: +49-8161-713717; fax: +49-8161-714143.
E-mail address: albrecht@wzw.tum.de (H. Albrecht).

0167-8809/$ – see front matter © 2003 Elsevier Science B.V. All rights reserved.
doi:10.1016/S0167-8809(03)00081-1

1. Introduction

More than 38% of the total area of Germany are arable fields and about 16% is grassland (Statistisches Bundesamt, 1998). The majority of the remainder is used for commercial production forest, settlement, and traffic. Thus, agriculture is the most important type of land use in Germany and the importance of agricultural land for recreation, wellbeing, and species diversity is evident. In the course of intensive farming over the last decades this diversity has suffered a severe decline (Meisel, 1984; Blab et al., 1989; Plachter, 1991; Albrecht, 1995). To counteract this development, most of the German federal states took measures to conserve the organism diversity in agro-ecosystems. Well known examples for these activities were the field margin strip program (Schumacher, 1980) and the program for birds breeding in meadows (Hutterer et al., 1993). On a global scale, governments as well as non-governmental organizations began to develop concepts to conserve and increase biodiversity in agricultural ecosystems, following the environmental summit in Rio de Janeiro in 1992 (Nellinger, 2000). Corresponding indicators proposed by OECD (1999) are the diversity of "domesticated" plants and livestock, as well as "wildlife" biodiversity. For particular agro-ecosystems, however, it was considered that further research is needed to find out representative "key indicator" wildlife species.

The aim of the present study is to examine the use of arable weeds as indicator organisms of biodiversity in agro-ecosystems to evaluate species conservation effects of management practices. The ecosystematic relevance, the relation to the management, the influence of temporal and spatial variation on the measurability and the time needed for inventory are examination criteria. In addition, the selection of appropriate methods to evaluate species diversity and socioeconomic effects of different conservation strategies are discussed.

2. Ecosystematic relevance

"Key species" are defined as those species, the loss of which leads to serious changes in the remaining biocoenosis (Calow, 1998). Arable weeds belong to the carbon autotrophic producers which provide food for the consumers in agro-ecosystems. The importance of this relation was impressively demonstrated by Heydemann (1983) who found 1200 phytophageous animal species feeding on the 100 most frequent arable weeds.

In search of suitable indicator groups for biodiversity in cultivated ecosystems Duelli and Obrist (1998) proposed to calculate the correlation coefficients between the number of species in every taxonomic group and the total number of species. In an exemplary investigation the number of flowering plant species showed—in contrast to most of the other organism groups tested—a highly significant correlation with the total number of arthropod taxa. The investigators concluded that plant species belong to the best indicators for biodiversity evaluation. In contrast, the correlations among the number of individuals of collembola, carabid beetles, skylarks, and weeds recorded in arable fields by Albrecht et al. (2001) were low. These authors contributed their results to evident differences in habitat and food requirements among the investigated groups. The only highly significant correlation was observed between the weeds and the carabid beetles. This correlation was caused by high number of phyto- and polyphageous beetles predominantly sampled in dense weed stands.

Generally, these results show that not all organism groups occurring in arable fields are closely related to each other. Nevertheless, they suggest that weeds belong to the group of species with a high ecosystematic relevance.

3. Relation to management practice

As a high density of arable weeds can cause severe problems by reducing the yield and the quality of cultivated crops, weed control is an essential element of arable farming. This obvious impact on farm management may rise the question if such species should be considered at all in nature conservation activities. One reason for such efforts is that only a small percentage of the arable weed species cause remarkable infestation damage. Of 306 plant species listed by Hofmeister and Garve (1998) which occur regularly in arable fields, only 26 are defined problematic. The remaining majority scarcely cause losses in yield and significantly contribute to the species diversity of arable landscapes. Consequently, it appears justified to consider the conservation of weeds in future strategies for a sustainable land use.

Operations like ploughing or herbicide spraying kill most of the weed plants growing on the soil surface thus suggesting that single management measures can cause a severe decline in population density. That these effects are less critical than it appears is demonstrated by a comparison of the aboveground flora to the weed seed bank in soil. Roberts and Ricketts (1979), Barralis and Chadoeuf (1980), and Albrecht and Pilgram (1997) observed that the number of individual plants on the soil surface represented only 1–10% of the total weed numbers including the seeds in soil. Thus, arable weed populations seem to have adapted to such management induced collapses through the development of persistent seed banks. They are characterized by a high percentage of seeds remaining dormant in soil while only a few germinate and establish seedlings (Thompson and Grime, 1979).

A canonical correspondence analysis (CCA; ter Braak, 1987–1992) which includes data on the actual and recent management (Fig. 1) shows the long term management practice to be an important factor determining the species composition in the weed seed bank. In ordination plots, the distance of two points to each other indicates the similarity of the sites in terms of species composition. It is evident that even

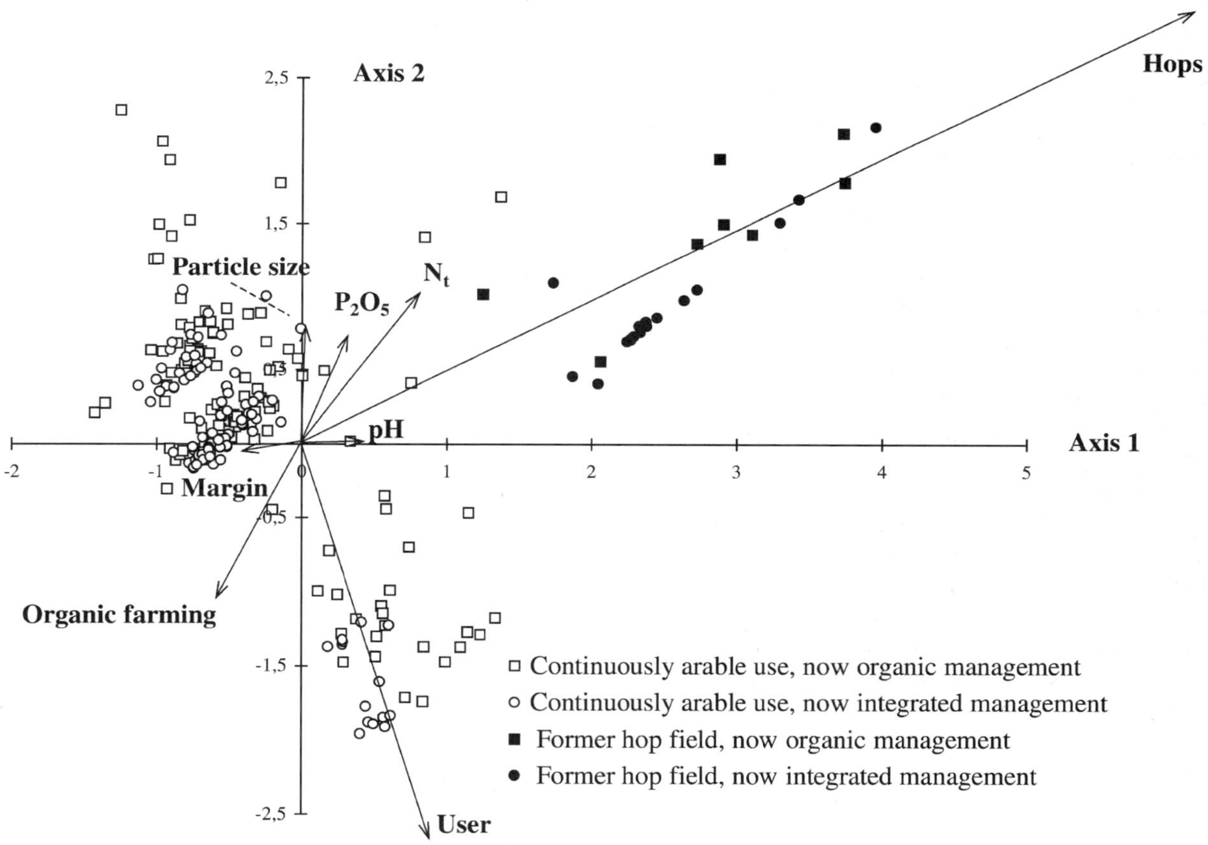

Fig. 1. CCA ordination of the weed seed bank composition at 256 sites on the FAM research station in Scheyern (southern Bavaria) with environmental variables. Particle size = median soil particle size, pH = pH level ($CaCl_2$), P_2O_5 = P_2O_5 concentration in soil (CAL), N_t = N_t content in soil, hop = previous cultivation of hops (cleared 5 years before sampling) or annual arable crops, user = former land use by different farmers (private farmers/Scheyern Abbey), margin = effect of field margins, organic farming = organic or integrated farming for 2 years. Sampling date: February 1995.

Table 1
The number of characteristic weed species in cereal and root crops under 4–5 years conventional and organic farming in North Rhine-Westphalia; size of sampling plots 100 m^2 (from Frieben, 1998, Table 36)

	Cereals				Root crops			
Management	Organic	Organic	Conventional	Conventional	Organic	Organic	Conventional	Conventional
Position in field	Margin	Center	Margin	Center	Margin	Center	Margin	Center
Median number of characteristic weed species	13	11	5	4	14	11	5	0

4 years after cleaning hop plants the weed seed bank composition of former hop fields still differs clearly from the one found in normal arable fields. Furthermore, the length and the proximity of the vector to the first axis indicates that the former cultivation of hops is still the most important variable to differentiate the weed species composition. Another factor obviously affecting the seed bank composition is the former land use of previous farmers. All sites were brought together under one management regime 4 years before the investigation began. In contrast, the change from conventional to organic or integrated farming which took place 2 years before sampling as well as soil factors seem to be of minor importance.

That the management practice affects not only the composition but also the number of species can be seen from results of Frieben (1998) in Table 1. Comparing the number of characteristic weed species in North Rhine-Westphalia fields values from plots which were under organic farming for 4–5 years ranged clearly above the corresponding numbers recorded in fields. In general terms, these results indicate that the arable weeds are an organism group which are significantly affected by the type and intensity of management. A high percentage of dormant seeds reduces the importance of single cultivation measures and emphasizes the significance of long term influences like cultivation systems, crop rotations, or field edges.

That chemical weed control impacts not only on plants but also on invertebrates and birds by modifying weed abundance and species assemblages is summarized by Boatman (2002) and Brown (2002).

4. Criteria to evaluate weed species diversity

The total number of species includes species characteristic for the living conditions in a certain habitat and ubiquitous generalists. As the characteristic species are greatly affected by changes in land use and intensification than the generalists, they should be considered with more importance in nature conservation issues. Another argument to select only the characteristic species for the evaluation of plant species diversity in arable fields is that highly noxious perennials like *Cirsium arvense*, *Elymus repens* and *Equisetum arvense* which frequently occur outside fields are not positively valued. A list of the “characteristic weeds” of German arable fields is given in Table 2.

Vegetation relevés from 130 arable fields in seven different landscapes in Bavaria with a standard sampling area of 100 m^2 give an impression of the variation in the number of such characteristic species in winter cereals. There, the median number per site was 11 and the 25 and 75% quartiles were 8 and 14, respectively. In total, the values ranged between a minimum of 2 and a maximum of 23. These numbers were recorded in two vegetation relevés, one before weed control in spring and one before harvest in summer. Compared with “normal” arable fields, management intensity on these sites was low and the spectrum of species well developed.

Apart from counting the number of species, diversity can also be estimated by calculating the evenness or combining the number of species with their relative abundance (Magurran, 1988). The Shannon and the Simpson index are the most frequently used indices of the latter type (Usher, 1994). To evaluate species diversity for practical nature conservation issues, these indices are disadvantageous for two reasons. The first is that an accurate record of plant numbers for each species is needed which is very time consuming. The second reason is that these indices favor communities with low number of species and individuals. A corresponding example is given in Table 3 where the Shannon index is calculated for two weed communities

Table 2
List of higher plant species assigned to the plant-sociological sub-class *Violenea asrvensis* (Hüppe and Hofmeister, 1990)[a]

Adonis aestivalis L.	*Galeopsis speciosa* Mill.	*Persicaria masculosa* Gray
Adonis flammea Jacq.	*Galinsoga ciliata* (Raf.) S.F.Blake	*Ranunculus arvensis* L.
Aethusa cynapium L.	*Galinsoga parviflora* Cav.	*Raphanus raphanistrum* L.
Ajuga chamaepytis (L.) Schreb.	*Galium spurium* L.	*Scandix pecten-veneris* L.
Allium vineale L.	*Galium tricornutum* Dandy	*Scleranthus annuus* L.
Alopecurus myosuroides Huds.	*Geranium dissectum* L.	*Setaria pumila* (Poir.) Roem. and Schult.
Anagallis arvensis L.	*Geranium rotundifolium* L.	*Setaria viridis* (L.) P. Beauv.
Anagallis foemina Mill.	*Iberis amara* L.	*Sherardia arvensis* L.
Anchusa arvensis (L.) M. Bieb.	*Kickxia elatine* (L.) Dumort.	*Silene linicola* C.C.Gmel.
Androsace maxima L.	*Kickxia spuria* (L.) Dumort.	Silene noctiflora L.
Anthemis arvensis L.	*Lamium amplexicaule* L.	*Sinapis arvensis* L.
Anthoxanthum aristatum Boiss.	*Lamium hybridum* Vill.	*Sonchus asper* (L.) Hill
Apera spica-venti (L.) P.B.	*Lamium purpureum* L.	*Spergula arvensis* L.
Aphanes arvensis L.	*Lathyrus aphaca* L.	*Spergularia segetalis* (L.) G. Don.
Aphanes inexpectata W. Lippert	*Lathyrus hirsutus* L.	*Stachys annua* L.
Arnoseris minima (L.) Schweigg. And Körte	*Lathyrus nissolia* L.	*Stachys arvensis* L.
Asperula arvensis L.	*Legousia hybrida* (L.) Delarbre	*Thlaspi alliaceum* L.
Avena fatua L.	*Legousia speculum-veneris* (L.) Chaix	*Thlaspi arvense* L.
Bifora radians M. Bieb.	*Linaria arvensis* (L.) Desf.	*Thymelaea passerina* (L.) Coss. and Germ.
Bunium bulbocastanum L.	*Lithospermum arvense* L.	*Torilis arvensis* (Huds.) Link
Bupleurum rotundifolium L.	*Matricaria recutita* L.	*Tulipa sylvestris* L.
Calendula arvensis L.	*Melampyrum arvense* L.	*Turgenia latifolia* (L.) Hoffm.
Caucalis platycarpos L.	*Mercurialis annua* L.	*Vaccaria hispanica* (Mill.) Rauschert
Centaurea cyanus L.	*Misopates orontium* (L.) Raf.	*Valerianella carinata* Loisel.
Chenopodium polyspermum L.	*Muscari neglectum* Guss. Ex Ten.	*Valerianella rimosa* Bastard
Chrysanthemum segetum L.	*Myagrum perfoliatum* L.	*Veronica agrestis* L.
Conringia orientalis (L.) Dumort.	*Myosotis arvensis* (L.) Hill	*Veronica arvensis* L.
Consolida regalis Gray	*Neslia paniculata* (L.) Desv.	*Veronica hederifolia* L.
Digitaria ischaemum (Schreb.) Muhl.	*Nigella arvensis* L.	*Veronica opaca* Fr.
Echinochloa crus-galli (L.) P. Beauv.	*Nonea pulla* (L.) DC.	*Veronica persica* Poir.
Erucastrum gallicum (Willd.) O.E. Schulz	*Orobanche ramosa* L.	*Veronica polita* Fr
Euphorbia exigua L.	*Odontites vernus* (Bellardi) Dumort.	*V. triphyllos* L.
Euphorbia helioscopia L.	*Orlaya grandiflora* (L.) Hoffm.	*Vicia angustifolia* L.
Euphorbia peplus L.	*Ornithogalum nutans* L.	*Vicia hirsuta* (L.) Gray
Fallopia convolvulus (L.) A. Löve	*Ornithogalum umbellatum* L.	*Vicia lutea* L.
Filago neglecta (Soy.-Will.) DC.	*Oxalis stricta* L.	*Vicia tetrasperma* (L.) Schreb.
Fumaria officinalis L.	*Papaver argemone* L.	*Vicia villosa* Roth ssp. *Villosa*
Fumaria rostellata Knaf	*Papaver dubium* L.	*Viola arvensis* Murr.
Fumaria vaillantii Loisel.	*Papaver rhoeas* L.	
Gagea villosa (M.Bieb.) Sweet	*Polycnemum arvense* L.	

[a] This sub-class comprises 118 plant species which have their main habitat in Germany in arable fields and vineyards (=characteristic arable weeds). It does not include therophytic ruderals as well as perennial species frequently occurring outside arable fields. The original list was given by Hüppe and Hofmeister (1990), several species were added according the sociological classification by Oberdorfer (1990) and Hofmeister and Garve (1998). Names of species accord with Wisskirchen and Haeupler (1998).

with different number of individual plants and species. Although plant community A has only 10 species, its Shannon diversity is as higher than the one in community B where 13 species were found. Obviously, the more even distribution of species abundance compensates the lack in the number of species. That number of individuals are more evenly distributed among species when the application of herbicides is intensively confirmed by the seed bank investigations of Squire et al. (2000). They observed that the number of species slightly more than doubled as herbicide applications were reduced while the total number of seeds increased by two orders of magnitude. Consequently, indices integrating the evenness of plant species cannot

Table 3
Shannon index of two differently structured arable weed communities

Species	Plant community	
	A (plants m^{-2})	B (plants m^{-2})
Capsella bursa-pastoris	1	2
Chenopodium album	1	1
Tripleurospermum perforatum	1	1
Taraxacum officinale	1	
Myosotis arvensis	1	1
Polygonum aviculare	1	2
Galinsoga ciliata	1	1
Stellaria media	2	2
Poa annua	1	2
Viola arvensis	1	14
Apera spica-venti		1
Matricaria recutita		17
Centaurea cyanus		1
Anthemis arvensis		1
Number of species	10	13
Number of individual plants	11	46
Shannon index	2.27	1.86
Evenness	0.99	

be recommended as a criterion to evaluate biodiversity in this ecosystem type.

Another useful criterion for a more sophisticated evaluation of single weed species is their degree of endangerment. As 42 of the 118 species from Table 2 are listed in the red data book of Germany and another 41 species are listed in the corresponding books of single federal states (Korneck et al., 1996), arable weeds belong to the vegetation types with the highest percentage of endangered plant species in Germany.

A second criterion for a differentiated evaluation on the species level may be the number of plant species which favor useful insects. Frieben (1998) prepared a list of such species named in literature and found that their numbers varied between 0 and 10 on an area of 100 m^2, depending on the management system.

Instead of using "static" criteria like the number of species occurring on a certain area, Pfadenhauer et al. (1997) recommended "dynamic" indicators to evaluate the efficacy of management measures for nature conservation issues. As lots of weeds possess no efficient strategies for an active dispersal over long distances and passive transport by management was greatly reduced by land use changes during the last decades, the dispersal rate could be such an indicator. Thus, farmers would not be paid for high number of species on their fields but for applying management practices which intensify the spread of species. Unfortunately, knowledge on the efficacy and workability of such measures is poor up to now.

5. Temporal variation

Great temporal variation in the apparent part of populations is a factor which makes it generally difficult to record organisms occurring in agricultural ecosystems.

One important reason for the differences between the number of weed species found in different arable crops under central European climate conditions is the seasonal variation of temperature. This is caused by cold spells in autumn and early spring favoring weeds with a strong adaptation to germination at low temperatures (Otte, 1996). As these species are unable to germinate in summer, they predominantly occur in winter annual stands. Thus, winter annual crops potentially inhabited by a higher number of species are recommended for sampling. When looking for well developed weed stands within the winter annual crops, autumn sown stands, where weed control is carried out in spring, are favorable as herbicide spraying at sowing time prevents an early development of weeds. In later stages the weeds are suppressed by light and nutrient competition of the crops. This especially applies to oil seed rape where weed sampling is significantly affected by the early herbicide application in autumn.

To get representative information on the composition of plant communities, repeated sampling is frequently needed. Fig. 2 illustrates the relation between the frequency of sampling and the number of weed taxa observed in the 130 arable fields in Bavaria, the same which were used for the analysis in Section 4. It can be seen that adding one relevé carried out in early spring to the usual sample made before harvest in summer increased the number of species by almost 30%.

Both natural and anthropogenic influences may be the reason behind this increase. Weeds may have faded in the course of weed control after the investigation in spring; thus a human influence. However, natural influences can cause withering of plants long before harvest time in summer. A corresponding

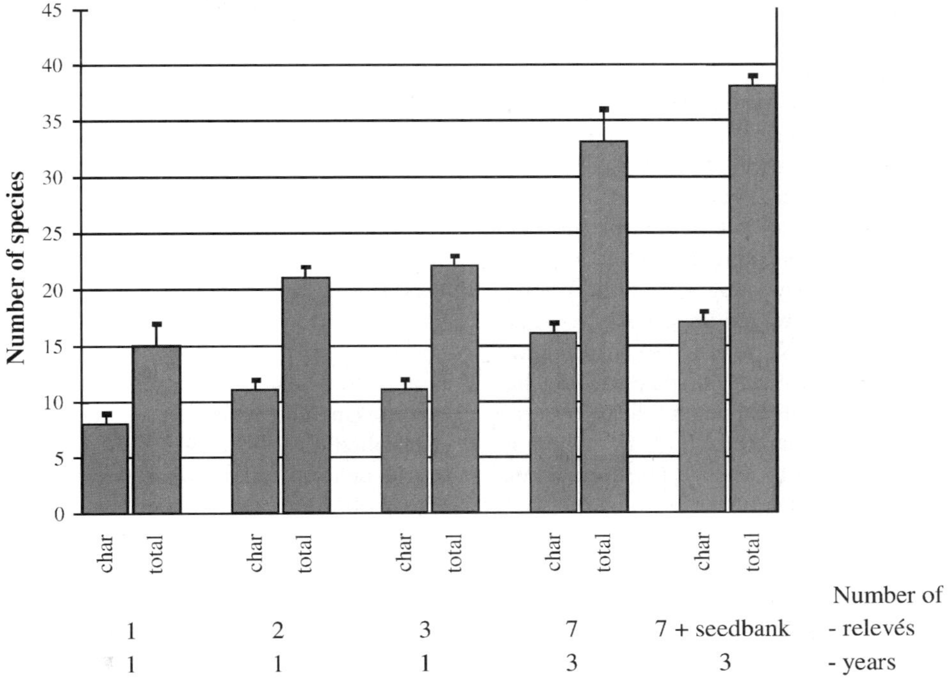

Fig. 2. Median numbers (with 95% intervals of confidence) of species recorded in 1, 2, 3 and 7 series of vegetation relevés in 130 winter cereal fields in seven different landscapes in Bavaria (see Fig. 3). The first series was carried out before harvest in summer, the second and third additionally comprise species observed before herbicide application in spring and during the flowering period in early summer, respectively. The right columns record the number of species found in seven relevés in different crops over 3 continuous years with or without an additional seed bank analysis. The "char" indicates characteristic species according to Table 2, while "total" includes all plant species found at the site.

example is *Veronica triphyllos* which occurs in German arable fields predominantly on coarse grained soils. This species escapes the summer drought on such substrates by early germination, flowering and fructification (Albrecht et al., 1999). These results demonstrate especially in herbicide treated fields that at least two records—one before the application and one before harvest—are necessary to give a general image of the total species spectrum.

As there were no additional species counted in a third series of relevés carried out during the flowering period of cereals suggests that this effort was not worthwhile. In contrast, another significant increase in the number of species was observed when the relevés were repeated seven times over 3 years in different crops and when additional seed bank analyses were made (columns 7–10 in Fig. 2). Thus, sampling in different years and crops in combination with investigations in the soil seed bank provide very detailed information on the weed flora. Unfortunately, economic circumstances limit such precision in large scale survey programs.

6. Spatial variation

Spatial variation is another factor which may aggravate an accurate evaluation of the species diversity in arable weed communities. At a field scale, this variation can be caused by differences in natural site conditions and former management. To overcome this problem using a sophisticated sampling method, Frieben (1998) recommended to record the species observed when crossing the field until no more new species can be found. Doubtlessly, this method is precise but also very time intensive.

To reduce this effort, most samples for the floristic description of the weed vegetation in arable fields in central Europe were made according to the method of the Zurich-Montpellier school (Braun-Blanquet, 1964). This method is based on the concept of minimal area and species–area curves. It involves doubling the size of the sampling area in a homogeneous stand until no new species are recorded. In practice, this instruction led to remarkable differences in the area researchers used for their vegetation relevés. Sampling areas reported in the corresponding literature vary from a minimum of $2\,m^2$ (Kulp, 1993) to a maximum of $1500\,m^2$ (Plakholm, 1989). As a consequence, the results cannot be compared with each other. To overcome this problem, a standardization of the sampling area is needed. According to Frieben (1998) most of the species occurring under homogeneous site and management conditions in arable fields can be found on an area of $100\,m^2$. Thus, this sampling area should be used to unify the method for vegetation records.

Conditions particularly favorable for high number of species and population densities can be found near field margins. Investigations of Marshall (1989), van Elsen (1994), and Wilson and Aebischer (1995) have shown that the majority of species are most abundant in the outermost meter and thin out sharply within the first 4 m from the field edge. A better supply of light (van Elsen, 1994, p. 157), lower crop yields and management intensity (Boatman and Sotherton, 1988), and the possibility to survive an to re-invade from neighboring habitats (Wilson and Aebischer, 1995) favor a high species diversity in the margin area. Using this effect for nature conservation issues, programs have been developed in which farmers were paid for limiting herbicide and fertilizer use in field margin strips (Schumacher, 1980). In Germany, these programs were introduced in most of the federal states and covered a total area of approximately 5000 ha in the beginning of the 1990s (Wicke, 1998). This area totaled to 0.04% of the arable land of the FRG. Since the sites included in these programs were continuously controlled and selected for their floristic inventory, many fields with well developed weed communities and with rare species listed in red data books could be included (e.g. Mattheis and Otte, 1994; Frieben, 1995). For farmers, however, managing the field margin and the field center differently is time consuming and can cause crop rotation and weed infestation problems (Boatman and Sotherton, 1988). Thus, the programs were only accepted in conjunction with a suitable compensation payment. As the expenses for the field margin strip programs of the German federal states were reduced since 1992, a severe set back in the area included was observed (Wicke, 1998). To

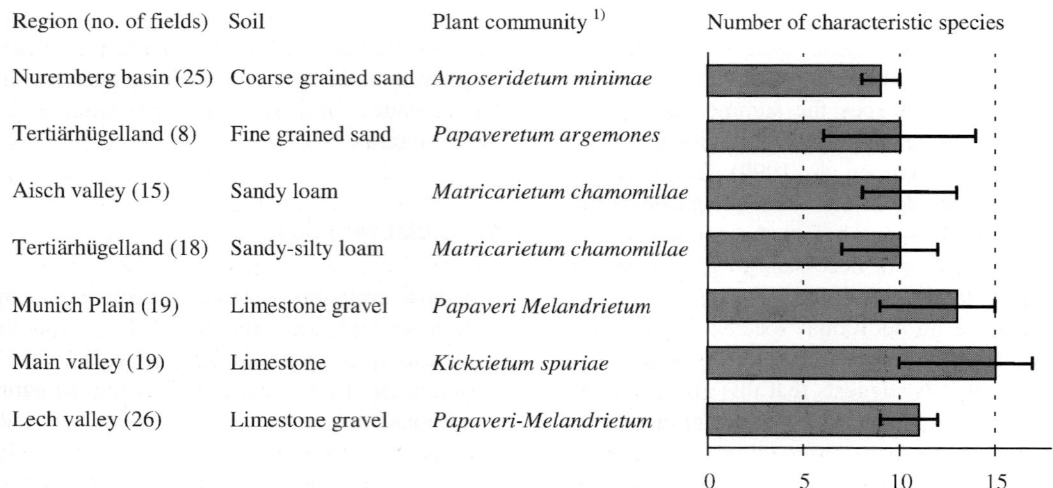

Fig. 3. Median numbers (with 95% confidence intervals) of characteristic arable weed species in seven different arable regions in Bavaria. Records were made in winter cereal fields with two relevés per site apart from the field margins. (1) Plant sociological communities as described by Hofmeister and Garve (1998).

increase the integration into the production process and make it more acceptable for the farmers, an extension of the protected area from the field margin strips to the whole fields could be advantageous.

At a large landscape scale arable weed communities are shaped by site conditions and—closely related to the site conditions—by the type and intensity of management. Fig. 3 shows that the median number of characteristic species observed in winter cereal crops in different regions of Bavaria varied between 9 and 15. That sandy and loamy soils did not significantly differ in their number of typical weeds (sandy sites in the Nuremberg basin and in the Tertiärhügelland were compared to the loamy soil type in the Aisch valley and in the Tertiärhügelland using the *U*-test by Mann–Whitney) is confirmed by the results of Frieben (1998), who investigated corresponding sites in northwestern Germany. The median value for the arable fields on limestone substrates (Lech valley, Main valley and Munich plain), however, significantly exceeded the number of species found on the other soil types. As this observation goes along with unpublished results cited by Frieben (1998), there could be a higher threshold recommended for arable fields on limestone substrates.

7. Economic implications

For the realization of nature conservation issues th costs and time expense for corresponding measures are of importance.

Programs which pay for the results of management like a high number of species or the presence of rare weeds necessitate control measures. To estimate the costs of such a control, an approximate calculation was drawn up for the Federal Republic of Germany. This country has an arable area of 11.8 million ha and an estimated average field size of 4 ha. Given a participation of 10% in this program, 300 000 arable fields must be controlled. Vegetation ecologists report that one vegetation relevé in a normal arable field takes about 30 min. Thus, the time needed for the two relevés proposed is 60 min. Further reduction of time is possible by leaving out the estimation of cover abundance, using pre-printed survey forms, and finishing the investigation when a certain number of plants is found. Thus, time to reach the sites is often greater than the time needed to record information. Calculating the price of two records per site (=field) at € 50, € 15 million have to be spent each year for the program.

Another possibility to increase agro-biodiversity not discussed in detail are programs which pay for the application of specific management practices (reduced fertilization, tillage, and weed control, measures of crop selection and rotation) or management systems like organic farming. As corresponding control is less time consuming such programs cause lower costs. Their positive effects on biodiversity are less specific and less reliable, however (e.g. Albrecht and Mattheis, 1998).

8. Conclusions

Both a high sensitivity to cultivation measures and a strong relation to other organism groups make weeds suitable indicators to evaluate management effec[illegible] wildlife diversity in arable fields. Consid-e[illegible] presented in Chapter 8 show that recording weed assemblages could feasibly be carried out even in a high number of fields. It remains however un[illegible]ain, whether nature conservation authorities are [illegible]ng to pay several million Euros each year for a system to control the maintenance of biodiversity at a certain level. Thus, a regular weed inventory may remain limited to areas with a high species conservation value (high number of species, occurrence of rare and endangered species). In fields where such a regular control is not possible, specific cultivation measures or management systems would have to take on an indicator function of wildlife biodiversity. This means that applying management systems like organic farming or cultivation patterns like minimum tillage should guarantee a certain level of species diversity. As management systems like organic farming comprise a broad spectrum of diverging cultivation measures, fields of different farms can show a great variation in species diversity (Becker and Hurle, 1998). Thus, single measures like the application of fertilizers or certain herbicides may show a stronger correlation to species diversity. Unfortunately, little is known about these effects on weed and wildlife diversity in arable fields at present Marshall et al. (2002) and should consequently be the focus of future arable ecosystem research. Proposals for evaluating the corresponding results were given in the present paper.

References

Albrecht, H., 1995. Changes in the arable weed flora of Germany during the last five decades. In: Proceedings of the Ninth European Weed Research Society Symposium on Challenges for Weed Science in a Changing Europe, Budapest, pp. 41–48.

Albrecht, H., Mattheis, A., 1998. The effect of organic and integrated farming on rare arable weeds on the Forschungsverbund Agrarökosysteme München (FAM) research station in southern Bavaria. Biol. Conserv. 86, 347–356.

Albrecht, H., Pilgram, M., 1997. The weed seed bank in a landscape segment in southern Bavaria. II. Relation to environmental factors and to the soil surface vegetation. Plant Ecol. 131, 31–43.

Albrecht, H., Mayer, F., Mattheis, A., 1999. *Veronica triphyllos* L. in the Tertiärhügelland landscape in southern Bavaria—an example for habitat isolation of a stenoeceous plant species in agroecosystems. Zeitschrift für Ökologie und Naturschutz 8, 219–226.

Albrecht, H., Kühn, N., Filser, J., 2001. Site effects on plant and animal distribution at the Scheyern experimental farm. Ecol. Stud. 147, 209–227.

Barralis, G., Chadoeuf, R., 1980. Etude de la dynamique d'une communauté adventice. I. Evolution de la flore adventice au cours du cycle végétatif d'une culture. Weed Res. 20, 231–237.

Becker, B., Hurle, K., 1998. Unkrautflora auf Feldern mit unterschiedlich langer ökologischer Bewirtschaftung. Zeitschrift für Pflanzenkrankheiten und Pflanzenschutz, Sonderheft XVI, 155–161.

Blab, J., Bless, R., Nowak, E., Rheinwald, G., 1989. Veränderungen und neue Entwicklungen im Gefährdungs- und Schutzstatus der Wirbeltiere in der Bundesrepublik Deutschland. Schriftenreihe für Landschaftspflege und Naturschutz 29, 9–37.

Boatman, N.D., 2002. Relationship between weeds, herbicides and birds. In: Marshall, E.J.P., Brown, V.K., Boatman, N.D., Lutman, P.J.W., Squire, G.R. (Eds.), The Impact of Herbicides on Weed Abundance and Biodiversity. DEFRA Desk Study PN0940, pp. 85–102.

Boatman, N.D., Sotherton, N.W., 1988. The agronomic consequences and costs of managing field margins for game and wildlife conservation. Aspects Appl. Biol. 17, 47–56.

Braun-Blanquet, J., 1964. Pflanzensoziologie, 3rd ed. Springer, Wien.

Brown, V.K., 2002. The impact of herbicides on invertebrates. In: Marshall, E.J.P., Brown, V.K., Boatman, N.D., Lutman, P.J.W., Squire, G.R. (Eds.), The Impact of Herbicides on Weed Abundance and Biodiversity. DEFRA Desk Study PN0940, pp. 59–84.

Calow, P. (Ed.), 1998. The Encyclopedia of Ecology and Environmental Management. Blackwell, Oxford.

Duelli, P., Obrist, M.K., 1998. In search of the best correlates for local organismal biodiversity in cultivated areas. Biodivers. Conserv. 7, 297–309.

Frieben, B., 1995. Effizienz des Schutzprogrammes für Ackerwildkräuter. Mitteilungen, Schriftenreihe der Landesanstalt für Ökologie, Landschaftsentwicklung und Forstplanung NRW 4/95, pp. 14–19.

Frieben, B., 1998. Verfahren zur Bestandsaufnahme und Bewertung von Betrieben des organischen Landbaues im Hinblick auf Biotop- und Artenschutz und die Stabilisierung des Agrarökosystems. D. Köster, Berlin.

Heydemann, B., 1983. Aufbau von Ökosystemen im Agrarbereich und ihre langfristigen Veränderungen. Daten und Dokumente zum Umweltschutz 35, 53–83.

Hofmeister, H., Garve, E., 1998. Lebensraum Acker, 2nd ed. Parey, Hamburg.

Hüppe, J., Hofmeister, H., 1990. Syntaxonomische Fassung und Übersicht über die Ackerunkrautgesellschaften der Bundesrepublik Deutschland. Berichte der Reinhold Tüxen-Gesellschaft 2, 61–61.

Hutterer, C.-P., Briemle, G., Fink, C. (Eds.), 1993. Wiesen, Weiden und anderes Grünland. Weitbrecht, Stuttgart.

Korneck, D., Schnittler, D.M., Vollmer, I., 1996. Rote Liste der Farn- und Blütenpflanzen (Pteridophyta et Spermatophyta) Deutschlands. Schriftenreihe für Vegetationskunde 28, 21–187.

Kulp, K.-G., 1993. Vegetationskundliche und experimentell-ökologische Untersuchungen der Lammkraut-Gesellschaft in Nordwestdeutschland. Dissertationes Botanicae 198. Cramer/Borntraeger, Stuttgart.

Magurran, A.E., 1988. Ecological Diversity and Its Measurement. Chapman & Hall, London.

Marshall, E.J.P., 1989. Distribution patterns of plants associated with arable field edges. J. Appl. Ecol. 26, 247–258.

Marshall, E.J.P., Brown, V.K., Boatman, N.D., Lutman, P.J.W., Squire, G.R., 2002. The Impact of Herbicides on Weed Abundance and Biodiversity. DEFRA Desk Study PN0940.

Mattheis, A., Otte, A., 1994. Ergebnisse der Erfolgskontrollen zum Ackerrandstreifenprogramm im Regierungsbezirk Oberbayern 1995–1991. Schriftenreihe der Stiftung zum Schutz gefährdeter Pflanzen 5, 56–71.

Meisel, K., 1984. Landwirtschaft und "Rote Liste"-Pflanzenarten. Natur und Landschaft 59, 301–307.

Nellinger, L., 2000. Politische Anforderungen und Vorgaben für Indikatoren im Bereich Landwirtschaft und Biodiversität. Symposium Indikatorenfindung für eine nachhaltige Landwirtschaft in den Bereichen Landschaft und Biodiversität, Freising, 1.

Oberdorfer, E., 1990. Pflanzensoziologische Exkursionsflora, 6th ed. Ulmer, Stuttgart.

OECD, 1999. Environmental Indicators for Agriculture, vol. 2, Issues and Design. In: Proceedings of the York Workshop. OECD Publications, Paris.

Otte, A., 1996. Populationsbiologischen Parameter zur Kennzeichnung von Ackerwildkräutern. Zeitschrift für Pflanzenkrankheiten und Pflanzenschutz, Sonderheft XV, 45–60.

Pfadenhauer, J., Albrecht, H., Auerswald, K., 1997. Naturschutz in der Agrarlandschaft - Perspektiven aus dem Forschungsverbund Agrarökosysteme München (FAM). Berichte Reinhold-Tüxen-Gesellschaft 9, 49–59.

Plachter, H., 1991. Naturschutz. Gustav Fischer UTB, Stuttgart.

Plakholm, G., 1989. Unkrauterhebungen in biologisch und konventionell bewirtschafteten Getreideäckern Oberösterreichs. Thesis Univ. Agricultural Sciences, Vienna.

Roberts, H.A., Ricketts, M.E., 1979. Quantitative relationships between the weed flora after cultivation and the seed population in the soil. Weed Res. 19, 269–275.

Schumacher, W., 1980. Schutz und Erhaltung gefährdeter Ackerwildkräuter durch Integration von landwirtschaftlicher Nutzung und Naturschutz. Natur und Landschaft 55, 447–453.

Squire, G.R., Rodger, S., Wright, G., 2000. Community-scale seedbank response to less intense rotation and reduced herbicide input at three sites. Ann. Appl. Biol. 136, 47–57.

Statistisches Bundesamt, 1998. Land- und Forstwirtschaft, Fischerei. Fachserie 3, Reihe 5.1 Bodenfläche nach Art der tatsächlichen Nutzung 1997. Metzler-Poeschel, Stuttgart.

ter Braak, C.J.F., 1987–1992. CANOCO—A FORTRAN Program for Canonical Community Ordination by [partial][detrended][canonical] Correspondence Analysis, Principal Components Analysis and Redundancy Analysis. Microcomputer Power, Ithaca, NY.

Thompson, K., Grime, J.P., 1979. Seasonal variation in the seed banks of herbaceous species in ten contrasting habitats. J. Ecol. 67, 893–921.

Usher, M.B., 1994. Erfassung und Bewertung von Lebnesräumen: Merkmale, Kriterien, Werte. In: Usher, M.B., Erz, W. (Eds.), Erfassen und Bewerten im Naturschutz. Quelle & Meyer/UTB, Heidelberg.

van Elsen, T., 1994. Die Fluktuation von Ackerwildkraut-Gesellschaften und ihre Beeinflussung durch Fruchtfolge und Bodenbearbeitungs-Zeitpunkt. Ökologie und Umweltsicherung 9/94.

Wicke, G., 1998. Stand der Ackerrandstreifenprogramme in Deutschland. Schriftenreihe Landesanstalt für Pflanzenbau und Pflanzenschutz, Mainz 6, 55–84.

Wilson, P.J., Aebischer, N.J., 1995. The distribution of dicotyledoneous arable weeds in relation to distance from the field edge. J. Appl. Ecol. 32, 295–310.

Wisskirchen, R., Haeupler, H., 1998. Standardliste der Farn- und Blütenpflanzen Deutschlands. Ulmer, Stuttgart.

ELSEVIER

Available online at www.sciencedirect.com

Agriculture, Ecosystems and Environment 98 (2003) 213–225

Agriculture Ecosystems & Environment

www.elsevier.com/locate/agee

Morphometric parameters: an approach for the indication of environmental conditions on calcareous grassland

Claus Mückschel[a,*], Annette Otte[b]

[a] Biometry and Population Genetics, Justus Liebig University, Heinrich-Buff-Ring 26-32 (IFZ), D-35392 Gießen, Germany
[b] Landscape Ecology and Landscape Planning, Justus Liebig University, Heinrich-Buff-Ring 26-32 (IFZ), D-35392 Gießen, Germany

Abstract

Morphological traits of plants often vary considerably in response to different environmental conditions, which may render them suitable as sensible indicators for conservation purpose.

To compare the plasticity of morphological traits of species of calcareous grassland under different environmental conditions (stands of different structure with respect to litter and vegetation height), morphometric investigations were carried out on calcareous grasslands differing in grazing regime. Populations of the calcareous grassland species *Plantago media* and *Scabiosa columbaria* subject to different grazing regimes showed considerable plasticity.

There were, e.g. relationships between shoot height, biomass, and fertility of individuals. Shoot height and the number of diaspores per inflorescence increased significantly with increasing cover of litter. Simultaneously there were considerable changes in the reproductive mode, from generative only to a combination of generative and vegetative reproduction.

The analysis of morphological plasticity showed that different environmental conditions could accurately and easily be monitored through the species-specific growth response.

Especially shoot height appears to play a central role, since the study species can vary this growth parameter to escape light deficiency of tall stands rich in litter. Due to its large variability, shoot height can be used as an indicator of growth conditions of the studied populations that are closely linked to land use and conservation management.

Keywords: Meadows; Grazing; Phenotypic plasticity

1. Introduction

In Central Europe low-productive meadows comprise diverse plant communities (Willems, 1982). Many of these have become fallow due to abandonment or too low intensity of management (e.g. Quinger et al., 1994). Biotope mosaics such as low-productive grasslands, especially Mesobromion-stands, can be regarded as among the most endangered habitats (Ellenberg, 1996; Korneck et al., 1996). Due to changes in agricultural techniques and land use changes only small remnants of the formerly widely distributed Mesobromion meadows are left in Central Europe (Willems, 1990; Ellenberg, 1996).

If the original use of these grasslands ceases, a succession series proceeds that after intermediate stages, best described as grassland–tall herb–scrub–woodland vegetation, leads to closed forests (Oberdorfer, 1993; Ellenberg, 1996). There is a rich literature concerning successional changes of vegetation structure of open calcareous grasslands (z. B. Schiefer, 1981; Dierschke, 1985; Bobbink and Willems, 1987; Wilmanns and

* Corresponding author.
E-mail address: claus.mueckschel@agrar.uni-giessen.de (C. Mückschel).

0167-8809/$ – see front matter © 2003 Elsevier Science B.V. All rights reserved.
doi:10.1016/S0167-8809(03)00082-3

Sendtko, 1995; Ellenberg, 1996; Sundermeier, 1999). These studies demonstrated changes in species abundance and composition during succession, though the rate of change in floristic composition was usually low. The extinction and new occurrence of species and corresponding changes in stand structure are often the result of long-lasting fallow processes. In most cases, changes occur slowly and are thus difficult and laborious to monitore. Morphological traits of populations and their individuals respond much faster than the whole phytocoenosis and can therefore often be observed earlier. Furthermore, quantitative measures of fitness of species can be deduced from morphometric parameters (cf. Noble and Slayter, 1980).

For example, Gluch (1980) used biomass production, shoot height and internode length of *Vaccinium myrtillus*, *Calamagrostis villosa* and *Deschampsia flexuosa* as indicators for the impact of herbicides. Individuals of *Cirsium acaule* responded to a cessation of grazing and mowing initially by increased length growth (Dierschke, 1985) and reached a greater height in taller vegetation. A similar response was also shown by *Antennaria dioica*; this species of low-productive meadows built no generative shoots under fallow conditions (Schwabe, 1990). Tamm (1972) showed that increased shading that also occurs on fallow land leads to reduced fertility and a reduction in leaf number and size in *Primula veris*, the number of plant individuals remained, however, constant for a relatively long time.

Phenotypic plasticity[1] as a response of plants to environmental change appears to be a quantitative trait that leads to a change in plant shape and morphology (e.g. Quinn, 1987; Primack and Kang, 1989).

The variability of morphological traits of a species or population may be related to the presence of different genotypes, that are adapted to specific environmental conditions. On the other hand, it may be related to the phenotypic plasticity of genotypes. The variation of a population trait in a habitat will thus depend on the combination of genetic variability and phenotypic plasticity (Bradshaw, 1965). The reaction norms of genotypes are only revealed under different environmental conditions. Different taxa but also different plant tissues will exhibit different degrees of plasticity (Primack and Kang, 1989; Briggs and Walters, 1997).

Within the biotope mosaic of a calcareous low-productive grassland we studied:

(a) how plastic are the morphologic characters of grassland species, populations and individuals in different grassland habitats within a relatively small area, and
(b) to what extend may morphometric parameters be used to indicate environmental conditions that are usually closely linked to land use regime and conservation management.

2. Study area

The study area is situated in the Western parts of the county Thuringia (Germany) within the landscape "Zechsteingürtel bei Bad Liebenstein" that lies at the south-western edge of the Thüringer Wald (Fig. 1). The study sites lie within the two adjacent nature protection areas (NSG), i.e. "Wacholderheide bei Waldfisch" (ca. 33 ha) and "Alte Warth bei Gumpelstadt" (ca. 86 ha), belonging to the country district of Wartburg. The altitudinal range is between 350 and 400 m a.s.l. The region has an annual mean temperature of 7.5 °C and a mean annual sum of precipitation of about 730 mm.

Fig. 1. Position of the study area within the landscape "Zechsteingürtel bei Bad-Liebenstein" (country district of Wartburg, Thuringia, Germany).

[1] According to Bradshaw (1965), phenotypic plasticity is defined as the ability of a genotype to adapt more than one form of morphology, physiological state or behaviour as a response to different environmental conditions.

Large mosaics of low-productive calcareous grasslands that are used for sheep grazing and—on a smaller area—cattle grazing exist here over shallow calcareous syrosem and rendzina soils. In areas grazed by sheep varieties of *Gentiano-Koelerietum* communities dominate, while transitions between drier forms of *Arrhenatheretum*-meadows and *Gentiano-Koelerietum*-communities occur on grassland grazed by cattle (cf. Mückschel, 2002).

The studied calcareous grasslands are regularly grazed since 15 years through a herd of 600 merino-sheep and 10–15 goats. Cattle grazing consists of herds of about 150 animals that are kept on paddocks of at least 10 ha. The timing of grazing is given in Table 1.

The land use variants differ mainly with respect to vegetation structure, i.e. the vertical and horizontal composition of vegetation with its living and dead components. Especially the accumulation of litter and the increasing stand height represent two important variables that may describe the state of fallow lands.

3. Material and methods

Within the two nature protection areas a total of four study plots were established (two in the NSG Wacholderheide and two in the NSG Alte Warth). These represent different types of grazing regimes in the area. Soil properties of different variants showed only minor differences, except for the proportion of skeleton (cf. Table 1). The grazing regimes comprised four types differing in the onset or timing of grazing (early, medium, late, (late-) sporadic). A summary of the characteristics of the different grazing regimes is given in Table 1. An early onset of grazing results in an increased removal of nutrients. Therefore, the study sites can be ranked according to decreasing grazing intensity (from early to (late-) sporadic) from A to D.

Table 1
Description of the study sites and land use variants[a]

	Site A	Site B	Site C	Site D
Structural site characteristics				
Number of plots	$n = 24$	$n = 24$	$n = 24$	$n = 24$
Land use/management	Early sheep grazed (from 30 April to 31 October)	Medium sheep grazed (from 15 July to 31 July)	Sporadically sheep grazed (irregularly from 1 August)	Late cattle grazed (from 1 September)
Vegetation	Grass-rich low-productive meadow	Orchid-rich low-productive meadow	Low-productive meadow with species of woodland margins and shrubs	Low-productive meadow with species of woodland margins
Exposure	SE	SW	S	W
Inclination (°)	10	15	9	22
Height of herb layer (cm)	18	20	25	35
Cover of herb layer (%)	85 (75–95)	85 (85–100)	95 (85–98)	85 (85–95)
Cover of cryptogams (%)	40	30	10	2.5
Cover of litter (%)	8	10	15	20
Number of species	56	69	66	72
Chemical and physical soil properties				
pH ($CaCl_2$)	7.0	7.1	6.8	7.0
Volume-ratio fine earth/soil skeleton	0.6	1.1	0.8	2.0
S (vol.%)	39.1	51.6	41.0	52.0
U (vol.%)	46.2	40.0	46.5	41.2
T (vol.%)	16.8	9.3	15.1	11.0

[a] For the structural site characteristics, data are median, arithmetic mean (for species number) and minimum, maximum (for cover of herb layer) of 24 subplots of 1 m^2. For chemical and physical soil properties, data are means of six replicate measurements per 24 subplots. S: sand, U: silt, T: clay.

Table 2
Frequency of *S. columbaria* and *P. media* on the studied site[a]

	Studied sites			
	A	B	C	D
S. columbaria	100^{p2-1-}	75^{r1-p4}	92^{a2-p4}	42^{r2-1-}
P. media	100^{p2-a4}	100^{p1-1+}	92^{r1-1-}	63^{r1-1+}

[a] Data are frequencies (%) in 24 subplots of $1\,m^2$, with the range of cover-abundance classes (LONDO-Scale) as superscripts.

Different grazing intensities are a result of the spatial position of the sites within the study areas. Grasslands in the edges of the nature protection areas are grazed later than the more central parts.

Permanent plots (24 subplots of $1\,m^2$ area within each site) were established in physiognomically uniform parts of sites A–D (Table 1). Morphometric measurements as well as biomass sampling were carried out on these plots.

3.1. Studied species

Two abundant and frequent species (*Plantago media*, *Scabiosa columbaria*) were selected for this study. These species reach relatively high population densities within the study area and therefore often dominate the vegetation (cf. Table 2). Important population biological traits of the studied species are given in Table 3.

3.2. Morphometric variables

To estimate the variation of mophological traits of the study species within different grazing regimes, the following measurements were carried out during the same phenological phase (peak flowering) and thus during the same phase of ontogenetic development during 1998 and 1999 (20 individuals per species per 24 subplots):

- *P. media*, shoot height, rosette diameter, number of rosettes per individual plant, number of diaspores per infructescence.
- *S. columbaria*, shoot height, number of rosettes per individual plant, number of flowerheads per plant.

It was not possible to assign individuals of the species to age classes (cf. Urbanska, 1992). The studied individuals within subplots were selected at regular distances along a tape-measure.

3.3. Biomass production

Additionally, the biomass of the two study species was estimated on all sites in 1999. The estimation is

Table 3
Population biological traits of the studied species (according to Grime et al., 1988; Ellenberg, 1992; Düll and Kutzelnigg, 1994; Oberdorfer, 2001 and own observations)

Trait	Information
P. media L. (Plantaginaceae)	
Life- and growth-form	Perennial rosette plant, hemikryptophyte, vegetative reproduction through root sprouting, height: 10–50 cm
Fruit/seed	Capsules with 1–4 seeds
Flowering period	Early summer
Fruiting period	July–August
Site conditions	Half-light plant, indifferent temperature response, on dry to mesic, mostly nitrogen-poor sites
Phytosociological behaviour	Anthropo-zoogenous heaths and grasslands
S. columbaria L. (Dipsacaceae)	
Life- and growth-form	Perennial half-rosette plant, hemikryptophyte, vegetative reproduction through root sprouting, height: 20–80 cm
Fruit/seed	Achaene
Flowering period	Mid-summer (autumn)
Fruiting period	Beginning of August/beginning of September
Site conditions	Light plant, indictor of intermediate temperatures, common on dry and nitrogen-poor sites
Phytosociological behaviour	Brometalia erecti

based on those individuals that were used for the morphometric study. Thus in 1999 morphometric variables can be directly related to biomass. After harvesting the aboveground biomass was dried to a constant weight at 105 °C, cooled in an exsiccator and weighed.

To achieve comparable values for biomass on all studied sites, small subplots were fenced to exclude grazers. Due to the presence of shallow soils rich in skeleton, roots could not be harvested quantitatively and all values for biomass relate to oven dry weight of aboveground biomass.

3.4. Statistics

Morphometric data were mostly normally distributed. In these cases, parametric tests were applied. Data were tested for homogeneity of variances using the Sen and Puri test. To achieve normal distribution and homogeneity of variances in all variables, some data had to be ln-transformed. The effect of site was tested statistically using analysis of variance (ANOVA) and significant differences between means were analysed using an a posteriori Tukey test (HSD). Differences were declared statistically significant at a level of α of 5%. Data on individual biomass showed a large variation and were only rarely normally distributed. Therefore, these data were analysed using non-parametric statistical methods.

Besides descriptive statistics, Spearman's rank correlation was applied to analyse the relationships between biomass and morphometric variables and traits. Statistical analyses were carried out using the programme Statistica 5.0 for Windows.

4. Results

4.1. Grazing regimes

The studied sites differed primarily in structural characteristics (cf. Table 1). Site A, grazed early in the year by sheep, showed a relatively low accumulation of litter (8%), a low stand height of 18 cm and a relatively large cover of cryptogams which may further indicate the open structure of this site. With only 56 species this plot showed the lowest species diversity. Species of woodland margins and woody species were infrequent here.

The highest species diversity with 72 species was recorded on site D grazed by cattle. Litter covered about 20% of the surface and stand height reached 35 cm. This site is characterised by a late onset of grazing which leads to an increased abundance of species of woodland margins and juvenile specimen of woody species. This may, in turn, explain the high species richness of this site. Plot B and the sporadically sheep grazed plot C take an intermediate position with respect to the accumulation of litter and stand height. With respect to the studied soil parameters sites differed primarily in the proportion of skeleton; site A showed the highest proportion of skeleton and site D the highest proportion of fine earth.

4.2. Morphometric variables

A comparison of morphometric data from 1999 and 1998 revealed that ratios relating the measured variables from both species to each other were rather constant, despite large variation in absolute values between years. Fig. 2a illustrates this for shoot height of *P. media*. A majority of variables, as also exemplified by shoot height, showed an increase within the study period over all sites. This may be related to weather variation. However, differences between years were rarely significant and only on site B differed shoot height significantly between years (Fig. 2a).

4.2.1. Shoot height

With a mean of 50 cm *P. media* showed significantly larger shoot heights on the cattle grazed grassland site D (1999), than in the other sites. The lowest values (20 cm) were recorded 1999 on the early mown, sheep grazed site A (Fig. 2a).

Similarly, the tallest plants of *S. columbaria* were found on cattle grazed site D (66 cm) (Fig. 3a), followed by the sporadically mown sheep grazed meadow. The lowest heights for this species were again recorded on the early mown, sheep grazed site A (33 cm).

4.2.2. Rosette diameter

Rosette diameter and thus leaf length of *Plantago* showed similar patterns to shoot height. The largest mean rosette diameters about 30 cm were found on the cattle grazed grassland site D. Plants of the early mown, sheep grazed site A had a small average rosette

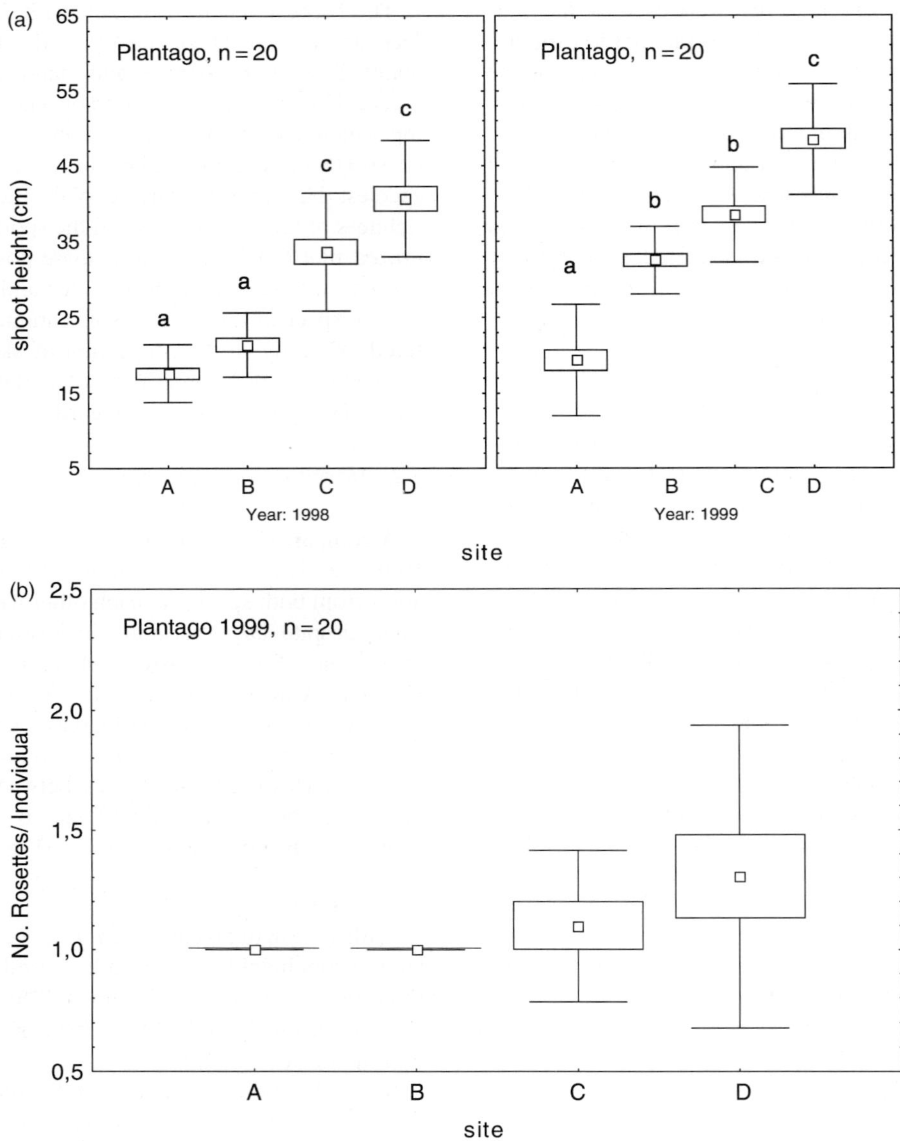

Fig. 2. (a) Shoot height (1998, 1999) and (b) number of rosettes per individual (1999) of *P. media* on sites of different grazing intensity. For a description of the sites see Table 1. Box and whisker plots give the mean (square), standard error (box) and standard deviation (whiskers). Significant differences among sites are denoted by different letters.

diameter of only 9 cm. Rosette diameters of the other two sites took an intermediate position. Leaf length appears to respond equally sensitive to changes in stand structure than shoot height.

Relatively small individuals with leaves suppressed on the ground occurred on open, litter-poor sites, while the largest shoot heights, the largest rosette diameters and therefore the largest leaves were found in plants of litter-rich or productive sites. Obvious were also changes in growth-form; in productive, tall stands rich in litter, leaves of *Plantago* enclosed the stalk of the inflorescence, giving the species the appearance of a

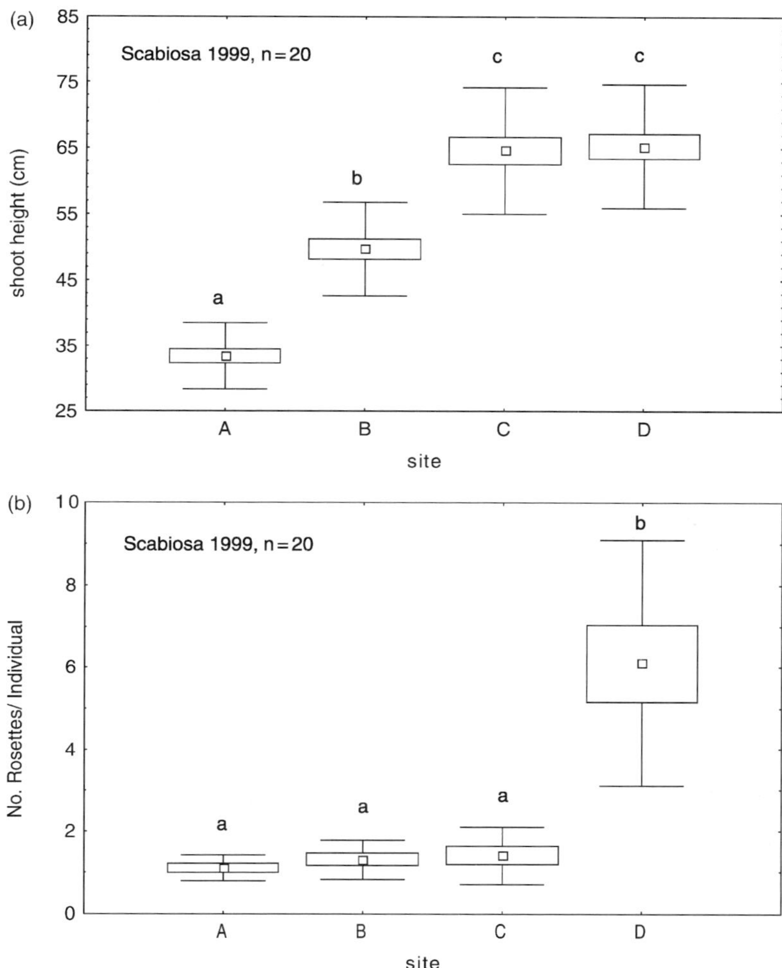

Fig. 3. (a) Shoot height and (b) number of rosettes per individual (1999) of *S. columbaria* on sites of different grazing intensity. For a description of the sites see Table 1. Box and whisker plots give the mean (square), standard error (box) and standard deviation (whiskers). Significant differences among sites are denoted by different letters.

tall herb and positioning the assimilating tissues in a favourable position with respect to light interception. Normally, leaves of this species are suppressed to the ground.

4.2.3. Number of rosettes per individual

The highest numbers of rosettes per individual of *Plantago* (Fig. 2b) were found on the cattle grazed site D (1.3 rosettes per individual). These number differed, however, not statistically significantly from those of the other study sites. It appears that *Plantago* is able to develop new ramets from the base of its root and thus may reproduce vegetatively. On the irregularly sheep grazed site (1.1 rosettes per individual), there was also a tendency to produce daughter rosettes. In individuals of the other sites that reached on average a smaller size, this behaviour was not recorded, only single rosettes were found.

Similarly, *Scabiosa* in most cases only developed one rosette per individual on sheep grazed sites, while on the site with tall vegetation and rich in litter, there were often several vegetative rosettes besides the

Table 4
Biomass per individual of *P. media* and *S. columbaria* at the time of peak flowering (1999) on sites of different grazing intensity ($n = 20$)[a]

Site	Biomass/individual (g dw) (*P. media*)					Biomass/individual (g dw) (*S. columbaria*)				
	Median (50%-quantile)	25%-quantile	75%-quantile	Minimum	Maximum	Median (50%-quantile)	25%-quantile	75%-quantile	Minimum	Maximum
A	0.46	0.33	0.60	0.19	0.79	0.35	0.27	0.37	0.20	0.45
B	0.64	0.58	0.79	0.39	1.18	0.57	0.44	0.70	0.32	1.16
C	0.91	0.77	1.19	0.64	1.66	1.10	0.85	1.25	0.52	2.50
D	2.11	1.65	3.17	0.97	5.22	1.82	1.63	2.16	1.08	5.50

[a] For description of study sites see Table 1.

flowering shoot. This differences were statistically significant (Fig. 3b). There were on average six daughter rosettes per individual in *Scabiosa* on site D. The lowest number of rosettes per individual was found on the early sheep grazed grassland. With an average of only 1.1 rosettes per individual, the production of daughter rosettes was very low.

4.2.4. Number of flowerheads per plant

In comparison to site A (an average of 2.0 flowerheads per plant) and B (2.7), the number of flowerheads per plant from *Scabiosa* was significantly larger on the cattle grazed site D (6.5) as well as on the irregularly sheep grazed grassland C (3.8) during 1999.

4.2.5. Number of diaspores per infructescence

Number of diaspores per infructescence in *Plantago* ranged from 75 to 365 and differed significantly among sites. The highest number of diaspores was found in plants of cattle grazed site D (354), the lowest average had plants of the early sheep grazed site A (75). The irregular and medium grazed sites B and C took an intermediate position with an average of 226 and 271 diaspores per infructescence, respectively.

4.3. Biomass

Table 4 shows that biomass of *Plantago* varied considerably among sites. The cattle grazed site (D) showed the highest values. The average biomass per plant on site D was fivefold larger that of plants from the early grazed site. It is interesting to note, that individuals of the cattle grazed site showed a much larger range of biomass than the other sites, viz. biomass of *Plantago* of site D ranged from approx. 1 to 5 g per

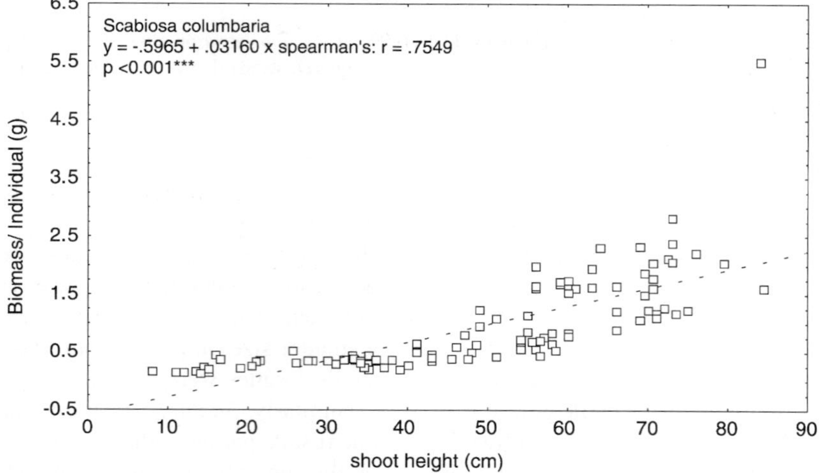

Fig. 4. Relationship between biomass per individual and shoot height of *S. columbaria* (1999).

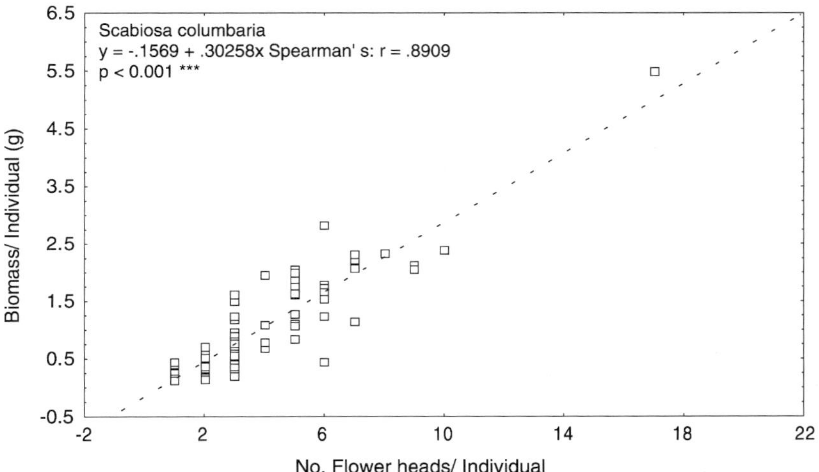

Fig. 5. Relationship between biomass per individual and number of flower heads per individual of *S. columbaria* (1999).

individual, while *Plantago* of the early grazed site varied only between 0.2 and 0.8 g per individual.

Biomass of *Scabiosa* showed the same patterns as in *Plantago*. The cattle grazed site harboured plants with the largest and early sheep grazed sites plants with the smallest biomass (Table 4). Average biomass of site D was approx. five times higher than on the early grazes site. Specimen with the lowest biomass were always recorded in open habitats poor in litter. As in *Plantago*, variation of biomass within sites in *Scabiosa* was highest in the cattle grazed site, while plants from litter-poor sites showed the lowest variation.

4.4. Relationships between aboveground biomass and morphometric variables

There were strong relationships between biomass and plant height of the studied species. Fig. 4 illustrates this for *Scabiosa*. Biomass of the latter was also correlated with the number of flowerheads per

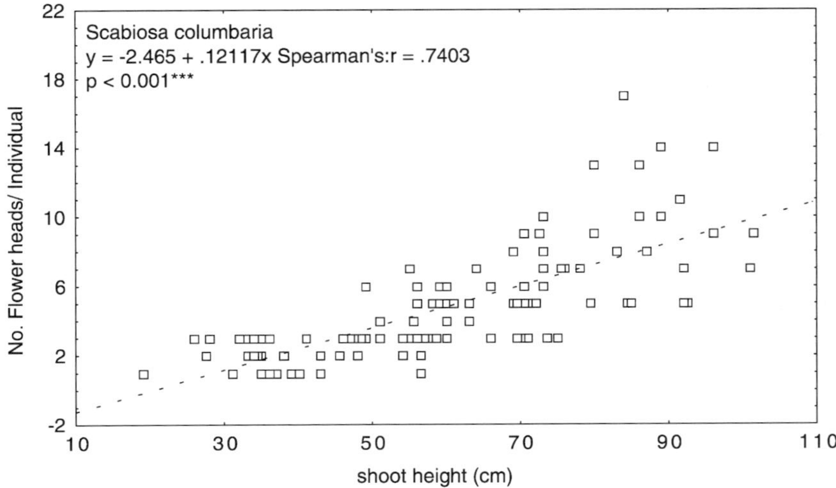

Fig. 6. Relationship between number of flower heads per individual and shoot height of *S. columbaria* (1999).

individual plant (Fig. 5). Similarly, the morphometric variables are significantly correlated with each other. Fig. 6 illustrates the increase in number of flowerheads and thus fertility in *Scabiosa* with shoot height.

5. Discussion

Individuals of both study species showed a high degree of plasticity of morphological traits along the different types of land use. It is very likely that this plasticity may be one reason explaining the occurrence of these species on all study sites. Both species responded rather similar.

On early grazed sites individuals of *Plantago* and *Scabiosa* showed reduced fitness, indicated by low production of daughter rosettes. Here only single rosettes (*Plantago*) or—exceptionally—plants with a few rosettes (*Scabiosa*) occurred, that were further characterised by short shoots and a low production of diaspores per infructescence (*Plantago*) or low numbers of flowerheads per individual (*Scabiosa*). The development of multiple rosettes and the "shift" towards vegetative reproductions appears to be a consequence of denser stands (taller stand height and larger accumulation of litter) and may be especially important for perennial species and their populations (cf. Urbanska, 1992) to be able to cope with changing environmental conditions, e.g. during succession.

Reduced irradiation in litter-rich tall stands may trigger the degree of vegetative reproduction as could be shown for *Arnica montana*, another species of low-productive meadows (Schwabe, 1990a). Vegetative reproduction leads to increased shoot density and thus an increased genet size, which will release competition for light in tall stands. Competition for light may increase the variability of morphological traits (e.g. Grubb et al., 1997; Huber, 1997). Along a gradient of increasing shade, an increase of shoot height may be interpreted as an indicator for stress (lack of light). The results of Grubb et al. (1997) in Mesobromion-stands in Southern England show that from a stand height of ca. 20 cm onwards competition for light is decisive for the aboveground success of (competitive) species. Applied to the calcareous meadows of the present study (cf. Table 1), the late grazed vegetation and species of sites C and D may be characterised by competition for light. Early grazing on sites A and B will increase the amount of light available and decrease competition for light. At the same time, an early onset of grazing will cause the removal of larger amounts of nutrients which will thus not be available later in the season. This may be one reason for smaller (lower stature, lower biomass) individuals on sites A and B.

Apart from generally lower illumination inside tall, litter-rich stands, light conditions within these stands may show larger variation as compared to early grazed sites since shape and orientation of plant parts shows large variation within tall stands (Sundermeier, 1999). The latter may also explain the generally larger heterogeneity of the measured traits and biomass on the late grazed sites (C, D).

The proportion of soil skeleton of the rendzina soils and thus the rooting space of plants appears to be important (Table 1). Site A showed the highest proportion of skeleton. Plants growing here may utilise less than half of the soil volume for the uptake of water. Site D that was inhabited by plants with the largest shoot heights, showed also the largest proportion of fine earth and thus presumably the best water supply. It may be assumed that lower amounts of water that are available on site A may be one reason for generally smaller individuals observed here.

Weiner (1988) regards plant size as a trait that may indicate variation of environmental conditions and that can be translated into variation of reproductive behaviour. Shoot height is considered to be an environmentally constrained variable that depends most strongly on the plant's growth conditions. In the present case these appear to be less favourable on early grazed meadows.

Our results are corroborated by the results of Schopp-Guth (1993) who found that growth-form of the rosette plant *Primula farinosa* changed with increasing mowing frequency. Plants decreased in frequently mown sites while the species showed considerable elongation growth under fallow conditions with increased litter accumulation. Studies of Miller (1998) on the restoration of calcareous low-productive meadows indicated that increasing stand height had a positive effect on vegetative growth of *Centaurea scabiosa*. Increasing land use intensity (mowing frequency), in turn, led to decreased plant size and reduced diaspore production. Furthermore, sexual reproduction of *Salvia pratensis* and *Leontodon incanus* increased with increasing height of the herb layer.

According to the ranking of sites in Table 1, these effects could also be observed in *Scabiosa* and *Plantago*.

Biomass production alike morphometric variables showed high intra-specific plasticity. On early grazed grasslands and thus litter-poor sites, species showed reduced "vitality" according to the above mentioned ranking of sites ("Vitality as the result of vegetative development" sensu Barkmann et al., 1964) leading to a decreased biomass production.

Individual biomass of *Plantago* and *Scabiosa* was mostly significantly higher when the matrix vegetation was higher. Similar results were obtained by Huber (1997) for *Trifolium fragiferum*.

In *Plantago* and *Scabiosa*, tight correlation between biomass of individuals and shoot height and number of flower heads existed. Therefore, morphometric parameters may be used as indirect estimators of biomass. In a literature review Horowitz (1976, as cited in Gluch, 1980) postulated that fresh weight, dry weight and plant height generally show tight correlation. Therefore, plant height, a variable that is very easy to measure in the field, appears to have a strong weight as indicator for certain habitats and environmental conditions (cf. Passarge, 1978).

However, since individuals may develop different growth-forms through their species-specific arrangement of leaves, stems and flowers with different (horizontal and vertical) distribution of biomass, the relationships between height and biomass should be evaluated for each study (Günter, 1997).

Due to their morphological plasticity plants of *Plantago* and *Scabiosa* may presumably endure in the vegetation for a certain period or adapt to the new environmental conditions after changes in land use regime such as, e.g. later grazing. They may respond through increased elongation growth, trying to compensate for deteriorated light conditions in litter-rich tall stands or through a shift in reproductive strategy (from generative to vegetative).

However, the observed trait differences among sites may also be related to differences in age structure or resource supply (water and nutrient budget) of individuals.

A high plasticity of morphological traits, i.e. number and size of leaves, was also found in *Plantago major* (Werner et al., 1989), closely related to *P. media*. Morphological plasticity of biomass production and especially reproductive plasticity (varying number of diaspores) enables *P. major* to adapt to the conditions of phytocoenoses of different weed communities. These results also emphases the fact that different environmental conditions may be characterised by differences in morphological traits of plants.

The knowledge of these species-specific responses and adaptations towards environmental changes is especially important with respect to the evaluation of low-productive grasslands from a management point of view, since these responses may be crucial for the fate of plant individuals or populations in the long run (Begon et al., 1997).

While the evaluation of management practice has traditionally been based on plant species and their abundance, the present study shows that especially the trait "shoot height" that is intimately linked to reproduction may be used as an indicator to answer various questions related to conservation management (e.g. the appropriate timing of grazing). This may be especially useful when directly comparing different types of land use and management within a narrowly delineated area. In contrast to an evaluation based on presence/absence and abundance of species, indicators based on morphological traits such as shoot height may not only depict the extremes but even demonstrate gradual differences between land use regimes (cf. Figs. 2 and 3).

This plasticity of response is exemplified by the rosette plant *Plantago* that builds low, small rosettes with short leaves suppressed to the ground. This reaction is typical for calcareous grasslands with a high conservation value that have a rather open structure with high light interception (sites A and B). Plants with such a growth-form appear also to be better adapted to browsing (cf. Walter, 1962).

In taller and less typical calcareous grassland stands with large litter accumulation (sites D and partly C), *Plantago* changed its reproductive strategy and approached the stature of a tall herb which may lead to improved light conditions for photosynthesis. Similar patterns were observed for *Scabiosa*.

Thus an evaluation of conservation value of different calcareous meadow stands could be done using, e.g. the reproductive strategy and/or shoot height of species.

The fact that morphological responses can be measured quantitatively (Hutchings, 1988) makes them a valuable tool for the evaluation of different environmental conditions that are closely linked to land use practice and management on low-productive grasslands. The main advantage is that management options can be evaluated immediately, based on quantitatively comparable data.

6. Conclusions

It can be concluded that morphometric parameters provide relatively easily applicable indicators for different environmental conditions.

Shoot height of the studied species was a valuable indicator for the studied land use regimes with their specific site and environmental conditions. Traits of species such as the individual shoot height reflect differences in stand structure (stand height, litter accumulation). Populations and their individuals respond to changes in their environment through changes in morphology. Different stand structure in the present study is the result of different types of land use.

Species that are widespread and common within a certain area and whose traits show a relatively large plasticity are especially suited for the comparison and evaluation of vegetation.

Generally, the use of morphometric characters makes it possible to efficiently compile large amounts of quantitative data that may be used for the evaluation of environmental conditions that are closely linked to land use and management for nature conservation.

Obviously, the above conclusions apply only to the habitats and species compared in the present study. Further investigations should address the question whether the above relationships are also valid for other (calcareous grassland) sites and other species. Further studies on this topic appear worthwhile since the tested morphometric variables are easy to measure in the field. They may present an alternative to destructive measurements such as the estimation of biomass.

Acknowledgements

The present study was conducted within the framework of a scientific project supported by the Bundesanstalt für Landwirtschaft und Ernährung (Frankfurt/Main).

References

Barkmann, J.J., Doing, H., Segal, S., 1964. Kritische Bemerkungen und Vorschläge zur quantitativen Vegetationsanalyse. Acta Bot. Neerlandica 13, 394–419.

Begon, M., Mortimer, M., Thompson, D.J., 1997. Populationsökologie. Spektrum Verlag, Heidelberg.

Bobbink, R., Willems, J.H., 1987. Increasing dominance of *Brachypodium pinnatum* (L.) Beauv. in chalk grasslands—a threat to a species-rich ecosystem. Biol. Conserv. 40, 301–314.

Bradshaw, A.D., 1965. Evolutionary significance of phenotypic plasticity in plants. Adv. Gen. 13, 115–155.

Briggs, D., Walters, S.M., 1997. Plant Variation and Evolution. Cambridge University Press, Cambridge.

Dierschke, H., 1985. Experimentelle Untersuchungen zur Bestandsdynamik von Kalkmagerrasen (Mesobromion) in Südniedersachsen. Vegetationsentwicklung auf Dauerflächen 1972–1984. In: Schreiber, K.F. (Hrsg.), Sukzession auf Grünlandbrachen. Münster. Geogr. Arbeiten 20, pp. 9–24.

Düll, R., Kutzelnigg, H., 1994. Botanisch-ökologisches Exkursionstaschenbuch. Quelle und Meyer, Heidelberg.

Ellenberg, H., 1992. Zeigerwerte der Gefäßpflanzen. In: Ellenberg, H., Weber, H.E., Düll, R., Wirth, V., Werner, W., Paulissen, D. (Eds.), Zeigerwerte von Pflanzen in Mitteleuropa. Scripta Geobotanica 18.

Ellenberg, H., 1996. Vegetation Mitteleuropas mit den Alpen in ökologischer Sicht. Ulmer Verlag, Stuttgart.

Gluch, W., 1980. Bioindikation mit produktionsbiologischen und morphometrischen Verfahren. Arch. Natursch. u. Landschaftsforsch. 20, 99–116.

Grime, J.P., Hodgson, J.G., Hunt, R., 1988. Comparative Plant Ecology: A Functional Approach to Common British Species. Unwin Hyman, London.

Grubb, J.P., Ford, M.A., Rochefort, L., 1997. The control of relative abundance of perennials in chalk grassland; is root competition or shoot competition more important? Phytocoenologia 27, 289–309.

Günter, G., 1997. Populationsbiologie seltener Segetalarten. Scripta Geobotanika XXII.

Huber, H., 1997. Architectural plasticity of stoloniferous and erect herbs in response to light climate. Ph.D. Thesis. Utrecht University, 120 pp.

Hutchings, M.J., 1988. Differential foraging for resources and structural plasticity in plants. Trends Ecol. Evol. 3, 200–204.

Korneck, D., Schnittler, M., Vollmer, I., 1996. Rote Liste der Farn und Blütenpflanzen (Pteridophyta et Spermatophyta) Deutschlands. Bundesamt für Naturschutz (Hrsg.). Schriftenreihe für Vegetationskunde (Bonn- Bad Godesberg) H. 28, 744 pp.

Miller, U., 1998. Renaturierung von Kalkmagerrasen: Demographische Differenzierung ausgewählter Kalkmagerrasenarten bei künstlicher Ansiedlung auf einer Ackerbrache. Herbert-UTZ-Verlag, München.

Mückschel, C., 2002. Zur Plastizität populationsbiologischer Merkmale ausgewählter Magerrasenarten Südthüringens unter Beweidungseinfluss. Herbert-UTZ-Verlag, München.

Noble, I.R., Slayter, R.O., 1980. The use of vital attributes to predict successional changes in plant communities subject to recurrent disturbances. Vegetatio 43, 5–21.

Oberdorfer, E., 1993. Süddeutsche Pflanzengesellschaften. Teil II. Fischer Verlag, Jena.

Oberdorfer, E., 2001. Pflanzensoziologische Exkursionsflora. Eugen Ulmer-Verlag, Stuttgart.

Passarge, H., 1978. Die Wuchshöhe, ein wichtiges Strukturmerkmal der Vegetation. Arch. Natursch. u. Landschaftsforsch 18, 31–41.

Primack, R.B., Kang, H., 1989. Measuring fitness and natural selection in wild plant populations. Annu. Rev. Ecol. Syst. 20, 367–396.

Quinger, B., Bräu, M., Kornprobst, M., 1994. Lebensraumtyp Magerrasen. 2. Teilbänd. Landschaftspflegekonzept Bayern, Bd. I und II.1. (Hrsg.), Bayerisches Staatsministerium für Landesentwicklung und Umweltfragen (StMLU) und Bayerische Akademie für Naturschutz und Landschaftspflege (ANL), München, 317 pp.

Quinn, J.A., 1987. Complex patterns of genetic differentiation and phenotpic plasticity versus an outmoded ecotype terminology. In: Urbanska, K.M. (Hrsg.), Differentiation Patterns in Higher Plants. Academic Press, London, pp. 95–113.

Schiefer, J., 1981. Bracheversuche in Baden-Württemberg. Vegetations- und Standortsentwicklung auf 16 verschiedenen Versuchsflächen mit unterschiedlicher Behandlung. Beih. Veröff. Naturschutz und Landschaftspflege Baden-Württemberg 22, 1–325.

Schopp-Guth, A., 1993. Einfluß unterschiedlicher Bewirtschaftu[illegible] auf populationsbiologische Merkmale von Streuwiesenpfla[illegible] und das Samenpotential im Boden. Diss. Bot. 204.

Schwabe, A., 1990. Veränderungen in montanen Borstgrasrasen durch Düngung und Brachlegung: *Antennaria dioica* und *Vaccinium vitis- idaea* als Indikatoren. Tuexenia 10, 295–310.

Schwabe, A., 1990a. Syndynamische Prozesse in Borstgrasrasen: Reaktionsmuster von Brachen nach erneuter Rinderbeweidung und Lebensrhythmus von *Arnica montana*. Carolinea 48, 45–68.

Sundermeier, A., 1999. Zur Vegetationsdichte der Xerothermrasen nordwestlich von Halle/Saale. Diss. Bot. 316.

Tamm, C.O., 1972. Survival and flowering of perennial herbs III. Oikos 23, 159–166.

Urbanska, K.M., 1992. Populationsbiologie der Pflanzen. Gustav Fischer-Verlag, Stuttgart.

Walter, H., 1962. Einführung in die Phytologie. Bd. II: Grundlagen der Pflanzenverbreitung. Eugen Ulmer-Verlag, Stuttgart.

Weiner, J., 1988. The influence of competition on plant reproduction. In: Lovett-Doust, J., Lovett-Doust, L. (Eds.), Plant Reproduction and Ecology—Patterns and Strategies. Oxford University Press, Oxford, pp. 228–245.

Werner, W., Wittig, R., Heimann, R., 1989. Biomasse und Reproduktion von *Plantago major* in verschiedenen ruderalen Pflanzengesellschaften. Verh. Ges. Ökol. XVIII, 671–681.

Willems, J.H., 1982. Phytosociological and geographical survey of Mesobromion communities in Western Europe. Vegetatio 48, 227[illegible].

Wi[illegible]H., 1990. Calcareous grassland in continental Europe. [illegible]ier, Walton, Wells (Eds.), Calcareous Grassland—Ecology and Management, Proceedings of the Joint British Ecological Society, vols. 3–10. Bluntisham Books, Bluntisham, [illegible]untigdon.

[illegible]nns, O., Sendtko, A., 1995. Sukzessionslinien in Kalkmagerrasen unter besonderer Berücksichtigung der Schwäbischen Alb. Beih. Veröff. Naturschutz Landschaftspflege Bad.-Württ. 83, 257–282.

ELSEVIER

Available online at www.sciencedirect.com

Agriculture, Ecosystems and Environment 98 (2003) 227–246

Agriculture Ecosystems & Environment

www.elsevier.com/locate/agee

Selecting target species to evaluate the success of wet grassland restoration

Gert Rosenthal*

Institute for Landscape Planning and Ecology, University of Stuttgart, Keplerstr. 11, 70174 Stuttgart, Germany

Abstract

The evaluation of restoration success needs the definition of goals and the set up of evaluation criteria. This will be done for wet grassland restoration on fen soils in NW Germany. These ecosystems lost their habitat function for co-adapted plant species due to melioration and intensive farming since the 1950s. The recovery of this function is a major concept deduced from environmental policies. In order to evaluate the success of restoration measures in wet grassland restoration target plant species are identified. Doing this, sets of required abiotic conditions are deduced and translated into calibrated mean indicator values for plant communities [Scripta Geobot. 18 (1992) 248]. These are arranged in a successional series which is expected to occur on fens when restoration measures are realised.

Thresholds of mean indicator values are used in order to select "target systems". Four hundred and seventy three plant species that occur in the successional fen series are reduced to 136 target species using different selection criteria. For the evaluation of restoration success it is necessary to establish assemblages of target species with different indication power, invasion probability and frequency. As an indicator integrating these demands, the potential of species to occur during different temporal stages (within successional lines) on fen soils and different habitats is classified. A further ranking of these target species therefore is based on their different niche breadths and temporal persistence during secondary succession. Six vulnerability classes are established which represent species groups with gradually changing vulnerability for intensification or abandonment, respectively. Species with a narrow niche and short persistence during succession are strongly endangered by supposed changes in land management. It is clear from restoration experiments in NW Germany and The Netherlands however that these species are hardly capable of re-establishing during restoration. Higher frequencies in intensive grasslands enhance the re-invasion into recovered grassland swards. Considering this low re-colonisation rate it has to be a priority concept to protect those relic habitats still existing, especially remnants of small sedge-, wet grassland- and stream valley communities.

Keywords: Vulnerability; Succession; Abandonment; Intensification; Nature protection; Open habitats

1. Introduction

Wet grassland ecosystems on fen soils in NW Germany have suffered strongly from intensive agricultural land use, reclamation, abandonment and fragmentation since the 1950s. Both their habitat function for co-adapted species and their retention and sink-capacity for water and nutrients have been lost. Only a few areas representative of the former widespread wet grassland ecosystems remain. Priority concepts of environmental policies like "sustainability" and "protection of resource functions" must, therefore, also be applied to fen grasslands (Halbritter, 1994). According to Fürst et al. (1989), Kaule (1991) and Plachter (1994), the specification of

* Tel.: +49-711-121-4143; fax: +49-711-121-3381.
E-mail address: gr@ilpoe.uni-stuttgart.de (G. Rosenthal).

0167-8809/$ – see front matter
doi:10.1016/S0167-8809(03)00083-5

such broad recommendations is necessary to ensure that they adapt to the demands of special habitats. In this paper the hydrological mire types "inundated mires" and "paludified mires" (after Succow and Joosten, 2001) are considered.

For wet grassland ecosystems on fen soils, the aforementioned priorities result in two major concepts. Concept 1: the restoration of the retention and sink-capacity which can be achieved by strong rewetting and cessation of the land use (Pfadenhauer and Kratz, 2001). Concept 2: the recovery of the habitat function for many (in particular light-demanding) fen species by means of land use extensification, nutrient depletion and raising the water table (Dierßen, 1998). Concept 2 will be discussed here as it highlights the importance of secondary, anthropogenically modified open habitats on fen soils for plant species. The main reason this is so important is that these secondary habitats are at least partly capable of compensating for the loss of primary habitats such as mires and flood plains. These primary habitats are unlikely to be restored under present environmental conditions (Pfadenhauer and Kratz, 2001). Secondary habitats must therefore (at least temporarily) provide a substitute for their habitat function. This is particularly true for species from oligo- and mesotrophic sites (Dierßen and Schrautzer, 1997). Abandoning the land use and rewetting (concept 1) does not automatically result in a high diversity of such species when—like in most cases—natural disturbances are excluded (Rosenthal, 1992; Müller and Rosenthal, 1998; Walter et al., 1998).

The major goal of this paper is to identify suitable target species in order to evaluate the success of wet grassland restoration. The most "popular" procedure for selecting target species is generally based on their vulnerability and rates of decrease, as summarised in the red data books (Korneck et al., 1998). These species are, however, often too rare to become (re)established, even if habitat conditions are suitable (Gigon et al., 1998). In order to get a ranking of species representing different frequencies within the landscape their habitat amplitude and persistence within a successional series will be considered (Pirkl and Riedel, 1991; Reck, 1996; Poschlod et al., 1999).

Species taxonomy follows Ehrendorfer (1973), and phytosociological taxonomy follows Meisel (1977) and Schrautzer (1988).

2. Selecting "target plant communities" and arranging a successional series

Targets for the restoration of degraded fen grasslands can be deduced from semi-natural fen landscapes which were, or still are, extensively used. Historical and actual surveys of these landscapes document the vegetation types that are potentially achievable (Meisel and Hübschmann, 1976; Rosenthal and Müller, 1988; Dierschke and Wittig, 1991). The results show that there are wide-ranging possibilities of different meadow, pasture, reed, sedge and forest communities related to the alliances of Calthion palustris, Caricion fuscae, Caricion davallianae, Molinion caeruleae, Magnocaricion, Phragmition australis, Filipendulion ulmariae, Agropyro-Rumicion, Cynosurion cristati, Alnion glutinosae and Alno-Ulmion minoris (Rosenthal et al., 1998).

Targeted abiotic conditions should be capable of both realising the habitat demands of light-demanding fen species and the necessities of a moderately intensive land use. The carrying capacity of fen peat for agricultural machinery during hay-making is adequate when the water table is 30–60 cm below ground-level (Hennings, 1996; Harter and Luthardt, 1997; Zeitz, 2000). Higher water levels (>30 cm) outside that period guarantees a saturation of the upper soil layers over most parts of the year (Hennings, 1996). This is a precondition for matching the demands of fen species and equates to mean moisture indictor values (calculated after Ellenberg et al., 1992) higher than 7 (Rosenthal, 2000).

Another precondition for establishing species-rich fen meadow communities is a reduction of productivity. The highest species diversities in wet grasslands can be expected at values less than 40 dt/ha (Grime, 1979; Vermeer and Berendse, 1983; Bakker, 1989; Oomes, 1992). According to the mean N-indicator values for the nutrient availability (calculated after Ellenberg et al. (1992)) high number of total species (>30) and red data book species (>10) are expected at mean N-values of less than 5 (Rosenthal, 2000). This equates to mesotrophic conditions, and is in accordance with the findings of Ellenberg (1989), Briemle et al. (1991), Bollens et al. (1998) and Korneck et al. (1998) from other grassland communities.

In the following section the potential vegetation types of inundated and paludified fens are classified according to the following criteria: management, moisture and trophic level. Two hundred and twenty eight vegetation types were classified into 15 vegetation groups (=Vg) (Table 1). The first 11 Vgs relate to progressive and regressive successional lines found on fen habitats. Vgs 12–15 represent other (also extensively used) grassland and heath communities in Northwest Germany that are not necessarily related to fens.

Using the threshold indicator values established above 38 vegetation types can be selected as "target communities". They belong to the alliances Molinion caeruleae, Calthion palustris and Caricion fuscae (meso- and eutrophic extensively managed wet grasslands) which are classified in Vgs 3 and 4. The mean *F*-values are higher than 7, the *N*-values range between 2.9 and 5.0 for mesotrophic conditions (Vg 4) and between 5.0 and 5.6 for more eutrophic conditions (Vg 3) (mean indicator values calculated after Ellenberg et al. (1992)). All associations of Vgs 3 and 4 are endangered (red data book) plant communities (Preising, 1997; Rosenthal et al., 1998).

The vegetation groups, Vgs 1–11, as distinguished in Table 1, are connected by different successional processes (Fig. 1). These include well-documented processes such as progressive secondary successions, from wet grasslands through to reed- and sedge communities, and regressive successions following eutrophication, drainage, land use intensification and peat destruction. Natural reforestation normally takes several decades, especially under wet conditions (Müller and Rosenthal, 1998). The "target processes" such as detrophication, rewetting, land use extensification or re-establishment of a mowing regime, peat production and invasion of species can only be partly realised within a few years (Hellberg, 1995; Hennings, 1996; Kundel, 1998; Rosenthal, 2000). From these findings it is clear that the aforementioned "target systems" are unlikely to be reached in a short time. New sets of abiotic and biotic conditions result rather in new "atypical" plant communities that have not previously been described. It is therefore more reasonable to use species rather than plant communities as target indicators. The main reason being that plant species are capable of creating new assemblages according to changing environmental conditions, only depending on their physiological, morphological and dispersal life traits which determine their invasion power and

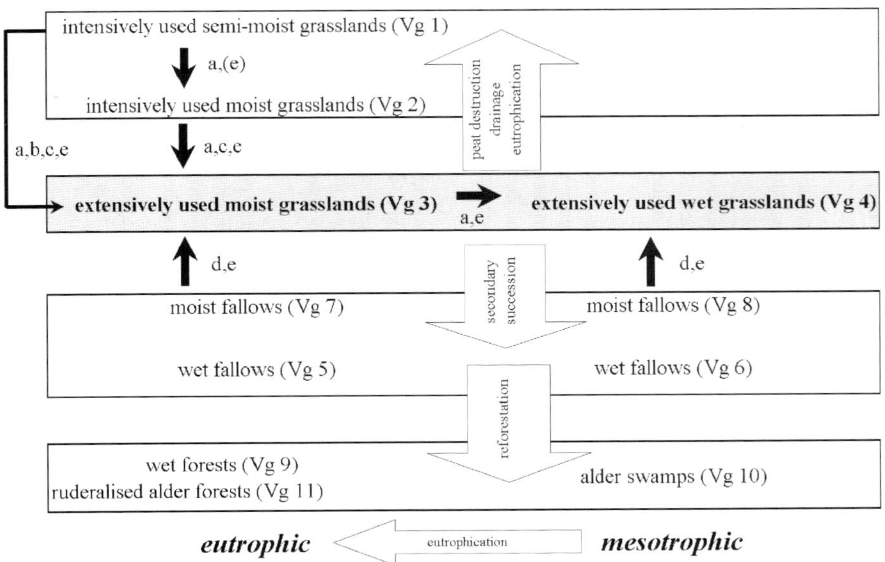

Fig. 1. "Target systems" and target processes (black arrows, see below) for the restoration of species-rich wet grassland systems within successional fen series. For the number of the Vgs refer to Table 1. Processes which are characteristic of intensification or abandonment are shown in open arrows. Target processes: (a) detrophication; (b) rewetting; (c) extensification of land use; (d) re-establishment of a (moderately intensive) management regime on fallow land; (e) re-invasion of species.

Table 1
Ecological integration of grassland plant communities and temporarily related successional stages on fen soils and spatially related communities on other wet habitats[a]

Vegetation groups (Vg)	Vegetation types	Management	Moisture	F-value	Trophic level	N-value
Intensively used grasslands						
1. Intensive semi-moist grasslands	Molino-Arrhenatheretea-rump communities	Intensive	Semi-moist	<6.5	Eutrophic	>5.0
2. Intensive moist grasslands	Rumici-Alopecurelum geniculati	Intensive	Moist	>6.5	Eutrophic	>5.0
Extensively used moist/wet grasslands						
3. Moist grasslands	Calthion communities	Extensive	Moist	7–8	Eutrophic	>5.0
4. Moist/wet grasslands	Calthion-, Mollinion-, Caricion fuscae communities	Extensive	Moist-wet	7–9	Mesotrophic	<5.0
Fallows						
5. Wet eutrophic fallows	Phragmition-, Magnocaricion-communities	Fallow	Wet	>8	Eutrophic	>5.0
6. Wet meso-trophic fallows	Phragmition-, Magnocaricion-communities	Fallow	Wet	>8	Mesotrophic	<5.0
7. Moist eutrophic fallows	Phragmition-, Magnocaricion-, Filpendulion-communities	Fallow	Moist	<8	Eutrophic	>5.0
8. Moist mesotrophic fallows	Magnocaricion-communities	Fallow	Moist	<8	Mesotrophic	<5.0
Forests						
9. Wet eutrophic forests	Alnetum glutinosae cardaminetosum et typicum, Pruno-Fraxinetum, Carici remotae-Fraxinetum	Forest	Moist-wet	–	Eutrophic	–
10. Mesotrophic alder swamps	Alnetum glutinosae betuletosum	Forest	Moist-wet	–	Mesotrophic	–
11. Ruderalised alder forests	Robus idaeus-Alnus glutinosa community	Forest	Semi-moist	–	Eutrophic	–
Other extensively used grasslands (stream valleys)						
12. Wet grasslands (streams)	Lathyrus palustris-com., Caricetum vulpinae	Extensive	Moist-wet	>7.0	Eutrophic	>5.0
13. Semi-moist grasslands (streams)	Sanguisorbo-Silaetum, Cnidio-Violetum, Dauco-Arrhena-theretum, Chrysanthemo-Rumicetum	Extensive	Semi-moist	5–7	Meso-, eutrophic	3–7
Other extensively used grasslands (pasture-, health vegetation)						
14. Eutrophic grasslands (pastures)	Lolio-Cynosuretum, Luzulo-Cynosuretum, Angelico-Cirsietum, Senecioni–Brometum	Extensive/intensive	Semi-moist	<6.5	Eutrophic	>5.0
15. Mesotrophic grasslands (health, pastures)	Callunetea-com, Junco-Molinietum, Lolio-Cynosuretum, Luzulo-Cynosuretum	Extensive/fallow	Semi-moist	<6.5	Mesotrophic	<5.0

[a] Two hundred and twenty eight vegetation types (originating from 4467 relevees) were classified into 15 vegetation groups (Vgs). Data collected from Meisel (1977), Schrautzer (1988), Zacharias et al. (1988) and Döring-Mederake (1991). F-value: mean indicator value for moisture, N-value: mean indicator value for nutrients (both calculated after Ellenberg et al., 1992).

competitive vigour (Ellenberg, 1956; Grime, 1979; Van der Valk, 1981; Keddy, 1992; Rosenthal, 1992; Gigon and Leutert, 1996; Kleyer, 1997).

Plant communities are not suitable for the representation of "target systems" in restoration ecology if a floristically fixed combination of species has to be reached. Yet, as representatives of a specific set of ecological conditions, they can be used to define target species sets and the expected frequencies of those species (Bakker et al., 2000). "Target systems" in this understanding are used as a link between abiotic thresholds and plant species that can be expected with increasing probability when conditions are adopted to meet the target conditions.

3. Selecting target species and assigning vulnerability values

Criteria used to select target species are:

1. The distribution of species over vegetation groups Vgs 1–15: only species with a significant positive deviation from expectation according to their distribution over Vgs were selected (Chi^2-test, 0.1% level).
2. Occurrence within the "target systems" of the vegetation groups Vgs 3 and 4.
3. "Decreasing tendency" in Germany during the last decades. Data refer to species surveys on topographical map grids (1:25.000) (Haeupler and Schönfelder, 1988). Species with a "decreasing tendency" from 2 = strongly decreasing to 4 = decreasing were selected (numbers according to Ellenberg et al. (1992)).
4. Red data book species and Molinietalia-species were added regardless of whether they were selected via steps 1–3.

Following this procedure the 599 vascular plant species that occur in the 15 vegetation groups was reduced to 136 target species.

For the evaluation of restoration success it is necessary to establish assemblages of target species with different indication power, invasion probability and frequency. As an indicator integrating these demands, the potential of species to occur during different temporal stages (within successional lines) on fen soils and different habitats is classified. For each species it consists of an assortment of different vegetation groups (Vgs). In order to complete the spectrum of vegetation types beyond fen communities additional information was taken from Korneck and Sukopp (1988). They considered the distribution of all plant species over all vegetation types that are present in Germany (Table 2).

The vulnerability of species is deduced from (1) their incidence in different successional stages and habitats, (2) the susceptibility to eutrophication, drainage and/or intensification of the vegetation types in which they occur (Table 2), and (3) the availability of vegetation types in a spatial distance that can normally be overcome by dispersal. The vulnerability is assessed individually for each species within each plant formation (A–F). Three cases are distinguished (Table 3): species restricted to endangered vegetation types (labelled as !!), species concentrated in, but not restricted to, endangered vegetation types (labelled as !), or species that are evenly distributed in endangered and non-threatened vegetation types (labelled as ×).

Fig. 2 provides a key to the integrative assessment of vulnerability values for individual species according to the aforementioned criteria. Species missing from non-threatened refuge habitats are considered to be strongly endangered (vulnerability value 1). That is especially true for species that depend on open wet grasslands like *Bromus racemosus*, *Taraxacum palustre*, *Pedicularis palustris* and *Gentiana pneumonanthe*. Species from oligotrophic, mostly calcareous fens like *Carex dioica*, *C. flava*, *C. hostiana*, *C. pulicaris* and *Pedicularis sylvatica* may also be included (Jensen and Schrautzer, 1999). These results emphasise the importance of protecting wet, low productivity grasslands while maintaining an extensive land use system. These species must be assumed to be extraordinarily sensitive against any kind of changing habitat conditions and successional trends towards forest communities. Their primary habitats in central Europe were lost long ago. Alternative habitats were created by human activities starting with the settlement period of the Younger Stone Age (Schwaar, 1990; Speier, 1996). This might have fostered dispersal and enlarged geographical ranges. Today, however, these secondary habitats are under threat or have even disappeared.

Species with the vulnerability value 2 are better protected against land abandonment because they are also able to grow in forests. If a landscape includes

Table 2
Classification of plant formations[a]

Plant formations (classes A–F)	Not endangered	Endangered or spatially remote
Non-forest systems		
(A) Anthropogenic plant formations	Intensively used semi-moist grasslands (Vg 1), intensively used moist grasslands (Vg 2), *grass pioneer communities*	Extensively used wet grasslands (Vgs 3, 4, 12), small-sedge communities (Vg 4), extensively used semi-moist grasslands (Vgs 13–15), heath- and mesotrophic pasture communities (Vg 15), *weed communities and communities of waste ground*
(B) Natural plant formations, expected with abandonment (except C)	*Pioneer vegetation of mud banks, nitrophylous ruderal communities, swards of pathways and flooded areas*	*Salt water vegetation, vegetation of raised bogs and wet heath, alpine und subalpine swards, xerothermic fringe communities*
(C) Natural plant formations, expected with abandonment (except B)	Eutrophic reeds, tall-sedge- and tall-herb communities (Vgs 5, 7)	Mesotrophic tall-sedge communities (Vgs 6, 8)
Fresh-water		
(D) Freshwater and springs (except C)	*Eutrophic freshwater communities, springs*	*Oligotrophic freshwater communities*
Forests		
(E) Wet forests	Ruderalised alder forest (Vg 11)	Alder swamp-, birch swamp woods, alder and broad-leaved woods of floodplains (Vgs 9, 10), *willow woods*
(F) Forests (except E)	*Broad-leaved woods on fertile soils, acid soil mixed oak woods and coniferous woods*	*Xerothermic woods*

[a] Vulnerability is evaluated within each of vegetation types of anthropogenic and natural origin (in italics: after Korneck and Sukopp (1988), otherwise see Table 1, figures for vegetation groups Vgs 1–15 in brackets).

(mesotrophic) wet forest communities they will find refuge habitats there. *Carex panicea*, *Veronica scutellata*, *Eriophorum angustifolium*, *Juncus acutiflorus* and *Carex echinata* belong to this group.

Plant species with refuge habitats in non-threatened reed communities are labelled with a vulnerability value 3. Species that also occur in non-threatened forest communities are classified with a vulnerability value 4. Species are supposed to be potentially endangered (vulnerability value 5) if they have their priority habitats in non-threatened grassland types but moreover only occur within threatened plant communities. With further intensification they may have to be put into a higher category of vulnerability (1 or 2). Species that occur in threatened, as well as in non-threatened habitats, are well protected against a further decline due to their wide range of potential habitats and successional stages (vulnerability value 6).

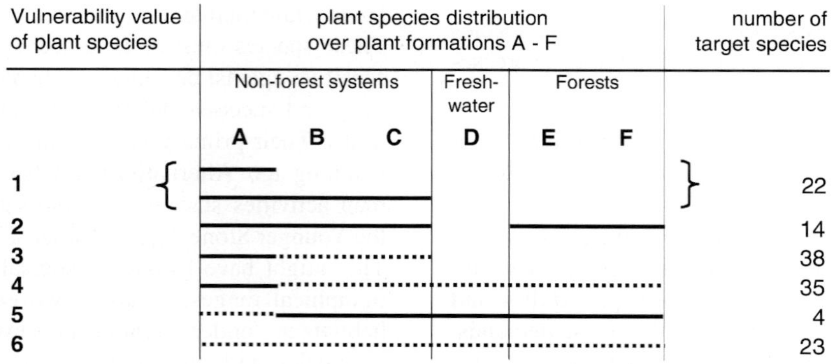

Fig. 2. Deducing "vulnerability values" that result from their incidence in endangered/spatially remote (full line) or non-endangered (dotted line) plant communities within plant formations A–F (after Table 2).

Table 3
Distribution of priority target species in moist and wet grasslands (vegetation groups Vgs 2–4) over plant formations A–F[a]

Decreasing tendency	Red data book Germany	Red data book Ni/Br		Vegetation groups: Intensive grasslands 1, sem, eu	2, moi, eu	Extensive wet grasslands 3, moi, eu	4, m/w, me	Fallows 5, wet, eu	6, wet, me	7, moi, eu	8, moi, me	Wet forests 9, m/w, eu	10, sem, eu	11, m/w, me	Other extensive grasslands 12, m/w, eu	13, sem, eu/me	14, sem, eu	15, sem, me	Vulnerability value	Plant formations: Non-forest systems A	B	C	Fresh water D	Forests E	F
4			*Alopecurus geniculatus*	56	**62**	14	16	4	2	11	3	–	–	–	34	5	10	–	6	×	×	×			
5			*Potentilla anserina*	40	**35**	5	5	2	3	11	6	–	–	–	20	2	13	7	6	×	×	×			
4			*Cardamine pratensis*	35	47	77	73	14	33	32	61	9	2	6	60	55	**74**	30	6	×	×	×	×	×	×
5			*Trifolium hybridum*	4	**14**	–	6	–	–	–	–	–	–	–	2	+	6	–	6	×	×				
4		3	*Carex vulpina*	4	**9**	–	1	+	–	–	–	–	–	–	22	+	+	–	6	×	×	×			
3			*Poa palustris*	4	**26**	32	6	7	1	5	3	–	–	–	66	2	4	–	6	×	×	×		!	
4			*Eleocharis palustris*	–	**16**	8	6	7	2	2	3	–	–	–	24	–	2	–	6	×		×			
2	3	3	*Oenanthe fistulosa*	–	**6**	2	4	2	+	+	2	–	–	–	9	–	+	–	3	!		×			
5			*Glyceria fluitans*	2	**38**	44	41	7	4	10	7	16	3	25	23	1	11	1	6	×	×	×	×	!	
4			*Galium palustre*	4	40	64	61	**49**	71	42	71	53	17	46	79	10	10	1	6	×		×		×	
5			*Phalaris arundinacea*	5	21	10	5	**28**	10	26	7	15	16	7	21	7	2	–	6	×		×		×	
4		3	*Caltha palustris*	6	17	**83**	72	24	37	48	73	29	5	3	58	1	10	1	4	!		×	×	!	
5			*Filipendula ulmaria*	4	6	**74**	53	19	30	58	62	46	18	10	34	21	17	8	4	!	×	×		×	
5			*Myosotis palustris*	–	14	**69**	33	18	18	29	46	22	4	3	49	3	7	2	4	!		×	×	!	
3		3	*Carex vesicaria*	–	2	**15**	11	6	4	+	3	3	2	3	6	–	+	–	3	!!		×		×	
2	3	2	*Bromus racemosus*	4	4	**24**	13	–	–	–	–	–	–	–	–	10	12	4	1	!					
2		3	*Veronica longifolia*	–	4	**12**	2	+	–	3	1	–	–	–	2	4	4	3	3	!		×			
3		3	*Rhinanthus serotinus* agg.	–	–	**11**	9	–	–	3	5	–	–	–	–	8	5	+	3	!!	!!	×			
2	3	2	*Carex cespitosa*	–	–	**11**	1	–	+	4	3	–	1	–	–	–	–	–	4	!!		×		×	
3			*Agrostis canina*	4	12	19	**47**	+	29	+	8	1	1	20	59	7	9	4	5	×	!!	!		!	
4			*Carex nigra*	5	17	17	**73**	1	30	3	59	2	5	43	5	2	14	12	5	×	!!	!		!	
4			*Anthoxanthum odoratum*	10	3	57	60	+	12	10	30	+	+	5	26	45	48	**75**	6	×	!!	×		×	×
5			*Lychnis flos-cuculi*	6	17	20	**61**	5	27	29	60	6	2	4	20	12	22	8	6	×		×		×	
4			*Ranunculus flammula*	4	14	4	**34**	2	8	2	8	6	+	2	20	+	5	2	6	×	×	×	!!	×	
4			*Carex gracilis*	+	14	13	**45**	16	23	21	16	6	3	5	15	–	10	+	6	×		×		×	
5			*Ajuga reptans*	+	+	10	**21**	1	+	10	21	16	7	+	–	19	6	6	4	!	!!	×		×	×
5			*Mentha arvensis*	4	5	3	**15**	3	1	1	–	2	1	+	3	+	7	–	6	×	×	×		×	
5			*Lotus uliginosus*	+	2	37	**63**	9	24	32	71	+	–	+	9	15	17	18	3	!!		×		!!	
3			*Achillea ptarmica*	4	6	2	**18**	+	2	6	22	+	–	–	2	10	4	5	3	!		×		!!	
4		(3)	*Galium uliginosum*	4	–	11	**35**	1	14	10	28	+	–	–	4	7	6	4	3	!	!!	×		!!	
4		3	*Carex panicea*	–	1	–	**21**	–	4	+	21	–	–	+	2	4	1	6	2	!!	!!	!		!!	
3		3	*Senecio aquaticus*	–	4	3	**14**	–	–	2	4	+	–	–	7	2	3	3	3	!		×		!!	
3		3	*Juncus filiformis*	–	2	35	**47**	2	–	4	2	–	–	+	11	–	3	2	3	!!	!!	×		!!	
2			*Veronica scutellata*	–	2	3	**4**	+	1	–	–	–	–	–	5	+	+	–	3	!		!!	×	!!	
5			*Juncus conglomeratus*	–	–	2	**7**	+	5	4	4	+	–	+	–	–	+	1	3	!!	×	×		!!	
3		3	*Polygonum bistorta*	–	–	3	**5**	–	–	+	–	+	–	–	–	+	–	–	3	!!	!!	×		!!	
3		3	*Dactylorhiza maculata*	–	–	6	**7**	–	–	–	–	–	–	+	–	+	2	+	4	!!	!!			!!	×
5			*Luzula multiflora*	–	–	3	**7**	–	3	–	–	–	+	–	–	1	+	8	4	!!	!!	!!		×	×
4		3	*Carex pallescens*	–	–	–	**4**	–	–	–	–	+	+	–	–	+	–	+	4	!!				×	×
4			*Hypericum maculatum*	–	–	–	**4**	–	–	–	–	–	–	–	–	3	–	2	1	!!	!!				

Table 3 (*Continued*)

Decreasing tendency	Red data book Germany	Red data book Ni/Br		Vegetation groups															Vulnerability value	Plant formations					
				Intensive grasslands		Extensive wet grass-lands		Fallows				Wet forests			Other extensive grasslands					Non-forest systems			Fresh water	Forests	
				1, sem, eu	2, moi, eu	3, moi, eu	4, m/w, me	5, wet, eu	6, wet, me	7, moi, eu	8, moi, me	9, m/w, eu	10, sem, eu	11, m/w, me	12, m/w, eu	13, sem, eu/me	14, sem, eu	15, sem, me		A	B	C	D	E	F
3		3	*Valeriana dioica*	–	–	4	**17**	2	9	5	22	9	+	6	–	3	+	–	3	!!	!!	×		!	
3			*Eriophorum angustifolium*	–	–	–	**25**	+	16	–	7	–	–	5	–	–	+	2	2	!!	!!	!		!!	
4			*Juncus acutiflorus*	–	–	–	**14**	+	5	1	6	+	–	4	–	–	–	+	2	!!	!!	!		!!	
3		3	*Carex echinata*	–	–	–	**9**	–	2	–	–	–	–	9	–	–	–	–	2	!!	!!	!!		!!	
3	3	1	*Juncus subnodulosus*	–	–	2	**11**	2	+	+	8	–	–	–	–	–	–	–	3	!!	!!	×			
2	3	1	*Parnassia palustris*	–	1	–	**7**	–	+	–	4	–	–	–	–	+	–	–	1	!!	!!	!!			
2	3	1	*Taraxacum palustre* agg.	–	–	2	**7**	–	–	–	–	–	–	–	–	–	–	–	1	!!	!!				
2	3	1	*Pedicularis palustris*	–	–	–	**13**	–	2	–	–	–	–	–	–	–	–	+	1	!!	!!	!!			
3		3	*Salix repens*	–	1	–	**3**	–	2	–	–	–	–	–	–	–	–	2	1	!!	!!	!!			
2	3	2	*Gentiana penumonanthe*	–	–	–	**4**	–	–	–	–	–	–	–	–	–	–	2	1	!!					

[a] The numbers indicate frequencies within each vegetation group (Vg) in %. Italics: significant positive deviation from expectation according to a Chi^2-test (0.1% level). Plant formations according to Table 2. Moisture: sem, semi-moist; moi, moist; m/w, moist/wet. Trophical status: eu, eutrophic; me, mesotrophic. Decreasing tendency according to changing frequencies in all of 10 km × 10 km grids in Germany; After Ellenberg et al. (1992): 5, stable; 4, between 3 and 5; 3, decreasing; 2, strongly decreasing. Red data book Germany: 3, threatened; 2, strongly threatened; 1, close to extinction. Red data book Niedersachsen/Bremen (Garve, 1994, evaluated for lowlands only): ditto. Vulnerability within plant formations: !!, restricted to endangered vegetation types; !, concentrated but not restricted to endangered vegetation types; ×, evenly distributed between endangered and non-threatened vegetation types. For vulnerability values see Fig. 2.

4. Description of target species

4.1. Priority species in managed wet grasslands on fen soils

This plant species group can be characterised by its preference for wet grassland systems (Vgs 2–4 in Table 3). *Alopecurus geniculatus* is representative of species with the widest range of habitats, particularly in intensive wet grasslands. Therefore, it can be supposed to be only slightly threatened, though missing in later successional stages such as reed and forest communities. The same is true for the *Trifolium hybridum* group.

The following species groups, however, have a priority distribution in extensive grasslands. The *Caltha palustris* group covers a wide range of different successional stages and is therefore well "buffered" against successional trends towards reed- and forest communities. In contrast the *B. racemosus* group is missing in forests. *Bromus* itself is strongly restricted to extensively managed grasslands. It suffers as a result of intensification, afforestation and abandonment and responds very quickly to changing environmental conditions due to its short life span (Lutz, 1996).

The *Agrostis canina* and *Lotus uliginosus* groups are frequently distributed in mesotrophic wet grasslands and mesotrophic reed, sedge or forest communities. The *Agrostis* group occupies a wide range of habitats and successional stages. It is therefore obvious that these species are not listed in the red data books (Garve, 1993; Korneck et al., 1996). The Lotus group is less frequent in forests, meaning that species thus get some higher vulnerability values.

Senecio aquaticus group resembles the aforementioned *B. racemosus* group, but is more restricted to mesotrophic wet grasslands. The habitat range of *S. aquaticus* itself is closely related to *Bromus*. Both species were therefore selected in order to characterise floristically the "Senecioni–Brometum racemosi" community (Lenski, 1953). Before the reclamation period it was the most widespread grassland community on fen soils in northern Germany (Rosenthal et al., 1998). During the last decades *Bromus* decreased much more than *Senecio*. The reason is that *Bromus* is more "gap-demanding" due to its annual life cycle. It was, therefore, more strongly restricted to the historical grassland management that integrated hay-making with a grazing period in late summer. *Senecio*, however, is a biennial plant that also grows in wetter habitats, where this management continues. Both are very sensitive indicators for this specific set of environmental conditions.

The *Dactylorhiza maculata* group occurs, not only in managed wet grasslands, but also in some subordinate sites such as mesotrophic heath, pasture and acidophilous forest communities. The following species groups have only very few non-threatened refuge habitats. The *Valeriana dioica* and *Juncus subnodulosus* groups occur in mesotrophic forest or reed communities, respectively. The *T. palustre* group is restricted to endangered open plant communities from oligo- and mesotrophic-sites such as wet grassland, sedge and heath communities. Their primary habitats (minerotrophic bog-edges and oligotrophic fens) have long since been destroyed. These species would disappear completely if their secondary (man-made) habitats were also destroyed. They are, therefore, strongly threatened by intensification, drainage and eutrophication. Another underestimated threat in Niedersachsen (NW Germany) is abandonment. For example, 30% of all (28) sites of *P. palustris* recorded between 1982 and 1992 disappeared exclusively as a result of abandonment (Rosenthal and Fink, 1996). The main determining factor is a dense layer of litter that inhibits the establishment of seedlings even in low productive grasslands.

Generally, short lived plants like *P. palustris*, *B. racemosus* and *S. aquaticus* are capable of rapidly responding to changing environmental conditions (particularly to deteriorating microclimatic conditions). There is almost no time-lag between impact and response (Matthies, 1991), because they regularly demand vegetation gaps in a narrow temporal window for germination and establishment.

4.2. Priority species in abandoned wet grasslands (reed and sedge communities)

Cessation of land management in wet grassland communities triggers a succession towards reed, tall sedge, tall herb, and forest communities. In this section priority species in unmanaged non-forest communities are considered (Vgs 5–8, Table 4). Increasing frequency and dominance of these species often indicates decreasing management intensity (Rosenthal, 1992).

Table 4
Distribution of priority target species in moist and wet fallow land (vegetation groups Vgs 5–8, Table 1; for further explanation see Table 3)

Decreasing tendency	Red data book Germany	Red data book Ni/Br		Vegetation groups: Intensive grasslands 1, sem, eu	2, moi, eu	Extensive wet grasslands 3, moi, eu	4, m/w, me	Fallows 5, wet, eu	6, wet, me	7, moi, eu	8, moi, me	Wet forests 9, m/w, eu	10, sem, eu	11, m/w, me	Other extensive grasslands 12, m/w, eu	13, sem, eu/me	14, sem, eu	15, sem, me	Vulnerability value	Plant formations: Non-forest systems A	B	C	Fresh water D	Forests E	F
5			*Gylceria maxima*	+	2	11	16	**23**	8	19	5	11	2	5	5	–	4		3	!		×		!	
5			*Mentha aquatica*	4	3	17	30	**26**	28	27	46	27	2	8	3	1	2	–	3	!		×		!	
5			*Iris pseudacorus*	–	4	9	11	**24**	16	13	6	28	14	22	50	–	+	–	4	!		×		×	
5			*Lycopus europaeus*	–	2	6	10	**20**	22	7	10	37	9	39	3	+	2	–	4	!		×		×	
5			*Angelica sylvestris*	5	2	15	21	9	12	**20**	31	18	8	2	–	5	6	+	4	!	×	×		×	
4			*Carex disticha*	4	13	24	13	6	13	**35**	62	–	1	–	21	2	4	–	6	×		×		×	
4			*Equisetum palustre*	5	9	11	24	12	18	**24**	41	7	7	9	9	6	10	2	6	×	!!	×		×	
5			*Cirsium oleraceum*	2	–	10	6	3	4	**25**	21	13	2	–	–	3	2	–	3	!		×		!	
5			*Scirpus sylvaticus*	–	1	18	18	2	2	**23**	7	16	5	6	–	+	3	–	4	!!		×		×	
4		3	*Geum rivale*	–	–	5	8	1	5	**17**	19	5	–	+	–	1	–	–	3	!!	!!	×		!!	
5			*Epilobium parviflorum*	–	–	3	1	2	–	**5**	8	+	–	+	–	–	–	–	3	!!	×	×		!!	
5			*Valeriana procurrens*	–	–	2	2	–	–	**3**	–	–	–	–	–	–	+	+	3	!!	×	×		!	
4			*Phragmites australis*	+	–	+	9	**30**	35	19	26	8	4	15	–	+	+	–	4	!!	!!	×		×	
5			*Carex paniculata*	–	–	1	12	**13**	17	7	17	13	8	27	–	–	–	–	4	!!		×	×		
4			*Berula erecta*	–	–	1	4	**9**	6	7	8	5	+	2	–	–	–	–	4	!!		×	×	×	
5			*Eupatorium cannabinum*	–	–	1	7	**13**	5	10	12	22	5	5	–	–	–	–	4	!!	×	×		×	
3			*Carex riparia*	–	+	2	6	10	12	**18**	6	4	+	+	–	–	–	–	4	!!		×		×	
5			*Carex acutiformis*	–	1	11	13	18	18	**41**	45	25	15	10	–	1	2	–	4	!!		×		×	
3			*Sium latifolium*	–	–	–	4	**11**	1	1	–	+	–	+	30	–	–	–	3	!!		×		!!	
4			*Sparganium erectum*	–	–	+	+	**10**	2	+	–	4	–	5	–	–	–	–	3	!!		×		!!	
4			*Rumex hydrolapathum*	–	1	–	2	**12**	7	3	1	3	–	+	–	–	–	–	3	!!		×		!!	
5			*Epilobium hirsutum*	–	+	3	7	**8**	3	9	8	+	–	–	–	–	3	+	3	!!		×		!!	
4			*Rorippa amphibia*	–	6	–	1	**11**	3	+	–	–	–	–	7	–	+	–	6	×	×	×			
3			*Myosotis laxa*	–	–	–	+	**1**	1	1	4	–	–	–	–	–	–	–	3	!!		×	×	!	
4			*Acorus calamus*	–	–	–	1	**6**	+	–	–	–	–	–	–	–	–	–	3	!!		×			
5			*Juncus effusus*	8	9	13	31	9	29	28	32	30	18	**65**	2	4	6	3	6	×	×	×		×	
5			*Cirsium palustre*	4	–	11	24	13	**39**	25	40	33	25	39	2	8	9	12	4	!	×	×		×	
5			*Lythrum salicaria*	4	2	2	14	17	**34**	13	18	26	4	23	5	1	3	–	3	!		×		!	

4			*Calamagrostis canescens*	–	1	5	3	25	**55**	15	15	24	37	55	–	–	+	–	4	!!		×		×
5			*Epilobium palustre*	–	+	3	21	8	**22**	9	23	11	5	12	6	1	2	–	4	!!	!!	×	×	×
5			*Equisetum fluviatile*	–	2	43	35	23	**40**	18	19	20	+	19	9	–	2	–	3	!!	!!	×		!
2			*Stellaria palustris*	–	14	6	21	5	**19**	5	19	+	–	–	9	+	2	–	5	×	!!	!		!!
4			*Peucedanum palustre*	–	–	+	13	16	**43**	5	13	22	9	59	–	–	–	–	4	!!		×		×
3	3	3	*Lysimachia thyrsiflora*	–	–	1	3	10	**26**	4	11	3	1	16	2	–	–	–	3	!!	!!	×		!
4	1	(3)	*Viola palustris*	–	–	1	16	3	**28**	2	10	11	29	32	–	–	+	–	4	!!	!!	!	×	×
3			*Potentilla palustris*	–	2	2	40	4	**59**	–	28	+	–	23	2	–	–	+	2	!!	!!	!		!!
4			*Carex rostrata*	–	–	–	19	+	**31**	1	23	+	+	25	–	–	–	1	2	!!	!!	!		!
4			*Hydrocotyle vulgaris*	–	1	–	11	5	**29**	4	20	+	–	12	–	–	+	2	2	!!	!!	!		!!
2	3	2	*Menyanthes trifoliata*	–	–	3	22	1	**23**	3	18	+	–	5	2	–	+	–	2	!!	!!	!		!!
2	3	2	*Dactylorhiza majalis*	–	–	1	12	+	6	1	**25**	+	–	–	–	+	2	–	2	!!	!!	!		!!
2	3	2	*Carex appropinquata*	–	–	–	1	–	**7**	+	–	–	–	–	–	–	–	–	2	!!		!		!
2	2	1	*Carex diandra*	–	–	–	+	–	**3**	–	–	–	–	–	–	–	–	–	1	!!	!!	!!		
2	3	2	*Carex lasiocarpa*	–	–	–	+	–	**1**	–	–	–	–	–	–	–	–	–	1	!!	!!	!!		
2	3	3	*Ranunculus lingua*	–	–	–	+	2	**4**	–	–	1	–	+	–	–	–	–	2	!!		×		!!
2	3	2	*Triglochin palustre*	–	+	–	+	–	2	**3**	6	–	–	–	–	–	–	–	3	!!	×	×		
2	2	2	*Calamagrostis stricta*	–	–	–	+	–	–	+	**3**	–	–	–	–	–	–	–	1	!!	!!	!		

Reduction beneath a critical value, however, results in a predominance of tall-growing species and hence strong light competition. Species diversity decreases to the disadvantage of many weak competitors such as rosette plants and many red data book species (e.g. *S. aquaticus*). Species of Vgs 5–8 are indicative of the sensitive balance between intensification and abandonment; many of them disappear with increasing intensity but become dominant with decreasing intensity.

Glyceria maxima, *Angelica sylvestris*, *Geum rivale*, *Phragmites australis*, *Sium latifolium* and *Rorippa amphibia* groups show a preference for eutrophic conditions. They are able to withstand eutrophication, as well as vegetation change during secondary succession. Most of them occur even in wet alder forests. Their wide range of habitats and successional stages could explain why these species are less endangered according to the red data books than those that depend on open grasslands (previous section).

The other species groups in Table 4 show a preference for mesotrophic wet sedge communities. Most of them cover a wide range of successional stages from managed grasslands to forests. Vulnerability values of the *Potentilla palustris*, *Carex appropinquata* and the *Triglochin palustre* groups indicate increasing sensitivity to eutrophication and intensification (cf. Table 4). It is plausible, therefore, that many red data book species are present in this group. The *Juncus effusus* group is comparatively well represented in intensive grasslands. An increasing contribution of *Juncus* indicates decreasing management intensity, especially in pastures.

4.3. Priority species in other grassland types

These species show a preference for habitats other than paludified or inundated fens (Table 5). The *Thalictrum flavum*, *Sanguisorba officinalis* and *Avenochloa pubescens* groups frequently occur in extensively managed grasslands along large stream valleys, such as the Elbe river valley in Northern Germany. The *Sieglingia* and *Succisa* groups occur mainly in low productivity pastures and heath communities. The *Cynosurus cristatus* group, on the other hand, has its main habitats in semi-moist, extensively used grasslands, but also tolerates some intensification.

Natural habitats of these species are supposedly flood plains (in the case of *Veronica serpyllifolia*, *Leontodon saxatilis* and *Carex leporina*), stream valley forests (*Silaum silaus*), acidophilous forests (*Calluna vulgaris*), alpine vegetation types (*A. pubescens*, *S. officinalis* and *Succisa pratensis*) and oligo-, meso-trophic heath vegetation (*Erica tetralix* and *Briza media*). For some species, like *Genista anglica*, *Rhinanthus minor* and *Sieglingia decumbens*, no natural habitats could be deduced from this database.

4.4. Priority species in forests

These species have demonstrated a preference for wet forests (vegetation groups Vgs 12–15, Table 6). Within the *Ranunculus auricomus* group only *Crepis paludosus* and *Scutellaria galericulata* occur frequently in wet grasslands. The *Molinia caerulea* group has its main habitats in mesotrophic alder forests within the successional line on fen soils. It is also, therefore, indicative of eutrophication in grasslands. The *Deschampsia cespitosa* group, however, also occurs in intensive grasslands. *Deschampsia* itself is a highly tolerant species. It strongly profits from decreasing management following grazing on semi-moist soils with fluctuating water tables (Rosenthal, 1992).

5. Evaluation of restoration experiments

There are several applications of this database in evaluating restoration experiments. First, threshold values for “target communities” on the type-level shall be defined. Here the proportion of target species with high vulnerability values (1–4) is used as a parameter to evaluate different vegetation groups within the successional fen series. The highest proportion (51%), realised in mesotrophic wet grasslands (vegetation group Vg 4), is treated as a maximum standard for calculating the coefficient E:

$$E = \text{number of target species with vulnerability values (1–4) total species number} \times 0.51$$

The lowest E-coefficients are characteristic of intensively used semi-moist grasslands (vegetation group Vg 1). Within the successional fen series (Fig. 3), E increases until reaching its maximum in vegetation group Vg 4. This vegetation group provides the most

Table 5
Distribution of priority target species in other extensively used grassland systems (vegetation groups Vgs 12–15, Table 1; for further explanation see Table 3)

Decreasing tendency	Red data book Germany	Red data book Ni/Br		Vegetation groups: Intensive grasslands		Extensive wet grasslands		Fallows				Wet forests			Other extensive grasslands				Vulner-ability value	Plant formations: Non-forest systems			Fresh water	Forests	
				1, sem, eu	2, moi, eu	3, moi, eu	4, m/w, me	5, wet, eu	6, wet, me	7, moi, eu	8, moi, me	9, m/w, eu	10, sem, eu	11, m/w, me	12, m/w, eu	13, sem, eu/me	14, sem, eu	15, sem, me		A	B	C	D	E	F
3		3	*Thalictrum flavum*	4	1	2	1	+	4	2	3	–	–	–	**29**	5	+	+	3	!		×			
2	3	2	*Lathyrus palustris*	–	–	+	1	+	+	3	7	–	–	–	**12**	+	–	–	3	!!		×			
5			*Stellaria uliginosa*	+	+	2	2	+	1	2	9	5	2	6	**49**	+	2	–	4	!	!!	×	×	×	
3		3	*Sanguisorba officinalis*	4	1	–	3	+	5	–	–	–	–	–	–	**14**	2	–	1	!	!!	!			
5		(3)	*Centaurea jacea*	4	1	2	+	–	–	+	–	–	–	–	–	**30**	13	29	3	!		×			
4			*Agrostis gigantea*	4	2	7	8	–	–	–	–	–	–	–	–	**10**	5	+	3	!	×				
3		2	*Silaum silaus*	–	2	1	–	–	–	–	–	–	–	–	–	**18**	4	1	4	!					×
3		2	*Avenochloa pubescens*	–	–	1	6	–	1	–	7	–	–	–	–	**18**	3	16	1	!!	!!	!!			
3		2	*Briza media*	–	–	–	3	–	–	–	–	–	–	–	–	**13**	3	12	1	!!	!!				
4		2	*Leontodon hispidus*	–	–	–	+	–	–	–	–	–	–	–	–	**7**	4	5	1	!!					
4		3	*Alchemilla vulgaris*	–	–	1	+	–	–	2	3	–	–	–	–	**4**	–	+	3	!!		×			
4		1	*Colchicum autumnale*	–	–	1	–	–	–	–	–	–	–	–	–	**11**	3	+	4	!!				!	×
4		(3)	*Cynosurus cristatus*	4	–	+	9	–	–	–	–	–	–	–	–	2	**45**	33	5	!					
4			*Veronica serpyllifolia*	13	6	4	5	–	–	–	–	–	–	–	–	9	**23**	14	6	×	×				
4			*Carex leporina*	4	7	4	13	–	–	–	–	–	1	2	–	4	18	18	6	×	×		×	×	
5			*Leontodon autumnalis*	39	29	22	20	–	–	2	–	–	–	–	53	34	**55**	36	6	×	×	×			
4		3	*Succisa pratensis*	–	–	–	11	–	2	–	–	–	–	–	–	4	1	**22**	2	!!	!!	!!		!	
4			*Luzula campestris*	–	–	–	2	–	+	1	4	–	1	–	–	11	9	**46**	1	!!		!			
4		3	*Rhinanthus minor*	–	–	–	2	–	–	–	–	–	–	–	–	2	4	**8**	1	!!					
4			*Leontodon saxatilis*	–	–	–	1	–	–	–	–	–	–	–	–	5	7	**23**	3	!!	×				
5			*Trifolium dubium*	–	4	–	6	–	–	–	–	–	–	–	–	16	9	**23**	1	!					
4			*Sieglingia decumbens*	–	–	–	+	–	–	–	–	–	–	–	–	1	–	**24**	1	!!					
3		(3)	*Nardus stricta*	–	–	–	3	–	–	–	–	–	–	–	–	–	1	**16**	1	!!	!!				
3			*Calluna vulgaris*	–	–	–	+	–	–	–	–	–	–	–	–	–	–	**16**	4	!!	!!				×
3			*Erica tetralix*	–	–	–	1	–	–	–	–	–	–	+	–	–	–	**6**	4	!!	!!			!!	×
3		3	*Genista anglica*	–	–	–	2	–	–	–	–	–	–	–	–	–	–	**4**	1	!!					
3		2	*Saxifraga granulata*	–	–	2	–	–	–	–	–	–	–	–	–	–	–	**2**	1	!!					
3		3	*Vaccinium oxycoccos*	–	–	–	+	–	–	–	–	–	–	–	–	–	–	**2**	1	!!	!!				

Table 6
Distribution of priority target species in wet forests (vegetation groups Vgs 9–11, Table 1; for further explanation see Table 3)

Decreasing tendency	Red data book Germany	Red data book Ni/Br		Vegetation groups															Vulner-ability value	Plant formations					
				Intensive grasslands		Extensive wet grasslands		Fallows				Wet forests			Other extensive grasslands					Non-forest systems			Fresh water	Forests	
				1, sem, eu	2, moi, eu	3, moi, eu	4, m/w, me	5, wet, eu	6, wet, me	7, moi, eu	8, moi, me	9, m/w, eu	10, sem, eu	11, m/w, me	12, m/w, eu	13, sem, eu/me	14, sem, eu	15, sem, me		A	B	C	D	E	F
5		(3)	*Ranunculus auricomus* agg.	5	–	4	2	–	–	1	6	**11**	1	–	9	12	4	–	4	!		!		!	×
4			*Scutellaria galericulata*	4	1	+	10	16	17	5	10	**32**	12	15	11	–	2	–	4	!		×		×	
4			*Crepis paludosa*	–	–	7	17	2	3	13	29	**28**	6	3	–	+	2	–	4	!!	!!	×	×	×	
4			*Cardamine amara*	–	–	4	1	1	+	6	5	**25**	5	6	–	–	–	–	3	!!		×	×	!	
5			*Stachys palustris*	–	–	2	3	1	+	2	3	**7**	3	+	–	–	–	–	4	!!		×		×	
5			*Valeriana officinalis*	–	–	2	2	5	6	7	6	**22**	4	–	–	1	–	–	3	!!		×		!	
4		3	*Primula elatior*	–	–	+	+	–	–	–	–	**7**	–	–	–	+	–	–	4	!!	!!			!!	×
6			*Deschampsia cespitosa*	25	9	6	22	1	2	8	13	66	**63**	29	6	13	26	14	6	×	!!	×		×	×
5			*Lysimachia vulgaris*	–	2	5	18	24	41	15	30	65	**53**	79	39	+	2	–	4	!!		×		×	
5			*Molinia caerulea*	–	–	–	5	–	6	+	–	3	17	**56**	–	6	–	6	4	!!	!!	!		×	×
4			*Carex canescens*	–	–	–	11	+	8	–	–	3	4	**44**	–	–	2	–	2	!!	!!	!		!	
3	3	3	*Myrica gale*	–	–	–	1	–	–	–	–	–	–	**6**	–	–	–	+	2	!!	!!			!!	

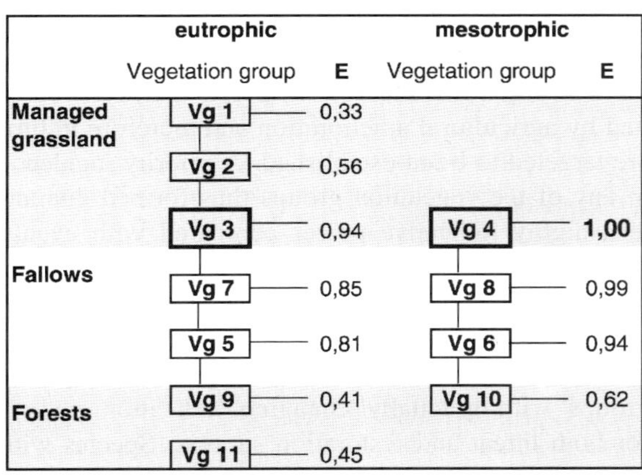

Fig. 3. Evaluation of vegetation groups (Vgs 1–11) within the successional fen series by coefficient *E*. *E* indicates the proportion of endangered species (with vulnerability values 1–4) from the total vascular plant species number. The highest value was standardised to 1 (=Vg 4).

suitable conditions for such species. However, the decline of *E* with a secondary succession towards fallow and forest indicates a decreasing proportion of endangered grassland species. Low values are found again in eutrophic wet forests (vegetation groups Vgs 9 and 11). The calculation of *E*-coefficients (Fig. 3) provides a framework on the type level for evaluating individual successional stages on the object level. They can be used to evaluate the success or failure of grassland restoration measures in an actual situation.

Results from a large assortment of experimental fen grassland restoration plots in NW Germany and The Netherlands indicate that, in general, there is a low invasion of target species with vulnerability values of 1–4 (Rosenthal, 2000). The balance of re-invasion and extinction of target species was much higher in species that regularly occur in intensively used grasslands (those with vulnerability values 5 and 6). This means that such species are preferential successful in-

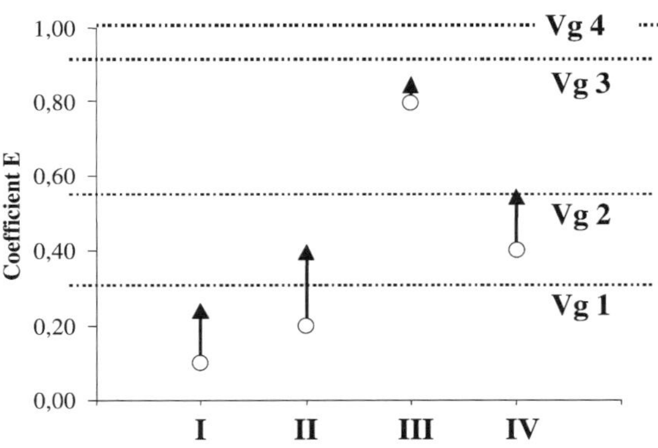

Fig. 4. Changing *E*-coefficients in different grassland communities (I–IV, see below) due to nutrient depletion by mowing (without fertilising). Circles: start of the experiment; triangles: end of the experiment. Dotted lines indicate threshold values for vegetation groups Vg 1 (intensively used semi-moist grasslands), Vg 2 (intensively used moist grasslands), Vg 3 (extensively used moist eutrophic grasslands) and Vg 4 (extensively used moist/wet mesotrophic grasslands); see Table 1. Data from 35 permanent plots in fen grasslands in NW Germany and The Netherlands (source: see Table 7).

vaders, whose habitat demands are met by intensively used grasslands, as well as in many other vegetation [illegible].

In Fig. 4 successional trends were evaluated individually for different plant communities following reduced management intensity (no fertiliser application and nutrient depletion due to mowing and removal of the hay). In general there are positive trends indicated by increasing *E*-coefficients. These increases, however, are small and there are some differences in detail. Initially species-poor communities (I, II and IV) show a stronger increase in target species than III which showed a higher diversity from the beginning. In II and IV the threshold values for intensively used semi-moist grasslands (Vg 1), and intensively used moist grasslands (Vg 2) respectively, are exceeded (Table 7).

Table 7
Balance of invasion and extinction rates of target species from fen grassland restoration experiments in NW Germany and The Netherlands (71 permanent plots studied over different time periods ranging from 2 to 19 years; Bakker, 1989; Schwartze, 1992; Rosenthal, 1992; Michels, 1993; Hellberg, 1995; Janhoff, 1996; Kundel, 1998)

	Vulnerability value (see Fig. 2)					
	1	2	3	4	5	6
Balance of target species invasion and extinction (% of species numbers in that category)	+4.5	+9.9	+11.5	+6.9	+32.0	+43.5

6. Discussion

It is proposed that targets be defined for the restoration of disturbed or destroyed ecosystems that match the fundamentals of environmental policies (e.g. Halbritter, 1994). Here sets of abiotic conditions are deduced for the recovery of wet grassland communities on fen soils and translated into calibrated mean indicator values for plant communities (calculation according to Ellenberg et al. (1992)). Threshold values (e.g. for groundwater levels and mean moisture-values, respectively) offer the opportunity to select "target systems" from potentially achievable ecosystems that are expected within a specific region and on a specific habitat type. The large data set used is highly representative of the variety of different conditions which are met in NW German fens (4467 relevees from 228 different plant communities). The "target systems" form part of the successional time series of this habitat type (either progressive or regressive) and of spatial complexes of different habitats in the landscape. Temporal series are modelled using the technique of "time by site substitution" (van der Maarel, 1996).

"Target systems" themselves, however, are not suitable tools for measuring restoration success: new sets of environmental conditions result in new communities that can hardly be predicted. Indeed, restoration experiments on different habitat types show that species assemblages do not reach the state of "typical" "target plant associations" even after many years (Kapfer, 1988; Bakker, 1989; Schwartze, 1992; Rosenthal, 1992, 2000; Kiefer, 1998; Kundel, 1998).

Therefore plant species that can be considered as the fundamental elements constituting plant communities are used as indicators for the evaluation of restoration success (Reck, 1996; Jensen and Schrautzer, 1999; Bakker et al., 2000). Here the question arises, which if not all, species may be suitable? Red data book species are unlikely to be expected in restoration experiments if they are rare in the vicinity (as is mostly the case, compare Gigon et al. (1998)). On the other hand, species should not be too common because their indication power is then low.

The proposed selection method starts with the whole species pool of the "target systems". As a first selection criteria their decreasing rates in Germany during recent decades (according to Haeupler and Schönfelder (1988) and Ellenberg et al. (1992)) are used to allow an evaluation of species from a priority point of view. This focuses on species that were inhibited by agricultural amelioration and therefore in turn are targeted to be re-established. A priority incidence in any of the vegetation groups thus formed guarantees highly indicative power compared with evenly distributed species. The allocation of different vulnerability values to selected species is based on different niche breadths and temporal persistence during secondary succession. This latter step creates species groups with gradually changing indication powers for both threat and restoration success. Species with vulnerability values 1–4 are strongly threatened by intensification of grassland management. Those with vulnerability values 1 and 3 are additionally threatened by abandonment and a progressive succession towards forests. Some are even obligatorily restricted to managed grasslands and fail within reed and tall sedge communities.

A brief evaluation of restoration experiments demonstrates that the re-invasion rate for target species is generally low. There does exist, however, a marked difference between species with vulnerability values 1–4 and 5–6. It has been proven that higher frequencies in intensive grasslands (species with vulnerability values 5 and 6) enhance the re-invasion into recovered grassland swards. Low frequencies, however, limit the invasion, even when restoration measures have been performed (compare Zobel et al. (1998)). Considering this low re-colonisation rate it has to be a priority concept to protect those relic habitats still existing, especially remnants of small sedge, wet grassland, stream valley, heath and nutrient poor pasture communities (Dierßen and Schrautzer, 1997). This can be realised by maintaining abiotic conditions (high water table, low nutrient status) and the traditional, moderately intensive management regime. The re-invasion of those species during restoration experiments in degraded plant communities, in turn, strongly indicates restoration success (in the sense of the recovery of the systems).

The restoration is much more successful when the traditional mowing regime becomes re-established in fallow communities which had been developed after abandoning the land use. The decrease of predominating tall-growing reed species results in a decrease of light competition. Some meadow target species such as *Myosotis palustris* and *L. uliginosus* which were out-competed during abandonment are capable of re-establishing from their persistent seed banks even after some decades of abandonment. Grassland species with transient seed banks however were lost (e.g. *S. aquaticus* and *V. dioica*) (Rosenthal, 2000).

The relevance of the proposed vulnerability values for evaluating restoration processes can be considered if one compares them with the rate of species decline defined by Ellenberg et al. (1992). There is a marked correlation between vulnerability values and the strength of their decline. The broader is the niche and the longer is the persistence within successional fen series, the less has been that decline in recent decades (Rosenthal, 2000). Species with a narrow niche and short persistence during succession were, however, strongly reduced. For restoration success this obviously still plays an important role; only species that are distributed frequently and which survive on refuge habitats are capable of recovery in a short time.

The proposed method for selecting target plant species is also applicable to other systems and other targets. Target species selected in this paper refer only to wet grassland systems on fen soils in NW Germany (paludified fens and inundated fens). They are partly comparable with species sets proposed by Jensen and Schrautzer (1999). These authors also stressed the threats of abandonment or further intensification for plant species and classified the risk of extinction. The system very clearly differentiates between these two reasons for extinction. For applying a target species system to restoration efforts it seems, however, more convenient to integrate the information in the condensed way discussed here. A set of indicators with a broad variety of different habitat demands guarantees that the restoration process is measurable and comparable for different sites and time periods.

7. Conclusions

The recovery of the habitat function of wet meadows for endangered plant species can be achieved by reversing the processes which caused the destruction of these systems. Long term investigations however show that there exist a lot of obstacles which are difficult to overcome such as high nutrient availability and slow depletion rates, irreversible loss of the peat structure and its water storage capacity, missing seed banks and limited seed dispersal of target species. Restoration needs the evaluation of restoration success.

Efficient evaluation measures which are capable of being transferred into nature conservation furthermore need evaluation criteria which evaluate the final outcome. This is the re-establishment of desired target species which is the most convincing argument for successful restoration.

In this paper a method is presented which allows a systematic deduction of those target species from priority restoration goals. These goals may either exist in terms of abiotic thresholds or desired target plant communities. Mean indicator values for plant species provide a suitable tool in "translating" abiotic guidelines into biotic parameters and vice versa. The proposed deduction method consists of several selection steps which reduce the large regional species pool in order to get a manageable number of target species. This target species set should be optimised in terms of its ability in indicating different levels of restoration success. Re-established species with a narrow habitat niche and limited temporal persistence during secondary succession indicate a high level of restoration success both in recovery of habitats and conditions for re-invasion and re-establishment of target species. Species with a broader niche and/or high persistence during succession on the other side indicate a lower level of success.

The re-invasion of species however is often inhibited by missing soil seed banks and/or limited dispersal of propagules from relic populations or simply the absence of those populations in close vicinity. Rare species are often too rare to be expected soon after restoration measures have been applied, even if the habitat conditions already fit. Restoration projects often suffer from those hysteresis effects. The indication of abiotic habitat conditions by plant species is then only partly possible. Restoration success however finally depends on both recovered abiotic habitat conditions and successful re-invasion of targeted species. An integrating evaluation approach by target species as it is proposed here evaluates the overall result and not single processes. This however could be a further research aim, to analyse the most limiting factors of unsuccessful restoration.

The target species set is only valid for the proposed restoration goal (wet grasslands) and for inundated and paludified fens in NW Germany. The procedure could however be applied to select comparable species sets for other fen types, ecosystems, regions or restoration goals. This would require corresponding data sets which must reflect the potential species pool that could be expected in realising that goal.

Acknowledgements

I would like to thank Dr. Peter Schwartze (Steinfurt), Dieter Janhoff, Eike Frese, Ele Lienkamp and Dr. Burghard Wittig (Bremen) and Dr. Ute Döring-Mederake (Göttingen) for providing vegetation relevees and ground water data. Many thanks to Wendy Bell (Adelaide) and Dr. Mark Morgan (Wales) who helped me to improve my English. I have also to thank an anonymous reviewer for improving the text.

References

Bakker, J.P., 1989. Nature management by grazing and cutting. Geobotany 14.

Bakker, J.P., Grootjans, A.P., Hermy, M., Poschlod, P., 2000. How to define targets for ecological restoration?—Introduction. Appl. Veg. Sci. 3 (1), 1–6.

Bollens, U., Güsewell, S., Klötzli, F., 1998. Zur relativen Bedeutung von Nährstoffeintrag und Wasserstand für die Biodiversität in Streuwiesen. Bull. Geobot. Inst. ETH 64, 91–101.

Briemle, G., Eickhoff. D., Wolf, R., 1991. Mindestpflege und Mindestnutzung unterschiedlicher Grünlandtypen aus landschaftsökologischer und landeskultureller Sicht. Veröffentlichungen für Naturschutz und Landschaftspflege in Baden-Württemberg, Supplement 6, 160 pp.

Dierschke, H., Wittig, B., 1991. Die Vegetation des Holtumer Moores (Nordwest-Deutschland). Veränderungein in 25 Jahren (1963–1988). Tuexenia 11, 171–190.

Dierßen, K., 1998. Zerstörung von Mooren und Rückgang von Moorpflanzen. In: Bundesamt für Naturschutz (Ed.), Ursache des Artenrückgangs von Wildpflanzen und Möglichkeiten zur Erhaltung der Artenvielfalt, pp. 229–240.

Dierßen, K., Schrautzer, J., 1997. Wie sinnvoll ist ein Rückzug der Landwirtschaft aus der Fläche? In: Bayerische Akademie für Naturschutz und Landschaftspflege (Ed.), Laufener Seminarbeiträge 1, pp. 93–104.

Döring-Mederake, U., 1991. Feuchtwälder im nordwestdeutschen Tiefland. Scripta Geobot. 19, 122.

Ehrendorfer, F., 1973. Liste der Gefäßpflanzen Mitteleuropas, 2 ed., Fischer Verlag Stuttgart.

Ellenberg, H., 1956. Aufgaben und Methoden der Vegetationsgliederung. Ulmer Verlag, Stuttgart.

Ellenberg Jr., H., 1989. Eutrophierung—das gravierendste problem im Naturschutz. NNA—Berichte 2 (1), 4–13.

Ellenberg, H., Weber, H.E., Düll, R., Wirth, V., Werner, W., Paulißen, D., 1992. Zeigerwerte von Pflanzen Mitteleuropas. Scripta Geobot. 18, 248.

Fürst, D., Kiemstedt, H., Scholles, F., Ratzbor, G., Gustedt, E., 1989. Umweltqualitätsziele für die ökologische Planung. In: Umweltbundesamt Berlin (Ed.), Forschungsbericht, 1989.

Garve, E., 1993. Rote Liste der gefährdeten Farn- und Blütenpflanzen in Niedersachsen und Bremen. Informationsdienst Naturschutz Niedersachsen 1, 1–37.

Garve, E., 1994. Atlas der gefährdeten Farn- und Blütenpflanzen in Niedersachsen und Bremen. Schriftenreihe für Naturschutz und Landschaftspflege in Niedersachsen 30 (1–2), Hannover.

Gigon, A., Leutert, A., 1996. The dynamic keyhole-key model of coexistence to explain diversity of plants in limestone and other grasslands. J. Veg. Sci. 7, 29–40.

Gigon, A., Langenauer, R., Meier, C., Niervergelt, B., 1998. Blaue Listen der erfolgreich erhaltenen oder geförderten Tier- und Pflanzenarten der Roten Listen, vol. 129. Veröffentlichungen des Geobotanischen Institutes der ETH, Stiftung Rübel, Zürich, pp. 3–137.

Grime, J.P., 1979. Plant Strategies and Vegetation Processes. Wiley, New York, 222 pp.

Haeupler, H., Schönfelder, P., 1988. Atlas der Farn- und Blütenpflanzen der Bundesrepublik Deutschland. Ulmer Verlag, Stuttgart, 768 pp.

Halbritter, G., 1994. Möglichkeiten der Umsetzung des Leitbildes einer dauerhaft-umweltgerechten Entwicklung in die praktische Umweltpolitik. Laufener Seminarbeiträge 4, 25–38.

Harter, A., Luthardt, V., 1997. Revitalisierungsversuche in zwei degradierten Niedermooren in Brandenburg—Eine Fallstudie zur Reaktion von Boden und Vegetation auf Wiedervernässung. Telma 27, 147–169.

Hellberg, F., 1995. Entwicklung der Grünlandvegetation bei Wiedervernässung und periodischer Überflutung. Dissertationes Bot. 243, 271.

Hennings, H., 1996. Zur Wiedervernäßbarkeit von Niedermoorböden. Dissertation. University of Bremen, 157 pp.

Janhoff, D., 1996. Dauerquadrate zur Untersuchung der Sukzession von Vegetation und Nährstoffverhältnissen im NSG Borgfelder Wümmewiesen. Gutachten im Auftrag der Umweltstiftung WWF Deutschland, Bremen, 54 pp.

Jensen, K., Schrautzer, J., 1999. Consequences of abandonment for a regional fen flora and mechanisms of successional change. Appl. Veg. Sci. 2, 79–88.

Kapfer, A., 1988. Versuche zur Renaturierung gedüngten Feuchtgrünlandes—Aushagerung und Vegetationsentwicklung. Dissertationes Bot. 120, 144.

Kaule, G., 1991. Arteh- und Biotopschutz. Stuttgart UTB–Eugen Ulmer Verlag, 2. erweiterte Auflage, 519 pp.

Keddy, P.A., 1992. A pragmatic approach to functional ecology. Funct. Ecol. 6, 621–626.

Kiefer, S., 1998. Wiederherstellung brachgefallener oder aufgeforsteter Kalkmagerrasen. Berichte des Institutes für Landschafts- und Pflanzenökologie der Universität Hohenheim, Beih., vol. 7, 309 pp.

Kleyer, M., 1997. Vergleichende Untersuchungen zur Ökologie von Pflanzengemeinschaften. Dissertationes Bot. 286, 201.

Korneck, D., Sukopp, H., 1988. Rote Liste der in der Bundesrepublik Deutschland ausgestorbenen, verschollenen und gefährdeten Farn- und Blütenpflanzen und ihre Auswertung für den Arten- und Biotopschutz. Schriften-Reihe für Vegetationskunde 19, 210.

Korneck, D., Schnittler, M., Vollmer, I., 1996. Rote Liste der Farn- und Blütenpflanzen (Pteridophyta et Spermatophyta) Deutschlands. Schriftenreihe für Vegetationskunde 28, 21–187.

Korneck, D., Schnittler, M., Klingenstein, F., Ludwig, G., Takla, M., Bohn, U., May, R., 1998. Warum verarmt unsere Flora? Auswertung der Roten Liste der Farn- und Blütenpflanzen Deutschlands. Schriftenreihe für Vegetationskunde, Heft 29, Bundesamt für Naturschutz (Ed.), Ursache des Artenrückgangs von Wildpflanzen und Möglichkeiten zur Erhaltung der Artenvielfalt, 299–350.

Kundel, W., 1998. Untersuchungen an Dauerbeobachtungsflächen im Grünland von Ausgleichsflächen des südlichen Niedervielandes im Zeitraum von 1986–1996. In: Landschaftsökologische Forschungsstelle Bremen (Ed.), Gutachten im Auftrag des Senators f. Umweltschutz, Bremen, 230 pp.

Lenski, H., 1953. Grünlanduntersuchungen im mittleren Ostetal. Mitteilungen der Floristisch-Soziologischen Arbeitsgemeinschaft, N.F. 4, 26–58.

Lutz, S., 1996. Soziologisches, ökologisches und populationsbiologisches Verhalten von *B. racemosus* L. im Bremer Raum. Diplomwork. University of Bremen, 83 pp.

Matthies, D., 1991. Räumliche und zeitliche Dynamik in Populationen der seltenen Art Melampyrum arvense L. In: Schmid, B., Stöcklin, J. (Eds.), Populationsbiologie der Pflanzen. Birkhauser, Basel, pp. 109–122.

Meisel, K., 1977. Die Grünlandvegetation nordwestdeutscher Flusstäler und die Eignung der von ihr besiedelten Standorte für einige wesentliche Nutzungsansprüche. Schriftenreihe für Vegetationskunde 11, 121.

Meisel, K., Hübschmann, A. von, 1976. Veränderungen der Acker- und Grünlandvegetation im nordwestdeutschen Flachland in jüngerer Zeit. Schriften-Reihe für Vegetationskunde 10, 109–124.

Michels, C., 1993. Grünlandextensivierung im Feuchtgebiet Saerbeck. Ergebnisse einer vegetationskundlichen Dauerflächenuntersuchung im Rahmen einer Effizienzkontrolle zum Feuchtwiesenschutzprogramm. LÖLF-Mitteilungen 18, 51–55.

Müller, J., Rosenthal, G., 1998. Brachesukzessionen—Prozesse und Mechanismen. Ber. Inst. Landschafts- u. Pflanzenökologie Universität Hohenheim 5, 103–132.

Oomes, M.J.M., 1992. Yield and species density of grasslands during restoration management. J. Veg. Sci. 3, 1–4.

Pfadenhauer, J., Kratz, R. (Eds.), 2001. Ökosystemmanagement für Niedermoore: Strategien und Verfahren zur Renaturierung. Ulmer, Stuttgart, 317 pp.

Pirkl, A., Riedel, B., 1991. Arten- und Biotopschutz für Deutschland. In: Henle, K., Kaule, G. (Eds.), Berichte aus der Ökologischen Forschung, vol. 4, pp. 343–346.

Plachter, H., 1994. Methodische Rahmenbedingungen für synoptische Bewertungsverfahren im Naturschutz. Ökologie und Naturschutz 3, 87–106.

Poschlod, P., Kiefer, S., Jackel, A.-K., Fischer, S., 1999. Populationbiologische Untersuchungen an Pflanzen der Trockenrasen—ein zönosenbezogener Ansatz der Gefährdung durch Fragmentierung und Isolation. In: Amler, K., Bahl, A., Henle, K., Kaule, G., Poschlod, P., Settele, J. (Eds.), Populationsbiologie in der Naturschutzpraxis, pp. 78–92.

Preising, E., 1997. Die Pflanzengesellschaften Niedersachsens—Rasen-, Fels- und Geröllgesellschaften. Naturschutz und Landschaftspflege in Niedersachsen 20/5, 148 pp.

Reck, H., 1996. Flächenbewertung für die Belange des Arten- und Biotopschutzes. In: Umweltakademie Baden Württemberg (Ed.), Bewertung im Naturschutz, pp. 71–112.

Rosenthal, G., 1992. Erhaltung und Regeneration von Feuchtwiesen—Vegetationsökologische Untersuchungen auf Dauerflächen. Dissertationes Bot. 182, 283.

Rosenthal, G., 2000. Zielkonzeptionen und Erfolgsbewertung von Renaturierungsversuchen in nordwestdeutschen Niedermooren anhand vegetationskundlicher und ökologischer Kriterien. Habilitation. University of Stuttgart, 230 pp.

Rosenthal, G., Fink, S., 1996. *Pedicularis palustris* L. im Bremer Gebiet—Verbreitung, Ökologie und Rückgangsursachen. Abhandlungen des Naturwissenschaftlichen Vereins Bremen 43 (2), 429–447.

Rosenthal, G., Müller, J., 1988. Wandel der Grünlandvegetation im mittleren Ostetal—Ein Vergleich. Tuexenia 8, 79–99.

Rosenthal, G., Hildebrandt, J., Zöckler, C., Hengstenberg, M., Mossakowski, D., Lakomy, W., Burfeindt, I., 1998. Feuchtgrünland in Norddeutschland—Ökologie, Zustand, Schutzkonzepte. In: Bundesamt für Naturschutz (Ed.), Angew. Landschaftsökolgie 15. Bad Godesberg, Bonn, 336 pp.

Schrautzer, J., 1988. Pflanzensoziologische und standörtliche Charakteristik von Seggenriedern und Feuchtwiesen in Schleswig-Holstein. Mitteilungen der Arbeitsgemeinschaft Geobotanik in Schleswig-Holstein und Hamburg 38, 189.

Schwaar, J., 1990. Natur und Vergangenheit—Bremen und sein Umland in den letzten 12.000 Jahren. Abhandlungen des Naturwissenschaftlichen Vereins Bremen 41 (2), 49–86.

Schwartze, P., 1992. Nordwestdeutsche Feuchtgrünlandgesellschaften unter kontrollierten Nutzungsbedingungen. Dissertationes Bot. 183, 204.

Speier, M., 1996. Paläoökologische Aspekte der Entstehung von Grünland in Mitteleuropa. Berichte der Reinhold-Tüxen-Gesellschaft 8, 199–219.

Succow, M., Joosten, H. (Eds.), 2001. Landschaftsökologische Moorkunde. Schweizerbart'sche Verlagsbuchhandlung, Stuttgart, 622 pp.

van der Maarel, E., 1996. Pattern and process in the plant community: Fifty years after A.S. Watt. J. Veg. Sci. 7, 19–28.

Van der Valk, A.G., 1981. Succession in wetlands: a Gleasonian approach. Ecology 62 (3), 688–696.

Vermeer, J.G., Berendse, F., 1983. The relationships between nutrient availability, shoot biomass and species richness in grassland and wetland communities. Vegetatio 53, 121–126.

Walter, R., Reck, H., Kaule, G., Lämmle, M., Osinski, E., Heinl, T., 1998. Regionalisierte Qualitätsziele, Standarts und Indikatoren für die Belange des Arten- und Biotopschutzes in Baden-Württemberg—Das Zielartenkonzept—ein Beitrag zum

Landschaftsrahmenprogramm des Landes Baden-Württemberg. Natur und Landschaft 73 (1), 9–25.

Zacharias, D., Janßen, C., Brandes, D., 1988. Basenreiche Pfeifengras-Streuewiesen des Molinietum caeruleae W.Koch, ihre Brachestadien und ihre wichtigsten Kontaktgesellschaften in Südost-Niedersachsen. Tuexenia 8, 55–78.

Zeitz, J., 2000. Befahrbarkeit von Niedermooren in Abhängigkeit von der Nutzungsintensität. http://www.agrar.hu-Berlin.de/pflanzenbau/tip/pages/h3online.

Zobel, M., van der Maarel, E., Dupre, C., 1998. Species pool: the concept, its determination and significance for community restoration. Appl. Veg. Sci. 1, 55–66.

ELSEVIER

Available online at www.sciencedirect.com

Agriculture, Ecosystems and Environment 98 (2003) 247–254

Agriculture Ecosystems & Environment

www.elsevier.com/locate/agee

Development and control of weeds in arable farming systems

B. Gerowitt*

Research Centre for Agriculture and the Environment, University of Göttingen, Am Vogelsang 6, D-37075 Göttingen, Germany

Abstract

It is investigated, whether the arable weed vegetation can perform as a characteristic to stimulate sustainable development of agriculture in Central Europe. The concept of sustainable development represents a dynamic process, influenced by ecological, economic and social aspects. Characteristics are required to integrate desired ecological aspects into farming concepts, in order to stimulate production systems which add to a sustainable development of the sector. The investigations rely on data of an arable farming system experiment (INTEX). Three farming systems called good-farming-practice (GFP), integrated-flexible and integrated-non-inversion were established from 1994 to 1998 on two sites. The systems differed in crop rotation, soil tillage, fertilisation and pesticide use. Thus, weed management comprised cultural or indirect control by plant husbandry and direct mechanical and chemical control. A residual weed vegetation after terminating all direct control in the arable crops was ex-ante intended in the integrated systems.

The spring weed densities were observed in winter wheat previous to any direct control treatments. The residual weed vegetation was investigated in all crops by monitoring species numbers and total ground cover after terminating all direct control.

The direct chemical control intensity was considerably lower in the integrated farming system, in which the crop rotation was extended and the soil was ploughed annually. Spring densities of problematic weeds like annual grass weeds and *Galium aparine* in the winter wheat crop were not higher in these systems.

Higher covers of the residual weed vegetation occurred more often in the productive arable crops of the two integrated farming systems, which additionally hosted higher number of species after terminating direct control. The different system strategies, including a higher input of herbicides in the system integrated-non-inversion, averted severe problems, including those arising from the dominating species being generally problematic in arable land use.

Keywords: Arable farming systems; Weed vegetation; Weed control

1. Introduction

Both, the political and the public discussion has been stimulated by the paradigm of sustainable development first presented by the Brundtland Commission (WCED, 1987). Agricultural development is part of this discussion (Becker, 1997). Without repeating details of the vast discussion about sustainable development, here two principles are accepted in general:

- Sustainable development is a dynamic process, stirred by economical, social and ecological aspects.
- Sustainable development is an overall concept. In order to come to subsidiary operational concepts it must be adapted to concrete situations and applications.

In Central Europe the ecological aspects for sustainable development of agriculture are to reduce the

* Tel.: +49-551-395538.
E-mail address: bgerowil@gwdg.de (B. Gerowitt).

0167-8809/$ – see front matter © 2003 Elsevier Science B.V. All rights reserved.
doi:10.1016/S0167-8809(03)00084-7

outputs with environmental relevance, e.g. leaching of nitrate, emission of greenhouse gases, erosion of fertile soil on the one hand and to maintain the biodiversity within the cultural landscape on the other (Gerowitt et al., 2000a). The ways to include these goals in sustainable development of agriculture are to negotiate with social and economic goals of the farming and the non-farming community.

The process requires characteristics, that are usable to stimulate desired transformations. The results of the process than need to be observed. Thus, both actions require indicators to ex-ante induce and to ex-post survey the transformation.

Different levels of actions can be distinguished for the processes, e.g. the global, the national, the regional, the farm and the field level. Different methods of inducing the desired development are adequate for these levels. To assess indicators on the basic level of fields in an arable farming system is attempted in this paper.

The spontaneous vegetation of arable fields is the exclusive contribution of the arable site itself to plant biodiversity. However, controlling weeds is one of the definite prerequisites for any successful arable production. Farmers can use various instruments, being either long-term regulative (e.g. crop rotation, tillage system), influencing the short-term competition of the arable crop (e.g. nitrogen fertilisation) or interfering directly (mechanical and chemical control). Almost all of the outlined actions touch the aim of reducing the environmentally relevant outputs of arable farming.

Thus, in arable farming systems weeds and their control could be an indicator to stimulate and survey changes in farming practice. The goal could be ex-ante integrated in weed management decisions, while an ex-post survey would concentrate on the existing weed vegetation. The way of obtaining such a weed vegetation, developed in terms of quality or quantity, is crucial for the concept. The questions on how this goal could be integrated into the production methods and if this fits into the general ideas of sustainable development of arable farming in Central Europe are important aspects together with data on such weed vegetation.

Results of the Göttingen INTEX project – an arable farming system project – are used to experimentally investigate these aspects.

2. Material and methods

The observations were embedded into the arable farming system Project INTEX in Göttingen, located in the northern German Federal State Lower Saxony.

The experimental fields were set up at two sites: Reinshof, 10 km south of Göttingen representing favourite arable conditions on flat fluvial soils, while Marienstein is sited in the hilly area 10 km north of Göttingen with heterogeneous heavy clayey and loamy soils. The farming system called good-farming-practice (GFP) represented intensive plant production in strict keeping with the German legal framework: fertiliser application and the use of pesticides are practised in accordance to official standards. A three-course crop rotation with oil seed rape (OSR) followed by winter wheat and winter barley was set up (Table 1).

Table 1
Overview of the farming systems during the experimental period 1995–1998, all crops were planted in all years

	Farming system		
	GFP	Integrated-flexible	Integrated-non-inversion
Soil cultivation	Plough	Plough (optional)	Chisel plough
Rotation	OSR, winter wheat, winter barley	OSR, cover crop[a], oats, winter wheat, spontaneous fallow	OSR, cover crop[a], oats, winter wheat, spontaneous fallow
N-fertilisation	Good practice	Balanced	Balanced
Plant protection	Good practice, using herbicides, fungicides, insecticides and growth regulators	Resistant varieties, mechanical weed control, pesticide use according to action thresholds, no growth regulators	Resistant varieties, pesticide use according to action thresholds, no growth regulators
Additional		3 m boundary strips	3 m boundary strips

[a] Volunteer OSR.

Since autumn 1994 two integrated systems called integrated-flexible and integrated-non-inversion were established. The crop rotation was changed so that the OSR was now followed by a spring-sown crop (oats). Winter wheat and a set aside period (with spontaneous vegetation) followed. The prevention of autumn losses of nitrogen is the key factor for this type of crop rotation. Healthy crops were established by choosing robust and resistant varieties and by adjusting the sowing dates in order to reduce the amount of agri-chemical applications in comparison to GFP. This framework for an integrated arable system was the basis for the establishment of the two subtypes: In the system integrated-flexible annual ploughing was deemed optional. Ploughing, in fact, was done in all 4 years. In the system integrated-non-inversion the soil was not ploughed.

Direct weed control was adapted to each system. In GFP decisions were applied according to the regional official advice, as associated with a well informed farmer. Accordingly, also reduced dosages for herbicides were considered. Weed control in the integrated systems accepted a certain amount of residual weeds, although long-term problems for arable production should be prevented. Detailed data on applied herbicides are given by Gerowitt et al. (2000b).

The INTEX-project relied on field-sized, long-term farming systems. All crops in the systems were planted each year in fields grouped according to the specific system rotation pattern. Single fields had a size between 1 and 3.7 ha. All agronomy treatments were carried out with normal-sized farming equipment (Steinmann and Gerowitt, 2000). Running since 1989, this paper refers to data of the INTEX research period from 1995 to 1998.

After terminating all weed control treatments, the residual weed vegetation was investigated in a vegetation survey on all fields in June–July. Three observation areas (10 m × 10 m) per field were permanently established during the experimental period (autumn 1994–harvest 1998). The total and the individual ground cover (in %) of all species was recorded. A simple factor ANOVA could be applied on the data of species numbers. Means were compared with the Tukey-test for equal sample sizes and the Scheffe-test for unequal sample sizes.

From autumn 1995 to 1998 three additional observation areas (10 m × 15 m) were set up in each field carrying a productive crop to investigate the spring weed densities before any direct control treatments were applied. Data on spring weed densities in the winter wheat crops, presented here, were observed in March–April in 10 random subplots of 0.1 m^2 per observation area. The spring weed density data of the winter wheat fields in 1995–1998 were compared using the Wilcoxon-test for summarised ranks.

In this paper, direct weed control treatments are characterised according to the part of the total weed vegetation they are focussing, being either monocotyledonous or dicotyledonous weeds. Because of their relevance for control two species are considered separately: *Galium aparine* L. and *Cirsium arvense* (L.) Scop. The relative frequency of direct chemical and mechanical treatments is calculated for each part (monocotyledonous weeds, dicotyledonous weeds or the single species *G. aparine* and *C. arvense*).

3. Results

3.1. Weed control

The control frequency of the weeds varied between the arable farming systems—being overall considerably lower in integrated-flexible than in GFP and integrated-non-inversion (Fig. 1). Mechanical treatments were used in the system integrated-flexible to control *G. aparine*, while no mechanical control was used in GFP, and only shopping of the set aside was applied in integrated-non-inversion. Grass weeds at the site Marienstein required incidental chemical control in integrated-flexible, but regular in GFP and in integrated-non-inversion. Additionally, in the latter system the chemical control frequency of *C. arvense* was increased compared to the other systems.

3.2. Spring weed densities

The spring weed densities in winter wheat differed between the systems (Fig. 2). At Reinshof grass weed densities were overall small and no significant difference occurred. Although more dicotyledonous weeds appeared in both integrated systems at Reinshof, the density of the species determining all control decisions, *G. aparine*, was not significantly different.

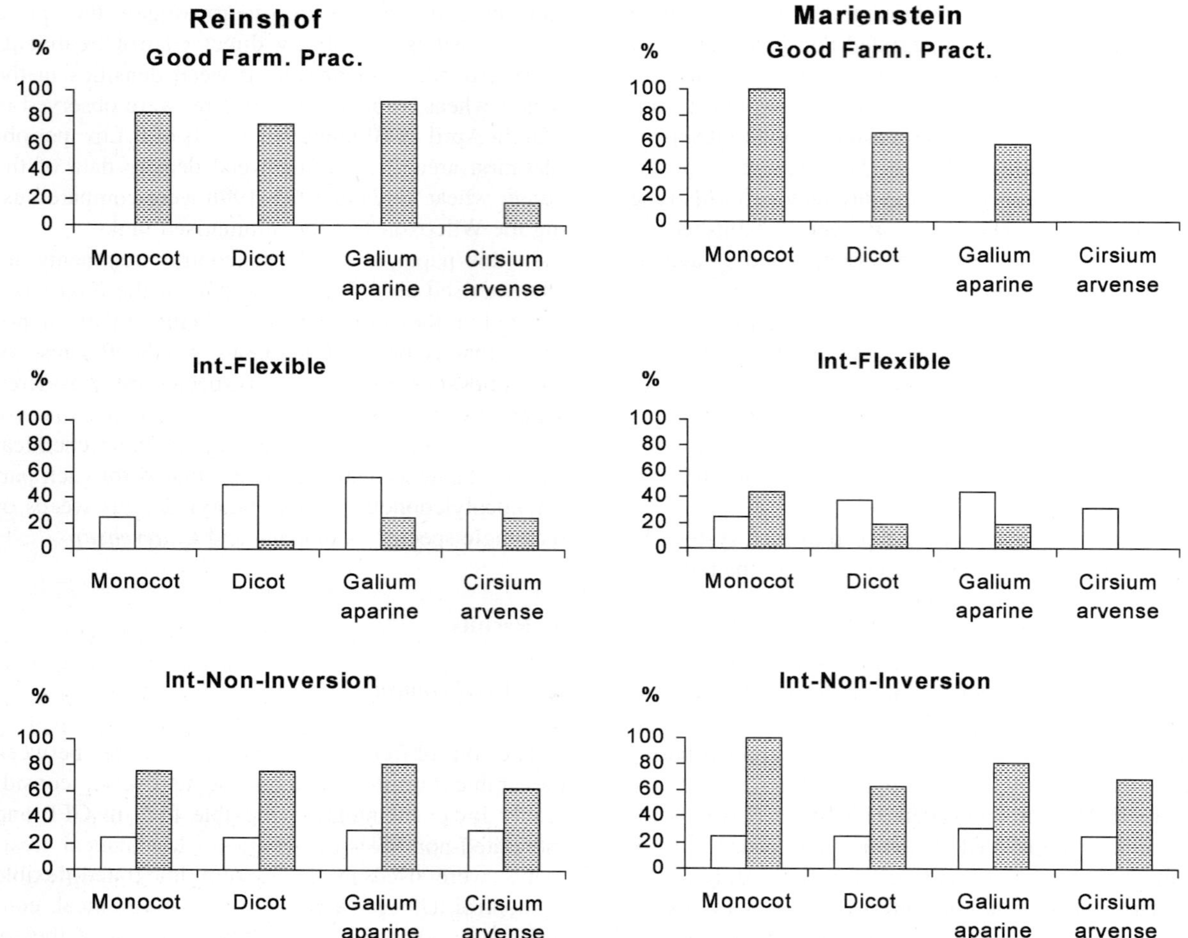

Fig. 1. Relative frequencies (%) of chemical and mechanical control of "monocotyledonous", "dicotyledonous", *G. aparine* and *C. arvense* in all crops 1995–1998, white bars = mechanical control, grey bars = chemical control (100% = always controlled, 25% mechanical control in the integrated systems by shopping the set-aside).

Grass weeds, which were mainly *Alopecurus myosuroides* Huds., had a major impact at Marienstein: In integrated-flexible their spring densities were effectively regulated by an expanded crop rotation (containing a spring-sown crop) and annual inversion tillage. A higher spring infestation with grass weeds in GFP indicates that, despite of the intensive chemical control in this system, the short rotation with only autumn sown crops favoured *A. myosuroides*. Dispensing with ploughing in integrated-non-inversion had the same effect on the spring densities of grass weeds.

3.3. Residual weed vegetation

Species numbers of the residual weed vegetation (when all direct weed control treatments were applied) in the productive crops OSR and cereals were higher in the integrated systems than in GFP (Fig. 3). In the integrated systems the rotational spontaneous set-aside additionally increased the mean number of species of the crop rotation. When the total number of species per farming system was used instead of the mean number, the ranking was the same, but on a higher level (Marienstein: GFP 37

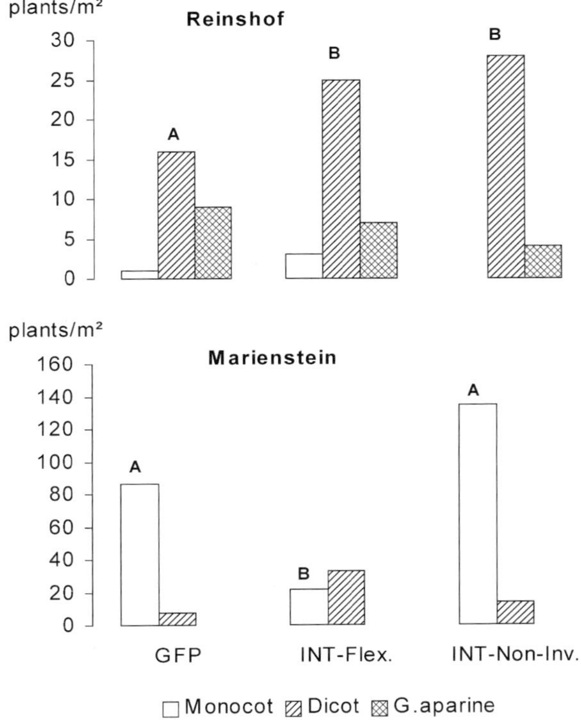

Fig. 2. Spring weed densities in the winter wheat crops of the farming systems, median 1996–1998, three observation areas per field (A, B = significant differences between the farming systems, Wilcoxon, $\alpha \leq 0.05$, $n = 9$).

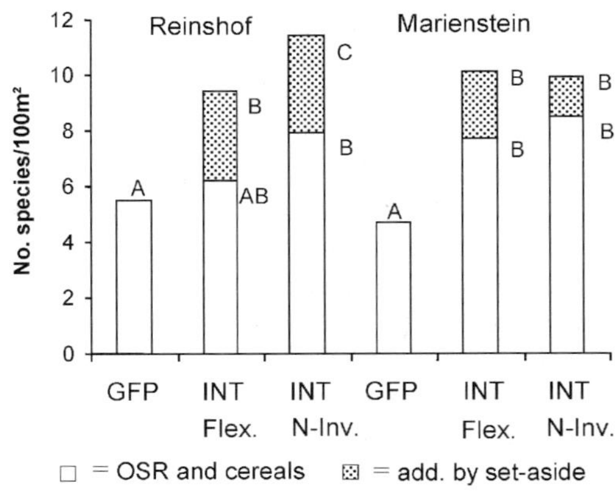

Fig. 3. Mean (arithmetic) number of weed species (without crops) – residual weed vegetation of the productive crops of the farming systems and the additional contribution of the set-aside fields in the integrated systems, 1995–1998.

species, integrated-non-inversion (set-aside included) 71 species, Reinshof: GFP 36 species, integrated-non-inversion (set-aside included) 73 species). The majority of all species were common segetal plant species (Gerowitt and Kirchner, 2000). Species dominating the residual weed cover were *C. arvense* and *G. aparine* (Reinshof) and *A. myosuroides* (Marienstein). With one exception of an evident increase due to farming practice (*Kickxia elatine* (L.) Dum. in Marienstein in integrated-non-inversion), rare arable weed species (listed by Garve (1993)) spontaneously occurred in all systems: *Centaurea cyanus* L., *Silene noctiflora* L., *Bromus arvensis* L. in Reinshof and *Consolida regalis* S.F. Gray, *Centaurea cyannus* L., *Myosurus minimus* L. in Marienstein. In Reinshof *Polygonum mite* Schrank and *Veronica agrestis* L. only emerged on the set-aside fields.

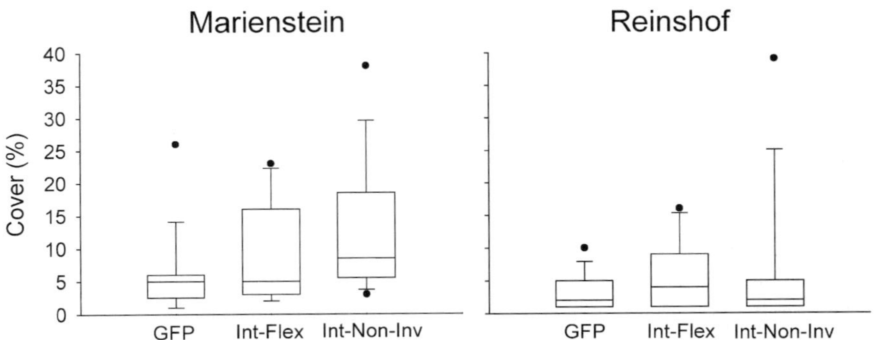

Fig. 4. Ground cover (%) of the residual weed vegetation in OSR/winter wheat/winter barley (farming system GFP), respectively, OSR/oats/winter wheat (integrated farming systems) – Box-and-Whisker plots, $n = 36$, 1995–1998.

Differences in the mean cover (median) of the residual weed vegetation were small, but the variation of the residual weed vegetation (plotted as Box-and-Whiskers in Fig. 4) indicates differences between the experimental sites and the farming systems. The cover of weeds surviving control tends to be higher at the heterogeneous heavy clay soils of Marienstein, whereas at Reinshof fields with almost no residual weed cover occurred in all farming systems. The deviation of the weed cover values above the median is always more extended in the integrated systems than in GFP. The set-aside fields in the integrated systems are excluded to obtain a balanced comparison of the residual vegetation in the productive crops. Including the cover of the spontaneous vegetation of the set-aside fields would shift the upper end of the distribution of the values in the integrated system up to 80%.

4. Discussion

This paper contributes to the question, whether the arable weed vegetation can be a characteristic to ex-ante stimulate and ex-post survey sustainable development of arable farming. Data of an arable farming system experiment enabled the investigation of various aspects of weeds and their control in case-studies with scientific and practical relevance.

In agricultural science, farming system approaches are used to fill the gap between disciplinary experimental studies and practical applications. Results of this type of experiments contribute to two demands on agricultural research: testing the results of analytical designs in more holistic approaches and being almost directly transferable into practice (Francis, 1994; Norman et al., 1994). Furthermore, the investigation of ecological and environmental effects of complete farming systems is an important motivation to use this experimental design (Edwards et al., 1993). To characterise farming system experiments, Vereijken (1992) introduced the term of experiments to synthesise whole systems rather than to analyse single factors. Generally, farming systems experiments are set up with not replicated whole fields as major experimental units combined with false replicates of observations in the fields (Norman et al., 1994). False replicates can be a few random, either durable or movable observation areas as used here. However, the number of false replicates can be systematically increased in form of observation grids on fields (Rew and Cousens, 2001).

Management factors, which are in classical field trials divided into single, combinable factors with fixed factor levels, are flexibly combined in farming systems in order to achieve ex-ante determined thematic goals. Thus, farming systems are dynamic and have a self-learning ability. Consequently, data on the external inputs required to manage the system under the thematic goal of a farming system represent results for the issue of sustainable development.

In arable farming weeds are generally controlled. Those plants surviving all control treatments may produce seeds or other dispersal units and thereby create objectives for further control. Arable farming can meet these interactions by regulating weed populations indirectly by crop rotation, tillage and further crop management supplemented by direct chemical and physical regulation (Gill et al., 1997; Doucet et al., 1999).

If the existence of weeds shall be stimulated their long-term control should be possible without increasing indications for direct weed control. The spring weed density, before any control in the current year took place, is relevant for this goal. The parameter represents the effect of the entire past management, without the strong superimposing influence of the direct control treatments in the current year (Doucet et al., 1999). Low spring densities may offer possibilities to skip or reduce chemical control. In the present study the spring densities of weeds conflicting with arable objectives do not indicate a severe population build-up running out of control. This is also true regarding the sequential development during the experimental time (Gerowitt and Kirchner, 2000).

Effects of a divers crop rotation in reducing spring infestations with weeds were shown by Claupein and Baeumer (1992), Dubois et al. (1998) and Pallutt (1999). Especially, high percentages of autumn sown crops favoured problematic weeds, e.g. like grass weeds (Pallutt, 1999).

Withdrawing inversion tillage in a farming system is almost always accompanied by chemical weed control (Hutcheon et al., 1998). Without an effective direct control, weed populations generally increase, if the produced seeds are not buried through ploughing (Schmidt et al., 1995). Therefore, it is remarkable, that weed control in the system integrated-non-inversion could be managed without a distinct increase in herbicide input compared to GFP. Further advantages for the system integrated-non-inversion in terms of energy consumption and the environmental risk associated with herbicides are underlined by Gerowitt et al. (2000b).

In the system integrated-flexible the combination of cultural control, including inversion tillage, resulted in an effective restriction of weeds and direct chemical control could be remarkably reduced. Albrecht and Mattheis (1996) investigated an arable organic farming system and found that weeds were effectively regulated by cultural methods. The implementation of cultural control methods was less far reaching in the system integrated-flexible compared to an organic farming system, but also effective. It could be less far reaching, since occasional chemical control was accepted. Thus, "conventional" farmers can also profit from cultural control by moderate changes in their crop husbandry. Investigations of other system parameters, like the gross margin, indicate no general disadvantage for both integrated systems (Steinmann, 1998; Steinmann and Gerowitt, 2000).

Actually, to have any weed vegetation is no primary agricultural goal neither in terms of quantity (ground cover) nor of quality (species numbers). However, if this goal should be stimulated, the residual weed vegetation is the most suitable variable, since it is determined by both, the original weed vegetation occurring in autumn or spring and the weed management. Therefore, the residual weed vegetation reflects the entire combination of all control methods, including cultural and direct control. In the presented study, the residual weed vegetation in the crops of the integrated systems had more species and occurred more often with higher covers. However, rare or endangered species did not occur regularly in these systems. If the seed bank of these species is depleted, the ability of arable weed species to infest fields depends on the species strategy for seed distribution, the distances to existing populations, the frequency and the amount of soil transport between fields (Marshall and Hopkins, 1990). For many rare weed species, the possibility of being distributed to other sites, will be rather low.

In order to discuss this result in context with those of other farming systems experiments, it is worthwhile to reflect their ex-ante determined thematic goal. The attribute "Integrated" is utilised in various farming systems approaches (Zadoks, 1989; El Titi, 1996; Albrecht and Mattheis, 1996; more examples were given by Holland et al. (1994)). However, this attribute promises a false comparability, since the combinations of the indirect and direct weed control instruments differ, serving different objectives. Objective of the integrated system described by Albrecht and Mattheis (1996) is an environmentally friendly weed control. It was not purposed to have a residual vegetation nor was it achieved. Like in our investigations, Zadoks (1989) and El Titi (1996) attempted a residual weed vegetation in their integrated systems and achieved this goal in terms of quality and quantity. Although, this does not necessarily mean that the residual vegetation always consists of botanically interesting and lush vegetation, it is confirmed in experiments with practical farming relevance, that the arable weed vegetation could be both, maintained and controlled in sustainable developed farming systems. However, this maintenance should not focus the weed vegetation of historical forms of arable land use, but of land use, adapted to contemporary techniques and production methods.

5. Conclusions

In this paper results of a farming system experiment are used to investigate how and to which degree the weed vegetation could be both, developed and controlled and whether the process meets the general idea of sustainable development in agriculture. In this experimental case-study on arable farming systems, the weed vegetation could perform as an ex-ante and ex-post relevant characteristic to promote sustainable developed farming systems. Consequently, this result demands for a more principal concept for applications (Gerowitt et al., 2003), which then can be implemented on a larger scale, e.g. regional and surveyed, whether the resulting practice still contributes to the overall concept of sustainable development of agriculture in Central Europe.

Promoting the arable weed vegetation on this level possibly can extend the overlapping of primary agricultural goals and sustainable development of arable farming systems. At the moment, this area of overlap is underdeveloped not only for the farmers in terms of management, profit and status but also for the non-farming community in terms of interest and value.

Acknowledgements

The INTEX farming systems are established with financial support of the State Ministry of Agriculture, Food and Forestry of Niedersachsen and the European Union. Investigating various ecological effects of the farming systems has been possible with financial support of the German Federal Ministry of Environment, Nature Conservation and Reactor Safety.

References

Albrecht, H., Mattheis, A., 1996. Die Entwicklung der Ackerwildflora nach Umstellung von konventionellen auf integrierten bzw. ökologischen Landbau. J. Plant Dis. Protect. XV, 211–224 (special issue).

Becker, B., 1997. Sustainability assessment: a review of values, Concepts and Methodological Approaches. Issues in Agriculture, p. 10.

Claupein, W., Baeumer, K., 1992. Einfluß von Fruchtfolge, chemischem Pflanzenschutz und Stickstoffdüngung auf die Segetalflora eines Dauerversuches auf Lößboden. J. Plant Dis. Protect. XIII, 243–251 (special issue).

Doucet, C., Weaver, S.E., Hamill, A.S., Zhang, J., 1999. Separating the effects of crop rotation from weed management on weed density and diversity. Weed Sci. 47, 729–735.

Dubois, D.E., Scherrer, C., Gunst, L., Jossi, W., Stauffer, W., 1998. Auswirkungen verschiedener Landbauformen auf den Bodenvorrat an Unkrautsamen in den Langzeitversuchen Chaiblen und DOK. J. Plant Dis. Protect. XVI, 67–74 (special issue).

Edwards, C.A., Grove, T.L., Harwood, R.R., Pierce Colfer, J.C., 1993. The role of agroecology and integrated farming systems in agricultural sustainability. Agric. Ecosys. Environ. 46, 99–121.

El Titi, A., 1996. Veränderung der Unkrautartenzusammensetzung nach 16 Jahren integrierter Bewirtschaftung auf dem Lautenbacher Hof. J. Plant Dis. Protect. XV, 201–210 (special issue).

Francis, C.A., 1994. Practical applications of agricultural systems research in temperate countries. J. Prod. Agric. 7, 151–157.

Garve, E., 1993. Rote Liste der gefährdeten Farm- und Blütenpflanzen in Niedersachsen und Bremen. Informationsdienst Naturschutz Niedersachsen 1/93.

Gerowitt, B., Kirchner, C., 2000. Entwicklung und Begrenzung der Ackervegetation in Ackerbausystemen. In: Steinmann, H.-H., Gerowitt, B. (Eds.), Ackerbau in der Kulturlandschaft—Funktionen und Leistungen. Ergebnisse des Göttinger INTEX-Projekts. Mecke-Verlag, Duderstadt, pp. 55–80.

Gerowitt, B., de Mol, F., Moerschner, J., Tremel, S., Steinmann, H.-H., 2000a. Merkmale einer nachhaltigen Entwicklung im Ackerbau—dargestellt am Beispiel des Göttinger INTEX-Versuchs. Landbauforschung Völkenrode SH 212, pp. 67–92.

Gerowitt, B., de Mol, F., Moerschner, J., Steinmann, H.-H., 2000b. Charakterisierung der Unkrautbekämpfung in Ackerbausystemen anhand der Kosten, des Energieeinsatzes, des Umweltrisikos durch Herbizide und der Ackervegetation. J. Plant Dis. Protect. XVII, 725–734 (special issue).

Gerowitt, B., Isselstein, J., Marggraf, R., 2003. Rewards for ecological goods—requirements and perspectives for agricultural land use. Agric. Ecosyst. Environ. 98, 541–547.

Gill, K.S., Arshad, M.A., Moyer, J.R., 1997. Cultural control for weeds. In: Pimentel, D. (Ed.), Techniques for Reducing Pesticide Use. Wiley, New York, pp. 237–275.

Holland, J.M., Frampton, G.K., Cilgi, T., Wratten, S.D., 1994. Arable acronyms analysed—a review of integrated arable farming systems research in western Europe. Ann. Appl. Biol. 125, 399–438.

Hutcheon, J.A., Stride, C.D., Wright, K.J., 1998. Manipulation of weed seedbanks in reduced tillage systems for sustainable weed control. Aspects Appl. Biol. 1, 249–254.

Marshall, E.J.P., Hopkins, A., 1990. Plant species composition and dispersal in agricultural land. In: Bunce, R.G.H., Howard, D.C. (Eds.), Species Dispersal in Agricultural Habitats. Belhaven Press, London, New York.

Norman, D.W., Frankenberger, T.R., Hildebrand, P.E., 1994. Agricultural research in developed countries: past, present and future of farming systems research and extension. J. Prod. Agric. 7, 124–131.

Pallutt, B., 1999. Einfluß von Fruchtfolge, Bodenbearbeitung und Herbizidanwendung auf Populationsdynamik und Konkurrenz von Unkräutern im Wintergetreide. Gesunde Pflanzen 51, 109–120.

Rew, L.J., Cousens, R.D., 2001. Spatial distribution of weeds in arable crops: are current sampling and analytical methods appropriate? Weed Res. 41, 1–18.

Schmidt, W., Waldhardt, R., Mrotzek, R., 1995. Extensivierungsmaßnahmen im Ackerbau: Auswirkungen auf Flora, Vegetation und Samenbank—Ergebnisse aus dem Göttinger INTEX-Projekt. Tuexenia 15, 415–435.

Steinmann, H.-H., 1998. Can integrated arable farming systems achieve long-term stability of crop yield and weed infestation? Med. Fac. Landbouww. Univ. Gent 63, 697–703.

Steinmann, H.-H., Gerowitt, B., 2000. Ackerbau in der Kulturlandschaft—Funktionen und Leistungen. Ergebnisse des Göttinger INTEX-Projekts. Mecke-Verlag, Duderstadt.

Vereijken, P., 1992. A methodic way to more sustainable farming systems. Neth. J. Agric. Sci. 40, 209–217.

WCED (World Commission on Environment and Development), 1987. Our Common Future. Oxford University Press, Oxford.

Zadoks, J.C. (Ed.), 1989. Development of Farming Systems. Pudoc Wageningen/NL.

PART III
BIODIVERSITY AT DIFFERENT SCALE LEVELS

Chapter 4 Biodiversity and Soil Habitat

M. Schloter, J.C. Munch

Synoptic Introduction

ELSEVIER

Available online at www.sciencedirect.com

Agriculture, Ecosystems and Environment 98 (2003) 255–262

Agriculture Ecosystems & Environment

www.elsevier.com/locate/agee

Indicators for evaluating soil quality

M. Schloter [a,*], O. Dilly [b], J.C. Munch [a]

[a] Institute of Soil Ecology, GSF-National Research Center for Environment and Health, Ingolstadter Landstr. 1, 85764 Neuherberg, Germany

[b] Institute for Plant Nutrition and Soil Science, University of Kiel, Olshausenstr. 40, 24118 Kiel, Germany

Abstract

Interactions between the diversity of primary producers (plants) and of decomposers (microbes and mesofaunal communities), the two key functional groups that form the basis of all ecosystems have major consequences on the functioning of agricultural ecosystems. Soil microorganisms control the transformation and mineralization of natural compounds and xenobiotics. The soil microbiota, existing in extremely high density and diversity, rapidly modify the energetic performance and activity rates to changing environmental conditions. Thus, the microbial consortium possesses the ability to accommodate environmental constraints by adjusting (i) activity rates, (ii) biomass, and (iii) community structure. These parameters are particularly important to take into consideration when evaluating soil quality. The present paper gives an overview about the possibilities to use bacterial and fungal populations as an indicator for soil quality. Furthermore also the applicability of nematodes for the determination of soil health will be discussed.

Keywords: Indicators for soil quality; Microbial biomass; Microbial diversity; Microbial activity; Nitrogen turnover; Nematodes

1. Selection of indicators

Soil quality is defined as the 'continued capacity of soil to function as a vital living system, within ecosystem and land use boundaries, sustain biological productivity, to promote the quality of air and water environments, and to maintain plant, animal and human health' (Doran and Safley, 1997). Since soil microorganisms can respond rapidly, they reflect a hazardous environment and are, therefore, considered when monitoring soil status. However, it is still unclear whether naturally occurring environmental factors can damage the genotypic ability of the soil microbiota to recover after harsh conditions and become healthy again (Sparling, 1997). Research on the resilience of soil microbiota may be a significant task of molecular approaches.

The ideal soil microbiological and biochemical indicator to determine soil quality would be simple to measure, should work equally well in all environments and reliably reveal which problems existed where. It is unlikely that a sole ideal indicator can be defined with a single measure because of the multitude of microbiological components and biochemical pathways. Therefore, a minimum data set is frequently applied (Carter et al., 1997). Thus, the basic indicators and the number of estimated measures are still under discussion. However, national and international programs for monitoring soil quality presently include biomass and respiration measurements but extended also to nitrogen mineralization, microbial diversity and functional groups of soil fauna (Bloem et al., 2003).

* Corresponding author. Tel.: +49-89-3187-2304; fax: +49-89-3187-3376.
E-mail address: schloter@gsf.de (M. Schloter).

0167-8809/$ – see front matter © 2003 Elsevier Science B.V. All rights reserved.
doi:10.1016/S0167-8809(03)00085-9

Irrespectively, it is essential to consider a set of abiotic and biotic properties and processes as soil indicators in ecosystems.

2. Soil microbial biomass

The soil microbial biomass can be defined as organisms living in soil that are generally smaller than approximately 10 μm. Most attention is given to fungi and bacteria, these two groups of microbes being the most important with reference to energy flow and nutrient transfer in terrestrial ecosystems (Richards, 1987). Fungi and bacteria are generally dominating within the biomass. However, most biomass estimates do not reliably exclude protozoa. The microbial biomass consists of dormant and metabolically active organisms. However, the presently widespread biomass estimates, either direct or indirect (biochemical) techniques, were not properly valid and checked for separating these fractions.

It has been suggested that the microbial biomass content is an integrative signal of the microbial significance in soils because it is one of the few fractions of soil organic matter that is biologically meaningful, sensitive to management or pollution and finally measurable (Powlson, 1994). With the development of the four now widespread indirect methods, fumigation-incubation (FI), substrate-induced respiration (SIR), fumigation-extraction (FE) and ATP content (Jenkinson and Powlson, 1976; Anderson and Domsch, 1978; Jenkinson and Ladd, 1981; Vance et al., 1987), a great deal of effort has gone into the measurement of the size of the microbial biomass and its associated nutrient pools. All these methods are designed to quantify the microbial biomass carbon in different soil samples, soil horizons, soil profiles and sites (Elliott, 1994). However, it must be realized that between different soil samples different biomass may occur without direct correlation to soil quality (Martens, 1995; Dilly and Munch, 1998).

Nevertheless the soil microbial biomass is the eye of the needle through which all organic matter needs to pass through (Jenkinson et al., 1987). As a susceptible soil component, the biomass may be therefore a useful indicator since pollution may reduce this pool as, e.g. demonstrated by Fritze et al. (1996) for heavy metals.

3. Structural microbial diversity

However the measurement of the microbial biomass is a black box approach, without differentiating the heterogeneity of the microbial community. With the rise of molecular genetic tools in microbial ecology it became apparent that we know only a very small part of the diversity in the microbial world. Most of this unexplored microbial diversity seems to be hiding apparently in the high amount of yet uncultured bacteria. New direct methods, independent from cultivation, based on the genotype (Amann et al., 1995) and phenotype (Zelles, 1996) of the microbes allow a deeper understanding of the composition of microbial communities. Using, e.g. the rDNA directed approach of dissecting bacterial communities by amplifying the 16S rDNA (rrs) gene from environmental samples by polymerase chain reaction (PCR), and studying the diversity of the acquired rrs sequences, almost exclusively new sequences became apparent which are only to a certain degree related to the well studied bacteria in culture collections (Amann et al., 1995). Frequently occurring, yet uncultured bacteria became visible microscopically by using fluorescently labelled rRNA-directed oligonucleotide probes. Based on molecular studies it can be estimated that 1 g of soil consists of more than 10^9 bacteria belonging to about 10,000 different microbial species (Ovreas and Torsvik, 1998). This huge amount of diversity makes it often difficult to handle the microbial community structure as an indicator for soil quality.

However there are some studies which could demonstrate clear effects of changes in the farming management or contamination of a site on the total microbial community structure. Ovreas and Torsvik (1998) compared the influence of crop rotation and organic farming on microbial diversity and community structure. They found a higher diversity in soils which were under organic farming management. However almost nothing is known about sustainability of measured microbial parameters. Only Smit et al. (2001) investigated the seasonal fluctuation of bacterial soil community in a wheat field. Mainly for the monitoring of contaminations microbial diversity parameter are often used, for the assessment of soil quality. Muller et al. (2001), for example, investigated the long-term effects of long-term exposure to mercury on the soil microbial community along a gradient of pollution. It could be shown that bacterial diversity was reduced in the contaminated soils, whereas there was no difference in fungal biomass. Most available information is about the effects of pesticides on microbes and their degradation by bacteria and fungi. Fantroussi et al. (1999) could demonstrate that due to the application of urea herbicides microbial diversity was decreased. Ibekwe et al. (2001) showed a clear impact of fumigants on the soil microbial community. Similar results were obtained for other pesticides (triadimefon) by Yang et al. (2000).

Due to the mentioned complicity of the whole microbial community it might be useful to look at indicator organisms only, which are correlated to soil quality, for example, beneficial microbes like *Rhizobium* or arbuscular mycorrhiza (AM).

AM are the most ancient and ubiquitous root symbioses, formed by fungi belonging to the order of Glomales (Zygomycetes) and 80% of terrestrial plants (Saif and Khan, 1975). AM fungi are obligatory biotrophic symbionts living in the roots of most terrestrial plants which positively affect on plant growth, and plant nutrition. The fungi involved act as biofertilizers, and are very important for agriculture (Gianinazzi and Schuepp, 1994). Furthermore, AM represent a direct interface between soil and roots, and a place of exchange not only of nutrient elements but also of toxic elements. Since Glomales and/or AM symbioses are sensitive to PAH (polycyclic aromatic hydrocarbons) and MTE (metallic trace elements), both could also be used as bioindicator of contaminated soils (Weissenhorn et al., 1995). The decline of AM occurrence and infectivity of AM in metal-polluted soils can be used as bioindicators of soil contamination.

Natural rhizobia populations are essential to increase the yield of leguminous crops. The importance of the interaction is based on the capacity of symbiotic *Rhizobium* strains to form nodules and fix atmospheric nitrogen. Some papers describe the influence of different farming systems on plant growth promoting *Rhizobium* (Miethling et al., 2000). The survival of *Rhizobium* on chickpea seeds, treated separately with one of the four commercial fungicides was improved by Kyei-Boahen et al. (2001) under laboratory conditions. Fungicide treatment in general decreased the viability of *Rhizobium* strains, forming capacity of nodulation, N_2 fixation, and plant growth. Also in other studies effect of various pesticides (insecticides, fungicides and herbicides) on growth and efficiency of symbiotic properties were found (Madhavi et al., 1993). Although only very little is known about the evolution of natural bacterial populations through the years in relation to a host plant diversity and abundance of rhizobia might be a good indicator for soil quality.

4. Microbial activity

Soil microbial activity leads to the liberation of nutrients available for plants but also to the mineralization and mobilization of pollutants and xenobiotics. Thus microbial activity is of crucial importance in biogeochemical cycling. Microbial activities are regulated by nutritional conditions, temperature and water availability. Other important factors affecting microbial activities are proton concentrations and oxygen supply. The group of methods on soil microbial activities embraces biochemical procedures revealing information on metabolic processes of microbial communities. To estimate the soil microbial activity, two groups of microbiological approaches can be distinguished.

First, experiments in the field that often require long periods of incubation (i.e. Hatch et al., 1991; Alves et al., 1993) before significant changes of product concentrations are detected, i.e. 4–8 weeks for the estimation of net N mineralisation. In this case, variations of soil conditions during the experiment are inevitable, i.e. aeration, and may influence the results (Madsen, 1996). Furthermore field measurements are often difficult to interpret, for example, soil respiration determined in the field suffers in separating the activity of microorganisms and other organisms such as plants, which vary significantly in different systems and throughout the season (Dilly et al., 2000).

In contrast short-term laboratory procedures that are usually carried out with sieved samples at standardized temperature, water content and pH value. Short-term designs of 2–5 h minimize changes in biomass structure during the experiments (Brock and Madigan, 1991). Such microbial activity measurements include enzymatic assays that catalyze substrate-specific transformations and may be helpful to ascertain effects of soil management, land use and specific environmental conditions (Burns, 1977). Laboratory methods have the advantage in standardizing environmental factors and, thus, allowing the comparison of soils from different geographical locations and environmental conditions and also results from different laboratories. They are frequently used to gain information on 'functional groups'. However, laboratory results refer to microbial capabilities, as they are determined under optimized conditions of one or more factors, such as temperature, water availability and/or substrate.

When measuring soil enzyme activity, it is important to understand what type of information is being collected and how can it be used. Taylor et al. (2002) mentioned two main reasons for measuring soil enzymes. First, as indicators of process diversity, which informs about the biochemical potential, possible resilience and potential for manipulation of the soil system. Second, as indicators of soil quality, in the sense that changes in key functions and activities can provide information about the progress of remediation operations or the sustainability of particular types of land management. Despite the obvious benefits of having these types of information, Pettit et al. (1977) pointed out that it is important to realize the restrictions on enzyme assays and the limitations on the interpretation. Soil enzyme assays generally provide a measure of the potential activity, i.e. that encoded in the "soil genotype", but this will rarely ever be expressed. However, it may represent the redundancy of the soil biochemical system and as such is an aspect of resilience. Some soil enzyme assays attempt to measure real activity, i.e. a phenotypic property, but are rarely successful. In considering soil enzymes as an indicator of soil quality, which enzymes are important? A case can be argued for at least 500 enzymes with critical roles in the cycling of C or N or both, but clearly this many cannot be measured routinely. If there is genuine redundancy in enzymatic functions in soil, the loss of activity of a specific "keystone" enzyme should not have a major effect. If, on the other hand, changes in the activity of some "benchmark" enzymes provide an early indication of changes in process diversity, soil enzymatic measurements have a clear role in the assessment of soil quality. The question that remains is which enzymes should be measured for this purpose?

5. Nitrogen turnover as an indicator for soil quality

Bioavailable nitrogen is one of the keys for plant growth in agriculture. At the same time nitrogen compounds like nitrate, nitrite or N_2O play an important role in environmental pollution. Therefore it is of great interest to understand the key processes in the nitrogen cycle in more detail, to define ways for a high productive agriculture which protects environment. On the one hand two main delivery processes (mineralisation and nitrogen fixation) are known. On the other hand nitrification and denitrification can cause significant losses of nitrogen from the bound pool.

The microbial mineralisation of proteins in terrestric ecosystems fulfils the key function to mobilize organically bound nitrogen (Ladd and Butler, 1972). Nitrogen is transformed in the cycle as ammonium. Extra cellular proteases are produced by microbes and secreted to the environment to hydrolyze macromolecular polypeptides into smaller molecules, which can be removed by the cell (Kalisz, 1988). They have in general a very low substrate specificity. Results from Bach and Munch (2000) indicate that differences in protease activities are not caused by different microbial populations but by variable expression rates of the same community. As proteases are exoenzymes they are after secretion no longer under the regulation of the cell and can stabilized by clay particles in the soil. Therefore the enzymatic activity for its own is no sensible indicator for the actual microbial proteolytic activity. Several attempts are made to identify ecophysiological conditions that cause an induction or repression of the peptidase expression in the habitat (Bach et al., 2001).

Nitrogen fixation is performed by phylogenetically and physiologically diverse groups of prokaryotic organisms and poses a challenge to microbial ecologists in terms of diversity and activity assessment. The ecology of free-living diazotrophs has received little attention due to the low fixation rates that are usually attributed to non-symbiotic nitrogen fixation. Nevertheless, recurring accounts of unusually high N inputs into studied systems in addition to quickly activated N_2-fixing activity in soil when energy sources are present indicate that this phenomenon plays an important role in natural systems and might have applications in land management. To provide better tools for the study of this group of organisms, molecular approaches have been developed based on PCR amplification of the *nifH* gene and its mRNA transcripts for the group-specific detection of free-living diazotrophs in soil (Widmer et al., 1999).

Nitrification is the chemolithoautotrophic oxidation of ammonium via nitrite to nitrate. Nitrification can be measured directly using labeled nitrogen. Another possibility to determine potential nitrification is based on the addition of chlorate to inhibit the nitrite oxidation (Kandeler, 1989). The potential nitrification can be measured as accumulation of nitrite after addition of ammonium in short-term experiments. This method is well suited for measuring potential nitrification in high number of samples. A reduction in diversity for ammonia oxidizers for tilled soils was found by Bruns et al. (1999) in comparison to the native plots.

Denitrification is one of the key processes in the global nitrogen cycle as nitrate is turned into gaseous products (Flessa et al., 1995). During the process nitrate is stepwise reduced via nitrite, NO and N_2O to N_2. Due to the action of denitrifying microorganisms, the global dinitrogen content in the atmosphere is largely in balance due to the formation of the dinitrogen gas from terrestrial nitrate. On the other hand, nitrogenous oxides released from soils and waters have several impacts on the atmosphere. Nitrous oxide is next to CO_2 and CH_4 in its importance as a potent greenhouse gas. Nitric acid and its chemical oxidation product NO_2 are major constituents of acid rain, and NO and also N_2O interact with ozone in complex reactions and are major causes of the destruction of the protective ozone layer in the stratosphere. Nitrate is the main N-source for the growth of plants in agriculture but can simultaneously be used also by microorganisms in soils. Denitrification is generally regarded as an anaerobic process, but there are indications that it may take place also in well-aerated soils with high contents of bioavailable organic matter. The conditions which favor denitrification in soils have not yet been elucidated in much detail. It is, however, clear that any use of nitrate by bacteria means a loss of N for the growth of plants. Thus, denitrification has also severe impact on agriculture. In addition, products of denitrification (nitrate respiration) have manifold other, mainly adverse effects on soils but also on the atmosphere and waters. While the denitrification product N_2O can be easily measured using gas chromatography, the determination of N_2 is not straight forward because comparatively small amounts of N_2 produced during denitrification have to be distinguished from a large background of 78% N_2 in the atmosphere. The methods available for measuring denitrification in the field are based on the use of the stable isotope ^{15}N or on acetylene for blockage of the enzyme N_2O-reductase. In a preliminary study Cheneby et al. (2000) investigated denitrifying bacteria in three agricultural soils using classical cultivation techniques. They found a good correlation between number and diversity of denitrifiers and soil type.

6. Faunal indicators: nematodes

The use of faunal groups as indicators for soil quality needs a choice of organisms, that (a) form a dominant group and occurs in all soil types, (b) have a high abundance and high biodiversity and (c) play an important role in soil functioning, e.g. in food webs.

Nematodes fulfill these conditions and seem to be at present state of knowledge the most promising group, also because different tests in ecotoxicology, realized for single species as well as for communities, shows the suitability (Traunspruger and Drews, 1996; Freeman et al., 2000; Peredney and Williams, 2000; Haitzer et al., 1999). The group, composed of more than 11,000 species (Andrassy, 1992) includes species with different degree of tolerance again stress. Furthermore, the most important species used for ecotoxicologic assessments, *Caenorhabditis elegans*, is one the few organisms with full genetic information. So, not only the classical toxicity parameter such as lethality, growth, reproduction and behavior can be realized, but also toxicity essays at molecular level, with the possibility to get information on the bioavailability of contaminants. Nematodes have life cycles over a broad range (a few days to over 2 years). This gives the possibility to integrate effects over different time scales.

Use of soils fauna as indicators offers different possibilities. Single species bioassays are important to assess effects of single stressors and bioconcetration studies. However, these tests are often realized in laboratory experiments, with soil samples transferred in experimental systems and spiked with contaminants. Experiments on community level are ecologically more relevant. They integrate interactions of all soil factors including management and pollutants effects. Effects recorded by nematode bioassays reflect, e.g. also the environmental conditions of the community, so effects of the physical habitat or the food availability. Community assays offer analysis of different features: abundance of individuals or species, biomass of species, species composition, feeding strategies, presence and abundance of key species. The date obtained are to be analyzed by different techniques, univariate or multivariate, depending on the required quality of response of the experimental device. Different measures are used. Shannon index gives the distribution of species abundance and also reveals rare species (higher index = higher diversity). Simpson index shows the distribution of species abundance, with more weight to common species (higher index = higher dominance). The evenness (value between 0 and 1) gives information on the distribution of species abundance (higher index = higher diversity). Feeding types are reflected by the index of tropic diversity. The maturity index (scales from 1 to 5) is an indicator for the persistence of colonizers or for the life strategies of nematodes (disturbance indicated by a low index).

Multivariate methods consider species or groups in combination with data an abundance or biomass. Similarities or dissimilarities between such assemblages are visualized by cluster analysis. Multivariate statistics tests the differences in community structure.

7. Aggregation of indicators

The aggregation of indicators for evaluating soil quality should consider the complexity of microbial life in soil. Multiple indicators can be regarded to refer to the 'driving forces' for C and N cycling in soils. As a minimum data set microbial biomass content and microbial activity rates including enzyme activities were often estimated together with measures on some basic soil components, i.e. organic C content (Carter et al., 1997). Data sets can be compared by designing sun ray plots (Dilly and Blume, 1998; Kutsch et al., 1998; Dilly and Kutsch, 2000). They show the pattern of the considered features and prospectively may evaluate with reference to both real and acceptable values of properties and processes and thus characteristics with respect to thresholds, limits or the window of viability. The lower the serration of the star as in case of the A horizon of a wet grassland in contrary to maize monoculture, the higher is the association between the microbial features and link between microbial processes (Dilly and Blume, 1998).

In contrast to this star approach, canonical component analyses that represent state-space orientation are frequently lacking in a clear explanation of ecological interrelations between dependent and controlling, e.g. biotic and abiotic factors. Furthermore, the rationale with respect to 'emergent' properties of soils and ecosystems (Müller, 1996; Dilly and Kutsch, 2000) will not be achieved.

In addition, microbial activities related to microbial biomass are used for evaluating environmental conditions calculating, i.e. the metabolic quotient, which is the ratio between CO_2 production under standardized conditions and microbial C content (Anderson and Domsch, 1993). Finally, soil microbial activities of C and N cycles should be related to soil C and N stocks providing information concerning transformation intensity in labile pools by looking at substrate transformation and product formation.

To evaluate soil quality, spatial heterogeneity of microbiological characteristics in ecosystems is important to take into account since microbiological features may vary scale-dependently (Stork and Dilly, 1998).

For holistic approaches, indicators may be displayed in hierarchical schemes for analyzing interactions and signal transfers in different subsystems (Dilly and Kutsch, 2000; Dilly et al., 2001). The abundance of specific populations and active components are probably more variable in contrast to the biomass. Particularly the activity of the whole biomass may change considerably with reference to environmental impact in contrast to biomass itself. These alterations may only slightly or slowly be affected in more stabile ecosystem components such as the soil organic C content.

8. Conclusions

The great abundance and diversity of microorganisms in soil have high metabolic potentials. Since microorganisms are generally growth-limited in soils, they may poorly exploit their capabilities. In contrast, soil microorganisms respond rapidly to stressors by adjusting (i) activity rates, (ii) biomass, and (iii) community structure. Combining soil microbiological estimates, e.g. in sun rays or quotients, seems to be of great relevance for evaluating soil quality. This is shown in four papers presented by: (1) Ruf et al. 'A biological classification concept for the assessment of soil quality'; (2) Anderson 'Microbial eco-physiological indicators to assess soil quality'; (3) Eckschmitt et al. 'On the quality of soil biodiversity indicators—three case studies at different spatial scales'; (4) Schloter et al. 'Influence of precision farming on the microbial community structure and selected functions in nitrogen turnover with indicator value for soil quality.

References

Alves, B.L.R., Urquiaga, S., Cadisch, G., Souto, C.M., Boddy, R.M., 1993. In situ estimation of soil nitrogen mineralization. In: Mulongoy, K., Merckx, R. (Eds.), Soil Organic Matter Dynamics and Sustainability of Tropical Agriculture. Wiley-Sayce Co-Publication, IITA/K.U. Leuven, pp. 173–180.

Amann, R., Ludwig, W., Schleifer, K.H., 1995. Phylogenetic identification and in situ detection of individual microbial cells without cultivation. Microbiol. Rev. 59, 143–149.

Anderson, J.P.E., Domsch, K.H., 1978. A physiological method for measurement of microbial biomass in soils. Soil Biol. Biochem. 10, 215–221.

Anderson, T.-H., Domsch, K.H., 1993. The metabolic quotient for CO_2 (qCO_2) a specific activity parameter to assess the effects of environmental conditions, such as pH, on the microbial biomass of forest soils. Soil Biol. Biochem. 25, 393–395.

Andrassy, I., 1992. A short census of free-living nematodes. Fundam. Appl. Nemat. 15, 187–188.

Bach, H.J., Munch, J.C., 2000. Identification of bacterial sources of soil peptidases. Biol. Fertil. Soils. 31, 219–224.

Bach, H.-J., Hartmann, A., Schloter, M., Munch, J.C., 2001. PCR primers and functional probes for amplification and detection of bacterial genes for extracellular peptidases in single strains and in soil. J. Microbiol. Meth. 44, 173–182.

Bloem, J., Schouten, T., Sørensen, S., Breure, A.M., 2003. Application of microbial indicators in ecological approaches to monitor soil quality. Ambio, (in press).

Brock, T.D., Madigan, M.T., 1991. Biology of Microorganisms. Prentice-Hall, New Jersey, pp. 874–890.

Bruns, M.A., Stephen, J.R., Kowalchuk, G.A., Prosser, J.I., Paul, E.A., 1999. Comparative diversity of ammonia oxidizer by 16S rRNA gene sequencing in native, tilled and successional soils. Appl. Environ. Microbiol. 65, 2994–3000.

Burns, R.G., 1977. Soil enzymology. Sci. Prog. 64, 275–285.

Carter, M.R., Gregorich, E.G., Anderson, D.W., Doran, J.W., Janzen, H.H., Pierce, F.J., 1997. Concepts of soil, quality and their significance. In: Gregorich, E.G., Carter, M.R. (Eds.), Soil Quality for Crop Production and Ecosystem Health. Elsevier, Amsterdam, pp. 1–19.

Cheneby, D., Philippot, L., Hartmann, A., Germon, J.C., 2000. 16S rDNA analysis for the characterization of denitrifying bacteria isolates from three agricultural soils. FEMS Microbiol. Ecol. 34, 121–128.

Dilly, O., Blume, H.P., 1998. Indicators to assess sustainable land use with reference to soil microbiology. Adv. GeoEcol. 31, 29–36.

Dilly, O., Kutsch, W.L., 2000. Rationale for aggregating microbiological information for soil and ecosystem research. In: Benedetti, A., Tittarelli, F., Bertoldi de, S., Pinazari, F. (Eds.) Biotechnology of Soil: Monitoring, Conservation and Remediation, EUR 19548. European Commission, Brussels, pp. 135–140.

Dilly, O., Munch, J.C., 1998. Ratios between estimates of microbial biomass content and microbial activity in soils. Biol. Fertil. Soils 27, 374–379.

Dilly, O., Bach, H.-J., Buscot, F., Eschenbach, C., Kutsch, W.L., Middelhoff, U., Pritsch, K., Munch, J.C., 2000. Characteristics and energetic strategies of the rhizosphere in ecosystems of the Bornhöved Lake district. Appl. Soil Ecol. 15, 201–210.

Dilly, O., Winter, K., Lang, A., Munch, J.C., 2001. Energetic eco-physiology of the soil microbiota in two landscapes of southern and northern Germany. J. Plant Nutr. Soil Sci. 164, 407–413.

Doran, J.W., Safley, M., 1997. Defining and assessing soil health and sustainable productivity. In: Pankhurst, C., Doube, B.M., Gupta, V. (Eds.), Biological Indicators of Soil Health. CAB International, Wallingford, pp. 1–28.

Elliott, E.T., 1994. The potential use of soil biotic activity as an indicator of productivity, sustainability and pollution. In: Pankhurst, C., Doube, B.M., Gupta, V. (Eds.) Soil Biota. Management in Sustainable Farming Systems. CSIRO, Australia, pp. 250–256.

Fantroussi, S., Verschuere, L., Verstraete, W., Top, E.M., 1999. Effects of phenylurea herbicides on soil microflora communities estimated by analysis of the 16S rRNA gene fingerprints and community level physiological profiles. Appl. Environ. Microbiol. 65, 982–988.

Flessa, H., Dörsch, P., Beese, F., 1995. Seasonal variation of N_2O and CH_4 fluxes in differently managed arable soils in southern Germany. J. Geophys. Res. 100, 115–124.

Freeman, M.N., Perdney, C.L., Williams, P.L., 2000. A soil bioassay using the nematode *Caenorhabditis elegans*. In: Henshel, D.S., Blakck, M.C., Harrass, M.C. (Eds.), Environmental Toxicology and Risk Assessment: Standardization of Biomarkers for Endocrine Disruption and Environmental Assessment, vol. 8. ASTM STP 1364. American Society for Testing and Materials, West Conshohocken, PA, pp. 305–318.

Fritze, H., Vanhala, P., Pietikäinen, J., Mälkönen, E., 1996. Vitality fertilization of Scots pine stands growing along a gradient of heavy metal pollution: short-term effects on microbial biomass and respiration rate of the humus layer. Fresen. J. Anal. Chem. 354, 750–755.

Gianinazzi, S., Schuepp, H., 1994. Impact of Arbuscular Mycorrhizas on Sustainable Agriculture and Natural Ecosystems, Advances in Life Sciences. Birkhauser, Basel, Switzerland.

Haitzer, M., Burnison, B.K., Traunspurger, W., Steinberger, E.W., 1999. Effects of quantity, quality and contact time of dissolved organic matter on bioconcentration of benzo[a]pyrene in the nematode *Caenorhabditis elegans*. Environ. Toxicol. Chem. 18, 459–465.

Hatch, D.J., Jarvis, S.C., Reynolds, S.E., 1991. An assessment of the contribution of net mineralization to N cycling in grass swards using a field incubation method. Plant Soil 138, 23–32.

Ibekwe, A.M., Paiernik, S.K., Gan, J., Yates, S.R., Yang, C.H., Crowley, D., 2001. Impact of fumigants on soil microbial communities. Appl. Environ. Microbiol. 67, 3245–3257.

Jenkinson, D.S., Ladd, J.N., 1981. Microbial biomass in soil: measurement and turnover. In: Paul, E.A., Ladd, J.N. (Eds), Soil Biochemistry, vol. 5. Dekker, New York, pp. 415–471.

Jenkinson, D.S., Powlson, D.S., 1976. The effect of biocidal treatment on metabolism in soil. V. A method for measuring soil biomass. Soil Biol. Biochem. 8, 209–213.

Jenkinson, D.S., Hart, P.B.S., Rayner, J.N., Parry, L.C., 1987. Modelling the turnover of organic matter in long-term experiments at Rothamsted. INTECOL Bull. 15, 1–8.

Kalisz, H.M., 1988. Microbial proteases. Adv. Biochem. Eng. Biotechnol. 36, 1–65.

Kandeler, E., 1989. Aktuelle und potentielle Nitrifikation im Kurzzeit-bebrütungsversuch. VDLUFA-Schriftreihe 28, Kongreßband Teil II, pp. 921–931.

Kutsch, W.L., Dilly, O., Steinborn, W., Müller, F., 1998. Quantifying ecosystem maturity—a case study. In: Müller, F., Leupelt, M. (Eds.) Eco Targets, Goal Functions, and Orientors. Springer, New York, pp. 209–231.

Kyei-Boahen, S., Slinkard, A.E., Walley, F.L., 2001. Rhizobial survival and nodulation of chickpea as influenced by fungicide seed treatment. Can. J. Microbiol. 47, 585–589.

Ladd, J.N., Butler, J.H.A., 1972. Short-term assays of soil proteolytic enzyme activities using proteins and dipeptide derivates as substrates. Soil Biol. Biochem. 4, 19–30.

Madhavi, B., Anand, C.S., Bharathi, A., Polasa, H., 1993. Effect of pesticides on growth of rhizobia and their host plants during symbiosis. Biomed. Environ. Sci. 6, 89–94.

Madsen, E.L., 1996. A critical analysis of methods for determining the composition and biogeochemical activities of soil microbial communities in situ. In: Stotzky, G., Bollag, J.M. (Eds.), Soil Biochemistry, vol. 9. Marcel Dekker, New York, pp. 287–370.

Martens, R., 1995. Current methods for measuring microbial biomass C in soil: potentials and limitations. Biol. Fertil. Soils 19, 87–99.

Miethling, R., Wieland, G., Backhaus, H., Tebbe, C.C., 2000. Variation of microbial rhizosphere communities in response to species, soil origin, and inoculation with *Sinorhizobium meliloti*. Microbiol. Ecol. 40, 43–56.

Müller, F., 1996. Emergent properties of ecosystems—consequences of self organizing processes? Senkenbergiana maritima 27, 151–168.

Muller, A.K., Westergaard, K., Christensen, S., Sorensen, S.J., 2001. The effect of long-term mercury pollution on the soil microbial community. FEMS Microbiol. Ecol. 36, 11–19.

Ovreas, L., Torsvik, V.V., 1998. Microbial diversity and community structure in two different agricultural soil communities. Microbiol. Ecol. 36, 303–315.

Peredney, C.L., Williams, P.L., 2000. Utility of *Caenorhabditis elegans* for assessing heavy metal contamination in artificial soil. Arch. Environ. Con. Toxicol. 39, 113–118.

Pettit, N.M., Gregory, L.J., Freedman, R.B., Burns, R.G., 1977. Differential stabilities of soil enzymes. Assay and properties of phosphatase and arylsulphatase. Acta Biochim. Biophys. 485, 357–366.

Powlson, D.S., 1994. The soil microbial biomass: before, beyond and back. In: Ritz, K., Dighton, J., Giller, G.E. (Eds.), Beyond the Biomass. Wiley, Chichester, UK, pp. 3–20.

Richards, B.N., 1987. The Microbiology of Terrestrial Ecosystems. Longman, Essex.

Saif, S.R., Khan, A.G., 1975. The influence of season and stage of development of plant on endogone mycorrhiza of field-grown wheat. Can. J. Microbiol. 21, 1020–1024.

Smit, E., Leeflang, P., Gommans, S., van Den Broek, J., van Mil, S., 2001. Diversity and seasonal fluctuations of the dominant members of the bacterial soil community in a wheat field as determined by cultivation and molecular methods. Appl. Environ. Microbiol. 67, 2284–2291.

Sparling, G.P., 1997. Soil microbial biomass, activity and nutrient cycling as indicators of soil health. In: Pankhurst, C., Doube, B.M., Gupta, V. (Eds.), Biological Indicators of Soil Health. CAB International, Wallingford, pp. 97–119.

Stork, R., Dilly, O., 1998. Maßstabsabhängige räumliche Variabilität mikrobieller Bodenkenngrößen in einem Buchenwald. J. Plant Nutr. Soil Sci. 161, 235–242.

Taylor, J.P., Wilson, M., Mills, S., Burns, R.G., 2002. Comparison of microbial numbers and enzymatic activities in surface soils and subsoils using various techniques. Soil Biol. Biochem. 34, 387–401.

Traunspruger, W., Drews, C., 1996. Toxicity analysis of freshwater and marine sediments with meio- and macrobenthic organisms: a review. Hydrobiologia 328, 215–261.

Vance, E.D., Brookes, P.C., Jenkinson, D.S., 1987. An extraction method for measuring soil microbial biomass C. Soil Biol. Biochem. 19, 703–707.

Weissenhorn, I., Mench, M., Leyval, C., 1995. Bioavailability of heavy metals and arbuscular mycorrhiza in a sewage-sludge-amended sandy soil. Soil Biol. Biochem. 27, 287–296.

Widmer, F., Shaffer, B.T., Porteous, L.A., Seidler, R.J., 1999. Analysis of *nifH* gene pool complexity in soil and litter at a Douglas fir forest site in the Oregon cascade mountain range. Appl. Environ. Microbiol. 65, 374–380.

Yang, Y., Yao, J., Hu, S., Qi, Y., 2000. Effects of agricultural chemicals on the DNA sequence diversity of microbial communities. Microbiol. Ecol. 39, 72–79.

Zelles, L., 1996. Fatty acid patterns of microbial phospholipids and lipopolysaccharides. In: Schinner, F., Öhlinger, R., Kandeler, E., Margesin, R. (Eds.), Methods in Soil Biology. Springer, Berlin, pp. 80–93.

PART III
BIODIVERSITY AT DIFFERENT SCALE LEVELS

Chapter 4 Biodiversity and Soil Habitat

M. Schloter, J.C. Munch

Original Papers

ELSEVIER

Available online at www.sciencedirect.com

Agriculture, Ecosystems and Environment 98 (2003) 263–271

Agriculture Ecosystems & Environment

www.elsevier.com/locate/agee

A biological classification concept for the assessment of soil quality: "biological soil classification scheme" (BBSK)

A. Ruf[a,*], L. Beck[b], P. Dreher[c], K. Hund-Rinke[c], J. Römbke[d], J. Spelda[b]

[a] *Institute for Ecology and Evolutionary Biology and UFT, University of Bremen, FB 2, D-28334 Bremen, Germany*
[b] *Staatliches Museum für Naturkunde Karlsruhe, D-76133 Karlsruhe, Germany*
[c] *Fraunhofer Institute for Environmental Chemistry and Ecotoxicology, D-57377 Schmallenberg, Germany*
[d] *ECT Oekotoxikologie GmbH, D-65439 Flörsheim/Main, Germany*

Abstract

The protection of soils as habitat for soil organisms which is ascertained in the German soil protection act calls for the development of a broad, holistic approach with biological objectives. As a first step towards establishing a system that meets these criteria, two pilot studies have been conducted. The scope of these studies was to investigate whether there were characteristic soil fauna communities for specific sites or for pedologically defined groups of sites. We sampled the macrofauna groups earthworms, chilopods, diplopods, and isopods and the mesofauna groups enchytraeids, predatory mites (Gamasina), and moss mites (Oribatida). We could show that there were typical soil fauna communities that belong to specific site groups, e.g. acid forests or agricultural sites. The recorded patterns were more distinct the more taxa were incorporated in the analyses. The macrofauna alone gave good results but did not differentiate within the main site groups. Earthworms separated the open sites from the forests, whereas the arthropods differentiated within the forest sites. Mesofauna taxa added valuable information to the macrofauna results. We concluded that macro- and mesofauna together form site specific species assemblages that may be used for defining typical soil fauna communities for specific soils. This site specific soil fauna community can be used as a reference for assessing biological soil quality.

Keywords: Habitat function; Biological classification; Soil fauna; Community

1. Introduction

In our modern society soil has to meet several functions, e.g. to buffer pesticides, nutrients, and metals, to enable agricultural production, or to ground houses, streets, and railroads. In addition to these functions that are directly useful to man, soil also has to perform natural functions like being the substrate for natural vegetation and the habitat for soil organisms. In the German Federal Soil Protection Act of 17 March 1998 (BBodSchG, 1998) these variety of functions of soils are explicitly addressed. For each of these functions soil quality may be defined by very different criteria and approaches. The following paper discusses soil quality criteria for the function of the soil as a habitat for soil fauna.

There is a variety of methods for investigating soil biological parameters and some of these are proposed for soil monitoring programmes (see Römbke and Kalsch, 2000) in addition to chemical and physical investigations. Contrasting this the assessment of the habitat quality of soils is still in a preliminary

* Corresponding author. Tel.: +49-421-218-7681; fax: +49-421-218-7654
E-mail address: aruf@uni-bremen.de (A. Ruf).

0167-8809/$ – see front matter © 2003 Elsevier Science B.V. All rights reserved.
doi:10.1016/S0167-8809(03)00086-0

stage, there is no commonly accepted procedure to evaluate the habitat function (Spurgeon et al., 1996; Dunger, 1998). It is evident after all that there is an urgent need for biological methods to assess the condition of the living system "soil". However, the habitat function can by no means measured by pedological properties but must be defined by biological parameters.

The concept we apply for that purpose is a biological soil classification scheme (BBSK), which relies on simple assumptions. They are: (1) That the composition of the soil fauna is mainly determined by the abiotic characteristics of sites. (2) That it is possible to find most important site parameters with the greatest influence on soil fauna. According to these assumptions sites with similar soils should have a similar soil fauna. So it should be possible to define a site specific reference soil fauna that can be used as baseline against an actual investigation. In a previous study (Römbke et al., 1997) we could show that it was possible to define site specific reference values (species composition or other community parameters) for many groups of the soil fauna.

Based on our assumptions and on the preliminary empirical results a four step procedure has been developed:

1. Site groups are defined by similar pedological and climatological parameters.
2. For each site group a specific soil fauna assemblage can be addressed that is the theoretical site specific reference community.
3. A site of interest is sampled for its soil fauna after it was classified to a site group due to its pedological and climatological characteristics.
4. The deviation between the reference and the actual sampled community is evaluated.

On the basis of single taxa, we have already tested our approach (Römbke et al., 1997, 2000, Dreher et al., 1999; Ruf et al., 1999). Comparable concepts are proposed for the UK (SOILPACS; Weeks, 1997) and are used in several Dutch investigations (Sinnige et al., 1992; Schouten et al., 2000).

We present the results of two surveys which were conducted to evaluate the concept of the BBSK. The approach was to include a variety of soil fauna taxa which belong to different size, life-form, and trophic groups. Therefore we studied the well known macrofauna taxa (macrosaphrophages and predators) and the poorly known and species rich taxa within the mesofauna (fungivors, microsaprophages, and predators).

2. Sites, material, and methods

2.1. Case study I

In a first pilot study we surveyed 11 forest sites in south-west Germany (Baden-Württemberg). All of these were included in a monitoring programme for investigating the impact of air borne pollutants on forest ecosystems. The chosen sites cover a broad spectrum of potential impact of air borne pollutants, from close to industrial centres to more pristine rural landscapes. Site parameters are published in LfU (1993) and Römbke et al. (1997). We sampled for 2 years, twice in spring and in autumn. Sample size for the microarthropods was 25 cm × 25 cm in four replicates for the organic layers and 25 cm^2 to a depth of 10 cm for the mineral soil in eight replicates per date. Extraction was done horizontalwise on a standard Berlese funnel system. Enchytraeids were sampled by a corer (25 cm^2 surface) and extracted by a modified O'Connor system. Earthworms were handsorted in the litter layer and expelled from the mineral soil by a formol solution. Macrofauna was caught in Barber traps during the vegetation period in two consecutive years. Taxa identified on species level were: Carabidae, Isopoda, Chilopoda, Diplopoda, Lumbricidae, Enchytraeidae, Oribatida, and Gamasina. They all cover different size classes and positions in the food web. The soil parameters were measured by standard methods (pH in $CaCl_2$; organic matter content as loss of ignition) or were made available by the "Landesanstalt für Umweltschutz Baden-Württemberg, Karlsruhe". Soil parameters used to classify sites to ecologically similar groups were pH-value, organic matter content, *C/N* ratio, soil moisture, and soil texture. All parameters themselves were classified into 4 or 5 groups, mainly according to the Kartieranleitung, 4th ed. (AG Boden, 1994), in the other cases according to the criteria given by Dreher et al. (1999).

This study served as a pilot project in which methods for classifying fauna and soil parameters were

Table 1
Site parameters for the 15 investigated sites in the second study (the first 10 are forests, the next four grasslands (last letter G) and the last an arable field (last letter A); SBB and CRM are the least acid forests)

	Texture	pH (in $CaCl_2$)	Precipitation (mm per annum)	Soil moisture "nFKWe"	*C/N*	SOM (%)
SBB	Lt2	5.1	940	93	25.7	21.7
CRM	Lts	5.9	800	139	13.6	12.5
NIB	Uls	3.9	1120	255	14.4	11.5
SCF	Lts	3.2	833	139	23.7	12.0
TAM	Us	3.1	784	257	17.9	17.0
MEM	Su3	3.5	950	171	19.0	14.0
LUB	Sl3	2.8	730	134	23.9	2.5
EHE	Sl3	3.1	720	137	16.2	7.1
BBK	S	3.4	600	95	20.8	9.7
BEK	Su3	3.2	596	89	25.9	7.6
SCG	Sl4	4.8	833	191	10.1	9.0
BRG	Tu3	4.6	828	122	11.2	7.9
AKG	Ut4	4.9	722	205	9.3	8.0
SBG	Lt2	5.7	940	76	7.6	3.8
SBA	Ls2	5.4	940	83	10.4	4.1

elaborated. Reference values were defined by the experts for each taxa separately. Deviations from the reference values were classified in "−" (clear deviation between reference and actual value), "±" (some differences, but not clear), and "+" (no deviation within 30% range). For more details see Römbke et al. (1997).

2.2. *Case study II*

The second study was conducted in a more realistic framework as it can be expected when a soil classification concept is used routinely by governmental agencies. Sampling was done only once in late autumn/early winter 1998 and we did not use Barber traps. All other methods were applied accordingly. We sampled 15 sites all over Germany, 10 of which were forests, 4 grasslands and 1 arable field (for more details see Römbke et al., 2000). Site parameters are given in Table 1. Taxa identified to species level were Isopoda, Chilopoda, Diplopoda, Lumbricidae, Enchytraeidae, Oribatida, and Gamasina. Nematoda were identified for three sites, only.

In the second study we invented a community approach, although each fauna taxon was also analysed separately (cf. Römbke et al., 2000). Community analysis was done by an indirect gradient analysis (correspondence analysis) with the software CANOCO and by cluster analysis with the software TWINSPAN.

3. Results

3.1. *Case study I: south-west Germany*

In our first pilot study we could show that soil fauna gives hints to the ecological condition of forest soils (Table 2). Most fauna taxa gave comparable results concerning the condition of the sites. For example, sites 350 and 520 showed major deviations for many taxa. Interestingly, these are the two sites which seem to be affected by atmospheric emissions from an industrial area. In the mineral soil there were slightly elevated (350) or clearly elevated (520) concentrations of Pb, Cd, and Zn. Epiphytic lichens were severely damaged at both sites (LfU, 1993). On the other hand, at sites 130 and 140 there was no indication for a disturbance for any taxa. These sites are located in the eastern part quite far from urban regions. Despite of that general pattern, there seem to be more sensitive taxa (e.g. the predatory Gamasina) and others that are more indifferent (e.g. Carabidae). At many sites earthworms gave no unequivocal results because some species that were expected could not be recorded. A complete species list and detailed information on the protocol for establishing reference values can be found in Römbke et al. (1997). Macro- and mesofauna were both suited within the framework of our concept. We were able to define site specific reference values

Table 2
Results of the taxawise evaluation of 11 forest sites in south-west Germany[a]

Taxon	Site numbers										
	130	140	292	310	350	380	400	410	450	470	520
Lumbricidae	+	±	+	+	−	±	±	±	+	+	−
Enchytraeidae	+	+	+	+	−	+	+	+	±	+	−
Diplopoda	n.d.	+	n.d.	+	+	+	−	+	+	±	−
Chilopoda	n.d.	+	n.d.	+	+	+	+	+	+	−	−
Isopoda	n.d.	+	n.d.	+	+	+	+	+	+	±	−
Carabidae	+	+	+	+	−	+	+	+	+	+	−
Oribatida	+	+	+	+	−	±	+	+	+	+	−
Gamasina	+	+	+	+	−	±	+	+	+	−	−

[a] −: Clear deviation between reference and actual value; ±: some differences, but not clear; +: no major deviation; n.d.: not determined.

for each taxa separately derived from published data which served as baseline for the fauna we actually found. In this stage it was not possible to define a common framework of evaluation for all investigated soil fauna groups. The reference values have been presence or absence of species for the macrofauna, only. For the species rich mesofauna groups integrated parameters like distribution of systematic groups (Beck et al., 1997) or life history tactics (Ruf, 1998) were defined.

3.2. Case study II: out of the forests

In the second project we focussed on the community analysis. The results for the macrofauna taxa Lumbricidae, Chilopoda, Diplopoda, and Isopoda are shown in Fig. 1a and b. Agricultural sites were well separated from the forests by the investigated macrofauna and within the forests the least acid ones form a group of their own. The correspondence analysis by the species reveals that the difference between agricultural sites and the forests is based on earthworms, but that the arthropods differentiated within the forests. There were no characteristic earthworm species for forest sites, except the two *Dendrobaena* species (*D. octaedra*: DENO and *D. rubida*: DENR). The earthworm species that are responsible for the mull humus form at CRM also occurred at the agricultural sites and are hence not diagnostic for weakly acid forest soils. A complete species list is given in Ruf et al. (2000).

After integration of the mesofauna taxa into the analysis the pattern did not change dramatically. Both the correspondence and the cluster analysis for all taxa show a distinct separation within the main sites groups (Fig. 2a and b). The structuring soil parameters were the pH-value and the *C/N* ratio. The prediction model using only pH-value yields 100% right results for the first step in the cluster analysis and for the second step 92% right prediction using both parameters, pH and *C/N* ratio. Comparing the results of the macrofauna analysis with the analysis done with all investigated taxa, it is evident that the patterns are more distinct the more taxa are included. More details are revealed when mesofauna taxa are incorporated. Especially within the forest sites there is more differentiation due to the microarthropods (mainly Oribatida). The coniferous forest sites BEK, BBK, and SCF form a very close subgroup within the other acid forest sites (Fig. 2a). The agricultural sites are more clearly separated from the forests due to the mite taxa Gamasina and Oribatida. In the Gamasina there are specialised species that only occur at the agricultural sites, whereas in the Oribatida there are only few species that could exist outside the forests. The two marshy pastures (AKG and BRG) again form a subgroup because they are inhabited by a hygophilous community of mites and enchytraeids.

We found site specific communities in the second investigation. The main separating parameter was the land use (forest vs. agricultural use). The second most important parameter was the pH-value and the third the *C/N* ratio. In grasslands the moisture regime was the major factor.

4. Discussion

In both studies we could show that soil fauna communities differentiate between different site types and

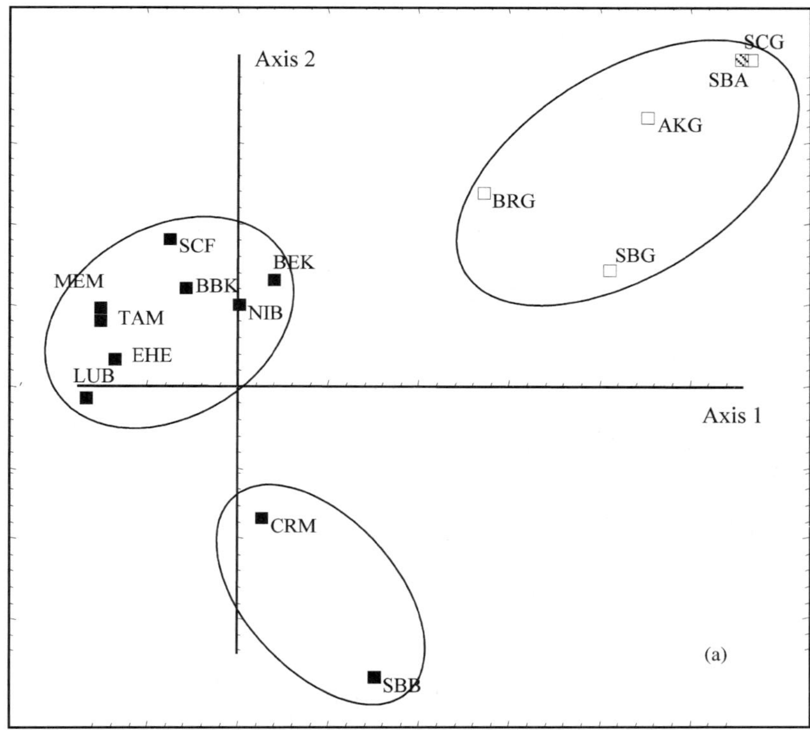

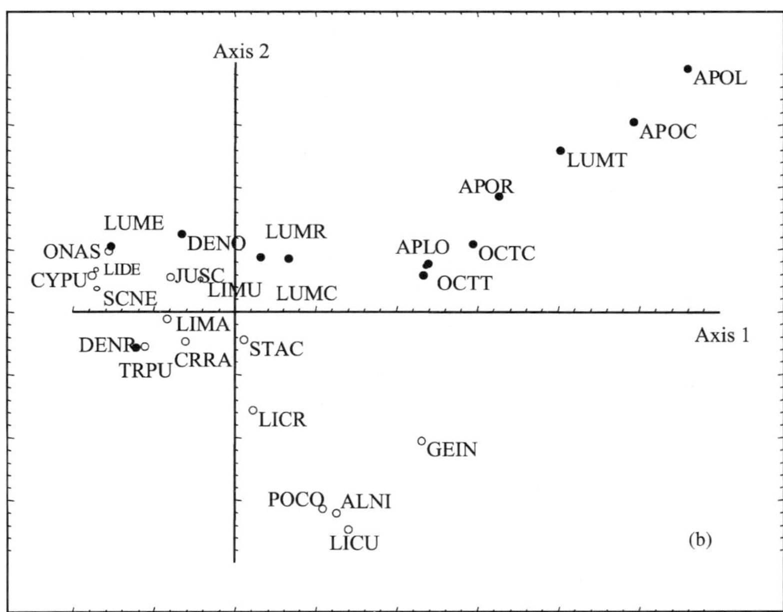

Fig. 1. Correspondence analysis for the macrofauna taxa Lumbricidae, Diplopoda, Chilopoda, and Isopoda. For site parameters refer to Table 1. (a) Sites, grouped according to their fauna assemblage; black squares: forest sites, open squares: pastures, hatched symbol: arable field. (b) Species, grouped according to the assemblage on each site; filled circles represent earthworms, and the empty circles are arthropods.

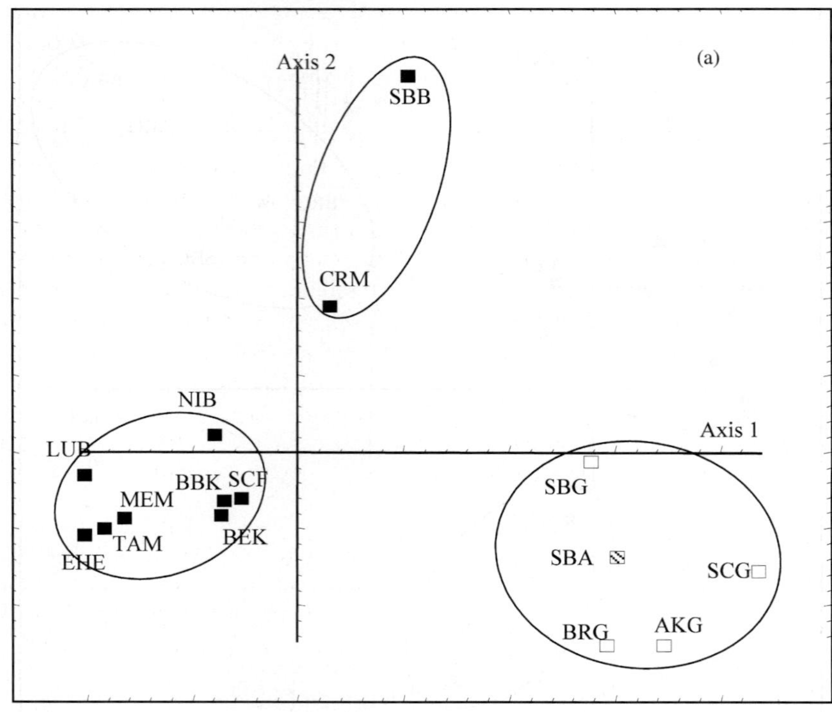

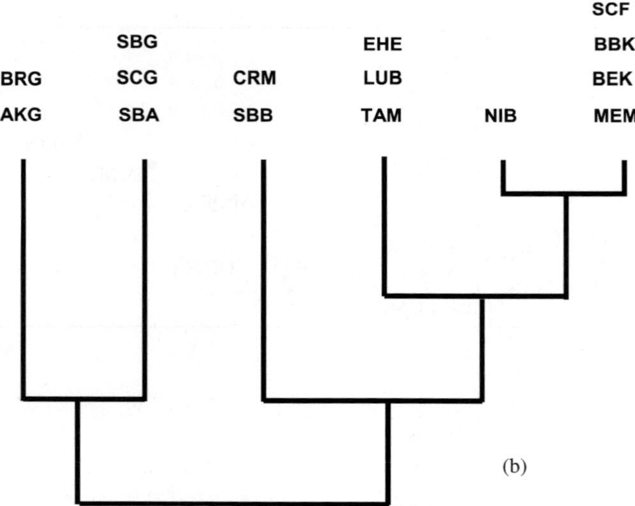

Fig. 2. Correspondence (a) and cluster analysis (b) for all taxa identified to species level. Black squares: forest sites, open squares: pastures, hatched symbol: arable field. For site parameters refer to Table 1.

different land use regimes. So groups of sites with similar features do have similar soil fauna, and on the other hand soil fauna can be used to characterise ecological site groups.

But still the definition of reference values is ambiguous. To predict the occurrence of a certain species at a definite site where it could dwell is a risky task, especially for species rich taxa like Oribatids, predatory mites or enchytaeids. Within these taxa there are very rare species that may not be discovered even in a favourable habitat simply because their numbers are too small to be found with reasonable effort. The difference between sites according to the species composition may appear larger than it really is ecologically. Species may react very specifically, differences are emphasised and shared characters neglected. Reasons for that are manifold: ecosystems represent singular unities which cannot occur exactly identical twice. Each ecosystem has its own history, its very specific characteristics, its surroundings, past and present disturbances. To draw general conclusions we have to find parameters that are not part of this singularity but belong to a whole class of sites or ecosystems and are not restricted to a very specific place and time. To meet this challenge one idea is to define parameters that integrate over species like taxonomic groups, life history tactics, dispersal strategies, or feeding guilds. These parameters are less variable and are more predictable on the level of pedologically similar site groups. Examples for the application of integrated taxonomical parameters are given by Beck et al. (1997) and Maraun and Scheu (2000) for Oribatid mites. Life history tactics for oribatids are distinguished and classified by Siepel (1995) and Behan-Pelletier (1999), for predatory mites by Ruf (1998), and for nematodes by Bongers (1990), and Yeates et al. (1993) classify nematodes to trophic groups.

The contrasting idea to this taxawise integration over species rather is to analyse the whole community of soil fauna and to classify this community. Each taxon gives its own valuable information but the community approach seems to be more differentiated (Sinnige et al., 1992; Spurgeon et al., 1996). Temporal variability in forest soil communities is shown to be low by Bengtsson (1994). This is especially true, when the analysis is related not to species level but to higher taxa. Therefore, a community approach like it is presented here is evidently well suited for bioindication and deserves further elaboration (see also Van Straalen, 1997).

A tiered or hierarchical procedure that relies on macro- as well as on mesofauna taxa has the advantage of incorporating the well known and in terms of energy turnover important species and not neglecting the species that are more sensitive to pollutants and other disturbances. In a first step, earthworms and macroarthropods are suited for characterising the community. Microarthropods and Enchytraeids can be used in a second step because these species rich but taxonomically difficult groups have a higher potential for differentiation. Site specific reference values on a community basis can serve as a benchmark for the habitat quality of soils. Further effort should be spent to the following items:

- Reduction of site parameters, not all are equally important. The main discriminating factor is the land use practise, then pH and *C*/*N* ratio which seem to be key factors structuring the soil fauna community. Further studies have to illuminate whether there is a land use specific hierarchy of the most important soil factors.
- Further definition of specific communities for each abiotically characterised site group. Each site group has to be identified in its spatial dimension and distribution in the landscape.
- Development of criteria for the assessment of a disturbance: What are threshold values at the community level, whose deviation has to be considered as being serious?
- Standardisation of guidelines for the practical performance of this concept for assessing biological soil quality.

5. Conclusion

In addition to the community approach the elaboration of taxa specific parameters to assess soil quality should also be continued. The definition of functional groups and life history tactics seems to be most promising (Siepel, 1994; Van Straalen, 1997). The community analysis is done by multifactorial statistics, indirect or direct gradient analysis. The distinguished groups can be defined as assemblages that occur under specific environmental conditions, similar to associations in plant sociology. This approach was applied in a comparable way by Graefe (1993) and Graefe and Belotti (1999) for the Annelida and can easily be transformed to the whole community. An example of how this can be accomplished is the RIVPAC classification and prediction system (Wright et al., 1993) for running waters. To transfer this system for assessing soil quality in an accordant manner, much more reference sites have to be studied. However our results indicate that this is a manageable task.

Acknowledgements

Both studies have been done by a larger group of soil biologists and soil scientists. Besides the authors they were Silvia Pieper and Werner Kratz from Terra Protecta, Berlin, Werner Kördel at the FHG IUCT, Schmallenberg, and Steffen Woas working at the Museum for Natural History, Karlsruhe. We acknowledge their contributions very much, without their sound expertise the studies could not have been done. Financial support was kindly granted by the German Federal Environmental Agency (UBA, F + E Vorhaben No. 207 05 006) and the Landesanstalt für Umweltschutz Baden-Württemberg.

References

BBodSchG (Bundes-Bodenschutzgesetz), 1998. Gesetz zum Schutz vor schädlichen Bodenveränderungen und zur Sanierung von Altlasten. BGBl I 502 vom, March 17, 1998.

Beck, L., Woas, S., Horak, F., 1997. Taxonomische Ebenen als Basis der Bioindikation—Fallbeispiele aus der Gruppe der Oribatiden. Abh. Ber. Naturkundemus. Görlitz 69, 67–86.

Behan-Pelletier, V., 1999. Oribatid mite biodiversity in agroecosystems: role for bioindication. Agric. Ecosyst. Environ. 74, 411–423.

Bengtsson, J., 1994. Temporal predictability in forest soil communities. J. Anim. Ecol. 63, 635–665.

Boden, A.G., 1994. Bodenkundliche Kartieranleitung, 4th ed. BGR, Hannover.

Bongers, T., 1990. The maturity index: an ecological measure of environmental disturbance based on nematodes species composition. Oecologia 83, 14–19.

Dreher, P., Kördel, W., Knoche, H., 1999. Grundlagen für die Erarbeitung eines Bewertungsrahmens für die Bodenfunktion "Lebensraum für Bodenorganismen". Teil I. Definition und räumliche Zuordnung von bodenkundlich/bodenbiologisch definierten Standorttypen. Mitteilgn. Dtsch. Bodenkundl. Gesellsch. 89, 173–176.

Dunger, W., 1998. Die Bindung zwischen Bodenorganismen und Böden und die biologische Beurteilung von Böden. Bodenschutz 3, 62.68.

Graefe, U., 1993. Die Gliederung von Zersetzergesellschaften für die standortökologische Ansprache. Mitteilgn. Dtsch. Bodenkundl. Gesellsch. 69, 95–98.

Graefe, U., Belotti, E., 1999. Strukturmerkmale der Bodenbiozönose als Grundlage für ein natürliches System der Humusformen. Mitteilgn. Dtsch. Bodenkundl. Gesellsch. 89, 181–184.

LfU (Landesanstalt für Umweltschutz Baden-Württemberg), 1993. Immissionsökologisches Wirkungskataster Baden-Württemberg. Jahresbericht 1990/91. Karlsruhe, 144 pp.

Maraun, M., Scheu, S., 2000. The structure of oribatid mite communities (Acari, Oribatida): patterns, mechanisms and implications for further research. Ecography 23, 374–383.

Römbke, J., Kalsch, W., 2000. Protokoll des Internationalen Fachgesprächs über "Ansätze für biologische Bewertungsstrategien und -konzepte im Bodenschutz". Bericht für das Bundesministerium für Umwelt, Naturschutz und Reaktorsicherheit, Bonn.

Römbke, J., Beck, L., Förster, B., Fründ, H.-C., Horak, F., Ruf, A., Rosciczewski, K., Scheurig, M., Woas, S., 1997. Boden als Lebensraum für Bodenorganismen und die bodenbiologische Standortklassifikation. Eine Literaturstudie. Texte und Berichte zum Bodenschutz, 4/97. Landesanstalt für Umweltschutz Baden-Württemberg, Karlsruhe.

Römbke, J., Dreher, P., Beck, L., Hammel, W., Hund, K., Knoche, H., Kördel, W., Kratz, W., Moser, T., Pieper, S., Ruf, A., Spelda, J., Woas, S., 2000. Bodenbiologische Bodengüte-Klassen, UBA Text, Berlin.

Ruf, A., 1998. A maturity index for predatory soil mites (Mesostigmata: Gamasina) as indicator of environmental impacts of pollution on forest soils. Appl. Soil Ecol. 9, 447–452.

Ruf, A., Beck, L., Hammel, W., Hund, K., Kratz, W., Römbke, J., Spelda, J., 1999. Grundlagen für die Erarbeitung eines Bewertungsrahmens für die Bodenfunktion "Lebensraum für Bodenorganismen". Teil II. Erste Ergebnisse zur Anwendung von bodenkundlich/bodenbiologisch definierten Standorttypen. Mitteilgn. Dtsch. Bodenkundl. Gesellsch. 89, 177–180.

Ruf, A., Beck, L., Römbke, J., Spelda, J., 2000. Standortspezifische Erwartungswerte für die Gemeinschaftsstruktur ausgewählter Taxa der Bodenfauna als Bodenqualitätskriterium. Ber. Nat.-med. Ver. Innsbruck 87, 361–380.

Schouten, T., Bloem, J., Didden, W.A.M., Rutgers, M., Siepel, H., Posthuma, L., Breure, A.M., 2000. Development of a biological indicator for soil quality. SETAC Globe 4, 30–32.

Siepel, H., 1994. Life-history tactics of soil microarthropods. Biol. Fertil. Soils 19, 263–278.

Siepel, H., 1995. Application of microarthropod life history-tactics in nature management and ecotoxicology. Biol. Fertil. Soils 19, 75–83.

Sinnige, N., Tamis, W., Klijn, F., 1992. Indeling van Bodemfauna in ecologische Soortgroupen. Centrum voor Milieukunde, Rijksuniversität Leiden.

Spurgeon, D.J., Sandifer, R.D., Hopkin, S.P., 1996. The use of macro-invertebrates for population and community monitoring of metal contamination—indicator taxa, effect parameters and the need for a soil invertebrate prediction and classification scheme (SIVPACS). In: Van Straalen, N.M., Krivolutsky, D.A. (Eds.), Bioindicator Systems for Soil Pollution. Kluwer Academic Publishers, Dordrecht, pp. 95–109.

Van Straalen, N.M., 1997. Community structure of soil arthropods as a bioindicator of soil health. In: Pankhurst, C.E., Doube, B.M., Gupta, V.V.S.R. (Eds.), Biological Indicators of Soil Health. CAB International, Wallingford, pp. 235–264.

Weeks, J.M., 1997. A demonstration of the feasibility of SOILPACS. Environment Agency, London.

Wright, J.F., Furse, M.T., Armitage, P.D., 1993. RIVPACS—a technique for evaluating the biological quality of rivers in the UK. Eur. Water Poll. Control 3, 15–25.

Yeates, G.W., Bongers, T., de Goede, R.G.M., Freckman, D.W., Georgieva, S.S., 1993. Feeding habits in nematode families and genera—an outline for soil ecologists. J. Nematol. 25, 315–331.

ELSEVIER

Available online at www.sciencedirect.com

Agriculture, Ecosystems and Environment 98 (2003) 273–283

Agriculture Ecosystems & Environment

www.elsevier.com/locate/agee

On the quality of soil biodiversity indicators: abiotic and biotic parameters as predictors of soil faunal richness at different spatial scales

Klemens Ekschmitt [a,*], Thomas Stierhof [a], Jens Dauber [a], Kurt Kreimes [b], Volkmar Wolters [a]

[a] IFZ—Department of Animal Ecology, Justus Liebig University, Heinrich Buff Ring 26-32, D-35392 Giessen, Germany
[b] Baden-Wuerttemberg Ministry of Environment and Transport, Kernerplatz 9, D-70182 Stuttgart, Germany

Abstract

Direct measurement of soil biodiversity is expensive, and therefore a substitution of measurement by indication is desirable. We analysed three large datasets for the potential to predict the diversity of soil faunal groups from other parameters. The datasets represent different spatial scales, namely grassland nematodes on an European scale, forest collembola on a regional scale, and grassland ants on a local scale. We tested two groups of parameters as possible surrogates for species richness: (1) environmental parameters, such as climate, soil and vegetation characteristics, and (2) community parameters, such as higher taxon richness, indicator taxa, and maximum dominance.

Climate and soil parameters were significantly correlated with biodiversity in all datasets. However, in spite of the large variety of measurement types analysed, prediction quality of environmental variables was weak and the explained proportion of variance ranged generally below 50%. Richness was subject to considerable stochastic variation among subsamples (CV = 20–60%) thereby evading a narrow correlation with environmental parameters. Higher taxon richness, based on taxa of intermediate hierarchical order, proved the best predictor of richness in collembola and nematodes explaining 55 and 89% of total variance, respectively. A combination of two ant species was the best predictor of ant richness explaining 59% of total variance.

The authors conclude that a rough guess of soil faunal diversity can be cost-effectively derived from environmental data while an estimate of moderate quality can be obtained with reduced taxonomic effort. The precise richness of a soil community, however, is subject to autogeneous community dynamics, to biotic interactions with other populations, and to conditions in the past, and can therefore only be retrieved by immediate investigation of the community itself. Criteria for the quality of indicator parameters are discussed.

Keywords: Diversity indicator; Soil biodiversity; Nematoda; Collembola; Formicidae

1. Introduction

To measure soil biodiversity requires high taxonomic expertise, takes considerable time-effort and is therefore expensive. With the singular exception of earthworms in temperate and northern soils, the identification of soil fauna poses challenges of extraction efficiency, sample processing, visual recognition, and

* Corresponding author. Tel.: +49-641-99-35712; fax: +49-641-99-35709.
E-mail address: klemens.ekschmitt@allzool.bio.uni-giessen.de (K. Ekschmitt).

0167-8809/$ – see front matter
doi:10.1016/S0167-8809(03)00087-2

difficult taxonomy. These technical hurdles result, for example, in an average performance of five samples per day for a collembola expert and of two samples per day for a nematode expert. In this paper, we investigate to which degree the diversity of soil faunal groups can be assessed from other parameters which are easier to retrieve than a direct quantification of diversity itself.

Various concepts how to indicate biodiversity have been proposed in the literature, such as indicator groups and higher taxon richness (Mikusinski et al., 2001; Chase et al., 2000; Gaston, 2000; Kerr et al., 2000; Carroll, 1998). Here, we consider the term *diversity* indicator in the broadest sense, we accept any variable that is sufficiently correlated with species richness to serve as a surrogate of richness. As a familiar paradigm of the indicator principle we consider photometry, where colour intensity is used as an indicator of substance concentration. Very much like in photometry, we judged indication quality by the co-linearity and by the scatter of the relation between indicator and target variable. We tried to improve the results by using multivariate measures, corresponding, for example to a differential measurement of two wavelengths in the photometry paradigm. We tested two groups of parameters as possible indicators of soil biodiversity: (1) environmental parameters which are expected to regulate soil fauna composition, such as climate, soil, and vegetation characteristics, and (2) measures inherent to the soil fauna community itself, such as higher taxon richness, indicator taxa, and maximum dominance. We analysed three large datasets representing different spatial scales, namely grassland nematodes on an European scale, forest collembola on a regional scale, and grassland ants on a local scale. Although we present the best richness indicator obtained for each dataset, our main focus is on evaluating the achievable quality of biodiversity indication, rather than on proposing optimal diversity indicators for specific soil faunal groups. The main question addressed is, whether soil biodiversity can efficiently be indicated by environmental parameters, or whether an analysis of the soil fauna itself is essentially inevitable.

2. Material and methods

In this paper, we present a concise description of study sites and methods. Detailed descriptions are provided in Ekschmitt et al. (1999), Frede and Bach (1999), Waldhardt et al. (1999), Wolters et al. (1999), LFU (1994), and Schick and Kreimes (1993). Fig. 1 presents an overview of the geographical location of sample sites, Table 1 summarises the measured parameters, and Table 2 gives an overview of sample sizes and other properties of the sampling designs applied. While sampling and identification of soil fauna followed acknowledged standard methods in each case, the environmental data differed largely among the three investigations. In the European project, actual environmental data were measured for each soil sample. In the regional project, long-term average environmental data were available from the regional authorities. In the local project, environmental information was largely calculated from topographical data present in a GIS database.

The sites of the nematode dataset lie on a climatic cross-gradient, ranging from oceanic to continental and from boreal to Mediterranean climate, and simultaneously representing six major types of European grassland: Northern tundra (Abisko, Sweden), Atlantic heath (Otterburn, United Kingdom), wet grassland (Wageningen, The Netherlands), semi-natural temperate grassland (Linden, Germany), East European steppe (Pusztaszer, Hungary), and Mediterranean garigue (Mt. Vermion, Greece). To extend the validity of the results beyond any chance effects of 1 year's weather, the microclimate at each site was manipulated using light reflection, shading, wind shelter, rain protection, and irrigation. From a full factorial table with six levels of temperature and six levels of moisture, the optimal set of 14 treatments was calculated as a D-optimal design using Fedorov's method (Cook and Nachtsheim, 1980). In Hungary, an additional treatment with high humidity and intermediate temperature was included. The Swedish experiment was done as a transplantation experiment due to the short vegetation period and the remoteness of the Abisco site.

At all sites, six samples were taken from each treatment during the period from June to December 1996. For each sample, 12 soil cores of 1.65 cm diameter and 10 cm depth were combined and sieved (2 mm mesh) prior to chemical analysis, and five similar cores were combined and used for nematode extraction. In total, 510 bulk samples from 85 treatments were taken for soil analysis, and for nematode analysis, respectively.

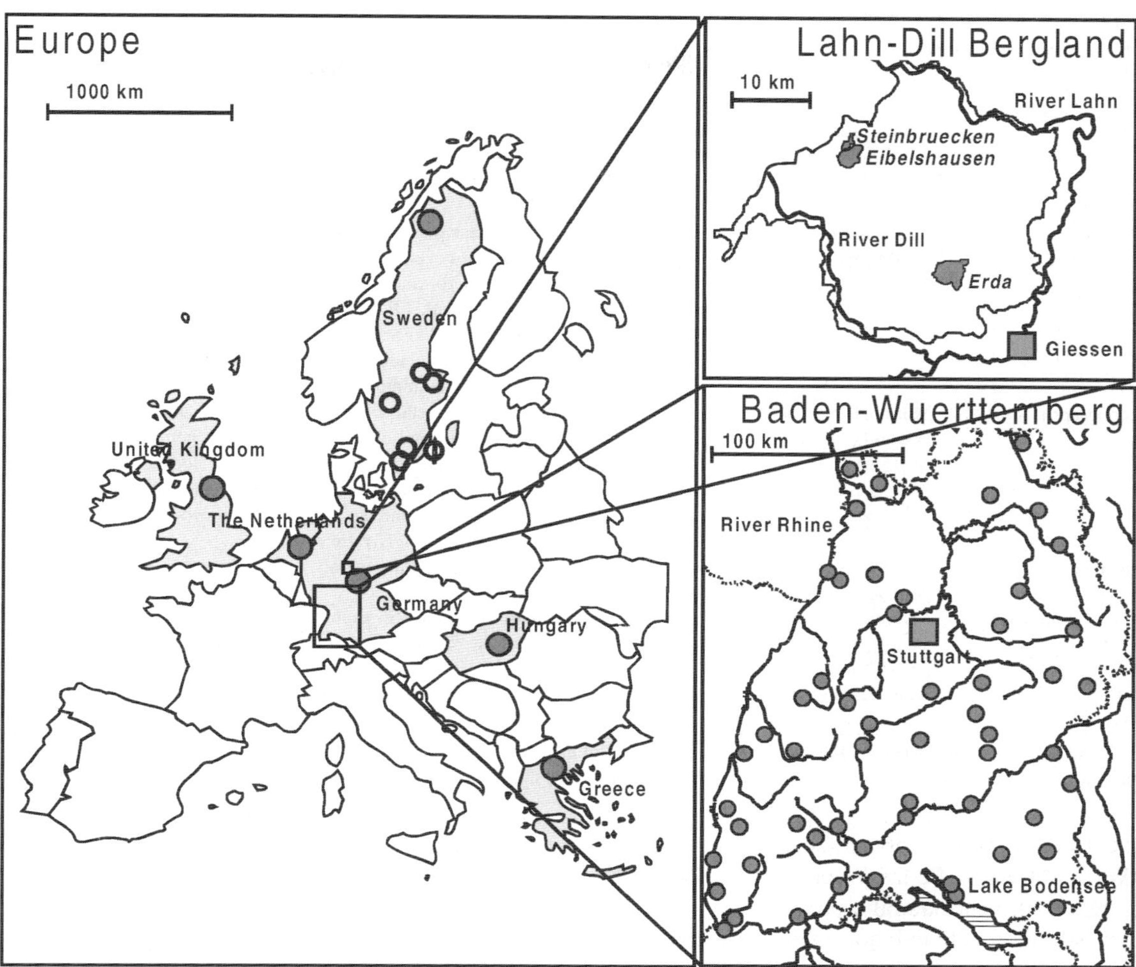

Fig. 1. Geographical locations of sampling sites. (a) Europe: solid circles indicate primary nematode sampling sites, open circles indicate transplantation sites in Sweden. (b) Lahn-Dill region: shaded areas indicate the three locations where ant nests were recorded. (c) State Baden-Wuerttemberg: solid circles indicate the forest sites sampled for collembola. Scales of the maps, and the location of the two smaller maps within Europe are indicated in the figure.

For each sample, soil microclimate was quantified by measuring soil temperature and soil water content. Nematodes were isolated using a modified Cobb's method (s'Jacob and Van Bezooijen, 1984) and were identified to genus level according to Bongers (1988) and Andrassy (1984). Microbial biomass-C was measured using fumigation-incubation (Ross, 1990; Jenkinson and Powlson, 1976). Soil ergosterol content was recorded as a measure of active fungal biomass (Djajakirana et al., 1996). Bacterial substrate utilisation was measured using a modified BIOLOG assay (Vahjen et al., 1995). Soil respiration was measured by absorption in alkali (Isermeyer, 1952). NH_4-N was measured by distillation using MgO to liberate ammonia, NO_3-N was determined by distillation using Devarda's alloy for reduction of NO_3^- to NH_4^+, and organic N of the soil was assessed from Kjeldahl analysis (Allen, 1974).

The regional dataset is based on samples from 57 permanent monitoring sites held by the Baden-Wuerttemberg agency for environmental protection (Landesanstalt fuer Umweltschutz Baden-Wuerttemberg,

Table 1
List of measured parameters[a]

Parameter group	Nematoda	Collembola	Formicidae
Climate	Actual precipitation (mm/day), soil temperature at 5 cm (°C), soil water content (% WHC)	Long-term annual precipitation (mm), long-term annual mean air temperature (°C)	Transport capacity after Moore and topographical wetness index (water balance), hillshade and reflectance map (insulation)
Vegetation	Plant production (dry matter) (g DM/m^2)	Species richness of cormophytes and of mosses, coverage of tree-, shrub-, herb- and moss-layer (%area), Ellenberg indicator values (*Zeigerzahlen*) for light, nitrogen, acidity, temperature and humidity	Plant richness, total plant coverage (%area), vegetation height (cm)
Soil/topography	Microbial biomass (mg C/g DM soil), soil ergosterol content (μg/g DM soil), BIOLOG AWCD (ext. 650/g DM soil), soil contents of NO_3, NH_4 and organic N (mg N/g DM soil)	Proportions of skeletal grain (%vol.), sand, silt and clay (%DM), water holding capacity (%vol.), soil bulk density (g/l), soil organic matter content (μg/g DM soil), soil C/N ratio	Height above drainage channel, slope, hemispherical dispersion after Hodgson (topographical roughness), maximum curvature after wood
Indicator groups	Dominance of feeding types, genera, families and orders	Dominance of species, genera and families	Dominance of species, genera and subfamilies
Maximum dominance	Maximum dominance in each sample of any feeding type, genus, family and order	Maximum dominance in each sample of any species, genus and family	Maximum dominance in each sample of any species, genus and subfamily
Higher taxon richness	Numbers of families and orders	Numbers of genera and families	Numbers of genera and subfamilies

[a] The table gives an overview of the measurements supplied in each of the three datasets. The measured variables are classified into abiotic (climate, vegetation, soil) and biotic parameters (indicator groups, dominance, higher taxon richness).

LFU). The sites are distributed across an area of ca. 230 km × 180 km in Baden-Wuerttemberg, Germany. They represent near-natural woodlands, i.e. they are neither strongly polluted nor intensively managed. Sampling plots are 20 m × 24 m large, surrounded by a buffer zone. No forestry measures are allowed in the plots and in the buffer zones. Most stands are deciduous and mixed forests dominated by beech. Some sites bear coniferous forests dominated by fir and spruce, and there are also a few riverine woods.

Table 2
Sampling designs and diversity dimensions[a]

Parameter	Nematoda	Collembola	Formicidae
Dimension of spatial scale	1000 km	100 km	10 km
Number of plots	85	57	20
Samples per plot	6	9	15
Average counts per sample	218 specimens	179 specimens	2 nests
Overall richness observed	121 genera	117 species	13 species
Higher taxon richness 1 observed	55 families	52 genera	4 genera
Higher taxon richness 2 observed	11 orders	5 families	3 subfamilies
Average richness per plot observed	35 genera	37 species	5 species
Average richness per plot, ACE extrapolation	40 genera	43 species	5 species
95%-confidence range of richness across samples	±38%	±71%	±126%

[a] The table presents summary data that characterise the quality of the datasets analysed. Spatial extent, number of plots and sample size are given for each dataset. The taxonomic diversity represented in each of the datasets is indicated by the average abundance and the total richness observed, followed by information on the richness of higher taxa and the richness within single plots.

Altitudes range from 98 m asl near the river Rhine up to 1260 m asl in the Black Forest. Soil acidity of the A-horizon varies between pH 3 and 7 across sites. Long-term annual temperature ranges from 5.5 to 9.5 °C and long-term annual precipitation ranges from 600 to 1700 mm across sites.

Collembola were sampled in the framework of a long-term biomonitoring programme (Oekologisches Wirkungskataster) from 1986 to 1988 and from 1990 to 1993. Half of the sites were visited every year in turns of 2 years. Sampling took place three times a year, i.e. in spring, summer and autumn. Three soil cores of 6.8 cm diameter and 8.0 cm depth were taken from each plot and combined. Soil arthropods were expelled from soil by means of a modified MacFadyen high gradient canister extractor and were stored in 70% ethanol. The Collembola were selected under a dissecting microscope and bleached in a mixture of lactic acid and glycerine, thereafter transferred onto slides. Identification to species level followed the nomenclature of Gisin (1960). Extensive data on vegetation, as well as on physicochemical soil properties were made available by previous investigations and are held in an environmental database by the LFU. Measurement methods are described in a dedicated handbook (LFU, 1994). Since single plant species and soil chemical elements did not show significant correlations with collembola in a previous analysis, a set of integrative plant parameters and mainly soil physical characteristics were selected for the present investigation, in addition to climate parameters (Table 1).

The local dataset is based on studies carried out in the rural districts of Steinbruecken, Eibelshausen and Hohenahr-Erda (Hesse, Germany) in the framework of the Collaborative Research Center "Land-use options for peripheral regions" (DFG, SFB 299). The region is a low mountain range at 270–385 m asl. Major soil types are regosols and cambisols, mean annual temperatures range from 6.5 to 8 °C, and mean annual precipitation varies from 700 to 1200 mm depending on the district. A small-scale mosaic of different land-use types characterises the cultivated area of the region. Poor physical soil conditions, a climate unsuitable for crops, and farm sizes too small to be economically viable resulted in a considerable decrease of cropland since 1950, while grassland and fallow land gradually increased. Much of the area is now covered by moderately managed grasslands, mainly by meadows mown once per year and occasionally grazed by sheep or cattle. *Festuca rubra—Agrostis tenuis* communities dominate the vegetation at the sites converted from cropland to grassland before 1980, whereas no typical plant communities of matured grassland have yet developed on the sites that were converted more recently.

Species richness and nest abundance of ants were determined from June to August 1999 on a number of randomly chosen subplots at each of 20 grassland sites. A minimum of 5 m was left between the subplots to avoid an overestimation of patchily distributed species in seemingly homogeneous habitats (Andersen, 1990). In addition, 2 m were left between the plots and the border zones of the sites to avoid edge effects. Ant nests without conspicuous mounds on the soil surface were recorded by raising the turf with a small chisel. Up to 10 worker ants were collected from each nest and fixed in 70% ethanol immediately. Taxonomic identification to the species level was based on Seifert (1996).

The grassland vegetation was recorded during May–August 1997, excluding a strip of 2 m width along the borders of sites to avoid possible edge effects. GIS-data based on a 20 m × 20 m grid were used from the project's database in order to indicate microclimate by topographic parameters related to insulation and water balance, and to indicate soil properties by parameters related to height, slope, and topographical roughness (Hodgson and Gaile, 1999; Wood, 1996; Moore and Wilson, 1992).

In order to eliminate sampling errors of richness estimates depending on sample size, extrapolated total richness, rather than observed richness was used as diversity parameter throughout this study. Richness extrapolations were calculated according to the ACE-estimator (Chazdon et al., 1998; Chao et al., 1993). Where ACE did not apply due to numerical restrictions, the CHAO1-estimator was used, as recommended by Colwell (1997). All calculations were programmed and conducted in MS Excel (Microsoft, Redmont, USA). The quality of indicators was tested by multivariate regressions comprising linear and quadratic terms. Regression models were confined to contain only significant variables by applying forward stepwise inclusion of independent parameters. Indicator quality was primarily assessed by the coefficient of determination R^2. If a regression model explained less than 50% of richness variance ($R^2 < 0.5$) then

Table 3
Overview of regression results[a]

Parameter	Nematoda	Collembola	Formicidae
(a) Environmental variables			
Climate			
Best single parameter	Average soil temperature, linear + quadratic ($R^2 = 0.24$, Rel.Err. = 32%, $P < 0.0001$)	Long-term annual precipitation, linear ($R^2 = 0.18$, Rel.Err. = 17%, $P = 0.0011$)	Reflectance map (insulation), linear + quadratic ($R^2 = 0.38$, Rel.Err. = 29%, $P = 0.0167$)
Best multiple model	Identical with univariate model above	Identical with univariate model above	Identical with univariate model above
Vegetation			
Best single parameter	Annual plant production, linear + quadratic ($R^2 = 0.10$, Rel.Err. = 37%, $P = 0.0289$)	Percentage coverage of tree layer, linear ($R^2 = 0.12$, Rel.Err. = 17%, $P = 0.0092$)	Plant richness, coverage, vegetation height (not significant)
Best multiple model	Not available	Coverages of tree layer and herb layer, linear + quadratic ($R^2 = 0.21$, Rel.Err. = 17%, $P = 0.0062$)	Plant richness, coverage, plant height (not significant)
Soil/topography			
Best single parameter	Soil microbial C, linear ($R^2 = 0.43$, Rel.Err. = 28%, $P < 0.0001$)	Soil organic matter, linear + quadratic ($R^2 = 0.16$, Rel.Err. = 17%, $P = 0.0189$)	Hemispherical dispersion (roughness), linear ($R^2 = 0.29$, Rel.Err. = 26%, $P = 0.0185$)
Best multiple model	Soil microbial C, ergosterol and BIOLOG, linear + quadratic ($R^2 = 0.54$, Rel.Err. = 25%, $P < 0.0001$)	Identical with univariate model above	Height above drainage channel, hemispherical dispersion, linear ($R^2 = 0.47$, Rel.Err. = 23%, $P = 0.0062$)
Climate, vegetation and soil combined			
Best multiple model	Soil microbial C, ergosterol and annual plant production, linear + quadratic ($R^2 = 0.57$, Rel.Err. = 26%, $P < 0.0001$)	Long-term annual precipitation, coverages of tree layer and herb layer, linear + quadratic ($R^2 = 0.34$, Rel.Err. = 15%, $P = 0.0002$)	Reflectance map, height above drainage channel and hemispherical dispersion, linear ($R^2 = 0.60$, Rel.Err. = 21%, $P = 0.0027$)
(b) Community parameters			
Indicator taxa			
Best single parameter	Dominance of family Paratylenchidae, linear ($R^2 = 0.61$, Rel.Err. = 23%, $P < 0.0001$)	Dominance of family Onychiuridae, linear + quadratic ($R^2 = 0.13$, Rel.Err. = 17%, $P = 0.0256$)	Dominance of *M. rubra*, linear ($R^2 = 0.44$, Rel.Err. = 27%, $P = 0.0014$)
Best multiple model	Paratylenchidae, Teratocephalidae and Diphtherophoridae, linear ($R^2 = 0.82$, Rel.Err. = 16%, $P < 0.0001$)	Isotomidae, Onychiuridae and Sminthuridae, linear ($R^2 = 0.15$, Rel.Err. = 17%, $P = 0.0324$)	*M. rubra* and *M. schencki*, linear ($R^2 = 0.59$, Rel.Err. = 24%, $P = 0.0006$)
Dominance			
Best regression	Maximum dominance of orders, linear ($R^2 = 0.51$, Rel.Err. = 25%, $P < 0.0001$)	Maximum dominance of families, linear + quadratic ($R^2 = 0.18$, Rel.Err. = 19%, $P = 0.0049$)	Maximum dominance at any taxonomic level (not significant)
Higher taxon richness			
Intermediate level taxon	Richness of nematode families, linear ($R^2 = 0.89$, Rel.Err. = 12%, $P < 0.0001$)	Richness of collembolan genera, linear ($R^2 = 0.55$, Rel.Err. = 12%, $P < 0.0001$)	Richness of ant genera, linear ($R^2 = 0.36$, Rel.Err. = 29%, $P = 0.0048$)
High level taxon	Richness of nematode orders, linear + quadratic ($R^2 = 0.38$, Rel.Err. = 29%, $P < 0.0001$)	Richness of collembolan families (not significant)	Richness of ant subfamilies, linear ($R^2 = 0.35$, Rel.Err. = 29%, $P = 0.0062$)

[a] The table lists the best regressions found in each category of potential richness indicators, together with an indication whether quadratic terms were significant. Where appropriate, multivariate models are included. Regression quality is specified by the coefficient of determination (R^2), the relative error of estimate (Rel.Err.) and the error probability (P).

indicator quality was judged as weak. The Statistica for Windows package (StatSoft Inc., Tulsa, USA) was used for all statistical tests.

3. Results

In total, 121 nematode genera, 117 collembola species and 13 ant species were recorded, corresponding to an average extrapolated total richness of 40 nematode genera, 43 collembola species and 5 ant species per sampling site (Table 2). Variation of richness estimates between replicate samples was high, leading to 95% confidence ranges of 25–55 genera for nematodes, 13–74 species for collembola and 0–10 species for ants, per sample on average.

Regression results are summarised in Table 3. The table depicts the indicators with the highest coefficient of determination (R^2) found among the tested variables in the categories climate, vegetation, soil, indicator taxa, maximum dominance and higher taxon richness.

Climate was a significant but weak ($R^2 < 0.25$) predictor of richness in all three datasets, showing a strongly humped regression line on the largest scale (Fig. 3b). Vegetation parameters were weak predictors of collembolan richness ($R^2 = 0.2$), poor predictors of nematode richness ($R^2 = 0.1$), and they were not significantly correlated with ant richness. Soil parameters and topographical parameters showed better but still weak correlations with nematode richness ($R^2 =$ 0.4–0.5) and ant richness ($R^2 =$ 0.3–0.5), and the correlation with collembola did only explain less than 20% of richness variance ($R^2 = 0.16$). A combination of climate, vegetation and soil parameters yielded R^2 values of ca. 0.6 for nematode and ant richness, and 0.35 for collembolan richness. To summarise, within the set of 44 environmental parameters tested, only six parameters and parameter combinations correlated with soil faunal richness by an R^2 higher than 0.4. The large majority of environmental parameters, including multivariate parameter combinations, explained less than 25% of variance in soil faunal richness, or were not significantly correlated with soil faunal richness at all.

In contrast to the environmental variables, parameters derived from community composition repeatedly exhibited R^2 larger than 0.5. Higher taxon richness, based on taxa of intermediate hierarchical order,

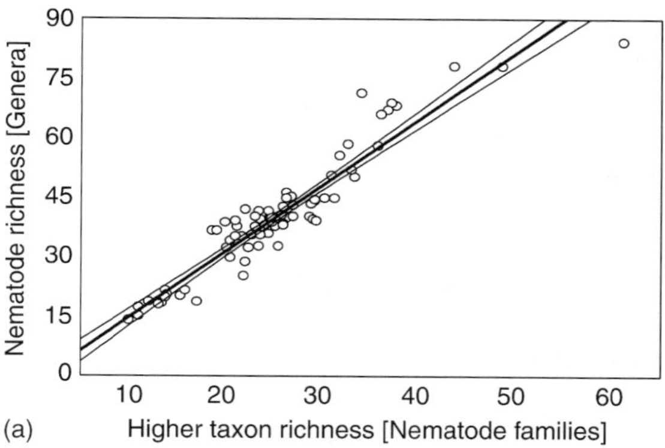

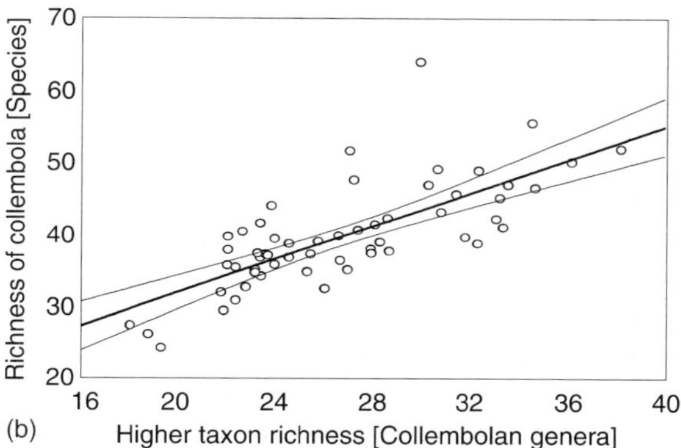

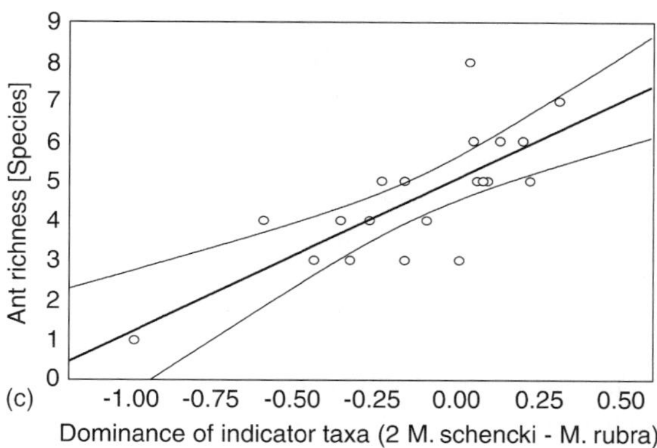

Fig. 2. The best community-based indicators found. Among all parameters tested, higher taxon richness proved the best predictor for nematode richness (a) and collembolan richness (b). Ant richness was well predicted by the combined dominance of the ant species *M. schencki* and *M. rubra* (c).

proved the best predictor of richness in collembola and nematodes explaining nearly 60 and 90% of richness variance, respectively ($R^2 = 0.55$ and 0.89). A combination of the ant species *Myrmica rubra* and *Myrmica schencki* was the best predictor of ant richness explaining nearly 60% of richness variance ($R^2 = 0.59$) (Fig. 2).

4. Discussion

In this paper, we investigate the potential of a range of parameters to serve as indicators of soil biodiversity. To this end, we analysed three datasets of a very different nature. If results share similarities across these datasets then we assume that the similarities point to a general rule and that comparable results can be expected in future investigations.

Among the three datasets analysed, the data on grassland nematodes presented the largest gradients covering a spatial scale of thousands of kilometres, while simultaneously providing actual environmental data from the immediate vicinity of each soil sample. This dataset represents a typical "researcher's" situation, where specific measurements are available for each single sample. The data on forest collembola, in contrast, represent an "environmentalist's" situation where data covering a regional scale have been collected over several years. Data from different categories, such as vegetation and soil originate from separate sampling campaigns and are loosely linked in time and space. They characterise the medium term average local situation. Finally, the data on grassland ants represent a modern perspective where environmental information is not based on direct measurement, but is indirectly inferred from topographical data and remote sensing data integrated into a GIS database.

Our results confirm the general hypothesis that soil biodiversity is driven by environmental factors, such as climate, vegetation and soil properties. At the same time, the results elucidate that soil biodiversity is only partly formed by the state of the environment. Environmental variables explained roughly half (34–60%) of the variance in soil animal richness, leaving the other half unexplained and therefore unpredictable from environmental factors alone. Soil animal richness varied considerably between replicate samples (CV = 20–60%) indicating a high degree of independence of richness from environmental conditions.

Fig. 3 illustrates three major causes why many environmental parameters did not prove good indicators of soil faunal richness: high scatter, non-linearity and non-representativeness. The most general cause was evidently a highly scattered relationship between indicator variable and richness (Fig. 3a). Furthermore, the relationship was strongly non-linear in many cases (Fig. 3b), as is demonstrated by the significance of quadratic terms in the regression results (Table 2). While a non-linear relationship between indicator and richness does not present a strict cut-off criterion, it complicates the indication of biodiversity because there is no unequivocal coincidence of increase, or decrease, of richness with an increase of the indicator value. A third and important deficiency of indicator quality is demonstrated in Fig. 3c. Here, the overall correlation was based on a few data points located to the right of the graph, while the majority of data, marked in the graph by a circle, exhibited a complete lack of correlation between indicator variable and richness. Thus, the overall correlation was not representative for most of the data and a prediction of richness from this data appears less justified. High scatter, non-linearity, non-representativeness, and combinations of these excluded many of the tested environmental variables from the list of potential biodiversity indicators.

However, there are more impediments to the prediction of soil faunal richness from environmental data than statistical properties of regressions. As has been forcefully argued by Salt and Hollick (1946), the observed distribution pattern of soil animals can bear little relation to presently measured environmental conditions, because the faunal distribution pattern is influenced: (1) by autogeneous dynamics of the population under consideration itself, (2) by interaction of this population with predators, parasites and competitors, as well as (3) by conditions and forces active in the past but indiscernible in the present. Depending on the quantitative importance of these mechanisms, distribution patterns of populations and therefore composition and diversity of communities can be largely independent from presently observed environmental patterns. Based on these arguments, we assume that correlations between environmental variables and soil faunal richness exhibiting R^2 values as high as 0.6,

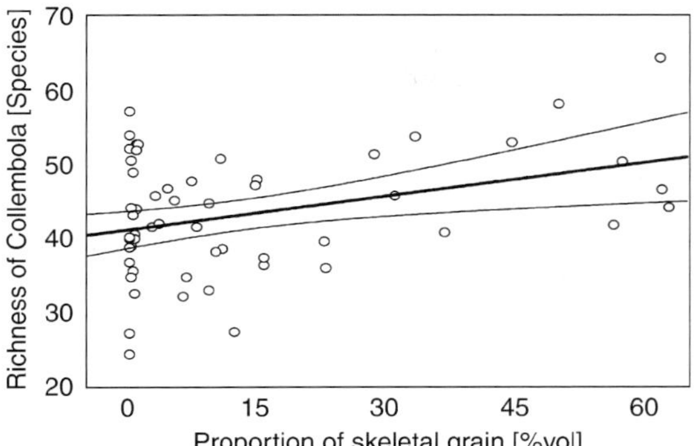

(a) Large scatter

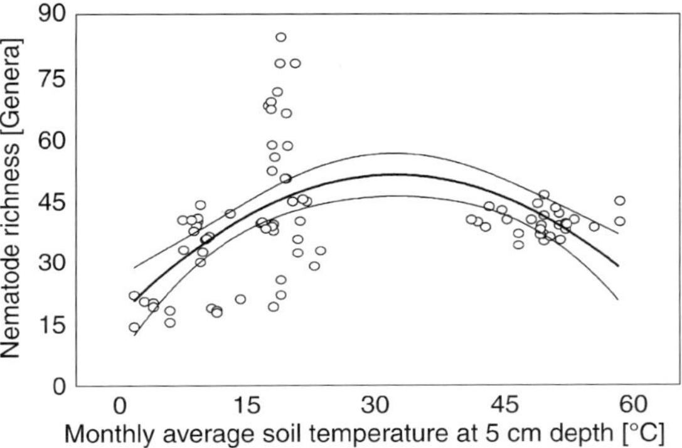

(b) Non-linearity

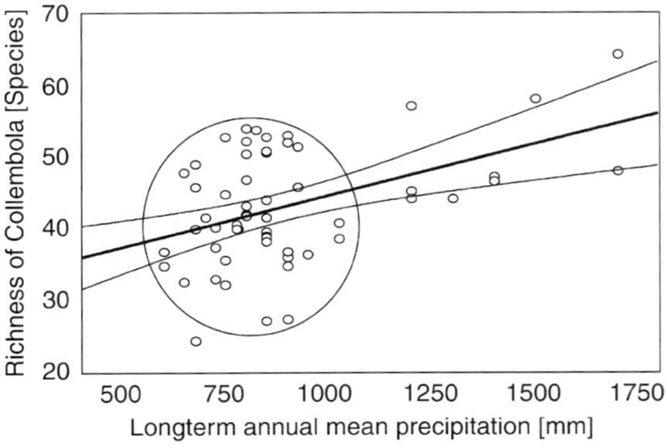

(c) Non-representativeness

Fig. 3. Examples of undesirable indicator properties. The figure demonstrates three types of relationships between indicators and target variables that deteriorate the quality of indication: (a) indicator variable and target variable show a scattered relation, (b) the relation between indicator variable and target variable is strongly non-linear, (c) the regression between indicator and target variable is coined by a minority of data points and is not generally valid in the majority of datapoints.

as they were observed in this study, represent a good achievement and will not generally be improvable by performing more or different measurements, nor by augmenting sampling effort. An inevitable consequence of such limited correlation strength is that prediction accuracy is also limited. In the present analysis, soil faunal richness as predicted from environmental variables deviated from measured richness by a coefficient of variation of ±15–26%, depending on the faunal group. This implies that 95% of predictions deviated up to ±30–51% from the true value, while 5% of predictions fell even beyond this range.

Given the partial independence of richness patterns from environmental conditions, it appears promising to retrieve information on biodiversity from the community itself and to save labour effort by reducing the amount of taxonomic identification. In fact, correlations based on community information such as higher taxon richness or indicator species yielded R^2 values of 0.55 up to 0.89. In nematodes and collembola, these correlations improved substantially over the best correlations with environmental variables, and in ants the same level of R^2 as with environmental variables was achieved. In all three faunal groups, the relationships between richness and the best community-based indicators were highly linear and representative (Fig. 2). However, even the best community-based indicators exhibited correlations, which would be seen far from acceptable as a calibration of a photometric measurement. The improvement of correlation quality meant that 95% of richness predictions ranged within ±24% of the true value in nematodes and collembola, and within ±46% of the true value in ants.

5. Conclusions

The authors conclude that a rough guess of soil faunal diversity can be cost-effectively derived from environmental data. Here, particularly the good predictions of ant richness provided by topographical data suggest that basic predictions of soil biodiversity can be developed in the future from GIS databases. The richness of a soil community is however likely to depend on autogeneous community dynamics, on biotic interactions with other populations, and on conditions in the past. Our results suggest that this fact is rather independent of the spatial extent and the spatial resolution at which measurements are made. Reliable richness estimates can therefore only be retrieved by immediate investigation of the community itself. Here, higher taxon richness proved a good diversity indicator in nematodes and collembola, where a large number of 40 taxa was present. In contrast, the richness of ant communities, which exhibited a low number of only five species per plot, was best predicted by a small set of indicator species. Indicator species may apply more generally to groups in which a relatively small number of species is present (Andersen, 1995). Even in these improved, community-based indicators, accuracy of richness predictions remained fairly low. We conclude that a full identification of the community at high taxonomic resolution is inevitable whenever precise knowledge on the diversity of a soil community is desired.

Acknowledgements

The nematode study was conducted within the project "Diversity Effects in Grassland Ecosystems of Europe", funded by the European Commission (DG XII, ENV4-CT95-0029). The collembola study was performed within the long-term biomonitoring programme (Oekologisches Wirkungskataster) held by the Baden-Wuerttemberg agency for environmental protection (Landesanstalt fuer Umweltschutz Baden-Wuerttemberg, LFU). The study on ants was conducted in the framework of the Collaborative Research Center "Land-use options for peripheral regions" funded by the Deutsche Forschungsgemeinschaft (DFG, SFB 299).

References

Allen, S.E., 1974. Chemical Analysis of Ecological Materials. Blackwell Scientific Publications, Oxford, p. 368.

Andersen, A.N., 1990. The use of ant communities to evaluate change in Australian terrestrial ecosystems: a review and a recipe. Proc. Ecol. Soc. Aust. 16, 347–357.

Andersen, A.N., 1995. Measuring more of biodiversity—genus richness as a surrogate for species richness in Australian ant faunas. Biol. Conserv. 73, 39–43.

Andrassy, I., 1984. Klasse Nematoda. Akademie-Verlag, Berlin.

Bongers, T., 1988. De Nematoden van Nederland. KNNV-Bibliotheek-uitgave 46, Pirola, Schoorl.

Carroll, S.S., 1998. Modelling abiotic indicators when obtaining spatial predictions of species richness. Environ. Ecol. Stat. 5, 257–276.

Chao, A., Ma, M.-C., Yang, M.C.K., 1993. Stopping rules and estimation for recapture debugging with unequal failure rates. Biometrika 80, 193–201.

Chase, M.K., Kristan, W.B., Lynam, A.J., Price, M.V., Rotenberry, J.T., 2000. Single species as indicators of species richness and composition in California coastal sage scrub birds and small mammals. Conserv. Biol. 14, 474–487.

Chazdon, R.L., Colwell, R.K., Denslow, J.S., Guariguata, M.R., 1998. Statistical methods for estimating species richness of woody regeneration in primary and secondary rain forests of NE Costa Rica. In: Dallmeier, F., Comiskey, J.A. (Eds.), Forest Biodiversity Research, Monitoring and Modeling. Conceptual Background and Old World Case Studies. Parthenon Publishing, Paris, pp. 285–309.

Colwell, R.K., 1997. Estimates: statistical estimation of species richness and shared species from samples. User's Guide. http://viceroy.eeb.uconn.edu/estimates.

Cook, R.D., Nachtsheim, C.J., 1980. A comparison of algorithms for constructing exact D-optimal designs. Technometrics 22, 315–324.

Djajakirana, G., Joergensen, R.G., Meyer, B., 1996. Ergosterol and microbial biomass relationship in soil. Biol. Fert. Soils 22, 299–304.

Ekschmitt, K., Bakonyi, G., Bongers, M., Bongers, T., Boström, S., Dogan, H., Harrison, A., Kallimanis, A., Nagy, P., O'Donnell, A.G., Sohlenius, B., Stamou, G.P., Wolters, V., 1999. Effects of the nematofauna on microbial energy and matter transformation rates in European grassland soils. Plant and Soil 212, 45–61.

Frede, H.-G., Bach, M., 1999. Perspektiven für periphere Regionen. Zeitschrift für Kulturtechnik und Landentwicklung 40, 193–196.

Gaston, K.J., 2000. Biodiversity: higher taxon richness. Prog. Phys. Geogr. 24, 117–127.

Gisin, H., 1960. Collembolenfauna Europas. Museum d'Histoire Naturelle, Genève, 312 pp.

Hodgson, M.E., Gaile, G.L., 1999. A cartographic modeling approach for surface orientation-related applications. Photogramm. Eng. Rem. Sens. 65, 85–95.

Isermeyer, H., 1952. Eine einfache Methode zur Bestimmung der Bodenatmung und der Karbonate im Boden. Zeitschrift fuer Pflanzenernaehrung und Bodenkunde 56, 26–38.

Jenkinson, D.S., Powlson, D.S., 1976. The effects of biocidal treatments on metabolism in soil—V. Soil Biol. Biochem. 8, 209–213.

Kerr, J.T., Sugar, A., Packer, L., 2000. Indicator taxa, rapid biodiversity assessment, and nestedness in an endangered ecosystem. Conserv. Biol. 14, 1726–1734.

LFU—Landesanstalt fuer Umweltschutz Baden-Wuerttemberg (Ed.), 1994. Methoden zu Wirkungserhebungen—ein Methodenhandbuch. Karlsruhe.

Mikusinski, G., Gromadzki, M., Chylarecki, P., 2001. Woodpeckers as indicators of forest bird diversity. Conserv. Biol. 15, 208–217.

Moore, I.D., Wilson, J.P., 1992. Length–slope factors for the revised universal soil loss equation: simplified method of estimation. J. Soil Water Conserv. 47, 423–428.

Ross, D.J., 1990. Measurements of microbial biomass C and N in grassland soils by fumigation-incubation procedures: influence of inoculum size and the control. Soil Biol. Biochem. 3, 289–294.

Salt, G., Hollick, F.S.J., 1946. Studies of wireworm populations. II. Spatial distribution. The J. Exp. Biol. 23, 1–46.

Schick, H., Kreimes, K., 1993. Der Einsatz von Collembolen als Bioindikatoren. In: Ehrnsberger, R. (Ed.), Bodenmesofauna und Naturschutz. Informationen zu Naturschutz und Landschaftspflege in Nordwestdeutschland, vol. 6. Wardenburg, pp. 309–323.

Seifert, B., 1996. Ameisen beobachten, bestimmen. Naturbuch Verlag, Augsburg.

s'Jacob, J.J., Van Bezooijen, J., 1984. A Manual for Practical Work in Nematology. Wageningen Agricultural University, Wageningen, The Netherlands, 77 pp.

Vahjen, W., Munch, J.C., Tebbe, C.C., 1995. Carbon source utilisation of soil extracted microorganisms as a tool to detect the effect of soil supplemented with genetically engineered and non engineered *Corynebacterium glutamicum* and a recombinant peptide at the community level. FEMS Microbiol. Ecol. 18, 317–328.

Waldhardt, R., Fuhr-Bossdorf, K., Otte, A., Schmidt, J., Simmering, D., 1999. Classification, localisation and regional extrapolation of vegetation potential in a peripheral cultural landscape. Zeitschrift für Kulturtechnik und Landentwicklung 40, 246–252.

Wolters, V., Dauber, J., Hirsch, M., Steiner, N., 1999. Diversität der Fauna im Landnutzungsmosaik einer peripheren Region. Zeitschrift für Kulturtechnik und Landentwicklung 40, 253–257.

Wood, J.D., 1996. The geomorphological characterisation of digital elevation models. Ph.D. Thesis. University of Leicester, Leicester.

ELSEVIER

Available online at www.sciencedirect.com

SCIENCE DIRECT®

Agriculture, Ecosystems and Environment 98 (2003) 285–293

Agriculture Ecosystems & Environment

www.elsevier.com/locate/agee

Microbial eco-physiological indicators to asses soil quality

Traute-Heidi Anderson*

Institute of AgroEcology, Federal Agricultural Research Centre, Bundesallee 50, 38116 Braunschweig, Germany

Abstract

The demand for quick and reproducible soil quality indicators in order to detect soil quality changes have increased over the last decade. Since soil functioning and sustaining soil fertility is governed largely by the decomposition activity of the microflora, there is particularly a need for microbial-based indicators. This short overview presents concepts and recent discussions found in the pertaining literature about the "microbial biodiversity" aspect of taxonomic species diversity in relation to environmental impact and taxonomic species diversity in relation to soil functioning, both of which seem at present inadequate for use as indicators. It also introduces the potential of ecological theory at the microbial community level and its use for indicator development. Here, own experiences on eco-physiological indicators are presented and their possible use for establishing eco-physiological profiles of soil sites are proposed.

Keywords: Soil quality; Soil microbiology; Biodiversity; Eco-physiological indicators; qCO_2; C_{mic}:C_{org} ratio

1. Introduction

Intensification of soil use during the last decades particularly in agriculture has caused fear that soil quality could be changed for long or irreversibly. A negative change would be seen in a reduction of soil fertility. On the other hand, subsequent proposals for sustainable land use have shown that there is a lack of suitable parameters for the assessment of "soil quality" from a soil microbiological point of view. An overview of suggested research and test strategies on biological soil quality indicators are given by Dighton (1997), Lynch and Elliott (1997), Bolinder et al. (1999) and Staddon et al. (1999). Following soil microbial parameters in a long field experiment (over 7 years), Wardle et al. (1999b) found no indication of detrimental effects on soil microbes because of agricultural intensification.

The term "soil quality" is used here in the sense as described recently in review articles by Doran and Safley (1997) and Doran and Zeiss (2000), who use it in context or synonymously to soil health: the "quality" of a soil being represented by a "suit of physical, chemical, and biological properties ..." (Doran and Safley, 1997, p. 6). Soil health, however, focuses more on the biotic components of a soil, reflecting, i.e., the maintenance of soil organisms and their proper functioning as regulators of nutrient cycling and therewith of soil fertility. In the pertaining literature both terms are intermingled and it is felt here that soil quality encompasses soil health.

Microorganisms play a leading role in soil development and preservation. Recent reviews on the interlinkage between the biotic compartment and biogeochemical cycling are given by Beare et al. (1995), Elliott (1997) and Pankhurst et al. (1997). The interrelationship of these organisms to their abiotic environment and the course of successions of microorganisms during the decomposition of dead

* Tel.: +49-531-5962541; fax: +49-531-5962699.
E-mail address: heidi.anderson@fal.de (T.-H. Anderson).

0167-8809/$ – see front matter © 2003 Elsevier Science B.V. All rights reserved.
doi:10.1016/S0167-8809(03)00088-4

plant material are all part of a self-regulatory process which determines to a great extent a site-specific soil fertility. In search of suitable soil quality indicators it would therefore be obvious to consider parameters where the biotic–abiotic interlinkages would find their expression. This leads to the controversial discussion on biodiversity when being related to soil microorganisms as a possible indicator of ecosystem function and sustainability, terms which encompass soil quality (Beare et al., 1995; Bengtsson, 1996, 1998; Giller et al., 1997; Kennedy and Smith, 1995; Wardle et al., 1999a) and to the also suggested holistic approach by consideration of ecological theory (Wiebe, 1984) which would include, for instance, eco-physiological indicators. Experience with eco-physiological indicators will be presented in this paper.

2. Current interest on microbial diversity as soil indicators

The observed loss of above-ground biodiversity of animal and plant species as a result of the expansion of farm production and intensification of land management has also initiated questions directed to the soil microbiologist. These critical questions come from national and international policy advisory boards, namely, do the causes for above-ground biodiversity loss affect as well below-ground (microbial) biodiversity and if so, to which degree? If the term biodiversity is used to mean taxonomic species diversity of bacteria and fungi in the way Atlas (1984) applied it to assess environmental stress, it could be an indicator of environmental change per se. Analogous to the loss of above-ground animal or plant species this would indicate that the lost species cannot cope with a new environmental situation. While, on the other hand, animal and plant ecologists can obviously better predict what a particular loss of an above-ground species would have for implications for an ecosystem, the soil microbiologist cannot do so, except for the few cases where the lost species is very well studied and its function is known.

It is accepted that the set of environmental conditions determines the degree of species diversity. The ability to adapt depends on the species genetic make-up. Forest soils, for instance, are by nature richer in species than agricultural soils (Domsch, 1982). Controlling factors of species richness for microbes were recently reviewed by Reber and Wenderoth (1997). It is further accepted that microbes and higher plants must react according to their own inherent ecological strategies (Andrews and Harris, 1986; review Wardle and Giller, 1996). The "hump-backed" response of species diversity seen for higher plants with respect to soil fertility, i.e. the highest soil fertility shows the lowest species richness (Marrs, 1993), may as well be true for the soil microflora, and would explain the low species richness at high soil fertility levels in agricultural soils (Domsch, 1982; Bardgett and Cook, 1998; Bardgett and Shine, 1999, p. 317).

The determination of species losses from a microbial soil community (due to environmental impacts) is, however, impossible to quantify for the reason that with classical microbiological cultural techniques only a fraction of the microorganisms have so far been identified (Hawksworth and Mound, 1991). From the authors review it can be extracted (considering only bacteria and fungi) that until now a total of $\sim$70 000 species have been described while an assumed number of 1 530 000 species remain undiscovered. Consequently, it is a very small window giving access to not more than 5% of microbes which could potentially be identified from a soil community. Even this is hardly possible, since identification to the species level is dependent on specialists knowledge. In addition, the tremendous cultural lab work involved to get all potentially identifiable bacteria and fungi to grow is hardly applicable for a quick, routine lab procedure. Further, in order to be sure if the loss of a species is just a transient or a permanent event (Domsch et al., 1983), it would be necessary to have the investigation repeated several times, a work unfeasible to cope with. At the end, the total loss of certain species from a community cannot be given after all.

Additional points must be considered when dealing with species diversity as a soil indicator. First, the species spectrum of any soil is never a constant entity. The composition of species within soil communities is shifting permanently, even under "normal" agricultural management as was exemplified in elaborate investigations on the fungal spectra of two crop rotation soils (Gams et al., 1969). Although the soils were comparable and under the same plant cover, the species spectra and abundances of fungal species differed significantly between the two soils. The authors concluded that the main determinant for the establishment of species composition has been the previous crop. Secondly, since natural stresses (freezing, draught) can temporarily decrease the organismal community by more than 50% it would be very difficult to determine man-made effects by taxonomic species diversity, since they could be masked by natural impacts (Domsch et al., 1983; Visser and Parkinson, 1992). Further, determination of species number and frequencies of a soil sample are relative values and only valid under the experimental conditions applied. It is for that reason impossible to use a single soil analysis for species diversity estimations. Already Domsch (1960) in his elaborate assay on "The fungal spectrum of a soil sample. Proof of homogeneity" states that a sufficient number of replicate soil analyses of a soil is necessary for statistical treatment of species number and dominance estimations. With a plot of species number over number of isolations a hyperbolic curve can be generated which allows to find the number of isolations necessary after which more isolations would not end up in more species (saturation point). In the study of Domsch (1960) a total of 13 soil analyses (of one soil) with over 6000 fungal isolates (which need to be identified to the species level) were necessary to reach this point. It produced 133 fungal species. Again, this shows the tremendous amount of work involved for a nearly accurate total species number assessment for a particular environmental situation or time span.

For that reason considerable hope was set in molecular approaches for an eventually rapid assessment of microbial community diversity. A detailed review on the state of the art is given by White and Macnaughton (1997). It seems that this powerful tool still has a lot of obstacles which need to be solved. One major drawback at present is the lack of quantitative recovery of nucleic acids. It is not possible for the time being to truly quantify the number of species present in a sample from clonal libraries. The importance of collections and databases on microorganisms becomes evident (Hawksworth and Mound, 1991).

A number of recent articles discuss the possible interdependency between biodiversity and functional parameters (Bengtsson, 1996, 1998; Wardle and Giller, 1996; Giller et al., 1997; Wardle et al., 1999a). The basic underlying question is: are microbial controlled soil functions, i.e. decomposition of organic material (a guarantee of soil fertility) depend on species diversity. This is even a more difficult question to answer than the previous one on biodiversity itself as indicator of an environmental change. The reason is, as mentioned above, we have only a poor quantitative measure to determine microbial species diversity. Early thorough investigations on pesticide side-effects on microorganisms indicated in general no detrimental effects on soil fertility, although shifts in species spectra could be assumed (Domsch, 1974). Reasons for this are seen in the great restitutional ability of the soil flora after detrimental (i.e. fungicidal) impacts (Domsch, 1968). Beare et al. (1995) in their review article on the significance of soil biodiversity and biogeochemical cycling come to the conclusion that the determination of species richness will contribute little if anything to our understanding of soil functioning. It is felt that the diversity of species in any soil, in general, harbors a high degree of functional redundancy. This notion is somewhat confirmed by the experimental work on decomposition processes and organismal diversity by Andrén et al. (1995), who found little or no indications that decomposition rates are controlled by the occurrence of species richness. However, here the whole decomposer community, fauna and flora, was considered. Also, experimentally modified microbial diversity by soil fumigation and reinoculation techniques showed no negative effects on straw decomposition rates (Degens, 1998). Our own investigations on the functional ability (diversity) of bacteria from acidic or neutral soil environments by application of the BIOLOG technique also showed only little decrease in the functional diversity of these two communities, although in the acidic soils we see less bacteria. There was, however, a difference in the utilization of substrate types between the two communities (Kreitz and Anderson, 1997). The existence of a true interlinkage between taxonomic, genetic to functional diversity (Zak and Visser, 1996) remains, however, unsolved. A detailed discussion (and hypotheses) on the possible coexistence of soil organisms and functional consequences of species losses is given by Ekschmitt and Griffiths (1998) which could apply as well to the microflora.

A more realistic approach would be to identify "keystone" species or "keystone microbial controlled processes" as indicator of environmental change or disturbance (Domsch, 1968; Hawksworth and Mound, 1991; Walker, 1992; Beare et al., 1995), since after a thorough literature survey Domsch et al. (1983)

Table 1
Overview of sensitivity groups with respect to losses of organisms or functions in decreasing order of sensitivity (extracted from Domsch et al., 1983)

Degree of sensitivity	Organisms/functions
High	Nitrifiers
	Rhizobium
	Actinomycetes
	Organic matter decomposition
	Nitrification
Medium	Bacteria
	CO_2 production/O_2 uptake
	Fungi
	Denitrification
	Ammonification
Low	N_2 fixation
	Azotobacter
	Ammonifiers
	Protein degraders

found that tolerance to stress is not evenly distributed but specific groups of organisms react more sensitive than others. Three sensitivity groups with respect to organisms or functions are compiled in Table 1.

A recent review of work done on functional groups of soil organisms as indicators of soil change is found by Roper and Ophel-Keller (1997).

3. Ecological theory and eco-physiological indices

Independent of the notion which tends to not imperatively link taxonomic species diversity with soil functional attributes, it is legitimate to raise the question if species diversity could be related to functional efficiency, i.e. fastest catabolic activity at lowest energy costs, lowest maintenance energy costs. Although functional redundancy seems to be widely spread in microbial communities, the question remains, how well organisms which are endowed with the same functional ability are able to do a job in terms of efficiency in comparison with each other (the jack of all trades aspect versus the specialist!). According to Odum (1969) ecological theory there exists a trend towards efficiency in energy utilization concomitant with an increase in diversity (primary producer) during ecosystem succession to maturity. The communities of successional stages for instance differ in primary production and community respiration, whereby at maturity the highest diversity of plant biomass but the lowest community respiration is expected.

In the following the concept of an eco-physiological approach in soil microbiology will be described which was introduced quite a while ago (Anderson and Domsch, 1985a,b, 1986a) which makes it possible to study this challenging question on the soil microbial community level. The term eco-physiology already implies an interlinkage between cell-physiological functioning under the influence of environmental factors. For a quantitative assessment of soil systems fundamental principles from autecological studies (pure culture systems) (Pirt, 1975; Slater, 1979) are extrapolated to the synecological level (cell community level). The microbial biomass (cell mass of fungi plus bacteria) are than taken as a collective, whereby specific metabolic activities of all single species within a community are reflected in the total metabolic capacity of the whole biomass. According to Odums theory, when extrapolated onto the microbial community, we should expect less community respiration with an increasing degree of below-ground species diversity. For the microbial community this could be measured by determining community respiration per biomass unit, the qCO_2. If community respiration is low, more carbon will be available for biomass production which than should be reflected in a high percentage microbial carbon (C_{mic}) to total organic carbon (C_{org}) (Anderson and Domsch, 1986b). So far, there are only few informations on a possible link between above-ground to below-ground taxonomic diversity (see review by Hooper et al., 2000). Zak and Visser (1996), cite Gochenaur (1975), who found an increased fungal taxonomic diversity with high resource heterogeneity. Anderson and Domsch (1990) observed a significant lower community respiration (qCO_2) with continuous crop rotation (standing for heterogeneous resource input) to continuous monocultures of agricultural plots in a quasi-steady state (at least 15 years). Wardle and Nicholson (1996) using a pot experiment with an increasing number of plant species did not find necessarily a lower microbial community respiration (qCO_2) with increased plant diversity. Here one should mention, however, that a 1-year pot experiment may be too short to bring the soils in a quasi-steady state, considering the pre-history of the soil with respect to carbon availability (Anderson and Domsch, 1986b). However, Bardgett and Shine (1999), using a litter bag approach were able to show a decrease in qCO_2 with the highest plant diversity in litter input. Studying chronosequences it could as well be observed that the qCO_2 decreased with time of successional stages of reclaimed soil systems (Insam and Domsch, 1988; Insam and Haselwandter, 1989) or age of agricultural soils (Anderson and Domsch, 1990) which is conform with the ecological theory. It is assumed since plants and soil organisms have coevolved that "above-ground species (plants) may be functionally linked to entire assemblages of below-ground species" (Wall and Moore, 1999). Early publications report adaptive responses of microbial organisms to certain plant residues (Domsch and Gams, 1968; Martyniuk and Wagner, 1978). Further, Degens et al. (2000) found a relationship between functional microbial diversity and pools of soil organic C. However, a direct evidence of below-ground microbial taxonomical diversity as a prerequisite for functional efficiency is until now a still unproven ecological question to be tackled. Already Wiebe (1984) in his review article encouraged microbiologists to test the relevance of microbes to ecological theory.

4. Eco-physiological indices

In order to be able to estimate if practices of land use or soil management would be more detrimental than others, the goal must be that such indicators quantify impacts for direct comparative purposes.

As put forth in the foregoing section, eco-physiological indices (metabolic quotients) are generated by basing physiological performances (respiration, carbon uptake, growth/death, etc.) on the total microbial biomass per unit time (Pirt, 1975; Slater, 1979; Anderson and Domsch, 1986a; Anderson, 1994). Any environmental impact which will affect members of a microbial community should be detectable at the community level by a change of a particular total microbial community activity, which can be quantified (i.e. qCO_2, qD, V_{max}, K_m). Fig. 1 shows a summary of own experiments on the development of microbial biomass-C per unit soil organic C, from which (%) C_{mic} in C_{org} can be calculated, called C_{mic}:C_{org} ratio, the indicator for growth together with qCO_2 values of long-term agricultural plots and mainly mature forest soils. The C_{mic}:C_{org} ratio of agricultural and forest soils at neutral pH is very similar and in the range between 2.0 and 4.4% C_{mic} of total C_{org}, depending on nutrient status and soil management. The metabolic quotient qCO_2 ranged between 0.5 and 2.0 mg CO_2-C g^{-1} C_{mic} h^{-1} in neutral soils. Values below 2.0 for the C_{mic}:C_{org} ratio or above 2.0 for the qCO_2 could be considered as critical for soils with a neutral soil pH. The microbial biomass (SIR) and CO_2 determinations were done using an automated infra-red-gas analyzer which measures hourly rates of CO_2 output (Heinemeyer et al., 1989). For accurate qCO_2 determination it is absolutely necessary to be able to follow CO_2 output in hourly rates (Anderson and Domsch, 1986a). The figure clearly demonstrates the interdependencies between maintenance and growth of soil microbial communities, whereby the qCO_2 reflects the maintenance energy requirement (Anderson and Domsch, 1985a,b) and the C_{mic}:C_{org} ratio the carbon available for growth (Anderson and Domsch, 1986b). Research on the maintenance energy requirement of organisms under in vitro conditions goes back to the 1960s (McGrew and Mallette, 1962; Mallette, 1963; Marr et al., 1963; Dawes and Ribbons, 1964; Pirt, 1965), testing the partitioning of the carbon substrate for energy and cell synthesis purposes. When this concept is extrapolated to soil systems, the total C input per annum from the primary producer must exceed the C requirements for maintenance in order to support growth (Anderson and Domsch, 1985a,b). As depicted in Fig. 1 microbial communities of long-term crop rotation systems are energetically more efficient (lower qCO_2) with a corresponding higher C_{mic}:C_{org} value (increased biomass) as compared to monoculture soil systems. The same could be verified with natural soil systems—pure beech stands to beech/oak stands. In addition, the figure shows the effect of soil pH (only forest soils). Under acidic conditions the qCO_2 is elevated since maintenance requirements are higher (Brown et al., 1980). This happens when the cell-internal pH (which needs to be kept around 6.0) diverges from the surrounding pH conditions and accordingly little biomass can be produced (low C_{mic}:C_{org} ratio). It can be easily deduced from this that both parameters together could be soil quality indicators.

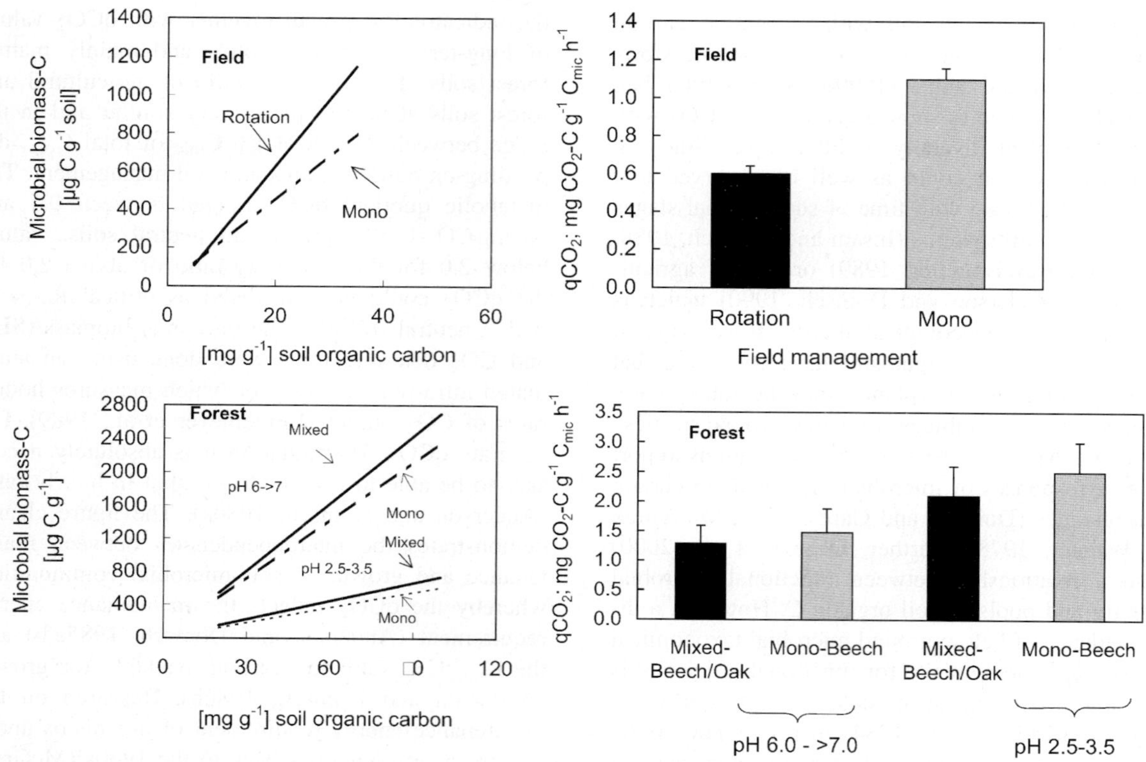

Fig. 1. Microbial biomass-C to soil organic C and qCO_2 of long-term field plots and mainly mature forest stands. Calculated $C_{mic}:C_{org}$ ratio: field rotation, 2.9, $n = 43$; field monoculture 2.3, $n = 34$; forest mixed beech/oak (pH 6 to >7.2), 2.7, $n = 66$; forest mono beech (pH 6 to >7.2), 2.3, $n = 147$; forest mixed beech/oak (pH 2.3–3.5), 0.8, $n = 103$; forest mono beech (pH 2.3–3.5), 0.6, $n = 103$ (data extracted from Anderson and Domsch, 1989, 1990; Anderson, 2003). Differences between variants (bars) are statistically significant at least at $P < 0.01$.

Using natural pH gradients of forest stands, (Anderson, 1999; Anderson and Domsch, 1993) this pH controlling function on microbial community respiration was more closely studied (Fig. 2). From this it can be seen that with an increase in soil pH, we have an increased $C_{mic}:C_{org}$ ratio and a decreased qCO_2.

Subsequent laboratory experiments which were carried out to study the pH impact on soil microbial communities under controlled conditions, verified the empirically obtained field and forest data of Fig. 1 (Blagodatskaya and Anderson, 1999). After an initial pH stress (shown by a high initial qCO_2 and immediate biomass loss), the survivors of a microbial community adapted to the new established soil pH. Adaptation went along (as expected) with a continuous decrease of the qCO_2. Interestingly enough, the qCO_2 values and $C_{mic}:C_{org}$ ratios of the final pH values obtained under laboratory conditions, correspond to those which are commonly found for natural sites of similar pH.

There is evidence in the literature (see Table 7.1 of Anderson, 1994; Turco et al., 1994; Visser and Parkinson, 1992; Sparling, 1997) which support these parameters as valuable soil indices. The qCO_2, as a cell-physiological entity, is a constant only under unchanging environmental conditions, however. Naturally, any impact on the cells (change in temperature, moisture, nutrient status, storage time, etc.) will be reflected in a change of the qCO_2 (Anderson, 1994). For that reason, experiments must obviously be designed with care. Not this parameter should therefore be criticized (Wardle and Ghani, 1995) but experimentators should be aware of pitfalls and limitations of this parameter (Anderson, 1994). Nevertheless, both pa-

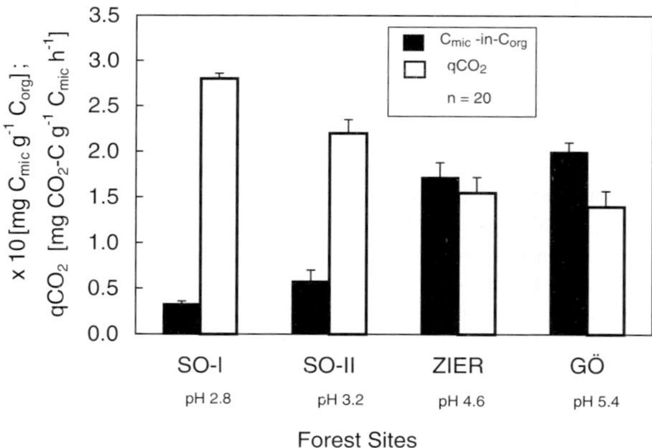

Fig. 2. Determined C_{mic}:C_{org} and qCO_2 values along a pH gradient of four forest sites in Lower Saxony (data extracted from Anderson, 1999). SO-I: solling no lime; SO-II: solling limed; ZIER: Zierenberg; GÖ: Göttinger Wald. Differences of C_{mic}:C_{org} and qCO_2 values between pH gradients are in general statistically significant at least at $P < 0.01$.

rameters are potential tools to study ecosystem theory, stress effects and adaptational processes. Ideally, both parameters should be used together and never alone.

5. Conclusions

Each metabolic activity of organisms is dependent on available carbon sources. Comparative investigations on agricultural and forest plots have demonstrated the very close quantitative relationship between microbial carbon and the total soil carbon (the so-called C_{mic}:C_{org} ratio). This C_{mic}:C_{org} ratio could be developed to a site-specific baseline value for different soil systems and could be used as a stability indicator for quick recognition of an environmental change. Physiological performances such as the specific respiration (qCO_2) could be employed in addition to the C_{mic}:C_{org} parameter for the characterization of the "baseline performance" of a microbial community of a particular soil category which could lead to an "eco-physiological profile" of a site. A necessary research demand lies in the determination of natural deviations of such parameters. However, a strong deviation from a site-specific baseline value would be indicative of a changing environment and the establishment of a new soil community.

References

Anderson, T.-H., 1994. Physiological analysis of microbial communities in soil: applications and limitations. In: Ritz, K., Dighton, J., Giller, K.E. (Eds.), Beyond the Biomass. Compositional and Functional Analysis of Soil Microbial Communities. Wiley, Chichester, UK, pp. 67–76.

Anderson, T.-H., 1999. Einfluss des Standortes auf die Gesellschaftsstruktur von Mikroorganismen und deren Leistungen unter besonderer Berücksichtigung der C-Nutzung. In: Bredemeier, M., Wiedey, G.-A. (Eds.), Verbundprojekt Veränderungsdynamik von Waldökosystemen, Teil 2. Forschungszentrum Waldökosysteme, Göttingen, pp. 385–394.

Anderson, T.-H., 2003. Microbial biomass carbon in Lower Saxony of Germany. In: Brumme, R. (Ed.), Human Impacts on Carbon and Nitrogen Cycles in Temperate Beech Forests. Ecological Studies. Springer, Heidelberg (in press).

Anderson, T.-H., Domsch, K.H., 1985a. Maintenance requirements of actively metabolizing microbial populations under in situ conditions. Soil Biol. Biochem. 17, 197–203.

Anderson, T.-H., Domsch, K.H., 1985b. Determination of ecophysiological maintenance carbon requirements of soil microorganisms in a dormant state. Biol. Fert. Soils 1, 81–89.

Anderson, T.-H., Domsch, K.H., 1986a. Carbon assimilation and microbial activity in soil. Zeitschrift Pflanzenernährung Bodenkunde 149, 457–468.

Anderson, T.-H., Domsch, K.H., 1986b. Carbon link between microbial biomass and soil organic matter. In: Megusar, F., Gantar, M. (Eds.), Perspectives in Microbial Ecology. Slovene Society for Microbiology, Ljubljana, Mladinska knjiga, pp. 467–471.

Anderson, T.-H., Domsch, K.H., 1989. Ratios of microbial biomass carbon to total organic-C in arable soils. Soil Biol. Biochem. 21, 471–479.

Anderson, T.-H., Domsch, K.H., 1990. Application of eco-physiological quotients (qCO_2 and qD) on microbial biomasses from soils of different cropping histories. Soil Biol. Biochem. 22, 251–255.

Anderson, T.-H., Domsch, K.H., 1993. The metabolic quotient for CO_2 (qCO_2) as a specific activity parameter to assess the effects of environmental conditions such as pH, on the microbial biomass of forest soils. Soil Biol. Biochem. 25, 393–395.

Andrén, O., Bengtsson, J., Clarholm, M., 1995. Biodiversity and species redundancy among litter decomposers. In: Collins, H.P., Robertson, G.P., Klug, M.J. (Eds.), The Significance and Regulation of Soil Biodiversity. Kluwer Academic Publishers, Dordrecht, The Netherlands, pp. 141–151.

Andrews, J.H., Harris, R.F., 1986. R- and K-selection and microbial ecology. Adv. Microb. Ecol. 9, 99–147.

Atlas, R.M., 1984. Use of microbial diversity measurements to assess environmental stress. In: Klug, M.J., Reddy, C.A. (Eds.), Current Perspectives in Microbial Ecology. American Society for Microbiology, Washington, DC, pp. 540–545.

Bardgett, R.D., Cook, R., 1998. Functional aspects of soil animal diversity in agricultural grasslands. Appl. Soil Ecol. 10, 263–276.

Bardgett, R.D., Shine, A., 1999. Linkages between plant litter diversity, soil microbial biomass and ecosystem function in temperate grassland. Soil Biol. Biochem. 31, 317–321.

Beare, M.H., Coleman, D.C., Crossley, D.A., Hendrix, P.F., Odum, E.P., 1995. A hierarchical approach to evaluating the significance of soil biodiversity to biogeochemical cycling. Plant and Soil 170, 5–22.

Bengtsson, J., 1996. What kind of diversity? Species richness, keystone species or functional groups? In: Wolters, V. (Ed.), Functional Implications of Biodiversity in Soil. Ecosystem Research Report No. 24, EUR 17659 EN, European Commission, pp. 59–85.

Bengtsson, J., 1998. Which species? What kind of diversity? Which ecosystem function? Some problems in studies of relations between biodiversity and ecosystem function. Appl. Soil Ecol. 10, 191–199.

Blagodatskaya, E.V., Anderson, T.-H., 1999. Adaptive responses of soil microbial communities under experimental acid stress in controlled laboratory studies. Appl. Soil Ecol. 11, 207–216.

Bolinder, M.A., Angers, D.A., Gregorich, E.G., Carter, M.R., 1999. The response of soil quality indicators to conservation management. Can. J. Soil Sci. 79, 37–45.

Brown, J.M., Mayes, T., Lelieveld, H.L.M., 1980. The growth of microbes at low pH values. In: Gould, G.W., Corry, J.E.L. (Eds.), Microbial Growth and Survival in Extremes of Environment. Academic Press, London, pp. 71–98.

Dawes, E.A., Ribbons, D.W., 1964. Some aspects of the endogenous metabolism of bacteria. Bacteriol. Rev. 28, 126–149.

Degens, B.P., 1998. Decreases in microbial functional diversity do not result in corresponding changes in decomposition under different moisture conditions. Soil Biol. Biochem. 30, 1989–2000.

Degens, B.P., Schipper, L.A., Sparling, G.P., Vojvodic-Vukovic, M., 2000. Decreases in organic C reserves in soils can reduce the catabolic diversity of soil microbial communities. Soil Biol. Biochem. 32, 189–196.

Dighton, J., 1997. Is it possible to develop microbial test systems to evaluate pollution effects on soil nutrient cycling? In: van Strahlen, N.M., Løkke, H. (Eds.), Ecological Risk Assessment of Contamination in Soil. Chapman & Hall, London, pp. 51–69.

Domsch, K.H., 1960. Das Pilzspektrum einer Bodenprobe. I. Nachweis der Homogenität. Archiv für Mikrobiologie 35, 181–195.

Domsch, K.H., 1968. Mikrobiologische Präsenz-und Aktivitätsanalysen von fungicidbehandelten Böden. Arbeiten der Universität Hohenheim, Band 44. Ulmer Verlag, Stuttgart, pp. 1–79.

Domsch, K.H., 1974. The maintenance of soil fertility and the use of pesticides. Zeitschrift für Pflanzenkrankheiten und Pflanzenschutz 81, 679–682.

Domsch, K.H., 1982. Beeinflussen Pflanzenschutzmittel das Bodenleben? Der aktuelle Stand der Forschung. DLG-Mitteilungen 14, 860–862.

Domsch, K.H., Gams, W., 1968. Die Bedeutung vorfruchtabhängiger Verschiebungen in der Bodenmikroflora. Phytopathologische Zeitschrift 63, 64–74.

Domsch, K.H., Jagnow, G., Anderson, K.-H., 1983. An ecological concept for the assessment of side-effects of agrochemicals on soil microorganisms. Residue Rev. 86, 65–105.

Doran, J.W., Safley, M., 1997. Defining and assessing soil health and sustainable productivity. In: Pankhurst, C.E., Doube, B.M., Gupta, V.V.S.R. (Eds.), Biological Indicators of Soil Health. CAB International, New York, pp. 1–28.

Doran, J.W., Zeiss, M.R., 2000. Soil health and sustainability: managing the biotic component of soil quality. Appl. Soil Ecol. 15, 3–11.

Ekschmitt, K., Griffiths, B.S., 1998. Soil biodiversity and its implications for ecosystem functioning in a heterogeneous and variable environment. Appl. Soil Ecol. 10, 201–215.

Elliott, E.T., 1997. Rationale for developing bioindicators of soil health. In: Pankhurst, C.E., Doube, B.M., Gupta, V.V.S.R. (Eds.), Biological Indicators of Soil Health. CAB International, New York, pp. 49–78.

Gams, W., Domsch, K.H., Weber, E., 1969. Nachweis signifikant verschiedener Pilzpopulationen bei gleicher Bodennutzung. Plant and Soil 31, 439–450.

Giller, K.E., Beare, M.H., Lavelle, P., Izac, A.-M.N., Swift, M.J., 1997. Agricultural intensification, soil biodiversity and agroecosystem function. Appl. Soil Ecol. 6, 3–16.

Gochenaur, S.E., 1975. Distributional patterns of mesophilous and thermophilous microfungi in two Bahamian soils. Mycopathol. Mycol. Appl. 57, 155–164.

Hawksworth, D.L., Mound, L.A., 1991. Biodiversity databases: the crucial significance of collections. In: Hawksworth, D.L. (Ed.), The Biodiversity of Microorganisms and Invertebrates: Its Role in Sustainable Agriculture. CAB International, Wallingford, UK, pp. 17–29.

Heinemeyer, O., Insam, H., Kaiser, E.-A., Walenzik, G., 1989. Soil microbial biomass measurements: an automated technique based on infra-red gas analysis. Plant and Soil 116, 191–195.

Hooper, D.U., Bignell, D.E., Brown, V.K., Brussard, L., Dangerfield, J.M., Wall, D.H., Wardle, D.A., Coleman, D.C., Giller, K.E., Lavelle, P., Van der Putten, W.H., De Ruiter, P.C., Rusek, J., Silver, W.L., Tiedje, J.M., Wolters, V., 2000. Interactions between aboveground and belowground biodiversity in terrestrial ecosystems: patterns, mechanisms, and feedbacks. BioScience 50, 1049–1061.

Insam, H., Domsch, K.H., 1988. Relationship between soil organic carbon and microbial biomass on chronosequences of reclamation sites. Microb. Ecol. 15, 177–188.

Insam, H., Haselwandter, K., 1989. Metabolic quotient of the soil microflora in relation to plant succession. Oecologia 79, 174–178.

Kennedy, A.C., Smith, K.L., 1995. Soil microbial diversity and sustainability of agricultural soils. Plant and Soil 170, 75–86.

Kreitz, S., Anderson, T.-H., 1997. Substrate utilization patterns of extractable and non-extractable bacterial fractions in neutral and acidic beech forest soils. In: Insam, H., Rangger, A. (Eds.), Microbial Communities. Functional Versus Structural Approaches. Springer, Berlin, pp. 149–160.

Lynch, J.M., Elliott, L.F., 1997. Bioindicators: perspectives and potential value for landusers, researchers and policy makers. In: Pankhurst, C.E., Doube, B.M., Gupta, V.V.S.R. (Eds.),

Biological Indicators of Soil Health. CAB International, New York, pp. 79–96.

Mallette, M.F., 1963. Validity of the concept of energy of maintenance. Ann. NY Acad. Sci. 102, 521–535.

Marr, G.A., Nilson, E.H., Clark, D.J., 1963. The maintenance requirement of *Escherichia coli*. Ann. NY Acad. Sci. 102, 536–548.

Marrs, R.H., 1993. Soil fertility and nature conservation in Europe: theoretical considerations and practical management solutions. Adv. Ecol. Res. 24, 241–300.

Martyniuk, S., Wagner, G.H., 1978. Quantitative and qualitative examination of soil microflora associated with different management systems. Soil Sci. 125, 343–350.

McGrew, S.B., Mallette, M.F., 1962. Energy of maintenance in *Escherichia coli*. J. Bacteriol. 83, 844–850.

Odum, E.P., 1969. The strategy of ecosystem development. Science 164, 262–270.

Pankhurst, C.E., Doube, B.M., Gupta, V.V.S.R., 1997. Biological indicators of soil health: synthesis. In: Pankhurst, C.E., Doube, B.M., Gupta, V.V.S.R. (Eds.), Biological Indicators of Soil Health. CAB International, New York, pp. 419–435.

Pirt, S.J., 1965. The maintenance energy of bacteria in growing cultures. Proc. R. Soc. London, Ser. B 163, 224–231.

Pirt, S.J., 1975. The Principles of Microbe and Cell Cultivation. Blackwell Scientific Publications, Oxford.

Reber, H., Wenderoth, D.F., 1997. Diversity and metabolic versatility of microbial communities in soil. In: Welling, M. (Ed.), Biologische Vielfalt in Ökosystemen—Konflikt zwischen Nutzung und Erhaltung. Schriftenreihe des BML, Reihe A: Angewandte Wissenschaften, Heft 465. Köllen Druck, Bonn, pp. 168–184.

Roper, M.M., Ophel-Keller, K.M., 1997. Soil microflora as bioindicator of soil health. In: Pankhurst, C.E., Doube, B.M., Gupta, V.V.S.R. (Eds.), Biological Indicators of Soil Health. CAB International, New York, pp. 157–177.

Slater, J.H., 1979. Microbial population and community dynamics. In: Lynch, J.M., Poole, N.J. (Eds.), Microbial Ecology. A Conceptual Approach. Blackwell Scientific Publications, Oxford, pp. 45–63.

Sparling, G.P., 1997. Soil microbial biomass, activity and nutrient cycling as indicators of soil health. In: Pankhurst, C.E., Doube, B.M., Gupta, V.V.S.R. (Eds.), Biological Indicators of Soil Health. CAB International, New York, pp. 97–119.

Staddon, W.J., Duchesne, L.C., Trevors, J.T., 1999. The role of microbial indicators of soil quality in ecological forest management. For. Chron. 75, 81–86.

Turco, R.F., Kennedy, A.C., Jawson, M.D., 1994. Microbial Indicators of Soil Quality. Defining Soil Quality for a Sustainable Environment. SSSA Special Publication No. 35, pp. 73–90.

Visser, S., Parkinson, D., 1992. Soil biological criteria as indicators of soil quality: soil microorganisms. Am. J. Alternative Agric. 7, 33–37.

Walker, B.H., 1992. Biodiversity and ecological redundancy. Conserv. Biol. 6, 18–23.

Wall, D.H., Moore, J.C., 1999. Interactions underground. Soil biodiversity, mutualism, and ecosystem processes. BioScience 49, 109–118.

Wardle, D.A., Ghani, A., 1995. A critique of the microbial metabolic quotient (qCO_2) as a bioindicator of disturbance and ecosystem development. Soil Biol. Biochem. 27, 1601–1610.

Wardle, D.A., Giller, K.E., 1996. The quest for a contemporary ecological dimension to soil biology. Soil Biol. Biochem. 28, 1549–1554.

Wardle, D.A., Nicholson, K.S., 1996. Synergistic effects of grassland plant species on soil microbial biomass and activity: implications for ecosystem-level effects of enriched plant diversity. Funct. Ecol. 10, 410–416.

Wardle, D.A., Giller, K.E., Barker, G.M., 1999a. The regulation and functional significance of soil biodiversity in agroecosystems. In: Wood, D., Lenné, J.M. (Eds.), Agrobiodiversity: Characterization, Utilization and Management. CAB International, New York, pp. 87–121.

Wardle, D.A., Yeates, G.W., Nicholson, K.S., Bonner, K.I., Watson, R.N., 1999b. Response of soil microbial biomass dynamics, activity and plant litter decomposition to agricultural intensification over a seven-year period. Soil Biol. Biochem. 31, 1707–1720.

White, D.C., Macnaughton, S.J., 1997. Chemical and molecular approaches for rapid assessment of the biological status of soils. In: Pankhurst, C.E., Doube, B.M., Gupta, V.V.S.R. (Eds.), Biological Indicators of Soil Health. CAB International, New York, pp. 371–396.

Wiebe, W.J., 1984. Some potentials for the use of microorganisms in ecological theory. In: Klug, M.J., Reddy, C.A. (Eds.), Current Perspectives in Microbial Ecology. American Society for Microbiology, Washington, DC, pp. 17–21.

Zak, J.C., Visser, S., 1996. An appraisal of soil fungal biodiversity: the crossroads between taxonomic and functional biodiversity. Biodiversity Conserv. 5, 169–183.

ELSEVIER

Available online at www.sciencedirect.com

Agriculture, Ecosystems and Environment 98 (2003) 295–304

Agriculture Ecosystems & Environment

www.elsevier.com/locate/agee

Influence of precision farming on the microbial community structure and functions in nitrogen turnover

M. Schloter*, H.-J. Bach, S. Metz, U. Sehy, J.C. Munch

GSF-National Research Centre for Environment and Health, Institute of Soil Ecology, Ingolstädter Landstraße 1, D-85764 Neuherberg, Germany

Abstract

In this paper the effects of precision farming were compared to conventional agricultural management estimating bacterial and fungal diversity and also microbial processes in the nitrogen cycle during the vegetation period of maize. The aim was to find parameters which can be used as indicators for sustainable soil quality. Two plots, a high yield and a low yield site, were selected in the two systems. It could be demonstrated that the microbial biomass and community structure of bacteria and fungi in the top soils were not influenced by precision farming. Microbial biomass was reduced during the summer month, due to the dryness and hot temperatures. The microbial community structure changed in late spring time probably due to the application of fertilizers and high amounts of root exudates in the rhizosphere.

The measured enzymatic activities however did not show only a seasonal variation with the highest measured values in spring and early summer, but also clear effects based on the investigated plot (high yield or low yield) and the used farming management system (conventional or precision). Proteolytic activity was significantly higher on the high yield plots at all measured times. Mainly on the low yield sites precision farming caused a higher proteolytic activity compared to the conventional management. Nitrification and denitrification activities were mainly effected after the application of N-fertilizer. Significant higher values were found on the low yield plots, where conventional farming was applied. The results indicate that the effectiveness of nitrogen turnover might be a good indicator for sustainable agriculture and soil quality.

Keywords: Microbial diversity; Denitrification; Nitrification; Proteolytic activity; Soil quality indicator

1. Introduction

Biodiversity describes in its entirety all forms of live in a given ecosystem (Myers, 1996). It can be measured on the one hand simply as the variety of organisms in the ecosystem, but also as the diversity of functional and structural genes on the other hand, which are expressed currently or potentially (Andersen and Rygiewicz, 1999). As nutrient turnover is mainly catalysed by microbes, the expression rates of functional genes of bacteria and fungi play an important role for the stability and productivity of agricultural soils (Schloter et al., 1998). Therefore for a sustainable and nature saving agriculture with high crop yields the genetic diversity of microbes (bacteria and fungi) may play an important role and could be used as an indicator for soil quality, which has to be determined for understanding turnover processes.

A simple isolation of characteristic microbes from the soils for functional analysis however is frequently impossible, as these organisms can not be cultured with classical techniques (Amann et al., 1995).

* Corresponding author. Tel.: +49-89-3187-2304; fax: +49-89-3187-3376.
E-mail address: schloter@gsf.de (M. Schloter).

0167-8809/$ – see front matter © 2003 Elsevier Science B.V. All rights reserved.
doi:10.1016/S0167-8809(03)00089-6

Furthermore some enzyme systems are not directly correlated with living organisms and are stabilized in soil (Bach and Munch, 2000; Bach et al., 2001). Therefore an determination of genes, enzymes and corresponding microbes is only possible by combining classical enzyme tests and modern biochemical tools.

In agricultural ecosystems, nitrogen is frequently the factor limiting crop yield (Ruser et al., 1998). However, the application of N compounds to enhance plant growth may pollute the environment, if fertilization is not temporally and spatially adjusted to the current plant requirements. If microbial ammonification of organically bound nitrogen or its application in mineral form provides more nitrogen than the plant is actually able to immobilize, residual nitrogen may be transformed by microbial processes depending on prevalent surrounding conditions. Nitrate, provided by fertilization or by nitrification may be leached into the groundwater or may get lost by denitrification as dinitrogen (N_2) and nitrous oxide (N_2O), the latter being involved in global warming and stratospheric ozone destruction (Singh et al., 1992). For sustainable agricultural management, the application of nitrogenous compounds must refer to the temporal, spatial and quantitative requirements of the crop, allowing maximum crop yields and minimized nitrogen loss.

Precision farming however summarizes cultivation practices that consider for spatial and temporal variability of soil attributes and crop parameters within an agricultural field. Distinct areas in a field may be managed by applying site-specific levels of input depending on the yield potential. Benefits are the reduction of the cost for crop production and thus, conserving resources while maintaining high yield and minimizing the risk of environmental pollution (Dawson, 1997).

The aim of the study presented in this paper is to investigate the applicability of indicators based on microbial community structure and function to assess soil quality. As nitrogen is a key compound both for high yields and environmental protection, key processes in the nitrogen cycle (proteolytic activity, nitrification and denitrification) were investigated in detail to improve the understanding of different farming management types on microbial catalysed processes. As a model a study was chosen, which compares conventional management practice with precision farming.

2. Materials and methods

2.1. Site

Investigations were carried out at the research farm of the agro-ecological research network 'Forschungsverbund Agrarökosysteme München' (FAM) from April to December 1999. The farm is located in Southern Germany in the 'Tertiärhügelland' (Tertiary Hill Land), approximately 40 km north of Munich (48°30.0′N, 11°20.7′E).

The soil characteristics are given in Table 1. Four study sites were established in a field fertilized according to precision farming practices. From maps integrating crop yield over 3 years, areas of high yield (>95% of average yield) and areas of low yield (<95% of average yield) were identified and fertilized differently (for details see Table 2). At the high and the low yield site, one plot was subjected to precision farming practices, and one plot to conventional agricultural practices, were monitored. In the investigated period maize was cropped.

2.2. Sampling

Bulk soil samples were taken from 0 to 20 cm by combining 5–10 soil cores. Samples were kept cool for transportation and extracted within 6 h after

Table 1
Soil characteristics of the top soil (0–20 cm)

	Site	
	HC, HP (high yield area)	LC, LP (low yield area)
Soil classification (USDA taxonomy)	Typic udifluvent	Dystric eutrochrept
pH (0.01 M $CaCl_2$)	5.9	6.1
C_{org} (%)	1.7	1.4
N_t (%)	0.17	0.15
Grain size fractions (%)		
Clay <2 μm	15	20
Silt 2–63 μm	49	51
Sand 63–2000 μm	36	29

HC: high yield area, fertilized according to conventional practices; HP: high yield area, fertilized according to precision farming practices; LC: low yield area, fertilized according to conventional practices; LP: low yield area, fertilized according to precision farming practices; data from own measurements and FAM database.

Table 2
Soil management in 1999

25 March	Herbicide durano (2.5 kg ha^{-1})
3 May	Fungicide mesurol
4 May	Sowing of maize Fertilizer ammonium-phosphate 40 kg N ha^{-1} (uniformly all sites)
5 May	Moluscizide metaldehyde (4 kg ha^{-1})
18 May	Herbicide stentan (1 kg ha^{-1}), starane (0.5 kg ha^{-1})
1 June	Fertilizer calcium-ammonium-nitrate HC: 110 kg N ha^{-1} HP: 135 kg N ha^{-1} LC: 110 kg N ha^{-1} LP: 85 kg N ha^{-1}
2 June	Herbicide cato (0.3 kg ha^{-1})
19 September	Harvest of maize
22 October	Tillage (0–20 cm) and sowing of winter wheat

HC: high yield area, fertilized according to conventional practices; HP: high yield area, fertilized according to precision farming practices; LC: low yield area, fertilized according to conventional practices; LP: low yield area, fertilized according to precision farming practices; data from FAM database.

sampling. Soil moisture was determined gravimetrically after drying for at least 24 h at 105 °C.

2.3. Microbial biomass and community structure

For the determination of microbial biomass and community structure phospholipids were used as a biomaker. For extraction of phospholipids and separation of fatty acids a modified one-phase procedure was performed (Zelles, 1996). The lipid material was fractionated into neutral lipids, glycolipids and phospho-(polar)-lipids. The phospholipid fraction was subjected to mild alkaline hydrolysis in order to liberate the ester-linked (EL) fatty acids. The products of hydrolysis were separated into unsubstituted fatty acids (UNSFAs), hydroxy substituted fatty acids (PLOH), and the remaining unsaponifiable lipids. Unsubstituted fatty acids were further separated into saturated (SATFA), monounsaturated (MUFA) and polyunsaturated fatty acids (PUFA) The unsaponifiable lipid fraction was subjected to a one-step acid hydrolysis. The liberated fatty acids of the non-ester-linked phospholipid fatty acids (NEL-PLFA) were separated into unsubstituted fatty acids and hydroxy substituted fatty acids (UNOH). The FAs and their derivatives were analysed by GC–MS, using a Hewlett-Packard 5971A MDS combined with a 5890 series II GC System.

2.4. Mineral nitrogen

Moist soil (80 g) was extracted with 160 ml of 0.01 M $CaCl_2$, shaken in a rotary shaker for 1 h, centrifuged at 4000 × *g* for 10 m, and filtered through a 0.45 μm membrane filter. Extract solutions were stored frozen until analysed for NH_4^+ and NO_3^- in a continuous flow analyser (Ruser et al., 1998).

2.5. Enzyme measurements

The overall activity or proteases in the soil samples was measured by the procedure of Ladd and Butler (1972), using casein as model substrate.

The activity of nitrite reductases (denitrification) was measured using soil suspensions anaerobic conditions and nitrate as substrate in closed flasks. N_2O reductases was inhibited by the addition of acetylene. Evolved N_2O is measured by gas chromatography (Ryden et al., 1979).

The measurement of the actual nitrification was performed according to Kandeler (1989). The oxidation of nitrite during the time of measurement for 24 h was stopped by sodium chlorate.

2.6. Statistics

All statistical analysis (mean values, normal distribution, standard deviation, principal component analysis) were calculated using the SPSS statistic program (SPSS, Germany). All analysis were performed with five (enzymes) or three (PLFA) independent replicates.

3. Results

3.1. Microbial biomass and community structure

The total amount of phospholipids extracted from the soil was used as a marker for microbial biomass. The biomass results during the investigation period on

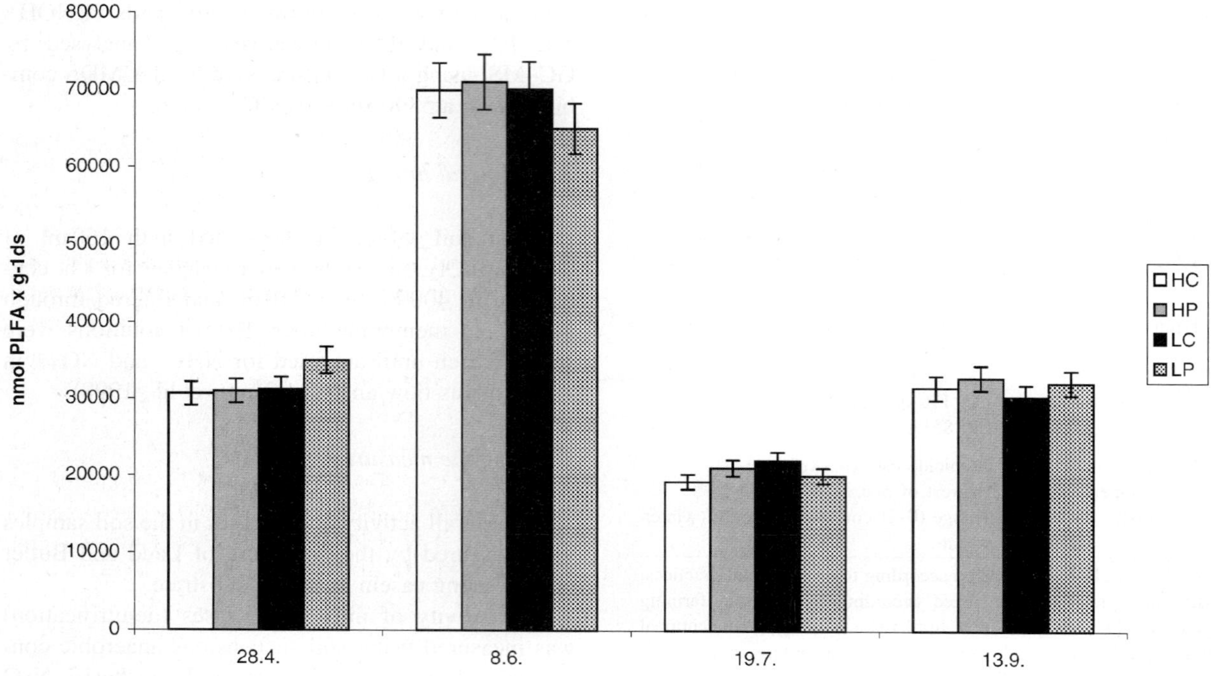

Fig. 1. Biomass results (μg PLFA g ds^{-1}) in the top from April to October 1999. HC: high yield area, fertilized according to conventional practices; HP: high yield area, fertilized according to precision farming practices; LC: low yield area, fertilized according to conventional practices; LP: low yield area, fertilized according to precision farming practices.

all four plots are shown in Fig. 1. The four plots did not show significant differences in their microbial biomass content before precisions farming starts (April 1999). The values ranged between 20,000 nmol PLFA g^{-1} dry weight of soil (July 1999, high yield plot, conventional farming) to more than 70,000 nmol PLFA g^{-1} dry weight of soil (June 1999, high yield plot, precision farming). A significant increase in biomass was found in June after the second application of fertilizers, due to the high amount of mineral nitrogen, the high temperatures and good water content of the soils. An significant influence of the differentiated amount of fertilizer on the amount of microbial was not visible. However a tendency to reduced microbial biomass values in the plot which was treated with lower amounts of nitrogen (LP) was visible. In contrast, the amount of microbial biomass in summer was reduced on all four plots. As nitrogen was not the limiting factor (at least on the high yield plots) (for details see Fig. 3), the low water content and the high temperatures in the top soil might be the reason. The decrease in microbial biomass had clear influences on the N-turnover processes in soil (see below). In September, the amount of biomass increased slightly on all four plots. As to the early sampling times also in summer and autumn no influence of the precision farming on microbial biomass was visible.

The composition of the microbial community in the top soils of the four plots under investigation over the vegetation period using the phospholipid fatty acid technique is shown in Fig. 2. The data were reduced using principal component analysis. The results clearly indicate, that the community structure of bacteria and fungi changed in dependence of the soil water content, the temperature and time of fertilization. A shift in the population structure was found after the application of fertilizer in June, but no influence of the different amounts of given nitrogen could be detected. In summer and autumn only slight changes of the community structure compared to June were found. Also a detailed analysis by different taxonomic relevant phospholipid groups did not give hints for an influence of precisions farming on the microbial community structure.

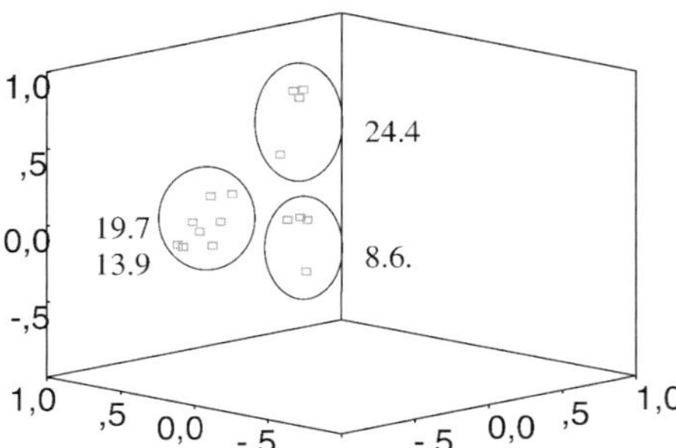

Fig. 2. Composition of the microbial community in the top soils over the vegetation period using the phospholipid fatty acid technique. The data were reduced using principal component analysis.

3.2. Microbial nitrogen turnover

Fig. 3 shows the nitrate concentrations in the top soils from April to October 1999. While the application of ammonium-phosphate in May had almost no influence on the nitrate concentrations in the soils of all plots, after the application of Calcium-ammonium-nitrate fertilizer in June a significant increase in NO_3-N concentrations almost by factor 10 could be measured on all four plots, in depended from the given amount of nitrate. Whereas the concentrations of nitrate dropped until July almost to the starting point (10 kg NO_3-N ha^{-1}) on the low yield plots (LC and LP), the amount of nitrate in soil reached a maximum on the high yield plots (HC and HP) in July with 150 kg NO_3-N ha^{-1} (HC), respectively 230 kg NO_3-N ha^{-1} (HP). Until October the amount of nitrate on the high yield plots was significantly higher compared to the low yield plots.

The potential proteolytic activity is shown in Fig. 4. During the whole period significant higher values could be measured in the "high yield" plots. Whereas the high yield plots show a run, which is influenced by the season, with a maximum peak (550 μg tyrosin g^{-1} dry weight) of soil in spring and a stepwise decrease of the potential proteolytic until winter (240 μg tyrosin g^{-1} dry weight) changes in activities in the low yield plots were quite constant over the investigation period with maximum values from 250 μg tyrosin g^{-1} dry weight and minimum values from 140 μg. The values were very similar in the high yield areas, which indicates no influence of the management type (conventional–precision). In contrast, clear differences were found in the proteolytioc activities in the two investigated low yield plots, mainly in the summer period (June–August) the values are significantly higher on the sites under precision farming. This might be an indication that if N-availability is reduced, protease activity increases. After harvesting

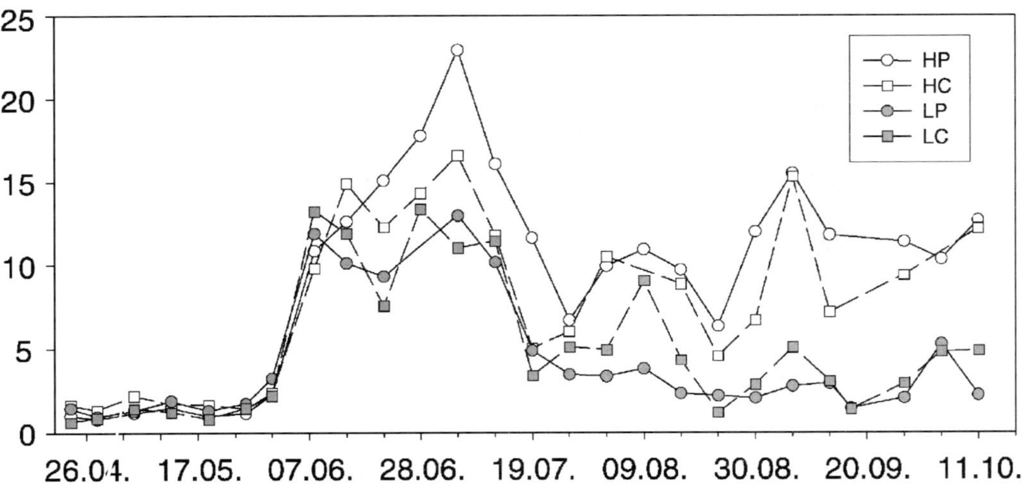

Fig. 3. Nitrate concentrations (kg NO_3-N ha^{-1}) in the top soils from April to October 1999. HC: high yield area, fertilized according to conventional practices; HP: high yield area, fertilized according to precision farming practices; LC: low yield area, fertilized according to conventional practices; LP: low yield area, fertilized according to precision farming practices.

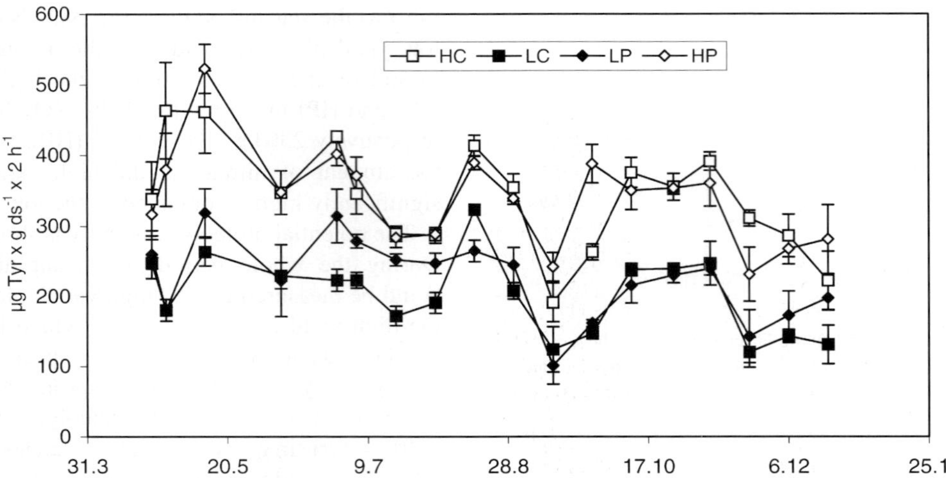

Fig. 4. Potential proteolytic activities from April to October 1999. HC: high yield area, fertilized according to conventional practices; HP: high yield area, fertilized according to precision farming practices; LC: low yield area, fertilized according to conventional practices; LP: low yield area, fertilized according to precision farming practices.

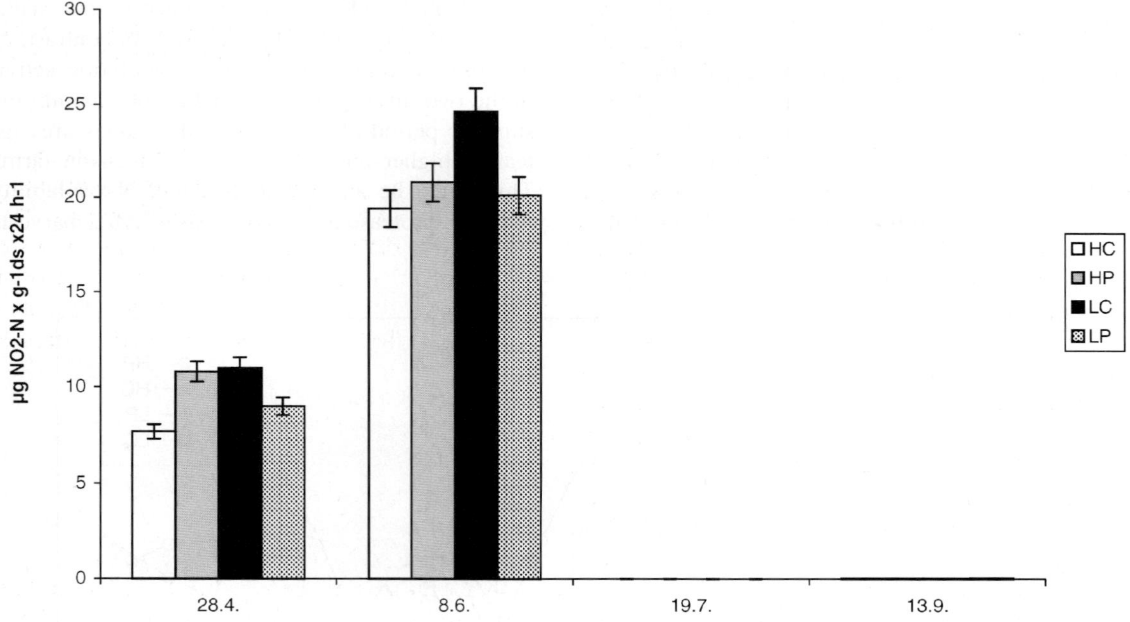

Fig. 5. Development of nitrification from April to October 1999. HC: high yield area, fertilized according to conventional practices; HP: high yield area, fertilized according to precision farming practices; LC: low yield area, fertilized according to conventional practices; LP: low yield area, fertilized according to precision farming practices.

the values for proteolytic activity on the low yield plots were similar. Overall it is obvious that the high yield plots showed a much higher potential for ammonification compared to the low yield sites.

The uniform fertilization in May (40 kg N ha^{-1}) resulted in an significant increase of activity on all investigated plots. The differentiated fertilization in June (HC and LC 111 kg N ha^{-1}, HP 135 kg N ha^{-1} and

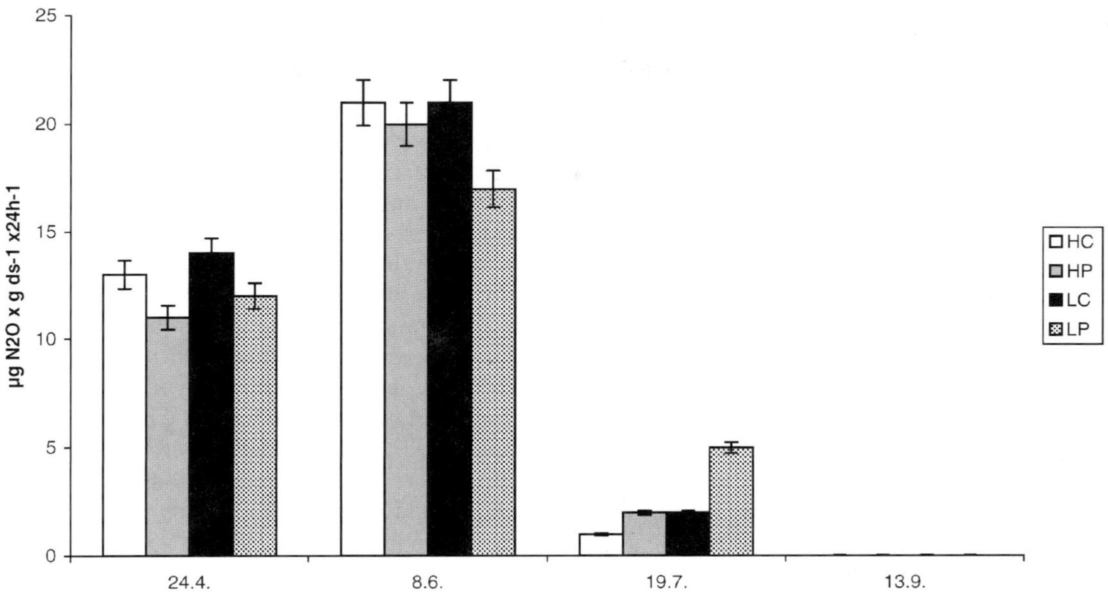

Fig. 6. Denitrification activity from April to October 1999; HC: high yield area, fertilized according to conventional practices; HP: high yield area, fertilized according to precision farming practices; LC: low yield area, fertilized according to conventional practices; LP: low yield area, fertilized according to precision farming practices.

LP 85 kg ha^{-1}) resulted in an increase on all the plots except LC.

The decreasing activities during summertime (June–September) may be a result of the decreasing biomass and microbial activity (see above). After harvest in general activities increase again, which may be an result of degraded harvest residues in the soil, which increase the amount of nitrogen.

A strong correlation between the application of the fertilizers in April and June and nitrification is obvious (Fig. 5). The highest amounts of NO_2-N were measured after the second application. The values measured on plot LC were significantly higher compared to the other sites. As the application of fertilizers in April was lower the same effect could be found, but with lower expression. In summer almost no nitrification could be measured on the four plots. The reasons for the missing nitrification at that time are similar to those discussed for the low proteolytic activity in summer.

The development of denitrification and nitrification is similar (Fig. 6). The highest amounts of N_2O were measured after the second fertilizer application (June 1999) on all four investigated plots. Due to high amounts of rain and the high soil temperatures increased microbial activity caused anoxic conditions, which resulted in a shift to denitrification. There is a trend that the reduced amount of fertilizer on plot LP resulted in a lower amount of expressed nitrite reductases. In summer, as for both other turnover processes, the denitrification potential is very low.

4. Discussion

4.1. Biomass and structural diversity

As microbial biomass has the roles as a pool of labile nutrients on the one hand and as the agent of decomposition of organic materials on the other hand, it has been proposed as an early indicator of changes in the total soil organic matter (Powlson et al., 1987). However complications may be taken into account, when microbial biomass is used as an indicator for active soil organic matter, as portions of the microbial community can turnover very rapidly. It has been reported that microbial biomass can fluctuate sharply over days following agricultural management like tillage or natural influences like drying and rewetting of a soil (Ocio et al., 1991; Wyland et al., 1995). Our results are in good accordance to the mentioned studies: Shifts in the microbial biomass were found after fertilizer application spring time and after reduction of the water content in summer. Detailed studies about the influence of nitrogen application on the microbial biomass were also done by Miller and Dick (1995). It could be shown that the amount of fertilizer has an influence on the amount of microbial biomass. Similar results were found in our study. Nevertheless the measured lower microbial biomass values (plot LP) were not significant. This might be more clear after some years with differentiated fertilizer application, when the nitrogen pools in the soil have changed.

Many studies have shown that the structure of the microbial community has a big impact on the yield (Kloepper et al., 1992). Plant growth is mainly influenced by an efficient microbial nutrient turnover and the abundance of plant pathogenic and plant beneficial microbes (Sorensen, 1997; Bever et al., 1997). However only a few studies have been published, which investigate the influence of different farming management on the microbial community structure in detail. In a former study from our group it could be shown, that significant shifts in the microbial population structure occurred when farming management was changed from conventional to organic farming (Schloter et al., 1998). Other authors (Buyer and Kaufman, 1996; Lupwayi et al., 1998) also found a high influence of the farming practice on the microbial community structure. In all the published papers, the effects were found on long term experimental field sites and the shifts in the management practice were quite distinct. In this study however changes in the community structure were followed in the first year after the farming practice was changed. Furthermore the differences in the amount of applied fertilizer were not very high and were the only change in farming management. Therefore the stability of the community structure under conventional and precision farming were expected. The community structure of bacteria and fungi showed a dynamic which is typical for agricultural soils, with maximal diversity values during flowering. High root exudates (Vancura, 1980) seem to provide sufficient available carbon. In contrast, in later plant development stages exudation is significantly reduced (Vancura, 1980) and easy accessible carbon might be running short, at least for a highly active microbial community. Furthermore the amount of bio available water is reduced in summer time. Both factors result in a reduced microbial biomass and diversity on all investigated plots, independent from the farming management.

In general mainly microbial community structure and diversity is very difficult to measure, to standardize and to interpret, as due to the high uptake rates of genetic information by microbes, mainly under selective pressure, structural microbial diversity is not directly correlated to functional microbial diversity. Therefore microbial community structure as its own is of no indicator value for soil quality and has to be combined with functional properties. Another possibility to use structural microbial diversity as an indicator would be to monitor groups of microbes, which are known to have a benefit for the ecosystem, like for example *Rhizobia* or *Mycorrhiza*.

4.2. Enzymatic activities

Although the influence of different farming systems on enzymatic activities in the nitrogen cycle are very important for agricultural management only several studies are published. Most of the work done so far is based on questions about soil specific differences. Kandeler et al., 1999 could demonstrate that after application of maize straw protease activity was two magnitudes higher than thew corresponding values of a control soil. Webb et al., 2000 measured fluxes of N-inputs and outputs at several sites over three seasons. The result of their study was, that in arable systems where no organic manures are applied, most of the N-losses can be correlated to nitrate leaching. Wheatley et al., 2001 studied the effects of combination of C and N on the potential nitrification rates in an arable soil over 22 months. They concluded from their data that subtle and temporally variable interactions between general soil heterotrophs and autotrophic nitrifiers are possibly stimulates by the addition of C and N. Rice and Pancholy, 1972 found that nitrification rates are correlated with the plant development.

In our experiments also after the short period of the application of precision farming the measured enzymatic activities from the nitrogen cycle showed a clear response to the differentiated fertilizer application. Lower application of fertilizer resulted in significantly higher proteolytic activities combined with lower nitrification and denitrification activities, mainly on the low yield plots; this indicates a more efficient turnover of nitrogen, with a better delivery of nitrogen by proteolytic activities from more complex substrates (proteins) combined with lower lost of gaseous nitrogenic compounds. This response is only restricted to the time of the application of differentiated fertilization. A sustainable effect of a more efficient nitrogen turnover by the application of lower amounts of fertilizer was not found. The measured enzymatic (denitrification and nitrification) activities are in good accordance to the measured gas fluxes.

5. Conclusion

In future more information about regulation of microbial functions will be available mainly by the recent progress in molecular microbial ecology. On the one hand much genetic information about key processes is available that specific primers can be designed to study potential, diversity and expression of genes (Bach et al., 2001), on the other hand, primary processes and their relation to transformation of nutrients are studied in detail in the running soil genome projects (Rondon et al., 2000). Furthermore stable-isotope probing (Radajewski et al., 2000) will be another suitable tool to identify microbes involved in nutrient turnover. This might help to develop suitable indicators for soil quality based on microbial nitrogen turnover.

Acknowledgements

We thank Cornelia Galonska, Christine Kollerbauer and Tarik Durkaya for skilful technical assistance. This study was supported by a grant of the BMBF, Germany.

References

Amann, R., Ludwig, W., Schleifer, K.H., 1995. Phylogenetic identification and in situ detection of individual microbial cells without cultivation. Microbiol. Rev. 59, 143–149.

Andersen, C.P., Rygiewicz, P.T., 1999. Understanding plant-soil relationships using controlled environment facilities. Adv. Space Res. 24 (3), 309–318.

Bach, H.-J., Munch, J.C., 2000. Identification of bacterial sources of soil peptidases. Biol. Fertil. Soils 31, 219–224.

Bach, H.-J., Hartmann, A., Schloter, M., Munch, J.C., 2001. PCR primers and functional probes for amplification and detection of bacterial genes for extracellular peptidases in single strains and in soil. J. Microbiol. Meth. 44, 173–182.

Bever, J.D., Wetover, K.M., Antonovics, J., 1997. Incorporation the soil community into plant population dynamics: the utility of the feedback approach. J. Ecol. 85, 561–573.

Buyer, J.S., Kaufman, D.D., 1996. Microbial diversity in the rhizosphere of corn grown under conventional and low-input systems. Appl. Soil Ecol. 5, 21–27.

Dawson, C.J., 1997. Management for spatial variability. In: Stafford, V. (Ed.), Precision Agriculture '97, vol. I. BIOS Scientific Publishers Ltd., Oxford, pp. 45–58.

Kandeler, E., 1989. Aktuelle und potentielle Nitrifikation im Kurzzeit-bebrütungsversuch. VDLUFA-Schriftreihe 28, Kongreßband Teil II, pp. 921–931.

Kandeler, E., Luxhoi, J., Tscherko, D., Magid, J., 1999. Xylanase, invertase and protease at the soil-litter interface of a loamy sand. Soil Biol. Biochem. 31, 1171–1179.

Kloepper, J.W., Wie, G., Tzun, S., 1992. Rhizosphere populations dynamics and internal colonization of cucumber by plant growth promoting rhizobacteria which induce systemic resistance to *Colletrichum orbiculare*. In: Tjamos, E.C. (Ed.), Biological Control of Plant Diseases. Progress and Challenges for the Future. NATO ASI Series, vol. 230, Plenum Press, New York, pp. 182–192.

Ladd, J.N., Butler, J.H.A., 1972. Short-term assays of soil proteolytic enzyme activities using proteins and dipeptide derivates as substrates. Soil Biol. Biochem. 4, 19–30.

Lupwayi, N.Z., Rice, W.A., Clayton, G.W., 1998. Soil microbial diversity and community structure under wheat as influenced by tillage and crop rotation. Soil Biol. Biochem. 30 (13), 1733–1741.

Miller, M., Dick, R.P., 1995. Dynamics of soil C and microbial biomass in whole soil and aggregates in two cropping systems. Appl. Soil Ecol. 2, 253–261.

Myers, N., 1996. Environmental services of biodiversity. Proc. Natl. Acad. Sci. U.S.A. 93 (7), 164–169.

Ocio, J.A., Brookes, P.C., Jenkinson, D.S., 1991. Field incorporation of straw and its effects on soil microbial biomass and soil inorganic N. Soil Biol. Biochem. 32, 171–176.

Powlson, D.S., Brook, P.C., Christensen, B.T., 1987. Measurement of soil microbial biomass provides an early indicator of changes in total soil organic matter due to straw incorporation. Soil Biol. Biochem. 19, 159–164.

Radajewski, S., Ineson, P., Parekh, N.R., Murrell, J.C., 2000. Stable isotope probing as a tool in microbial ecology. Nature 403, 646–649.

Rice, E.L., Pancholy, S.K., 1972. Inhibition of nitrification by climax ecosystems. Am. J. Bot. 59, 1033–1040.

Rondon, M.R., August, P.R., Bettermann, A.D., Brady, S.F., Grossmann, T.H., Liles, M.R., Osburne, M.S., Clardy, J., Handelsmann, J., Goodmann, R.M., 2000. Cloning the soil metagenome: a strategy for accessing the genetic and functional diversity of uncultured micro organisms. Appl. Environ. Microbiol. 66, 2541–2547.

Ruser, R., Flessa, H., Schilling, R., Steindl, H., Beese, F., 1998. Soil compaction and fertilization effects on nitrous oxide and methane fluxes in potato fields. Soil Sci. Soc. Am. J. 62, 1587–1595.

Ryden, J.C., Lund, L.J., Focht, D.D., 1979. Direct measurement of denitrification loss from soils. Soil Sci. Soc. Am 43, 104–110.

Schloter, M., Zelles, L., Hartmann, A., Munch, J.C., 1998. New quality of assessment of microbial diversity in arable soils using molecular and biochemical methods. Plant Nutr. Soil Sci. 161, 425–431.

Singh, H.B., Herlth, D., O'Hara, D., Zahnle, K., Bradshaw, J.D., Sandholm, S.T., Talbot, R., Crutzen, P.J., Kanakidou, M., 1992. Relationship of peroxyacetyl nitrate to active and total odd nitrogen at northern high latitudes: influence of reservoir species on NO_x and O_3. J. Geophy. Res. 97, 16523–16530.

Sorensen, J., 1997. The rhizosphere as a habitat for soil microorganisms. In: VanElsas, J.D. (Ed.), Manual of Environmental Microbiology. Marcel Dekker, New York, pp. 233–260.

Vancura, V., 1980. Fluorescent pseudomonads in the rhizosphere of plants and their relation to root exudates. Folia Microbiol. 25, 168–173.

Webb, J., Harrison, R., Ellis, S., 2000. Nitrogen fluxes in three arable soils in the UK. Eur. J. Agron. 13, 207–223.

Wheatley, R.E., Ritz, K., Crabb, D., Caul, S., 2001. Temporal variations in potential nitrification dynamics in soil related to differences in rates and types of carbon and nitrogen inputs. Soil Biol. Biochem. 33, 2135–2144.

Wyland, L.J., Jackson, L.E., Schulbach, K.F., 1995. Soil-plant nitrogen dynamics following incorporation of a mature rye cover crop in a lettuce production system. J. Agric. Sci. 124, 17–25.

Zelles, L., 1996. Fatty acid patterns of microbial phospholipids and lipopolysaccharides. In: Schinner, F., Öhlinger, R., Kandeler, E., Margesin, R. (Eds.), Methods in Soil Biology. Springer, Berlin. pp. 80–93.

PART III
BIODIVERSITY AT DIFFERENT SCALE LEVELS

Chapter 5 Biodiversity and Landscape
R. Waldhardt

Synoptic Introduction

ELSEVIER

Available online at www.sciencedirect.com

Agriculture, Ecosystems and Environment 98 (2003) 305–309

Agriculture Ecosystems & Environment

www.elsevier.com/locate/agee

Biodiversity and landscape—summary, conclusions and perspectives

Rainer Waldhardt*

Division of Landscape Ecology and Landscape Planning, Justus-Liebig-University of Giessen, IFZ, Heinrich-Buff-Ring 26-32, D-35392 Giessen, Germany

Abstract

Based on the contributions in this special issue and further references approaches to indicate biodiversity at the landscape scale are summarized and discussed. The complexity of biodiversity, its levels of aggregation and its specific relations to "natural" and "anthropogenic" environmental conditions depending on spatio-temporal scales are emphasized. The necessity of further research is stressed and research fields are differentiated.

Keywords: Managed landscape; Spatio-temporal scale; Landscape indicator; Landscape structure; Matrix effect

1. Introduction

By signing the Convention on Biological Diversity in 1992, European countries, as well as more than 160 nations worldwide, have pledged themselves to the conservation of biological diversity. Even though the conservation of particularly valuable habitats is still important—this goal dominated nature conservation politics until ca. 1980—it is more and more recognized that the preservation of biodiversity is only to achieve through the (re)establishment of a mosaic of suitable habitat patches at the landscape scale. Since biodiversity of terrestrial ecosystems is expected to be mainly affected by land-use change within the next 100 years (Sala et al., 2000) and virtually the entire land area of Central Europe has been subject to anthropogenic land use, we have to focus on land use management for the preservation of biological diversity.

The assessment of biodiversity in managed landscapes remains a problem for mainly two reasons: (i) Diversity measures strongly depend on the chosen spatio-temporal scale of the prevailing assessment, and unfortunately there are no satisfying scaling functions applicable to transfer results to another scale. (ii) Relations between biodiversity and land-use are generally very complex (Szaro and Johnston, 1996). Besides of "natural" (e.g. geological and climatic conditions) and "anthropogenic" (e.g. specific management practices, habitat fragmentation) environmental conditions, ecological processes and socio-economic factors have to be taken into account.

The total species richness at the landscape scale could serve as a criterion for the evaluation of sustainable land use, but is impossible to assess (Duelli, 1997). Thus, in analogy to environmental indicators (Stein et al., 2001), there is a need to *search for indicators* that show either quantitative relationships to overall biodiversity (i.e., correlates), or which are at least qualitatively connected to biodiversity measures

* Tel.: +49-641-9937163; fax: +49-641-9937169.
E-mail address: rainer.waldhardt@agrar.uni-giessen.de (R. Waldhardt).

0167-8809/$ – see front matter © 2003 Elsevier Science B.V. All rights reserved.
doi:10.1016/S0167-8809(03)00090-2

(i.e., surrogates; cf. Duelli and Obrist, 1998) at various spatio-temporal scales.

The elaboration of such indicators on a regionally differentiated scale is a goal of various research projects. In this chapter "Biodiversity and landscape" the contributions of Jeanneret et al. (2003) (Swiss Federal Research Station for Agroecology and Agriculture) as well as of Dauber et al. (2003), Hirsch et al. (2003), Steiner and Köhler (2003) and Waldhardt and Otte (2003) (all part of the Collaborative Research Centre 299 "Land use concepts for marginal regions" at the University of Giessen, Germany), may exemplify this kind of research. Four further contributions present *concepts and procedures for the regionally differentiated indication* of biodiversity at the landscape scale that are currently being developed. Within the concepts of "structural indication by man made objects" (Osinski, 2003), of "mosaic indicators" (presented in two contributions by Hoffmann and Greef (2003) and Hoffmann et al. (2003)), and in the approach of "ecological area sampling" (Hoffmann-Kroll et al., 2003) regionally differentiated patterns of biotopes and selected species groups as indicators of animal and plant diversity are discussed.

Based on the contributions in this chapter and further references related to the topic "biodiversity and landscape" approaches and possibilities of indication of biodiversity at the landscape scale will be summarized and discussed in the following. Especially, it will be dealt with the questions: (i) how far scientific research on indication of landscape biodiversity considers landscape complexity and dynamics (cf. Chapter 1 of this special issue concerning the term biodiversity), (ii) whether the indicators found, allow valid conclusions on landscape biodiversity and (iii) which gaps in scientific research exist and have to be filled.

2. Indication of biodiversity at the landscape scale

2.1. Approaches presented in this special issue

The research programs, concepts and case studies on the indication of landscape biodiversity that are presented here deal with various aspects of biodiversity on different spatio-temporal scales. As often in biodiversity research, floristic and/or faunistic species diversity are the main focus (cf. Waldhardt and Otte, 2000). Further components of biodiversity such as the diversity of biocoenoses or the spatio-temporal patterns of habitats are evaluated with respect to their ability to indicate species diversity on the one hand, on the other hand, they are understood and discussed as important components of biodiversity themselves. It is fair to say that various aspects of biodiversity are not considered in this chapter: this is true for structural and functional aspects of diversity which are, according to Hobbs (1997) and Bunnell and Huggard (1999), poorly understood and it is also true for genetic diversity.

In accordance with Jeanneret et al. (2003) it must be emphasized that there are no general models relating components of biodiversity such as overall species diversity to landscape characteristics. The relationships strongly depend on the demands of the prevailing component of biodiversity (e.g. vascular plant species, zoocoenoses) on environmental conditions. Furthermore, both a wide spatial (local habitat up to landscape) and temporal scale must be taken into consideration.

Only few species groups are considered in the papers by Jeanneret et al. (2003), Dauber et al. (2003), Hirsch et al. (2003) and Waldhardt and Otte (2003). Furthermore, the reference areas of these contributions present more or less small landscape sections. Therefore, we cannot be sure whether the results of these studies may be validly applied in general. However, biodiversity indicators that may be suitable for, e.g., monitoring or evaluating management practices within the regions being described are given in the papers. One further aim of Dauber et al. (2003), Hirsch et al. (2003) and Waldhardt and Otte (2003) is to quantify relations between specific components of biodiversity and land-use. From these relations rules are derived, which are intended to serve as integrated parameters in a landscape biodiversity model (cf. the contribution of Steiner and Köhler (2003)).

Jeanneret et al. (2003), Dauber et al. (2003) and Hirsch et al. (2003) deal with the importance of both (i) the "internal" properties of the plots investigated, such as management-type and site conditions, and (ii) their matrix (i.e. the surroundings, the plots are embedded in) for species richness within the plots themselves. The results highlight that—depending on the species groups considered and the size of the matrix—matrix effects have to be included in modeling routines of landscape biodiversity. This also holds for effects of land use dynamics as shown by Waldhardt and Otte (2003). The latter study also demonstrates that effects on the diversity of flora and vegetation vary with the level of aggregation of diversity (α- to γ-diversity of plant species and phytocoenoses). Finally, the model developed by Steiner and Köhler (2003) reveals the importance of connectivity of habitat types for landscape biodiversity.

More strongly focussed on direct application are the concept and case study by Osinski (2003) and Hoffmann and Greef (2003), Hoffmann et al. (2003) and the routines for spatially differentiated indication of species and habitat diversity presented by Hoffmann-Kroll et al. (2003). Based on digital landscape information Osinski (2003), Hoffmann and Greef (2003) and Hoffmann et al. (2003) separate spatial units whose suitability as habitats for animal and plant species is then evaluated. Thus a spatially differentiated indication of landscape biodiversity and the evaluation of landscapes with respect to their suitability as habitats for plants and animals appear possible. The evaluation of habitats described by Hoffmann and Greef (2003) and Hoffmann et al. (2003) is based on an inventory of (groups of) indicator species with known habitat preferences. Therefore, the accuracy of predictions will crucially depend on the selection of appropriate (groups of) indicator species, though. The suitability of these selected (groups of) indicator species for the indication of total biodiversity within biotopes, landscape sections or landscapes has to be tested.

The approach of the ecological area sampling (EAS), introduced by Hoffmann-Kroll et al. (2003), is as the British countryside survey (Haines-Young et al., 2000), aiming at a representative country-wide inventory and evaluation of structural states and changes of nature and landscapes. By the means of frequently repeated field inventories on relatively few, representative study plots, the consequences of changes in land use for landscape quality will be quantified for all of Germany. Biotope diversity and species composition are recorded as indicators of landscape quality. The selection of plots is based on a geostatistical analysis of remote sensing data on land use. Experiences gained in a pilot study testing this methodology was conducted by the authors and revealed satisfying results.

2.2. Conclusions and perspectives

The approach of combining organismic and landscape indicators—this approach unites all contributions of this chapter with their different foci—is regarded as straightforward for the analysis and indication of biodiversity at the landscape scale (Debinski and Humphrey, 1997; Duelli and Obrist, 1998; Margules and Pressey, 2000; Wagner et al., 2000; Wagner and Edwards, 2001). However, currently there is hardly any set of regionally applicable biotic indicators (in the sense of correlates and surrogates) available. Further efforts are needed for identifying biotic groups of indicators as carried out by Duelli and Obrist (1998) and Sauberer (2001) to quantify overall species diversity, respectively, by Waldhardt et al. (2000) to quantify plant species diversity within grasslands. Apart from this, further studies should try to quantify to which extend various components of landscape biodiversity depend on landscape structure (Lindenmayer et al., 2000). Analyses of the importance of composition and fragmentation of habitats (e.g. Hansen et al., 1988; Poschlod et al., 1996; Fagan et al., 1999; Steffan-Dewenter and Tscharntke, 1999; Kurki et al., 2000), their heterogeneity (e.g. Turner and Gardner, 1991; Wrbka et al., 1999; Thies and Tscharntke, 1999; Romero-Alcaraz and Avila, 2000; Weibull et al., 2000; Atauri and Lucio, 2001; Kie et al., 2002), connectivity (e.g. Clergeau and Burel, 1997; Mönkkönen and Reunanen, 1998; Hanski, 1999) and their total area at the landscape scale (e.g. Waldhardt and Otte, 2001) supply further information important within this context. Furthermore, land use intensities (e.g. Bjorklund et al., 1999; Dumanski and Pieri, 2000; MacDonald et al., 2000; Zechmeister and Moser, 2001), dynamics and processes affecting biodiversity (e.g. Turner, 1989; Jones et al., 1994, 1997; Petit and Burel, 1998; Loreau et al., 2001; Parody et al., 2001; Simmering et al., 2001) must be taken into account. Finally, the significance of interactions among the above mentioned and further possible landscape characteristics at different spatial and temporal scales is largely unknown.

Biodiversity of landscapes, even when focussing on single components such as species diversity, will depend on numerous landscape characteristics related to land use. The impact of these land use characteristics will probably vary in relation to spatio-temporal scale and level of aggregation (α- to γ-diversity; cf. Whittaker, 1972). Consequently, it appears to be very unlikely that there should be one single indicator for landscape biodiversity. Therefore, sets of indicators must be determined and valid models of landscape biodiversity must include a sufficient number of indicator-response rules concerning the consequences of land use and land use changes on biodiversity.

References

Atauri, J.A., Lucio, J.V., 2001. The role of landscape structure in species richness distribution of birds, amphibians, reptiles and lepidopterans in Mediterranean landscapes. Landsc. Ecol. 16, 147–159.

Bjorklund, J., Limburg, K.E., Rydberg, T., 1999. Impact of production intensity on the ability of the agricultural landscape to generate ecosystem services: an example from Sweden. Ecol. Econ. 29, 269–291.

Bunnell, F.L., Huggard, D.J., 1999. Biodiversity across spatial and temporal scales: problems and opportunities. For. Ecol. Manage. 115, 113–126.

Clergeau, P., Burel, F., 1997. The role of spatio-temporal patch connectivity at the landscape level: an example in a bird distribution. Landsc. Urban Plan. 38, 37–43.

Dauber, J., Hirsch, M., Simmering, D., Waldhardt, R., Otte, A., Wolters, V., 2003. Landscape structure as an indicator of biodiversity: matrix effects on species richness. Agric. Ecosyst. Environ. 98, 321–329.

Debinski, D.M., Humphrey, P.S., 1997. An integrated approach to biological diversity assessment. Nat. Areas J. 17, 355–365.

Duelli, P., 1997. Biodiversity evaluation in agricultural landscapes: an approach at two different scales. Agric. Ecosyst. Environ. 62, 81–92.

Duelli, P., Obrist, K., 1998. In search of the best correlates for local organismal biodiversity in cultivated areas. Biodivers. Conserv. 7, 297–309.

Dumanski, J., Pieri, C., 2000. Land quality indicators: research plan. Agric. Ecosyst. Environ. 81, 93–102.

Fagan, W.F., Cantrell, R.S., Cosner, C., 1999. How habitat edges change species interactions. Am. Nat. 153, 165–182.

Haines-Young, R.H., Barr, C.J., Black, H.I.J., Briggs, D.J., Bunce, R.G.H., Clarke, R.T., Cooper, A., Dawson, F.H., Firbank, L.G., Fuller, R.M., Furse, M.T., Gillespie, M.K., Hill, R., Hornung, M., Howard, D.C., McCann, T., Morecroft, M.D., Petit, S., Sier, A.R.J., Smart, S.M., Smith, G.M., Stott, A.P., Stuart, R.C., Watkins, J.W., 2000. Accounting for nature: assessing habitats in the UK Countryside. DETR (Dept. of the Environment, Transport and the Regions), London, UK, pp. 1–134.

Hansen, A.J., di Castri, F., Naiman, R.J., 1988. Ecotones: what and why? In: Di Castri, F., Hansen, A.J., Holland, M.M. (Eds.), A New Look on Ecotones. Biol. Int. 17, 9–45 (special issue).

Hanski, I., 1999. Habitat connectivity, habitat continuity, and metapopulations in dynamic landscapes. Oikos 87, 209–219.

Hirsch, M., Pfaff, S., Wolters, V., 2003. The influence of matrix type on flower visitors of *Centaurea jacea* L. Agric. Ecosyst. Environ. 98, 331–337.

Hobbs, R., 1997. Future landscapes and the future of landscape ecology. Landsc. Urban Plan. 37, 1–9.

Hoffmann, J., Greef, J.M., 2003. Mosaic indicators—theoretical approach for the development of indicators for species diversity in agricultural landscapes. Agric. Ecosyst. Environ. 98, 387–394.

Hoffmann, J., Greef, J.M., Kiesel, J., Lutze, G., Wenkel, K.-O., 2003. Practical example of the mosaic indicators approach. Agric. Ecosyst. Environ. 98, 395–405.

Hoffmann-Kroll, R., Schäfer, D., Seibel, S., 2003. Landscape indicators from ecological area sampling in Germany. Agric. Ecosyst. Environ. 98, 363–370.

Jeanneret, P., Schüpbach, B., Luka, H., 2003. Quantifying the impact of landscape and habitat features on biodiversity in cultivated landscapes. Agric. Ecosyst. Environ. 98, 311–320.

Jones, C.G., Lawton, J.H., Shachak, M., 1994. Organisms as ecosystem engineers. Oikos 69, 373–386.

Jones, C.G., Lawton, J.H., Shachak, M., 1997. Positive and negative effects of organisms as physical ecosystem engineers. Ecology 78, 1946–1957.

Kie, J.G., Bowyer, R.T., Nicholson, M.C., Boroski, B.B., Loft, E.R., 2002. Landscape heterogeneity at differing scales: effects on spatial distribution of mule deer. Ecology 83, 530–544.

Kurki, S., Nikula, A., Helle, P., Lindén, H., 2000. Landscape fragmentation and forest composition effects on goose breeding success in boreal forests. Ecology 81, 1985–1997.

Lindenmayer, D.B., Margules, C.R., Botkin, D.B., 2000. Indicators of biodiversity for ecologically sustainable forest management. Conserv. Biol. 14, 941–950.

Loreau, M., Naeem, S., Inchausti, P., Bengtsson, J., Grime, J.P., Hector, A., Hooper, D.U., Huston, M.A., Raffaelli, D., Schmid, B., Tilman, D., Wardle, D.A., 2001. Ecology—biodiversity and ecosystem functioning: current knowledge and future challenges. Science 294, 804–808.

MacDonald, D., Crabtree, J.R., Wiesinger, G., Dax, T., Stamou, N., Fleury, P., Lazpita, J.G., Gibon, A., 2000. Agricultural abandonment in mountain areas of Europe: environmental consequences and policy response. J. Environ. 59, 47–69.

Margules, C.R., Pressey, R.L., 2000. Systematic conservation planning. Nature 405, 243–253.

Mönkkönen, M., Reunanen, P., 1998. On critical thresholds in landscape connectivity—management perspective. Oikos 84, 302–305.

Osinski, E., 2003. Operationalisation of a landscape oriented indicator. Agric. Ecosyst. Environ. 98, 371–386.

Parody, J.M., Cuthbert, F.J., Decker, E.H., 2001. The effect of 50 years of landscape change on species richness and community composition. Global Ecol. Biogeogr. Lett. 10, 305–313.

Petit, S., Burel, F., 1998. Effects of landscape dynamics on the metapopulation of a ground beetle (Coleoptera, Carabidae) in a hedgerow network. Agric. Ecosyst. Environ. 69, 243–252.

Poschlod, P., Bakker, J, Bonn, S., Fischer, S., 1996. Dispersal of plants in fragmented landscapes. In: Settele, J., Margules, C.R., Poschlod, P., Henle, K. (Eds.), Species Survival in Fragmented Landscapes, pp. 123–127.

Romero-Alcaraz, E., Avila, J.M., 2000. Landscape heterogeneity in relation to variations in epigaeic beetle diversity of a Mediterranean ecosystem. Imp. Conserv. Biodivers. Conserv. 9, 985–1005.

Sala, O.E., Chapin, F.S., Armesto, J.J., Berlow, E., Bloomfield, J., Dirzo, R., Huber-Sanwald, E., Huenneke, L.F., Jackson, R.B., Kinzig, A., Leemans, R., Lodge, D.M., Mooney, H.A., Oesterheld, M., Poff, N.L., Sykes, M.T., Walker, B.H., Walker, M., Wall, D.H., 2000. Biodiversity—global biodiversity scenarios for the year 2100. Science 287, 1770–1774.

Sauberer, N., 2001. Surrogate taxa in biodiversity assessment—working better than thought. Predicting Biodiversity in European Landscapes, Vienna, November 18–20, 2001, conference abstracts.

Simmering, D., Waldhardt, R., Otte, A., 2001. Syndynamik und Ökologie von Besenginsterbeständen des Lahn-Dill-Berglands unter Berücksichtigung ihrer Genese aus verschiedenen Rasengesellschaften. Tuexenia 21, 51–89.

Steffan-Dewenter, I., Tscharntke, T., 1999. Effects of habitat isolation on pollinator communities and seed set. Oecologia 121, 432–440.

Stein, A., Riley, J., Halberg, N., 2001. Issues of scale for environmental indicators. Agric. Ecosyst. Environ. 87, 215–232.

Steiner, N.C., Köhler, W., 2003. Effects of landscape patterns on species richness—a modelling approach. Agric. Ecosyst. Environ. 98, 353–361.

Szaro, R.C., Johnston, D.W. (Eds.), 1996. Biodiversity in Managed Landscapes. Oxford University Press, Oxford.

Thies, C., Tscharntke, T., 1999. Landscape structure and biological control in agroecosystems. Science 285, 893–895.

Turner, M.G., 1989. Landscape ecology: the effect of pattern on process. Ann. Rev. Ecol. Syst. 20, 171–197.

Turner, M.G., Gardner, R.H. (Eds.), 1991. Quantitative Methods in Landscape Ecology: The Analysis and Interpretation of Landscape Heterogeneity. Springer, New York.

Wagner, H.H., Edwards, P.J., 2001. Quantifying habitat specificity to assess the contribution of a patch to species richness at a landscape scale. Landsc. Ecol. 16, 121–131.

Wagner, H., Wildi, O., Ewald, K.C., 2000. Additive partitioning of plant species diversity in an agricultural mosaic landscape. Landsc. Ecol. 15, 219–227.

Waldhardt, R., Otte, A., 2000. Zur Terminologie und wissenschaftlichen Anwendung des Begriffs Biodiversität. Wasser Boden 52, 10–13.

Waldhardt, R., Otte, A., 2001. Abschätzung der zur Erhaltung einer lokaltypischen Ackerflora benötigten Fläche am Beispiel der Gemarkung Erda (Lahn-Dill-Bergland, Hessen). Peckiana 1, 101–108.

Waldhardt, R., Otte, A., 2003. Indicators of plant species and community diversity in grasslands. Agric. Ecosyst. Environ. 98, 339–351.

Waldhardt, R., Simmering, D., Otte, A., 2000. Standortspezifische Surrogate und Korrelate der alpha-Artendichten in der Grünland-Vegetation einer peripheren Kulturlandschaft Hessens. Ber. ANL 24, 79–86.

Weibull, A.C., Bengtsson, J., Nohlgren, E., 2000. Diversity of butterflies in the agricultural landscape: the role of farming system and landscape heterogeneity. Ecography 23, 743–750.

Whittaker, R.H., 1972. Evolution and measurement of species diversity. Taxon 21, 213–251.

Wrbka, T., Szerencsits, E., Moser, D., Reiter, K., 1999. Biodiversity patterns in cultivated landscapes: experiences and first results from a nationwide Austrian survey. In: Maudsley, M., Marshall, J. (Eds.), Heterogeneity in Landscape Ecology: Pattern and Scale, in: Proceedings of the IALE (UK) Conference, Bristol, 9/1999.

Zechmeister, H.G., Moser, D., 2001. The influence of agricultural land-use intensity on bryophyte species richness. Biodivers. Conserv. 10, 1609–1625.

PART III
BIODIVERSITY AT DIFFERENT SCALE LEVELS

Chapter 5 Biodiversity and Landscape

R. Waldhardt

Original Papers

ELSEVIER

Available online at www.sciencedirect.com

Agriculture, Ecosystems and Environment 98 (2003) 311–320

Agriculture Ecosystems & Environment

www.elsevier.com/locate/agee

Quantifying the impact of landscape and habitat features on biodiversity in cultivated landscapes

Ph. Jeanneret[a,*], B. Schüpbach[a], H. Luka[b]

[a] Swiss Federal Research Station for Agroecology and Agriculture, Reckenholzstr. 191, CH-8046 Zurich, Switzerland
[b] Research Institute of Organic Agriculture, Ackerstrasse, CH-5070 Frick, Switzerland

Abstract

Determining habitat and landscape features that lead to patterns of biodiversity in cultivated landscapes is an important step for the assessment of the impact of extensification programmes in agriculture. In the context of an assessment of the effect of national extensification programme on biodiversity in agriculture, field data of three regions (7 km^2 each) were collected according to a stratified sampling method. A distribution model of three taxa (spiders, carabid beetles, and butterflies) is related to influencing factors by means of multivariate statistics (canonical correspondence analysis (CCA), partial CCA). Hypothetical influencing factors are categorised as follows: (1) habitat (habitat type, plant species richness) and (2) landscape (habitat heterogeneity, variability, diversity, proportion of natural and semi-natural areas). The correlative model developed for the spider assemblages revealed that the most important local habitat factors are those directly influenced by management practices. Landscape variability, heterogeneity and diversity in the surroundings are not significant factors. Carabid beetle assemblages show a positive reaction to landscape features and respond to particular cultivated surroundings. The model developed for butterflies shows that species assemblages are sensitive to landscape features. Surrounding land use in particular, has a major influence.

There are no general models relating overall species diversity to landscape diversity. The relationship strongly depends on the organism examined. Therefore, biodiversity response to landscape and habitat changes (e.g. the extensification programme) has to be identified by means of multi-taxon concept.

Keywords: Landscape diversity; Species diversity; Arthropods; Canonical correspondence analysis; Biotic indicators

1. Introduction

1.1. Biodiversity measurement

Although biodiversity is conceptually defined (genetic diversity, organismal diversity, ecological diversity, cultural diversity; Heywood and Baste, 1992), its measurement remains controversial and is often the consequence of financial resources and observation techniques. Numerous surrogates and correlates have been proposed to quantify biodiversity (Gaston, 1996). As they make up about 65% of the species of all multicellular organisms (Hammond, 1992), arthropods are good candidates to represent biodiversity. Investigating a large set of faunistic groups and vascular plants, the advantages and disadvantages of reducing the number of indicators to a representative set of organismal diversity have been studied by Duelli and Obrist (1998), Duelli et al. (1999). They found correlations of high significance between species richness

* Corresponding author. Tel.: +41-1-377-72-28;
fax: +41-1-377-72-01.
E-mail address: philippe.jeanneret@fal.admin.ch (Ph. Jeanneret).

0167-8809/$ – see front matter © 2003 Elsevier Science B.V. All rights reserved.
doi:10.1016/S0167-8809(03)00091-4

of 19 of the 25 the groups studied with the overall number of species.

Nevertheless, species richness of one single group or of a small set of organisms will not provide sufficient information in several research areas:

- Any kind of comparison between habitats, ecosystems and landscapes. Because the same species richness in different habitats, ecosystems and landscapes usually represents different species compositions.
- Analysis of the species diversity function in ecosystems. Because one single taxonomic group can hardly represent more than one or two functions.
- Analysis of the impact of environmental explanatory factors on biodiversity. Because one single group is unlikely to react the same way as any other group.

As mentioned by Huston (1994), in addition to the simple number of species present, biodiversity has many other components. Which species are present and which are most abundant are important aspects of biological diversity that cannot be summarised in a simple figure. In the past, controversy surrounding the statistical quantification of diversity was largely a result of the unreasonable expectation that a single statistic should contain all the information about the assembly of objects that it represents. This is clearly an unrealistic expectation.

Approaches taking into account species composition, which is an essential biodiversity component, and looking at species assemblages, constitute a more appropriate method of quantifying the influence of environmental explanatory factors.

Saying that biodiversity and its response to environmental conditions cannot be measured by means of a single taxonomic group, indicators primarily adapted to the objectives (i.e. specific habitat restoration, agricultural practices) should be defined. Beside reaching specific goals, indicators should meet general criteria, e.g. stable taxonomy, response to environmental changes, etc. (Reck, 1990; Riecken, 1992; Reid et al., 1993; Pearson, 1995; Stork and Samways, 1995). The indicators should represent all major functional guilds and take into account several spatial and temporal scales (Noss, 1990; Hunsaker, 1993; Hammond, 1995).

1.2. Mosaic of cultivated landscape

A cultivated landscape is usually a mosaic-like structure, composed of patches, implying discreteness of elements and the existence of boundaries between neighbouring patches (Forman, 1995). The possible mosaic structure is unlimited and most probably, there is no single mosaic pattern which allows for a theory to be generated (Wiens, 1995). Nevertheless, generalisations should be based on the grouping of landscape types according to various indices of structure (e.g. O'Neil et al., 1988). Moreover, species guilds or groups should be studied according to their ecological characteristics. This approach will focus on the functionality of landscape structure.

A cultivated landscape is heterogeneous and provides several types of habitats. Because most of the time, single habitats usually prove to be unsuitable for a large number of species, the spatial and temporal distribution of suitable habitats is a key factor for biodiversity in cultivated landscapes. For certain species, isolation, a primary characteristic of island biogeography, is a major problem (Mac Arthur and Wilson, 1967). On land, however, most species can cross the mosaic, at least at low rates (Knaapen et al., 1992). Because cultivated landscapes exhibit a high quantity of patches which are not suitable for most species with regard to time and space, the incorporation of a high proportion of natural and semi-natural areas, functioning as refuges, is very important (Duelli, 1997). In addition to the proportion of natural and semi-natural areas, habitat heterogeneity and variability per unit of area are pertinent factors for the prediction and evaluation of biodiversity. Isolation probably has no negative impact on species diversity in cultivated patches. It does, however, have a negative impact on refuges. Interactions between the matrix (disturbed patches, including cultivated land), and environmental, remnant or regenerated patches (sensu Forman, 1995), and particularly species movement in this context (permeability, connectivity, and fractal dimension), is very important for species distribution and diversity in cultivated landscapes.

Studies focussing on the relationship between species diversity and landscape diversity in cultivated landscapes describe the theoretical links (Baudry and Baudry-Burel, 1982), were based on restricted groups of organisms (Burel, 1989, 1992; Miller et al.,

1997), on landscapes of particular structure (Burel and Baudry, 1995), or on gradients of agricultural intensity (Burel et al., 1998).

1.3. Ecological compensation: a particular case of landscape transformation

In countries where intensive agriculture has led to an alarming level of ecological degradation, an increasing number of efforts are now being made to restore agricultural landscapes and enhance biodiversity by implementing agri-environmental programmes, such as introducing semi-natural habitats and field margins into farmland.

According to the Swiss Federal Law on Agriculture, only those farmers adhering to specific ecological production rules are entitled to receive subsidies. Parallel to measures which apply directly to the field management, 7% of the land must be set aside for ecological compensation and managed according to special guidelines. Fifteen types of ecological compensatory areas (ECA) were defined, six of which are subsidised by the government. The six categories comprise extensively used meadows (no fertilisation, late mowing), low intensity meadows (restricted fertilisation, late mowing), humid meadows used for litter, wild flower strips, standard fruit trees in traditional orchards, hedgerows and groves.

Since 1992, the total increase in ECA in Swiss cultivated landscape has risen to 89,000 ha (8.6% of the arable land) in 1999 (Swiss Federal Office of Agriculture, 2000). The Swiss Federal Office for Agriculture (SFOA) is responsible for the assessment of the effects of ecological measures in agriculture on environmental indicators. One of these is biodiversity.

The present study is part of a project in which biodiversity and landscape patterns are assessed in three regions of approximately 7–8 km^2. A number of variables known to have an impact on biodiversity on the local (management, soil, exposition) and landscape level (e.g. the distribution and arrangement of crops and ECAs, connectivity) are measured in these regions and used to explain diversity patterns of five taxa: plants, butterflies, spiders, carabid beetles, and birds. We concentrate our analysis on observation points of biodiversity (sampling points = sites) situated in the mosaic of cultivated land and calculate landscape parameters around these observation points. The purpose is to identify and quantify the impact of the main environmental variables having an influence on biodiversity and specify the role of the ECAs.

In this context, the influence of ECAs is supposed to act at two scales:

(i) on a local scale, by habitat influencing variables (management, soil, exposition, etc.);
(ii) on a landscape scale, in interaction with other habitats, by landscape influencing variables (distribution, variability and arrangement of cultures, ECAs, connectivity, etc.).

This study shows the results of spider, carabid beetle and butterfly assemblages collected and observed in one of the three regions in 1997 and 1998. The relationship between habitat, landscape explanatory variables and species distribution is quantified by means of multivariate statistics. The results are then examined, with regard to the Swiss cultivated landscape restoration programme.

2. Method

2.1. Region, sampling methods

The area chosen to carry out the study was Ruswil (20 km NW of Lucerne). It comprised a total surface of 8 km^2, consisting of grassland (44%), arable land (19%), ECAs (7%), forests (20%), and villages (10%). Four ECA habitat types, usually small areas of approximately 400 m^2 size can be found in the perimeter, namely extensively used meadows (no fertilisation, late mowing), low intensity meadows (restricted fertilisation, late mowing), hedgerows and standard fruit trees in traditional orchards. Each field in the perimeter was visited, categorised according to its use and digitised by means of a geographical information system (GIS) (Fig. 1).

Spiders, carabid beetles and butterflies were recorded according to a stratified sampling method. ECAs, cultivated areas and forest edges were defined as strata. The number of samples per ECA type was determined in proportion to the number of elements in each type occurring in the study area (Table 1). Seven highly intensive meadows were chosen to serve as references for the cultivated area, because they are predominant of the landscape in the region. Seventeen

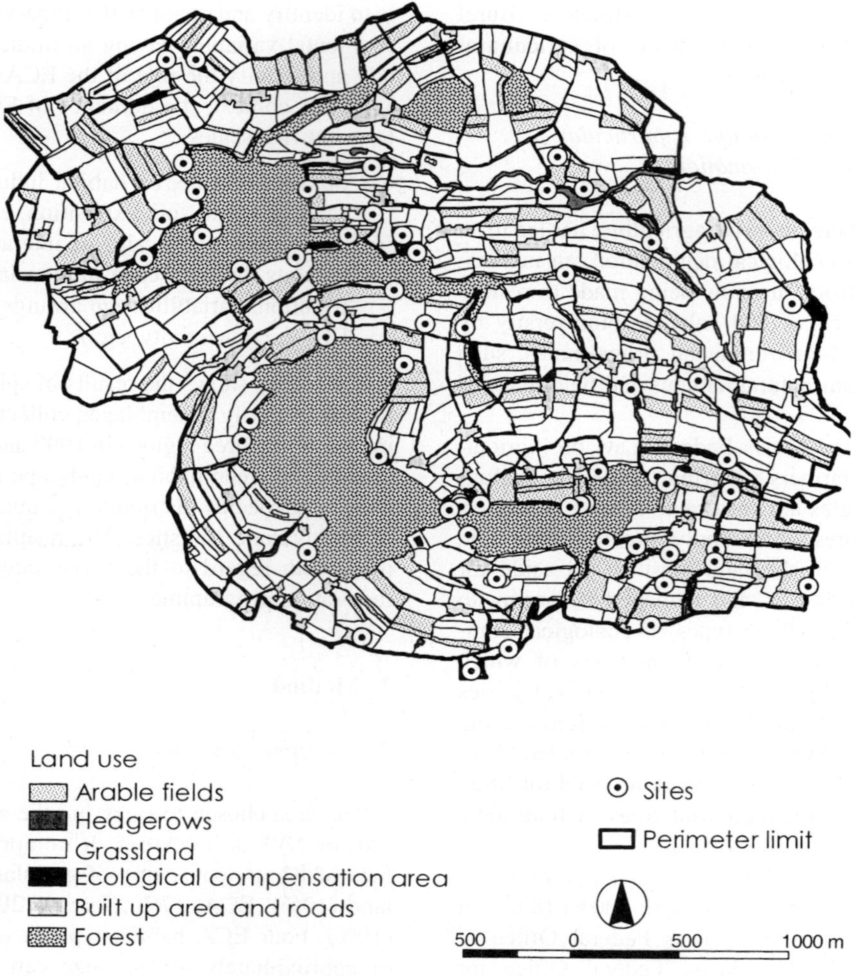

Fig. 1. Land use map of the study region (Ruswil, 20 km NW of Lucerne).

Table 1

Strata, habitat types, and the number of sampling sites in the 7 km^2 area of Ruswil (for explanation about the ECAs, see Section 1.3)

Strata	Habitat type	Number of sites
Cultivated areas	High intensity meadows	7
ECAs	Extensively used meadows	16
	Low intensity meadows	7
	Hedgerows	3
	Standard fruit trees in traditional orchards	8
Forest edges		17

observation sites were set up along the forest edge belonging to three forest plots.

Spiders and carabid beetles were collected in 1997 and butterflies observed in 1998 on the 58 sites. Spiders and carabid beetles were collected with three pitfall traps per site, during 5 weeks (during the first 3 weeks of May and last 2 weeks of June), as proposed by Duelli (1997) to optimise the number of species caught compared to the sampling effort. During the season, across a 0.25 ha large area per site, butterflies were observed five times during 10 min each. At forest edges, butterflies were recorded along the edge. At each of the observation sites, the vegetation was

Table 2
Characterisation of the habitats and landscape acting as explanatory variables on biodiversity

Scale	Environmental variables	Land use types
Habitat descriptors	Plant species richness	
	Habitat type	Six types: high intensity, extensively used and low intensity meadow, hedgerow, standard fruit trees in traditional orchard, forest edge
Landscape descriptors	Surrounding habitat variability	Eighteen types: habitat type + cereal fields, root crops, corn, pasture, artificial meadow, grove, rape, nursery, slope, brook, built up area, forest
	Surrounding habitat heterogeneity	Eighteen types: habitat type + cereal fields, root crops, corn, pasture, artificial meadow, grove, rape, nursery, slope, brook, built up area, forest
	D_1 index of landscape pattern	Eighteen types: habitat type + cereal fields, root crops, corn, pasture, artificial meadow, grove, rape, nursery, slope, brook, built up area, forest
	Surrounding land use	Four classes: cultivated land, ECA, built up area, forest

assessed over a single area of $100\,m^2$ according to the Braun–Blanquet method.

2.2. Habitat and landscape characteristics acting as explanatory variables on spider, carabid beetle and butterfly assemblages

The environmental influence on spider, carabid beetle and butterfly assemblages was tested using six descriptors (Table 2), introduced as explanatory variables in CCA. The first set of descriptors defines the habitats and the second, the landscape surrounding the habitats.

2.2.1. Habitat descriptors

1. Plant species richness is supposed to have a major influence on the spider and butterfly assemblages. For spiders and carabid beetles, higher plant species richness offers more diverse habitat structure (architecture) and more niches for prey (Strong et al., 1984). Higher plant species richness represents more host and feeding plants in time and space that should influence butterfly assemblages.
2. Depending on their management, habitat types will have a major influence on epigeal arthropod and butterfly communities. Actually, habitat types summarise variables like time and number of harvests per season, fertiliser applications, etc. Habitats were assigned to the six types listed in Table 2.

2.2.2. Landscape descriptors

On a broader scale, landscape characteristics are relevant explanatory variables for plant and animal communities, because they define ecosystem arrangement and interactions (Forman and Godron, 1986; Forman, 1995). For landscape characterisation, a 200 m radius circle around the observation points was defined using a GIS. Within this circle, four landscape descriptors were defined (Table 2).

First, patches (a patch = a relatively homogeneous nonlinear area that differs from its surroundings) were assigned to 18 land use types. Three landscape pattern indices, probably the simplest ones, were calculated on the basis of the percentage cover of the different land use types within the circle:

1. Surrounding habitat variability = number of surrounding land use types (Duelli, 1997).
2. Surrounding habitat heterogeneity = number of patches of different land use types (Duelli, 1997).
3. D_1 index of landscape pattern (O'Neil et al., 1988); D_1 is a measure of dominance.

   $D_1 = \ln n + \sum P_i \ln P_i$, where n is the total number of land use types and Pi the proportion of patches in land use type i.

   Second, a qualitative measure of landscape diversity was introduced. The 18 land use types were assembled in four surrounding land use classes to record information about the landscape quality (types of habitats):
4. Surrounding land use classes: cultivated land, ECA, forest and built up area.

2.3. Statistical analysis

To conserve the whole information on the observed indicators, we defined species diversity as composed of species variety and relative abundance of the species. Species–environment relationship was then analysed with the help of ordinations and multivariate statistics, instead of summarising the biotic information in one single value such as species richness or a diversity index where interpretation would be difficult and the loss of information too substantial.

To identify the main environmental variables having an effect on the species diversity, a canonical correspondence analysis (CCA) and a partial CCA, were carried out by means of the CANOCO programme (Ter Braak and Smilauer, 1998). In CCA, the significance of a particular environmental variable can be assessed by Monte Carlo testing (bootstrapping) of the axis associated with that variable, using the axis eigenvalue as the test statistic (Ter Braak, 1987).

With GIS, main landscape descriptors were analysed and introduced in the CCA and partial CCA. The detailed model describing the use of CCA with each separate variable prior to a forward selection being carried out, to be followed by CCA involving all the variables, is described in Jeanneret et al. (1999) and Jeanneret (2000). The goal is to establish a hierarchy between explanatory variables and eliminate those which do not explain any variance significantly. Partitioning of variance is then performed through partial CCA (e.g. Borcard et al., 1992). The fraction of the variance explained (and its significance, obtained by means of a Monte Carlo permutation test) by each of the environmental descriptors is given separately, after eliminating the variance due to the other (partialed) variables, which are used as covariables.

3. Results

3.1. Faunistical description of the sites

Altogether, 16,057 spiders belonging to 135 species and 9325 carabid beetles belonging to 79 species were collected from the 58 sites. Published results showed that forest edges are well characterised by the spider and carabid beetle communities and represent a particular habitat where typical forest species were found together with species of adjacent meadows (Jeanneret et al., 2000; Pfiffner et al., 2000). Looking at the species richness, forest edges are significantly richer than grassland habitats for both groups of epigeal arthropods.

Altogether, 892 butterflies belonging to 17 species were observed on the 58 sites. Butterfly species richness was significantly higher in the extensively used and low intensity meadows than in the high intensity meadows (Jeanneret et al., 2000). According to the correspondence analysis, studied sites are not clearly differentiated by the butterflies. Nevertheless, some extensively managed meadows have a particular species set.

3.2. Environmental explanatory variables having an influence on spider, carabid beetle and butterfly assemblages

Within the scope of separate CCA and forward selection procedures, environmental variables and classes able to explain a significant share of variance were recognised and then introduced in partial CCA.

3.2.1. Habitat descriptors

Plant species richness explained a significant part of the variance of spider, carabid beetle and butterfly assemblages (Table 3). Habitat type significantly determines spider and carabid beetle assemblages. Habitat type has no significant influence on butterfly assemblages, for which plant species richness is a decisive factor.

3.2.2. Landscape descriptors

Surrounding habitat variability and heterogeneity as well as the D_1 index of dominance do not explain a significant part of the epigeal arthropod variance (Table 3). Logically, landscape variables have a strong influence on butterfly assemblage, which need landscape structure to move from one suitable habitat to the other. Butterflies are then significantly influenced by the heterogeneity of the surrounding habitats. Surrounding land use, all classes put together, is a significant landscape explanatory variable for each indicator. Nevertheless, the percentage of forest and cultivated land were negatively correlated (more cultivated land = less forest; Pearson $r = -0.75$). After separate CCA and forward selection of variables, the percentage of forest only remains significant for spiders, the percentage of cultivated land for carabid beetles, and the forest and ECA for butterflies (Table 3).

Table 3
Summary of the percentage variance explained and P-values (Monte Carlo permutation test) by environmental variables for spiders, carabid beetles, and butterflies (n.s.: $P \geq 0.05$)

Scale	Environmental variables	Spiders		Carabid beetles		Butterflies	
		%	P	%	P	%	P
Habitat descriptors	Plant species richness	4	<0.05	3.6	<0.05	4.2	<0.05
	Habitat type	24	<0.05	18.4	<0.05	11.4	n.s.
Landscape descriptors	Surrounding habitat variability	1.9	n.s.	1.8	n.s.	2.8	n.s.
	D_1 index of landscape pattern	2.1	n.s.	2	n.s.	2.3	n.s.
	Surrounding habitat heterogeneity	2.3	n.s.	2.4	n.s.	3.5	<0.05
	Surrounding land use classes:						
	Cultivated land	5.2	n.s.[a]	3.4	<0.05	2.9	n.s.[a]
	Forest	5.5	<0.05	6.1	n.s.[a]	3.4	<0.05
	ECA	2.4	n.s.	1.6	n.s.	3.9	<0.05
	Built up area	1.5	n.s.	1.4	n.s.	2.3	n.s.

[a] Following forward selection.

3.3. Variance partitioning

Following the procedure explained in Jeanneret (2000), explanatory variables which did not have significant effects on the communities were eliminated before performing a partial CCA.

The significant habitat and landscape variables (plant species richness and habitat type; surrounding habitat heterogeneity and the significant surrounding land use classes) were introduced in a partial CCA in order to rank them according to explained variance. The fraction of variance explained and its significance (Monte Carlo permutation test) by each of the environmental variables is given separately, after eliminating variance due to the other (partialed) variables, which are used as covariables. For spiders, the share of variation attributed to the habitat type is greater than any other explained variation (Fig. 2). The remaining variance is attributed to plant species richness. No landscape variables explained a significant part of variance in the spider species assemblage. The presence of forest in the surroundings is eliminated after the partitioning, because it is correlated with the descriptor forest edge (habitat descriptor; Pearson $r = 0.53$) and is no longer significant.

Compared with spiders, habitat type has a weaker impact on carabid beetles (12.4% vs. 19.7%, Figs. 2 and 3). Plant species richness remains a significant variable having an effect on carabid beetles. Landscape represented by the surrounding land use class "cultivated land" significantly influences carabid beetle assemblages (2.7%).

For butterflies, plant species richness, the surrounding habitat heterogeneity, the forest and the ECA explained a significant part of the variance and were therefore selected. In a partial CCA, heterogeneity does not explain a significant part of variance any longer (Fig. 4). Landscape quality, expressed as surrounding land use classes, is sufficient to explain the difference in those butterfly assemblages with habitat plant species richness (Fig. 4).

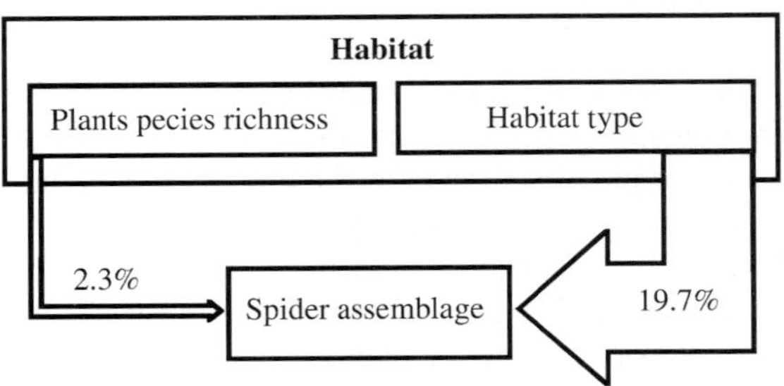

Fig. 2. Synthetic model of correlative relations between environmental variables and spider assemblages, based on partial CCA. Landscape descriptors do not explain any significant part of the variance. The arrows are proportional to the percentage of the explained variance and can be compared with the percentages stated for carabid beetle and butterfly assemblages in Figs. 3 and 4.

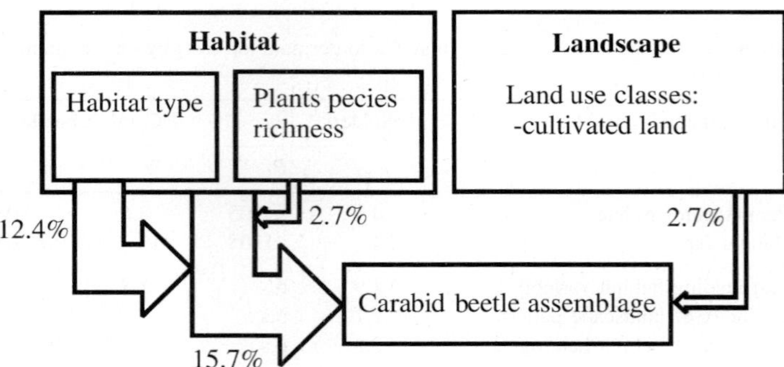

Fig. 3. Synthetic model of correlative relations between environmental variables and carabid beetle assemblages, based on partial CCA. Both habitat and landscape descriptors explain a significant part of the variance. The arrows are proportional to the percentage of the explained variance and can be compared with the percentages stated for spider and butterfly assemblages in Figs. 2 and 4.

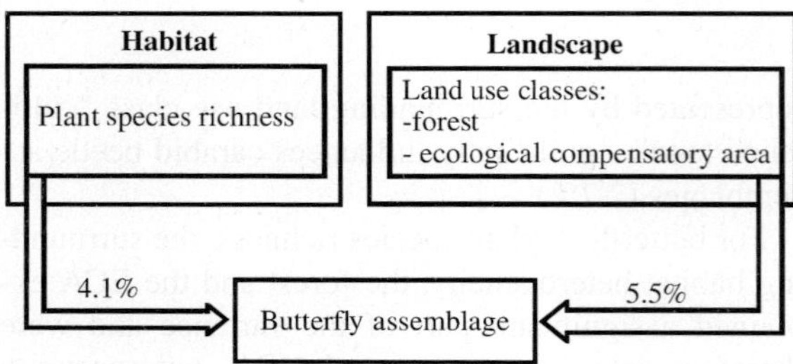

Fig. 4. Synthetic model of correlative relations between environmental variables and butterfly assemblages, based on partial CCA. Both habitat and landscape descriptors explain a significant part of the variance. The arrows are proportional to the percentage of the explained variance and can be compared with the percentages stated for spider and carabid beetle assemblages in Figs. 2 and 3.

4. Discussion

The correlative model developed for the spider assemblages revealed that local habitat variables (habitat type), that affect the site directly are the most important ones. The habitat type, which is directly influenced by management practices and especially by the extensification programme, is the most affecting variable, followed by the plant species richness of the site. Management effect of cultivated fields on spider assemblages is not astonishing and has been demonstrated in different situations (e.g. Alderweireldt, 1989; Dennis et al., 2001; Pozzi and Borcard, 2001). Landscape variability, heterogeneity, and diversity in the surroundings are not significant explanatory variables. These results confirm precedent studies carried out by Asselin and Baudry (1989), Burel and Baudry (1995) and show that spiders move easily among habitats by ballooning (Blandenier and Fürst, 1997). They can reach any habitat and are subsequently influenced by local habitat conditions.

Results for carabid beetles showed that species assemblages are significantly sensitive to both habitat and landscape characteristics. Habitat type is the variable with the greatest influence, followed by plant species richness. As for spiders, the characterisation of habitats with carabid beetles has been described (e.g. Turin et al., 1991; Kramer, 1996). Among the habitat types, forest edge has a strong influence, because typical forest species were also collected in this habitat. Landscape has a significant influence and additional analysis have stressed the effect on carabid beetle assemblages of the presence of root crops and pastures in the adjacent surroundings.

In spite of the low species richness of the study area, the model developed for butterflies showed that species assemblages are sensitive to landscape features. The surrounding land use type in particular has a major influence. As emphasised by Dower et al. (1992), field boundaries are very important for butterflies as vital links between different habitats. In our study, forest edges and ECA in the surroundings determine the species diversity of an observed site. Plant species richness at the site is not sufficient to explain the difference in the species assemblages. Most butterfly species, need structure to move and often require several habitats to complete their life-cycles. In the landscape studied, habitat arrangement and plant species richness at the site had the greatest influence. Nevertheless, observation of the breeding sites and the larval stages would be necessary to complete information on population of particular species.

On the basis of our results, habitat and landscape parameters have varying impacts, depending on the indicator. This is also true of the particular case of ECAs. On one hand, spider, carabid beetle and butterfly assemblages are influenced by the habitat type, which in part, is a result of the extensification rate, but on a very local scale. On the other hand, ECAs in the surrounding landscape have a significant influence on butterfly assemblages observed on a given site, which is a result of the extensification rate at landscape level.

Because of the differentiated response, it is important to approach the role of the extensification programme by examining different species assemblages, as recommended by Noss (1990) and Huston (1994). Biodiversity response to the extensification programme cannot be summarised by one single indicator because every taxon may react differently to the environmental conditions that are influenced by this programme.

Examining species assemblages allows for a comprehensive appreciation of the impact of habitat and landscape parameters on biodiversity. Reduction of biodiversity to species richness or to a diversity index conducts to a substantial loss of information that is crucial in examining the effect of environmental factors (Huston, 1994). The species richness of spiders, e.g. does not make it possible to differentiate sufficiently between extensively used meadows and intensively used meadows (Jeanneret et al., 2000). Species assemblages, however, make a very significant distinction possible.

5. Conclusion

The purpose and value of using indicator organisms is to observe the biotic response to environmental stress or restoration efforts. Such information is useful for the development or refining of restoration and management plans as well as for monitoring the success of these efforts. From our study it is clear that the habitat management (fertilisation and cutting regime) has to be taken into account to optimise the restoration program for the three indicator organisms. For carabid beetles the close neighbourhood of a given habitat must be examined. For butterflies, habitat management is not at all sufficient to explain their distribution. The restoration programme must then act at the landscape scale.

Furthermore, other typical landscape features, e.g. the isolation of semi-natural areas, connectivity, permeability and fractal dimension will be tested. Another approach that will be chosen, is the analysis of landscape structure and quality, as applied to landscape windows (Burel et al., 1998).

Acknowledgements

This study was partially financed by the Swiss Federal Office for Agriculture. The authors wish to thank J. Steiger, G. Blandenier and H. Hänggi for spider identification, F. Herzog, Th. Walter and L. Pfiffner for comments on the manuscript, S. Bosshard, M. Waldburger and D. Schrag for their assistance in field work.

References

Alderweireldt, M., 1989. An ecological analysis of the spider fauna (Aranae) occurring in maize fields, Italian ryegrass fields and their edge zones, by means of different multivariate techniques. Agric. Ecosyst. Environ. 27, 293–306.

Asselin, A., Baudry, J., 1989. Les araneides dans un espace agricole en mutation. Acta Oecol. 4, 143–156.

Baudry, J., Baudry-Burel, F., 1982. La mesure de la diversité spatiale. Relations avec la diversité spécifique. Utilisation dans les évaluations d'impact. Acta Oecol. Appl. 3, 177–190.

Blandenier, G., Fürst, P.-A., 1997. Ballooning spiders. In: Selden, P.A. (Ed.), Proceedings of the 17th European Colloquium of Arachnology, Edinburgh.

Borcard, D., Legendre, P., Drapeau, P., 1992. Partialling out the spatial component of ecological variation. Ecology 73, 1045–1055.

Burel, F., 1989. Landscape structure effects on carabid beetles spatial patterns in western France. Landscape Ecol. 2, 215–226.

Burel, F., 1992. Effect of landscape structure and dynamics on species diversity in hedgerow networks. Landscape Ecol. 6, 161–174.

Burel, F., Baudry, J., 1995. Species biodiversity in changing agriculture landscapes: a case study in the Pays d'Auge, France. Agric. Ecosyst. Environ. 55, 193–200.

Burel, F., Baudry, J., Butet, A., Clergeau, P., Delettre, Y., Le Coeur, D., Dubs, F.L., Morvan, N., Paillat, G., Petit, S., Thenail, C., Brunel, E., Lefeuvre, J.-C., 1998. Comparative biodiversity along a gradient of agricultural landscapes. Acta Oecol. 19, 47–60.

Dennis, P., Young, M.R., Bentley, Ch., 2001. The effects of varied grazing management on epigeal spiders, harvestmen and pseudoscorpions of *Nardus stricta* grassland in upland Scotland. Agric. Ecosyst. Environ. 86, 39–57.

Dower, J.W., Clarke, S.A., Rew, L., 1992. Habitats and movement pattern of satyrid butterflies (Lepidoptera: Satyridae) on arable farmland. Entomol. Gazette 43, 29–44.

Duelli, P., 1997. Biodiversity evaluation in agricultural landscapes: an approach at two different scales. Agric. Ecosyst. Environ. 62, 81–91.

Duelli, P., Obrist, M.K., 1998. In search of the best correlates for local organismal biodiversity in cultivated areas. Biodivers. Conserv. 7, 297–309.

Duelli, P., Obrist, M.K., Schmatz, D.R., 1999. Biodiversity evaluation in agricultural landscapes: above-ground insects. Agric. Ecosyst. Environ. 74, 33–64.

Forman, R.T.T., 1995. Land Mosaics, The Ecology of Landscapes and Regions. Cambridge University Press, Cambridge.

Forman, R.T.T., Godron, M., 1986. Landscape Ecology. Wiley, New York.

Gaston, K.J., 1996. Biodiversity. A Biology of Numbers and Difference. Blackwell, London.

Hammond, P.M., 1992. Species inventory. In: Groombridge, B. (Ed.), Global Biodiversity, Status of the Earth's Living Resources. Chapman & Hall, London, pp. 17–39.

Hammond, P.M., 1995. Practical approaches to the estimation of the extent of biodiversity in speciose groups. In: Hawksworth, D.L. (Ed.), Biodiversity, Measurement and Estimation. Chapman & Hall, London, pp. 119–136.

Heywood, V.H., Baste, I., 1992. Introduction. In: Heywood, V.H., Watson, R.T. (Eds.), Global Biodiversity Assessment. UNEP, Cambridge University Press, Cambridge, pp. 1–19.

Hunsaker, C.T., 1993. New concepts in environmental monitoring: the question of indicators. The Science of the Total Environment Supplement, pp. 77–95.

Huston, M.A., 1994. Biological Diversity. Cambridge University Press, Cambridge.

Jeanneret, Ph., 2000. Interchanges of a common pest guild between orchards and the surrounding ecosystems: a multivariate analysis of landscape influence. In: Ekbom, B., Irwin, M.E., Robert, Y. (Eds.), Interchanges of Insects Between Agricultural and Surrounding Landscapes. Kluwer Academic Publishers, Dordrecht, pp. 85–107.

Jeanneret, Ph., Schüpbach, B., Lips, A., Harding, J., Steiger, J., Waldburger, M., Bigler, F., Fried, P.M., 1999. Biodiversity patterns in cultivated landscapes: modelling and mapping with GIS and multivariate statistics. In: Maudsley, M., Marshall, J. (Eds.), Heterogeneity in Landscape Ecology: Pattern and Scale. Colin Cross Printers Ltd., Garstang, pp. 85–94.

Jeanneret, P., Schüpbach, B., Steiger, J., Waldburger, M., Bigler, F., 2000. Evaluation Ökomassnahmen: Biodiversität, Tagfalter und Spinnen. Agrarforschung 7, 112–116.

Knaapen, J.P., Scheffer, M., Harms, B., 1992. Estimating habitat isolation in landscape planning. Landscape Urban Plan. 23, 1–16.

Kramer, I., 1996. Biodiversität von Arthropoden in Wanderbrachen und ihre Bewertung durch Laufkäfer, Schwebefliegen und Stechimmen. Agrarökologie 17, Bern.

Mac Arthur, R.H., Wilson, E.O., 1967. The Theory of Island Biogeography. Princeton University Press, Princeton, NJ.

Miller, J.N., Brooks, R.P., Croonquist, M.J., 1997. Effects of landscape patterns on biotic communities. Landscape Ecol. 12, 137–153.

Noss, R.F., 1990. Indicators for monitoring biodiversity: a hierarchical approach. Conserv. Biol. 4, 355–364.

O'Neil, R.V., Krummel, J.R., Gardner, R.H., Sugihara, G., Kackson, B., Deangelis, D.L., Milne, B.T., Turner, M.G., Zygmunt, B., Christensen, S.W., Dale, V.H., Graham, R.L., 1988. Indices of landscape pattern. Landscape Ecol. 1, 153–162.

Pearson, D.L., 1995. Selecting indicator taxa for the quantitative assessment of biodiversity. In: Hawksworth, D.L. (Ed.), Biodiversity, Measurement and Estimation. Chapman & Hall, London, pp. 75–79.

Pfiffner, L., Luka, H., Jeanneret, Ph., Schüpbach, B., 2000. Evaluation der Ökomassnahmen: Biodiversität, Effekte ökologischer Ausgleichsflächen auf die Laufkäferfauna. Agrarforschung 7 (5), 212–217.

Pozzi, S., Borcard, D., 2001. Effects of dry grassland management on spider (Arachnida: Araneae) communities on the Swiss occidental plateau. Ecoscience 8, 32–44.

Reck, O., 1990. Zur Auswahl von Tiergruppen als Biodeskriptoren für den tierökologischen Fachbeitrag zu Eingriffsplanungen. Schriftenreihe für Landschaftspflege und Naturschutz 32, 99–119.

Reid, V.W., McNeely, J.A., Tunstall, D.B., Bryant, D.A., Winograd, M., 1993. Biodiversity Indicators for Policy-makers. World Resources Institute, New York.

Riecken, U., 1992. Planungsbezogene Bioindikation durch Tierarten und Tiergruppen. Schriftenreihe für Landschaftspflege und Naturschutz 36. Bundesforschungsanstalt für Naturschutz und Landschaftsökologie.

Stork, N.E., Samways, M.J., 1995. Inventorying and monitoring of biodiversity. In: Heywood, V.H., Watson, R.T. (Eds.), Global Biodiversity Assessment. UNEP, Cambridge University Press, Cambridge, pp. 453–544.

Strong, D.R., Lawton, J.H., Southwood, R., 1984. Insects on Plants. Blackwell Scientific Publications, Oxford.

Swiss Federal Office of Agriculture, 2000. Evaluation des mesures écologiques et des programmes relatifs à la garde d'animaux de rente, synthèse des rapports spécifiques à chacun de domaines de recherche. Office fédéral de l'agriculture, Berne.

Ter Braak, C.J.F., 1987. Unimodal models to relate species to environment. Doctoral Thesis. Agricultural Mathematics Group-DLO, Wageningen, The Netherlands.

Ter Braak, C.J.F., Smilauer, P., 1998. CANOCO Reference Manual and User's Guide to Canoco for Windows: Software for Canonical Community Ordination, Version 4. Microcomputer Power, Ithaca, NY.

Turin, H., Alders, K., Boer, P.J. Den, Essen, S. van, Heijerman, Th., Laane, W., Peterman, E., 1991. Ecological characterization of carabid species (Coleoptera, Carabidae) in The Netherlands from thirty years of pitfall sampling. Tijdschfrift voor Entomologie 134, 279–304.

Wiens, J.A., 1995. Landscape mosaics and ecological theory. In: Hansson, L., Fahrig, L., Merriam, G. (Eds.), Mosaic Landscapes and Ecological Processes. Chapman & Hall, London, pp. 1–26.

ELSEVIER

Available online at www.sciencedirect.com

Agriculture, Ecosystems and Environment 98 (2003) 321–329

Agriculture Ecosystems & Environment

www.elsevier.com/locate/agee

Landscape structure as an indicator of biodiversity: matrix effects on species richness

Jens Dauber[a,*], Michaela Hirsch[a], Dietmar Simmering[b], Rainer Waldhardt[b], Annette Otte[b], Volkmar Wolters[a]

[a] *Department of Animal Ecology, Justus-Liebig-University, IFZ, Heinrich-Buff-Ring 26-32, D-35392 Giessen, Germany*
[b] *Division of Landscape Ecology and Landscape Planning, Justus-Liebig-University, IFZ, Heinrich-Buff-Ring 26-32, D-35392 Giessen, Germany*

Abstract

Sustainable conservation management in cultivated landscapes urgently needs indicators that provide quantitative links between landscape patterns and biodiversity. As a contribution to this aim, the influence of the matrix surrounding managed grassland sites on species richness of ants, wild bees and vascular plants was investigated at two different scales (50 and 200 m radius). In addition, patch variables describing habitat quality were included in the analyses. Species richness of the three taxa was not significantly inter-correlated. Multiple linear regression analysis revealed significant predictor variables for the species richness of the different taxa at both matrix scales. The variation of the matrix radius had no impact on the variance explanation of the regression models. The degree of variance explained by the regression models varied between taxa (bees > plants > ants). Moreover, the predictive variables were different for the taxa, with the regression model for wild bees including both patch and matrix variables, that for plants richness including patch variables only, and that for ants including matrix variables only. We conclude that landscape diversity and percentage cover of certain land-use types might serve as useful indicators for species richness at the landscape scale. However, the specific response patterns revealed in our study suggest that a variety of taxa must be included in this type of approach.

Keywords: Apoidea; Formicidae; Plants, matrix; Conservation management; Agricultural landscape

1. Introduction

Biodiversity has become a general and basic conservation value over the past few decades (Gaston, 1996; Duelli et al., 1999). Recent advances in metapopulation theory, landscape ecology and macroecology (e.g. Rosenzweig, 1995; Hanski, 1999; Lawton, 2000) have revealed the strong impact of landscape configuration on local diversity and community structure. Though habitat quality may be the most important factor determining the presence of a species at a given site (Duelli, 1997), diversity within a patch additionally depends on the structure of the surrounding landscape. These so-called 'matrix effects' have been demonstrated by various authors (e.g. Jonsen and Fahrig, 1997; Miller et al., 1997; Burel et al., 1998; Thies and Tscharntke, 1999; Weibull et al., 2000). Moreover, the composition of a landscape is one of the key factors explaining species richness at the regional scale (Dunning et al., 1992; Dale et al., 2000; Wagner et al., 2000). Fahrig (2001) has shown that an improvement in

* Corresponding author. Tel.: +49-641-9935717; fax: +49-641-9935709.
E-mail address: jens.dauber@allzool.bio.uni-giessen.de (J. Dauber).

0167-8809/$ – see front matter © 2003 Elsevier Science B.V. All rights reserved.
doi:10.1016/S0167-8809(03)00092-6

matrix quality can significantly reduce extinction thresholds when the amount of habitat is decreasing. Conservation strategies for sustaining biodiversity must consider that species richness and ecological processes are controlled by parameters operating at a wide array of scales (Baudry et al., 2000). However, the measurement of these parameters at large spatial scales is, if practicable at all, costly and time consuming. We thus urgently need simple indicators of biodiversity, providing quantitative links between landscape patterns and species richness.

Our previous investigations carried out in different land-use types (i.e. arable land, grassland, and fallow land) had shown that the strong variability in species richness between study sites could not sufficiently be explained by differences in internal factors like habitat quality (e.g. Dauber and Wolters, 2000). It seems obvious that external factors such as spatio-temporal dynamics (Purtauf et al., 2001; Waldhardt and Otte, 2003), boundary characteristics (e.g. Fagan et al., 1999) or neighbourhood effects (e.g. Tilman, 1994) also contribute to variations in species richness and community composition at the patch level. The study presented here thus focussed on the impact of matrix variables on the species richness of selected taxa inhabiting managed grassland sites. In particular, we addressed the question as to whether parameters that are easily measured at large spatial scales are suitable predictors of species richness. The study is part of a multidisciplinary research project on the impact of land-use on ecological and economical conditions of a marginal landscape (Frede and Bach, 1999; Waldhardt et al., 1999; Wolters et al., 1999).

Matrix effects are due to a variety of processes including source–sink dynamics or neighbourhood effects (Dunning et al., 1992; Schluter and Ricklefs, 1993; Wiens et al., 1993). These in turn are modulated by the mobility, the migratory behaviour and the specific demands of the species involved. Therefore, we selected three taxa differing in both their mode of dispersal and their habitat requirements: vascular plants, ants and wild bees. Plants are sessile and disperse through seeds or vegetative organs (e.g. rhizomes). The number of dispersal units per plant, as well as the longevity of dispersal units (e.g. in the soil seed bank), the mode and distance of dispersal vary between plant strategy types (Grime and Hillier, 1992; Poschlod et al., 1996; Thompson et al., 1997; Bonn and Poschlod, 1998). Ant colonies are relatively sessile. Some species move their colonies at most every 2 weeks and some do not move at all (Smallwood, 1982). Foragers move from the core of the nest through the environment searching for food resources (Kaspari, 2000). Dispersal of most species results from mating flies of alates. Ant colonies resemble plants in many ways (Andersen, 1991). Bees provide for their broods by central place foraging from their nest (Cresswell et al., 2000). Westrich (1989) estimates an average flight radius of about 50 m for most European wild bees. Foraging distances of bumblebees are considerably greater, reaching up to 600 m for some species (Walter-Hellwig and Frankl, 2000).

2. Study sites and methods

2.1. Study sites

The study was carried out at 20 managed grassland sites situated in the Lahn-Dill-Bergland (Hesse, Germany). This region is a low mountain range, with soil conditions and climate that are unsuitable for modern crop production. The resulting increase in the proportion of managed grassland and fallow land, and the accompanying decrease in the proportion of arable land over the past decades has led to a major change of landscape structure (Waldhardt et al., 1999). Fourteen sites were located in the rural district of Erda (270–385 m asl, 8 °C mean annual temperature, 700–800 mm mean annual precipitation), and six in Steinbrücken (320–420 m asl, 6.6 °C mean annual temperature, 1100–1200 mm mean annual precipitation).

2.2. Landscape analysis

Habitat quality depends on two different sets of variables: (i) intra-patch variables (e.g. field-size, soil type, aspect, vegetation cover), and (ii) matrix variables (e.g. heterogeneity of the surrounding landscape, portions of surrounding land-use types) (Landis and Marino, 1999; Wagner et al., 2000). Intra-patch variables used in this study were soil type, aspect, and age (i.e. time since last ploughing) (see Appendix A). Soil types were lumped into two categories: regosols as well as shallow cambisols representing drier soils (d) and pseudogley (planosol) representing more humid soils (h). The slopes were characterised as south-facing (S) or north-facing (N). Age of the grassland sites was derived from a GIS-supported stereoscopic analysis of black and white aerial photographs from the period of 1945–1997 (Fuhr-Boßdorf et al., 1999).

The matrix variables used in our study were derived from the same digital maps as mentioned above by means of a GIS (ArcView 3.2). The matrix of each study site was analysed at two different scales. A 50 m radius comprised the immediate surroundings and a 200 m radius covered a more general pattern of the vicinity of the sites (Fig. 1). Five different land-use types were identified within these two matrix scales: arable land (A), managed grassland (G), fallow land (FL), forest (F) and various urban land-use types (U). The latter category includes gardens, orchards, settlements and other forms of urban land-use. The proportion of the single land-use types in the matrices was calculated. The only linear structures were farm roads and country roads. Though the percentage cover of theses structures was implicitly included in the calculation via their proportionate contribution to the total area, roads were not used as separate independent variables because of their high variability in habitat quality (e.g. grassy field paths or tarred roads). Within the 50 m radius around the study sites, the proportion of different land-use types ranged from 0 to 76% for arable land, 0 to 96% for grassland, 0 to 35% for fallow land, 0 to 28% for forest and 0 to 18% for urban land-use. Within 200 m radius around the study sites the corresponding proportions ranged from 0 to 72% for arable land, 0 to 89% for grassland, 0 to 38% for fallow land, 0 to 52% for forest and 0 to 32% for urban land-use.

2.3. Biotic data collection

Ants were surveyed on all of the 20 sites, wild bees on 15 sites, and plants on 13 sites (see Appendix A). Ants were sampled using pitfall traps. A transect of five traps (min. 5 m spacing) was placed at each of the 20 sites. The pitfall traps were plastic cups 10 cm in diameter and with a 0.5 l volume, dug in level to the ground surface. Traps were filled with 0.2 l of a 70% solution of an ethanol/glycerine mixture as a preservative. One-half of the study sites traps were operated for a 4-week period in May 1997. The other half was operated for the same time period in May 1998. All ant individuals trapped were identified to species level.

Wild bees were sampled by pan traps of three different colours, i.e. yellow, blue and white. Pan traps were circular plastic bowls 18 cm in diameter and 12.5 cm deep. Traps were half-filled with a 1:1 mixture of water saturated with salt and 70% ethanol. A detergent was added to reduce surface tension. One pan trap of each colour was placed at the centre of 16 of the 20 study sites. Traps were operated for two 1-week periods in August 1998. All bees were identified to species

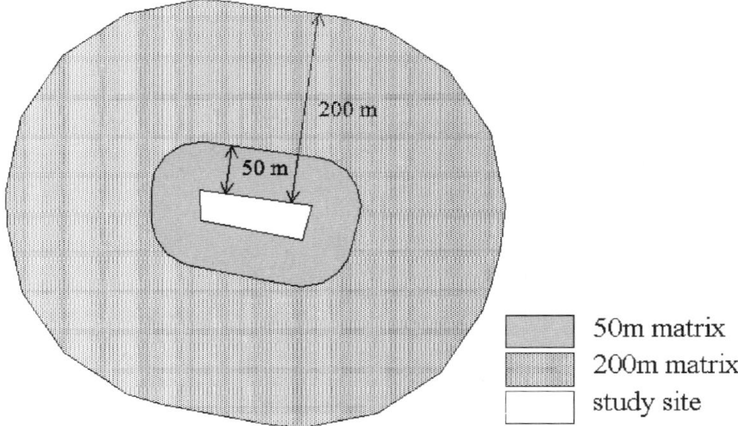

Fig. 1. Example of the two matrices surrounding a study site. The buffered area within a distance of 50 m comprises the immediate surroundings and the area within a distance of 200 m describes a more general pattern in the vicinity of the study site.

level. Honeybees (*Apis mellifera*) were excluded from further analyses.

Numbers of vascular plant species were recorded on 13 of the 20 study sites in May–August 1997. The sites were walked up and down until no new species were found at all within the last 20 min. A strip of 2 m width along the edges was excluded to avoid possible edge effects.

2.4. Data analysis

Correlations in species richness of the three taxa were tested using Pearson's product moment correlation. Backward stepwise multiple linear regressions were conducted to determine whether matrix variables were significant predictors of species richness at the two matrix scales (50 and 200 m). Predictor variables were sequentially omitted according to their relative reduction of R^2s. Intra-patch variables were included in all analyses (Jonsen and Fahrig, 1997). Aspect and soil type were included in the regressions as dummy-variables (Brosius, 1988). Inter-correlation among explanatory variables was investigated with Pearson's product moment correlation. Strong negative correlations were detected between percentage cover of arable land and grassland as well as between percentage cover of forest and grassland at both matrix scales. Therefore, percentage cover of grassland was not included as a matrix variable in the multiple regressions. Plant data had to be corrected for a potential bias caused by the well established correlation between sample area and species richness (e.g. Rosenzweig, 1995) because plants were the only taxon in our study for which sampling depended on the area of sampling. This correction was done by fitting the species–area equation $S = bA^z$ (S the species richness of plants, A the sampling area (0.1 ha up to 1.0 ha); cf. MacArthur and Wilson, 1967). We then extracted the residuals and used these for further multiple regression analyses. All statistical analyses were performed using the Statistica 5.0 for Windows package (StatSoft Inc., Tulsa, USA).

3. Results

In total 18 ant species, 16 wild bee species, and 153 vascular plant species were recorded. Species richness of the individual sites is shown in Appendix A. Pearson's product moment correlation did not reveal

Table 1
Results of the multiple regression analyses for all three taxa for the 50 m matrix and the 200 m matrix scale[a]

Taxa			Patch variables			Matrix variables		
Matrix radius (m)	Model R^2	Model P	Predictor	Beta	P	Predictor	Beta	P
Vascular plants								
50	0.67	0.0041	Aspect	−0.771	0.0023	–	–	–
			Soil type	0.557	0.0151	–	–	–
200	0.67	0.0041	Aspect	−0.771	0.0023	–	–	–
			Soil type	0.557	0.0151	–	–	–
Ants								
50	0.45	0.0062	–	–	–	%FL	0.412	0.037
			–	–	–	%F	0.463	0.021
200	0.51	0.0004	–	–	–	%F	0.715	0.0004
Wild bees								
50	0.89	0.0000	Soil type	−1.175	0.0000	%A	0.363	0.004
			Age	0.758	0.0002	–	–	–
200	0.89	0.0000	Soil type	−1.291	0.0000	%A	0.356	0.0049
			Age	0.837	0.0001	–	–	–

[a] All variables that contributed significantly to the regression model for a given taxa are shown. Table includes beta (relative importance of the predictor) and P (significance level) for each variable in the models, as well as the significance level and R^2 for the overall models. %: percentage cover, A: arable land, F: forest, FL: fallow land.

any significant correlation of the species richness among the three taxa, suggesting that the species number of one taxon cannot be predicted from the species number of another taxon.

The multiple linear regression analysis revealed significant predictor variables for the species richness of all taxa (Table 1). Results were either not or only slightly affected by increasing matrix radius. The degree of variance explanation varied between taxa (bees > plants > ants). Furthermore, the variables included in the regression models varied between the taxa. The models for plant species richness only included two intra-patch variables (aspect, soil type), with both variables together explaining 67% of total variance (Fig. 2). The models for the species richness of ants, in contrast, included matrix variables only: percentage cover of fallow land and forests. The former variable had a slightly lower explanatory power than the latter in the 50 m radius model and was eliminated from the 200 m radius model (Table 1). Under these conditions, the variance explanation slightly increased with increasing matrix scale but never exceeded 50% of the total variance (Fig. 2). The model for the species richness of wild bees included both patch (soil type, age) and matrix (percentage cover of arable land) variables (Table 1). The two patch variables had a higher explanatory power than the matrix variable. The variance of the species richness of wild bees explained by the regression models was particularly high (89%), with the matrix variable adding between 12 and 13% to the variance explanation (Fig. 2).

The regression models indicate that plant species richness was higher on south-facing slopes and on more humid soils than on north-facing slopes and dryer soils. Ant species richness increased with increasing percentage cover of forest and fallow land within the 50 m matrix and with an increase of percentage cover of forest within the 200 m matrix. Species richness of wild bees was higher on dry soils, increased with patch age and increased with increasing percentage cover of arable land within the matrices.

4. Discussion

Because of the enormous problems associated with the determination of total species richness, clear choices concerning the features of the parameters used as surrogates are required (Margules and Pressey, 2000). One strong temptation is to use one group of species to indicate the richness of other or even all species (Margules and Pressey, 2000). The fact that we could not establish any correlation in species richness suggests that this approach might be misleading: none of the taxa investigated had an indicative potential for any of the other taxa—at least in managed grassland habitats. The fact that this contradicts the finding that these taxa are among the best indicators of overall species richness in a Swiss agricultural

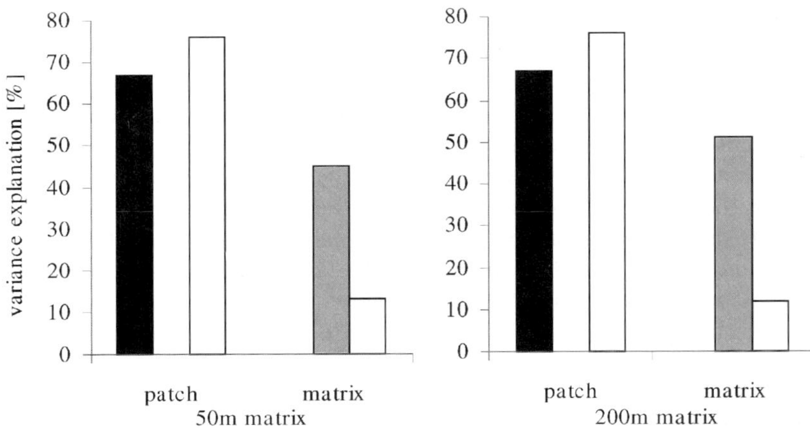

Fig. 2. Partitioning of the variance explanation between patch and matrix variables of the regression models for both matrix scales. Black columns represent vascular plants, grey columns ants and white columns wild bees.

landscape (Duelli and Obrist, 1998) confirms that tests of taxonomic surrogacy produce contradictory results (Prendergast, 1997; Duelli and Obrist, 1998; Osborn et al., 1999). Nevertheless, searching for combinations of taxa complementing one another in their indicative potential or testing higher taxon surrogacy (Gaston, 2000) remains an essential task for the identification of biodiversity indicators.

A further approach, to be combined with taxonomic surrogacy, is to understand and quantify the relationship between species groups and landscape characteristics controlling species richness (Dunning et al., 1992; Barbault, 1995). Landscape features might be useful predictors for the management of species richness at the landscape level (Franklin, 1993; Baudry et al., 2000; but see Araújo et al., 2001), but their impacts on species richness at any one point in a landscape act at multiple scales (Debinski et al., 2001). Therefore, we analysed the influence of matrix composition on species richness at two different scales. We expected the matrix radius to have an influence on species richness of all three taxa. Due to the small home ranges of most ant and wild bee species (Smallwood, 1982; Westrich, 1989) and to the restricted dispersal of grassland plants via seeds and rhizome growth (Bonn and Poschlod, 1998), we assumed the 50 m matrix to have a greater influence on species richness than the 200 m matrix. Yet, no striking differences in variance explanation were found between the two scales in the cases where matrix variables were part of the regression models. The question as to whether variance explanation would increase with further increase in the matrix scale must be answered in future studies. These studies should also address the question of the existence of optimal matrix scales.

Precision of the prediction of species richness obtained by the regression models varied between the taxa. Most interestingly, the degree of variance explanation corresponded to differences in mobility and habitat requirements of the taxa. About 89% of the variation in bee species richness was explained by the regression model including both patch and matrix variables. This points to the complex habitat requirements of bees, which need a core habitat for nesting and a sufficient foraging habitat in the vicinity. The explanatory value of patch variables describing soil conditions and continuity of management was particularly high, suggesting that availability of brood habitats is a critical factor determining the diversity of wild bees. Most species of wild bees nest in the soil (Cane, 1991) and nearly all of these are known to be thermophilous and to prefer fairly well-drained sites (Linsley, 1958; Westrich, 1989). The modulating effect of the matrix variable 'percentage cover of arable land' indicates the positive influence of arable land on the availability of food resources during the time of the year the sampling was done. The low-input farming systems of the study region obviously guarantee a high number of flowering plants in the crop fields at times when the availability of flowering plants in grassland is low due to mowing.

Plant species richness was solely explained by patch variables, reflecting the fact that the actual state of the established grassland vegetation largely depends on patch quality. This contrasts to the finding that neighbouring habitats can affect plant species richness by providing a source of rhizomes and diaspores over a short distance (Zonneveld, 1995). However, matrix effects are particularly important for the plant colonisation of a patch after disturbance (e.g. Kotanen, 1997). Following the mass effect hypothesis of Shmida and Wilson (1985), even mature grasslands being grazed during parts of the year have lots of small-scale disturbances that can serve as regeneration foci for propagules of the patch context. Yet, such matrix effects were obviously not strong enough or were not sufficiently represented by the matrix variables used to influence the regression models of our study.

Ants were the only taxon in which species richness was solely described by the landscape composition within the matrices. This contradicts our initial expectation that both patch and matrix variables should contribute to the prediction of ant species richness, as this group is strongly linked to environmental conditions, like temperature (e.g. Kaspari et al., 2000). However, the regression models revealed in our study explained only around 50% of the variance. It might thus be concluded that the quality and variability of the patch variables used in our study was not appropriate to precisely predict ant species richness. Local ant community structure and diversity are well known to be mediated by the regional context (Bestelmeyer and Wiens, 1996). Serrano et al. (1993) have shown that ant succession in Mediterranean fallow land depends on the surrounding landscape, with the proportion of mature species being greater in areas with a more mature matrix. The proximity of source habitats like undisturbed forest patches affect ant diversity in agriculturally disturbed areas in wet forests of Costa Rica (Roth et al., 1994). Though this might suggest that forests of our study region also serve as a source habitat, the species composition of the grassland patches does not support this assumption, as no forest species were collected at the study sites. We thus hypothesise that the impact of forest cover found in our study rather reflects the impact of other landscape qualities that were not taken into account in our study. These could include, e.g., shallow and less productive soils and associated vegetation features, because the percentage cover of forest is highest close to hilltops.

5. Conclusions

The aim of this study was to establish indicators of species richness at the regional scale. We conclude that the missing correlation between species richness of the taxa included in our investigation and the species-specific differences in the regression models makes it difficult to find appropriate indicators of overall species richness of the landscape. Patch variables describing habitat quality were proven to be essential predictors for two of the taxa studied. For some of the taxa, matrix variables like percentage cover of one or several land-use types can be used as predictors of species richness. Nevertheless, the applicability of these predictors may be restricted to landscapes where landscape composition is closely correlated to environmental conditions and cultivation practices, as for instance in the low intensity farming systems of the region studied.

Acknowledgements

We are indebted to Katja Fuhr-Boßdorf for providing access to her data on landscape history. Jan Bengtsson and Dagmar Schroeter made valuable comments on early drafts of the manuscript. We thank Klemens Ekschmitt for his help with statistical analyses and Thorsten Behrens for GIS assistance. We are very grateful to the landowners who allowed us to use their fields. The German Research Foundation (DFG) provided financial support for this study.

Appendix A

Species richness of ants, bees and plants per site and site characterisation

Site	Species number			Patch variables				Matrix variables of 50 m matrix					Matrix variables of 200 m matrix				
	Ants	Bees	Plants	Aspect	Soil	Field-size (ha)	Age (years)	%A	%FL	%G	%F	%U	%A	%FL	%G	%F	%U
1	10[a]	11	54	S	d	0.29	53	59.8	21.9	0.1	9.4	1.0	35.8	4.6	9.0	44.1	1.2
2	11[a]	5	52	S	d	0.23	25	44.2	19.0	0.0	24.1	3.9	26.0	2.4	4.0	52.0	8.8
3	5[a]	8	n.d.[b]	S	d	0.39	31	74.3	11.7	3.6	0.0	0.0	72.2	3.0	17.7	0.2	0.0
4	1[c]	2	n.d.	S	h	0.55	53	27.0	0.0	65.0	0.0	0.0	20.4	5.3	66.4	0.6	0.0
5	3[c]	3	68	S	h	0.74	53	18.1	0.0	58.0	0.0	11.0	27.6	0.4	51.2	0.6	10.8
6	7[c]	n.d.	n.d.	N	d	0.26	53	25.6	0.0	35.1	27.9	0.0	27.5	0.5	38.8	24.0	0.0
7	6[c]	n.d.	41	N	d	1.00	53	39.4	0.0	35.6	0.0	17.6	21.7	1.7	31.1	0.0	32.5
8	5[c]	1	45	N	h	0.27	53	12.1	0.0	75.1	0.0	0.9	17.3	0.0	55.5	0.0	12.4
9	4[c]	4	33	N	d	0.28	9	20.6	0.0	52.1	0.0	16.2	32.5	0.2	54.7	0.7	4.0
10	6[a]	7	n.d.	S	d	0.14	25	75.8	0.0	0.0	18.2	0.0	55.9	3.8	0.0	34.6	0.0
11	4[c]	n.d.	n.d.	N	h	0.57	53	8.8	15.3	65.6	0.0	0.0	56.0	5.8	28.7	0.0	0.8
12	4[c]	1	53	N	h	0.78	53	44.7	0.0	42.8	0.0	0.0	54.5	0.7	32.5	0.5	0.0
13	6[c]	n.d.	32	S	d	0.16	31	48.5	22.0	13.8	7.0	0.0	32.8	2.6	24.0	25.1	4.5
14	7[c]	n.d.	50	S	h	0.14	53	28.4	0.0	45.5	0.0	16.8	35.1	0.0	49.9	0.0	5.7
15	8[a]	6	48	S	d	0.32	26	2.4	34.8	56.2	0.0	0.0	4.5	38.1	46.3	5.5	0.0

Appendix A (*Continued*)

Site	Species number			Patch variables				Matrix variables of 50 m matrix					Matrix variables of 200 m matrix				
	Ants	Bees	Plants	Aspect	Soil	Field-size (ha)	Age (years)	%A	%FL	%G	%F	%U	%A	%FL	%G	%F	%U
16	7[a]	3	46	S	d	0.12	26	0.0	0.0	96.4	0.0	0.0	4.1	15.7	68.9	4.8	0.1
17	5[a]	8	43	S	d	0.14	53	0.0	0.0	93.0	0.0	0.0	0.6	6.0	64.7	3.3	2.2
18	6[a]	3	n.d.	S	d	0.10	19	0.0	21.4	59.3	9.3	0.0	0.0	22.7	43.9	23.5	2.1
19	6[a]	4	n.d.	S	d	0.13	26	0.0	0.0	88.1	0.0	0.0	0.0	14.1	58.3	11.4	6.5
20	6[a]	4	61	S	d	0.65	25	0.0	0.0	88.3	0.0	0.0	0.9	0.0	88.5	3.0	0.0

[a] 1998.
[b] No data.
[c] 1997.

References

Andersen, A.N., 1991. Parallels between ants and plants: implications for community ecology. In: Huxley, C.R., Cutler, D.F. (Eds.), Ant–Plant Interactions. Oxford University Press, Oxford, pp. 539–558.

Araújo, M.B., Humphries, C.J., Densham, P.J., Lampinen, R., Hagemeijer, W.J.M., Mitchell-Jones, A.J., Gasc, J.P., 2001. Would environmental diversity be a good surrogate for species diversity? Ecography 24, 103–110.

Barbault, R., 1995. Biodiversity dynamics: from population and community ecology approaches to a landscape ecology point of view. Landscape Urban Plan. 31, 89–98.

Baudry, J., Burel, F., Thenail, C., Le Cæur, D., 2000. A holistic landscape ecological study of the interactions between farming activities and ecological patterns in Brittany, France. Landscape Urban Plan. 50, 119–128.

Bestelmeyer, B.T., Wiens, J.A., 1996. The effects of land use on the structure of ground-foraging ant communities in the argentine chaco. Ecol. Appl. 6, 1225–1240.

Bonn, S., Poschlod, P., 1998. Ausbreitungsbiologie der Pflanzen Mitteleuropas. Grundlagen und kulturhistorische Aspekte. Quelle and Meyer, Wiesbaden.

Brosius, G., 1988. SPSS/PC+ Basics und Graphics. McGraw-Hill, Hamburg.

Burel, F., Baudry, J., Butet, A., Clergeau, P., Delettre, Y., Le Coeur, D., Dubs, F., Morvan, N., Paillat, G., Petit, S., Thenail, C., Brunel, E., Lefeuvre, J.-C., 1998. Comparative biodiversity along a gradient of agricultural landscapes. Acta Oecol. 19, 47–60.

Cane, H.C., 1991. Soils of ground-nesting bees (Hymenoptera: Apoidea): texture, moisture, cell depth and climate. J. Kans. Entomol. Soc. 64, 406–413.

Cresswell, J.E., Osborne, J.L., Goulson, D., 2000. An economic model of the limits to foraging range in central place foragers with numerical solutions for bumblebees. Ecol. Entomol. 25, 249–255.

Dale, V.H., Brown, S., Haeuber, R.A., Hobbs, N.T., Huntly, N., Naiman, R.J., Riebsame, W.E., Turner, M.G., Valone, T.J., 2000. Ecological principles and guidelines for managing the use of land. Ecol. Appl. 10, 639–670.

Dauber, J., Wolters, V., 2000. Species richness of ants in the land use mosaic of a marginal landscape. Mitt. Dtsch. Ges. Allg. Angew. Entomol. 12, 281–284.

Debinski, D.M., Ray, Ch., Saveraid, E.H., 2001. Species diversity and the scale of the landscape mosaic: do scales of movement and patch size affect diversity? Biol. Conserv. 98, 179–190.

Duelli, P., 1997. Biodiversity evaluation in agricultural landscapes: an approach at two different scales. Agric. Ecosyst. Environ. 62, 81–91.

Duelli, P., Obrist, M.K., 1998. In search of the best correlates for local organismal biodiversity in cultivated areas. Biodivers. Conserv. 7, 297–309.

Duelli, P., Obrist, M.K., Schmatz, D.R., 1999. Biodiversity evaluation in agricultural landscapes: above-ground insects. Agric. Ecosyst. Environ. 74, 33–64.

Dunning, J.B., Danielson, B.J., Pulliam, H.R., 1992. Ecological processes that affect populations in complex landscapes. Oikos 65, 169–175.

Fagan, W.F., Cantrell, R.S., Cosner, C., 1999. How habitat edges change species interactions. Am. Nat. 153, 165–182.

Fahrig, L., 2001. How much habitat is enough? Biol. Conserv. 100, 65–74.

Franklin, J.F., 1993. Preserving biodiversity: species, ecosystems, or landscapes. Ecol. Appl. 3, 202–205.

Frede, H.-G., Bach, M., 1999. Perspectives for peripheral regions. Z. Kulturtech. Landentwicklung 40, 193–196.

Fuhr-Boßdorf, K., Waldhardt, R., Otte, A., 1999. Effects of land use dynamics (1945–1998) on the potential of plant communities and species in a marginal cultural landscape. Verh. Ges. Ökologie 29, 519–530.

Gaston, K.J. (Ed.), 1996. Biodiversity: A Biology of Numbers and Differences. Blackwell Scientific Publications, Oxford.

Gaston, K.J., 2000. Biodiversity: higher taxon richness. Prog. Phys. Geogr. 24, 117–127.

Grime, J.P., Hillier, S.H., 1992. The contribution of seedling regeneration to the structure and dynamics of plant communities and larger units of landscape. In: Fenner, M. (Ed.), Seeds.

The Ecology of Regeneration in Plant Communities. CAB International, Wallingford, UK.

Hanski, I., 1999. Metapopulation Ecology. Oxford University Press, UK.

Jonsen, I.D., Fahrig, L., 1997. Response of generalist and specialist insect herbivores to landscape spatial structure. Landscape Ecol. 12, 185–197.

Kaspari, M., 2000. A primer on ant ecology. In: Agosti, D., Majer, J.D., Alonso, L.E., Schultz, T.R. (Eds.), Ants. Standard Methods for Measuring and Monitoring Biodiversity. Smithsonian Institution Press, Washington, pp. 9–24.

Kaspari, M., Alonso, L., O'Donnel, S., 2000. Three energy variables predict ant abundance at a geographical scale. Proc. R. Soc. Lond. B 267, 485–489.

Kotanen, P.M., 1997. Effects of gap area and shape on recolonization by grassland plants with differing reproductive strategies. Can. J. Bot. 75, 352–361.

Landis, D.A., Marino, P.C., 1999. Landscape structure and extra-field processes: impact on management of pests and beneficials. In: Ruberson, J.R. (Ed.), Handbook of Pest Management. Marcel Dekker, New York, pp. 79–104.

Lawton, J.H., 2000. Community Ecology in a Changing World. Ecology Institute, Oldendorf/Luhe.

Linsley, E.G., 1958. The ecology of solitary bees. Hilgardia 27, 543–599.

MacArthur, R.H., Wilson, E.O., 1967. The Theory of Island Biogeography. Princeton University Press, Princeton, NJ.

Margules, C.R., Pressey, R.L., 2000. Systematic conservation planning. Nature 405, 243–253.

Miller, J.N., Brooks, R.P., Croonquist, M.J., 1997. Effects of landscape patterns on biotic communities. Landscape Ecol. 12, 137–153.

Osborn, F., Goitia, W., Cabrera, M., Jaffé, K., 1999. Ants, plants and butterflies as diversity indicators: comparisons between strata at six forest sites in Venezuela. Stud. Neotrop. Fauna Environ. 34, 59–64.

Poschlod, P., Bakker, J., Bonn, S., Fischer, S., 1996. Dispersal of plants in fragmented landscapes. In: Settele, J., Margules, C., Poschlod, P., Henle, K. (Eds.), Species Survival in Fragmented Landscapes. Kluwer Academic Publishers, Dordrecht.

Prendergast, J.R., 1997. Species richness covariance in higher taxa: empirical tests of the biodiversity indicator concept. Ecography 20, 210–216.

Purtauf, T., Dauber, J., Wolters, V., 2001. Effects of changes in landscape patterns on carabid diversity and community structure. In: Mander, Ü., Printsmann, A., Palang, H. (Eds.), Development of European Landscapes. Publicationes Instituti Geographici Universitatis Tartuensis 92, Tartu, pp. 70–75.

Rosenzweig, M., 1995. Species Diversity in Space and Time. Cambridge University Press, New York.

Roth, D.S., Perfecto, I., Rathke, B., 1994. The effects of management systems on ground-foraging ant diversity in Costa Rica. Ecol. Appl. 4, 423–436.

Schluter, D., Ricklefs, R.E., 1993. Species diversity: an introduction to the problem. In: Ricklefs, R.E., Schluter, D. (Eds.), Species Diversity in Ecological Communities: Historical and Geographical Perspectives. The University of Chicago Press, Chicago, pp. 1–10.

Serrano, J.M., Acosta, F.J., Lopez, F., 1993. Belowground space occupation and partitioning in an ant community during succession. Eur. J. Entomol. 90, 149–158.

Shmida, A., Wilson, M.V., 1985. Biological determinants of species diversity. J. Biogeogr. 12, 1–20.

Smallwood, J., 1982. Nest relocations in ants. Insect. Soc. 29, 138–147.

Thies, C., Tscharntke, T., 1999. Landscape structure and biological control in agroecosystems. Science 285, 893–895.

Thompson, K., Bakker, J., Bekker, R., 1997. The Soil Seed Banks of North West Europe: Methodology, Density and Longevity. Cambridge University Press, Cambridge.

Tilman, D., 1994. Competition and biodiversity in spatially structured habitats. Ecology 75, 2–16.

Wagner, H.H., Wildi, O., Ewald, K.C., 2000. Additive partitioning of plant species diversity in an agricultural mosaic landscape. Landscape Ecol. 15, 219–227.

Waldhardt, R., Fuhr-Boßdorf, K., Otte, A., Schmidt, J., Simmering, D., 1999. Classification, localisation and regional extrapolation of vegetation potentials in a peripheral cultural landscape. Z. Kulturtech. Landentwicklung 40, 246–252.

Waldhardt, R., Otte, A., 2003. Indicators of plant species and community diversity in grasslands. Agric. Ecosyst. Environ. 98, 339–351.

Walter-Hellwig, K., Frankl, R., 2000. *Bombus* spp. (Hym., Apidae), in an agricultural landscape. J. Appl. Entomol. 124, 299–306.

Weibull, A.-C., Bengtsson, J., Nohlgren, E., 2000. Diversity of butterflies in the agricultural landscape: the role of farming system and landscape heterogeneity. Ecography 23, 743–750.

Westrich, P., 1989. Die Wildbienen Baden-Württembergs. Bd. 1 Allgemeiner Teil: Lebensräume, Verhalten, Ökologie und Schutz. Ulmer, Stuttgart.

Wiens, J.A., Stenseth, N.C., Van Horne, B., Ims, R.A., 1993. Ecological mechanisms and landscape ecology. Oikos 66, 369–380.

Wolters, V., Dauber, J., Hirsch, M., Steiner, N., 1999. Fauna in a mosaic landscape at the peripheral region. Z. Kulturtech. Landentwicklung 40, 253–257.

Zonneveld, I.S., 1995. Vicinism and mass effect. J. Veg. Sci. 5, 441–444.

ELSEVIER

Available online at www.sciencedirect.com

Agriculture, Ecosystems and Environment 98 (2003) 331–337

Agriculture Ecosystems & Environment

www.elsevier.com/locate/agee

The influence of matrix type on flower visitors of *Centaurea jacea* L.

Michaela Hirsch*, Sabine Pfaff, Volkmar Wolters

Department of Animal Ecology, IFZ, Justus-Liebig-University, Heinrich-Buff-Ring 26-32, D-35392 Giessen, Germany

Abstract

The structure of invertebrate communities is impacted by landscape variables. Here we present a study on the influence of the surrounding matrix on pollinating flower visitors of the knapweed *Centaurea jacea* in a small-scale agricultural landscape (Central Hesse, Germany). The study was carried out in late summer 1998 by monitoring visits of 24 insect taxa at 15 *C. jacea* patches. The following matrix types were studied: (i) arable land only, (ii) arable land close to grassland (<50 m), (iii) a mosaic of arable land, grassland and forests, and (iv) grassland only.

More than half of the flower visitors were bees, with honeybees (*Apis mellifera*) and bumblebees being the dominant taxa. The matrix type did not affect either the mean frequency of total flower visits or the mean richness of all taxa. The same applied to different size classes. Significant matrix effects were confined to five large taxa of the Apoidea: *Bombus lapidarius*, *B. pascuorum*, the *B. terrestris* group, large wild bees (other than bumblebees) and *A. mellifera*. *A. mellifera* significantly preferred patches surrounded by grassland, while large wild bees preferred patches surrounded by a mosaic of arable land, grassland and forests. The flower visiting frequency of all three bumblebee species was high in patches surrounded by grassland, but only the *B. terrestris* group showed a clear-cut preference for this matrix type.

It is concluded that matrix effects on flower visitors of *C. jacea* are taxon- and body size-specific. Strong matrix effects on large pollinators suggest that these taxa are able to discriminate between patches. Considering the ecological services provided by pollinators, the preservation of large areas covered by interconnected grassland sites as well as by a mosaic of different land use forms should have high priority in future management strategies.

Keywords: Pollinators; Apoidea; Matrix effects; *Centaurea jacea*; Landscape ecology

1. Introduction

The composition of invertebrate communities is affected by a variety of landscape variables including patch conditions, length of ecotones and quality of the surrounding matrix (Aizen and Feinsinger, 1994; Webb, 1989; Ås, 1999; Duelli et al., 1999; Golden and Crist, 1999; Dauber and Wolters, 2000). However, results on the direction of these effects are conflicting. For example, while decreasing patch size reduced both the diversity and density of insects (Aizen and Feinsinger, 1994; Matter, 1997), patch size had no influence on the visitation rates of the pollinator communities investigated by Schmalhofer (2001). Such contradictory results can partly be explained by the fact that landscape effects on community structure are taxon specific, i.e. depend on habitat-specific responses of the organisms involved and on specific ecological processes that link landscape

* Corresponding author. Tel.: +49-641-9935653; fax: +49-641-9935709.
E-mail address: michaela.hirsch@allzool.uni-giessen.de (M. Hirsch).

0167-8809/$ – see front matter © 2003 Elsevier Science B.V. All rights reserved.
doi:10.1016/S0167-8809(03)00093-8

variation to population dynamics (Dunning et al., 1992).

The impact of landscape fragmentation on pollinators has been strongly underestimated in the past (Walther-Hellwig and Frankl, 2000). For this reason, we studied the influence of the surrounding matrix on pollinating flower visitors of the knapweed *Centaurea jacea* L., 1753. Agricultural management has a very strong isolating effect on this species, which makes it particularly well suited for the analysis of matrix effects. By selecting *C. jacea* patches that differ in the proportion of arable land in the surrounding area it is possible to systematically vary the degree of isolation. The following hypotheses were tested: (i) matrix effects on pollinating flower visitors of *C. jacea* are taxon- and body size-specific, and (ii) taxonomic richness and density of pollinating flower visitors decrease when the matrix of *C. jacea* patches is dominated by arable land, whereas both parameters increase when the matrix is dominated by grassland.

2. Methods

The study was carried out in late summer 1998 in the Lahn-Dill-Bergland (Central Hesse, Germany). All sites are situated in the rural districts of Hohenahr (Erda) and Biebertal (Frankenbach). A total of 15 *C. jacea* patches were selected by means of a GIS supported analysis (ArcInfo). Each patch was situated on sites that could be assorted to one of the following matrix types (Fig. 1): (A) arable land only (distance from the next grassland at least 200 m, $n = 4$), (AG) arable land close to grassland (distance from the next grassland less than 50 m; $n = 3$), (AGF) a small-scale mosaic of arable land, grassland and forest ($n = 4$), and (G) grassland only (distance from the next arable land at least 200 m, $n = 4$).

Invertebrates foraging at flowers of *C. jacea* were determined seven times during the period of investigation. All observation campaigns were performed under weather conditions that have proven appropriate for the community analysis of flower visitors (e.g. Schwenninger, 1992; Erhardt, 1985; Teräs, 1976; Pollard et al., 1975; Witsack, 1975; Matthews and Matthews, 1971). According to the method suggested by Aizen and Feinsinger (1994), the number and taxonomic composition of insects visiting flowers of *C. jacea* patches were recorded for 15 min on 15 flower heads per patch by one of five experienced entomologists at each date. Each study site was sampled

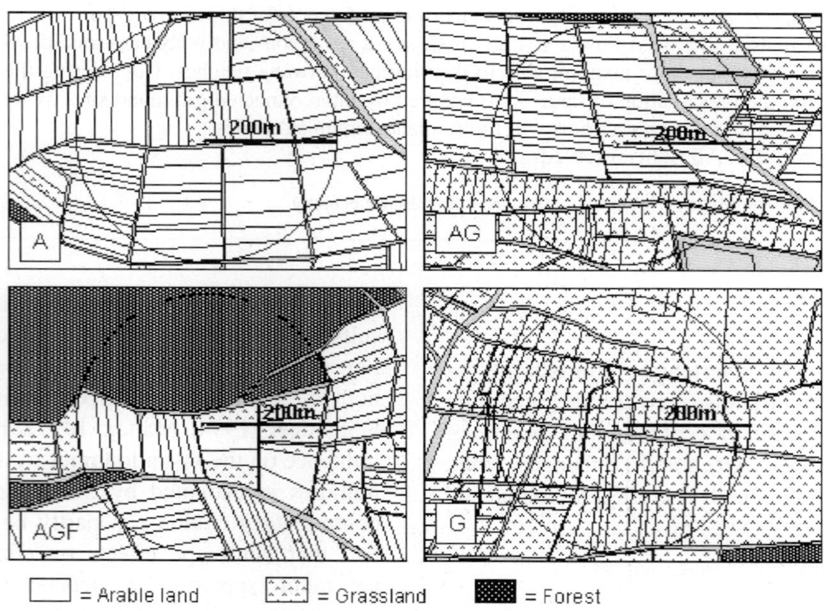

Fig. 1. Examples of the four matrix types.

Table 1
Mean frequency of flower visits and taxonomic richness monitored at *C. jacea* patches surrounded by four different matrix types in late summer 1998 (A: arable land only ($n = 4$), AG: arable land close to grassland (<50 m; $n = 3$), AGF: mosaic of arable land, grassland and forest ($n = 4$), and G: grassland only ($n = 4$))[a]

	A			AG			AGF			G		
	Small	Medium	Large	Small	Medium	Large	Small	Medium	Large	Small	Medium	Large
Coleoptera	2.00 (1.15)	1.00 (2.00)	1.50 (2.38)	1.67 (2.08)	0.67 (1.15)	1.67 (2.89)	1.00 (0.82)	0.25 (0.50)	3.25 (2.75)	0.75 (0.50)	1.00 (1.41)	0.25 (0.50)
Diptera												
Syrphidae	0.00 (0.00)	1.50 (1.73)	1.00 (1.15)	0.33 (0.58)	0.67 (1.15)	2.67 (2.52)	0.00 (0.00)	3.50 (1.29)	3.00 (3.37)	0.00 (0.00)	1.50 (0.58)	5.75 (5.32)
E. balteatus			4.25 (3.30)			4.67 (3.21)			3.75 (1.71)			3.50 (1.00)
Other Diptera	0.50 (1.00)	0.75 (0.50)	2.25 (0.96)	2.00 (2.65)	7.00 (9.64)	0.33 (0.58)	0.25 (0.50)	1.50 (2.38)	2.75 (1.26)	0.00 (0.00)	1.75 (2.22)	1.00 (1.41)
Hymenoptera												
A. mellifera			11.50 (5.26)			12.67 (21.08)			6.00 (8.49)			66.25 (44.62)
B. lapidarius			7.25 (4.79)			4.00 (4.00)			6.00 (8.83)			5.75 (3.50)
B. terrestris association			0.25 (0.50)			1.33 (2.31)			0.50 (1.00)			4.25 (2.22)
B. pascuorum			6.75 (6.55)			1.67 (2.08)			1.25 (1.26)			5.75 (4.92)
Other *Bombus*			7.0 (5.72)			7.00 (7.27)			3.25 (1.71)			4.50 (3.11)
Wild bees (other than *Bombus*)	0.50 (0.58)	0.50 (0.58)	2.25 (2.22)	0.00 (0.00)	1.00 (1.00)	2.67 (1.15)	0.00 (0.00)	0.00 (0.00)	23.50 (21.49)	0.25 (0.50)	0.50 (0.58)	15.50 (15.44)
Other Hymenoptera	4.00 (2.94)	1.00 (0.82)	3.00 (0.82)	2.67 (2.08)	0.33 (0.58)	0.33 (0.58)	2.75 (3.59)	1.25 (1.50)	1.50 (1.29)	3.75 (2.87)	1.75 (1.71)	1.00 (2.00)
Lepidoptera									4.50 (5.26)			4.00 (3.16)
Maniola jurtina			0.25 (0.50)			0.33 (0.58)			1.50 (2.38)			2.50 (1.73)
Argynnis papia			0.00 (0.00)			0.00 (0.00)			1.00 (2.00)			1.00 (2.00)
Other Lepidoptera			0.81 (1.66)			0.33 (0.50)			2.00 (2.45)			0.50 (1.00)
Taxonomic richness	2. 50 (1.29)	1.26 (2.75)	12.25 (1.26)	2.67 (0.58)	2.67 (1.53)	8.00 (2.65)	1.75 (0.50)	3.25 (0.50)	11.00 (3.37)	2.00 (0.82)	3.75 (0. 96)	11.00 (2.58)
Total flower visits	7.00 (4.19)	4.75 (4.57)	48.06 (15.10)	6.67 (3.06)	9.67 (12.77)	39.67 (31.94)	4.00 (4.08)	6.50 (3.30)	63.75 (32.99)	4.75 (2.99)	6.50 (1.26)	121.50 (56.65)

[a] The species of each taxon where assorted to different size classes: small (<5 mm), medium (5–10 mm), and large (>10 mm). Standard deviation is given in parenthesis.

only once or twice by the same expert to reduce individual bias (Haeseler and Ritzau, 1998; Hermann, 1996). The following taxa were recorded: Coleoptera (pollinating taxa only), Syrphidae, other Diptera (pollinating taxa only), Hymenoptera (other than Apoidea), Apoidea (other than *Bombus* (Latr., 1802) and *Apis mellifera* L., 1758), *Bombus*, *A. mellifera*, and Lepidoptera. The genus *Bombus* was further separated into five taxonomic categories (four species and *Bombus* spp.), Lepidoptera were separated into seven taxonomic categories (six species and 'other Lepidoptera'), and the species *Episyrphus balteatus* (DeGeer, 1776) was separated as an extra taxonomic unit from the Syrphidae. Moreover, all taxa were assigned to one of three size classes: small (<5 mm), medium (5–10 mm), and large (>10 mm). Because not all taxa cover all size classes, a total of 24 taxonomic units could be differentiated (Table 1). Several coarse parameters of community composition were additionally determined: (i) total frequency of flower visits per hour, (ii) total taxonomic richness (i.e. the sum of all taxonomic units), and (iii) frequency of flower visits of the three size classes, and (iv) taxonomic richness of the three size classes. Thus, a total of 32 dependent variables were available for further analyses.

Statistical analyses were carried out using the program STATISTICA for Windows 5.0 (Statsoft, 1995). Small Syrphidae had to be excluded because of their low density. A one-way analysis of variance (ANOVA) was applied to test the effect of matrix type (independent variable) on the flower visiting frequency of the dependent variables listed above. Data were log-transformed prior to analysis. To eliminate a bias caused by patch-specific variations in the density of flower heads, the number of flower heads within a radius of 10 m of both *C. jacea* and other flowering plants was included in the ANOVAs as a co-variable. Differences between means were tested using the Tukey HSD test ($P < 0.05$).

3. Results

A total of 1245 flower visitors belonging to 24 taxa were recorded (Table 1). The most abundant groups were Apoidea (>50%), followed by other Hymenoptera and Diptera. Honeybees and bumblebees had the highest share of Apoidea with 46 and 33%, respectively. The matrix type did not affect either the mean frequency of total flower visits ($F = 0.93$) or the mean richness of all taxa ($F = 1.01$). The same applied to the mean frequency of flower visits and the mean taxonomic richness of the different size classes (Table 2).

Table 2
Results of the one-way ANOVA on the effect of matrix type on the frequency of flower visits per hour by and taxonomic richness of pollinators of *C. jacea* (significant results are marked with an asterisk, $P < 0.05$; data transformation: $x' = \ln(x + 1)$)

	F-value		
	Small	Medium	Large
Coleoptera	1.33	0.46	1.56
Diptera			
E. balteatus	–[a]	–	0.28
Other Syrphidae	–	0.87	1.34
Other Diptera	0.68	2.65	1.81
Hymenoptera			
A. mellifera	–	–	3.49*
B. lapidarius	–	–	2.83*
B. terrestris s. l.	–	–	3.10*
B. pascuorum	–	–	3.01*
Other *Bombus*			1.14
Other Apoidea	1.19	2.09	2.91*
Other Hymenoptera	0.47	0.84	1.04
Lepidoptera	–	–	
Maniola jurtina	–	–	2.23
Argynnis papia	–	–	0.72
Other Lepidoptera			0.54
Taxonomic richness	1.20	0.10	2.13
Total flower visits	0.90	0.89	0.95

[a] Not found or less than 10 individuals recorded.

Significant matrix effects were confined to five large taxa of the Apoidea: *Bombus lapidarius* (L., 1758), *B. pascuorum* (Scop., 1763), the *B. terrestris* group (L., 1758), large wild bees and *A. mellifera* (Table 2). Though the significant effects revealed by the ANOVA suggest a decline of *B. lapidarius* in patches surrounded by the matrix type AG as well as a decline of *B. pascuorum* in patches surrounded by the matrix types AG and AGF, no significant difference between means could be established by the Tukey test (Fig. 2). The particularly high frequency of flower visits by individuals belonging to the *B. terrestris* group in patches surrounded by the matrix

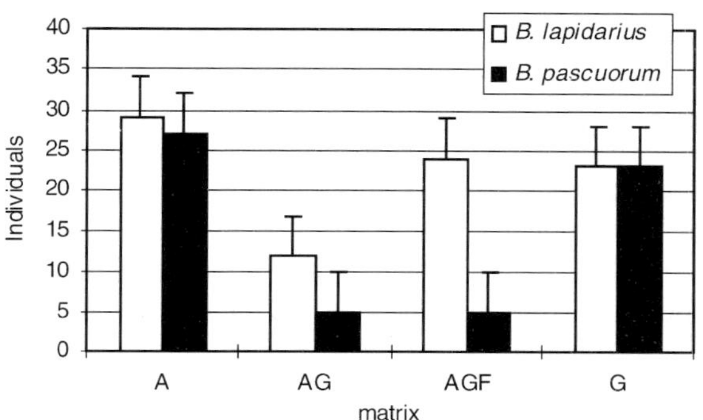

Fig. 2. Density of *B. lapidarius* and *B. pascuorum* (non-transformed data) in *C. jacea* patches surrounded by four different matrix types (A: arable land only ($n = 4$), AG: arable land close to grassland (<50 m; $n = 3$), AGF: mosaic of arable land, grassland and forest ($n = 4$), and G: grassland only ($n = 4$)).

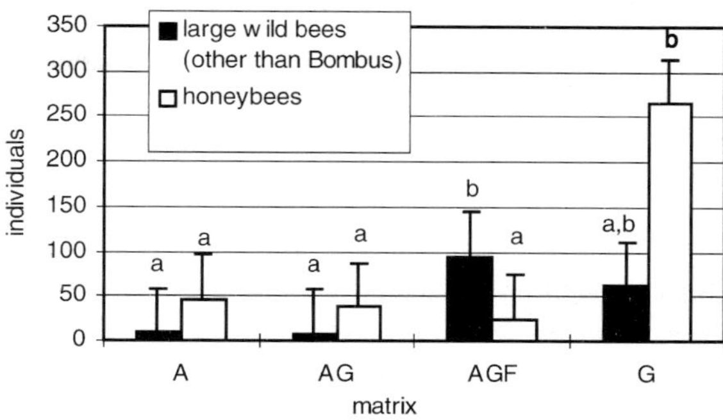

Fig. 4. Density of large wild bees and *A. mellifera* (non-transformed data) in *C. jacea* patches surrounded by four different matrix types (see Fig. 2 for abbreviations; columns of the same shading sharing identical letters are not significantly different according to the Tukey test ($P < 0.5$)).

type G, in contrast, was confirmed by significant differences between means (Fig. 3).

The flower visiting frequency of large wild bees (other than bumblebees) was significantly higher in patches surrounded by the matrix type AGF than in those surrounded by A and AG, while patches surrounded by G had an intermediate position (Fig. 4). Significantly more *A. mellifera* were recorded in patches surrounded by the matrix type G than in the other patches, without any significant difference between the matrix types A, AG, and AGF, respectively (Fig. 4).

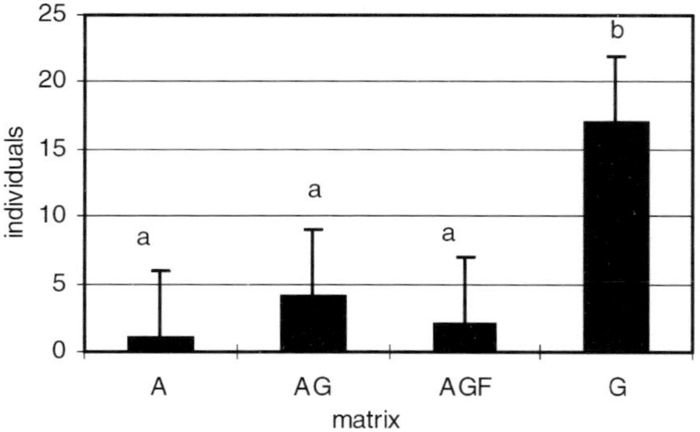

Fig. 3. Density of individuals belonging to the *B. terrestris* association (non-transformed data) in *C. jacea* patches surrounded by four different matrix types (see Fig. 2 for abbreviations; columns sharing identical letters are not significantly different according to the Tukey test ($P < 0.5$)).

4. Discussion

The significant response of five large Apoidea taxa to differences in matrix type revealed in this study confirms our first hypothesis: matrix effects on pollinating flower visitors of *C. jacea* are taxon- and body size-specific. However, our results also show that coarse parameters of community structure such as total frequency of flower visits or taxonomic richness do not provide suitable indicators of matrix effects on flower visitors of *C. jacea*. Moreover, most taxonomic units investigated in our study did not respond to changes in matrix composition. This contrasts results reported in the literature (e.g. Jeanneret and Charmillot, 1995; Pfaff and Wolters, 1999 for Lepidoptera; Knecht et al., 2000 for flower visiting Coleoptera). Considering that the matrix types chosen in our study can be related to different degrees of isolation (A > AG > AGF > G), our data on the other hand confirm the findings of Steffan-Dewenter and Tscharntke (1999) and Schmalhofer (2001), who could not establish any relationship between the degree of isolation and the total number of flower visitors.

The strong matrix effect on large bees found in our study seems to contradict the finding of a positive correlation between the degree of isolation and the mean body size of insects visiting flowers (Gathmann et al., 1994; Steffan-Dewenter and Tscharntke, 1999). Because of the fact that large bees are able to forage much greater ranges than small- and medium-sized bees (Gathmann et al., 1994), one might conclude that matrix effects on large taxa are leveled out by their high flight capacity. However, the studies cited above were carried out in landscapes with widely separated flower patches (up to 1000 m), while our study focused on a small-scale mosaic landscape with distances about 200 m. We hypothesize that, under conditions in which the ability to overcome long distances does not play a role, large flower visitors use their high flight capacity to discriminate between patches and are thus able to select the most appropriate resources. Large social insects such as bumblebees and honeybees, which have much higher energetic costs than small solitary bees (Heinrich, 1976), can best fulfill their nutritive requirements by foraging in comparatively large, connected *C. jacea* patches situated in grassland sites (Agren, 1996; Sih and Baltus, 1987; Pleasants, 1981). Solitary bees, in contrast, depend much more on the availability of a small-scale mosaic of microhabitats suitable for both nesting and food provision (Westrich, 1989). This would explain why the highest number of large wild bees was found in patches surrounded by a mosaic of grassland, arable land and forests.

The hypothesis that taxonomic richness and density of flower visitors decreases when the matrix of *C. jacea* patches is dominated by arable land, but increases when the matrix is dominated by grassland is not confirmed by our results. However, we did find a positive effect of a grassland matrix on honeybees and on the *B. terrestris* group. These bees show forage constancy not only for certain flowers, but also for the foraging area (e.g. Osborne and Williams, 2001). Especially honeybees have large foraging ranges and preferentially recruit discrete vegetational patches of high nutritional quality (Ginsberg, 1983; Aizen and Feinsinger, 1994). Honeybees thus tend to explore other plant species or patches than wild bees (Aizen and Feinsinger, 1994; Ginsberg, 1983). Though the density of all three investigated bumblebee species was high in *C. jacea* patches surrounded by grassland, only the individuals of the *Bombus terrestris* group showed a clear-cut preference for this matrix type. The species-specific response patterns confirm that bumblebees differ in foraging strategies and range (Hedtke and Schricker, 1996; Walther-Hellwig and Frankl, 2000).

5. Conclusion

All the flower visitors monitored in our study are pollinators (Allen-Wardel et al., 1998; Kevan, 1999), with bees probably being particularly effective (Batra, 1995; Kevan, 1999). Though our results suggest that the pollinator community of *C. jacea* as a whole is quite robust against matrix effects, the significant response of large Apoidea points to the important role of the surrounding landscape in modulating the association between pollinating insects and flowering plants. Considering the ecological services provided by pollinators in agricultural landscapes, our study indicates that the preservation of large areas covered by interconnected grassland sites as well as by a mosaic of different land use forms should have high priority in future management strategies.

Acknowledgements

We are grateful to the experienced entomologists Doris Nothaft, Björn Thissen and Silvia Rank. The study was supported by the German Research Foundation.

References

Agren, J., 1996. Population size, pollinator limitation, and seed set in the self-incompatible herb *Lithrum salicarina*. Ecology 77, 1779–1790.

Aizen, M.A., Feinsinger, P., 1994. Habitat fragmentation, native insect pollinators, and feral honey bees in Argentine "Chaco Serrano". Ecol. Appl. 4, 378–392.

Allen-Wardel, G., Bernhard, P., Bitner, R., Burquez, A., Buchmann, S., Cane, J., Cox, P.A., Dalton, V., Feinsinger, P., Ingram, M., Inouye, D., Jones, C.E., Kennedy, K., Kevan, P., Koopowitz, H., Medellin, R., Medellin-Moralis, S., Nabham, G.P., 1998. The potential consequences of pollinator declines on the conservation of biodiversity and stability of food crop yields. Conserv. Biol. 12, 8–18.

Ås, S., 1999. Invasion of matrix species in small habitat patches. Conservation Ecology [online] 3: Internet. http://www.consecol.org/vol3/iss1/art1.

Batra, S.W.T., 1995. Bees and pollination in our changing environment. Apidologie 26, 361–370.

Dauber, J., Wolters, V., 2000. Diversität der Ameisenfauna im Landnutzungsmosaik einer peripheren Region. Mitt. Dtsch. Ges. Allg. Angew. Ent. 12, 281–284 (in German, with English abstract).

Duelli, P., Obrist, M.K., Schmatz, D.R., 1999. Biodiversity evaluation in agricultural landscape: above-ground insects. Agric. Ecosyst. Environ. 74, 33–64.

Dunning, B.D., Danielson, B.J., Pulliam, H.R., 1992. Ecological processes that effect populations in complex landscapes. OIKOS 65, 169–175.

Erhardt, A., 1985. Wiesen und Brachland als Lebensraum für Schmetterlinge. Eine Feldstudie im Tavetsch (GR). Birkhäuser, Basel.

Gathmann, A., Greiler, H.-J., Tscharnktke, T., 1994. Trap nesting bees and wasps colonizing set-aside fields: succession and body size, managed by cutting and sowing. Oecologia 98, 8–14.

Ginsberg, H.S., 1983. Foraging ecology of bees in an old field. Ecology 64, 165–175.

Golden, D.M., Crist, T.O., 1999. Experimental effects of habitat fragmentation on old-field canopy insects: community, guild and species responses. Oecologia 118, 371–380.

Haeseler, V., Ritzau, C., 1998. Zur Aussagekraft wirbelloser Tiere in Umwelt- und Naturschutzgutachten—was wird tatsächlich erfasst? Z. Ökologie u. Naturschutz 7, 45–66 (in German, with English abstract).

Hedtke, C., Schricker, B., 1996. Homing in *Apis mellifera* and four Bombus species in comparison. Apidologie 27, 320–323.

Heinrich, B., 1976. The foraging specialisations of individual bumblebees. Ecol. Monogr. 42, 105–128.

Hermann, G., 1996. Zur Bearbeiterabhängigkeit faunistischer Beiträge am Beispiel von Heuschreckenerhebungen und Konsequenzen für die Praxis. Laufener Seminarbeitr. 3, 143–154 (in German).

Jeanneret, P., Charmillot, P.J., 1995. Movements of Tortricid moths (Lep. Tortricidae) between apple orchards and adjacent ecosystems. Agric. Ecosyst. Environ. 55, 37–49.

Kevan, P.G., 1999. Pollinators as bioindicators of the state of enviromen:species, activity and diversity. Agric. Ecosyst. Environ. 74, 373–393.

Knecht, C., Hirsch, M., Wolters, V., 2000. Blütenbesuchende Käfer in einem Landnutzungsmosaik. Agrarspectrum 31, 180–188 (in German, with English abstract).

Matter, S.F., 1997. Population density and area: the role of between- and within-patch processes. Oecologia 110, 533–538.

Matthews, R.W., Matthews, J.R., 1971. The malaise trap. Its utility and potential for sampling insect populations. Michigan Entomol. 4, 117–122.

Osborne, J.L., Williams, I.H., 2001. Site constantly of bumble bees in an experimentally patchy habitat. Agric. Ecosyst. Environ. 83, 129–141.

Pfaff, S., Wolters, V., 1999. The impact of agricultural management on diurnal lepidopteran communities in a mosaic landscape. In: Proceedings of the Conference on Sustainable Landuse Management 1999, Salzau, Germany, vol. 28. EcoSys Beiträge zur Ökosystemforschung, pp. 159–167.

Pleasants, J.M., 1981. Bumblebee response to variation in nectar availability. Ecology 62, 1648–1661.

Pollard, E., Elias, D.O., Skelton, M.J., Thomas, J.A., 1975. A method of assessing the abundance of butterflies in Monks Wood Natural Nature Reserve in 1973. Ent. Gaz. 26, 79–88.

Schmalhofer, V.R., 2001. Tritrophic interactions in a pollination system: impacts of species composition and size of flower patches on the hunting success of a flower-dwelling spider. Oecologia 129, 292–303.

Schwenninger, H.R., 1992. Untersuchungen zum Einfluss der Bewirtschaftungsintensität auf das Vorkommen von Insekten in der Agrarlandschaft, dargestellt am Beispiel der Wildbienen (Hymenoptera, Apoidea). Zool. Jb. Syst. 119, 543–561 (in German, with English abstract).

Sih, A., Baltus, M.S., 1987. Patch size, pollinator behavior, and pollinator limitation in catnip. Ecology 68, 1679–1690.

Statsoft, Inc. 1995. STATISTICA for Windows (Computer Program Manual). Statsoft, Tulsa, OK.

Steffan-Dewenter, I., Tscharntke, T., 1999. Effects of habitat isolation on pollinator communities and seed set. Oecologia 121, 432–440.

Teräs, I., 1976. Flower visits of bumblebees, *Bombus* Latr. (Hymenoptera, Apidae), during one summer. Ann. Zool. Fenn. 13, 200–232.

Walther-Hellwig, K., Frankl, R., 2000. Foraging habitats and foraging distances of bumblebees, Bombus spp. (Hym., Apidae), in an agricultural landscape. J. Appl. Entomol. 124, 299–306.

Webb, N.R., 1989. Studies on the invertebrate fauna of fragmented heathland in Dortset, UK, and the implications for conservation. Biol. Conserv. 47, 153–165.

Westrich, P., 1989. Die Wildbienen Baden-Württembergs. Bd. 1. Allgemeiner Teil, Lebensräume, Verhalten, Ökologie und Schutz. Ulmer, Stuttgart (in German).

Witsack, W., 1975. Eine quantitative Keschermethode zur Erfassung der epigäischen Arthropodenfauna. Ent. Nachr. Dresden 8, 121–128 (in German).

ELSEVIER

Available online at www.sciencedirect.com

Agriculture, Ecosystems and Environment 98 (2003) 339–351

Agriculture Ecosystems & Environment

www.elsevier.com/locate/agee

Indicators of plant species and community diversity in grasslands

Rainer Waldhardt*, Annette Otte

Division of Landscape Ecology and Landscape Planning, Justus-Liebig-University of Giessen, IFZ, Heinrich-Buff-Ring 26-32, D-35392 Giessen, Germany

Abstract

Parameters which are directly related to both land use change and biodiversity may be useful tools to indicate biodiversity in marginal landscapes. In these landscapes for about five decades abandonment of cultivation, especially in favour of extensive grassland use and succession on abandoned fields, has led to considerable changes in the landscape structure. Regions such as the Lahn-Dill Highlands (Hesse, Germany), formerly characterized by small-parcelled crop and grassland rotation, increasingly feature old grassland communities over large areas.

Impacts of changes in the landscape structure on the floristic–phytocoenotic diversity have been studied in two landscape tracts of this region that today are mainly used as grassland. Reconstruction of land use dynamics based on multitemporal aerial photograph interpretations of the period from 1945 to 1997 confirm the predominance of cultivation until ca. 1960 in both areas. On the basis of phytosociological surveys in one stand abandoned 3 years before the survey and in each three grassland stands of different age classes (11–27, 28–38, 39–46 and over 46 years), the floristic–phytocoenotic diversity of these stands is characterized as follows:

(i) Flora and vegetation are clearly differentiated in relation to stand age.
(ii) The vegetation of the older (>38 years) stands is more comparable among one another than is the vegetation of younger stands.
(iii) 19- to 33-year-old stands have the highest number of exclusive species.
(iv) Old stands (>46 years) have the highest α-species richness.
(v) The stands can be classified into different vegetation types in relation to age.

The floristic–phytocoenotic diversity is associated with site differences of the above age classes. Older stands are more frequent at upper slopes and the pH values of their soils are lower.

With a small methodological outlay, grassland stands of different age and species diversity can be differentiated by red–green–blue (RGB) colour tonal values from false-colour infrared (FCIR) aerial photographs.

The results open up possibilities for the qualitative and quantitative indication of floristic–phytocoenotic diversity in grasslands on the basis of stand age, site factors and also green and blue tonal values from the respective FCIR aerial photographs.

Furthermore the results indicate that it is necessary to retain old grassland stands, as well as a mosaic of extensively used grassland stands of different ages to retain plant species and community diversity in the study region.

Keywords: Marginal agricultural landscape; Land use change; Vegetation dynamics; *Festuca rubra–Agrostis capillaris* community; Aerial photographs

* Corresponding author. Tel.: +49-641-9937163; fax: +49-641-9937169.
E-mail address: rainer.waldhardt@agrar.uni-giessen.de (R. Waldhardt).

0167-8809/$ – see front matter © 2003 Elsevier Science B.V. All rights reserved.
doi:10.1016/S0167-8809(03)00094-X

1. Introduction

With the signing of the Convention on Biological Diversity in 1992, Germany has pledged itself to the conservation of biological diversity. Since virtually the entire land area in Germany is managed, this goal must be realized by sustainable land use over as large an area as possible (BMU, 1998). This especially applies to agriculturally used landscapes, whose biological diversity has been in decline for several decades, owing to overly intensive land use or to land use abandonment (Burel et al., 1998; MacDonald et al., 2000).

A crucial aspect of the total biodiversity of a landscape is its floristic diversity, which is discussed as a correlate thereof (Duelli and Obrist, 1998). As with the closely associated diversity of plant communities, floristic diversity in agricultural landscapes is essentially dependent on the former and current land use forms, intensities, patterns and dynamics present in these landscapes (Waldhardt et al., 2000).

Until ca. 1950, crop/grassland rotation, often in small-parcelled mosaics, was a widespread practice in marginal agricultural landscapes in Germany with climatic and/or edaphic conditions unfavourable for cultivation. There, as in similarly disadvantaged regions of Europe, cultivation has been largely abandoned within the last five decades in favour of grassland and succession on abandoned fields (Baldock et al., 1996). The landscape structure of these regions has fundamentally changed. Increasingly older plant communities have been able to develop, often over wide areas.

One aim of our research group is to quantify the importance of both forms of land use dynamics—rotation and long-term land use change—for the floristic–phytocoenotic α-, β- and γ-diversity (Whittaker, 1972) of marginal agricultural landscapes. On the basis of these and other results, we are developing ecologically and economically sustainable land use concepts within a research cooperative (Frede and Bach, 1999). Our research is being carried out in the Lahn-Dill Highlands region, which can be regarded as an especially marginal agricultural landscape. In order to quantify the importance of grassland age for the floristic–phytocoenotic diversity, the research presented here was conducted in two typical hill slope areas that have been successively taken out of cultivation since ca. 1950.

In developing sustainable, landscape-orientated land use concepts, we are devoting particular attention to current hot spots of floristic diversity. From the analysis of site factors and from the reconstruction of the land use history of these areas, important indications of the origins of the present-day diversity can be derived, as well as demands on future land use quality in order to retain this diversity. If necessary, however, it is possible—at great expense—to conduct a comprehensive field survey of the floristic–phytocoenotic diversity at a landscape scale in order to determine hot spots. Aerial photograph analysis may be an efficient procedure for determining hot spots. Hence, we have examined the question of whether false-colour infrared (FCIR) aerial photographs at a scale of 1: 5000—with a small methodological outlay—permit quantitative conclusions to be drawn about the local floristic–phytocoenotic diversity.

Possibilities for the indication of the floristic–phytocoenotic diversity of grassland vegetation will be discussed on the basis of stand age, site factors and colour spectra of the stands in FCIR-photographs.

2. Study region and sites

The Lahn-Dill Highlands cover about 700 km^2 in the highlands of Hesse, Germany. Owing to the following conditions:

(i) topography and soils (altitudinal range: 200–600 m a.s.l.; acidic regosols to moderately deep brown earths dominating over Devonian claystone slates, gravelstone slates and greywackes on arable sites on slopes of up to 20° or more; Schotte and Felix-Henningsen, 1999);
(ii) climate (mean annual temperature 5–8 °C, annual precipitation 700–1200 mm);
(iii) agrarian-structure (mean farm size 14 ha; 78% of farms managed with a secondary means of income; low levels of fertilizer and pesticide application; Frede and Bach, 1999).

Cultivation, especially, is not sufficiently profitable on much of the agriculturally used land area (Frede and Bach, 1999) anymore. Hence, the visual landscape character has fundamentally changed in comparison with that of ca. 1950. In many places, extensive grassland use has replaced the traditional, small-parcelled crop production and crop/grassland rotation (Kohl, 1978). Sheep grazing and annual mowing predominate.

The research presented here was conducted in two areas 1 (18 ha) and 2 (12 ha) of the Steinbrücken district (340–400 m a.s.l.). On nutrient- and base-poor regosols and brown earths, the upper and mid-slopes of these south-facing tracts feature land use dynamics that are typical of the landscape. On slopes of about 15°, the soils of area 1 are shallower and more skeletal in character than those of area 2, present on slopes of only about 7°.

3. Methods

3.1. Land use dynamics

The reconstruction of the land use followed Fuhr-Boßdorf et al. (1999). Stereoscopic interpretation of black-and-white aerial photographs from the years 1945, 1953, 1961, 1972, 1989 and 1997 at a scale of approximately 1: 10,000, with a resolution to the level of the allotment, was applied. The following agricultural land use forms: arable land, grassland and abandoned fields with woody plant succession were differentiated. Further information about land use was acquired through discussions with farmers and local officials in the study region.

3.2. Floristic–phytocoenotic diversity

Thirteen grassland allotments (plots) of different age were chosen for the floristic–phytocoenotic studies in area 1 (Fig. 1). Cultivation was first abandoned on one allotment: (i) 3 years before the survey was carried out in 1999 (age class A). Three allotments each were last cultivated: (ii) 11–27 (age class B), (iii) 28–38 (age class C), (iv) 39–46 (age class D) or (v) more than 46 years (age class E) ago.

Five vegetation surveys were carried out on each plot in June, with a standard size of 5 m × 5 m (subplots) estimating the percentage of individual species cover and total vegetation cover. Additionally, the respective growth heights of the vegetation were recorded. The nomenclature of the botanical names is in accordance with Wisskirchen and Haeupler (1998). Statements concerning ecological characteristics of species are in accordance with Ellenberg et al. (1992).

For a comparison of the age classes B–E, the mean α-species richness (species count/25 m^2), total ground cover and growth heights of the vegetation of each of the 13 plots were initially calculated. To quantify the similarity of the vegetation between subplots (β-diversity of the plots) Renkonen coefficients (Renkonen, 1938) were calculated. The vegetation of the five subplots of each plot were compared in pairs and the mean similarity was determined. The mean Renkonen coefficients of the plots, as well as their total numbers of species documented within the five subplots (γ-species diversity of the plots; species number × 125 m^2), were taken to calculate mean values of each age class. Finally, a phytosociological analysis of the stands according to Braun-Blanquet (1964) was applied.

Using the described method, botanical surveys and analyses were carried out for three further allotments of differing age (one plot from each of the age classes B–D) in area 2. This was done to determine whether the vegetation characteristics in area 1 were also present on similar sites in the region.

3.3. Site diversity

In autumn 1999, soil samples from a depth of 5–10 cm were taken from all 65 subplots of area 1 to determine relations between the vegetation and soil conditions.

The proportion of coarse material (>3 mm) of the soil was determined, together with total carbon (C_t) and nitrogen (N_t) levels, levels of available phosphorous (CAL method; P_{CAL}) and pH (in $CaCl_2$) of the fine soil (analyses according to Schlichting et al., 1995). Considering the coarse material, carbon, nitrogen and phosphorous levels of the soil of each subplot were calculated.

Furthermore the topographic position (tp) of the plots within the slopes was determined. Seven classes of topographic positions from the mid-slope (tp1) to the upper slope (tp7) were differentiated (Fig. 1).

3.4. Colour spectra in FCIR aerial photographs

Non-georeferenciated FCIR aerial photographs at a scale of 1: 5000 were available from a flight in

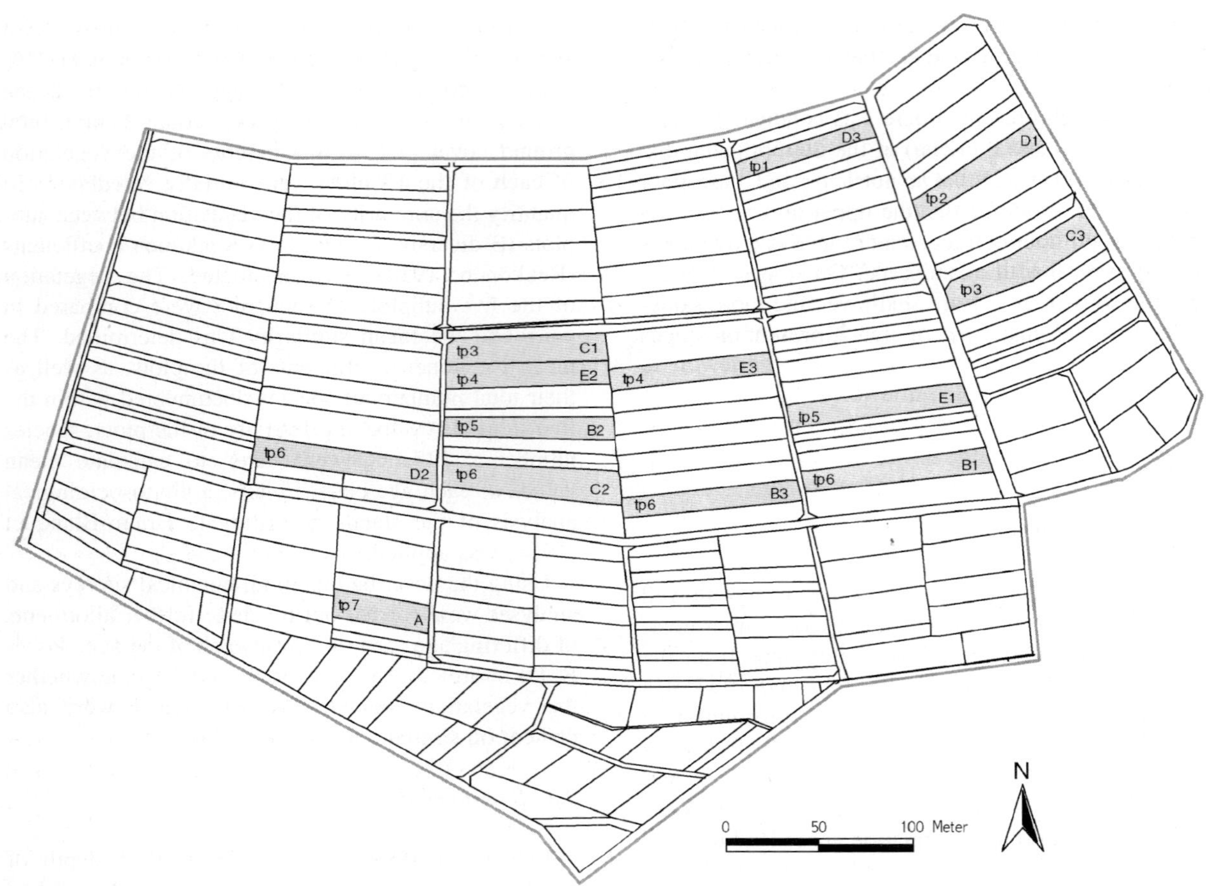

Fig. 1. Study plots in area 1. Age (in years since last in cultivation) according to aerial photographs: (A) 3; (B) 11–27; (C) 28–38; (D) 39–46; (E) >46. Classes of topographic position: tp1–tp7. In each age class three allotments (plots 1–3) were investigated. Five vegetation surveys (subplots a–e) were carried out on each plot, with a standard size of 5 m × 5 m.

May 1997. The aerial photographs were scanned in the red–green–blue (RGB) mode at a resolution of 300 dpi. The colour spectra of the photographs were not calibrated. The analysis of the colour spectra was carried out with the computer-program Adobe Photoshop 5.5 (Adobe Systems Inc., 1999).

The former allotment boundaries recognizable in the aerial photographs enabled a spatial classification of the study sites to be made. The allotment of age class A was not included, since it was still in cultivation at the time the aerial photograph was taken. Within the five subplots of each site standard-sized part-images (144 pixels) were cut out of the aerial photographs and their spectra of red, green and blue tones were analysed. The mean tonal values of the red, green and blue components were calculated, together with their standard deviations. A tonal value of between 0 and 255 was assigned to each component in every pixel, higher values representing brighter colour tones. Mean tonal values and standard deviations were subsequently calculated for each plot, and finally compared between age classes. Relations between mean α-species diversity and both the mean tonal values and their standard deviations of the subplots were tested.

3.5. Statistics

We performed a detrended correspondence analysis (DCA) with the species data (main matrix) of all 60 subplots of the age classes B–E from area 1 using the PC-ORD 4 software (McCune and Mefford, 1999). Axes were rescaled and rare species were downweighted. Method of detrending was by segments. To see how well the distances in the ordination spaces represent the distances in the original, unreduced spaces, it was proved in "after-the-fact" evaluations of how well distances in the ordination spaces match the relative Euclidean distances in the main matrices. The "after-the-fact" evaluation is performed by calculating the coefficient of determination (r^2) between distances in the ordination space and distances in the original space.

We created joint plots by having our environmental variables in second matrices. The angles and lengths of environmental lines tell the direction and strength of the relationships. In order to detect correlations between ordination axes and site conditions Pearsons-r (Krebs, 1999) was calculated based on the sample scores of the first and second ordination axes and environmental variables.

The effect of time (age classes B–E) on total ground cover, growth heights and similarity of the vegetation, as well as on both the mean tonal values and their deviations of the RGB images, was tested by means of a one-way ANOVA following Kolmogorov–Smirnov- and Sen & Puri tests. Whenever necessary, data were ln-transformed ($\ln x+0.1$) prior to statistical analyses. For multiple comparisons of mean values the Tukey HSD-test was applied. To quantify the correlation between two parameters we calculated Pearsons-r.

4. Results

4.1. Land use dynamics of the study sites

The land use in area 1 ca. 1950 was present on about 60 allotments with an average area of only 0.3 ha. About 70% of the total area was in cultivation (Table 1). In the period between 1961 and 1972, cultivation was abandoned over large parts of the area, and during the last 10 years on nearly all remaining allotments as well. On 2 ha of the former arable land, succession with woody species established, today overgrown with *Cytisus scoparius*. The remaining area was kept open as grassland. Since 1973, this grassland has been used for extensive sheep grazing and mulched once annually. There has been no fertilizer application since the abandonment of cultivation.

Table 1
Proportion of land use forms in areas 1 and 2 during the period 1945–1998[a]

	Proportion of area (percentage of total area)					
	1945	1953	1961	1972	1989	1997
Area 1 (total area: 18.2 ha)						
Arable land	70	67	58	17	16	2
Grassland	23	24	27	66	60	72
Abandoned fields[b]	0	2	8	8	13	13
Other land use[c]	7	7	7	9	11	13
Area 2 (total area: 11.8 ha)						
Arable land	73	65	68	30	20	2
Grassland	6	14	8	42	46	59
Abandoned fields[b]	0	0	2	6	6	7
Other land use[c]	21	21	22	22	28	32

[a] Black-and-white aerial photographs (scale 1: 10,000 approximately) were interpreted (Fuhr-Boßdorf et al., 1999).

[b] These are overgrown with *C. scoparius* in both landscape tracts (cf. Simmering et al., 2001).

[c] Other land uses are mainly roads and tracks.

The land use change followed a similar course in area 2 (Table 1). Again, a large proportion of the arable land was converted to grassland between 1961 and 1972, or between 1989 and 1997. Like area 1, area 2, is also mainly used for extensive sheep grazing. Some allotments are grazed by cattle or horses, or mown once or twice annually. Individual allotments are fertilized; here, more exact informations are not available.

4.2. Floristic–phytocoenotic and site diversity of grassland of differing age

The DCA of the surveyed vegetation in area 1 (Fig. 2)—site A, with considerably different vegetation, has not been included here—permits the recognition of a distinct species diversity of the stands. A strong correlation between ordination distance and distance in the original n-dimensional space was found ($r^2 = 0.78$). The first axis (DCA 1) represents the shift in community structure with increasing grassland age (correlation between ordination axis DCA-1 and age of vegetation: $r = 0.86$; $P < 0.001$). Moreover the older stands are more frequent at the

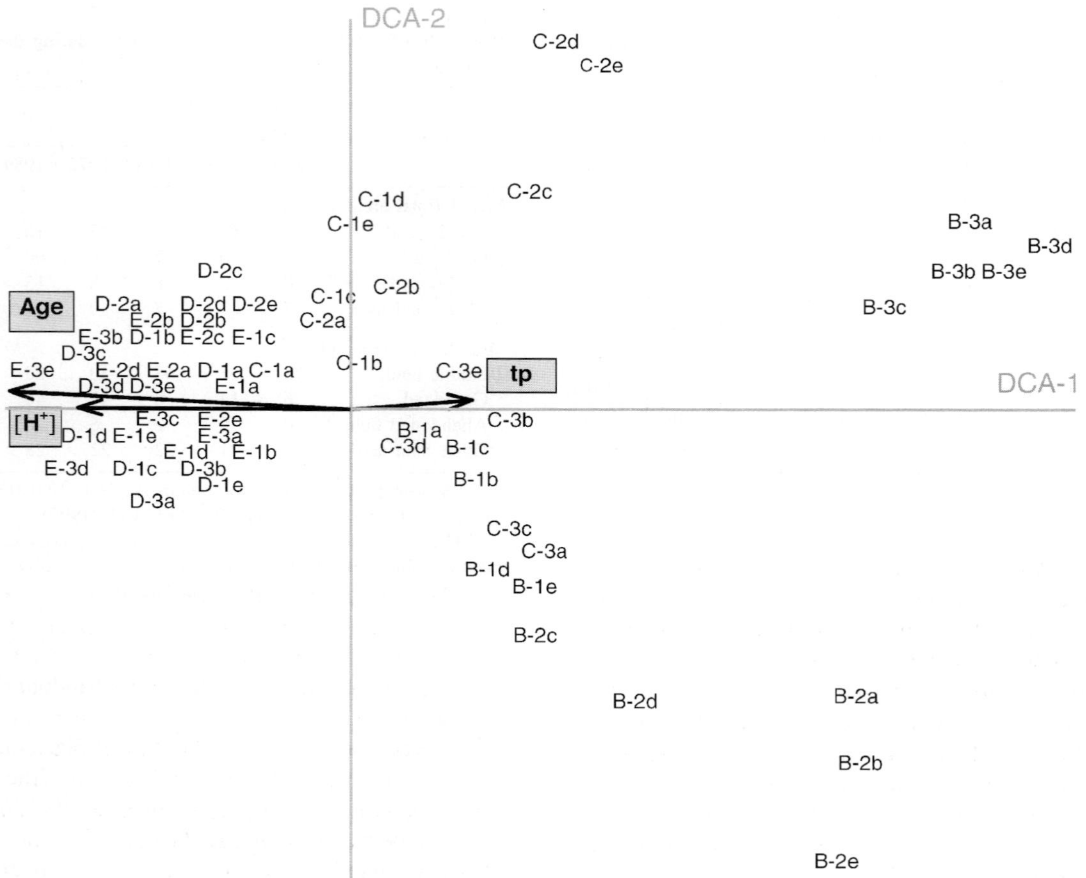

Fig. 2. Joint plot with sites of grassland in area 1. Age (in years since last in cultivation) according to aerial photographs: (A) 3; (B) 11–27; (C) 28–38; (D) 39–46; (E) >46. [H^+]: concentration of protons (see Table 2). tp: topographic position (see Fig. 1). In each age class three allotments (plots 1–3) were investigated. Five vegetation surveys (subplots a–e) were carried out on each plot, with a standard size of 5 m × 5 m.

upper slope and the pH values of their soils are lower (see Table 2).

The vegetation of the age classes D and E is much more similar within these classes than is the vegetation of younger stands within their respective age classes. On the contrary, the stands of the age classes B and C are clearly differentiated along ordination axis DCA-2.

A total of 80 plant species was recorded in the 65 subplots of area 1. A total of 27 species are found in one respective age class only. Each age class features four to seven exclusive species. Most of those species were recorded just in one or two subplots. With a presence in five respectively seven subplots *Convolvulus arvensis* and *Heracleum sphondylium* are common in age class C. Both species prefer to grow on base and nutrient rich soils.

In Table 3, for age classes A–E, plant species that essentially contribute to a differentiation of the vegetation are listed, together with their mean cover values. The table also includes total species counts together with the mean cover values, growth heights and Renkonen coefficients of the vegetation. With a dominance of the sown species *Lolium perenne* and *Trifolium repens*, 65% total ground cover and a growth height of less than 40 cm, the vegetation of

Table 2
Site characteristics of the topsoil (soil depth 5–10 cm) of grassland allotments of differing age in area 1[a]

Plot	Coarse material (wt.%)	C_t (%)	N_t (%)	P_{CAL} (mg × 100 g^{-1})	pH
A	22 ± 4	1.44 ± 0.09	0.16 ± 0.01	13 ± 2	4.4 ± 0.2
B-1	23 ± 4	1.72 ± 0.05	0.18 ± 0.01	4 ± 1	4.3 ± 0.1
B-2	21 ± 6	1.74 ± 0.14	0.18 ± 0.01	8 ± 2	4.4 ± 0.1
B-3	24 ± 5	1.67 ± 0.12	0.16 ± 0.01	6 ± 2	4.4 ± 0.2
C-1	24 ± 8	1.70 ± 0.17	0.17 ± 0.01	10 ± 2	4.3 ± 0.1
C-2	23 ± 6	1.72 ± 0.15	0.17 ± 0.02	6 ± 1	4.2 ± 0.1
C-3	21 ± 5	1.78 ± 0.15	0.18 ± 0.08	7 ± 2	4.1 ± 0.1
D-1	23 ± 7	1.76 ± 0.21	0.17 ± 0.02	10 ± 2	4.1 ± 0.1
D-2	27 ± 4	1.69 ± 0.09	0.16 ± 0.01	7 ± 2	4.0 ± 0.1
D-3	25 ± 4	1.72 ± 0.11	0.17 ± 0.01	7 ± 1	4.0 ± 0.1
E-1	27 ± 8	1.61 ± 0.21	0.16 ± 0.02	4 ± 1	3.8 ± 0.1
E-2	26 ± 5	1.69 ± 0.15	0.16 ± 0.01	7 ± 2	3.9 ± 0.2
E-3	27 ± 9	1.67 ± 0.21	0.17 ± 0.02	7 ± 2	3.9 ± 0.1

[a] Age (in years since last in cultivation) according to aerial photographs: (A) 3; (B) 11–27; (C) 28–38; (D) 39–46; (E) >46.

site A is distinctly more sparse and of lower canopy height than that of the other stands. Twenty further species attain a total cover of around 10%. The presence of several annual weeds of arable land (including *Capsella bursa-pastoris*, *Lamium purpureum* and *Viola arvensis*) indicates that the vegetation of this allotment is still at a very early stage of secondary succession. With a mean Renkonen coefficient of 0.87, the vegetation characteristics of the five subplots are very similar, while the mean α-species richness (12.0 species/25 m^2) is the lowest in comparison with the other age classes.

In age class B, the sown species *L. perenne* and *T. repens* also attain high cover values, yet these values are distinctly lower than those for the 3-year-old stand. While *L. perenne* is absent from older sites (C–E), *T. repens* is still present in these, although with a low cover. Age class B is characterized by comparatively high cover values of the ruderal species *Taraxacum officinale* and *Tanacetum vulgare*, but also by typical species of nutrient-poor grassland sites, such as *Bromus hordeaceus* and *Agrostis capillaris*. In contrast, more nutrient-demanding grassland species, such as *Arrhenatherum elatius*, *Galium mollugo* and *H. sphondylium*, attain the highest mean cover values in age class C.

Festuca rubra dominates in age classes D and E, with more than 65% cover. Low-growing forbs, such as *Rumex acetosella* and *Hypochoeris radicata*, reflect nutrient- and base-poor conditions. *Anthoxanthum odoratum* and *Teucrium scorodonia* characterize the oldest stands (age class E).

In all age classes B–E about 41–48 species/125 m^2 (mean γ-species richness) were recorded. The mean α-species richness is lower in age class B (17.8 species/25 m^2) than is in age class E (24.2 species/25 m^2). The variability of the α-species richness, as well as of the total ground cover, the growth heights and the Renkonen coefficients, is comparatively high in age classes B and C.

Hence, owing to its land use history, area 1, which today is predominantly used as grassland, features a differentiated floristic composition over a small area. In an aggregative fashion, the following vegetation types can be distinguished: with increasing stand age, there is (i) an early successional stage on abandoned fields (age class A); (ii) ruderalized grassland vegetation (age class B); (iii) an oat-grass meadow (*Arrhenatheretum elatioris* Br.-Bl. 1915) (age class C) and also (iv or v) red fescue–bent-grass meadows (*F. rubra–Agrostis tenuis* community) with a lesser (age class D) or greater number (age class E) of acid-tolerant, respectively. Hence, the γ-diversity of the grassland vegetation of this area increases in relation to its age structure.

The vegetation of the three additional allotments studied in area 2 is similarly differentiated with regards to stand age: the age class B allotment is characterized by a ruderalized grassland stand. The age class C allotment features an oat-grass meadow,

Table 3
Characteristics of the vegetation of grassland of differing age and differentiating species in area 1[a]

	Class A, 3 years	Class B, 11–27 years	Class C, 28–38 years	Class D, 39–46 years	Class E, >46 years
Number of plots	1	3	3	3	3
Species count × 25 m^{-2}					
Plot 1	12.0 ± 3.4	15.8 ± 0.8a	22.0 ± 3.0a	23.0 ± 2.5a	24.2 ± 1.1a
Plot 2		20.2 ± 1.5b	20.6 ± 3.3b	20.8 ± 2.8a	24.6 ± 1.8a
Plot 3		17.4 ± 1.8a	15.8 ± 1.8a	22.0 ± 1.9a	23.8 ± 2.9a
Age class	12.0	17.8 ± 2.2a	19.5 ± 3.3a	21.9 ± 1.1a	24.2 ± 0.4 b
Species count × 125 m^{-2}	22	41 ± 3.5a	46 ± 4.3a	48 ± 4.2a	45 ± 1.5a
Total ground cover (%)					
Plot 1	65.0 ± 4	91 ± 2a	87 ± 8a,b	86 ± 2a	93 ± 3a
Plot 2		74 ± 10b	92 ± 3a	89 ± 6a	85 ± 4a
Plot 3		94 ± 2a	80 ± 5b	85 ± 4a	85 ± 8a
Age class	65.0	86 ± 11a	86 ± 6a	86 ± 2a	87 ± 5a
Growth height (cm)					
Plot 1	37.0 ± 5	66 ± 2a	47 ± 7a	59 ± 7a	53 ± 6a
Plot 2		70 ± 8a	82 ± 6b	52 ± 4a	50 ± 4a
Plot 3		69 ± 6a	62 ± 3c	54 ± 4a	54 ± 2a
Age class	37.0	68 ± 2a	64 ± 18a	55 ± 4a	52 ± 2a
Similarity (Renkonen coefficient)					
Plot 1	0.87 ± 0.06	0.85 ± 0.06a	0.84 ± 0.05a	0.87 ± 0.03a	0.86 ± 0.05a
Plot 2		0.72 ± 0.08b	0.69 ± 0.12b	0.90 ± 0.03a	0.87 ± 0.02a
Plot 3		0.89 ± 0.04a	0.82 ± 0.06a	0.89 ± 0.04a	0.84 ± 0.03a
Age class	0.87	0.82 ± 0.09a	0.79 ± 0.08a	0.89 ± 0.01a	0.86 ± 0.02a
Mean ground cover (%)					
L. perenne	38	16.6 ± 23.0			
T. repens	29	16.0 ± 14.8	11.3 ± 6.1	11.3 ± 9.3	6.3 ± 3.1
C. bursa-pastoris	0.2	0.1 ± 0.1			
V. arvensis	0.2				
L. purpureum	0.1				
T. officinale	2.6	1.1 ± 1.0	0.9 ± 0.2		
T. vulgare		1.3 ± 2.0	0.1 ± 0.1		
B. hordeaceus		2.6 ± 4.3			
A. capillaris		22.7 ± 3.8a	13.8 ± 9.4a,b	6.1 ± 2.5b	5.2 ± 1.4b
A. elatius		0.2 ± 0.2	3.1 ± 3.9	0.6 ± 0.4	0.7 ± 0.6
G. mollugo			1.6 ± 2.1		
H. sphondylium			1.2 ± 2.1		
F. rubra	2.8	32.0 ± 15.7a,b	45.0 ± 10.4a,b,c	71.0 ± 1.0c,d	67.7 ± 2.5b,c,d
R. acetosella				6.0 ± 3.5	3.8 ± 3.1
Luzula campestris				2.3 ± 1.6	2.3 ± 1.3
Galium saxatile				0.4 ± 0.5	0.3 ± 0.4
H. radicata					
A. odoratum				0.1 ± 0.0	1.6 ± 1.2
T. scorodonia					0.4 ± 0.4

[a] Mean values of plot A1 (age class A), of plots 1–3 of each age classes B–E and of age classes are given with standard deviations (location of plots: see Fig. 1). The mean values of the plots and of the age classes B–E were compared. Values with different letters are significantly different ($P < 0.05$; Tukey-HSD).

Table 4
RGB colour values and standard deviations of the colour tonal value spectra of grassland allotments of differing age in areas 1 and 2[a]

	Class B, 11–27 years	Class C, 28–38 years	Class D, 39–46 years	Class E, >46 years
Area 1				
Number of plots	3	3	3	3
Red				
Mean colour tonal value	200.0 ± 7.7	199.0 ± 13.9	202.6 ± 2.3	200.2 ± 6.1
Mean standard deviation	3.8 ± 0.7	4.2 ± 0.6	4.3 ± 0.7	4.3 ± 0.3
Green				
Mean colour tonal value	116.8 ± 10.2a	123.9 ± 27.6a	168.7 ± 9.0b	156.2 ± 2.5b
Mean standard deviation	5.5 ± 1.3a	8.9 ± 2.8a	16.3 ± 1.8b	16.1 ± 0.6b
Blue				
Mean colour tonal value	168.8 ± 14.2a	158.4 ± 20.5a	198.5 ± 1.4b	186.0 ± 6.4a
Mean standard deviation	7.3 ± 1.1a,b	9.7 ± 1.4a,b,c	12.7 ± 1.9b,c,d	13.3 ± 0.6c,d
Area 2				
Number of plots	1	1	1	
Red				
Mean colour tonal value	227.5	227.2	221.3	
Mean standard deviation	2.2	2.8	3.3	
Green				
Mean colour tonal value	173.3	181.6	183.5	
Mean standard deviation	4.1	6.8	8.9	
Blue				
Mean colour tonal value	215.9	219.6	215.9	
Mean standard deviation	5.1	7.3	8.2	

[a] FCIR aerial photographs from 1997 were analysed using the image-processing program Adobe Photoshop 5.5. The aerial photographs were scanned in the RGB mode at a resolution of 300 dpi. Within the zone of five subplots of each allotment, standard-sized part-images (144 pixels) were cut out of the aerial photographs and their spectra of red, green and blue tones analysed. The mean tonal values of the R, G and B components were calculated, together with their standard deviations. Values with different letters are significantly different ($P < 0.05$; Tukey-HSD). The values in area 2 refer to one study site each.

and the age class D allotment a red fescue–bent-grass meadow with few acid-tolerant forbs. Similar to the situation in area 1, the mean α-species richness is lower in age class B (18.3 species/25 m^2) than is in age class C, respectively, D (22.6, respectively, 22.3 species/25 m^2).

4.3. Colour spectra of the survey plots in FCIR aerial photographs in relation to the age of the grassland vegetation

In the FCIR aerial photographs, there are differences in the colour spectra of the green and blue tones in relation to the age of the grassland vegetation (Table 4). In the aerial photograph of area 1, both the mean tonal values and their deviations are distinctly lower in the age classes B and C than the corresponding values in the age classes D and E. Thus the stands of the red fescue–bent-grass meadow (age classes D and E) can be differentiated by the analysis of the colour spectra from younger stands. Furthermore, positive quantitative relations between the mean α-species richness and both the mean green respectively blue tonal values and their deviations were found (Fig. 3).

In the allotments of area 2, the mean green tonal value and the deviations of the green and blue tones in the FCIR aerial photograph of the youngest grassland allotments (age class B) are the lowest, while those of the oldest grassland allotments (age class D) are the highest. In contrast to the situation in area 1 the mean blue tonal value is not different between stands of different age. Besides, in comparison with area 1, the mean red, green and blue tones of the three allotments of area 2 have distinctly higher values.

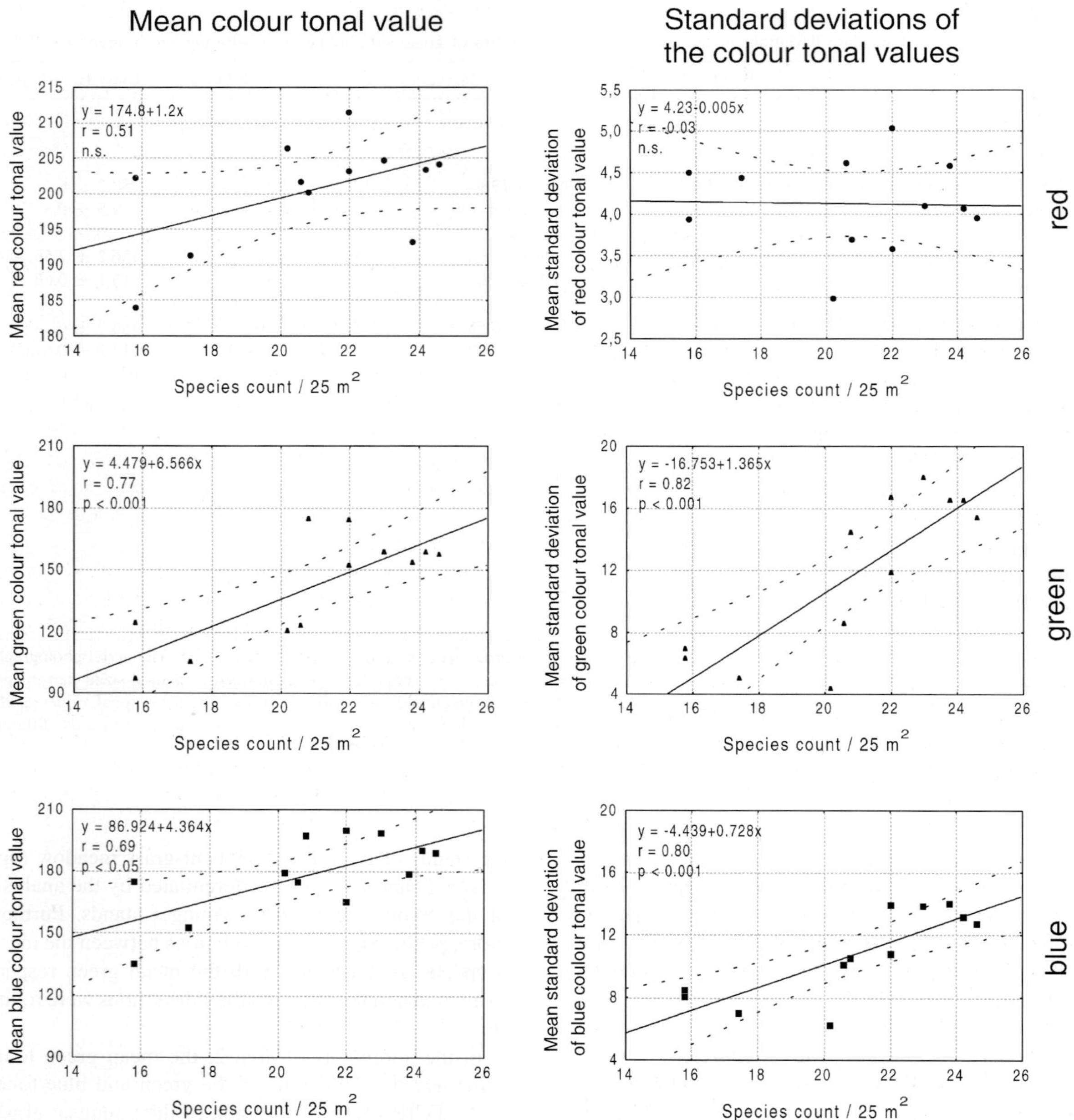

Fig. 3. Correlations between α-diversity of plant species and RGB colour spectra of FCIR aerial photographs. Mean species counts per $25\,m^2$ of 12 grassland allotments in area 1. The analysis of the colour spectra of the study sites in FCIR aerial photographs was carried out with the aid of the computer-program Adobe Photoshop 5.5. The aerial photographs were scanned in the RGB mode at a resolution of 300 dpi. r = Pearsons-r; 95%—confidence intervals.

5. Discussion

The hill slope areas presented here feature land use dynamics that are typical of the Lahn-Dill Highlands. Interpretation of aerial photographs from the period 1945–1997 from other parts of this landscape (Waldhardt et al., 2000) indicates that there, too, the proportion of arable land on upper and mid-slopes decreased markedly between ca. 1960 and 1970, and again after ca. 1990, particularly in favour of grassland use. This development can be attributed to several causes: These site types are often so steep that they cannot be safely cultivated with modern agricultural machines. In addition, the high coarse material content of the shallow soils renders ploughing more difficult. Hence, increasing mechanization of agriculture from ca. 1960, led to a rapid abandonment of cultivation on such arable land sites (Kohl, 1978). Meanwhile, increasingly unfavourable economic conditions for agricultural land use, also due to the EU agricultural reforms of 1993, have led to an almost complete abandonment of cultivation.

The land use dynamics presented here can be regarded as typical of marginal, agriculturally used regions of central Europe. Similar trends are shown by multitemporal aerial photograph interpretations by Gloaguen et al. (1994). Comparable developments in northern Europe can be found where agricultural land use was present in former decades. This is documented by aerial photograph interpretations for a rural area in central Finland (Ruuska, 1996).

According to Diquelou and Rozé (1997), land use extensification at a landscape scale leads to increased biodiversity, since new habitats, such as abandoned fields with woody plant succession, increase habitat diversity. In contrast, these authors hold that extensification at a local (village) scale leads to a greater uniformity with decreasing biodiversity. This is not the case in the present study. Although a large proportion of both investigated areas 1 and 2 has been grazed over large areas for some 30 years, the diversity of land use forms has increased during this period. Today, a considerable proportion of the total area is overgrown with *C. scoparius* succession. In the Lahn-Dill Highlands, according to our research group's studies, these successional areas have a surprisingly diverse floristic–phytocoenotic inventory (Simmering et al., 2001). At the same time, owing to the gradual abandonment of cultivation, as shown in this study, there is still a small-scale floristic–phytocoenotic diversity within the grassland today.

The heterogeneity of grassland in relation to age raises both the floristic and the phytocoenotic β- and γ-diversity of the studied hill slopes. A similar situation was found by Austrheim and Olsson (1999), who studied grassland in Norway with a comparable age spectrum. There, intermediate phase stands also have the highest number of exclusive species. Similarly in both studies, the oldest stands, with little variation in species composition, are richest in species. However, this only applies to the aboveground flora. Following the abandonment of cultivation, the soil seed reserve of the arable land species is more or less quickly depleted and many grassland species do not develop a persistent seed reserve (Bekker et al., 1997). In the Lahn-Dill Highlands, the soil seed reserve of arable land species is almost completely depleted within about 20 years (Waldhardt et al., 2001).

Stand age, topographic position of stands and pH values of their soils are found to be important site conditions to explain grassland diversity. Effects of further site conditions (e.g. nitrogen mineralization rates) on the vegetation could not be analysed in our investigation. We do not know, if the sheep grazing had been spatially homogenous within the investigated slopes. Spatial patterns of grazing are known to result in small-scale patterns of different nitrogen mineralization rates (Afzal and Adams, 1992) and of the vegetation itself (Semmartin and Oesterheld, 1996; Adler et al., 2001).

Several species typical of the species-rich old grassland vegetation (age classes D and E) indicate nutrient- and base-poor conditions. Species losses from the vegetation must be assumed in response to fertilizer application (Bakker, 1989; Burel et al., 1998). In this case changes in vegetation structure and species losses are likely to occur in grasslands of the Lahn-Dill Highlands, too. Hence, a low trophic level is of importance for maintaining the α-diversity of plant species in old grasslands.

Differences of the flora and vegetation of grasslands of different age as presented here should not be understood as a successional process. To us, this appears to be inadmissible for several reasons. It has been demonstrated that the land use pattern, as well as that of the entire landscape, changed considerably within the last five decades. Furthermore, it may be assumed that, at the time the land use was abandoned, arable sites that had been managed for a longer period had higher trophic levels than those that had already been converted to grassland around 1950. Since plant succession in abandoned fields and grasslands depends on the history of the allotments and also on their biotic environments (Gloaguen et al., 1994; Bakker et al., 1996), it would be wrong to assume that the present vegetation succession is the same as it was earlier.

Despite methodological simplification, the analysis of colour tonal values and their deviations from two FCIR aerial photographs scanned in the RGB mode permitted a differentiation of grassland stands of different age and species richness. Since the tonal values were not calibrated before the analysis, as described by Holopainen and Wang (1998), differences of the absolute tonal values of both aerial photographs cannot be attributed to floristic–phytocoenotic differences. However, data derived from one aerial photograph permit conclusions to be drawn on the floristic–phytocoenotic diversity. In particular, the quantitative positive relationships between the small-scale variability of green and blue tonal values of the RGB images and the α-species richness of the vegetation are suitable for an exact indication of hot spots of species diversity of the studied site at the level of the allotment.

6. Conclusions

As our investigation shows, the age of grassland stands is of varying importance, if focussing on specific components and spatial scales of their biological diversity (in sensu Noss, 1990; cf. Waldhardt and Otte, 2000). The α-species richness within grassland stands of the studied site is higher in older than is in younger stands. But there is no difference of the γ-species richness between stands of different age classes. The β-species diversity within and between stands and the structural diversity of the vegetation are comparatively high in the intermediate phase. The γ-species and vegetation diversity within a landscape tract, as well as its structural diversity, are positively effected through stands of different age within a landscape tract. Therefore, differentiated analyses are necessary to assess the importance of stand age of plant communities on their biological diversity. Those analyses will need to be considered in the development of sustainable land use concepts at the landscape scale.

In the present study, surrogates and correlates (Duelli and Obrist, 1998; Gaston and Spicer, 1998) of the floristic–phytocoenotic diversity of the grassland of two typical hill slope areas in a marginal agricultural landscape were determined using the qualitative and quantitative indicators “stand age”, “topographic position”, “pH value of the soil” and “green and blue tonal values” in FCIR aerial photographs. After validation, hot spots of diversity in the region can be identified without the need for a vegetation survey in the field. At the same time, the elucidated indicators provide indications for an explanation of the spatially differentiated diversity in the studied region.

The floristic α-diversity of the grassland of the studied site types (south-facing upper to mid-slopes) is highest in older stands with, at the same time, a moderate nutrient supply at best. A mosaic of stands of different age increases the floristic–phytocoenotic β- and γ-diversity of the slopes. Both these aspects will be taken into account in developing ecologically and economically sustainable land use concepts within the research cooperative mentioned earlier: it is necessary to retain old grassland stands, as well as a mosaic of extensively used grassland stands of different ages, on south-facing upper and mid-slopes of the region.

Acknowledgements

We would like to thank the German Research Foundation (DFG) for financial assistance.

References

Adler, P.B., Raff, D.A., Lauenroth, W.K., 2001. The effect of grazing on the spatial heterogeneity of vegetation. Oecologia 128, 465–479.

Adobe Systems Inc., 1999. Adobe Photoshop 5.5. User Guide. Adobe Systems Inc, Uxbridge, UK.

Afzal, M., Adams, W.A., 1992. Heterogeneity of soil mineral nitrogen in pasture prazed by cattle. Soil Sci. Soc. Am. J. 56, 1160–1166.

Austrheim, G., Olsson, E.G., 1999. How does continuity in grassland management after ploughing affect plant community patterns? Plant Ecol. 145, 59–74.

Bakker, J.P., 1989. Nature Management by Grazing and Cutting. Geobotany, vol. 14. Kluwer Academic Publishers, Dordrecht.

Bakker, J.P., Olff, H., Willems, J.H., Zobel, M., 1996. Why do we need permanent plots in the study of long-term vegetation dynamics? J. Veg. Sci. 7, 147–155.

Baldock, D., Beaufoy, G., Brouwer, F., Godeschalk, F., 1996. Farming at the Margins. Institute for European Environmental Policy, London.

Bekker, R.M., Verweij, G.L., Smith, R.E.N., Reine, R., Bakker, J.P., Schneider, S., 1997. Soil seed banks in European grasslands: does land use affect regeneration perspectives? J. Appl. Ecol. 34, 1293–1310.

BMU - Bundesministerium für Umwelt, Naturschutz, Reaktorsicherheit (Eds.), 1998. Bericht der Bundesregierung nach dem Übereinkommen über die biologische Vielfalt. Nationalbericht biologische Vielfalt. Neusser Druckerei und Verlags GmbH, Neuss.

Braun-Blanquet, J., 1964. Pflanzensoziologie: Grundzüge der Vegetationskunde. Springer, Wien.

Burel, F., Baudry, J., Butet, A., Clergeau, P., Delettre, Y., Le Coeur, D., Dubs, F., Morban, N., Paillat, G., Petit, S., Thenail, C., Brunel, E., Lefeuvre, J.C., 1998. Comparative biodiversity along a gradient of agricultural landscapes. Acta Oecol. 19, 47–60.

Diquelou, S., Rozé, F., 1997. Effets de l'intensification de l'agriculture et de la déprise agricole sur la dynamique paysagère en Bretagne intérieure. Ecol. Mediterr. 23, 91–106.

Duelli, P., Obrist, K., 1998. In search oft the best correlates for local organismal biodiversity in cultivated areas. Biodiv. Conserv. 7, 297–309.

Ellenberg, H., Weber, H.E., Düll, R., Wirth, V., Werner, W., Paulißen, D., 1992. Zeigerwerte von Pflanzen in Mitteleuropa. Scr. Geobot. 18, 1–258.

Frede, H.-G., Bach, M., 1999. Perspectives for peripheral regions. Z. Kulturtech. Landentw. 40, 193–196.

Fuhr-Boßdorf, K., Waldhardt, R., Otte, A., 1999. Auswirkungen der Landnutzungsdynamik auf das Potential von Pflanzengemeinschaften und Pflanzenarten einer peripheren Kulturlandschaft (1945–1998). Verh. Ges. Ökol. 29, 519–530.

Gaston, K.J., Spicer, J.I., 1998. Biodiversity—An Introduction. Blackwell, Oxford.

Gloaguen, J.-Cl., Rozé, F., Touffet, J., Clément, B., Forgeard, F., 1994. Etude des successions après abandon des pratiques culturales en Bretagne. Acta Bot. Gallica 141, 691–706.

Holopainen, M., Wang, G.X., 1998. Accuracy of digitized aerial photographs for assessing forest habitats at plot level. Scand. J. For. Res. 13, 499–508.

Kohl, M., 1978. Die Dynamik der Kulturlandschaft im oberen Lahn-Dillkreis. Wandlungen von Haubergswirtschaft und Ackerbau zu neuen Formen der Landnutzung in der modernen Regionalentwicklung. Giessener Geogr. Schr. 45.

Krebs, C.J., 1999. Ecological Methodology. Addison-Welsey, Menlo Park, CA.

MacDonald, D., Crabtree, J.R., Wiesinger, G., Dax, T., Stamou, N., Fleury, P., Lazpita, J.G., Gibon, A., 2000. Agricultural abandonment in mountain areas of Europe: environmental consequences and policy response. J. Environ. Manage. 59, 47–69.

McCune, B., Mefford, M.J., 1999. PC-ORD Multivariate Analysis of Ecological Data, Version 4. MjM Software Design, Glenedoen Beach, OR, USA.

Noss, R.F., 1990. Indicators for monitoring biodiversity: a hierarchical approach. Conserv. Biol. 4, 355–364.

Renkonen, O., 1938. Statistisch-ökologische Untersuchungen über die terrestrische Käferwelt der finnischen Bruchmoore. Ann. Zool. Soc. Zool.-Bot. Fenn. Vanamo 6, 1–231.

Ruuska, R., 1996. GIS analysis of change in an agricultural landscape in central Finland. Agric. Food Sci. Finland 5, 567–576.

Schlichting, E., Blume, H.-P., Stahr, E., 1995. Bodenkundliches Praktikum: eine Einführung in pedologisches Arbeiten für Ökologen, insbesondere Land- und Forstwirte und für Geowissenschaftler. Blackwell, Berlin.

Schotte, M., Felix-Henningsen, P., 1999. Application of ground-penetrating radar to survey the distribution and properties of periglacial layers in the Lahn-Dill-Bergland. Z. Kulturtech. Landentw. 40, 220–227.

Semmartin, M., Oesterheld, M., 1996. Effect of grazing pattern on primary productivity. Oikos 75, 431–436.

Simmering, D., Waldhardt, R., Otte, A., 2001. Syndynamik und Ökologie von Besenginsterbeständen des Lahn-Dill-Berglands unter Berücksichtigung ihrer Genese aus verschiedenen Rasengesellschaften. Tuexenia 21, 51–89.

Waldhardt, R., Otte, A., 2000. Zur Terminologie und wissenschaftlichen Anwendung des Begriffs Biodiversität. Wasser Boden 52, 10–13.

Waldhardt, R., Simmering, D., Fuhr-Boßdorf, K., Otte, A., 2000. Floristisch-phytocoenotische Diversitäten einer peripheren Kulturlandschaft in Abhängigkeit von Landnutzung, Raum und Zeit. Schriftenreihe Agrarspectrum 31, 121–147.

Waldhardt, R., Fuhr-Boßdorf, K., Otte, A., 2001. The significance of the seed bank as a potential for the reestablishment of arable-land vegetation in a marginal agricultural landscape. Web Ecol. 2, 83–87.

Whittaker, R.H., 1972. Evolution and measurement of species diversity. Taxon 21, 213–251.

Wisskirchen, R., Haeupler, H., 1998. Standardliste der Farn- und Blütenpflanzen Deutschlands. Ulmer, Stuttgart.

ELSEVIER

Available online at www.sciencedirect.com

Agriculture, Ecosystems and Environment 98 (2003) 353–361

Agriculture Ecosystems & Environment

www.elsevier.com/locate/agee

Effects of landscape patterns on species richness—a modelling approach

Nathalie Céline Steiner*, Wolfgang Köhler

Department of Biometry and Population Genetics, Justus-Liebig-University, Heinrich-Buff-Ring 26-32, 35392 Gießen, Germany

Abstract

Ecologists have suspected for a long time that landscape composition and landscape pattern are highly significant for species diversity. However, the way in which species diversity behaves in spatially different arranged landscapes is largely unexplained. This is currently the special focus of our investigation quantifying the relationship between landscape pattern and species richness. Therefore, we built a model which examines the effects of artificially structured landscape patterns on species richness (α-, β-, γ-diversity). By modifying habitat availability and spatial pattern of the habitat types we obtained different α-, β- and γ-diversity values. Additionally, we tested the effects of changing species community structure (ratio of generalists versus specialists) on diversity in combination with varying landscape patterns. The results showed that there was always a significant effect of landscape pattern on species richness. However, depending on the species community structure, we observed different strong relationships between habitat aggregation and species diversity. The higher is the rate of generalists in the species pool, the higher is the species richness at the local and regional scale. With increasing landscape aggregation, species richness decreased at all scales in all scenarios, whereas landscapes with only specialists occurring had always the lowest α-, β- and γ-diversity.

Keywords: Modelling; Spatial pattern; Landscape aggregation; Species diversity; Community structure

1. Introduction

Landscapes are never static; their elements are in permanent temporal and spatial flux. In agricultural areas, spatial dynamics on several time scales arise from cultural management of the landscape. Crop rotation shifts plant species, resulting in habitat changes for local animal populations (Merriam et al., 1991).

Thus species richness in agricultural landscapes depends not only on environmental factors such as climate, soil, topography and natural succession, but also on the spatial and temporal pattern of agricultural practices (Wagner et al., 2000). Studies which have dealt with species diversity and landscape structure focussed mainly on relations between species richness and area (e.g. Forman, 1995; Durrett and Levin, 1996; Losos and Schluter, 2000), habitat fragmentation (see Debinski and Holt, 2000) or habitat availability (Wagner et al., 2000), but little on the spatial context, i.e. the nature of the landscape surrounding a patch (see Debinski et al., 2001; Fahrig, 2001). At the same time many ecologists realised the importance of the spatial configuration of habitat for population dynamics (e.g. Cantrell and Cosner, 1991; Tilman et al., 1997), metapopulation dynamics (Wiens, 1997),

* Corresponding author. Tel.: +49-641-9937545; fax: +49-641-9937549.
E-mail address: nathalie.c.steiner@bio.uni-giessen.de (N.C. Steiner).

0167-8809/$ – see front matter © 2003 Elsevier Science B.V. All rights reserved.
doi:10.1016/S0167-8809(03)00095-1

community structure (Holt, 1997) and dispersal (e.g. Taylor et al., 1993; Gustafson and Gardner, 1996).

The major aim of this study was to elucidate the relationship between habitat aggregation and species richness at different scales. Habitat aggregation, as a quantitative measure of aggregation levels of spatial patterns in a landscape, is an important tool in landscape ecology when relating patterns to ecological processes (He et al., 2000). Therefore, we adapted the technique of Hiebeler (2000) to produce, via a landscape-generation algorithm, artificially structured landscapes with varying proportions of different habitat types which were differently aggregated. Additionally, these various landscape patterns were combined with different scenarios of species community structure. We assumed that every patch, here a single cell in the cellular automaton, as a discrete entity in a landscape has its own species inventory depending on environmental, regional and historical constraints (Ricklefs and Schluter, 1993) and land use. These patches interact with neighbouring patches in the way that species which can survive in a variety of habitats, so-called habitat generalists (Holt, 1997), disperse into two or more habitats. So the number of species in a patch (α-diversity) may be affected by the species inventory surrounding the patch and this in turn may influence the dissimilarity between habitat patches regarding species inventory (β-diversity) and that again will affect the overall diversity of a landscape (γ-diversity). Local extinction was disregarded in our simulation runs because of the very diverse reproductive and dispersal attributes of the different species and the very small predicted effect of habitat pattern on extinction (Fahrig, 2001).

Our aim was to quantify the impacts on the local diversity, the between diversity, and the overall diversity of a landscape when changing the spatial pattern of a landscape plus varying the proportions of generalists to specialists inhabiting the landscape. We obtained results for the following questions:

(a) What happens to α, β and γ when habitat availability, landscape aggregation and community structure are changed?
(b) Are there any thresholds, and do the diversity measures react in the same way to changes in spatial pattern and community structure?

2. Methods

The landscape model consists of a cellular automaton with different habitat types filling two-dimensional space like contiguous tessellations (Haydon and Pianka, 1999). Here, a habitat type corresponds to a type of land use and a patch to a management unit, in this case one cell. Each patch was characterised by a value indicating its habitat type. Only three habitat types (arable land, meadow and forest) were used because of the increasing number of possible neighbourhoods and states when using more than three habitat types. Applying the mechanisms of von Neumann neighbourhood relations (four neighbouring cells), the number of possible neighbourhood arrangements shot up from 15 combinations when using three habitat types to 210 combinations when having four habitat types (see Köhler et al., 1996).

The hypothetical landscapes used in this investigation were simulated on a 100×100 quadratic lattice. The scenarios differed in the spatial pattern of the landscape, the habitat available and the proportions of generalists to specialists (Fig. 1, Tables 1 and 2). Initially, each of the three habitat types was assigned a given proportion of the total area (Table 1) and was distributed randomly throughout the matrix. Then we modified in each step the overall aggregation of the virtual landscape precisely by using the aggregation index (AI) from He et al. (2000). The AI, proposed

Table 1
Proportions of habitats used in our model (%)[a]

Scenario	Arable field	Meadow	Forest
A	33	34	33
B	40	40	20
C	70	20	10

[a] Total sum = 100%.

Table 2
Proportions of generalists in each habitat (%)

Scenario	Arable field	Meadow	Forest
1	100	100	100
2	50	50	50
3	0	0	0
4	100	50	0

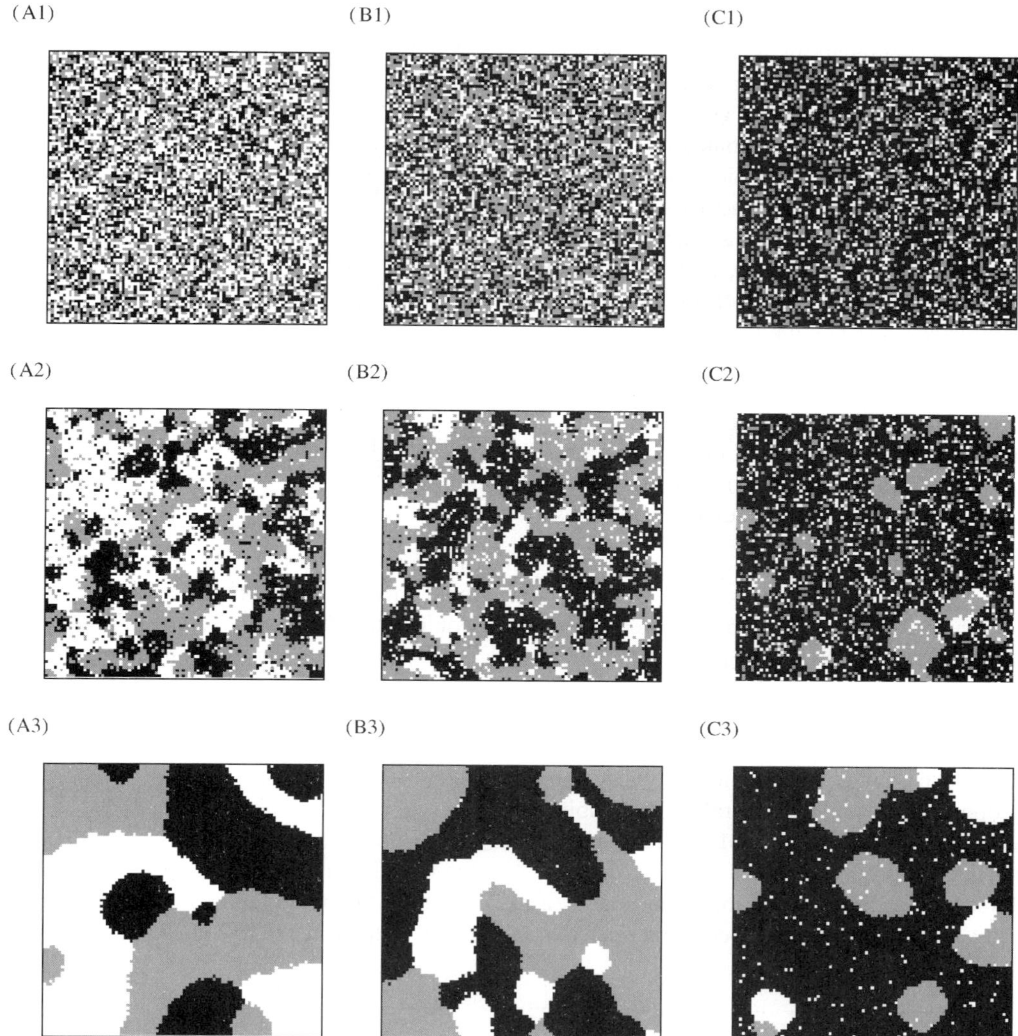

Fig. 1. Several artificially generated landscapes and their spatial pattern are shown. In each landscape the three habitat types a, f and m occur but with varying amounts of habitat and clustering. The overall aggregation of the landscape AI_L increases within each column from 33%, via 66–95%. In the first column the three habitat types occupy the same amount of habitat (33%, see Table 1). In B land composition is as follows: $a = 40\%$, $m = 40\%$, $f = 20\%$ (B1–B3). In the last column (C1–C3) arable land is the dominating land use: $a = 70\%$, $m = 20\%$ and $f = 10\%$. Arable land (a) is coloured black, meadow (m) grey and forest (f) white.

by He et al. (2000), is calculated as:

$$AI_i = \frac{e_{i,i}}{\max\_e_{i,i}}$$

where $e_{i,i}$ is the number of edges shared by habitat type i itself and $\max\_e_{i,i}$ the maximum possible number of edges shared by habitat type i.

The overall landscape aggregation (AI_L) is derived as:

$$AI_L = \sum_{i=1}^{n} AI_i \times A_i\%$$

where n is the total number of habitat types present in the landscape, and $A_i\%$ the percentage of the landscape of A_i. AI_i has a range from 0 to 1, with 0 indicating the lowest level of aggregation.

At each aggregation level for each cell, the within-area diversity (α), measured as the number of species occurring in one cell (Whittaker, 1972), and the between-area diversity (β), which measures the average changes in species between two sites (Cody, 1993), were calculated and then averaged for the entire landscape. For the calculation of the overall diversity of the hypothetical landscapes, we implemented Whittaker's (1977) multiplicative approach of linking α- and β-diversity.

Local diversity (α) and species spill-over (β) were affected by the states of the four neighbouring cells, i.e. the habitat type and the set of common species previously determined (see Table 2). The increase in species richness of each cell is calculated as follows:

$$S_{\mathrm{cl}} = S_{\mathrm{cl}} + S_{n1\mathrm{g}} + S_{n2\mathrm{g}}$$

where S_{cl} is the habitat-specific species inventory of the central cell, and $S_{n1\mathrm{g}}$ and $S_{n2\mathrm{g}}$ the number of habitat generalists of the neighbouring cells which immigrate in the central cell. The values $n1$ and $n2$ symbolise two different habitat types, i.e. two different land uses, each with its own species inventory with unequal proportions of habitat generalists and habitat specialists (Table 2). The number of habitat generalists of one habitat type was added only once, even if this habitat type occurred more than one time as a neighbouring cell. No increase in species number was made when all four surrounding cells had the same land use as the central cell. Also we ignored the cells at the edge of the lattice in our calculations. For the comparability of our simulation runs, we always used the same finite number of patches (100 × 100 cells) but manipulated the intrinsic species inventory of each habitat type, keeping the total number of species in each landscape constant (Table 3). Additionally, we modified the number of species in common between the three different habitats by shifting the proportions of habitat generalists, who could disperse into two or three different habitat types, to habitat specialists, who would settle down only in one habitat type (Table 2). Finally, we set the proportions of generalists and specialists subject to the land use (scenario 4, Table 2).

Table 3
The number of species intrinsic to each of the habitat types[a]

Scenario	Arable field	Meadow	Forest
s1	33	34	33
s2	20	40	40

[a] Total sum = 100 species.

3. Results

Over 5760 simulation runs were done to compare the variation in α-, β- and γ-diversity in relationship to habitat availability, landscape heterogeneity and community structure. Fig. 2 shows the regression curves

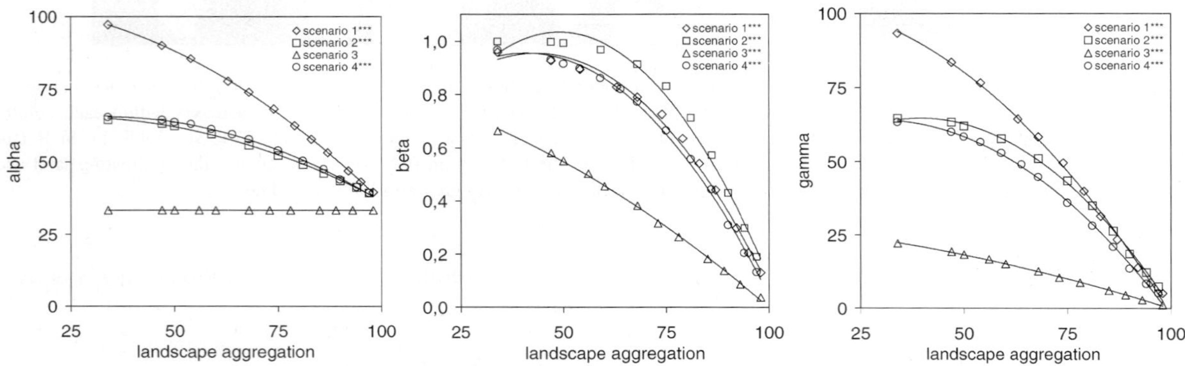

Fig. 2. Effect of landscape aggregation (AI_L) on α-, β- and γ-diversity (*** $P < 0.001$). Scenario (1) only generalists occur in the landscape; scenario (2) specialists and generalists occupy the landscape at a ratio 1:1; scenario (3) only specialists inhabit the landscape; scenario (4) here community structure depends on land use: only generalists occupy arable land, specialists and generalists occur in meadows at a ratio 1:1, and the forest is inhabited by specialists (cf. Table 2).

of the simulation runs when having an equal amount of habitat for each habitat type (33%, Table 1), equal proportions of species in each habitat (Table 3), and varying proportions of habitat generalists (cf. Table 2). The three graphics in the row represent the three diversity measures (α, β and γ) in relationship to the overall aggregation of the landscape. In each graphic the four different groups according to the predefined community structure are summarised (cf. Table 2).

All diversity measures in all scenarios were significantly affected by landscape aggregation except for α in one scenario (scenario 3) where species number per patch stayed constant at all levels of aggregation (Fig. 2).

The average species number per patch (α) decreased in three of the four scenarios when landscape aggregation increased. When only generalists occurred in the virtual landscapes (scenario 1) α reached the highest values but declined sharply when landscape aggregation increased. Also in scenarios 2 and 4 the loss of species richness per patch was notable when landscape aggregation increased. Local diversity at different aggregation levels was not affected when the landscape was inhabited solely by specialists (scenario 3), but the number of species per patch (α) was always smaller than those of the other scenarios (Fig. 2).

The second diversity measure β was significantly affected by the overall aggregation of the landscape. In all four scenarios β decreased when landscape aggregation increased. Here, scenario 2 reached the highest values and again declined notably with increasing landscape aggregation. Once more the regression curve of scenario 3 had the lowest β-values. In three of the four scenarios β decreased very slowly until an aggregation level of 60% and then faded rapidly. Only in scenario 3 did the curve drop in a linear way.

These previous results were also reflected in the last diagram of Fig. 2 where the overall diversity (γ) of the landscapes was the highest when the overall aggregation was the lowest and only generalists inhabit the virtual landscape (scenario 1). The curves declined in three cases in a quadratic manner, and in scenarios 2 and 4 showed a threshold effect.

Then the intrinsic species inventory was manipulated in the way that one habitat type hosted lesser species than the other two (scenario s2, Table 3). In addition, habitat availability, landscape aggregation and community structure were changed as before. The results of these simulation runs restate the previous findings: diversity was strongly negatively correlated with increasing landscape aggregation in all scenarios with the exception of α in scenario 3 (Fig. 3). There was only little variation compared to the simulations with a homogeneous distribution of species over the three habitat types, whereby the s2 scenarios always attained lower values compared to the s1 regression curves. In one case, in scenario 4-s2, the regression curve for α and γ is considerably lower than in scenario 4-s1.

Habitat availability played an important role when one habitat type notably dominated (arable land = 70%, Table 1). In these cases the diversity achieved in all scenarios were significantly lower compared

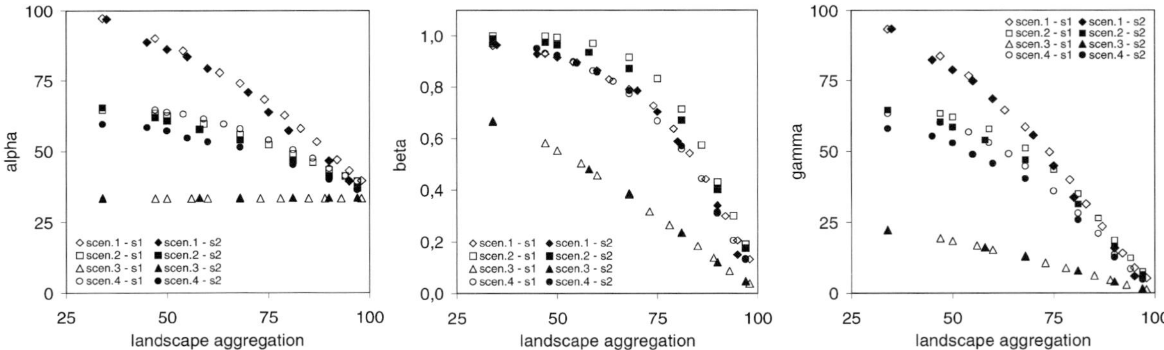

Fig. 3. Effect of landscape aggregation on α-, β- and γ-diversity when changing the intrinsic species inventory of each habitat (cf. Table 3). s1 indicates the scenario of homogeneous distribution of all 100 species over the three habitat types ($a = 33, m = 34, f = 33$), and in s2 one habitat type hosts lesser species than the other two ($a = 20, m = 40, f = 40$). Scenarios 1–4 signify the proportions of habitat generalists in the landscape (cf. Table 2).

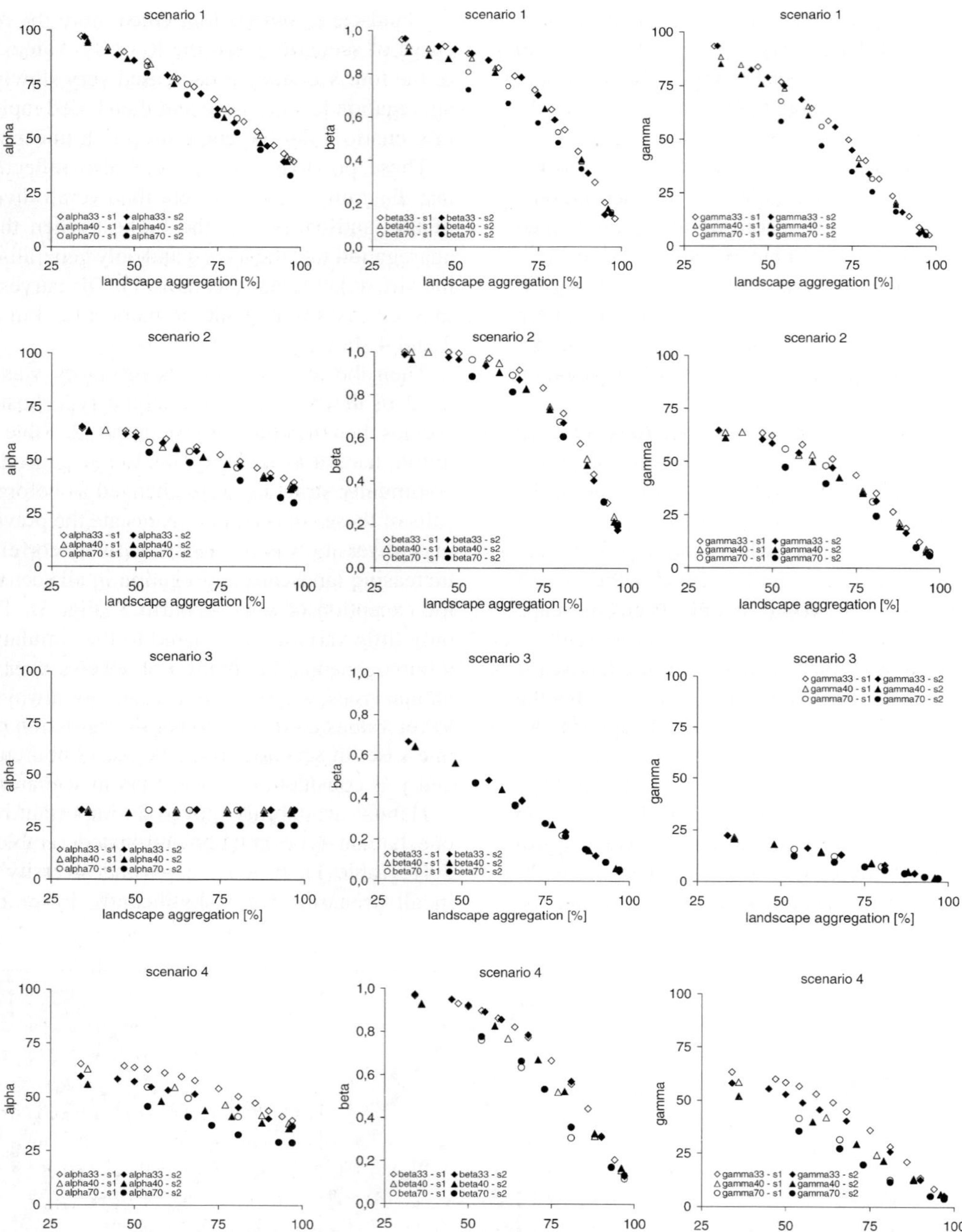

Fig. 4. Effect of landscape aggregation on α-, β- and γ-diversity when having different proportions of habitat (arable land: 33, 40 or 70% of the land surface) and a varying number of species per habitat (s1 and s2-scenarios, cf. Table 3).

to the ones where the three habitats had equal proportions (Fig. 4). When the dominating land use was assigned lesser species (s2-scenarios) the diversity values were consistently very low. The biggest variations of α, β and γ appeared when community structure was determined subject to land use (scenario 4). Again the s2-scenarios attained lower values than the s1-scenarios.

4. Discussion

The goal of our investigations was to study the effects of landscape structure and community structure on species richness at different scales. Using simulation runs with a cellular automata we demonstrated that the spatial heterogeneity of a landscape has a strong significant effect on local diversity (α), the between diversity (β) and the overall diversity of the landscape (γ). Diversity (α, β and γ) was the highest when landscape aggregation was the lowest. The community structure also played a major role in diversity at all scales: the greater is the number of common species the bigger is the local, the between and the overall diversity. Specialists were not affected by the landscape's heterogeneity at the local level, but the loss of between diversity (β) and overall diversity (γ) when landscape aggregation increased was even bigger compared to the other scenarios. The reasons were that the specialists did not move into the other habitats and so there was no increase of species richness per patch. In addition, the dissimilarity between the habitats (β) in the specialist scenario decreased in a linear way in contrast to the other scenarios where β remained constant until a threshold and then dropped rapidly.

Interestingly, diversity was also low when, supplementary to landscape aggregation, the intrinsic inventory and the proportions of generalists depended on land use. The impact was strengthened when one habitat type was dominating the landscape. The task now is to verify if species richness is dependent of the land use type and if community structure (ratio generalists to specialists) also depends on the land management type.

We do not deny that environmental variability is clearly the dominant factor in determining species richness (Palmer, 1992), i.e. species richness as a function of the number of different habitats, but in a given number of habitats the spatial distribution of the environmental variability is also extremely important as our results showed.

The simulation results support the findings of the empirical studies from Harrison (1999) who compared 48 patchy and continuous sites with respect to local diversity, regional diversity and among-site differentiation. For all herb species together, regional diversity and average local diversity were higher on small patches, mainly because of a significantly higher number of alien species. For the endemic species local diversity was lower and among-site differentiation higher on small patches. Debinski et al. (2001) also showed in their empirical and model investigations that patterns of species diversity and abundance differ according to the scale of the landscape mosaic. Simulation results suggest that both specialist and generalist butterflies can sample a much more diverse array of habitat types in a more fine-grained landscape. This trend may benefit species with restricted movement that require multiple resources to complete a complex life cycle. Wagner et al. (2000) as well observed a significant difference in diversity patterns between the edge and the core area for arable fields and meadows, with the edge being generally more diverse than the core.

In these simulation runs we first neglected local extinction in isolated habitats because of the manifold ways through which species disperse. A more integrated advance would be to link this basic model approach with investigations where species are grouped according to their dispersal abilities, as Tischendorf and Fahrig (2000) or Vos et al. (2001) did, in order to find the optimum level of aggregation for the maximum diversity (cf. With and Crist, 1995). The major conclusions derived from this model could also be combined with other spatially explicit information affecting species richness such as historical land use, soil properties, and information derived from digital elevation models in order to map species diversity and its distribution at a local and regional level.

5. Conclusions

Over the last few decades agriculturally used areas have experienced pronounced changes triggered by the development of new, highly productive agricultural techniques which was prompted by the increased demand for agricultural goods. Due to the complex interactions of the ecological processes present in agricultural environments, the effects on these environments are also multifaceted and not easily comprehensible. Presently, experts warn that the biggest threat to global diversity results from land use changes (Sala et al., 2000). Empirical studies on the effects of land use changes to species diversity are often extremely time consuming and cost-intensive. Ecological modelling is one approach to reduce such costs and provide quickly some preliminary results. The model presented in this paper was able to obtain some valuable insights into the effects of spatial landscape patterns on species richness and quantify them.

In addition we found, in agreement to Fahrig (2001), that conservation and land management strategies should always consider the quality of the whole landscape, especially the number of different habitats, their relative proportions and their spatial arrangement.

Acknowledgements

This research is part of the collaborative research programme "Land use conceptions for peripheral regions" (SFB 299) and we have to thank the German Research Foundation (Deutsche Forschungsgemeinschaft, DFG) for the financial support. We are particularly thankful to Marcus Hoffmann for his great support within the project. Also we would like to thank Klemens Ekschmitt, Gabriel Schachtel and Rob O'Neill for their helpful comments and Kerstin Wiegand for reviewing the paper.

References

Cantrell, R.S., Cosner, C., 1991. The effects of spatial heterogeneity in population dynamics. J. Math. Biol. 29, 315–338.

Cody, M.L., 1993. Bird diversity components within and between habitats in Australia. In: Ricklefs, R.E., Schluter, D. (Eds.), Species Diversity in Ecological Communities—Historical and Geographical Perspectives. Chicago Press, Chicago, pp. 147–158.

Debinski, D.M., Holt, R., 2000. A survey and overview of habitat fragmentation experiments. Conserv. Biol. 14 (2), 342–355.

Debinski, D.M., Ray, C., Saveraid, E.H., 2001. Species diversity and the scale of the landscape mosaic: do scales of movement and patch size affect diversity? Biol. Conserv. 98, 179–190.

Durrett, R., Levin, S., 1996. Spatial models for species–area curves. J. Theor. Biol. 179, 119–127.

Fahrig, L., 2001. How much habitat is enough? Biol. Conserv. 100, 65–74.

Forman, R.T.T., 1995. Land Mosaics: The Ecology of Landscapes and Regions. Cambridge University Press, Cambridge, pp. 54–65.

Gustafson, E.J., Gardner, R.H., 1996. The effect of landscape heterogeneity on the probability of patch colonization. Ecology 77 (1), 94–107.

Harrison, S., 1999. Local and regional diversity in a patchy landscape: native, alien, and endemic herbs on serpentine. Ecology 80 (1), 70–80.

Haydon, D.T., Pianka, E.R., 1999. Metapopulation theory, landscape models, and species diversity. Ecoscience 6 (3), 316–328.

He, H., DeZonia, B.E., Mladenoff, D., 2000. An aggregation index (AI) to quantify spatial patterns of landscapes. Landscape Ecol. 15, 591–601.

Hiebeler, D., 2000. Populations on fragmented landscapes with spatially structured heterogeneities: landscape generation and local dispersal. Ecology 81 (6), 1629–1641.

Holt, R.D., 1997. From metapopulation dynamics to community structure: some consequences of environmental heterogeneity. In: Hanski, I., Gilpin, M.E. (Eds.), Metapopulation Dynamics: Ecology, Genetics and Evolution. Academic Press, San Diego, pp. 146–165.

Köhler, W., Schachtel, G., Voleske, P., 1996. Biostatistik. Springer, Heidelberg, pp. 241–243.

Losos, J.B., Schluter, D., 2000. Analysis of an evolutionary species–area relationship. Nature 408, 847–850.

Merriam, G., Henein, K., Stuart-Smith, K., 1991. Landscape dynamics models. In: Turner, M., Gardner, R. (Eds.), Quantitative Methods in Landscape Ecology. Springer, New York, pp. 399–416.

Palmer, M.W., 1992. The coexistence of species in fractal landscapes. Am. Nat. 139 (2), 375–397.

Ricklefs, R.E., Schluter, D. (Eds.), 1993. Species diversity: regional and historical influences. In: Species Diversity in Ecological Communities—Historical and Geographical Perspectives. Chicago Press, Chicago, pp. 350–364.

Sala, E.S., Chapin III, F.S., Armesto, J.J., Berlow, E., Bloomfield, J., Dirzo, R., Huber-Sanwald, E., Huenneke, L.F., Jackson, R.B., Kinzig, A., Leemans, R., Lodge, D.M., Mooney, H.A., Oesterheld, M., Poff, N.L., Sykes, M.T., Walker, B.H., Walker, M., Wall, D.H., 2000. Global biodiversity scenarios for the year 2100. Science 287, 1770–1774.

Taylor, P.D., Fahrig, L., Henein, K., Merriam, G., 1993. Connectivity is a vital element of landscape structure. Oikos 68 (3), 571–573.

Tilman, D., Lehman, C.L., Kareiva, P., 1997. Population dynamics in spatial habitats. In: Tilman, D., Kareiva, P. (Eds.), Spatial Ecology—The Role of Space in Population Dynamics and Interspecific Interactions. Princeton University Press, New Jersey, pp. 3–20.

Tischendorf, L., Fahrig, L., 2000. How should we measure landscape connectivity? Landscape Ecol. 15, 633–641.

Vos, C.C., Verboom, J., Opdam, P.F.M., Ter Braak, C.J.F., 2001. Toward ecologically scaled landscape indices. Am. Nat. 183 (1), 24–41.

Wagner, H.H., Wildi, O., Ewald, K.C., 2000. Additive partitioning of plant species diversity in an agricultural mosaic landscape. Landscape Ecol. 15, 219–227.

Whittaker, R.H., 1972. Evolution and measurement of species diversity. Taxon 21, 213–251.

Whittaker, R.H., 1977. Evolution of species diversity in land communities. In: Hecht, M.K., Steere, B.W.N.C. (Eds.), Evolutionary Biology, vol. 10, pp. 1–67.

Wiens, J.A., 1997. Metapopulation dynamics and landscape ecology. In: Hanski, I., Gilpin, M.E. (Eds.), Metapopulation Biology: Ecology, Genetics, and Evolution. Academic Press, San Diego, pp. 43–60.

With, K.A., Crist, T.O., 1995. Critical thresholds in species' responses to landscape structure. Ecology 76 (8), 2446–2459.

ELSEVIER

Available online at www.sciencedirect.com

Agriculture, Ecosystems and Environment 98 (2003) 363–370

Agriculture Ecosystems & Environment

www.elsevier.com/locate/agee

Landscape indicators from ecological area sampling in Germany

Regina Hoffmann-Kroll*, Dieter Schäfer, Steffen Seibel

Federal Statistical Office, D 65180 Wiesbaden, Germany

Abstract

Up to now there is a great lack of statistically reliable information on biodiversity and its changes in the landscape of Germany on a national level. Research was required to develop appropriate methods for monitoring biodiversity. The ecological area sampling (EAS) is such a new statistical approach to provide indicators on the state and the development of landscapes, ecosystems and species. EAS does not focus on protected areas, biotopes of high value or rare species but on "normal landscape". It was designed in a way similar to the British Countyside Survey by the German Federal Statistical Office and the Federal Agency for Nature Conservation with support of private companies. Data collection in sample units is done by field survey after a previous interpretation of satellite images and aerial photographs. Detailed analysis of these primary data as well as the computation of main biodiversity indicators which are representative for the whole of Germany fill a great gap from a national point of view. The results are broken down by land classes; the accuracy is quantified by the standard error. EAS was tested successfully in a pilot study in 1995/1996 in a restricted test area.

Keywords: Biodiversity; Ecological area sampling; Indicator; Land class; Landscape quality

1. Introduction

Ecological area sampling (EAS) is part of a more comprehensive research project on the state of the environment which is integrated into the environmental economic accounting (EEA; cf. Radermacher and Stahmer, 1994). In EEA much work has already been done to provide information from a national point of view both for the *pressures* of economy on the environment and for *responses* to improve the environmental conditions. A description of the *state* of the environment on a national scale however is still missing. In order to fill this gap, land cover information for the whole of Germany has been derived from satellite data (project CORINE land cover; cf. Deggau, 1995), and a research project on the development of a system of indicators of the quality of environment in Germany has been carried out. The EAS is part of this latter project. It is a new tool for the collection of data on the nature and landscape structures and their development. These data can serve to quantify aggregated indicators on landscape structure from the point of view of nature protection or natural capital. The other aspects of the indicator system which are not covered by EAS are impacts of pollution and the functionality of ecosystems (cf. Radermacher et al., 1998).

Information on the state of the environment in Germany that is currently collected on a regular basis focuses mainly on the impact on media and organisms caused by harmful substances ("impacts" aspect). For this there exists a number of monitoring systems covering the area of Germany. Data on the state and quality of landscapes and nature from "physical structure"

* Corresponding author. Tel.: +49-611-752676;
fax: +49-611-753971.
E-mail address: regina.hoffmann-kroll@statistik-bund.de
(R. Hoffmann-Kroll).

0167-8809/$ – see front matter © 2003 Elsevier Science B.V. All rights reserved.
doi:10.1016/S0167-8809(03)00096-3

aspects, which provide just as important information on the state of the environment, are however available only for specific cases, limited spatial sections or do not meet the requirements of an indicator system on the state of the environment in a satisfying way. For this reason EAS is the precondition for systematic, periodic, representative and nation-wide data collection in this field (Barr et al., 1993; Haines-Young et al., 2000).

Up to now EAS has been tested in a pilot study in agricultural landscapes in three Länder of Germany. For this reason EAS can be seen as an interesting instrument in developing landscape indicators in the context of agri-environmental indicators, too.

2. Methods

For EAS, data on the landscape quality, the biotope quality, and the occurrence of species in biotopes are collected in periodically monitored sites that were selected at random. The concept was developed in cooperation between the Federal Statistical Office and the Federal Agency for Nature Conservation with the support of private companies (cf. Back et al., 1996; Statistisches Bundesamt and Bundesamt für Naturschutz, 2000). The survey design and programme are shown in Fig. 1.

Since the structure of landscapes and the occurrence of species heavily depend on the local land conditions, a classification of Germany into 28 land classes was developed by means of cluster analysis; each of these land classes is characterised by a largely homogeneous natural composition (regarding geology, climate, soil, hydrology and morphology). In addition to this classification of natural conditions the European CORINE land cover project provides information on land cover for the whole of Germany (on the basis of satellite data, aerial photographs and topographical maps; cf. Deggau, 1995). The intersection of both layers—the land classes and the CORINE land cover units—form the basic geometry for the stratification of the sample which in the following is organised in two steps:

- At the first level, indicators of landscape quality and of biotope quality are covered for sample units of the size of $1\,km^2$ each selected at random within the different strata. For this purpose, aerial photographs are evaluated to cover the biotopes existing in a given sample area. The scale of biotopes is quite appropriate to describe the state of the environment because it permits an analysis in terms of ecological theory, too. Subsequently, the landscape is examined (through a field survey) and the biotopes checked for their coverage or, where necessary, further specified by means of a biotope classification comprising some 500 items. Moreover, the field survey allows the coverage of small biotopes that are not visible on the aerial photograph. The pattern of biotopes allows one to derive indicators of landscape quality. In addition, for important biotope types, the field survey also serves to cover additional variables on the biotope quality. The results of aerial photograph interpretation and field survey then are digitalised and stored in a geographical information system (ARC/INFO). The statistical approach permits to subsequently raise the sample data to higher levels such as land classes or biotope types in Germany. Theoretical work on the concepts of this level is described in detail in Back et al. (1996).
- At the second level, these results are supplemented by an analysis of the species (plants and some groups of species of animals) existing in randomly selected smaller subsample units within the sample areas of the first level. The concept for this second level is described in detail in Statistisches Bundesamt and Bundesamt für Naturschutz (2000).

Only when landscape quality and biotope quality (level I) are linked with the stock of species in biotopes (level II) it will be possible to arrive at a satisfactory assessment of the ecosystem quality with regard to the physical structure, as is planned for the indicator system. The present article however is restricted to the further explanation of the landscape quality aspects, i.e. the analysis of how the landscape is composed by biotopes of different types (as part of level I). Methods and results concerning biotope quality or species analysis are not considered here.

The concepts were tested to a limited extent in a pilot study in summer of 1995 and 1996. Generally, any landscape—agricultural landscapes, urban areas, woodland and land cover types close to nature—has to be included in such a sample for the Federal Republic of Germany. For reasons of pragmatism, however, the pilot project focused on agricultural areas (type of land cover according to CORINE) in Brandenburg,

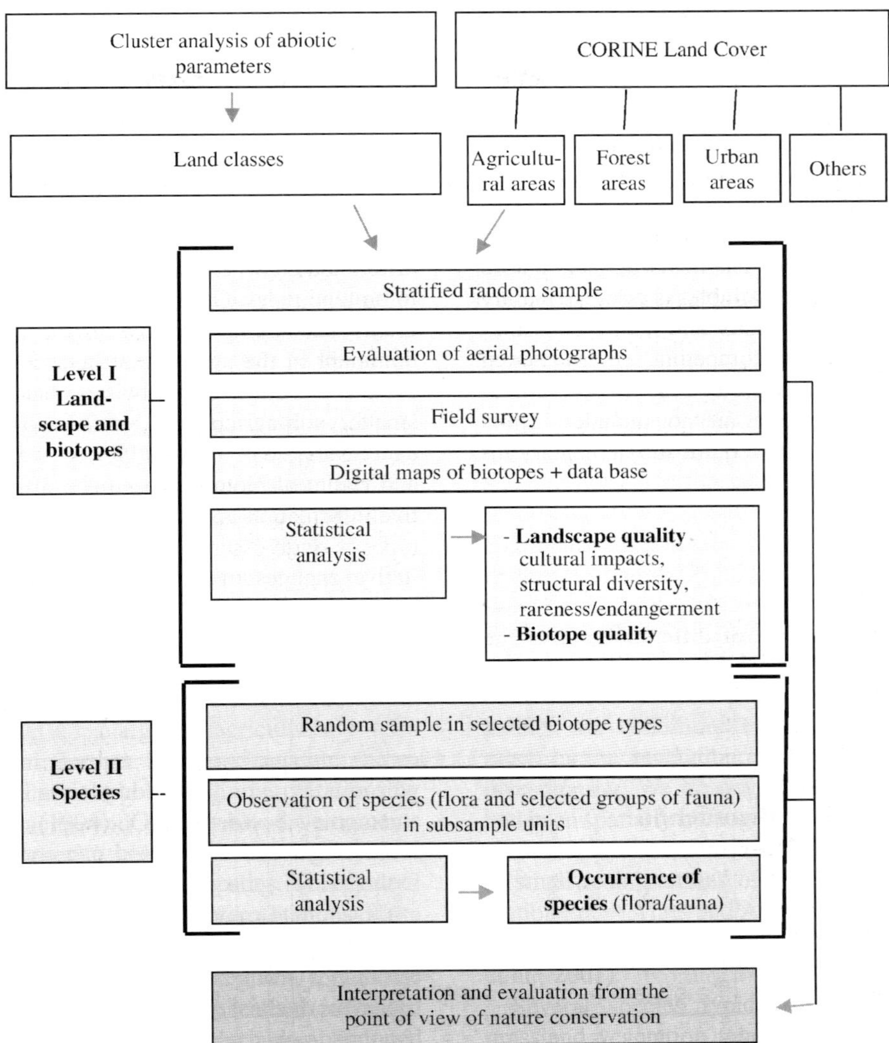

Fig. 1. Survey design and programme of EAS.

Berlin and Thuringia. Because of methodological and financial reasons only 7 of the total of 28 land classes were selected in the test area of the pilot study for the coverage of landscape and biotope quality. Sample size was limited to 70 sample units of 1 km^2 each.

This pilot study focused in particular on finding solutions to issues of methodology (e.g. sampling concept, operationalisation of indicators, sampling errors) and survey organisation. Especially for an entirely new survey tool such as EAS, a methodological pilot study is indispensable. For analysing the contents, the results of the pilot study are suitable only to a limited degree, which is due to two factors. First, the mainly methodological character of the pilot study and, second, the small sample size which, in turn, is due to limited funds. The consequences of these restrictions are the delimitation of the pilot study test area and the sampling random errors that occurred in the pilot study; because of the small sample size, these errors are considerably more serious than would be expected for the main survey. However, since the pilot study is a rather large survey if compared with other data available for the ecological field, the standards applied to the representation of results should be less exacting than usual in official statistics—all the more so, since there is a great national demand for data which at least give an impression of magnitudes.

The indicators mentioned in the following are tailored to the prevailing type of agricultural area. Nevertheless the indicators partly seem to be applicable for other non-urban areas, too. The landscape and nature quality based on structural variables is assessed mainly from the aspect of biodiversity and nature conservation. Other, and possibly competing, environmental aims such as groundwater protection, climate protection, recreation and the like, are not included in this project. This would in part require supplementary indicators.

3. Results

The much higher degree of differentiation of the EAS compared with the coverage of land cover (the lower limit of coverage for biotopes is 400 m^2, which is just 0.16% of the lower limit of coverage for the evaluation of land cover as part of CLC) becomes obvious simply by comparing the biotope types covered by the EAS with the land cover units. Fig. 2 shows how a landscape unit that looks homogeneous on a small scale (the CORINE agricultural area in Brandenburg, Berlin and Thuringia) becomes more differentiated when changing from land cover types to biotope types. Of course, arable land and managed grass—accounting for 85% of the area—are highly dominant in the agricultural land cover types. However, it becomes also obvious that there do occur smaller non-agricultural biotope types to a significant extent, e.g. forest (3%) as well as settlement and technical biotope types (also 3%). This quantitative information about the shares of different biotope types is quite essential for the interpretation of qualitative changes over time (cf. Krack-Roberg et al., 1995).

The data on landscape quality will be evaluated with regard to three items:

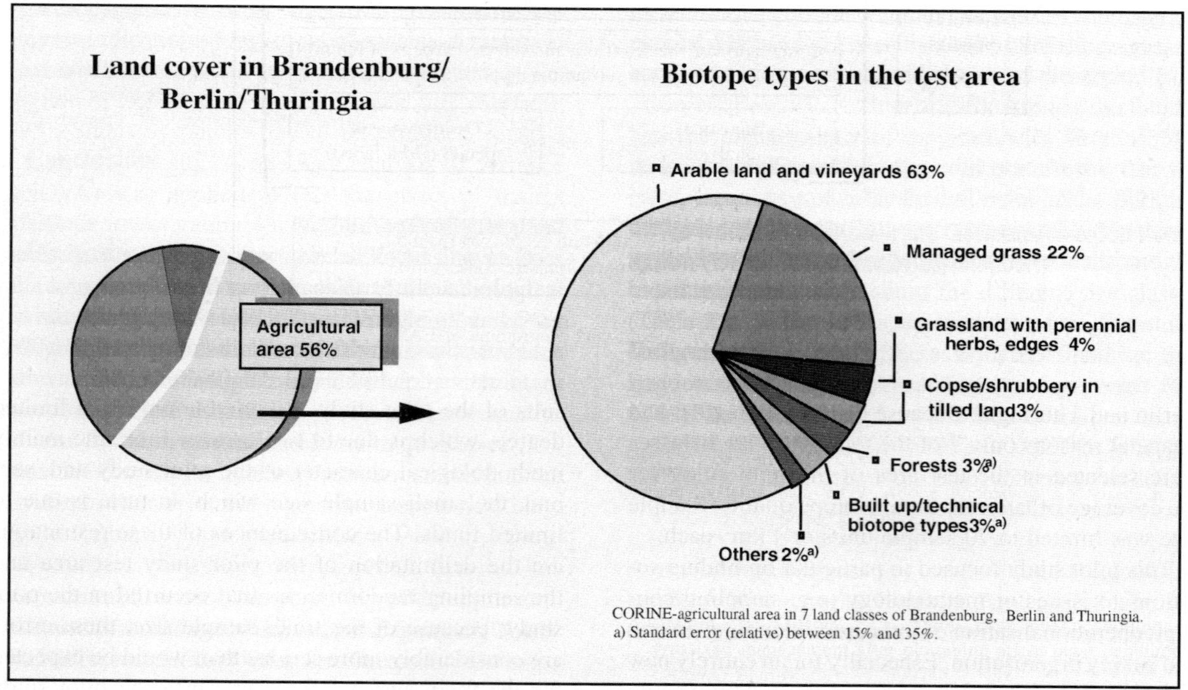

Fig. 2. Stock of biotope types 1995 in the test area.

Table 1
Indicators of landscape quality in agricultural pattern

Item	Indicator
Intensity of use	Artificiality Soil sealing Risk of erosion of arable land Fragmentation
Structural diversity	Biotope diversity Plot size of agricultural areas Elements with linear features Small biotopes Spatial distribution of structural elements Spatial distribution of endangered biotopes
Rareness/endangerment	Occurrence of endangered biotopes

- cultural impact/intensity of use;
- structural diversity;
- rareness/endangerment.

The indicators to be covered in the sphere of cultural impact/intensity of use are artificiality/naturalness, soil sealing, risk of erosion, and fragmentation (Table 1). The degree of artificiality/naturalness of the sample area is shown by the shares of the area of biotope types with low artificiality levels.

Generally, the biotope types may be allocated to different levels of naturalness on the type level. This work was carried out by prior technical discussion. The three groups distinguished here are natural biotope types (and biotope types close to nature); semi-natural biotope types; biotope types characterised by cultural impact and artificial biotope types (to simplify matters, the latter are referred to as "biotope types determined by culture" in the following). Table 2 shows the corresponding area shares, broken down by land classes. The extent to which the natural environment is already characterised by human impact in the agricultural landscape is shown by the large share of biotope types determined by culture (68%) and the small share of natural biotope types (and biotope types close to nature) (4%) in the test area. An examination of the individual land classes reveals considerable variations depending on the natural situation. Marshland and river plains, plain land near to groundwater and palaeozoic low mountain ranges are covered by semi-natural biotope types up to a degree of about a third. For the remaining four land classes the shares of biotope types determined by culture rise to about 80% or even more. Among the agricultural areas, plain land near to groundwater has the largest share (5%) of natural biotope types (and biotope types close to nature). These are essentially reeds. The fertile, and thus intensively used plain land distant from groundwater and the basin landscapes on loess show the lowest figures (3 and 2%) for natural biotope types (and biotope types close to nature).

Table 2
Artificiality of biotope types by land classes 1995 in the test area[a]

Land class	Degree of artificiality[b]			
	1 natural biotope types and biotope types close to nature (% area)	2 semi-natural biotope types (% area)	3/3–4/4 biotope types determined by culture (% area)	Without allocation (% area)
Total	4[c]	25[c]	68	3[c]
Bogs and river marshes	4[d]	31[d]	61[c]	5[d]
Lowland, near to groundwater	5[c]	35[c]	57[c]	3[c]
Lowland, far from groundwater	3[c]	14[d]	82	2[d]
Basin landscapes on loess	2[c]	10[d]	86	2[d]
Hilly landscapes on triassic sediments	4[d]	8[c]	83	4[d]
Low mountain range on schluff-, sand- and claystone	4[d]	14[c]	79	3[d]
Low mountain range, palaeozoic	4[c]	34[c]	52	10[d]

[a] CORINE-agricultural area in seven land classes of Brandenburg, Berlin and Thuringia.
[b] Allocation by Naturnah/BfN 1995.
[c] Standard error (relative) between 15 and 35%.
[d] Standard error (relative) 35% or more.

To assess the overall degree of soil sealing of the sample area, the degree of soil sealing of built-upon biotope types is roughly estimated as part of the field survey. As regards the risk of erosion, it will only be possible to develop a rough indicator which will mainly be derived from the slope inclination, possibly including precipitation. The presentation of fragmentation is to focus on the length of the paved road network per hectare in the areas outside the settlement area.

In the field of structural diversity, attempts are currently made to cover the following indicators: biotope diversity, biotope size of arable land, the occurrence of elements with linear features and of small biotopes, the spatial distribution of such elements and of biotopes of the red data book. To show biotope diversity, the number of biotopes and of biotope types within the sample area is to be determined.

Biotope types of the settlement sphere and technical biotope types such as dumping grounds are not taken into account here. Fig. 3 shows the average number of different biotope types by land classes. On average, there are 32 biotope types/km^2. The fertile basin landscapes on loess (23 biotope types/km^2) are the most monotonous ones, while the greatest number of different biotope types per area unit can be found in the palaeozoic low mountain ranges and in plain land near to groundwater (35). This result corresponds quite clearly to the distribution of biotope types with different levels of naturalness (Table 2). The results thus show quite clearly how aspects of cultural characterisation of the landscape are interlinked with aspects of structural impoverishment.

For arable and viticultural land, the biotope size as an important indicator is estimated. The occurrence of elements with linear features is determined by calculating the overall length of such elements within the sample area. For small biotopes such as ponds, isolated trees, rocks and the like, the frequency of occurrence in the sample area is determined on the basis of the results of the field survey.

As regards the sphere of rareness/endangerment, it will be attempted to cover the biotopes of the red data book and determine the shares of their areas.

As mentioned above, results on biotope quality, flora and fauna are not included in this paper but already published in another context (cf. Hoffmann-Kroll et al., 1997, 1998).

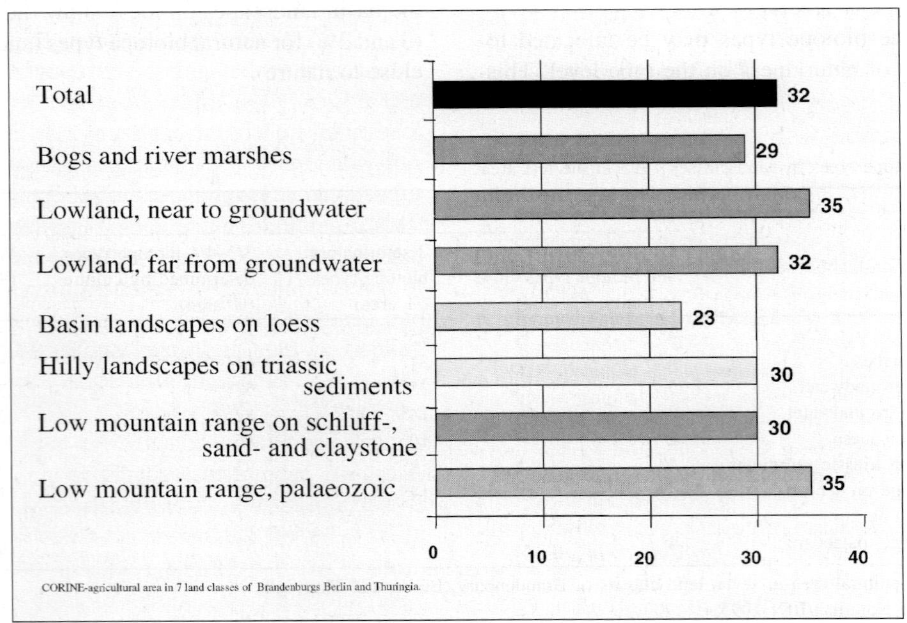

Fig. 3. Diversity of biotope types by land classes 1995 in the test area—mean number of biotope types per km^2.

4. Discussion

As a result of the pilot survey, there will be refinements or slight changes in indicator construction, especially in computing the indicators of spatial distribution of structural elements. In general however the indicators proved a success.

The results shown as examples will certainly not provide enough differentiation for many interesting issues. Although it would be possible also as part of the pilot study to obtain results for other survey characteristics, providing tables with a more detailed breakdown, the random error caused by the small sample size would in some cases reach a dimension that would involve a considerable risk of overinterpreting the results. A routine EAS main survey for the representation of trends would have to include forests and land cover types close to nature rather than being limited to the CORINE agricultural area; due to a larger sample size and the inclusion of the other land cover types, a considerable reduction of errors would be expected.[1] Thus a main survey for Germany would provide a more comprehensive and more differentiated programme which nevertheless could be evaluated with a justifiable degree of errors. Also, it would be possible to include in that programme additional survey contents (e.g. with regard to the fauna) and more comprehensive analysis (e.g. regarding the relation of different survey contents such as the fauna of birds in relation to the structural diversity of the landscape). Despite its focus on methodological aspects, the pilot study successfully demonstrated by means of examples the possible information value also in terms of contents and the potential of the EAS to provide highly important information on ecological structural situations which so far has not been available in Germany.

5. Conclusion

The EAS concept will be a reliable basis for information and a useful tool for decision-making in the field of nature conservation, which so far has not been available at the level of the Federal Republic:

- The EAS results permit supplying information on the state and changes of nature and the landscape in current land use areas.
- The EAS will provide for the first time data on the frequency and quality of ecosystem types which will be reliable in terms of statistics for the whole of Germany. This means considerable support for political action in the field of nature conservation because "large-area nature conservation", i.e. sustainable use also outside protected areas, requires information on the entire territory, i.e. in "normal landscape", too.
- By means of EAS, area-related changes of the biological diversity in Germany may be represented and evaluated for all-German nature conservation policies.
- EAS results will permit to measure the success or failure of nature protection policies at the federal level because the extent of changes can be observed.
- EAS results will provide a basis for comparison with regard to purposeful area protection measures as part of nature conservation, e.g. the efficiency of large-scale nature conservation projects of the federal government can be represented in comparison with the state of the overall landscape.

[1] First, there are still considerable random errors at present with regard to information on biotope types which occur mainly outside the CORINE agricultural area (e.g. forest). Second, in the case of areas continuously monitored, the errors of estimation occurring during trend estimation are clearly smaller than for stocktaking at a specific point in time.

Acknowledgements

EAS concepts are part of a research project which was funded by the German Ministry of Education and Research. Many persons contributed to the design of the concept, in particular Dr. U. Bohn, R. Dröschmeister, U. Riecken, Dr. G. Wolf (Federal Agency of Nature Conservation), W. Bihler, Dr. H. Heidrich-Riske, W. Radermacher (Federal Statistical Office) and the members of the private companies which were involved in the project by sub-contracts.

References

Back, H.-E., Rohner, M.-S., Seidling, W., Willecke, S., 1996. Konzepte zur Erfassung und Bewertung von Landschaft und Natur im Rahmen der "Ökologischen Flächenstichprobe". Beiträge zur Umweltökonomischen Gesamtrechnung, Heft 6, Statistisches Bundesamt, Wiesbaden.

Barr, C.J., Bunce, R.G.H., Clarke, R.T., Fuller, R.M., Furse, M.T., Gillespie, M.K., Groom, G.B., Hallam, C.J., Hornung, M., Howard, D.C., Ness, M.J., 1993. Countryside Survey 1990—Main Report. Institute of Terrestrial Ecology and Institute of Freshwater Ecology.

Deggau, M., 1995. Statistisches Informationssystem zur Bodennutzung. In: Wirtschaft und Statistik 12/1995, 895 ff.

Haines-Young, R.H., Barr, C.J., Black, H.I.J., Briggs, D.J., Bunce, R.G.H., Clarke, R.T., Cooper, A., Dawson, F.H., Firbank, L.G., Fuller, R.M., Furse, M.T., Gillespie, M.K., Hill, R., Hornung, M., Howard, D.C., McCann, T., Morecroft, M.D., Petit, S., Stier, A.R.J., Smart, S.M., Smith, G.M., Stott, A.P., Stuart, R.C., Watkins, J.W., 2000. Countryside Survey 2000—Accounting for Nature Assessing Habitats in the UK Countryside. Department of the Environment, Transport and the Regions, London.

Hoffmann-Kroll, R., Schäfer, D., Seibel, S., 1997. Land use and biodiversity indicators from ecological area sampling—results of a pilot study in Germany. Statist. J. United Nations ECE 14, 379–395.

Hoffmann-Kroll, R., Schäfer, D., Seibel, S., 1998. Biodiversität und Statistik—Ergebnisse des Pilotprojekts zur Ökologischen Flächenstichprobe. In: Wirtschaft und Statistik 1/1998, pp. 60–75.

Krack-Roberg, E., Riege-Wcislo, W., Wirthmann, A., 1995. Konzept einer Gesamtrechnung für Bodennutzung und Bodenbedeckung. Beiträge zur Umweltökonomischen Gesamtrechnung Heft 4. Statistisches Bundesamt, Wiesbaden.

Radermacher, W., Stahmer, C., 1994. Vom Umwelt-Satellitensystem zur Umweltökonomischen Gesamtrechnung in Deutschland. In: Zeitschrift für angewandte Umweltforschung Jg. 7, Heft 4, pp. 531–541 (Part 1), Jg. 8, Heft 1, pp. 99–109 (Part 2).

Radermacher, W., Zieschank, R., Hoffmann-Kroll, R., v. Nouhuys, J., Schäfer, D., Seibel, S., 1998. Entwicklung eines Indikatorensystems für den Zustand der Umwelt in der Bundesrepublik Deutschland mit Praxistest für ausgewählte Indikatoren und Bezugsräume. Beiträge zu den Umweltökonomischen Gesamtrechnungen Bd. 5, Wiesbaden.

Statistisches Bundesamt, Bundesamt für Naturschutz, 2000. Konzepte und Methoden zur Ökologischen Flächenstichprobe—Ebene II. Monitoring von Pflanzen und Tieren. In: Bearbeitet von Schmidt, H., Herrmann, T., Foeckler, F., Deichner, O., Klein, M., Nuss, I., Schellhorn, F., Türk, W. (Eds.), Angewandte Landschaftsökologie Heft 33.

ELSEVIER

Available online at www.sciencedirect.com

Agriculture, Ecosystems and Environment 98 (2003) 371–386

Agriculture Ecosystems & Environment

www.elsevier.com/locate/agee

Operationalisation of a landscape-oriented indicator

Elisabeth Osinski*

Chair of Agricultural Economics, Center of Life and Food Sciences, Technical University of Munich, Agricultural Economics, Alte Akademie 14, D-85350 Freising, Germany

Abstract

Several indicators have been suggested in the field of landscape protection. However, is there a need to operationalise such indicators. According to the [Environmental Indicators for Agriculture, vol. 3, Methods and Results, OECD, Paris, 2001, p. 368], the structural indicator 'man-made objects' is illustrated by a method that includes the analysis of parameters for the operationalisation of the indicator. The method also includes a geographical information system (GIS)-aided spatial integration of landscape element distribution and surrounding land use types, for building the so-called landscape complexes (LCs). An LC is used to characterise natural units (a type of landscape-oriented classification of the entire test area), with regard to their landscape type qualities.

Carrying out this so-called operationalisation can assist in a more specific and improved region-oriented structuring (regionalisation) of agri-environmental programmes, which are currently negligibly space- and target-oriented.

The investigation area was the entire area of the German federal state Baden-Wuerttemberg, as well as single natural units comprising the state. The operationalisation method was firstly aimed at setting priorities for landscape protection on the spatial level of the state, which has the political and administrative power to enforce them. Secondly the method assisted in the identification of special packets of target-oriented measures, which can be used on the level of natural units.

Keywords: Baden-Wuerttemberg; Landscape protection; Landscape complex; Agri-environmental programme; GIS tool

1. Introduction

The protection of habitats for wild species is recorded in national law, programmes and rules that are targeted at protecting existing animal stocks and/or supporting the spread of endangered species. A common strategy is to designate nature reserves (e.g. national parks, biosphere reserves) to protect natural and semi-natural habitats. In agricultural landscapes where decreased numbers of wild species are demonstrated, the situation is different since these cannot be declared be nature reserves. Moreover, the number of rare wild species is decreasing and nature reserves are able to sustain only 35–40% of the wild species existing in Germany (see SRU, 1985, p. 296). This situation has remained unchanged as it is actually stated by the German National Committee of Environmental Experts (SRU, 2000). Instead of protecting single species in agricultural landscapes, a systematic approach focussing on landscapes is needed.

Protection of landscapes plays a role e.g. in the environmental sensitive areas (ESA) concept (MAFF, 1993), but also in theory of environmental sciences (Dolman et al., 2001; MacFarlane, 1998; O' Riordan et al., 2000). Hence the manner in which the system is regarded must be changed from the single element to the landscape scale. Particularly in agricultural landscapes, the interaction between unused landscape

* Tel.: +49-8161-71-4461; fax: +49-8161-71-4426.
E-mail address: osinski@wzw.tum.de (E. Osinski).

0167-8809/$ – see front matter © 2003 Elsevier Science B.V. All rights reserved.
doi:10.1016/S0167-8809(03)00097-5

elements and surrounding farmland should be considered as a quality influencing factor.

From an administrative and political viewpoint, the protection of landscapes is steered by the state. In a first step, valuable landscapes must be defined prior to the allocation of finances and concepts for maintaining these landscapes must be subsequently developed. The protection of landscapes first requires methods for their registration and evaluation; for biota this is possible evaluating landscape structures (see Forman and Godron, 1986). In order to utilise finances as efficiently as possible, protection strategies should be developed subsequent to landscape evaluation. Efficiency can be defined in this context as using the right measures at the adequate places. Therefore, the present aim for landscape protection involves defining measures which are regional-oriented (regionalisation).

Currently, the protection of agricultural landscapes is mainly supported by agri-environmental programmes (sponsored partly by the EU and partly by the member states). These programmes are under criticism due to their missing space-oriented ecological targets. According to the EU, these programmes should be more target-oriented, the offered measures more spatially focussed and the programmes better related to the landscapes (EU-COM, 1998).

In parallel to developing programmes, ways are sought for their testing, such that the eventual successes should be measurable. Agri-environmental indicators should be practical in this case, and several sets of indicators have been already suggested by different organisations (EEA, 2001; OECD, 2001) and states (MAFF, 2000; UBA, 1999).

In reference to landscapes, defined indicators should fulfil several tasks: (1) implicate used and unused parts of the landscape in addition to requirements relevant to each kind indicator type (Policy relevance, analytical soundness, measurability, see OECD, 1997; pp. 19–21); (2) reflect existing networks in landscapes, rather than single elements; and (3) provide recommendations on measures to protect the landscapes.

According to the example of the OECD indicator, man-made objects, a method for operationalisation of this specific indicator will be demonstrated. The indicator will be subsequently used for defining landscape protection targets.

2. Area description

The study was carried out in Baden-Wuerttemberg, one of the German federal states. Baden-Wuerttemberg is characterised by a large variance in both site and climate conditions of its various regions, which is reflected by different agricultural land use types. These have in turn resulted in different main landscape types related to their productivity.

The climatic-favoured regions are situated mainly in the western part in the Rhine valley, in the north-west (Kraichgau and Neckar valley) and in the southern part bordering with the Lake Constance. Regions having very poor land use compatibility are the Black Forest, areas of the Swabian Jura, the so-called Albvorland and the Keuper region. In regions partly covered by loess (Kraichgau, Neckar valley, Keuper region), agricultural land use suitability varies dependent upon the height of the loess cover and climatic conditions.

A visual comparison of a land use compatibility map (see MLR, 1990) with information on real existing land use type in Baden-Wuerttemberg results in a high grade of spatial conformity of both factors. Regions highly suited for agricultural purposes are mainly used as arable land or speciality crops; regions poorly suited due to climatic or edaphic conditions are characterised by grassland mixed with arable land, predominantly grassland and/or forest. In most cases, declining land use compatibility is accompanied by a lower land use intensity. In these landscapes, the number of unused landscape structures are often higher than in intensively used landscapes.

3. Methods

The spatial unit used in the present analysis will be first described, followed by an explanation of the geographical information system (GIS)-based method employed for the recognition of patterns in landscape structures. This method leads to operationalisation of the selected indicator for each of the landscapes in Baden-Wuerttemberg. Lastly, the regionalisation procedure will be described, whereby a brief description of the theoretical background will be first elucidated followed by a demonstration of the practical approach by the example of landscape protection in Baden-Wuerttemberg.

3.1. Spatial reference unit

In order to subdivide the entire Baden-Wuerttemberg area for both recognition of spatial patterns as well as for the regionalisation process, a suitable spatial reference is required. Such a spatial reference can either be newly constructed or an already existing spatial classification can be used. The present analysis is aimed at differentiating the complete area in respect to its land use characteristics and landscape elements closely associated with the landscape's history. In the case of administrative-delimited areas, i.e. Nomenclature of Territorial Units for Statistics (NUTS)—European regions comprised mostly of landscape units are divided (Eiden et al., 2000; EU-COM, 2000). The landscape classification system 'classification of natural units' of Meynen and Schmithuesen (1953–1962) has been considered better suited to reflect landscape characteristics in Germany. This classification is based on methodology developed in the second half of the last century aimed at mapping homogenous landscapes and differentiating among them on the basis of geological–edaphical and vegetation specific characteristics (see Schmithuesen, 1953). This approach, based on geographers view is still currently used for planning purposes (Heinl and Heck, 2000) since it represents more adequately better those areas with similar land use types and land use suitability than administrative units. This classification is available for the whole of Germany. Thus a potential basis for an analysis in other federal states is possible, comparable to that presented here. Fig. 1 presents the fourth level of the hierarchically organised classification of natural units (defined here as 'natural unit') and main landscapes as larger units; the latter are thus named to provide orientation on Baden-Wuerttemberg.

3.2. Indicator 'man-made objects'

The regionalisation process is targeted at spatial specific aims in landscape protection by utilising the indicator 'man-made objects'. Although there is no existing general method for deriving this indicator (OECD, 2001), the method is dependent upon the regional and/or state specific existence of peak, linear or flat landscape elements. By utilising an adequate data base, the indicator is then operationalised, and includes carrying out a target-oriented analysis of parameters. A parameter can be developed into an indicator by defining its respective threshold, target and/or baseline values (Riley, 2001).

In this study, these landscapes worth preserving and those with development potential as habitat for wild species were analysed, and comparisons of space relationships were carried out with the assistance of a data base containing the distribution of landscape elements and land use data.

3.3. Data base

The first data base used is the result of a mapping process of valuable habitats for wild species (Hoell and Breunig, 1995), which was commissioned in the 1980s by the landscape protection agency of Baden-Wuerttemberg. The selective so-called 'biotope mapping' was carried out by various individuals using the same mapping key and methodology, on a field scale of 1:25,000 and digitising the data base thereafter. Due to individual differences in using the mapping key, the resulting data base lacks homogeneity. A newer mapping approach conducted in the nineties, mapped only protected biotopes as defined by the German nature protection law, but did appear not to be sufficient for finding valuable landscapes. Thus, the results of the previous mapping approach are used in the current analysis.

For characterisation of agricultural landscapes, landscape element groups are selected whose development and maintenance are strongly related to agricultural practices. This selection includes hedgerow/copses, oligotrophic grassland, wet/medium grassland and orchards (Table 1), all of which contribute in the characterisation process. Additionally, classified satellite LANDSAT-5-TM images of land use types have been used. In order to adapt the data resolution of both landscape element and satellite image data, the 30 m grid of the satellite image was re-classified to a 100 m grid and the number of land use classes reduced from 16 to 6 (see Table 1).

3.4. Pattern analysis by GIS and cluster methodology

3.4.1. GIS-aided analysis

A GIS is an computer-aided system for digital recording, handling, storing, re-organising, modelling,

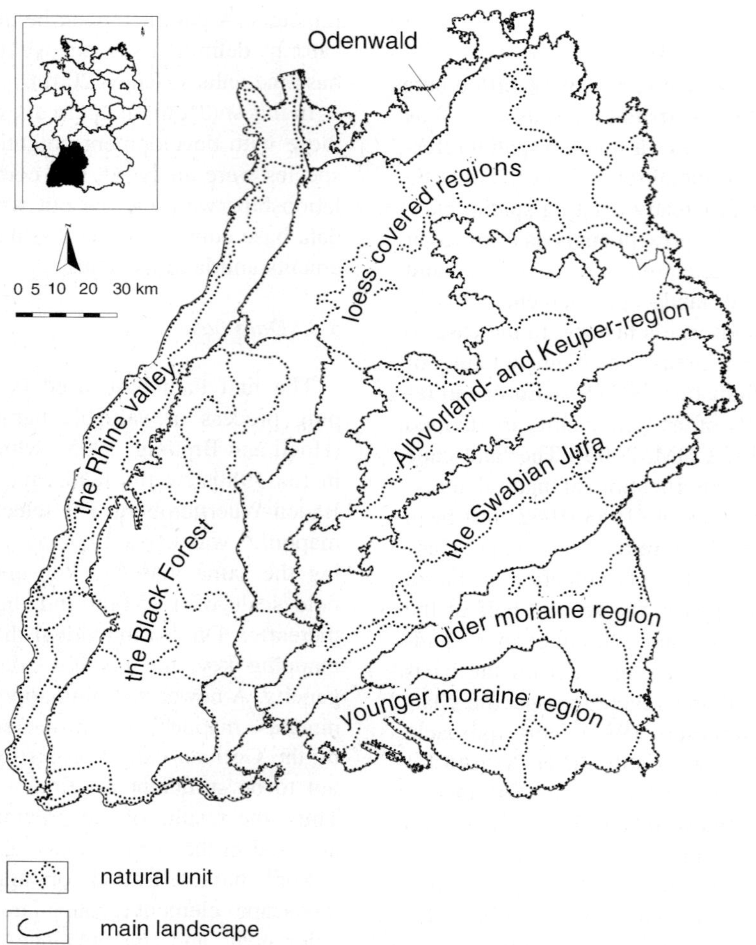

Fig. 1. Main landscapes and natural units in Baden-Wuerttemberg (own figure).

analysing and visualising geographical data (after Bill and Fritsch, 1994). As a prerequisite for GIS usage, a precise notion on the target under analysis is required. A scheme of the calculation can be then formalised. Fig. 2 is a schematic representation of the process for defining parameters from the data base containing landscape elements and land use, for deriving distribution patterns of landscape elements.

After selecting those landscape element types that are strongly related to agricultural practices, a grouping into functional sets called 'landscape element class' was done. This was followed by a spatial analysis, in which areas belonging to one and the same class were associated to a landscape element complex (LEC) if they were situated within a defined distance away from each other. The distance depends on the habitat function of these landscape elements. The target species for which these elements are regarded as 'valuable' is related to the distance between landscape elements that must be overcome. Blab (1993, p. 185) adduces a distance of 400–800 m as surmountable for hedgerow organisms such as birds, small mammals and flying insects. Knauer (1988, p. 62) describes distances from 100 to 400 m as a range not to be exceeded within an ecological network in agricultural landscapes. The selected distance of 300 m can be regarded as a mean value, which is also oriented at data resolution. If a more targeted analysis of the

Table 1
Data base used for the analysis of valuable landscapes in Baden-Wuerttemberg

Data base	Resolution/reference	Data management
Biotopkartierung Baden-Wuerttemberg (biotope mapping)	1:25,000/Landesanstalt fuer Umweltschutz (Hoell and Breunig, 1995)	Selection of four landscape element groups: *Hedgerows/copses*; copses, hedgerows, mesophytic bushes, humid *Oligotrophic grassland*; calciphile grassland, mat-grass grassland *Wet/medium grassland*; fertile meadow, pasture, litter meadow, wet meadow *Orchards*
Satellite image classification (LANDSAT-5-TM)	30 m pixel, 16 land use classes/Institut fuer Photogrammetrie und Fernerkundung, Karlsruhe, 1993[a]	Reclassification from 16 to 6 classes including grassland, arable land, forest, settlement, fruit production, others Remapping from 30 to 100 m pixel

[a] Institut für Photogrammetrie und Fernerkundung (IPF) Karlsruhe, 1993. Erstellung einer Landnutzungskarte des Landes Baden-Wuerttemberg, Unveroeff. Abschlussbericht im Auftrag des MLR Baden-Wuerttemberg.

differentiated habitat function of landscape elements regarding selected species occurred, a need for additional information on habitat quality would arise than what is available on the spatial level of the entire area of a federal state (i.e. Baden-Wuerttemberg). The analysis is constructed such that new distance values may be introduced without causing disruptions should corresponding new values were available.

Landscape elements situated within the defined distance of 300 m are merged with the area in between them and form LECs. Combining LEC with information on type of land use results in landscape complexes (LC). After conducting a spatial overlay with the

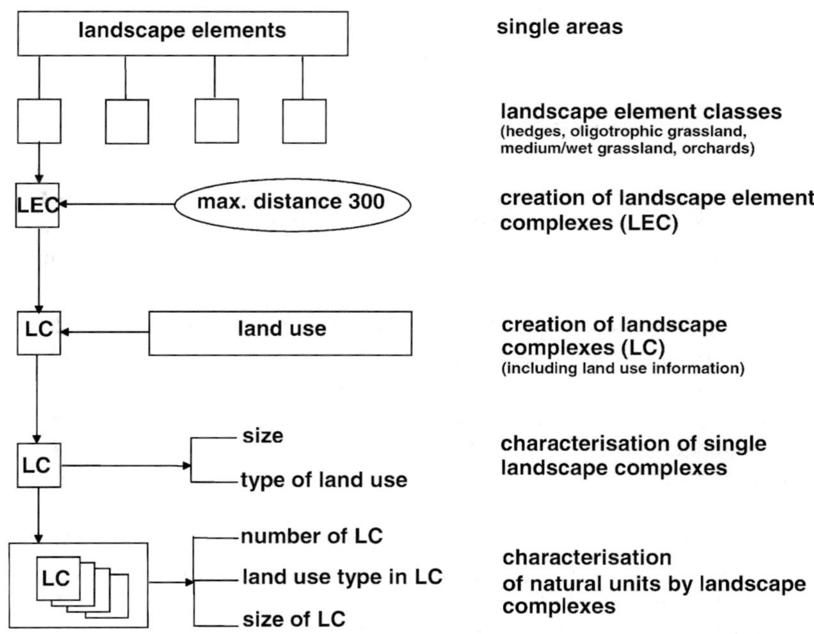

Fig. 2. Parametrisation process of the indicator 'man-made objects' (own figure).

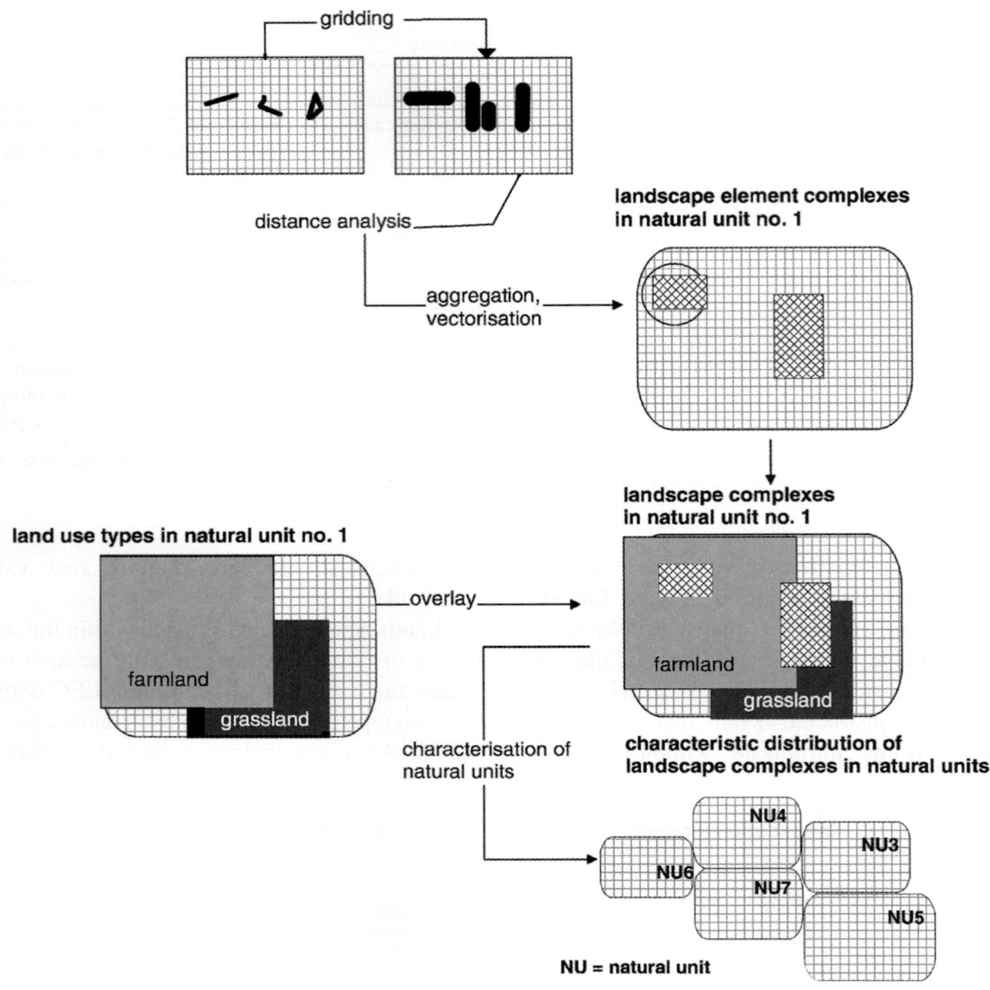

Fig. 3. Pattern analysis for characterisation of different landscape element sets in natural units (after Osinski, 2000).

borders of the natural units (see above), a data base was created, which included data sets containing the number and size of LCs per natural unit for each type of landscape element.

The technical execution of the analysis follows logical rules for data combination (see Fig. 3). Due to the heterogeneity of the data set, a homogenisation process of the entire set per landscape element type was necessary. A GIS-based distance transformation (see Bill, 1996) was therefore performed. The transformation was targeted at separating single landscape elements from those grouped together. The basis of this analysis is the transformation of the landscape elements stored as polygons into the grid format with the size of 1 ha. The grids are tested in the next step of the analysis according to the rule if there is another biotope area in the distance of up to 300 m.

Grid fields (grids) having more than a pre-defined number of neighbours (the limiting value is the third quartile of the maximum number of neighbours) containing landscape element areas are merged together into LECs. For characterisation of the complexes regarding land use, information on the data set of LECs is combined with the land use information obtained from satellite imaging. Thus it is possible to estimate the degree of potential influence on the landscape elements by type of land use but not by the intensity of land use. This step results in LCs.

For characterisation of natural units, the distribution of LCs per unit becomes more relevant than the single complexes. Specification of the indicator 'man-made objects' requires calculation of the following parameters, allowing the comparison of natural units according to their set of landscape elements:

- Statistical parameters of LCs per natural unit (medium, standard deviation).
- Relative total sum of size area of LCs per natural unit related to the best equipped natural unit; the natural unit having the largest area is valued at 100.

3.4.2. *Finding similar natural units by clustering*

The statistical parameters mentioned above are used as factors in the next step of the analysis for finding natural units with similar configuration of landscape elements. The analysis is conducted according to the Ward–Cluster method. This type of clustering yields groups with the highest intra-group similarity of elements and the highest inter-group dissimilarity and results in similar sized groups (Bahrenberg et al., 1992) and aids in the sorting of natural units. In turn, this sorting is interpreted as acting in natural unit groups formation. Each of these groups contains LCs of similar size and number, and forms the basis for the regionalisation process.

3.5. *The regionalisation approach*

The term regionalisation is used here in a manner similar as by geographers. This process primarily involves the subdividing of an area into smaller sections with further purpose-oriented classification of these (Grigg, 1967). This classification is aimed at the generalisation and the simplification of problem handling. The steps of the regionalisation process can be recognised in Fig. 4, showing that the planning area is the entire region of Baden-Wuerttemberg. This area is further subdivided by using the criterion 'equipment of LCs' in a defined spatial constellation. The process of generalisation is understood here as the transfer of information from the single landscape element to the natural unit within the larger spatial reference area, and allows the integration of analysis results of the different landscape element types on the same spatial reference.

In the following step, natural units having identical LC configurations can be combined for constructing target areas (Gebietskulissen) as a spatial basis for a common problem management. In addition to target development, these areas are useful as reference for

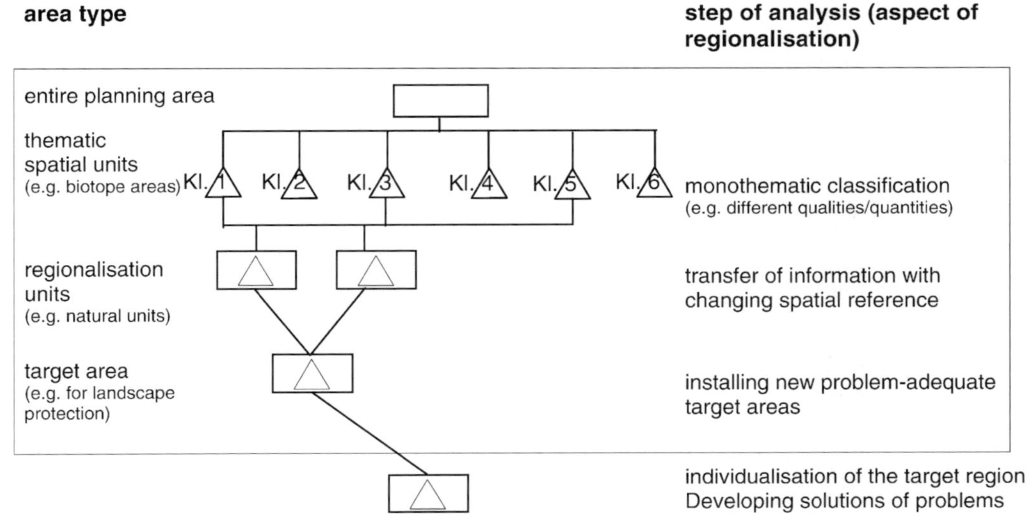

Fig. 4. Regionalisation approach for generalisation of landscape protection targets in Baden-Wuerttemberg (own figure).

defining packets of measures for reaching the defined goals.

4. Results

4.1. Pattern of LCs

For each group of landscape elements, the number and size of LCs present in each of the natural units were estimated. This analysis yielded a data set for each natural unit and each landscape element type. As example the hedgerow/copses are shown in Table 2 with a subset of the LCs with their number, the minimum and maximum size per natural unit. This data set was obtained by the GIS analysis and was processed by a spreadsheet program.

Each of the 58 natural units contain hedgerow/copses complexes. Their number and size varied markedly across the different natural units. A large variation in the number of LCs was observed. Between one and over 200 LCs were found in one natural unit, ranging in size between 0.01 to nearly 1900 ha. This variability in size and number demonstrates the necessity of characterising each of the natural units according to their different particular properties.

In a first step, the statistical distribution of LCs was characterised by the mean and standard deviation which are both needed as factors for the cluster analysis (see Table 3). In addition, a relative weight was introduced into the analysis by calculating the area sum of all LCs per type in each natural unit. The highest area sum observed was set at value 100 and all others were calculated relative to this value (see column 'relsize' in Table 3).

The standard deviation, mean value and sum of relative area size are the basic factors for the cluster

Table 2
Number and size (minimum–maximum) of hedgerow/copses complexes in the natural units in Baden-Wuerttemberg (extract of the whole data set)

Number of natural unit	Number of LC	Minimum size of LC (ha)	Maximum size of LC (ha)
160	1	3.00	3.00
151	3	7.00	13.37
203	4	1.00	11.91
212	4	1.00	20.00
161	6	4.00	90.00
152	7	6.89	25.00
154	8	5.85	44.00
106	9	1.00	42.00
⋮	⋮	⋮	⋮
1021	93	0.05	256.00
32	94	0.78	156.00
1081	105	0.01	224.00
128	112	0.86	1357.67
40	128	0.15	360.00
961	140	0.18	170.00
122	140	1.00	1408.00
1291	152	0.85	1148.00
123	155	1.00	312.17
95	196	0.92	1891.01
94	220	0.15	1218.99
125	227	0.02	1628.00
Total	3107	0.01	1891.01

Table 3
Distribution weights of hedgerow/copses complexes per natural unit in Baden-Wuerttemberg (extract)

No. of natural unit	Standard deviation	Mean	Relsize
30	11.22	13.64	0.70
31	55.08	27.75	5.05
32	34.05	30.97	13.59
33	22.07	17.54	5.81
34	6.24	9.45	0.49
44	45.64	52.56	2.70
45	64.56	40.91	3.25
92	55.78	37.33	8.37
93	255.37	136.61	29.97
94	165.20	97.37	100.00
⋮	⋮	⋮	⋮
128	169.26	60.56	31.66
141	101.21	44.06	3.50
150	51.43	62.03	8.11
151	2.80	9.46	0.13
152	6.32	15.27	0.50
153	21.56	21.37	1.70
154	14.36	20.76	0.78
155	33.98	27.69	4.14
160	0.00	3.00	0.01
161	29.25	26.00	0.73
200	63.46	40.13	11.61
201	31.98	22.58	3.27
202	20.01	20.82	1.65
203	4.03	5.97	0.11
210	32.75	26.42	5.67
1441	40.85	22.36	1.77
2241	93.51	59.07	8.55
Total	53.77	34.08	12.04

Table 4
Clusters of natural units based on similar distribution of hedgerow/copses complexes in Baden-Wuerttemberg

Cluster no.	Relative area sum (relsize)	Mean (ha) size of complexes	Standard deviation of complexes	Number of natural units in the cluster	Characteristic of hedgerow complex distribution and size
1	0.1–0.7	3–15	2–11	9	Local, small
3	6–10	17–32	14–34	17	Local, small
7	3–9	44–59	93–106	3	Local, medium sized
2	2–21	22–62	34–64	20	Local-scattered, medium
9	40–48	45–56	115–127	2	scattered, medium, big single complexes
8	31–47	61–72	164–169	2	scattered, big, very big single complexes
4	30	137	255	1	scattered, very big single complexes
5, 6	98–100	97–108	165–238	2	Predominant, big, very big single complexes

analysis (after Ward). The example 'hedgerow/copses complexes' resulted in a solution with nine clusters that appeared to be the best differentiated solution. This result was interpreted by describing the natural unit groups within the clusters according to characteristics. The cluster number (left column in Table 4) is not related to the size of LCs but depends rather on the sequence of cluster processing.

Table 4 presents the range of factor values used in the cluster analysis, the number of natural units per cluster and a short characteristic description of the hedgerow complex distribution for each cluster. Grouping the clusters according to relative area size, a difference in the sum arises between local (<10%), scattered (10–50%) and predominant (>50%) appearance. The size of the single complexes per natural unit are described in relation to the mean value on state level; this is 34 ha in the case of hedgerow complexes. Complexes smaller than 34 ha were defined as 'small', 34–68 ha (double mean) as 'medium', and complexes exceeding 68 ha were characterised as 'large' or 'very large' in case the size is much beyond 68 ha (>200%). Cases in which the standard deviation exceeded the mean value by more than 300% the existence of single big complexes are mentioned here.

The clusters 1, 2, 3 and 7 show local to scattered complexes, whereas clusters 4, 5, 6, 8 and 9 show scattered to predominant appearance of hedgerow complexes, with markedly large single complexes within clusters 5, 6 and 8.

In the same manner, LECs of oligotrophic grassland, medium/wet grassland and orchards were analysed. Combining both pattern and cluster analyses leads to operationalisation of the landscape indicator 'man-made objects'. By adding GIS information a concrete spatial reference of the LC patterns is produced. Thus, localisation of target priorities (protection and/or development) and activities for fulfilling the targets are possible.

4.2. Regionalisation of target categories

The grouping of natural units into clusters, as well as the descriptive classification of the single natural units allow for visualisation of the distribution of each landscape element type relative to the natural unit with the highest ranked LC set in Baden-Wuerttemberg. In each of the four landscape element distributions, similar groups could be defined:

Group 1. Only local presence, small to average sized LCs.
Group 2. Scattered presence, small to average sized LCs.
Group 3. Scattered presence, big to very big sized single LCs.
Group 4. Predominant presence, small to very big sized LCs.

In Fig. 5 (panels A–D), the natural units and their typical landscape element distribution is defined according to the previously mentioned groups. Natural units in Group 1 are shaded light grey; Group 2 in medium grey; Group 3 in dark grey and Group 4 are shaded black.

For each of the landscape elements different focal distribution points exist, so that identical signatures do not represent identical sizes and numbers of complexes of the different types. For each landscape element type there is a maximum single complex size or area sum per natural unit (see Table 5). Thus, the

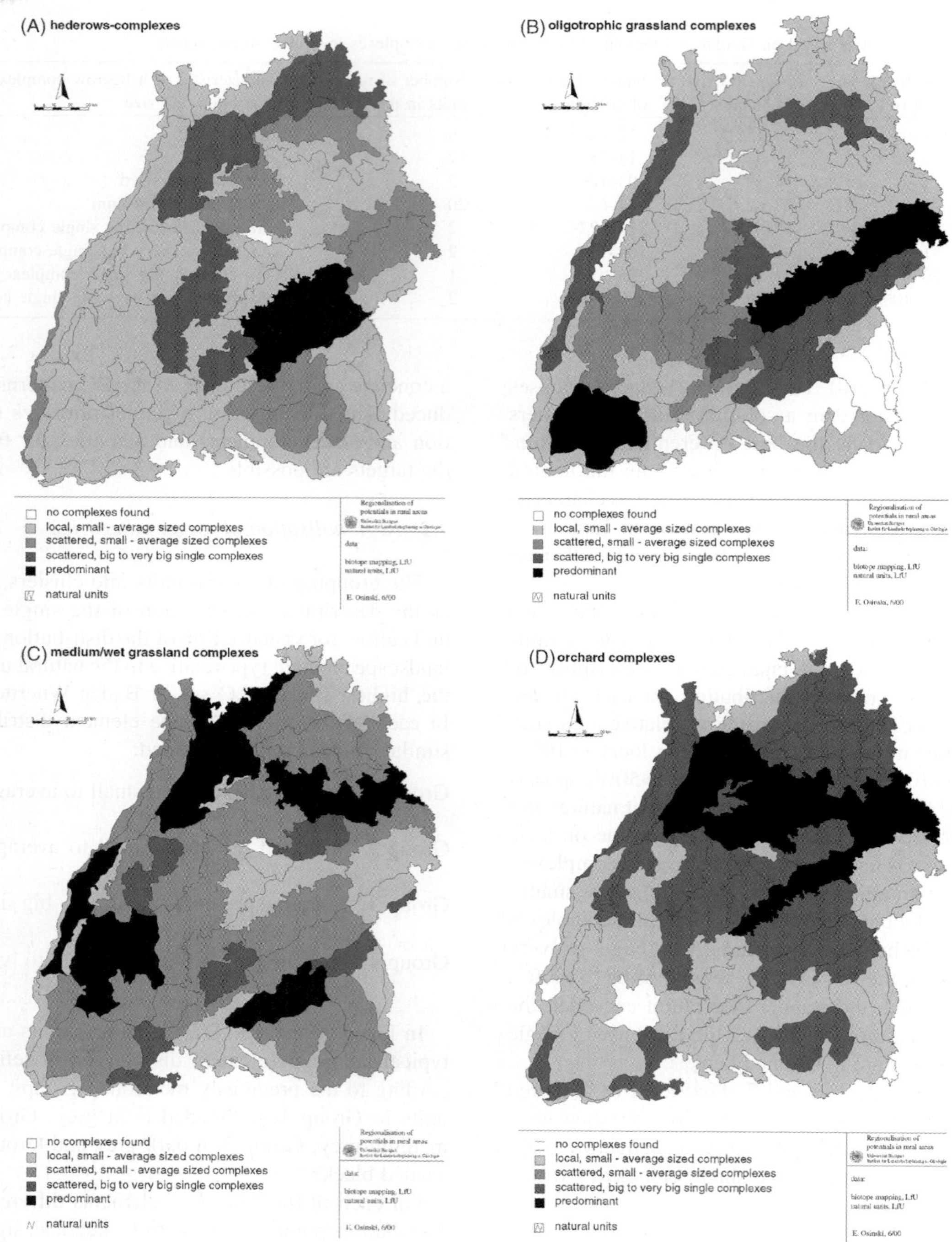

Fig. 5. Characterisation of natural units according to distribution of LECs; hedgerows, oligotrophic grassland, medium/wet grassland and orchards (own figure).

Table 5
Characteristics of LC types in Baden-Wuerttemberg

Landscape element type	Smallest complex (ha)	Biggest complex (ha)	Mean size of complexes (ha)	Number of natural units including this type of complex (max. 58)
Hedgerows	0.01	1891	34	56
Oligotrophic grassland	0.02	4383	48	49
Medium/wet grassland	0.02	3942	65	55
Orchards	0.02	2466	52	53

scaling is related to the state of Baden-Wuerttemberg, and specific for each landscape element.

Hence, the highest density of hedgerow complexes can be found in the Middle Swabian Jura (Fig. 5(A)). Oligotrophic grassland can be found in the central and eastern part of the Swabian Jura (limestone grassland) and in the Black Forest (acid grassland) (Fig. 5(B)). These landscape elements depend either on animal grazing activities and the by-products of a nutritional export (oligotrophic limestone grassland), or the product of altered soil use in forest to pasture transition with low nutritional contents (poor acid grassland) (Hoell and Breunig, 1995). The group of landscape elements 'medium/wet grassland' (Fig. 5(C)) summarise several grassland types under different management and typical species for wet grassland. The focal distribution point is situated in natural units having mainly only low percentage of grassland. The Odenwald region in the north, the Schwäbisch-Fränkische-Waldberge in the east and the Middle Black Forest in the west are predominantly forest landscapes. The Offenburger Rheinebene in the west and the Donau-Ablach-Platten belong to Riparian regions. However, the area mainly covered by grassland (Oberschwaben and Westallgäuer Hügelland in the south-east) show only scattered small to average sized complexes. Orchards (Fig. 5(D)) are predominant landscape elements in Baden-Wuerttemberg. Especially at the foothills of the Swabian Jura, in the Neckar valley and in Kraichgau, these landscape elements are characterised as large and highly dense.

5. Discussion

First we shall discuss the validity and transferability of the method, followed by a review of whether or not the goals of the study were met.

5.1. Pattern recognition

Working with the structural aspect of landscapes is comparable to the matrix-patch theory (see Forman and Godron, 1986), which defines 'matrix' as the mainly covered and best connected section of a landscape (Forman and Godron, 1986, p. 186). Furthermore the 'patches' are non-linear areas, differing in appearance from the surrounding matrix (Forman and Godron, 1986, p. 83). The LCs formed in this study are comparable to the patches, but they do not result from mapping of natural elements. Rather these result from the distance rule (see Section 3.4).

Although the approach presented here analyses structural characteristics of landscapes, a difference to methods dealing with structure indices such as patch density, edge density or Shannon diversity index (see Gallego et al., 2000; Blaschke, 1999) also exists. These authors attempt to describe the variety of land use types and structures by mostly grid-based analyses or by using species related distances, sizes or networks of single habitat structures. In contrast, we suggest here a more structural-quantitative determination of structure areas, related to the reference area of LCs, thus presenting an advantage since it considers the interrelation of landscape elements and potential interaction with the farmland in between. Therefore, the landscape level as an important scale level for sustainable development (Selman, 2000) is reached to a higher degree.

5.2. Operationalisation of the indicator 'man-made objects' and transferability of the method

By combining recognition patterns and finding similarities in the landscape element properties of natural units, it becomes possible to describe landscape structures (here from the viewpoint of habitat protection), and to estimate their function leading to establishing of values. Merging of natural units with similar values can result in main thematic emphases for region protection or region development targets.

The approach presented allows to ascertain the OECD indicator 'man-made objects', which itself comprises the area of a state through the assistance of a standardised method.

GIS-based approaches similar to those presented here are currently used in the US–mid-Atlantic region (Jones et al., 1997). Data sets on grid basis are used and transferred to new reference areas (water catchment areas) for solving resource protection problems in spatial differentiation. The grid analysis is here also used for demonstrating the interrelationship between farmland and surrounding areas (e.g. streams).

The possibility of applying this method to other German states is subject to the availability of a data base containing characteristic landscape elements and a data base containing land use information. However, not only selective habitat mapping data bases can be used here but also extensive land use mapping such as coordination of information on the environment (CORINE). This land use and biotope data are useful for this sort of analysis. There is currently a Research + Development project (sponsored by the Federal Office for nature protection in Germany) dealing with CORINE data, for the extraction of landscapes worth protecting. A broadening of the approach used in that study through consideration of characteristic landscapes by means of the method presented here seems plausible.

In addition, the method presented here is related to the state level and presents the advantage of a consistent evaluation scheme that considers regional peculiarities of landscapes. This is possible by scaling the evaluation with reference to the comparison between the natural units (see also Jones et al., 1997). Since the dilemma of evaluation or weight schemes often results in the omission of incongruent data (see Riley, 2001), no strictly formulated targets or thresholds are produced (rather regionalised at the state scale-oriented protection targets).

In other German states, a sort of natural unit classification exists which could be also instrumental in the analysis presented here. A hierarchical nested landscape related system has been established in Canada (CanSIS, 2001) using nearly identical terminology to that already developed in The Netherlands (Klijn, 1997) with varying ecozones, ecoregions and ecodistricts. This target region formation is different from landscape descriptions by the British English Nature organisation (English Nature, 1998–2002), which emphasises more individual character of each 'natural area'. In the present paper, natural units are not only described but also compared to each other. Thus an evaluation is made for setting preferences for landscape protection and for setting up adapted measure packets for more efficient protection programmes.

5.3. Regionalisation of targets and practical use of the results

In the case of agri-environmental programmes, which are targeted at agricultural landscapes, financial compensation is made to farmers who conduct measures having positive environmental impact. Especially the protection and maintenance of landscapes as targets for agricultural landscapes has not been very successful. Further the biotic properties have been decreasing for years (SRU, 2000).

Which effect can result in a new development of a programme? Considering the factor landscape protection on a regional scale can improve both the financial and ecological efficiency of programmes. The process of regionalisation is focussed on a more precise targeting of measures.

The financial efficiency is a criterion for evaluating agri-environmental programmes, and increases when finances are spent at the right place. From the viewpoint of landscape protection those landscapes with a still-existing value due to their landscape element qualities should receive preferential protection.

Areas having a good to very good quality (see Fig. 5, dark grey to black signature) may be regarded as target areas with the aim to protect preferentially the respective landscape element. Thus, natural units having extraordinary high qualities merit the highest priority from the state.

Though some critique may arise that this approach may appear overly conservative, it appears necessary to protect the still-existing characteristic landscapes with high priority. Under the continuing trends of landscape intensification with high suitability for agriculture, and the abandoning of agricultural poorly suitable land, characteristic landscapes are endangered. These could be protected by agri-environmental programmes if such existed and were well adapted. In contrast to the ESA (environmentally sensitive areas in England (MAFF, 1993)), a selective approach (delimiting the ESA from non-ESA areas) it was not followed but rather one more extensive for setting protection goals. By that, each of the natural units can regard landscape protection as their specific job for maintaining its individuality.

Areas with only local or scattered LCs can be regarded as areas for development. But is there a need for a target in which the elements should be developed, e.g. orienting at the history or at habitat requests?

Extrapolating these findings to the already present agri-environmental programme in Baden-Wuerttemberg, a need for landscape related protection arises as target priority. Through assistance of this programme, approximately 93% of the orchard area can be cultivated (LEL, 1998, p. 103). However, for grassland, only on 50% of the area is supported. In this case, only a slight directive function of the programme occurs due to regional dependent premium. Stronger and clearer target- and space-oriented agricultural guidelines are still lacking.

Concerning protection of oligotrophic grassland or hedges, there is no spatial protection target established indicating which landscape element should be predominantly protected and in which region. Using the method presented here allows landscapes with various qualities to be differentiated and preferential protection targets to be formulated on this basis.

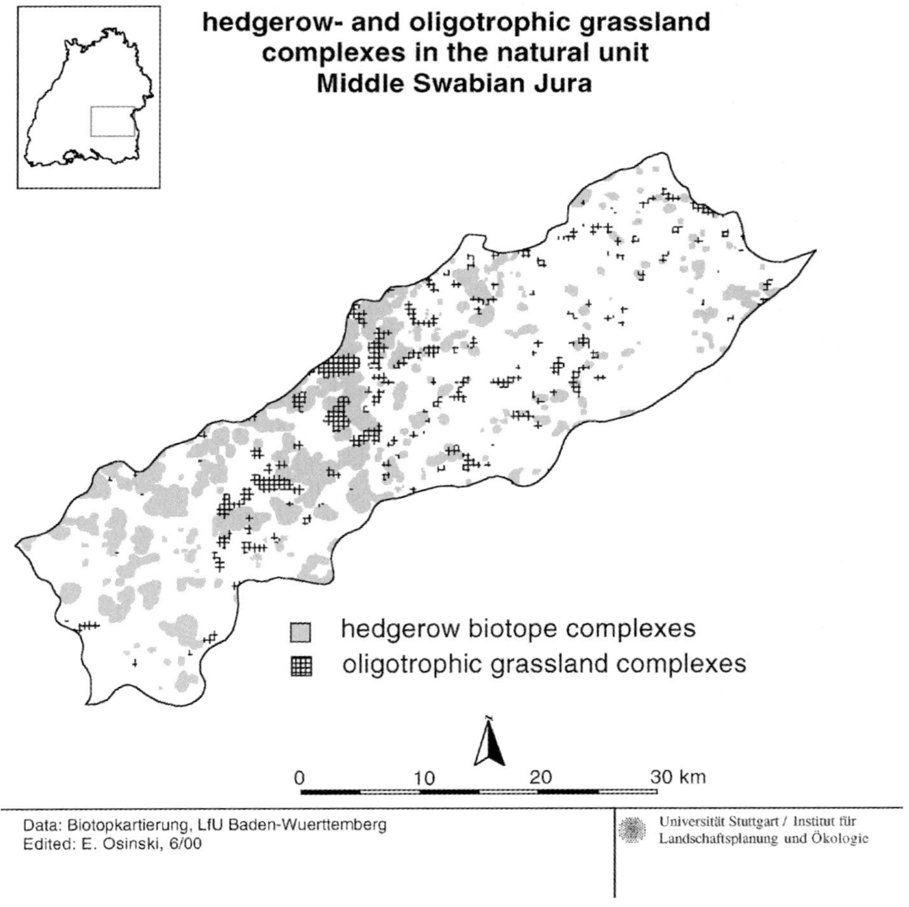

Fig. 6. Hedgerow and oligotrophic grassland complexes in the natural unit 'Middle Swabian Jura' (own figure).

5.4. Regionalisation of measures

Through consideration of land use types, the method presented here allows the formulation of measure packets on the level of the state Baden-Wuerttemberg. Also these measures can be applied on single natural units.

As example, the LCs including hedgerows and oligotrophic grassland, for the natural unit 'Mittlere Flächenalb' are shown in Fig. 6.

On the level of the natural unit, hedgerow complexes contain 32% arable land, 30% grassland and 24% forest on the area surrounding the landscape elements. The oligotrophic grassland complexes comprise 22% arable land, 29% grassland and 34% forest, and have therefore a higher proportion of forest area. This information, together with data on e.g. the amount of reforested area in this region, a set of measures on protection target can be defined. The target to be protected can be redefined by 'keeping landscape elements free from forest'. A measure packet for protection of this landscape type should also include a check of each planned reforestation process, due to its potential collision with landscape protection targets. Focussing on the Middle Swabian Jura it becomes clear that hedgerow and oligotrophic grassland complexes are often spatially correlated. Hence, recommendations can be given for appropriate management of both landscape types. If there is popular acceptance of landscape protection targets, measures should be introduced for (a) stabilising the different areas of arable land, grassland and forest; and (b) minimising the impact of farming practices, such as input of nutrients and pesticides, on the landscape elements.

Placing such measures into practice must be first planned on smaller spatial units. Other concepts must also work on this spatial level including habitat requests of defined species (see Hoffmann and Greef, 2003). For specifying measures, target species must be defined their abundance must be investigated and the habitat quality of the landscape elements themselves must be ascertained. Thus, such approaches as the mosaic-indicator concept (Hoffmann and Greef, 2003) and the regionalisation approach can complement each other.

An alternative approach, combining measures targeted at more general and more special aims successful for establishing agri-environmental programmes, is proposed in the evaluation report of these programmes (EU-COM, 1998).

Considering the role of farms, not only single strategies but more landscape-oriented and more than 'single farm-including' concepts are discussed (MacFarlane, 1998, p. 583). Different from other programmes the ownership of land should not play a decisive role in this type of landscape protection approach, since there is a need for more than a site related approach for ensuring biotic qualities of landscapes (Dolman et al., 2001).

6. Conclusions and outlook

This study attempted firstly to demonstrate a method for deriving regionalised landscape protection targets. Secondly, it shows the necessity for the operationalisation of universally formulated indicators, i.e. the structure indicator 'man-made objects', in other words, placing in concrete terms according to the OECD (2001) by a special method for each country.

While doing this, the planning framework of the region or state should be considered. Whereas present example is more conservative in terms of protection, another country (i.e. The Netherlands) would be more creative in dealing with landscapes. In this respect, The Netherlands have incorporated a landscape development plan into law (see MLNV, 1992).

Beside considering landscape protection targets from the biotic viewpoint, the analysis was extended to other landscape and land use influencing factors, such as erosion and groundwater control. Focal points of erosion control and groundwater protection can be derived in order to formulate measures, including protection of abiotic resources (Osinski, 2002). Thus, measures must be defined which should consider both kinds of targets the abiotic and the biotic by estimating their interactions.

Acknowledgements

The basis for this study has been worked out in the framework of an EU-Project (AIR-CT-1294-96). The data have been used by kind permission of the project co-ordinator Prof. Dr. G. Kaule, Stuttgart.

I want to thank Dr. Ursula Olazabal for carefully checking English syntax and Dr. Sylvia Herrmann for valuable remarks on the text.

References

Bahrenberg, G., Giese, E., Nipper, J.,1992. Statistische Methoden in der Geographie. Bd. 2. Teubner, Stuttgart.

Bill, R., 1996. Grundlagen der Geo-Informationssysteme. Bd. 2. Analysen, Anwendungen und neue Entwicklungen. Wichmann, Heidelberg.

Bill, R., Fritsch, D., 1994. Grundlagen der Geo-Informationssysteme. Bd. 1. Hardware, Software und Daten. 2. Aufl. Wichmann, Heidelberg.

Blab, J., 1993. Grundlagen des Biotopschutzes fuer Tiere. Schriftenreihe fuer Landschaftspflege und Naturschutz. H. 34. Bonn-Bad Godesberg.

Blaschke, T., 1999. Quantifizierung der Struktur einer Landschaft mit GIS: Potential und Probleme. In: Walz, U. (Ed.), Berichte des Institutes für ökologische Raumentwicklung Dresden, pp. 9–24.

CanSIS, 2001. Canadian Soil Information System. URL: http://sis.agr.gc.ca/cansis/.

Dolman, P.M., Lovett, A., O' Riordan, T., Cobb, D., 2001. Designing whole landscapes. Landscape Res. 26 (4), 305–335.

EEA (European Environment Agency), 2001. Towards agri-environmental indicators. Topic Report No. 6, Copenhagen.

Eiden, G., Kayadjanian, M., Viadal, C., 2000. Capturing landscape structures: tools. In: From land cover to landscape diversity in the European Union. URL: http://www.europa.eu.int/comm/agriculture/publi/landscape.

EU-COM (European Commission), 1998. State of Application of Regulation (EEC) No. 2078/92: Evaluation of Agri-Environment Programmes. Arbeitsdokument der Kommission—DG VI Commission Working Document, VI/7655/98.

EU-COM (European Commission), 2000. Commission Notice to the Member States of 14 April 2000 laying down guidelines for the Community initiative for rural development (LEADER+) (2000/C 139 05).

English Nature, 1998–2002. Natural Areas. URL: http://www.englishnature.org.uk/pubs/gis/tech_na.htm.

Forman, R.T.T., Godron, M., 1986. Landscape Ecology. Wiley, New York.

Gallego, F.J., Escribano, P., Christensen, S., 2000. Comparability of landscape diversity indicators in the European Union. In: From land cover to landscape diversity in the European Union. URL: http://www.europa.eu.int/comm/agriculture/publi/landscape.

Grigg, D.B., 1967. Regions, models and classes. In: Chorley, R.J., Hagget, P. (Eds.), Models in Geography, Methuen, London, pp. 461–509.

Heinl, T., Heck, T., 2000. Aufbau eines Informationssystems für die ökologisch orientierte Planung im Maßstabsbereich 1:200000. In: COMPUTERGESTÜTZTE RAUMPLANUNG, Manfred SCHRENK (Ed.), Beiträge zum 5. Symposion zur Rolle der INFORMATIONSTECHNOLOGIE in der und für die RAUMPLANUNG, Selbstverlag des Instituts für EDV-gestützte Methoden in Architektur und Raumplanung der Technischen Universität Wien, Wien, 2000.

Hoell, T., Breunig, T., 1995. Biotopkartierung Baden-Wuerttemberg. Ergebnisse der landesweiten Erhebungen 1981–1989. Beih. Veroeff. Naturschutz Landschaftspflege Bad.-Wuertt. 81, Karlsruhe.

Hoffmann, J., Greef, J.M., 2003. Mosaic indicators—a concept for the development of indicators for species diversity in agricultural landscapes. 1. Theoretical approach. In: Buechs, W. (Ed.), Biotic Indicators for Biodiversity and Sustainable Agriculture. Agric. Ecosyst. Environ. 98, 387–394.

Jones, K.B., Riitters, K.H., Wickham, J.D., Tankersley, R.D., O'Neill, R.V., Chaloud, D.J., Smith, E.R., Neale, A.C., 1997. An Ecological Assessment of the United States Mid-Atlantic Region: A Landscape Atlas. United States Environmental Protection Agency (EPA)/600/R-97/130, Washington, DC.

Klijn, F., 1997. A hierarchical approach to ecosystems and its implications for ecological land classification with examples of ecoregions, ecodistricts and ecoseries of The Netherlands. Thesis. Leiden University.

Knauer, N., 1988. Konzept eines Netzes aus oekologischen Zellen in der Agrarlandschaft und Bedeutung für das Agraroekosystem. In: ANL (Akademie für Naturschutz und Landschaftspflege) (Hrsg.), Biotopverbund in der Landschaft. Laufener Seminarbeiträge 10/86.

LEL (Landesanstalt fuer Entwicklung der Landwirtschaft und der Laendlichen Raeume), 1998. Evaluierung von Programmen nach der Verordnung (EWG) Nr. 2078/92 des Rates vom 30. Juni 1992 für umweltgerechte und den natuerlichen Lebensraum schuetzende landwirtschaftliche Produktionsverfahren in Baden.Wuerttemberg. Teil I Marktentlastungs- und Kulturlandschaftsausgleich (MEKA).

MacFarlane, R., 1998. Implementing agri-environment policy: a landscape ecology perspective. J. Environ. Plann. Manage. 41 (5), 575–596.

Meynen, E., Schmithuesen, J. (Eds.), 1953–1962. Handbuch der naturraeumlichen Gliederung Deutschlands, 1.-3. Lieferung, Bundesanstalt fuer Landeskunde, Remagen.

MAFF (Ministry of Agriculture, Fisheries and Food), 1993. Our living heritage. Environmentally sensitive areas. MAFF Environment Matters. MAFF Publications, London.

MAFF (Ministry of Agriculture, Fisheries and Food), 2000. Towards Sustainable Agriculture. A pilot set of Indicators. URL: http://www.maff.gov.uk.

MLNV (Ministerie van Landbouw, Natuurbeheer en Visserij), 1992. Nota Landschap—Regeringsbeslissing Visie Landschap, Nederlands.

MLR (Ministerium Laendlicher Raum Baden-Wuerttemberg) (Ed.), 1990. Oekologische Standorteignungskarte fuer den Landbau. 1:250,000.

OECD (Organisation for Economic Co-operation and Development), 1997. Environmental Indicators for Agriculture. OECD, Paris.

OECD (Organisation for Economic Co-operation and Development), 2001. Environmental Indicators for Agriculture, vol. 3, Methods and Results. OECD, Paris.

O' Riordan, T., Lovett, A., Dolman, P., Cobb, D., Suennenberg, G., 2000. Designing and implementing whole landscapes. Ecos 21 (1), 57–68.

Osinski, E., 2000. Ermittlung landesweit bedeutender Biotopschutzziele unter Einsatz eines Geographischen Informationssystems. In: Oesterreichische Zeitschrift für Vermessung und Geoinformation, Heft 1, pp. 32–37.

Osinski, E., 2002. GIS-gestuetzte Regionalisierung von Agraraumpotentialen—Anwendung zur effektiven Ausgestaltung von Agrar-Umweltprogrammen in zwei europaeischen Regionen. Der Andere Verlag, Osnabrueck (English abstract).

Riley, J., 2001. Multidisciplinary indicators of impact and change—key issues for identification and summary. Agric. Ecosyt. Eviron. 87 (2), 245–259.

Schmithuesen, J., 1953. Einleitung. In: Meynen, E., Schmithuesen, J. (Eds.), Handbuch der Naturraeumlichen Gliederung Deutschlands. 1. Lieferung. Bundesanstalt fuer Landeskunde, Remagen, pp. 1–34.

Selman, P., 2000. Landscape sustainability at the national and regional scales. In: Benson, J.F., Roe, M. (Eds.), Landscape and Sustainability. SPON Press, London, pp. 97–110.

SRU (Der Rat von Sachverständigen für Umweltfragen), 1985. Umweltprobleme der Landwirtschaft. Sondergutachten. Kohlhammer, Stuttgart.

SRU (Der Rat von Sachverständigen für Umweltfragen), 2000. Umweltgutachten 2000, Schritte ins nächste Jahrtausend. Metzler-Poeschel, Stuttgart.

UBA (Umweltbundesamt), 1999. Entwicklung von Parametern und Kriterien als Grundlage zur Bewertung ökologischer Leistungen und Lasten der Landwirtschaft—Indikatorensysteme. Texte 42/99 (English abstract).

ELSEVIER

Available online at www.sciencedirect.com

Agriculture, Ecosystems and Environment 98 (2003) 387–394

Agriculture Ecosystems & Environment

www.elsevier.com/locate/agee

Mosaic indicators—theoretical approach for the development of indicators for species diversity in agricultural landscapes

J. Hoffmann*, J.M. Greef

Federal Agricultural Research Centre, Institute of Crop and Grassland Science, Bundesallee 50, Braunschweig D-38116, Germany

Abstract

Changed land use conditions, especially the intensification of agricultural production, have lead to a significant reduction in species diversity over the past several decades. In order to limit the impact of this reduction, indicators were developed and practical measures put forth which should serve to maintain biological diversity, but generally pay too little attention to different natural area conditions. In this light, an indicator concept for area species diversity on the basis of regional particularities has been developed which is based on a clear delineation of natural areas according to their unique landscape conditions. From the large number of species which appear in one area, a small number of key-species typical to the landscape, for which good ecological information is available, are selected as indicators. These species serve as the current status indicators for each area. With the goal of promoting regionally specific species diversity, the composition of indicators varies from area to area, according to each unique landscape. On the basis of an anticipated population change (target indicators) a comparison was made of habitat requirements and deficits in the existing habitats. In this manner, regionally targeted measures to improve habitats could be derived and implemented. So, the abundance of indicator species can be positively influenced and the regionally characteristic species diversity can thus be promoted. Such a set of indicators can serve as the basis for biodiversity management in completely agricultural landscapes and provide a picture of the total of all the single mosaic landscape units.

Keywords: Biodiversity indication; Landscape peculiarities; Key-species; Agricultural landscapes

1. Introduction

The reduction of species diversity is currently considered one of the greatest ecological problems worldwide (Wilson, 1985, 1992; Heywood and Watson, 1995). Agriculture is to a large extent responsible for this development. While the number of species and habitats in the middle European landscape has historically expanded, in the past several decades the numbers have gone down dramatically, to a large extent due to increasing intensification of agriculture and the cessation of use of marginal locations in almost all agricultural areas (Haber, 1997). Intensive agriculture is currently considered one of the main causes of the reduction of biological diversity in many areas of the world. This situation is documented in the Red Lists for plants, animals and habitats (e.g. Bundesamt für Naturschutz, 1996; Jedicke, 1997).

In the future, the EU surveys on agriculture will be linked more closely to special ecological achievements, which can be expected to extend beyond the recommendations of the best management practices (Cross-Compliance Modulation). For this reason, new methods and practical solutions are required, independent of special advantages of each farmer under the typical natural conditions at his location, which

* Corresponding author. Tel.: +49-531-596-2312; fax: +49-531-596-2399.
E-mail address: joerg.hoffmann@fal.de (J. Hoffmann).

0167-8809/$ – see front matter © 2003 Elsevier Science B.V. All rights reserved.
doi:10.1016/S0167-8809(03)00098-7

provide him with the possibility of making ecological achievements, and accordingly of receiving funding. Agricultural indicators to promote farming in which the maintenance of biological diversity plays an important role in sustainable production systems are being worked upon (e.g. Breitschuh et al., 1996; Wascher, 1997; Piorr, 1998; Conacher, 1998; Paoletti, 1999b). While the development of indicators for the abiotic areas has made progress, indicators usable for biological diversity are lacking (OECD, 1999), although many indicators have been suggested for individual groups of species. An indicator approach is needed which takes into account natural conditions, historical developments and economic structures in the various regions (KOM, 2000). This requires regionalisation and the implementation of indicators characterising local, landscape-typical diversity in each area and which make an effective contribution to the sustenance of biological diversity in global terms as well. The indicators should not just be used for observation and evaluation, but should also serve as the basis for practical measures in order to make the largest possible impact in the securing and improvement of species diversity in agricultural areas.

In order to develop a suitable indicator concept, theories are being developed and discussed, and an indicator approach is being designed, in which typical regional species are used as indicators.

2. Theses for the development of an indicator approach for species diversity

There are different natural conditions with different species and communities of species, the constellation of which is different from area to area.

Due to the uniqueness of geographical spaces (landscapes, regions, states), for example, the relief and surface conditions, the climate, the hydrology and vegetation, different natural areas can be identified—about 900 areas in Germany (Meynen et al., 1962). In relation to the changing natural site conditions, different distribution patterns of species and species communities can be seen on distribution maps, for example, for ferns and flowering plants (Benkert et al., 1996) and birds (Rheinwald, 1993). Expansion is limited to certain areas due to natural conditions and differing landscape developments, resulting in different regional specific demands on the indicators.

Agricultural landscapes have developed differently over relatively small areas due to natural and socio-cultural conditions, with different landscape structures, different specialised species and different diversity.

In agricultural landscapes which differ widely (Fig. 1) a generally significant higher diversity versus minimally structured areas (Fig. 1, landscape C) can be observed (Kretschmer et al., 1995; Dierschke, 2000; Hoffmann et al., 2001). This depends on the

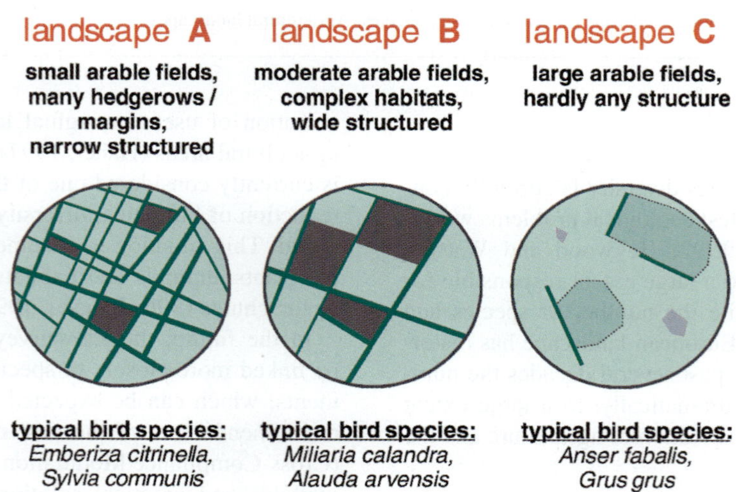

Fig. 1. Agricultural landscapes in northeast Germany with a different structure of habitats and of specific significance for species diversity (selected bird species).

heterogeneity and particularities of the natural conditions, the portion of characteristic habitat structures, for example, wooded areas, hedgerows or areas with small bodies of water, for example, small ponds and lakes. For this reason, a variety of ecological groups of species can thrive over relatively small geographical areas, for example, open spaces, forest perimeters, wooded areas as well as terrestrial species or aquatic species (Hoffmann and Kretschmer, 2001). A general evaluation and the development of individual areas with the help of generalised indicators, for example, the recommendation of a standard minimum length for hedges per square kilometre of arable land, or unspecific indicator types, would contribute to uniform, regional, naturally-linked conditions and structures grown out of historical context. This has for the most part, however, had as a consequence more of a reduction effect than an expansion effect on the diversity. A mosaic of differing structures through natural and semi-natural biotopes appropriate to the natural space conditions and various land uses in agricultural landscapes is advantageous.

Maintenance of biological diversity does not mean a local maximisation of number of species, but rather the sustenance of diversity typical for the landscape and the location.

A nature- and usage-dependent mosaic of changing species exists on cultural landscapes. The diversity of individual habitats is called α-diversity (Whittaker, 1972). Together with the habitats, this comprises the meaning of landscape diversity (Whittaker, 1972). The γ-diversity is not comprised of the sum of all individual α-diversities together, but rather from the total number of species in the area, independent of the fact if they, or how often they, appear in different biotopes. An area with few floral species, for example, a raised bog or moor area with highly specialised indigenous plant species, can serve as an important mosaic stone in the biological diversity of a landscape if these species do not appear in any other biotope type. In contrast, in species rich biotopes, for example, some garbage dumps and ruderal communities, a large number of generalists and alien species can be found which belong neither to the typical landscape inventory (many instable or some problematic neophyte populations), or which appear as generalists in many other areas (Hoffmann, 1999a). In comparison to the moor, which is low in species, this shows that high species diversity is relative. This is why unspecific species numbers give no information about the ecological value of an area (Bastian and Schreiber, 1994). As high a number as possible can therefore not be set as the basic goal. It appears to be more important to the maintenance of biological diversity to consider species and biotopes typical to the location and region (Plachter, 1991; Schumacher, 1998).

The majority of species are only partially suitable as indicators. A selection of indicators is necessary, the presence of which are representative for other species.

In Germany alone, there are over 72,000 animal and plant species and more than 500 types of biotopes (Bundesamt für Naturschutz, 1999). In richly structured agricultural areas in Middle Europe, up to 2500 species can be found on 100 ha (Kretschmer et al., 1995). In richly structured cultural landscapes (200 km^2) between 15,000 and 20,000 (Hoffmann, 1999b, unpublished). Many species types, for example, micro-organisms, insects, spiders and mosses can only be documented and counted with a high time investment and locally only for a short time. Also, for the majority of the species, only minimal knowledge is available on the population dynamics in dependence on natural and anthropogenic location changes. The hardly overseeable species diversity also makes it clear that a limit to fewer key indicators, for which good biological knowledge is available, seems to be required (Mühlenberg, 1993; Ellenberg, 1997; Welling, 1998; Saunders et al., 1998; OECD, 1998). The selection of several key-species, the presence of which characterise an area, and which are simultaneously representative of other species which are more difficult to document would be advantageous (Mühlenberg, 1993).

Species about which solid knowledge about the distribution and ecology are available are most suitable as indicators.

The best researched species types include at present birds and ferns and flowering plants, which for this reason appear to be suitable as indicators (Hoffmann, 1996, 1998; Niemi et al., 1997; Whitford et al., 1998; Ormerod and Watkinson, 2000). But other species groups, such as invertebrates, can also be used regionally as indicators (e.g. Favila and Halffter, 1997; Andersen, 1997; Churchill, 1997; Bohac, 1999; Duelli, 1999; Kevan, 1999; Paoletti, 1999a).

The habitat conditions required by a given population can be concluded with the help of the habitat requirements of the species.

If one assumes a species approach in the finding of indicators—"the idea of indicator species is a relatively old concept" (Hall and Grinnell, 1919)—then one can draw conclusions on the requirements of the species used as indicators on their habitat structure (e.g. birds, ferns and flowering plants). For example, relevant habitat types, qualitative characteristics of the habitats and minimum size for reproduction, resting and nutritional supply. It is assumed in an indicator concept, that through the promotion of representative, delicate species in need of protection, that other species in the same biotope can be protected as well (Walter et al., 1998).

Targeted changes in the habitat conditions can influence the size of species populations.

In the framework of the potential of a natural location, the biotope structure (quantitative, qualitative) can be influenced and changed in a targeted manner with practical measures (landscape management, maintenance of semi-natural biotopes, different use of land parcels, the new establishment of habitats) (among others, Jedicke et al., 1996; Glemnitz et al., 2000). The habitat requirements of the species or communities of species and the desired population development are the basis for such measures.

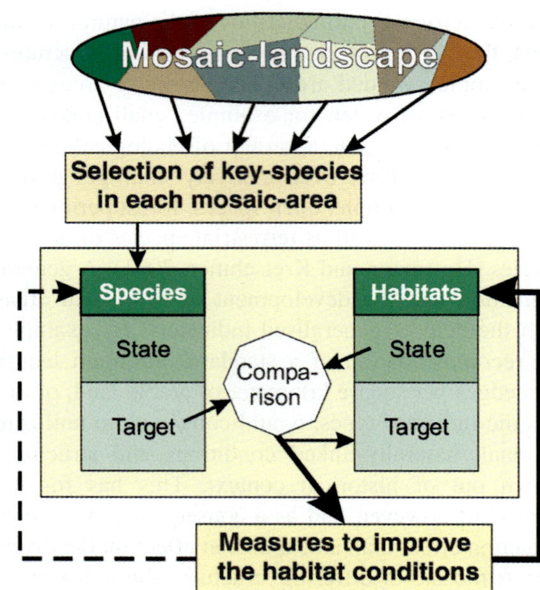

Fig. 2. Scheme of the mosaic-indicator concept.

3. The mosaic indicator approach

From the above-mentioned theses, the following requirements for an indicator concept for promoting species diversity in agricultural landscapes can be concluded.

- The definition of landscape mosaics with similar natural spatial conditions.
- The selection of indicator species typical of a natural area which make possible the regionally differentiated measures for the improvement of habitat conditions.
- Aggregation of regional indicators for supra-regional use of political relevance.

Dependent on the unique characteristics of the landscape (e.g. geomorphology, climate, hydrology), a mosaic landscape is assumed in a mosaic indicator approach. This is comprised of differing natural space units which differ in terms of landscape uniqueness. For each of these mosaic areas (cf. Fig. 2, above) a selection of only a few, but regionally typical indicator species (mosaic indicators), their presence and abundance in the given area are characterised. The types of indicators are selected on the basis of their ecological characteristics, which define them as typical species (key-species) for the natural area being studied. Here a distinction is made between national species with high protection priority and natural area species with special regional significance (Walter et al., 1998) as well as in the hierarchy between globally, regionally and locally endangered taxa (Breininger et al., 1998). The current abundance is used as a status indicator. The indicators (species composition and abundance) vary from landscape to landscape mosaic due to the uniqueness of each area.

In order to succeed in taking the step from passive monitoring and simple evaluation of the existing situation to finding an effective possibility for influencing the regional biological situation, target quantities of indicator species are defined from which development goals and measures for improving the habitat situation can be derived. These should not be fixed to rigid figures, but move within a range of tolerance (Plachter et al., 2003).

On the basis of the demands of the indicator species on their environments (literature, expert knowledge), the abundance is compared with the current habitat situation of the area and possible deficits in the habitat structure (qualitative and quantitative) as well as the form and intensity of use can be determined. The analysis of the discrepancy between the habitat requirements of an indicator and the existing habitat situation makes the derivation of differentiated, required practical measures for each area possible as well as their implementation, for example, the new establishment of small structures (e.g. hedges, margins, grassland), the care of available semi-natural biotopes or the extensification of partial areas still farmed intensively.

With this approach, targeted measures can be taken with the expectation of appropriate improvements of the habitat conditions. The extent of the measures can be drawn from the set goals and in which order the abundance of individual or multiple species shall be changed (population ecological criteria) and which changes are potentially sensible under consideration of the natural space conditions. The steps taken in the process of the mosaic indicator approach can be seen in Fig. 2.

As a result, with targeted habitat changes, a regeneration of indicators which can react with a population change in the direction of the targeted abundance.

4. Discussion

The need for suitable indicators for biological diversity results above all from the need to establish effective observation methods and practical measures to maintain diversity. Since biological diversity varies dramatically due to the varying natural conditions and historic development of individual regions, landscapes and partial landscapes, a regionalisation of the indicators (e.g. Piorr, 1998; KOM, 2000) and a regionally adapted target species is necessary (Mühlenberg, 1993; Walter et al., 1998).

In contrast to global and supra-regional goals of maintaining a maximum of species, at the local level in many cases this does not hold true. This is the result of specific differing species rich environments, for example, for oligotrophic raised bogs, deciduous forests, sandy dry grassland and arable areas with adapted diversity. The absolute species number or a count of how many hedges are in an area reveal little about the quality of an arable area or an agricultural landscape. Both species and habitat must be analysed in relation to location and natural site conditions and be evaluated (Bastian and Schreiber, 1994). Thus, some arable area with almost no hedges and hardly any semi-nature habitats can serve as a resting place for Nordic gooses (e.g. *Anas fabalis*, *Anas albifrons*) or cranes (*Grus grus*) in the fall and the spring, and thus serve as an important mosaic element for the maintenance of biological diversity in the entire landscape mosaic.

Often the ecological needs of the species are in conflict (Hoffmann et al., 2001) so that a decision must be made which species will be favoured in the changes or improvements for the habitat structure, and which will not. Due to the differing presence and endangerment of open land and forest species in individual agricultural landscapes, it does not, for example, seem to be sensible to use narrow structured hedge systems (cf. landscape A in Fig. 1) in all agricultural landscapes. In areas not so narrowly structured by hedges and groups of trees and shrubs between the fields (cf. landscapes B and C in Fig. 1), an increase in hedges is not necessarily required to maintain a typical regional species diversity, but rather to improve the stronger qualitative habitat characteristics of the hedges, for example, broader hedges, increase in the structural diversity within the hedges and widening of adjacent margins. In this way the negative effects of arable farming on semi-nature small structures can be reduced, without limiting the open character of the landscape and its function as a biosphere for specialised "steppe species" such as the field birds *Alauda arvensis* and *Miliaria calandra*.

The regional selection of typical natural space key-species as indicators can be made in dependence of the endangerment level of species (global, regional, local) and their representation of other species for which population changed represent similar changes for many other species (Mühlenberg, 1993; Walter et al., 1998). However, the selection leaves room for a subjective decision which species or ecological groups of species best mirror the typical natural space diversity. This is largely determined by the regional level of knowledge (literature, expert knowledge) for an inventory of species in the individual landscape mosaics.

Between areas a shifting in the choice of indicators is necessary to those for which local meaning and space needs of the population (minimum requirement, optimum) which differ from species to species, are influenced. It must also be considered, that many mobile species need different habitats in the cycle of their live, for example, for reproduction, for wintering and as a source of food (Duelli, 1997). Here the aspect of linkage of the biotope and the securing of suitable habitat structures between the landscape mosaics is significant.

5. Conclusion

With the mosaic indicator approach exists the possibility for defining concrete status and target indicators according to the regional particularities of agricultural landscapes and to plan regional measures. With the help of status indicators, the effects of farming on the biological diversity can be analysed and farming practices can be tested and evaluated according to location with the help of indicators (positive or negative effect). The target indicators (species) present an envisioned abundance of species, which shall be strived for though changes in the habitat conditions. But it cannot be assumed that the appropriate measure will lead to the desired change in abundance in species. An iterative approach, with the motto "learning by doing" is needed, and one must always be prepared to cope with unexpected developments.

Also regional measures can change habitat conditions on a small scale, but supra-regional and global factors, such as climate changes, also have an impact. Especially animals that are mobile, for example, many species of birds, experience disturbances in the areas where they spend the winter due to their migration patterns, which limit their ability to serve as indicators regionally (Breininger et al., 1998).

Overall, with the help of the suggested mosaic indicator approach, a broad biodiversity management system with strong regional character could be established. The Red Lists could conceivably serve as a super-ordinate indicator of political relevance. Here it would be necessary to update and expand the lists regularly every few years. Differing endangerment levels for species and habitats as well as lower numbers of animals plants and biotopes in the Red Data books can easily be evaluated statistically and in comparison to randomly distributed tests (Seibel et al., 1997; Hoffmann-Kroll et al., 1998) of high ecological relevance. In comparison to relatively complicated calculations taking into account changes in abundance (Brink, 1999) evaluation of the Red Lists can be easily used as indicator aggregates on the state and national level to document effects. Changes within the Red Lists are only partially influenced by the effects of agricultural management. To differentiate the effects of farming, a more specific approach in the Red Data books would be desirable. However, it is also conceivable that the mosaic indicator approach could be checked for its suitability not only for agricultural landscapes but also as an indicator concept with complete relevance.

References

Andersen, A.N., 1997. Ants as indicators of ecosystem restoration following mining: a functional group approach. In: Hale, P., Lamb, D. (Eds.), Conservation Outside Nature Reserves. Centre for Conservation Biology, University of Queensland, pp. 319–323.

Bastian, O., Schreiber, K.F. (Eds.), 1994. Analyse und ökologische Bewertung der Landschaft. Fischer, Jena.

Benkert, D., Fukarek, F., Korsch, H. (Eds.), 1996. Verbreitungsatlas der Farn- und Blütenpflanzen Ostdeutschlands. Fischer, Jena.

Bohac, J., 1999. Staphylinid beetles as bioindicators. Agric. Ecosystems Environ. 74, 357–372.

Breininger, D.R., Barkaszi, M.J., Smith, R.B., Oddy, D.M., Provancha, J.A., 1998. Prioritizing wildlife taxa for biological diversity conservation at the local scale. Environ. Manage. 22 (2), 315–321.

Breitschuh, G., Eckert, H., Roth, D., 1996. Effiziente und umweltverträgliche Landnutzung (EULANU). Ein Konzept für eine marktwirtschaftlich organisierte Landbewirtschaftung. Thüringer Landesanstalt für Landwirtschaft, Jena.

Brink, B.T., 1999. Case Study on Agro-biodiversity Indicators in The Netherlands. OECD Workshop on Expert Meeting on Biodiversity, Wildlife Habitats and Landscape, 3–5 May 1999, Paris.

Bundesamt für Naturschutz, 1996. Rote Liste gefährdeter Pflanzen Deutschlands. Schriftenreihe für Vegetationskunde 28, 1–744.

Bundesamt für Naturschutz, 1999. Daten zur Natur 1999. Landwirtschaftsverlag, Münster.

Churchill, T.B., 1997. Spiders as ecological indicators: an overview for Australia. Mem. Mus. Vict. 56, 331–337.

Conacher, A., 1998. Environmental quality indicators: where to from here? Aust. Geographer 29 (2), 175–189.

Dierschke, H., 2000. Kleinbiotope in botanischer Sicht—ihre heutige Bedeutung für die Biodiversität in Agrarlandschaften. Pflanzenbauwissenschaften 4 (1), 52–62.

Duelli, P., 1997. Biodiversity evaluation in agricultural landscapes: an approach at two different scales. Agric. Ecosystems Environ. 62, 81–91.

Duelli, P., 1999. Biodiversity evaluation in agricultural landscapes: above-ground insects. Agric. Ecosystems Environ. 74, 33–64.

Ellenberg, 1997. Biologische Vielfalt auf der Art-Ebene und ihre Gefährdung als Kriterium und Indikation für ein Monitoring der Nachhaltigkeit von Waldbewirtschaftung—Ein Diskussionsbeitrag zum "Helsinki-Prozess", Berichte über Landwirtschaft 76, 127–137.

Favila, M.E., Halffter, G., 1997. The use of indicator groups for measuring biodiversity as related to community structure and function. Acta Zoologica Mexicana (n.s.) 72, 1–25.

Glemnitz, M., Wurbs, A., Jacobsen, M., 2000. Ansätze zur Regionalisierung von Zielen für die Lebensraumfunktion in Agrarlandschaften. Agrarspectrum 31, 62–73.

Haber, W., 1997. Umweltprobleme der Pflanzenproduktion—Ursachen und Lösungsansätze. In: Diepenbrock, W., Kaltschmitt, M., Nieberg, H., Reinhard, G. (Eds.), Umweltverträgliche Pflanzenproduktion. Zeller, Osnabrück.

Hall, H.H., Grinnell, J., 1919. Life-zone indicators in California. Proc. Calif. Acad. Sci. 9, 37–67.

Heywood, V.H., Watson, R.T. (Eds.), 1995. Global Biodiversity Assessment. University Press, Cambridge.

Hoffmann, J., 1996. Patterns of plant species in the landscape as potential indicators of environment changes. Acta Agranomica Ovariensis 38 (1/2), 23–39.

Hoffmann, J., 1998. Assessing the effects of environmental changes in a landscape by means of ecological characteristics of plant species. Landscape Urban Plann. 41, 239–248.

Hoffmann, J., 1999a. Analyse und Bewertung regionaler Artenvielfalt als Grundlage für Schutzkonzepte. Landschaftökologie und Umweltforschung, Geoökologiekongress 99, TU Braunschweig 33, 138.

Hoffmann, J., 1999b. Species Diversity in a Cultural Landscape in Middle Europe in the "Nature Park Märkische Schweiz", unpublished.

Hoffmann, J., Kretschmer, H., 2001. Zum Biotop- und Artenschutzwert großer Ackerschläge in Nordostdeutschland. Peckiana 1, 17–31.

Hoffmann, J., Kretschmer, H., Pfeffer, H., 2001. Effects of Patterning on Biodiversity in Northeast German Agro-landscapes, Ecological Studies, vol. 147. Springer, New York, pp. 325–340.

Hoffmann-Kroll, R., Schäfer, D., Seibel, S., 1998. Biodiversität und Statistik—Ergebnisse des Pilotprojektes zur ökologischen Flächenstichprobe. StBA, Wirtschaft und Statistik 1/1998, 60–75.

Jedicke, E., Frey, W., Hundsdorfer, M., Steinbach, E., 1996. Praktische Landschaftspflege. Ulmer, Stuttgart.

Jedicke, E. (Ed.), 1997. Die Roten Listen—Gefährdete Pflanzen und Tiere, Pflanzengesellschaften und Biotoptypen in Bund und Ländern. Ulmer, Stuttgart.

Kevan, P.G., 1999. Pollinators as bioindicators of the state of the environment: species, activity and diversity. Agric. Ecosystems Environ. 74, 373–393.

KOM, 2000. Indikatoren für die Integration von Umweltbelangen in die gemeinsame Agrarpolitik. Mitteilung der Kommission an den Rat und das europäische Parlament. Kommission der Europäischen Gemeinschaften, Brüssel, 26.01. 2000 KOM (2000) 20, endgültig.

Kretschmer, H., Pfeffer, H., Hoffmann, J., Schrödl, G., Fux, I., 1995. Strukturelemente in Agrarlandschaften Ostdeutschlands—Bedeutung für den Biotop- und Artenschutz. ZALF-Bericht 19, 164.

Meynen, E., Schmithüsen, J., Gellert, J.F., Neef, E., Müller-Miny, H., Schultze, J.H., 1962. Handbuch der naturräumlichen Gliederung Deutschlands. Bundesanstalt für Landeskunde und Raumforschung, Bad Godesberg.

Mühlenberg, M., 1993. Freilandökologie. Quelle & Meyer, Heidelberg Wiesbaden.

Niemi, G.J., Hanowski, J.M., Lima, A.R., Nicholls, T., Weiland, N., 1997. A critical analysis on the use of indicator species in management. J. Wildlife Manage. 61 (4), 1240–1252.

OECD, 1998. Workshop on Agri-environmental Indicators, Breakout Session Group 2—Agriculture and Biodiversity, 22–25 September 1998. York, United Kingdom.

OECD, 1999. Environmental Indicators for Agriculture: Methods and Results—the Stocktaking Report Greenhouse Gases, Biodiversity, Wildlife Habitats, Paris, 13–15 October.

Ormerod, S.J., Watkinson, A.R., 2000. Editors' introduction: birds and agriculture. J. Appl. Ecol. 37, 699–705.

Paoletti, M.G., 1999a. The role of earthworms for assessment of sustainability and as bioindicators. Agric. Ecosystems Environ. 74, 137–155.

Paoletti, M.G., 1999b. Special issue—invertebrate biodiversity as bioindicators of sustainable landscapes—foreword. Agric. Ecosystems Environ. 74, 9–11.

Piorr, H.P., 1998. Development of an indicator framework for the analysis of agricultural landscapes. Bornimer Agrarwissenschaftliche Berichte 21, 70–76.

Plachter, H., 1991. Naturschutz. Thieme, Stuttgart.

Plachter, H., Müssner, R., Stachow, U., Werner, A., 2003. Leitlinien einer neuen Land- und Forstwirtschaft im Sinne des Nachhaltigkeitstheorems von Rio. Angewandte Wissenschaft, (in press).

Rheinwald, G., 1993. Atlas der Verbreitung und Häufigkeit der Brutvögel Deutschlands—Kartierung um 1985. Schriftenreihe des DDA 12.

Saunders, D., Margules, C., Hill, B., 1998. Environmental Indicators for National State of the Environment Reporting—Biodiversity, Australia: State of the Environment (Environmental Indicator Reports). Department of the Environment, Canberra.

Schumacher, W., 1998. Ziele des Naturschutzes für agrarisch genutzte Flächen—biotischer Ressourcenschutz. In: Bundesumweltministerium, Ziele des Naturschutzes und einer nachhaltigen Naturnutzung in Deutschland. Geographische Institute Rheinische Friedrich-Wilhelms Universität, Bonn, 133–138.

Seibel, S., Hoffmann-Kroll, R., Schäfer, D., 1997. Land use and biodiversity indicators from ecological area sampling—results of a pilot study in Germany. Stat. J. United Nations ECE 14, 379–395.

Walter, R., Reck, H., Kaule, G., Lämmle, M., Osinski, E., Heinl, T., 1998. The target species concept—regionalised quality goals, standards and indicators of species and biotope protection for the environmental development plan of Baden-Württemberg. Natur und Landschaft 73 (1), 9–25.

Wascher, D.M., 1997. Biodiversität des ländlichen Raumes in Europa—Bioindikatoren und Pilotstudien zur Entwicklung von Bewertungskonzepten. In: Umweltverträgliche Pflanzenproduktion—Indikatoren, Bilanzierungsansätze und ihre Einbindung in Ökobilanzen, Zeller, Osnabrück, 81–98.

Welling, M., 1998. Biologische Vielfalt in genutzten Ökosystemen. Berichte über Landwirtschaft 76, 598–614.

Whitford, W.G., Soyza, A.G., v. Zee, J.W., Herrick, J.E., Havstad, K.M., 1998. Vegetation, soil and animal indicators of rangeland health. Environ. Monit. Assess. 51, 179–200.

Whittaker, R.H., 1972. Evolution and measurement of species diversity. Taxon 21, 213–251.

Wilson, E.O., 1985. The biological diversity crisis. Bioscience 35, 700–706.

Wilson, E.O., 1992. The Diversity of Life. Penguin, London.

ELSEVIER

Available online at www.sciencedirect.com

Agriculture, Ecosystems and Environment 98 (2003) 395–405

Agriculture Ecosystems & Environment

www.elsevier.com/locate/agee

Practical example of the mosaic indicators approach

J. Hoffmann[a,*], J.M. Greef[a], J. Kiesel[b], G. Lutze[b], K.O. Wenkel[b]

[a] *Federal Agricultural Research Centre, Institute of Crop and Grassland Science, Bundesallee 50, Braunschweig D-38116, Germany*

[b] *Centre for Agricultural Landscape and Land Use Research, Institute of Landscape System Analysis, Eberswalder Str. 84, Müncheberg D-15374, Germany*

Abstract

The theoretical concept for the use of mosaic indicators for species diversity in agricultural landscapes is illustrated on the basis of a practical example. A method to organise data on the specific regional natural conditions on the basis of an existing organisational chart of natural areas in Germany, digital results of biotope mapping and aerial views was developed for a selection of indicators (species, ecological groups) appropriate to given natural conditions on the regional scale. The use of the mosaic indicator concept was demonstrated for the example of bird species in a selected mosaic landscape characterised by a typical regional biotope structure—a special density of hedges and small bodies of water. On the basis of the indicator species, measures for the improvement of the habitat situation were derived with the goal of maintaining the landscape peculiarities and promoting specific diversity adapted on these special site conditions.

The methods developed to regionalise the landscapes while taking into account spatial variations make it possible for farmers throughout Germany, no matter what natural characteristics each individual farm may have, to participate in programs to promote biological diversity and make a contribution to natural conservation.

Keywords: Mosaic indictors; Species diversity; Landscape peculiarities; Birds; Agricultural landscapes

1. Introduction

One of the goals of the use of indicators in agricultural landscapes is to establish effective methods for observing, evaluating and taking practical measures to maintain biological diversity. Another goal is to use the indicators for all parts of the landscape, independent of a special diversity but related to the landscape peculiarities, to help all the farmers to make a contribution toward the maintenance of landscape typical diversity in their own areas.

To achieve these goals, an ecological space division (state, region, landscape) into landscape units is required to express the differences in the unique landscape characteristics from one area to the other. The natural landscape conditions and differences in the historical agricultural landscape structures are of significance in such a landscape structuring (Meynen et al., 1962; Klijn, 1991; Rennings, 1994). A relevance to the agricultural production units (farm, field, plot) is needed to recommend concrete measures for the farmers or other land users.

On the basis of the theoretical mosaic indicator approach (cf. Hoffmann and Greef, 2003, 5.9 in this publication) the following contribution will develop a method for regionalised ecological landscape structuring. With the species group of birds, the practical approach for choosing indicator species in a sample area will be demonstrated, and regionally specific measures for the improvement of

* Corresponding author. Tel.: +49-531-596-2312; fax: +49-531-596-2499.
E-mail address: joerg.hoffmann@fal.de (J. Hoffmann).

0167-8809/$ – see front matter © 2003 Elsevier Science B.V. All rights reserved.
doi:10.1016/S0167-8809(03)00099-9

habitat conditions for the indicator species will be derived.

2. Materials and methods

2.1. Natural space divisions for the differentiation of landscape uniqueness

The geomorphological and geological conditions, the soil conditions, the climate, hydrology, vegetation and soil use will be considered as characteristics to differentiate between varying landscapes. This approach identifies 892 different natural units in Germany (Meynen et al., 1962). Although these areas are not homogenous units, as seen in Fig. 1, as a rule they are too heterogeneous to select indicator species representative for the whole area of a single natural unit. This is especially true in the division of natural and semi-natural habitats, for example wooded areas like hedges and small forest islands in agricultural landscapes and areas with small bodies of water. Each natural unit must therefore be analysed on the basis of its individual natural heterogeneity before indicator species are selected (natural unit specific biotope analysis), and, if necessary, subdivided into regionalised natural units—the mosaic-landscape units.

The example of two differently structured agricultural landscapes, the areas Hasenholz and Eggersdorf, in the natural unit "East Brandenburg Platte" in east Germany, was used for a regionalised natural unit division into mosaic-landscape units. The small wooded areas (hedges, forest islands within agricultural fields) are of significance in the landscape of both areas. But the portion of land with such wooded areas is very different. The extremes of the two areas are the highly wooded areas with a high density of hedges and forest islands versus very minimally structured, vacant agricultural steppes (Kretschmer et al., 1995; Hoffmann et al., 2001). Small bodies of water (cettel holes[1] and ponds) also define the landscape but appear in very different spatial distribution and density (Kalettka, 1996; Kalettka et al., 2001).

Using a geographical information system (GIS) and the ArcView/ArcInfo program packet, as well as digital results of the biotope mapping (Landesumweltamt Brandenburg, 1994, 2000), the wooded areas and small water bodies were calculated and evaluated with the help of a grid density model and by using the "Moving-Window-Method" (Silvermann, 1986; Bailey and Gatrell, 1995). In 5 m steps and a radius of 400 m, each rasterised biotope cell was accumulated, generalised and smoothed by a modified shifting mean function. By this process, mosaic-landscape units, in this case, units with similar density of wooded areas and small bodies of water were identified.

A classification of the densities was made step-wise, in a rough scale "high", "medium" and "low" (Table 1), to establish a practicable number of mosaic-landscape units with differing wooded and small water body densities. The density of habitats were recorded as a percent of the area portion per area unit of 1 km^2.

In order to bring the regionalised natural area types in accordance with available administrative borders, calculated maps available on the biotope type density (wooded areas, small water bodies) were overlaid with aerial photos behind them clearly showing the streets, paths, fields and other characteristics. Then an adjustment of the regionalised areas with varying densities of wooded areas and small water bodies on important administrative borders, for example, streets and field borders, was made. For these regionalised areas, an analysis of the area of all biotopes, e.g. hedges, forest islands, fields and grassland, was carried out.

Table 1
Classification of the density areas of wooded areas (high >2.7%, medium 1.3–2.7%, low >1.3%) and small bodies of water (high >5.1%, medium 2.6–5.1%, low >2.5%) in the East Brandenburg Platte for the definition of the mosaic-landscape units

Density of wooded areas	Density of small bodies of water	Regionalised natural areas
High	High	1
High	Medium	2
High	Low	3
Medium	High	4
Medium	Medium	5
Medium	Low	6
Low	High	7
Low	Medium	8
Low	Low	9

[1] Cettel hole—glacially shaped small lentic waters.

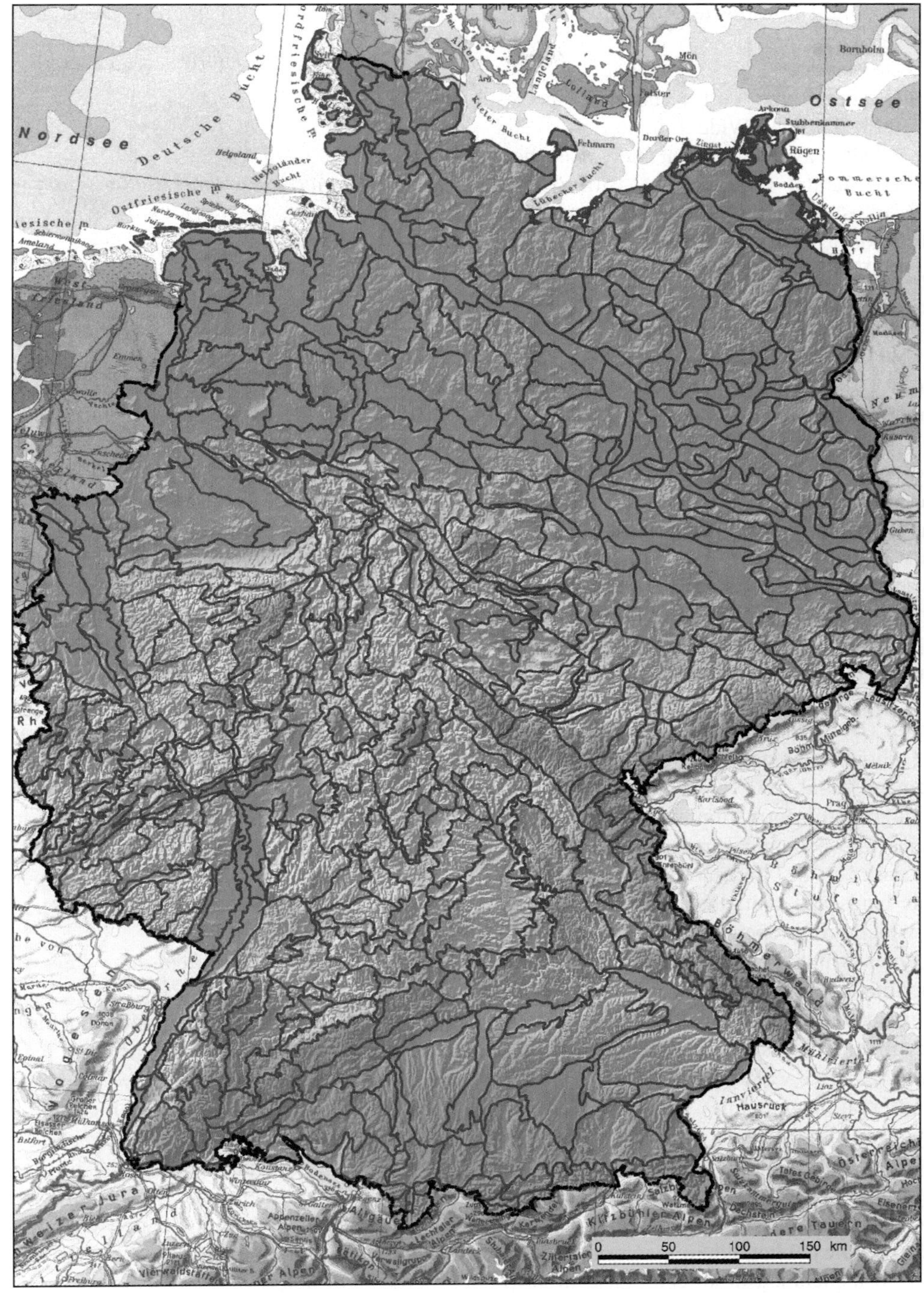

Fig. 1. Landscape units of Germany according to Meynen et al. (1962).

2.2. Selection of indicator species on the basis of the species group of birds

Birds count among the best researched species group in Central Europe. Wide ranging basic and expert knowledge is available about their existence, distribution and habitat requirements. Bird species are often used as indicators to show the ecological condition of agricultural areas (Ormerod and Watkinson, 2000). With regard to their habitat requirements, birds can be easily classified, for example in character species in particular habitats (Flade, 1994) or summarized in ecological groups (Zenker, 1982; Kretschmer et al., 1995) which contain species with similar habitat requirements.

Taking the Eggersdorf area, which falls under the regionalised area type 7 with minimal wooded areas and a high concentration of small water bodies (cf. Table 1), as an example, bird species were recorded on the basis of field survey (species inventory and abundance according to Dornbusch et al., 1968, two observation years). For the selection of the set of the indicator species, the species inventory in the superordinated areas, in this case the state of Brandenburg (Rutschke, 1983; Hoffmann and Haase, 2001), the county district (Hoffmann and Koszinski, 1993) and the directly adjoining areas of the mosaic-landscape units (Hoffmann et al., 2000) were taken into account as background information and a classification of the species into ecological groups was made according to Zenker (1982). In accordance with the defined mosaic-landscape unit, the bird species could be counted which prefer open agricultural landscapes as well as types which prefer the small water areas and their perimeters as habitats. In order to keep the number of indicator types manageable, a total of 13 bird species in order of species with special European significance (EG-VSRL—species of the European bird conservation guideline), species of Germany with high conservation priority (species of the German Red Data Book according the Bundesamt für Naturschutz, 1996; Jedicke, 1997) and species of particular regional importance (BRB) were chosen as indicators. The specific habitat requirements of each bird species were then included in an ecological description on their habitat requirements. These description contained essential information, e.g. about necessary habitats in the breeding time, nesting site, food and possible measures to improve the abundance of the populations.

3. Results

3.1. Regionalised natural area division with the example of the crop farming areas of Hasenholz and Eggersdorf

The densities of wooded areas and small water bodies in the two investigation areas Hasenholz and Eggersdorf calculated with the help of the grid model are presented in Fig. 2 in a selected land area of about 10 km^2. While the Hasenholz area (above) has a high level of wooded areas and minimal number of small water bodies, in Eggersdorf (below) the wooded areas are minimal, but there are many small bodies of water. Because of the different structures, both of these areas have a different meaning as living space and for the species diversity in agricultural areas.

The classification of the wooded areas and small bodies of water into the stages high, medium and low yielded a total of nine mosaic-landscape units within the larger natural unit "East Brandenburg Platte" (see Table 1). The calculated biotope density areas showed a very irregular area distribution which serves as a disadvantage for the practical measures. With the combination of the biotope type maps and the aerial photos, the borders of the regionalised spaces is adjusted to the clearly recognisable administrative borders (paths, streets, wood perimeters, field borders), see Fig. 3.

3.2. Bird species as indicators in Eggersdorf

In the Eggersdorf area, 37 breeding bird species and an additional 39 species identified as feeding or resting guests, were identified. Five of the breeding bird types—*Alauda arvensis*, *Coturnix coturnix*, *Perdix perdix*, *Miliaria calandra*, and *Emberiza hortunala*—are considered typical open area birds. Thirteen of the species are identified as water or water perimeter birds, sometimes in combination with adjoining open landscapes, for example *Motacilla flava*, *Saxicola rubetra* and *Emberiza schoeniclus*, as well as 10 of the forest and forest perimeter birds. The birds identified as forest and forest perimeter birds are in the superordinated areas the most frequently occurring birds, for

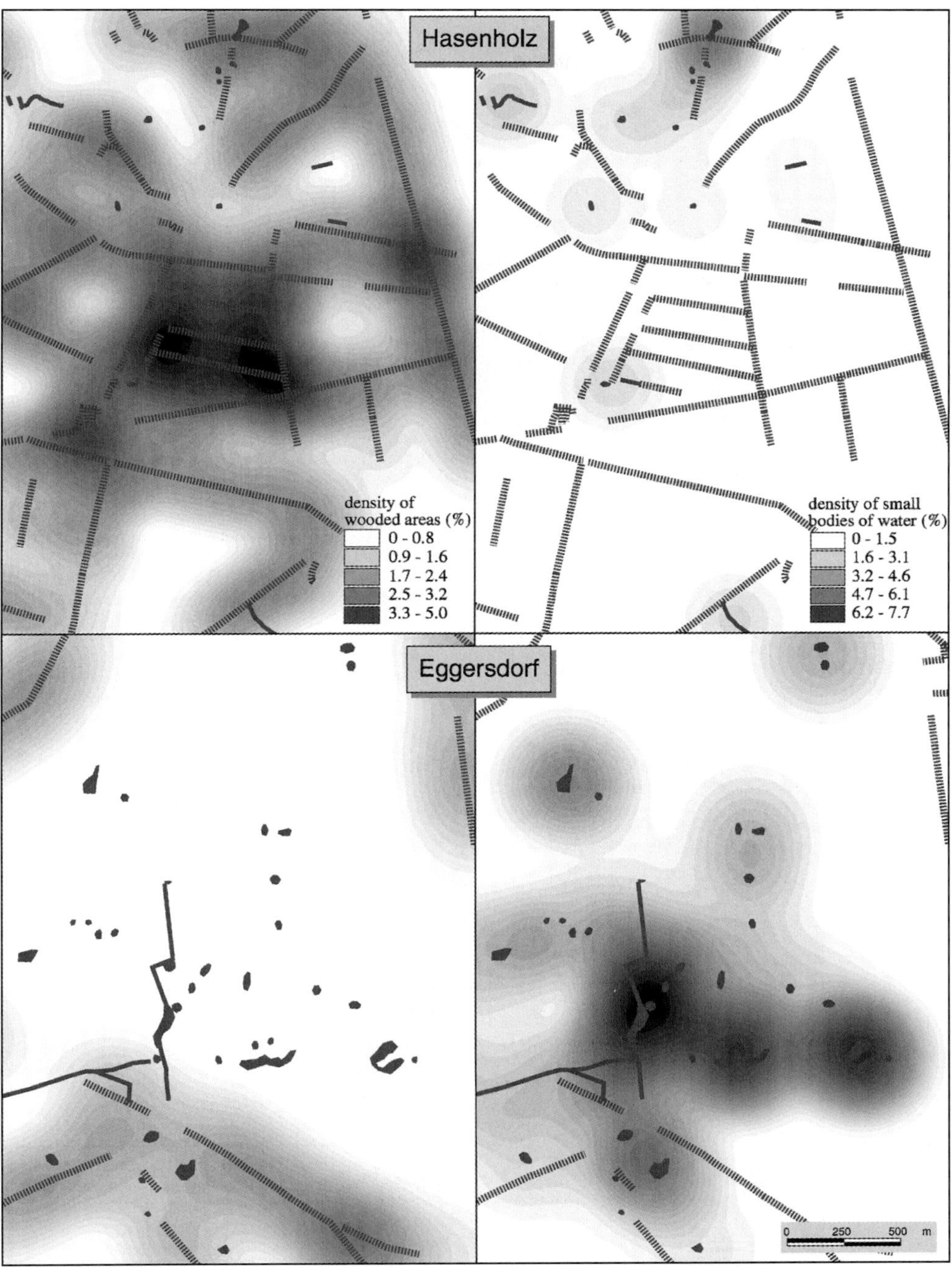

Fig. 2. Wooded areas (left) and small bodies of water (right) in the areas Hasenholz and Eggersdorf, above: high wooded area and low small water body density, below: low wooded area density and high density of small bodies of water.

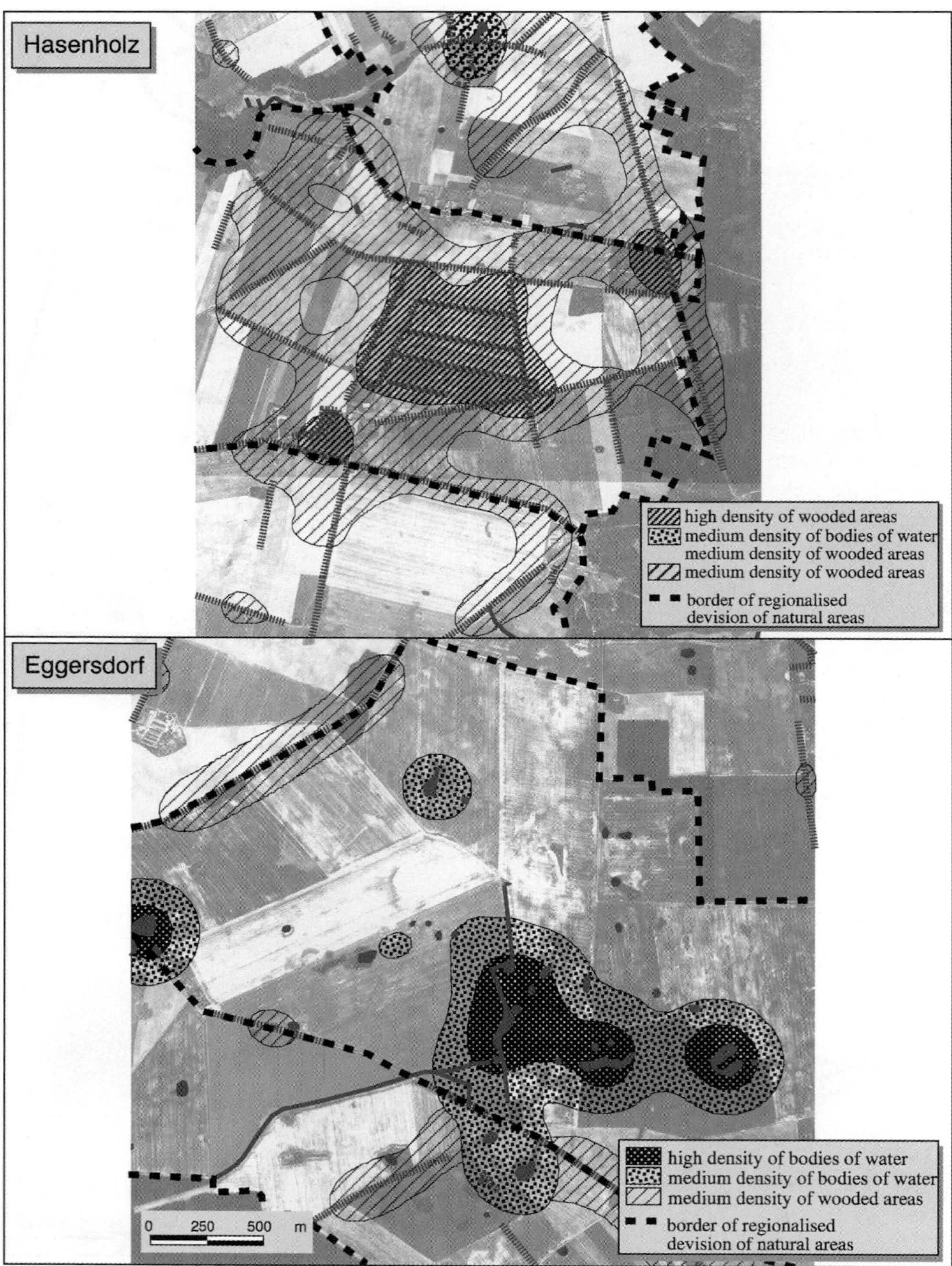

Fig. 3. Regionalised division landscape units with under-laid aerial views to show mosaic-landscape units with consideration of administrative borders, above: Hasenholz with high density of wooded areas and low density of small water bodies, below: Eggersdorf with low wooded areas and a high density of small water bodies.

example *Turdus merula* and *Parus major*, which have a relatively broad habitat spectrum. In contrast, the majority of the open area and water or water periphery birds are more strongly specialised species and adapted for the most part to agricultural structures with minimal wooded areas and more bodies of water. For this reason, birds from this ecological group were chosen as indicators typical to the landscape peculiarities in Eggersdorf (Table 2).

The current abundance of these indicator birds is used as status-indicator. Through a comparison with the endangerment situation (regional, supra-regional) and the current habitat conditions in the area, the target abundance for regional development targets are estimated. In this manner, it is possible to differentiate between species which are present in optimal or nearly optimal numbers with the target of maintaining the status quo and endangered species with the goal of increasing their population.

The abundance was estimated for each species on the basis of regional population data (status). An area of approximately 3.6 km^2 of the selected mosaic-landscape unit was used to develop specific targets for the population development for the indicator species. These are based on known abundance of the concerned indicator species in positively structured areas of the superordinated areas (Rutschke, 1983; Rheinwald, 1993; Hoffmann and Koszinski, 1993; Hoffmann et al., 2000). A division of the status and target indicators was made into groups based on frequency: 0—extinct, 1—very rare, 2—rare, 3—scattered, 4—widespread, and 5—frequent. These numbers were given as values for the current status and the development target for each indicator species. For example, for *Miliaria calandra*: (1,3), see Table 2.

An ecological description on the habitat requirements was made for each indicator species based on data from literature and regional expert knowledge (Table 3) which serves as the basis for measures for the improvement of habitat conditions for the species concerned.

In order to better visualise the indicators, they have been presented in a "Spider Web diagram" (Fig. 4). Behind every type, the status and target is given with the appropriate frequency value.

3.3. Derivation of regional measures to improve the habitat conditions

With the help of a GIS supported biotope type analysis in the mosaic-landscape units the available portion of areas of the natural and semi-natural biotopes were analysed and presented in a manner similar to the indicator species (Abb. 5). A distinction was made between the current status and the requirements drawn out of knowledge of the habitat requirements

Table 2
Indicator species (breeding birds) in the area of Eggersdorf (wooded area density low, small water body density high), hierarchy: EG-VSRL (species of the European bird conservation guideline), D (species of Germany with high conservation priority), BRB (species of particular regional importance), abundance (status, target) and aims for the population development

Indicator species	Ranking			Abundance		Aim
	EG-VSRL	D	BRB	State	Target	
Alauda arvensis	No	No	Yes	5	5	Maintain status quo
Cortunix cortunix	No	Yes	Yes	2	3	Increase population
Perdix perdix	Yes	Yes	Yes	1	2	Increase population
Miliaria calandra	No	Yes	Yes	1	3	Increase population
Lanius collurio	Yes	Yes	Yes	2	3	Increase population
Vanellus vanellus	No	Yes	Yes	0	1	Increase population
Grus grus	Yes	Yes	Yes	1	2	Increase population
Circus aeruginosus	Yes	Yes	Yes	2	3	Increase population
Saxicola rubetra	No	Yes	Yes	1	3	Increase population
Motacilla flava	No	No	Yes	3	3	Maintain status quo
Emberiza citrinella	No	No	Yes	3	3	Maintain status quo
Emberiza hortulana	Yes	Yes	Yes	1	3	Increase population
Passer montanus	No	No	Yes	2	3	Increase population

Table 3
Ecological description of one indicator species in the area of Eggersdorf with the example of the breeding bird *Emberiza calandra* (Corn bunting)

Miliaria calandra (Corn bunting): endangered in Germany, seriously endangered in Brandenburg
Habitat: wide, open landscapes with low density of wooded areas, avoids forests, forest perimeters and narrow structured hedgerow landscapes
Habitat during the breeding season: prefers grassland and ruderal areas in the edge of arable fields, self greened annual or perennial fallow fields, dry and semi-dry grasslands, minimum territory approximately 0.5–2 ha
Singing locations: single trees or bushes, perimeters of very small forest islands within arable fields or grassland, grassy and bushy areas on small bodies of water and paths
Nesting site: on the ground in grassy, weedy or herby vegetation
Food: insects (breeding season), cereal and wild plant seeds (winter), seeks food on fallow fields, field peripheries, on ruderal and grassy areas, and along sandy paths
Possibilities for population increase: an increase of self-greened fallow fields, ruderal areas with low density of the plant structure, dry and semi-dry grassland, an increase of breadth of margins along paths, reduction or halt in the use of herbicides and insecticides on portions of crop lands

for changes in the biotope structure with the goal of biotope development. As a practical measure for an increase in the population of *Miliaria calandra*, the population of which should increase from the frequency 1 to 3, a significant increase in the spontaneously greened fallow areas is needed (see Tables 2 and 3). Also, margins on the small wooded areas, on paths and small water bodies, and an increase in the grassland areas, and, if possible, a higher proportion of non-paved field paths which are bordered with fruit trees, would positively affect the population development.

The proportion of biotope for each biotope type is shown in Fig. 5 as an area measure as a percent to the total areas. The digitalised results of the biotope type mapping serve as a relatively good data base for the multiplicity of relevant biotopes, for example for wooded areas, water and dry grasslands. They are relatively simple to evaluate for the required area related analyses. Some small biotopes, for example the width of the margins on hedges and field paths, have not yet been documented sufficiently with the biotope mapping (Landesumweltamt Brandenburg, 2000). They must be expanded with terrestrial surveys if necessary.

4. Discussion

The regionalisation of the landscapes within the framework of the mosaic indicator approach is targeted

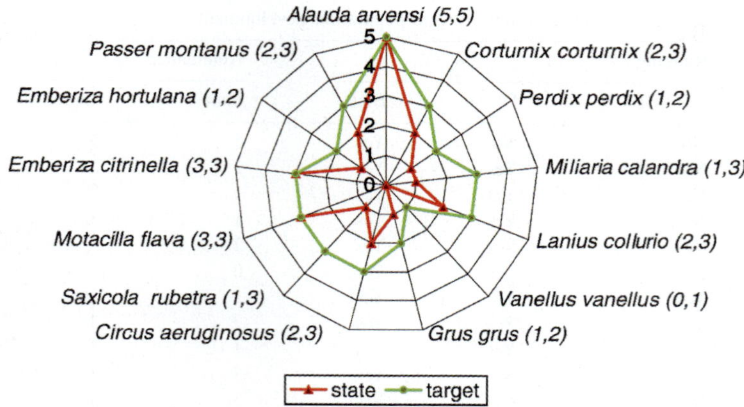

Fig. 4. Selected indicator species (breeding birds) in the areas of Eggersdorf with status and target values according the classification of their abundance.

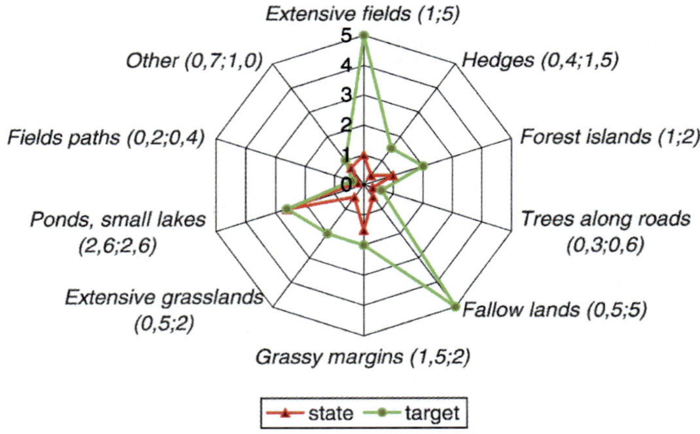

Fig. 5. Habitat conditions in the Eggersdorf area, status and target of each biotope as a percent of the total area of the mosaic-landscape unit.

at finding a practical method to considerate the main landscape peculiarities with their related species diversity to find the regionally typical key species in each area to serve as indicators. Through targeted promotion of regional particularities, landscape typical diversity of species and their habitats shall be maintained and their development promoted. At the same time, possible negative influences through agricultural practices which are uniformly implemented regardless of the special local conditions shall be compensated for.

The methodological approach of the mosaic indicator approach was tested in one area as a first step and as an example, with one species group, the birds selected, as landscape typical indicators. The use of species as indicators for one area should not be limited to one species group (Bastian and Schreiber, 1994; Breininger et al., 1998) but rather include multiple species groups if possible. Due to the large number of water bodies in the study area Eggersdorf, in addition to birds, amphibians have a particular ecological significance (Berger et al., 1998), so that some species from this group could also be suitable as indicators. Depending on the regional knowledge level in other areas, landscape typical species of other animal and plant groups, for example ferns and flowering plants (Hoffmann, 1996, 1998) and ground beetles (Duelli, 1999), are conceivable.

For the further development of regionalised natural area organisation, further biotopes in addition to the characteristic habitat structures of wooded areas and small bodies of water in the studied area should be considered. In other areas, for example, in pre-mountainous areas mountain meadows or on the sea coast, salt marshes which mark the landscape must also be considered. Also, other parameters, for example certain relief characteristics, can be included.

With about 900 different natural units in Germany, which must be broken down into smaller areas for a regionalised indicator selection, the question emerges if the mosaic indicator approach is too complicated. The following arguments speak for its practicability:

- there is a good data base available for important species groups, particularly for birds and plants (ferns and flowering plants) for all of Germany and some other countries in Europe, too;
- data on the natural division of the land areas is available for all of Germany and is being refined for some areas;
- the results of the biotope mapping needed for regionalised natural area division are also almost completely available and are being to a large extent digitalised and can be analysed relatively simply;
- aerial photos are available for all parts of Germany plus digital administrative data;
- the available computer technology (including software) seems to be relatively uncomplicated for complete data processing;
- one can build upon an existing system of well educated farmers, biologists and landscape planners

as well as a large number of available workers in agricultural areas who could assume the workload.

5. Conclusion

Briefly, it is possible to determine that the mosaic indicator approach is a data intensive process, which however can be clearly overseen in the regional, individual landscape mosaic units and seems to be able to be regulated with the help of modern computer technology. It can also be anticipated that due to the strong regionally targeted measures for species and biotope protection presented here, with consideration of limited financial resources, better effects on the landscape and local typical species diversity can be achieved.

Presently a strong shift in the decision making competence for agricultural tasks can be seen at the regional level. From an ecological perspective, an appropriate tool is necessary. Practical natural protection as a regional approach (Piorr, 1998; Plachter and Korbun, 2003) is necessary for the different natural space variations and quality goals with different extents and targeted achievement (Werner and Plachter, 2000). The complete integration of natural protection goals in farm production and how it is being tackled with the mosaic indicator approach, does not mean a division into production and protection areas, but rather the implementation of regionally different concepts in the structuring of agricultural landscapes (Kretschmer et al., 1995; Hoffmann et al., 2001; Hoffmann and Kretschmer, 2001). Also, more strongly targeted are land uses adapted to local conditions (Haber, 1998).

Through the combination of the regionalised mosaic-landscape units with aerial views, administrative structures (e.g. streets, field plots) land areas are easy to organise into the individual areas. On this basis, each farmer has the possibility to be included in programs to promote biological diversity.

As further steps are required:

- the indicator concept must be further developed, for example, the regionalised natural space division and the linked selection of appropriate criteria as well as methods for regional differentiated selection of indicator species must be improved;
- the mosaic indicator approach must be tested in different sample areas;
- participants, especially farmers, must be included from the outset.

References

Bailey, T., Gatrell, A., 1995. Interactive Spatial Data Analysis. Longman, Harlow.

Bastian, O., Schreiber, K.F. (Eds.), 1994. Analyse und ökologische Bewertung der Landschaft. Fischer, Jena.

Berger, G., Dürr, S., Kretschmer, H., 1998. Integration von Zielen des Biotop- und Artenschutzes in ackerbauliche Nutzungssysteme am Beispiel von Amphibien. Strukturwandel der Landnutzung 1998, 81–97.

Breininger, D.R., Barkaszi, M.J., Smith, R.B., Oddy, D.M., Provancha, J.A., 1998. Prioritizing wildlife taxa for biological diversity conservation at the local scale. Environ. Manage. 22 (2), 315–321.

Bundesamt für Naturschutz, 1996. Rote Liste gefährdeter Pflanzen Deutschlands. Schriftenreihe für Vegetationskunde. vol. 28, pp. 1–744.

Dornbusch, M., Grün, G., König, H., Stephan, B., 1968. Zur Methode der Ermittlung von Brutvogelsiedlungsdichten auf Kontrollflächen. Mitt. IG Avifauna DDR, pp. 7–16.

Duelli, P., 1999. Biodiversity evaluation in agricultural landscapes: above-ground insects. Agric. Ecosyst. Environ. 74, 33–64.

Flade, M., 1994. Die Brutvogelgemeinschaften Mittel- und Norddeutschlands. IHW-Verlag, Eching.

Hoffmann, J., 1996. Patterns of plants species in the landscape as potential indicators of environment changes. Acta Agronomica Ovariensis 38 (1/2), 23–39.

Hoffmann, J., 1998. Assessing the effects of environmental changes in a landscape by means of ecological characteristics of plant species. Landsc. Urban Plan. 41, 239–248.

Hoffmann, J., Greef, J.M., 2003. Mosaic indicators—theoretical approach for the development of indicators for species diversity in agricultural landscapes. Agric. Ecosyst. Environ. 98, 387–394.

Hoffmann, J., Koszinski, A. (Eds.), 1993. Die Vogelwelt im Landkreis Strausberg. Tastomat, Eggersdorf.

Hoffmann, J., Koszinski, A., Mittelstädt, H., Köhn, K.H., Fiddicke, M., Leue, J., Leue, M., Türschmann, H., Büxler, O., Lange, K., Haase, G., Kage, J., 2000. Die Vogelwelt des Naturparks Märkische Schweiz. Tastomat, Eggersdorf.

Hoffmann, J., Haase, G., 2001. Grauammer—*Miliaria calandra* (Linnaeus 1758). In: Die Vogelwelt Brandenburgs, pp. 619–622.

Hoffmann, J., Kretschmer, H., 2001. Zum Biotop- und Artenschutzwert großer Ackerschläge in Nordostdeutschland. Peckiana 1, 17–31.

Hoffmann, J., Kretschmer, H., Pfeffer, H., 2001. Effects of patterning on biodiversity in Northeast German agro-landscapes. Ecological Studies, vol. 147. Springer, Berlin, pp. 325–340.

Jedicke, E. (Ed.), 1997. Die Roten Listen—Gefährdete Pflanzen und Tiere, Pflanzengesellschaften und Biotoptypen in Bund und Ländern. Ulmer, Stuttgart.

Kalettka, T., 1996. Die Problematik der Sölle (Kleinhohlformen) im Jungmoränengebiet Nordostdeutschlands. Naturschutz und Landschaftspflege in Brandenburg S1996, 4–12.

Kalettka, T., Rudat, C., Quast, J., 2001. Potholes in Northeast German agro-landscapes: functions, land use impacts, and protection strategies. In: Tenhunen, J.D., Lenz, R., Hantschel, R. (Eds.), Ecosystem Approaches to Landscape Management in Central Europe. Ecological Studies, vol. 147. Springer, Berlin, pp. 291–298.

Klijn, F., 1991. Hierarchical classification of ecosystems: a tool for susceptibility analysis and quality evaluation for environmental policy. In: Ravera, O. (Ed.), Terrestrial and Aquatic Ecosystems, Perturbation and Recovery, New York, pp. 80–89.

Kretschmer, H., Pfeffer, H., Hoffmann, J., Schrödl, G., Fux, I., 1995. Strukturelemente in Agrarlandschaften Ostdeutschlands—Bedeutung für den Biotop- und Artenschutz. ZALF-Bericht 19. Selbstverlag, Müncheberg.

Landesumweltamt Brandenburg, 1994. Biotopkartierung Brandenburg—Kartierungsanleitung. Unze, Potsdam.

Landesumweltamt Brandenburg, 2000. Digitalisierte Ergebnisse der Biotopkartierung des Landes Brandenburg. Unze, Potsdam.

Meynen, E., Schmithüsen, J., Gellert, J.F., Neef, E., Müller-Miny, H., Schultze, J.H., 1962. Handbuch der naturräumlichen Gliederung Deutschlands. Bundesanstalt für Landeskunde und Raumforschung, Bad Godesberg.

Ormerod, S.J., Watkinson, A.R., 2000. Editors' introduction: birds and agriculture. J. Appl. Ecol. 37, 699–705.

Piorr, H.P., 1998. Development of an indicator framework for the analysis of agricultural landscapes. Bornimer Agrarwissenschaftliche Berichte 21, 70–76.

Plachter, H., Korbun, T., 2003. A methological primer for the determination of nature conservation targets in agricultural landscapes. In: Flade, M. (Ed.), Nature Conservation in Agricultural landscapes. Quelle and Meyer (in press).

Rennings, K., 1994. Indikatoren für eine dauerhaft-umweltgerechte Entwicklung. Metzler-Poeschel, Stuttgart.

Rheinwald, G., 1993. Atlas der Verbreitung und Häufigkeit der Brutvögel Deutschlands—Kartierung um 1985. Schriftenreihe des DDA 12.

Rutschke, E., 1983. Die Vogelwelt Brandenburgs. Fischer, Jena.

Silvermann, B.W., 1986. Density Estimation for Statistics and Data Analysis. Chapman and Hall, New York.

Werner, A., Plachter, H., 2000. Integration von Naturschutzzielen in die landwirtschaftliche Landnutzung—Voraussetzungen, Methodenentwicklung und Praxisbezug. Agrarspectrum 31, 44–61.

Zenker, W., 1982. Beziehungen zwischen dem Vogelbestand und der Struktur der Kulturlandschaft. Beiträge zur Avifauna des Rheinlandes 15, 1–249.

PART IV
EXPERIENCES AND APPLICATIONS

Chapter 6 Experiences of Applications

E. Osinski, W. Büchs,
B. Matzdorf, J. Weickel

Synoptic Introduction

ELSEVIER

Available online at www.sciencedirect.com

Agriculture, Ecosystems and Environment 98 (2003) 407–421

Agriculture Ecosystems & Environment

www.elsevier.com/locate/agee

Application of biotic indicators for evaluation of sustainable land use—current procedures and future developments

Elisabeth Osinski [a,*], Uwe Meier [b], Wolfgang Büchs [b], Joerg Weickel [c], Bettina Matzdorf [d]

[a] *Chair of Agricultural Economics, Center of Life and Food Sciences, Technical University Munich, Alte Akademie 14, Freising 85350, Germany*

[b] *Federal Biological Research Centre for Agriculture and Forestry, Messeweg 11/12, Braunschweig 38104, Germany*

[c] *Landesanstalt fuer Pflanzenbau und Pflanzenschutz Rheinland-Pfalz, Essenheimer Straße 144, Mainz 55128, Germany*

[d] *Bettina Matzdorf, Zentrum fuer Agrarlandschafts- und Landnutzungsforschung (ZALF), Institut fuer Soziooekonomie, Eberswalder Street 84, Muencheberg D-15374, Germany*

Abstract

Indicators in the field of bio-diversity and landscape are applied on various levels, including the continental field as well as the individual agricultural enterprise. Apart from the ecological evaluation of agricultural enterprises and agrarian policy measures, indicators are also used in environmental reporting and evaluation as well as in planning or simulation models in administrative and scientific fields. Already for longer period of time, indicators have been used as assessment criteria in landscape planning to support decisions regarding land use. Due to the standards the EU commission requires from the member states in this regard, the application of indicators to assess the effects of agri-environment programs have gained prominence. The EU requires proof of the achievement of the promotion aims such as soil (erosion, nutrients, plant-protective agents), water, bio-diversity and landscape (EU/VO 12004/00).

This commitment can be met by means of a functioning environmental reporting. However, only an insufficient number of suitable indicators exist in the fields of bio-diversity and landscape. In a bottom-up approach, control systems to assess ecological farming achieved on an operational level for a future development were recently developed. Here, a future development is seen in the alignment referring to results or goals, respectively, and in the regionalisation. The article gives an overview of the indicator application on different spatial levels and for different purposes.

Keywords: Application of indicators; Monitoring; Farm assessment; Landscape planning; Agri-environment program

1. Application objectives of indicators in the fields of bio-diversity and landscape

The necessity to develop indicators in the fields of bio-diversity and landscape has already been sufficiently documented (OECD, 1997, 2001; KOM, 2001a,b). The common goal is a standardised assessment leading to world-wide comparable results. When developing indicators, their different applications must be taken into consideration. Apart from the objectives to be obtained, the application of indicators vary with regard to their spatial reference. Attempts to provide and apply indicator systems already exist on a European (respectively continental), national as well as regional and local level, the latter also including the scale of agricultural enterprises.

* Corresponding author. Tel.: +49-8161-71-4461; fax: +49-8161-71-4426.
E-mail address: osinski@wzw.tum.de (E. Osinski).

0167-8809/$ – see front matter © 2003 Elsevier Science B.V. All rights reserved.
doi:10.1016/S0167-8809(03)00100-2

By means of indicators, the compliance of development processes with sustainable development on a national and international level are tested (Muessner et al., 2002). In larger regional units, the use of indicators is mostly limited to the evaluation of time rows of land use changes, of potential stress factors or the development of a specific species population. In large areas, an average has to be found between extremely aggregated characteristic values belonging mostly to pressure indicators, and the sectorial view of various species or groups of species with an indicative character. The smaller the observed area, the more characteristic values can be surveyed and a description of the cause and effect relations is more likely. Besides the influence on the biotic resources caused by an enterprise, the actual condition can also be registered immediately. This control can not only be used to verify agricultural practice with regard to its environmental relevance, but also to estimate the effect of the measures of agri-environment programs. The application fields stated include an ex-post-evaluation.

Another field for the application of indicators refers to a future-oriented assessment of planned measures. Due to the fact that the entire system observed cannot be shown in the field of planning, the use of indicators here is also of necessity.

Therefore, indicators must comply with the problems shown and the spaces for which they must be valid. Basically, the application fields of indicators can be defined as follows:

- Assessment of agrarian enterprises.
- Assessment of efficiency of political measures.
- Environmental reporting/environmental monitoring.
- Assessment within a planning framework.

The contributions of this volume for the application of indicators can be classified, with regard to the above-mentioned application of assessment of environmental effects of agricultural enterprises (Roth and Schwabe, 2003; Heyer et al., 2003; Oppermann, 2003; Braband et al., 2003) or the evaluation of agri-environmental measures (Marggraf, 2003; Menge, 2003) as a specific form of political intervention into agricultural action. The contributions stated, deal primarily with agri-environmental indicators, since agriculture, due to its superficial extension, heavily influences bio-diversity and landscape. Additionally, an overview of approaches on the integration of nature protection aims into operational management is given.

Even though the knowledge of agri-environmental indicators is still incomplete and a continuous development of the methods for the derivation of indicators is necessary, there already exists the necessity of using operational indicators for bio-diversity and landscape (comp. Larsson and Esteban, 2000).

Depending on the objective and the spatial reference of the application, the claims vary, i.e. to the transferability, the extent of investigation and the availability of data. In the following section, the application of indicators according to the requirements of the OECD (1997) in the various, above-mentioned fields is examined more closely with regard to the measurability of indicators, the policy relevance and transferability of application and the problems, namely the political soundness of landscape protection and bio-diversity.

2. Assessment of agrarian and forestal enterprises

2.1. On-farm assessment

With regard to environmental reporting, changes of environment state indicators or changes of pressure indicators, respectively, are documented. Through this approach, the cause factors remain anonymous. On the assumption that an improvement of the environmental state can be initiated by environment observation, the inclusion of the cause factors is necessary. In the field of the a-biotic resource protection, legal regulations for the reduction of emissions (fertiliser regulations, plant protection law, water protection regulations) have already been determined. In the field of the biotic resource protection in Germany there have been improvements, for one in the specification of legal provisions (republication of the Federal Nature Conservation Law which, for example, should regulate the minimum provision of landscapes with biotopes), secondly it has been attempted to improve agricultural and industrial operations through self-commitment or certifying procedures. The impact on a-biotic and biotic resources is coming from a single farm and can be classified according to farming intensity and the extent of measures.

The interaction of measure and location implies a certain effect on the respective protected good. If one wants more than an overview on the efficiency of agricultural measures on biotic resources, one has to focus on the agricultural enterprise. From this idea, operational control procedures have been developed in order to determine the certain degree of environmentally friendly farming of the enterprises. In some of these control systems biotic aspects have already been implemented (see compilation by Braband et al., 2003), while in others a-biotic factors are controlled. The systems are implemented either by experts or by the farmers themselves. The purpose of most of the procedures is to document the existing farm stock, wild species, biotopes, landscape elements and land use types. However, it is not enough to require the documentation of species or structures when they are not connected to a quantitatively or a qualitatively evaluating statement. An evaluation of the actual state can only be effected if corresponding reference values are determined. The contributions from Roth and Schwabe (2003) as well as from Oppermann (2003) consider specific regional characteristics and compare the stock with a target value.

As a relevant indicator, Roth and Schwabe (2003) present 'the ecologically and culturally important surfaces of the agrarian region' (OELF). A modified form of this structure indicator is also applied in other systems (see contribution of Braband et al., 2003). Instead of remaining at a statement regarding the actual state, the authors put it in relation to the set target by evaluating plans for the use and cultivation of the respective agrarian region. Through this way, deficits can be determined and compensated. With the above-mentioned plans, the State of Thuringia has, together with some states (Bavaria, Saxony-Anhalt), possess secure environmental planning approaches. Most states in Germany do not have information about the desired surface values for ecologically relevant structures. These objectives, however, are prerequisite for an effective application of indicators.

Oppermann's contribution (2003) surmounts the documentation and determination of the deficits in the field of ecologically relevant areas. The 'nature balance scheme for farms' (Oppermann, 2003) contains four assessment fields which can be carried out on an operational level without large-scale data surveying. Apart from structure-related fields, the indicators also describe the fields of bio-diversity (richness of species) and field management (i.e. farming intensity). Besides the documentation, points for the evaluation of the condition of the enterprise with regard to the various indicators are awarded. Even here, the actual objective comparison is formulated regionally specific. However, here it also has to have a qualitative value due to fact that in natural green space a specific list of species potentially occurring in this region from which a minimum of species should exist.

The amount of effort required for the data collection is a common criticism of operational assessment procedures. In order to lower this amount of work, Oppermann (2003) checks if the number of indicators, 47, may be reduced. He therefore identifies the results concerning the richness of species as characteristic for the remaining results. This means that the total number of given points for the enterprise is best correlated with the percentage of species rich areas. The question being, if the surveying of species is sufficient for an operational assessment. This would simplify the operational assessment process.

The relevance of the collected data for the correct image of the agrarian eco-systems cannot be evaluated by applying the presented operational assessment systems. However, there exists a significant tendency for the systems to determine action-related indicators, which are easier to collect, than result-oriented indicators. Nevertheless, one must take into consideration that the paid actions do not necessarily lead to the expected results. If on the other hand only results are paid, the procedure leading to these results cannot be clearly determined and thus they are not payable in the strict sense of payment.

Generally, the assessment of the state of bio-diversity on an operational level rising above simple indicator types such as single target species is considered to be too complicated. Since the varying quality and quantity of seasonally and annually appearing biotic indicators has been criticised in the past, the collection of easily measurable a-biotic factors was proposed. Measurements try to prove that a connection exists between the farm management, the subsequently measured changes of a-biotic factors and the consequences for indicator organisms (Heyer et al., 2003). Therefore, one tries to unify the effect of certain agricultural implementation procedures on biotic aspects. A verification of this thesis at different positions is still pending. However, in individual cases it is necessary to clarify and to secure the reference of the a-biotic indicator to the respective biotic indicator before been able to use the first one.

The section "Application of indicators" is rounded up by a contribution from Braband et al. (2003) which analyses various operational assessment procedures and cites points of criticism also relevant for other indicator systems. This also confirms that mostly action-oriented indicators are rewarded, although from a scientific point of view, the effects of these actions cannot be entirely predicted. On the other hand, actions that are not rewarded, would therefore only show an effect after a long period of time, even though they would be of use for environmental policy. However, a better transparency of the action-oriented indicators is significant.

Only in a few systems is the quality of non-cultivated agricultural areas taken into consideration. In most cases, only the area size is taken into account.

When characterising the agricultural areas either actions or, in the case of result compensation, the number of field products or the species of the accompanying flora and fauna are characterised.

The assessment systems compared in this abstract can partly be developed even further towards more result-oriented indicators. A prerequisite, however, is the concretion of the objectives and above all public discussions.

Furthermore, the integration of farmers into the assessment procedure seems to be of importance in order to become conscious of the effects the measures can have on the bio-diversity of an agricultural enterprise. For this, the farmers' knowledge in the area of bio-diversity must be improved (also see Kleijn et al., 2001). Oppermann (2003) indicates that the farmers in the Black Forest were open for ecological procedures on their farms and were prepared to accept criticism after noticing unexpected positive aspects. However, a change in thinking is necessary to not only include a fixed implementation of indicators, but also to reflect together with farmers on how to achieve optimal effects for the environment.

Another development direction pursues a stronger formalisation of control procedures. As an enhancement of the procedure KUL (criteria of environmentally justified agriculture), a certification procedure (USL: Environmental securing system agriculture) has been proposed to give the farmers a market advantage as it certifies ecological compatibility (VdLufa, 2001).

The certification procedures, so far only used in ecological farming, would therefore be extended with regards to environmental compatibility. A prerequisite is the registration by the indicators of the strains and the utility of the agriculture and the acceptance of the defined tolerance ranges.

2.2. Criteria systems and environmental management to assess performance in nature conservation by agrarian and forestal enterprises

Trust and credibility are always interlinked with an assessment. In this respect, existing structures, for example in agriculture, which as such are credible or not, can only be assessed with regard to previously determined evaluation parameters, like test criteria. In the 1990s, criteria-oriented evaluation of agricultural enterprises has become increasingly established in practice. Practical criteria systems are briefly introduced and compared. None of the criteria systems uses bio-indicators to prove that a conservation measure enhanced bio-diversity. Rather it is assumed that measures specifically directed at protecting biotic resources will "per se" have positive effects on bio-diversity. This is a pragmatic and controllable approach which should be maintained and promoted as long as there are no reasonable economic and ecological alternatives.

2.2.1. Environmental management systems and nature conservation

The existing environmental management systems, Regulation EC (2001) and ISO 14001 (1996), do not provide for biological indicators to measure the starting situation and the success of measures in securing bio-diversity. When the protection of biotic resources is integrated into the operating structures and processes of an agricultural enterprise, it is assumed that measures to preserve bio-diversity, such as integrated land management or habitat networks are in fact serving to protect biotic resources.

Under the two environmental management systems mentioned above, integrated crop growing could help to permanently improve the protection of biotic resources in cultivation areas if clear provisions and criteria are defined for that purpose. Integrated crop cultivation is a minimum standard suitable for the integration into environmental management schemes. It aims for a continuous improvement of the environmental situation, and allows the measurement of adherence to the principles, and accounting of the site conditions. Connected aquatic and terrestrial succession zones can be set up outside the growing areas. They would serve to maintain bio-diversity in line with the intention of the agrarian environmental management systems. Bio-diversity on the actual cultivation area could be furthered by ecological-oriented cultivation, but is not necessarily required by standardised environmental management systems.

2.2.2. Environmental criteria systems in plant growing

Environmental criteria systems are based on environment-oriented performance criteria. On that basis, a company may allow inspectors to evaluate it for its environmental performance. If the evaluation is successfully, the company may use this fact for publicity, for example, by displaying this on a specific label.

Protection of biotic resources is anchored in each system, but with varying binding functions, and with varying solution approaches, depending on the production direction. The basic assumption is that a reduced input of substances, environmentally gentle cultural measures and straight nature-conservative measures in and outside the actual productive area are sufficient. None of the criteria systems require any scientific proof of the success of each individual measure, for instance on the basis of bio-indicators. The reason being that there are no general bio-indicators for all the different cultivation systems in the different climatic regions. Even if any were available, the effort required for inspection and verification would be too high in terms of time and money. Therefore, any measures of nature conservation implemented and documented will count as a success for nature conservation.

Euro-Retailer Produce Working Group of Good Agricultural Practice (EUREP GAP) (2002) is a co-operative by a number of leading European retail companies. The EUREP system consists of 'required' and 'encouraged' criteria. 'Required' means the respective criteria must be fulfilled if an enterprise wants to participate in the program, 'encouraged' means that the criteria specify an objective which is desirable to be achieved.

A key aim must be to enhance the environmental bio-diversity on farms through a conservation management plan.

Required: Has a conservation management plan been established and have the growers considered how they can enhance the environment for the benefit of the local community and the flora and fauna?

Encouraged: Do the growers understand and assess the impact their farming activities have on the environment and do they pursue nature conservation policy plans on their farms? Are the policies compatible with sustainable commercial production, and do they minimise influences on the environment as far as possible?

The performance of a company with regard to the protection of biotic resources cannot be measured by EUREP criteria. It is the strategy of the EUREP system to first increase awareness of problems of nature conservation and protection of biotic resources, and then motivate farmers to enhance their understanding of these fields.

In the Flower Label Program (FLP) (2002), environmental criteria are based on the environmental criteria list by Meier and Feltes (1996) and on a relevant Internet publication by the Federal Biological Research Centre of Agriculture and Forestry (BBA, 1996). It is only possible to a very limited extent to protect biotic resources in areas of highly intensive protected flower cultivation. Yet FLP criteria include the requirement that healthy crop residues are composted, and that the compost is recycled into the production system. Apart from these soil conservation measures, FLP seeks solutions outside the production areas. This contributes to maintaining biological diversity outside the production areas and enhances the quality of living of residents near flower plantations, especially in developing countries, where workers and other people often live between the plantations.

The Forest Stewardship Council (FSC) (2000) wants to achieve, on a global scale, forest management which is at the same time economically buoyant and compatible with nature and the human society. FSC certification is guided by 10 principles with 56 criteria. Some of the relevant criteria for biotic resource protection shall be presented here. Environmental impacts shall be assessed according to the scale and intensity of forest management and the uniqueness of the affected resources, and assessment results shall be adequately integrated into management systems. Safeguards shall exist which protect rare, threatened and endangered species and their habitats (e.g. nesting and feeding areas). Conservation zones and protection areas shall be established, appropriate to the scale and intensity of forest management and the uniqueness of the affected resources. Inappropriate hunting, fishing, trapping and collecting shall be controlled.

Forest management shall conserve biological diversity and its associated values, water resources, soils, and unique and fragile eco-systems and landscapes, and, by doing so, maintain the ecological functions and the integrity of the forest.

The following ecological functions and values shall be maintained intact, enhanced, or restored:

- Forest regeneration and succession.
- Genetic, species, and eco-system diversity.
- Natural cycles that affect the productivity of the forest eco-system.

The Pan European Forest Certificate (PEFC) (2002) was introduced in 2000 as a product marketing label. The PEFC certificate is only awarded to forest companies following sustainable forest management guidelines and allowing inspections of the fulfilment of environmental and social criteria.

The wildlife and conservation policy is: planning of forest management and mapping of forest resources, including forest biotopes of ecological importance and taking into account protected, rare, sensitive or typical forest eco-systems, such as meadowlands, wet biotopes, habitats of endemic or endangered species, or endangered or protected genetic in situ resources.

The indicators for conservation are:

- Protection and promotion of rare tree and bush species.
- Historical ways of management which have produced valuable eco-systems, for instance, coppice forest, should be promoted in suitable places if economically viable.
- Maintenance and harvest measures shall be carried out so that they do not cause lasting damage to the eco-system. Wherever possible, steps to enhance bio-diversity shall be undertaken.

Rainforest Alliance (2002). A number of agricultural producers (bananas, coffee, citrus, sugar cane, cocoa) in Latin America voluntarily allowed their farms to be evaluated by the US-based environmental organisation 'Rainforest Alliance' for the fulfilment of social and environmental criteria.

The indicators for natural habitats are:

- A plan to conserve and recuperate different eco-systems within the farm that including a mapping of the boundaries of critical habitats such as wetlands, lagoons and wildlife roosting or nesting areas.
- Indiscriminate or unjustified deforestation is prohibited on farms.
- Incorporation of forested areas into private conservation easements, private reserves or government protection programs.
- Buffer zones around protected areas.

The indicators for reforestation are:

- Rivers should have a 10 m buffer in flat areas and 50 m on inclines greater than 30% (measured horizontally).
- Public roads that border or cross plantations should have a 10 m buffer of native vegetation.
- Areas not in production or not suited for production must be reforested.
- Vegetative barriers of 30 m required around schools, clinics and housing areas. These buffers should be designed to reduce pesticide drift. Production-related buildings (packing stations, showers and storage areas) should have a 10 m buffer zone.

The indicators for wildlife conservation are:

- Establishment of policies prohibiting hunting, collecting and trafficking.
- Consultation of national environmental legislation, specifically legislation related to wildlife protection and establishment of biological corridors.
- Company support of local research, particularly those projects recommended by the certification program.

3. Assessment of the effectiveness of political measures

The European Union, seeing the necessity of preservation measures for biological variety, has promoted the finding of so-called headline indicators as well as indicators for specific strategic instruments (KOM, 2001b). Therefore, the outcome of all strategies in total (headline indicator) and of single strategies can be checked. Additionally, there has been the desire that environmental issues should play a greater role in agricultural policies (COM, 2000). Since the 1980s there have already been efforts to include environmental aspects into the common agricultural policies. The 1992 EU agricultural policy reform enabled agri-environment programs to play an important role in agricultural development as the so-called second pillar. The EU stated the overall objectives with the guideline EEC 2078/92. Member states, however, could decide on the actual procedure (see Marggraf, 2003).

An evaluation of the individual member state programs is mandated by the EU guideline (EEC 746/96 Art. 16–20). An assessment is of importance for the further continuance of the programs as well as for the further inclusion of environmental issues in agricultural policies. The EU Commission expects the evaluations to provide insight into the effects of the implemented measures and suggests improvements (COM, 1998).

A common assessment method for these programs has not yet been developed, in part due to the lack of experience with these new programs. However, the measures prescribed by these programs are quite similar. This has enabled Wilhelm (1999) to evaluate the programs with regard to their ecological effectiveness. Three different location types with different agricultural suitability were to be assessed by ecologists on the program effectiveness. The measures, being quantifiable through the assignation of proportion factors, became functions of indicators. The measure sets of each program were linked together with the implementation area. This allowed a comparison between the different programs and an assessment of the ecological relevance. The programs have to be classified according to their scope, either broad and not specific or narrow and very specific, and the size of the implementation area, for example, either an agricultural area for a whole state or by certain criteria selected regions (e.g. specific wetland). The ecological assessment concentrates exclusively on the implemented measures and their effects. This assessment is then used for an economic assessment of the programs. This is done by comparing the sum of the ascertained ecological points with the expenditures required for these. The ecological points are derived from the sum of the indices of the individual measures. By using a theoretical approach for economic assessment, Marggraf (2003) does not establish the actual effects of the implemented measures. The official evaluation report also stated that the implementation area usually depicted the most used indicators (COM, 1998).

This approach was also used in the State of Saxony-Anhalt for the assessment of an agri-environment program for environmentally friendly cultivation (Menge, 2003). The distribution of the measures acceptance was determined with help of a descriptive statistical method. The surveyed farms were classified according to biotic resource preservation indicators. These classifications included the following: species diversity, resistance factors, percentage of legumes and grains, fallow duration and field size. The indicators are ascertained and averaged out through all farms. An indicator development over many years may show developments with regard to environmentally related effects. A defining beforehand of objectives or goals is not recommended. An assessment of these should be done after documentation of the development process. This declaration remains indirect since a monitoring of the immediate changes in the area of bio-diversity was not undertaken. The use of these indicators, however, is not equivalent to the assessment of the actual effectiveness of the measures on the implementation areas in relation to the sensitivity of the areas to a certain impact (Osinski, 2002). Through this way, measures with positive effects can target the right locations.

Because of the immense effort and time required for these measures to show positive environmental effects, assessment of agri-environment programs are dependent on the indirect approach. However, studies in The Netherlands have shown that a direct evaluation is necessary. A comparison of the biotic makeup of areas with and without management agreements showed no positive effects on the avian and plant population due to this form of agriculture (Kleijn et al., 2001). The authors therefore suggest a scientific component for the implementation of agri-environment programs. The EU Commission (KOM, 2000) handed out an indicator criteria catalogue for the assessment of these programs within Europe. This catalogue should then prove the success of the implemented measures. The indicators included, for the areas of bio-diversity and landscapes, concentrated on data of the implementation area and the surveying of species population (mammals, birds) (KOM, 2000, Section D). The actual assessment, however, is responsibility of each member state. In the case of a species survey this must be done on the basis of a local analysis. Because of this, the assessments concentrate on individual local areas or specific farms. For this approach, a mixture of measure-oriented and result-oriented indicators was chosen, similar to the operational assessment approach.

A key issue in the assessment of these programs is which qualities and what quantities of bio-diversity and landscapes are of relevance. What results are desired from 'good agricultural practice' and which results are going beyond? It is hoped that the establishing of a national framework with minimum requirements and control criteria would solve this issue in Germany (Knickel et al., 2001). However, the determining of these minimum requirements is difficult due to the lack of a common reference system and because of the immense variety on the regional, nature and agricultural level.

Thus, the discussion of 'good agricultural practice' and its development through the use of indicators is far from over. Action-oriented and result-oriented approaches have also been discussed (Knickel et al., 2001). An example of result-oriented system is the 'target species concept' of the State of Baden-Wuerttemberg's landscape program (Walter et al., 1998). This concept provides the currently most comprehensive approach for the determination of species-related minimum requirements at the natural units' level. This concept determines the type of indicators as well as the desired population sizes.

Such an approach would also allow the separation of the requirements for good agricultural practice from ecological services. This would avoid the so-called 'carry-on effect' and increase the 'ecological accuracy' of agri-environment programs (Hampicke, 2000).

There have already been suggestions on how a separation with regard to economic assessment could be implemented. Therefore, normally operated conventional agriculture would not receive government subsidies in the long run. However, if nature conservancy aspects are included, a limited compensation is possible. This would mean that traditional, integrated and ecological-oriented agriculture would only receive medium compensation, while landscape preservation services would receive the highest level of compensation (Knickel et al., 2001). These services however, have to be specified with help of indicators.

4. Environmental reporting/environmental monitoring

There have been political efforts, both on the national level, e.g. in Australia (Pearson et al., 1998) or England (MAFF, 2000) and European level (COM, 1999) to develop indicators that primarily concentrate on reporting and monitoring. They should control the environmental condition and evaluate the effectiveness of the implemented measures. National and international environmental indicator systems were developed in order to improve the information and communication of environmental conditions, the prioritisation within environmental policy and to determine environmental objectives (see Walz, 1997). The political importance of this was emphasised on the Helsinki Summit, where also an indicator system for the European level was designed. The architecture of this system is made up of so-called environmental-specific core indicators (these allow only individual indicators from each of the core areas) and of sector-specific indicators (COM, 1999). One of these core areas is 'Nature and Bio-diversity' whose indicator 'Species Catalogue of the basis of genetical variety and variety of biotopes' is still to be developed. Advances in the agricultural sector have been immense, especially under the efforts of the OECD Commission and the individual member states. A list of agri-environmental indicators exists, some of which have already been partly implemented. The achieved results have also been documented (OECD, 2001).

In many countries (for example, The Netherlands, Sweden, Norway, Canada) the development of indicator systems was based on the OECD pressure-state-response approach (OECD, 1997). The emphasis is either on pressure indicators (The Netherlands) or on state indicators (Sweden) (Walz, 1997). The biggest difficulty in the development of indicator systems, however, is the determination of the correct parameters of the effect systems. It is common to characterise stress parameters through the use of indirect indicators. Factors that influence the decline of species and biotopes include the increased agricultural intensity, the resulting destruction of habitats and changes in the living conditions in habitats (OECD, 2001). Thus, a change in the areas of bio-diversity and landscapes can be derived from taking into account the development of these impact indicators. This indirect derivation, however, does not allow a documentation of clear cut causalities. A documentation of the changes in nature and landscapes in relation to the implemented measures can only be achieved through monitoring. Monitoring would not only determine individual stress or biotic parameters, but would also document both of them and place them in relation. A method for environmental random area sampling was developed for Germany (Statistisches Bundesamt and BfN, 2000). This method aims to survey the changes in so-called "normal landscapes". This would allow agrarian and forestal areas in addition to conservation areas to be incorporated into a national test framework. This allows the collection of structural data as well as of data about the quality of the biotopes through a surveying of the species makeup. This comprehensive indicator system is currently the most advanced proposal for a monitoring approach in Germany. However, as it appears to be highly cost intensive it has not been put into practice.

There has been a strong demand for changes in the area of conservation of landscapes. Database surveys, that would allow a documentation of these changes on a national or European level, are even less available than for the area of bio-diversity. Proposals are available for respective indicators that would include natural coherence, diversity, and features as well as cultural identity, diversity and features (Wascher, 2000). As a prerequisite for successful monitoring, a documentation of European landscapes is needed first (see EEA, 1995). There have also been efforts on the German national level to categorise landscapes in order to better document their development. A research project financed by the national office for nature protection (BfN) is targeted at characterising and standardising of high valued landscape types. Additionally, impacts are to be documented and suggestions for protection are to be made.

The main problems of a European indicator system lie in the incoherence of the available databases, especially for the areas of bio-diversity and landscapes (Brouwer, 1999). There are European institutions that are attempting to construct an integrated database. However, since they also rely on national surveys, the database is incomplete for certain regions and sometimes portraits European nature and bio-diversity insufficiently (Delbaere, 1998).

5. Other areas for indicator implementation

Besides the use in operational assessment, in the evaluation of agri-environment programs or environmental monitoring guidelines, indicators can also be of use in planning. Thus, the regional objectives of the concept of 'Ecological and Cultural Significant Areas' (OELF) is based on regional planning (Roth and Schwabe, 2003). These objectives, however, have to be first determined by defined methods. The plans used in that case are called 'plans on the use and care of agrarian spaces (ANP) (Roth, 1996). These are comparable to landscape plans which are documenting the makeup of landscape elements and give target values for future development in agricultural landscapes.

Landscape planning per se and the construction of this kind of plan is unthinkable without the use of indicators. Landscape planning in Germany is the sector planning for nature protection and landscape care. It aims to provide a land use that enables to best preserve the potential of the land to maintain the a-biotic resources and the plants and animals (Spitzer, 1995). Therefore, the abstract concept of protecting 'variety, peculiarity and beauties of nature' (German Nature Protection Law, BNatSchG, 2002) have to be put in concrete terms with help of indicators.

The potential has to be first determined and then evaluated in order to then design respective measures for the protection, development and preservation. Each of these steps contain an abstraction of the reality. In Germany it is common to develop planning frameworks on the state level for the areas of species and biotope conservation. Such a planning framework already exists in the target species concept of the State of Baden-Wuerttemberg. This concepts lists not only the indicator types but also objective limits for the desired population size per usable area (minimum standard) and per protected biotopes (special population protection) (Walter et al., 1998).

Another area for the use of indicators is the environmental impact assessment (EIA, after BNatSchG, §18, 2002). EIA guidelines prescribe an assessment of the environmental compatibility of an implemented measure. The method of evaluation is still an experimental field (Koeppel et al., 1998) although there is a need for providing species-related data as an evaluation basis. Species are analysed following the criteria scarcity, degree of threat, number of species and individuals, dominance structure, diversity and evenness. Apart from this the impact evaluation includes the estimation of the threat which is put on the functionality of the area affected by the planned impact. For orientation a list of biotopes and impact-specific species groups suitable as indicators is provided (Koeppel et al., 1998).

The prerequisite for the determination of the conservation status of an area, for example, by the Nature Conservation Law (BNatSchG), is the occurrence of rare species. The occurrence of rare species following the so-called Red List of threatened plants and animals (e.g. Korneck and Sukopp, 1988) can be understood as an indicator to help in the assessment of the placing of the area under protective status. These indicators only represent the total number of species which are to be protected by law (BNatSchG, §1). The use of Red Lists is sometimes criticised because of different degrees of threats in different parts of the country. Thus, the Red List looses its character as tool for differentiating endangered and not endangered species. Another argument dealing with the use of Red Lists and its limit is mentioned by Garrelts and Krott (2002). By use of a so-called flagship species as single indicator it can be tried to influence political decision making. The example of the Great Bustard (*Otis tarda*) in Brandenburg demonstrated a negative effect on the use as an indicator. That species was given the role of the countries' development enemy. Only a few species living in that area had enough power to prevent building a motorway. The public could not accept the weighting of a few species at the limit of there habitat against the very important motorway. The authors of the article suggest instead of making decisions with the help of 'Red List Species' they should be taken as 'Early-warning System' (Garrelts and Krott, 2002).

Models are becoming even more important for the planning of measures and the prediction of the effects of the implemented measures. The modelling by Herrmann et al. (2003) can be used as an example on how with the help of indicators frameworks for landscape development can be developed. These can then be used for the design of scenarios. This prognostic aspect of indicator implementation is also present in habitat modelling. Individual biotope characteristics are used here as key factors in the determination of habitat suitability of biotopes (Kleyer et al., 1992). Weber and Koehler (1999) also demonstrate this approach in modelling the dispersion of field larks on a defined landscape. Other modelling approaches simultaneously use sets of ecological and economic indicators to predict the economic and ecological effects of agricultural measures (see Meyer-Aurich and Zander, 2003).

Thus, many indicators have already been used in various development processes, in fact their use is not been questioned. Indicators are used in all types of sectorial plannings that influence changes in landscapes and bio-diversity, for example, in landscape, urban, transport and forestal planning. While assessment systems with included indicators exist, the methods to land use development must be determined following. The evaluation in the framework of landscape planning processes are lacking in acceptance. Not-standardised and not-comprehensible methods play a significant role (Muessner and Plachter, 2002). But it is not sufficient to standardise methods because they are depending on justified indicators. Both have to be developed parallel.

6. Conclusion

The use of indicators was introduced in this article for the areas of assessment of agrarian and forestal enterprises, assessment of the efficiency of political measures, environmental reporting, planning and modelling. In general, the demands placed on indicators concern the analytical soundness, the measurability and their political relevance. Of great importance is their transferability and effort required for the data collection and availability of data. Due to the different forms of implementation, some of these criteria are more problematic than others.

6.1. Assessment of agrarian and forestal enterprises

The assessment of the environmental compatibility of agricultural enterprises using a variety of point systems has been further developed in the past years, as shown in the summary by Braband et al. (2003). The preservation of bio-diversity and landscapes has been introduced here. The database is collected on site and is dependent on the availability of experts. The applied methods are partly implemented regionally in order to correctly assess the area of unused structures and the existence of different species. However, some experts are critical about the extent of the data collection. Therefore, approaches to reduce indicators by determination of correlations, as proposed by Oppermann (2003), will be of even greater importance in the future. In other views, farming systems (Knickel et al., 2001) and especially ecological farming is considered to be advantageous for the development of bio-diversity (Koepke, 2002). This would mean that the type of farming would be used as indicator also for bio-diversity and landscape and not only for soil and water protection (see KOM, 2001a). This assumption is supposed to be verified by a current research project at the University of Kassel, Germany.

The political relevance of these farm-related indicators is given by the fact that decision makers and farmers receive information regarding the condition of the enterprises. From this information, recommendations for other enterprises or help with decisions concerning the own enterprise (compare Girardin et al., 1999) can be derived. Future developments of result-oriented indicator systems are possible.

Several experiences exist within the Dutch 'Yardstick for bio-diversity on farms-program' (Buys, 1995) showing an effect on initiative of farmers' own (Oosterveld and Guldemond, 1999).

Since some objectives can only be achieved after a long time, in certain areas measures must still be compensated. Independent of this, the objectives of the implemented measures should be known and their achievement grade evaluated in order to avoid a carry-on effect by the farmers. If the effects of individual measures are not clearly known, the necessity of research should be indicated (see evaluation of agri-environmental programs).

Apart from single farms, an increasing pressure is being placed on bigger agricultural and forestal enterprises also from abroad to prove ecological and, partially, social contributions. Commerce and consumers do not rely anymore only on governmental normative regulations and control procedures. Commerce and producers of agricultural products together determine production standards, production criteria in the ecological and social sector including the product quality.

All control systems have the same procedures and the same objective, but with different interest-oriented intention and motivation. The common aim is to prove, by documenting their actions, that the risks of plant production for humans and for the environment can be minimised to controllable criteria and that the voluntary control of these objectives shall complement governmental controls. In this regard, the intensity of biotic resource protection is represented varyingly. Biotic resources in forests and, in part, also in plantations of bananas, coffee, sugarcane, cacao and citrus fruits can be preserved as shown in the criteria of the Rainforest Alliance. In the protected cultivation of ornamental plants or vegetables under glass or foil, the possibilities of appropriate nature conservation of the production areas are limited. Therefore, it is recommended to implement measures outside the greenhouses. Bio-indicators that prove an increased bio-diversity beyond the measures of nature conservation are not used in any criteria system (see Section 2.2). On the contrary, it is assumed that biotic resource protection measures will per se have positive effects on bio-diversity. These pragmatic and controllable measures should be maintained and supported as long as no other economically and ecologically significant alternatives are available.

At the moment, neither the involved NGOs nor producers or commerce discuss the implementation of bio-indicators in criteria systems. On the contrary, there are still doubts if agrarian enterprises should undertake measures to improve bio-diversity. A future implementation of bio-indicators as control criteria to prove successful measures for improving bio-diversity should be connected to the following conditions.

The bio-indicator must

1. be, as far as possible, scientifically unquestioned;
2. be accepted by all participants of the control system;
3. be useful for all comparable agrarian eco-systems;
4. always exist with implementation of the measure;
5. be recognisable without using a special control procedure;
6. also be able to show the quality of the implemented measures;

The implementation of criteria systems to assess ecological performances of production enterprises as presently discussed and introduced on the international level, is always an intervention on the freedom of choice of the production and is, therefore, connected to a higher production cost. Since the implementation of the systems is voluntary, the economic effects on the enterprise must be taken into consideration as otherwise they will not be accepted in practice. In order to lower the cost of the operational assessment, key indicators should be developed. This would allow the simultaneous assessment of numerous control criteria by means of key indicators and thus reduce costs.

6.2. Evaluation of programs

The European Union requires a stronger control of already implemented agri-environmental programs (KOM, 2000). This assessment is not only of the implementation of the measures but also on their success (appendix D). It is therefore necessary to prove the efficiency of the promoted measures in a clear scientific manner. Although the control requirements were formulated EU-wide, the indicators will not be transferable since the implemented programs are differently oriented. Even here, the objective cannot be the determination of a uniform indicator set, but of a reliable assessment of these programs. This assessment, however, must be more than a statistic of the applied measures (compare Marggraf, 2003). Indicators which compare the situation before the introduction of the programs to the situation of the present evaluation and which are able to separate the effects of the programs still have to be determined. In this area of indicator application, the availability of data is still insufficient. However, the EU member states will be forced to comply with the requirements of the EU commission. For this purpose, the inclusion of local surveying results is planned (compare Part D, KOM, 2000).

An evaluation of such agri-environment programs would be easier to accomplish, if in the biotic field, environmental observation systems would be more widespread. Maximum requirements, like "environmental random area sampling" (Statistisches Bundesamt and BfN, 2000) proposed in Germany, must be discussed with regard to a multiple use. Parallel to this, more scientific efforts should be made to determine the actual "necessary" number of groups of species to be examined for the characterisation of bio-diversity. Reporting in the area of protected habitats is better than the reporting of species groups. A national biotope surveying and, for example, the identification of FHH areas prescribed by law lead to these results. Therefore, the political relevance is very high. Alone for agriculturally used landscapes (which this article focuses on), the so-called "normal landscape" there exists a demand for action. Approaches to connect agrarian statistical data from the farmers' subsidies applications (INVEKOS) with land use data (CORINE) have been made on the EU level (EEA, 2001) to gain a better understanding of environmental state. However, these approaches will only remain on the structural level and will not reach the bio-diversity level. Here national approaches are necessary. In Baden-Wuerttemberg, an inventory of all available target species groups including their potential distribution has built the basis which provided the target species concept (Walter et al., 1998).

However, with regard for landscape conservation, a uniform meaning of the expressions "intensively used", "extensively used" and "not used" should be determined (compare OECD, 2001).

For the inclusion of interactions between bio-diversity and landscapes, new aggregated indicators, Natural Capital index (NCI) (OECD, 2001, p. 316) should be developed, where the quality (change of the number of wildlife species emanating from a basic line) is multiplied by the quantity of the eco-system.

In order to secure structure indicators and their significance to bio-diversity, more research is necessary to find out the interaction of agricultural activity and habitats (fragmentation, heterogeneity, vertical vegetation structures) (OECD, 2001).

Whereas the term "bio-diversity" has a clear scientific meaning, the term "landscape" has a much broader scope. Landscape as a concept (compare Wascher, 2000), the relation of structure, function and value of landscapes must be better understood (OECD, 2001). The definition of the adapted indicators must then follow. These aspects must be included in documentation about bio-diversity and landscape.

Because of the fascinating possibility of seeing into the future, the use of indicators in planning and modelling is of special political relevance. Due to the connection with the development of plans and/or visual scenarios, a strong decision-supporting character is given to the indicators. Their communicative function comes to the foreground.

Due to their relationships within the models, new contents, modifying their analytic consistency related to the problem, are given to indicators. In this case, the models extend the indicative expressiveness aiming at the connection of relevant aspects of a problem and the presentation of possible solutions.

The demonstration of the different applications of indicators shows that they are connected to different requirements. In general, indicators are used to describe complex systems difficult to survey (Girardin et al., 1999). They can diagnose or assess future developments. Girardin et al. (1999) propose a way to develop an indicator which, in the first step, defines its targets, determines the user in the second step and provides the actual development of the indicator in the third step. This always results in a compromise between the existing information, the state of science and the user's requirements regarding the indicator's grade of simplicity (Girardin et al., 1999). Therefore, the indicators can never be developed without considering the users' demands. The objective of science should be to work on a securitisation of indicators.

References

Biologische Bundesanstalt (BBA), 1996. Check for ecological production on flower farms. URL: http://www.bba.de (English version). Institute for Plant Protection in Horticulture.

Braband, D., Geier, U., Koepke, U., 2003. Bio-resource evaluation within agri-environmental assessment tools in different European countries. In: Buechs, W. (Ed.), Biotic indicators for biodiversity and sustainable agriculture. Agric. Ecosyst. Environ. 98, 423–434.

Brouwer, F., 1999. Agri-environmental indicators in the European Union: policy requirements and data availability. In: Brouwer, F., Crabtree, B. (Eds.), Environmental Indicators and Agricultural Policy. CABI Publish., Wallingford, pp. 57–72.

Bundesnaturschutzgesetz (BNatSchG), 2002. Gesetz ueber Naturschutz und Landschaftspflege v. 25. Maerz 2002. URL: http://jurcom5.juris.de/bundesrecht/bnatschg_2002/.

Buys, J.C., 1995. Towards a Yardstick for Biodiversity on Farms. Centre for Agriculture and Environment (CLM), The Netherlands (in Dutch, English summary).

Commission of the European Communities (COM), 1998. Evaluation of agri-environment Programmes. State of Application of Regulation (EEC) No. 2078/92. DGVI Commission Working Document VI/7655/98.

Commission of the European Communities (COM), 1999. Report on environment and integration indicators to Helsinki summit. SEC'1999, 1942 final.

Commission of the European Communities (COM), 2000. Indicators for the integration of environmental concerns into the common agricultural policy. Communication from the Commission to the Council and the European Parliament COM'2000, 20 final.

Delbaere, B.C.W. (Ed.), 1998. Facts and Figures on Europe's Biodiversity. ECNC, Tilburg.

European Environmental Agency (EEA), 1995. Europe's Environment—The Dobris Assessment. Copenhagen.

European Environmental Agency (EEA), 2001. Towards agri-environmental indicators. Integrating Statistical and Administrative Data with Land Cover Information. Topic Report No. 6, Copenhagen.

Euro-Retailer Produce Working Group (EUREP), 2002. Europaeisches Handelsinstitut e.V. URL: http://www.eurep.org, http://www.ehi.org.

Flower Label Program (FLP), 2002. Internationaler Verhaltenskodex fuer die sozial- und umweltvertraegliche Produktion von Schnittblumen. URL: http://www.flower-label-program.org, 25 April 2002.

Forest Stewardship Council (FSC), 2000. URL: http://www.fsc-deutschland.de, 24 March 2000.

Garrelts, H., Krott, M., 2002. Erfolg und Versagen der Roten Listen—wann ist deren Einsatz ratsam? (Success and failure of red lists—in which cases is their use appropriate?). Nat. Landsc. 77 (3), 110–115 (English summary).

Girardin, P., Bockstaller, C., Van der Werf, H., 1999. Indicators: tools to evaluate the environmental impacts of farming systems. J. Sust. Agric. 13 (4), 5–21.

Hampicke, U., 2000. Moeglichkeiten und Grenzen der Bewertung und Honorierung oekologischer Leistungen in der Landwirtschaft. Schr.-R. d. Deutschen Rates fuer Landespflege 71, 43–49.

Herrmann, S., Dabbert, St., Schwarz-von-Raumer, H.G., 2003. Threshold values for nature protection areas as indicators for bio-diversity—a regional evaluation of economic and ecological consequences. In: Buechs, W. (Ed.), Biotic indicators for biodiversity and sustainable agriculture. Agric. Ecosyst. Environ. 98, 493–506.

Heyer, W., Huelsbergen, K.J., Papaja S., Wittmann, C., 2003. Field related organisms as possible indicators for evaluation of land use intensity. In: Buechs, W. (Ed.), Biotic indicators for biodiversity and sustainable agriculture. Agric. Ecosyst. Environ. 98, 453–461.

International Standardisation Organisation (ISO) 14001, 1996. Environmental management systems. Specification with Guidance for Use. URL: http://www.iso.ch.

Kleijn, D., Berendse, F., Smit, R., Gilissen, N., 2001. Agri-environment schemes do not effectively protect biodiversity in Dutch agricultural landscapes. Nature 413, 723–725.

Kleyer, M., Kaule, G., Henle, K., 1992. Landschaftsbezogene Oekosystemforschung fuer die Umwelt- und Landschaftsplanung. Z. Oekologie Nat. 1, 35–50.

Knickel, K., Janssen, B., Schramek, J., Kaeppel, K., 2001. Naturschutz und Landwirtschaft: Kriterienkatalog zur ‚Guten fachlichen Praxis. Schr. R. Angew. Landschaftsoek. 41, Bundesamt fuer Naturschutz (Ed.), Bonn-Bad Godesberg (English summary).

Koepke, U., 2002. Umweltleistungen des Oekologischen Landbaus. Oekologie und Landbau 122, 6–18.

Koeppel, J., Feickert, U., Spandau, L., Strasser, H., 1998. Praxis der Eingriffsregelung. Ulmer, Stuttgart.

Kommission der Europaeischen Gemeinschaften, Generaldirektion Landwirtschaft (KOM), 2000. Zweck und Anwendung des Katalogs gemeinsamer Bewertungsfragen mit Kriterien und Indikatoren VI/12004/00. Teil A-D.

Kommission der Europaeischen Gemeinschaften (KOM), 2001a. Statistischer Informationsbedarf fuer Indikatoren zur Ueberwachung der Integration von Umweltbelangen in die Gemeinsame Agrarpolitik. Mitteilung der Kommission an den Rat und das Europaeische Parlament, COM'2001, 144 final.

Kommission der Europaeischen Gemeinschaften (KOM), 2001b. Aktionsplaene zur Erhaltung der biologischen Vielfalt fuer die Gebiete Erhaltung der natuerlichen Ressourcen, Landwirtschaft, Fischerei sowie Entwicklung und wirtschaftliche Zusammenarbeit. Mitteilung der Kommission an den Rat und das Europaeische Parlament, COM'2001, 162 final.

Korneck, D., Sukopp, H. 1988. Rote Liste der in der Bundesrepublik Deutschland ausgestorbenen, verschollenen und gefaehrdeten Farn- und Bluetenpflanzen und ihre Auswertung fuer den Arten- und Biotopschutz. Schr.R. f. Veg.kunde. Heft 19, Bonn-Bad Godesberg (English summary).

Larsson, T.B., Esteban, J.A. (Eds.), 2000. Cost-effective indicators to assess biological diversity in the framework of the convention on biological diversity—CBD. URL: http://www.gencat.es/mediamb/bioind/bioind.pdf.

Marggraf, R., 2003. Comparative assessment of agri-environment programmes in federal states of Germany. In: Buechs, W. (Ed.), Biotic indicators for biodiversity and sustainable agriculture. Agric. Ecosyst. Environ. 98, 507–516.

Meier, U., Feltes, J., 1996. Bewertung von Blumenbetrieben in Nicht-EU-Laendern nach oekologischen Standards. Nachrichtenbl. Deut. Pflanzenschutzd. 48, 80–82.

Menge, M., 2003. Experiences with the application, recordation and valuation of agri-environmental indicators in agricultural practice. In: Buechs, W. (Ed.), Biotic indicators for biodiversity and sustainable agriculture. Agric. Ecosyst. Environ. 98, 443–451.

Meyer-Aurich, A., Zander, P., 2003. Consideration of biotic environmental quality targets in agricultural land use—a case study from the Biosphere Reserve Schorfheide-Chorin. In: Buechs, W. (Ed.), Biotic indicators for biodiversity and sustainable agriculture. Agric. Ecosyst. Environ. 98, 529–539.

Ministry of Agriculture Fisheries and Food (MAFF), 2000. Towards sustainable agriculture. A Pilot Set of Indicators. URL: http://www.maff.gov.uk.

Muessner, R., Plachter, H., 2002. Methodological standards for nature conservation: case-study landscape planning. J. Nat. Conserv. 10, 3–23 (Urban and Fischer).

Muessner, R., Jebram, J., Schmidt, A., Wascher, D., Bernotat, D., 2002. Derzeitiger Entwicklungsstand. In: Plachter, H., Bernotat, D., Muessner, R., Riecken, U., 2002. Entwicklung und Festlegung von Methodenstandards im Naturschutz. Schriftenreihe fuer Landschaftspflege und Naturschutz. Heft 70, pp. 35–53.

Organisation for Economic Co-operation and development (OECD), 1997. Environmental Indicators for Agriculture, Paris.

Organisation for Economic Co-operation and Development (OECD), 2001. Methods and results. Environmental Indicators for Agriculture, vol. 3. Paris.

Oosterveld, E.B., Guldemond, J.A., 1999. Measuring the Biodiversity Yardstick—A Three Year Trial on Farms. Centre for Agriculture and Environment (CLM) (in Dutch, English summary).

Oppermann, R., 2003. Nature balance scheme for farms—evaluation of the ecological situation. In: Buechs, W. (Ed.), Biotic indicators for biodiversity and sustainable agriculture. Agric. Ecosyst. Environ. 98, 463–475.

Osinski, E., 2002. GIS-gestuetzte Regionalisierung von Agrarraumpotentialen—Anwendung zur effektiven Ausgestaltung von Agrar-Umweltprogrammen in zwei europaeischen Regionen. Der Andere Verlag, Osnabrueck (English abstract).

Pan European Forest Certificate (PEFC), 2002. URL: http://www.pefc.de, vom 17 March 2002.

Pearson, M., Johnston, D., Lennon, J., McBryde, I., Marshall, D., Nash, D., Wellington, B., 1998. Environmental indicators for national state of the environment reporting. Natural and Cultural Heritage, Australia: State of the Environment (Environment Indicator Reports). Department of the Environment, Canberra.

Rainforest Alliance (RA), 2002. URL: http://www.rainforest-alliance.org, vom 28 January 2002.

Regulation EC, 2001. No. 761/2001 of the European Parliament and of the Council of 19 March 2001 allowing voluntary participation by organisations in a Community eco-management and audit scheme (EMAS). Amtsblatt der Europaeischen Gemeinschaft L 114/1.

Roth, D., 1996. Agrarraumnutzungs- und pflegeplaene—ein Instrument zur Landschaftsplan-Umsetzung. Nat. Landsc. 28 (8), 237–249.

Roth, D., Schwabe, M., 2003. Method for assessing the proportion of ecologically, culturally and provincially significant areas ("OELF") in agrarian spaces used as a criterion for environmental friendly agriculture. In: Buechs, W. (Ed.), Biotic indicators for biodiversity and sustainable agriculture. Agric. Ecosyst. Environ. 98, 435–441.

Spitzer, H., 1995. Einfuehrung in die raeumliche Planung. UTB-Ulmer, Stuttgart.

Statistisches Bundesamt and BfN (Bundesamt fuer Naturschutz) (Eds.), 2000. Konzepte und Methoden zur Oekologischen Flaechenstichprobe. Ebene II. Monitoring von Pflanzen und Tieren. Schr.-R. Angew. Landsc. 33.

Verband deutscher landwirtschaftlicher Untersuchungs- und Forschungsanstalten (VdLufa), 2001. USL-Projektstelle, Informationen zu USL (Umweltsicherungssystem Landwirtschaft) URL: http://www.vdlufa.de/kul/usl_01.htm.

Walter, R., Reck, H., Kaule, G., Laemmle, M., Osinski, E., Heinl, T., 1998. Regionalisierte Qualitaetsziele, Standards und Indikatoren fuer die Belange des Arten- und Biotopschutzes in Baden-Wuerttemberg: das Zielartenkonzept—Ein Beitrag zum Landschaftsrahmenprogramm des Landes Baden-Wuerttemberg. Nat. Landsc. 73, 9–25.

Walz, R., 1997. Grundlagen fuer ein nationales Umweltindikatorensystem. Umweltbundesamt, Texte 37.

Wascher, D. (Ed.), 2000. Agri-environmental indicators for sustainable agriculture in Europe. Technical report series. European Centre for Nature Conservation (ECNC), Tilburg.

Weber, A., Koehler, W., 1999. Modellierung der Verbreitung und Ausbreitung von Indikatorarten—ein erster Ansatz der Charakterisierung der Biodiversitaet (Modeling of Spread and Dispersal of Keyspecies—a first approach to characterise biodiversity). Z. F. Kulturtechnik Landentwicklung 40, 207–212 (Blackwell, English summary).

Wilhelm, J., 1999. Oekologische und oekonomische Bewertung von Agrarumweltprogrammen. Delphi-Studie, Kosten-Wirksamkeits-Analyse und Nutzen-Kosten-Betrachtung. Peter Lang, Frankfurt am Main.

PART IV
EXPERIENCES AND APPLICATIONS

Chapter 6 Experiences of Applications

E. Osinski, W. Büchs,
B. Matzdorf, J. Weickel

Original Papers

ELSEVIER

Available online at www.sciencedirect.com

Agriculture, Ecosystems and Environment 98 (2003) 423–434

Agriculture Ecosystems & Environment

www.elsevier.com/locate/agee

Bio-resource evaluation within agri-environmental assessment tools in different European countries

Dorothee Braband[a,*], Uwe Geier[b], Ulrich Köpke[b]

[a] *Department of Organic Farming and Cropping, University of Kassel, Nordbahnhofstr. 1a, D-37213 Witzenhausen, Germany*

[b] *Institute of Organic Agriculture, University of Bonn, Katzenburgweg 3, D-53115 Bonn, Germany*

Abstract

In order to achieve environmentally sound agriculture as well as a remuneration of ecological achievements on the farm level, practicable instruments for accurate measurement have to be developed. Whereas there are suitable indicators to assess agricultural impacts on abiotic resources, there is a lack of indicators for the assessment of biotic resources on the farm level. Proposals for and attempts at as well as first-hand experience in assessment methods of agricultural impacts on biotic resources exist in various European countries. A few of these methods require intensive measurements with a lot of indicators, but a sufficient assessment and evaluation of the biotic as well as the aesthetic resources (landscape structure/image) on the farm level has not yet been carried out.

Biotic resource indicators of seven assessment methods (Ecopoints Lower Austria, Solagro, Halberg, Nature Balance Scheme, KUL, Frieben, Biodiversity Yardstick) from five European countries are compared with each other. Some of the methods register biotic resources only marginally by indirect measurements (indicators), respectively, or action-oriented indicators such as time of cutting/mowing, frequency of cutting/mowing, age of grassland, percentage of unsprayed area. Other methods are very complex, so that their implementation on a wide scale does not seem possible.

There is a need to further develop and improve practicable and efficient tools in order to assess the agricultural impact on bio-resources on farm level and consequently in order to promote an environmentally sound agriculture.

Keywords: Bio-resource evaluation; Agricultural impacts; Agri-environmental indicators; Agri-environmental assessing tools; Productive farm area; Non-productive farm area

1. Introduction

Since the mid eighties an increasing environmental awareness has been encouraging people to deal with the negative effects of agriculture on the environment. In Germany, in 1985 for the first time, the German Council of Environmental Advisors (Rat für Sachverständigen für Umweltfragen; SRU, 1985) recognized the negative effects of modern farming methods on the environment. National as well as international governmental and non-governmental organizations are paying more attention to agricultural impacts on our resources, and debates about relevant and reliable agri-environmental indicators (e.g. OECD, 1994) have been started. After the UN Conference of Rio in 1992 "sustainability" should be a leitmotiv for every sort of

* Corresponding author. Tel.: +49-5542-98-1634; fax: +49-5542-98-1568.
E-mail address: braband@wiz.uni-kassel.de (D. Braband).

0167-8809/$ – see front matter © 2003 Elsevier Science B.V. All rights reserved.
doi:10.1016/S0167-8809(03)00101-4

management including agriculture. Against this background, payments to farmers for agri-environment purposes have become part of a variety of policy schemes all over Europe (e.g. see Council Regulation (EEC) no. 2078/92 (European Union, 1992); Deblitz and Plankl, 1998; European Union, 1999; Agenda, 2000). In Germany, intentions to assess, quantify and evaluate the agricultural impact on resources are constantly growing. Consequently much effort has been and is still being put into developing reliable sets of indicators which could be used in the field of production analysis (Ceuterick, 1998; Geier, 2000) as well as in the field of reports on the environment (Walz, 1997; Rudloff et al., 1999) without mentioning various agri-environmental policy schemes. Because of the complexity of the agricultural impact on "living" (bio-) resources most indicators deal with the abiotic resources, but there are attempts to put more emphasis on the bio-resources such as biodiversity and landscape (structure/image).

An environmentally sound agriculture might be achieved or supported via "rewarding the services of agriculture for nature conservation and landscape management" (see Deutscher Rat für Landespflege, 2000), but therefore suitable methods which make the assessment on farm level possible still have to be developed. Proposals for and attempts at as well as first-hand experience in assessment methods of agricultural impacts on resources exist in various European countries. As deficits have been mentioned from various sides concerning the assessment and evaluation of the agricultural impact on bio-resources (e.g. Geier and Köpke, 2000; Frieben, 1998), the aim of this review is to give a short comparison of different assessing methods on the farm level, focussing on the assessment of the agricultural impact on bio-resources. Indicators related to the effects on abiotic resources will be taken into account as far as there is also an obvious connection with biotic resources. The focus is laid on bio-resource indicators.

Assessing and evaluating the agricultural impact might contribute to the "environmental" awareness as well as promote a real integration of environmental issues (such as nature conservation issues as well) into farming practice. Assessing and evaluating the farmer's ecological contribution or ecological services would mean the first step towards a monetary remuneration.

2. The aim of agri-environmental assessing tools

2.1. Comparison and analysis of the various assessing and evaluation methods

A general tool for farm level assessment of the agricultural impact on the environment should meet the below mentioned requirements:

- The agricultural impact on abiotic as well as on biotic resources should be taken into account.
- The used indicators should very well reflect the state of the abiotic and biotic resources on the farm level.
- The handling should be as easy as possible, i.e. practice must not be too complex and work expenditure must not be too high.
- The method should be transferable into different regions and different farm management systems.
- There should exist clear definitions concerning goals and standards by taking into account local or regional differences.
- There should exist a high transparency especially in the field of evaluation.
- The method/tool should be compatible or take into account agricultural policy instruments, therefore control must be possible.

Regarding these requirements seven assessment methods from five European countries are analysed in which way the above-mentioned demands are met by the indicators *with reference to the impact on bio-resources*. Table 1 represents the analysed assessing methods.

All chosen methods are based on a farm level assessment:

- Ecopoints Lower Austria (Mayrhofer, 1997; Niederösterreichische Agrarbezirksbehörde, 1999): this assessment method is recognized as an agri-environmental policy scheme by the European Community (regulation no. 2078/92). Nine environmental criteria (indicators) are listed by which an environmentally sound management on the farm level is evaluated. According to the obtained "ecopoints"—by meeting the environmental criteria—every farm gets a remuneration (payment). About 1800 farms are taking part in this programme.
- "Criteria for environmentally sound agriculture" (KUL, Kriterien umweltverträglicher Landbe-

Table 1
Agri-environmental assessing and evaluation methods on farm level[a]

Method	Number of analysed farms	Aim	Evaluation level		Assessing	
			Farm	Plot/ area	Abiotic resources	Biotic resources
Ecopoints Lower Austria (A)	1800	Assessment, evaluation, remuneration recognized as regional agri-environmental policy scheme	+	+	+	+
KUL (D)	100	Assessment, evaluation, sustainability assessment of agricultural systems	+[a]	−	+	(+)
Halberg (DK)	20	Assessment, decision aid for farmers when dealing with resources	No evaluation[a]	−	+	(+)
Solagro (F)	300	Assessment, evaluation, sustainability assessment of agricultural systems	+	−	+	(+)
Frieben (D)	8	Assessment, evaluation, management advice for agricultural practice regarding further integration of nature conservation issues into *organic* farming practice	+	+	(+)	+
Nature Balance Scheme (D) (Naturbilanz)	16	Assessment, evaluation, management advice for agricultural practice regarding further integration of nature conservation issues into farming practice	+	(+)	+	+
Biodiversity Yardstick (NL)	22	Assessment, valuation tool to make farmers to become more aware of the biodiversity on their farms and to encourage the integration of nature conservation issues into farming practice	+	(+)	−	+

(+) Weak or very few indicators.
[a] Grassland is not taken into account.

wirtschaftung, Eckert and Breitschuh, 1997; Eckert et al., 1999) is an assessment method developed by the Thuringian Governmental Institution for Agriculture (Thüringer Landesanstalt für Landwirtschaft). There are five categories with 22 criteria. A defined scale for each criterion gives evidence of its fulfilment. This method has been tested on about 100 farms in different German regions.
- At the Danish Institute for Agriculture a method has been developed by Halberg (1997, 1999) which relies upon 13 criteria in five categories. This method is supposed to be a decision aid when dealing with resources on the farm level. It is not yet an evaluation tool.
- The French approach Solagro (Pointereau et al., 1999) which assesses and evaluates the contribution of agriculture to the environment with 16 criteria has already been applied to 300 farms all over France.
- The method developed by Frieben (1998) measures the contribution to the environment of organic farms by using ten criteria with a complex set of indicators. The stress is laid on the improvement of organic farming practice in regard of the integration of nature conservation goals on the farm level. The assessment tool is meant to expose performances and deficits in the field of agricultural impacts on the environment as well as provide management advice for agricultural practice.
- The Nature Balance Scheme: Naturbilanz started as a scheme launched by Naturschutzbund Deutschland e.V. (German Society for Nature Conservation, a German NGO) in south-western Germany in order to promote and help farmers integrating nature conservation aims into farming practice. There are about 42 criteria which help evaluating the farm (Naturschutzbund Deutschland e.V., 1999; Oppermann et al., 2000).

- The Biodiversity Yardstick developed at the Centre for Agriculture and Environment in Utrecht (Buys, 1995; Oosterveld and Guldemond, 1999) is a measuring method in order to reflect the state of bio-resources on the farm level. This method could be performed by farmers themselves at least for some important measurements. Lists of the chosen indicator species support them.

All methods listed above want to contribute to the development of environmentally sound agriculture. Beside this common aim Table 1 reveals that the different methods put their emphasis more or less on assessing either the abiotic or the biotic resources. KUL, Halberg and Solagro are tools that mainly consider the agricultural impact on the abiotic resources whereas Frieben and the Biodiversity Yardstick focus on the biotic resources. Ecopoints Lower Austria and the Nature Balance Scheme take into account both abiotic and biotic resources.

Until now, there do not exist very many tools with the aim to assess and evaluate the agricultural impact—especially on the biotic resources—on the farm level. These seven methods represent the most developed instruments in this field and although not always stressing the bio-resources that is one reason why they should be considered when discussing agri-environmental indicators and improving assessment methods.

The focus of the review will be laid on the comparison of these seven tools regarding the following questions. On which level is the agricultural impact on bio-resources assessed? On which level are these impacts evaluated? Which kind of indicators are used? Are the productive and non-productive farm area appropriately reflected?

2.2. Levels of assessing and evaluation

Assessing and evaluation are supposed to be two different stages when dealing with the mentioned tools. Against this background assessing is seen as the stage of collecting data without putting a value on these data. Nevertheless the recorded data are already in some kind value-weighted by choosing a set of indicators. Putting value on these data and measuring them against a yardstick with the aim of a "weighted assessment" is seen as the stage of evaluation. All of the analysed tools except for the method Halberg are developed in order to evaluate (see Table 1).

In order to describe appropriately the resources' state on farm level it is very important to have a look at the accuracy with which the data are collected, i.e. on which spatial level the indicators are used. The analysed tools show three different levels where assessment takes place. One level is represented by the farm itself (e.g. percentage of unsprayed area, number of cultivated plants); another level is represented by the plot—the smallest spatial level—and again another level is represented by biotope types (e.g. presence of indicator species in arable land, coverage of weeds in grain). Aggregation of data on one level might result in obtaining an indicator of the next higher level, for example, aggregating the data of all grain plots (coverage of weeds in the plot) may result in the assessment level "biotope types" (coverage of weeds in grain).

Aggregation often means simplification accompanied by a loss of information—the state of the single plot is not any more reflected in detail. But especially this detailed information is very important when ecological effects on the farm level have to be documented.

Different levels of assessing determine the accuracy of the description of the state and therefore finally determine the accuracy of the evaluation. This leads to the question: are assessment tools without indicators on the plot level able to describe the agricultural impact on biotic resources appropriately?

The final evaluation process is a very difficult task. The evaluation as well as the assessment may be performed on different spatial levels. Whereas the methods KUL and Solagro evaluate only on farm level (see Table 1) all other presented methods use indicators on the plot level although those are not always obvious in the final result of the evaluation (Nature Balance Scheme, Biodiversity Yardstick). That is because assessment on the one level is not necessarily succeeded by evaluation (weighted assessment) on the same level.

Two of the presented tools—Ecopoints Lower Austria and the method Frieben—evaluate on the farm level by evaluating single plots at a first stage whereas the Biodiversity Yardstick and the Nature Balance Scheme aggregate plots to biotope types. The "value" of a particular plot is not clear in the final result although every plot is checked.

2.3. Action- or results-oriented indicators?

When dealing with or discussing indicators another problem has to be faced. Most of the indicators which have already been put into practice evaluate by taking into account the way of the farmer's action; the action itself is assessed, the assessment is *action-oriented*. The other kinds of indicators do not assess the action but the result of the action or a state—these indicators should be called *results-oriented indicators*. Results-oriented indicators show more than a state but reflect a goal at the same time (see above "presence of indicator species").

Most of the approaches list action-oriented indicators (see Table 2). Despite the existing knowledge and scientific research about the interactions and relations between and within ecosystems, we do not yet know enough to forecast the results of a certain action in any case. In some cases, when a loss of (structural) biodiversity has already taken place, for example, the disappearance of plant species, these plant species are probably not going to reappear even if the "response"-action is considered to be environmentally sound (e.g. loss in the sense of extinction). On the other hand there are actions which are recognized as environmentally sound, but which do not immediately show effects on the "living" resources, for example, the positive effects of hedges; after having planted a hedge, it takes several years before its effect will be reflected for instance by the number and species of breeding birds in the area.

But a very important advantage of action-oriented indicators is their transparency and therefore its advantage concerning their implementation. Action-oriented indicators are often considered to be easier to handle in practice and fairer, most of the seven analysed methods are based upon such indicators. But as the results of actions have effects not just in one direction, either affecting abiotic or biotic resources, it is not possible to subdivide these action-oriented indicators properly in a satisfactory way into indicators for bio-resources on the one hand and indicators for abiotic resources on the other hand. Indicators cannot be seen in a one-dimensional way. Nevertheless, because of focussing on bio-resources, the authors have tried to "subdivide" the indicators in the above-mentioned way. Consequently, against this background, only the pesticide indicator is regarded as a relevant criterion for bio-resources because pesticide use is one of the main reasons for the decline of the typical weed flora (Schumacher and Schick, 1998). Although fertilizers (assessed by indicators such as surpluses and efficiencies of N and P) and especially N-fertilizers influence the agro-ecosystem in a considerable way too, they should be considered here as more closely related to abiotic resources. Table 2 shows the comparison between the analysed tools regarding the action- or results-orientation of bio-resource relevant indicators.

Having discussed the different kinds of indicators and their "value" for assessing abiotic and bio-resources, the focus has to be laid on the farm level assessment itself. In order to have a more detailed look at the assessment of bio-resources one should be aware of the fact that farm area can be subdivided into the cultivated (productive) farm area and uncultivated (non-productive) farm area. A way to compare the different approaches is by analysing where or in which of these areas the agricultural effects are assessed. In what way are the productive and the non-productive areas dealt with? In which way is the productive farm area (cultivated land: grassland (pasture/meadows)) and arable land assessed with regard to bio-resources? In which way is the non-productive farm area assessed and evaluated with regard to bio-resources?

3. Assessing and evaluating the bio-resources on the cultivated land

As already mentioned most of the indicators used are action-oriented (see Table 2), but their validity with regard to the state of bio-resources is not always obvious. Only a few action-oriented indicators seem to be relevant regarding bio-resources on farm level. Table 3 lists all at first sight relevant indicators used when assessing and evaluating farmland by the different methods. In order to compare the different assessment methods with each other, we have chosen an example: the assessment and evaluation of arable land. How is a plot assessed and evaluated? What kind of indicators are used, how many indicators are used, are these reliable?

Every presented method is based on action as well as on results-oriented indicators, except for the Dutch Biodiversity Yardstick, which relies only on results-oriented indicators in the form of indicator species.

Table 2
Assessing the cultivated and the non-cultivated farm area

Method	Number of criteria/indicators	Assessing the cultivated farm area (productive)		Assessing the uncultivated farm area (non-productive)	
		Action-oriented indicators	Results-oriented indicators	Quantity (indirect)	Quality (direct)
Ecopoints Lower Austria (A)	9	Exclusively	–	Area (%)	–
KUL (D)	22	(Almost) exclusively	(Diversity of crops)[a]	Area (%)	–
Halberg (DK)	13	Almost exclusively	Coverage of weeds in grain	Area (%)	–
Solagro (F)	16	(Almost) exclusively	(Number of crops)[a]	Density (m/ha farm area)	–
Frieben (D)	10 criteria with a complex set of indicators	Yes	Yes	Density (m/ha farm area)	Yes
Nature Balance Scheme (D) (Naturbilanz)	41	Yes	Yes	Area (%)	with restrictions
Biodiversity Yardstick (NL)	Lists with indicator species	–	Exclusively	–	Yes

[a] Considered results-oriented indicator: only one.

Table 3
Assessing the arable land

Method	Action-oriented indicators	Results-oriented indicators	Assessing input/expenditure	Evaluation level
Ecopoints Lower Austria (A)	Pesticide use (number or treatments/ha) Plot size	–	Plot card index	Farm
KUL (D)	Pesticide use (costs/ha) Median of plot size	Diversity of crops, (Shannon–Weaver index)[a]	Plot card index	Farm
Halberg (DK)	Unsprayed area (% farmland)[b]	Coverage of weeds in grain	Plot card index Inspection	Farm
Solagro (F)	Unsprayed area (% farmland)	Number of crops[a]	Plot card index	Plot Farm
Frieben (D)	Crop rotation[a] NBI (no running beneficial arthropod-indicator)	Number of typical weed species Number of endangered weed species Presence of endangered plant communities Flower supplies Number of plant species advancing beneficial insects Structure of vegetation	Plot card index Inspection Profound botanical knowledge is required (e.g. taxonomy of plant sociology)	Plot Farm
Nature Balance Scheme (D) (Naturbilanz)	Crop rotation[a] Management: no mineral fertilizer, no pesticides resp. ecological agriculture, kind of mower	Presence and number of indicator weed species	Plot card index Inspection Basic botanical knowledge is required	Farm resp. cultivated area
Biodiversity Yardstick (NL)	–	Presence and frequency of indicator species: plants, birds, mammals, reptiles, amphibians, butterflies	Inspection Botanical and faunistic knowledge is required	Farm resp. biotopes

[a] Only with restrictions seen as biotic resource indicator.
[b] The amount of pesticide use is considered by the TFI (treatment frequency index), but main impact (by the author) on ground water—abiotic resources.

Pesticide treatment is seen as a biotic resource relevant criterion (see above). This has been taken into account by five of the seven presented tools. Whereas the Ecopoints Lower Austria, Halberg, Solagro and the Nature Balance Scheme assess arable land by the criterion "area without pesticide treatment" among other criteria, the assessing method KUL uses a monetary parameter. Frieben ignores pesticide treatments because her method is a tool to assess organic farms only which do not use pesticides at all.

The KUL-method assesses the costs of the pesticide treatment on the farm level (per hectare) in relation to the regional costs. The optimum of pesticide use, i.e. the optimum, is reached when the costs are 30% lower on the farm level than regional level. Organic farming methods, for example, are not being taken into account at all when assessing arable land by this indicator or rather by using the proposed scale. The use of this indicator can only be justified by defining a new goal or by varying the evaluation scale. By the criterion "unsprayed area" (used by four of the seven methods) farms with little or no pesticide use are rewarded but, as it is only assessed on the farm level, it is not obvious which plots are treated with pesticides and the contribution to nature conservation cannot really be traced.

Another management action which affects bio-resources is the crop rotation or the number of crops. Used as an action-oriented indicator for bio-resources which is supposed to give evidence of the quality of bio-resources, it is not supposed to be very strong, although crop rotations do have effects on the biodiversity (e.g. van Elsen, 1994; Frieben, 1998; Prescher and Büchs, 2000) and although these positive effects seem to be generally recognized. But on the other hand it can be a very useful indicator when evaluating the crop plant diversity on farm level. A high number of different crop species and varieties should be considered as an ecological achievement in the field of maintenance and conservation (agro-) biodiversity (see the UNEP, 1998 Convention on Biological Diversity). Both KUL and Solagro take into account the number of crops. Whereas Solagro just assesses the number of crops, KUL considers in addition to that their distribution on the farmland by applying the Shannon–Weaver index. By using the Shannon–Weaver index the "crop plant diversity-indicator" has links to the landscape image (see KUL). But both criteria should not be used as strong "bio-resource indicators" on farm level. The ecological state of arable land is not reflected in a satisfactory way by naming the crops or crop sequences.

The plot size is taken into account by indicators such as "plot size", "average plot size". The NBI index (=Theoretisch nicht besiedelbare Innenfläche = no running beneficial arthropod-indicator, for details see Frieben (1998)) recognizes even the shape of a plot as a relevant indicator which is very important when taking into account the colonization of arable land by arthropods. But nevertheless these management indicators do not describe the ecological state of the plot—although there are obviously close linkages to it. Besides the size of a plot, other aspects, too, have to be taken into account. The KUL-method uses, e.g. the indicator plot size to evaluate the impact on the soil in the field of erosion risks.

But it seems to be obvious that the above mentioned and listed indicators do not allow conclusions concerning the quality of or the site-specific agricultural impact on the bio-resources.

Four out of seven methods are more or less based on results-oriented indicators beside action-oriented ones. Two of them work with just one result-oriented indicator when assessing arable land. The Danish method by Halberg uses, e.g. coverage of weeds in grain, and the Nature Balance Scheme uses a set of indicator plant species whose presence has to be documented on the arable land to obtain scores. The Dutch Biodiversity Yardstick uses "indicator species". Five groups of indicator species are listed: plants (199 species of vascular plants), birds (77 species of nesting birds, 14 species of wintering birds), mammals (17 species), amphibians (7 species), reptiles (2 species) and butterflies (26 species) with fixed ecological and for plants even perception value indices. Besides the presence of a specie its ecological importance (for example, rarity, population trends) and scenic value, i.e. the species' contribution to the attractiveness of the scenery (for example, height, colours of the flowers, flowering period) is included in the yardstick. According to this yardstick *Viola arvensis* obtains six points as ecological rating score and 53 as scenic value rating score, *Legousia speculum-veneris* obtains 63 points as ecological rating score and 51 as scenic value rating score. Twenty-two indicator species of vascular plants are chosen to describe arable land (for details see Buys (1995), pp. II–III, 83–89). The method developed by Frieben (1998) reveals the biggest number of results-oriented indicators: number of species of the typical weed flora (characteristic species of plant communities of arable land, for details see Frieben (1998), pp. 89–93) number of endangered species (according to red data lists), the presence of endangered plant communities (red data lists), amount of flowering weeds, number of species which advance beneficial insects (Frieben, 1998, pp. 96–98, 259–260), structure of vegetation. The amount of flowering weeds is estimated and assessed in three categories. Beneficial insects (only Syrphidae and Chrysopidae are considered) advancing weeds do not include grasses and sedges as well as very competitive weeds (e.g. *Cirsium arvense*, *Galium aparine*). A plot assessed by using the set of indicators proposed by Frieben is very well reflected concerning its ecological state at the moment. But taking into account the time input when assessing each plot in this way, soon leads to the conclusion that, despite its validity, it cannot really be put into a wide range of practice. The more the assessment method is based on results-oriented indicators, the more time is needed, with the result that it becomes less easy and more expensive to implement.

4. Assessing and evaluating bio-resources on non-productive farm area, uncultivated or very extensively cultivated farm land

Concerning the assessment of cultivated and non-cultivated farm area (Table 2) there generally do not exist very great differences between the different tools. All of the examined assessment methods except for the Dutch one assess more or less the quantity of uncultivated land on farm level. But unfortunately, the reference point is not always clearly defined: total farm area or cultivated farm area? Solagro neglects the area but quantifies the length of margins expressed in meters per hectare farm area. Frieben quantifies the density of hedges and the "presence of biotopes of special conservation interest". The Nature Balance Scheme considers "special" biotopes (e.g. ponds or slopes) by their presence or/and number. Indicators concerning the presence and number of "special" biotopes have got linkages to species diversity; a diverse landscape structure effects species diversity positively (Duelli et al., 1999).

By using the criterion length per hectare, the compound or the connection of structural elements in the landscape is reflected (as ecological validity) on the one hand and on the other hand it could be used as a criterion for landscape structure.

In order to better understand how non-productive farm area is dealt with, Table 4 shows the assessment of a "special" biotope—a field margin.

Whereas six out of the seven methods (see Table 3) quantify the non-productive farm area it should be worth comparing the assessment of a field margin with these different tools.

The area of field margins is assessed in four methods. Solagro works with the indicator length (see above). It does not quantify a field margin in general. Its assessable structures are restricted to hedges and edges of forests so that a herbal field margin will simply be neglected. Frieben does as-

Table 4
Assessing a field margin

Method	Action-oriented indicators	Results-oriented indicators	Assessing input	Evaluation level
Ecopoints Lower Austria (A)[b]	Area (% farmland)	Value dependent on adjacent structures	Maps/aerial pictures Inspection	Farm
KUL (D)[b]	Area (% farmland region)	–	Maps/aerial pictures	Farm
Halberg (DK)[b]	Area (% farmland)	–	Maps/aerial pictures	Farm
Solagro (F)[c]	Length (m/ha)	–	Maps/aerial pictures	Farm
Frieben (D)[a]	Breadth (m) Presence of a fallow margin	Plant coverage Flower supply after harvest Vegetation structure after harvest Number of endangered plants Presence of bushes, trees or hedges	Maps/aerial pictures Inspection Botanical knowledge is required	Margin itself Farm
Nature Balance Scheme (D) (Naturbilanz)[b]	Area (% farmland)		Maps/aerial pictures Inspection	Farm
Biodiversity Yardstick (NL)[c]	–	Field margins are not separately assessed—no special biotope module: presence and frequency of indicator species: plants, birds, mammals, reptiles, amphibians, butterflies	Inspection Botanical and faunistic knowledge is required	Farm resp. biotopes

[a] Field margins are proper assessable elements.
[b] Field margins are considered as non-cultivated area.
[c] Field margins are no assessable elements.

sess the density of hedges on farm level, but does not assess the density of herbal field margins. The breadth of the margin and the presence of a fallow margin are the quantifying indicators. Although herbal field margin are not quantified via area of farmland, field margins in general represent 1 criterion out of 10 when assessing a farm. The Dutch Biodiversity Yardstick relies on a set of biotope modules. Here as well herbal field margins are not yet considered as a special biotope module, but they probably might be integrated in the module arable land. There is no transparent assignment. Whereas the quantity seems to be reflected in a mostly satisfactory way, the quality is not really taken into account. The Ecopoints Lower Austria evaluates marginal structures in relation to adjacent structures: a tarred road, for example, reduces its value. It is evident that the assessment method by Frieben is the only method which considers the ecological quality of the field margin itself. There are five indicators which are supposed to reflect its ecological state (see Table 4). The quality of a field margin itself is reflected by using seven indicators which are listed in Table 4.

Five out of the seven methods do consider field margins, but only one out of the seven methods tries to reflect its “ecological state”.

5. Conclusion

All presented methods are instruments to assess/quantify and evaluate the agricultural impact on the environment (the method Halberg does not yet evaluate). Whereas the impacts on and the state of abiotic resources seem to be sufficiently reflected, there are evident lacks concerning the assessment of the impact on and the state of biotic resources. This review reveals deficiencies despite the aim to assess agricultural impacts comprehensively on the farm level. The ecological quality, the state, of the cultivated farmland is not really reflected. Action-oriented indicators are not suitable to assess and evaluate a state even less with regard to a clear defined goal concerning the quality of biotic resources. Indicators such as crop diversity and plot size have linkages to bio-resource but should not be used as strong bio-resource indicators. Much more emphasis should be laid on the ecological quality of cultivated land. Useful concepts seem to be the indicator species system of the Biodiversity Yardstick and the Nature Balance Scheme. Indicator species systems seem to be very valuable if not too complex (i.e. not too many species, species with a diagnostic value, easily recognizable species) and adapted to different regions. The method Frieben seems to be too complex to be put into practice. Nevertheless this method might be a very valuable basis if cut back in a most sensible way with regard to the essential indicators. In order to come to a most transparent evaluation on the farm level, a plot specific assessment or even evaluation seems to be indispensable. This has already been achieved in the Ecopoints Lower Austria and the method of Frieben. The non-productive farm area is well documented by its area and the lengths of its adjacent structures. Their quality on the farm level might be reflected by an indicator such as structure diversity (see Nature Balance Scheme) or at a much lower level again by indicator species. Here again, the method of Frieben seems to be too complex, despite its excellent reflection of the state. The area alone does not give evidence of the quality.

Promoting environmentally or bio-resource sound agriculture may be achieved via establishing this aim in policies and policy schemes. These should be based on sets of reliable bio-resource indicators. The comparison shows that there do exist many valuable approaches and attempts which need further developing.

Besides promoting environmentally sound agriculture via policies, there should also be the aim of raising the awareness concerning agricultural impacts on bio-resources on the farm level. Assessment tools such as the Biodiversity Yardstick, where farmers themselves are involved in assessing their farm area (presence, and frequency, of indicator species) or agri-environmental policy schemes in Baden-Württemberg, Germany (MLR, 1999) or Switzerland (Schweizerischer Bundesrat, 2001) are important steps towards *environmentally aware* agriculture.

Discrepancies between action- or results-oriented assessment and evaluation as well as the easiest possible way to act without neglecting important issues are problems which still have to be solved.

Acknowledgements

The authors would like to thank Anthony Alcock for language correction on the manuscript.

References

Buys, J.C., 1995. Naar een natuurmeetlat voor landbouwbedrijven. Centrum voor landbouw en milieu (CLM) Utrecht, No. 169.

Ceuterick, D. (Ed.), 1998. Proceedings of the International Conference on Application of Life Cycle Assessment in Agriculture, Agro-Industry and Forestry, Brussels, December 3–4, 1998.

Deblitz, C., Plankl, R., 1998. EU-wide Synopsis of Measures According to the Regulation REG (EEC) 2078/92 in the EU. Federal Agricultural Research Centre Braunschweig Völkenrode (FAL), Braunschweig.

Deutscher Rat für Landespflege (DRL)—German Council for Land Stewardship (Ed.), 2000. Honorierung von Leistungen der Landwirtschaft für Naturschutz und Landschaftspflege. Schr. d. Dt. Rats f. Landespflege, Nr. 71 (Juli 2000).

Duelli, P., Obrist, M.K., Schmatz, D.R., 1999. Biodiversity evaluation in agricultural landscapes: above-ground insects. Agric. Ecosyst. Environ. 74, 33–64.

Eckert, H., Breitschuh, G., 1997. Kritische Umweltbelastungen Landwirtschaft (KUL): Ein Verfahren zur Erfassung und Bewertung landwirtschaftlicher Umweltwirkungen. In: Diepenbrock, W., Kaltschmitt, M., Nieberg, H., Reinhardt, G. (Eds.), Umweltverträgliche Pflanzenproduktion: Indikatoren, Bilanzierungsansätze und ihre Einbindung in Ökobilanzen. Initiativen zum Umweltschutz 5, Deutsche Bundesstiftung Umwelt, Osnabrück, pp. 185–195.

Eckert, H., Breitschuh, G., Sauerbeck, D., 1999. Kriterien umweltverträglicher Landbewirtschaftung (KUL)—ein Verfahren zu ökologischen Bewertung von Landwirtschaftsbetrieben. Agribiol. Res. 52, 57–76.

European Union, 1992. Council Regulation (EEC) No. 2078/92 of June 30, 1992 on agricultural production methods compatible with the requirements of the protection of the environment and the maintenance of the countryside (July 30, 1992). Off. J. L215, pp. 0085–0090.

European Union, 1999. Europe's Agenda 2000—Strengthening and widening the European Union. http://europa.eu.int/comm/agenda2000/public_en.pdf.

Frieben, B., 1998. Verfahren zur Bestandsaufnahme und Bewertung von Betrieben des Organischen Landbaus im Hinblick auf Biotop- und Artenschutz und die Stabilisierung des Agrarökosystems. Schriftenreihe Institut für Organischen Landbau, Bd. 11, Berlin, Verlag Dr. Köster.

Geier, U., 2000. Anwendung der Ökobilanz-Methode in der Landwirtschaft. Schriftenreihe Institut für Organischen Landbau, Bd. 13, Berlin, Verlag Dr. Köster.

Geier, U., Köpke, U., 2000. Analyse und Optimierung des betrieblichen Umweltbewertungsverfahrens "Kriterien umweltverträglicher Landbewirtschaftung" (KUL). Berichte über Landwirtschaft Bd. 78 (1), 70–91.

Halberg, N., 1997. Farm level evaluation of resource use and environmental impact. In: Isart, J., Llerena, J.J. (Eds.), Proceedings of the Third ENOF Workshop on Resource Use in Organic Farming, Ancona, June 5–6, 1997, pp. 213–224.

Halberg, N., 1999. Indicators of resource use and environmental impact for use in a decision aid for Danish livestock farmers. Agric. Ecosyst. Environ. 76, 17–30.

Mayrhofer, P., 1997. Das Ökopunktemodell Niederösterreich-Aufbau und Umsetzung in der Agrarumweltpolitik. In: Diepenbrock, W., Kaltschmitt, M., Nieberg, H., Reinhardt, G. (Eds.), Umweltverträgliche Pflanzenproduktion: Indikatoren, Bilanzierungsansätze und ihre Einbindung in Ökobilanzen. Initiativen zum Umweltschutz 5, Deutsche Bundesstiftung Umwelt, Osnabrück, pp. 197–208.

MLR (Ministerium Ländlicher Raum Baden-Württemberg—Ministry for Rural Area Baden-Württemberg), 1999. Artenreiches Grünland. Anleitung zur Einstufung von Flächen für die Förderung im MEKA II. MLR-59-99.

Naturschutzbund Deutschland e.V.—German Society for Nature Conservation (Ed.), 1999. Naturschutz mit der Landwirtschaft. Eine Übersicht zu Naturschutzelementen im landwirtschaftlichen Betrieb.

Niederösterreichische Agrarbezirksbehörde, Niederösterreichischer Landschaftsfonds (Eds.), 1999. Regionalprogramm Ökopunkte Niederösterreich—Informationsheft, überarbeitete Fassung, Wien.

OECD (Organisation for Economic Co-operation and Development), 1994. Environmental Indicators. OECD Core Set, Paris.

Oosterveld, E.B., Guldemond, J.A., 1999. De natuurmeetlat gemeten. Centrum voor landbouw en milieu (CLM) Utrecht, No. 407.

Oppermann, R., Nürnberger, M., Kunz, S., 2000. Kriterien zur Messung ökologischer Leistungen in der Landwirtschaft. agrarspectrum Bd. 31: Entwicklung nachhaltiger Landnutzungssysteme in Agrarlandschaften, pp. 31–43.

Pointereau, P., Dimkic, C., Mayrhofer, P., Backhausen, J., Bochu, J.-L., Doublet, S., Meiffren, I., Schumacher, W., 1999. Umweltbewertungsverfahren für die Landwirtschaft. Drei Verfahren unter der Lupe. SOLAGRO Toulouse.

Prescher, S., Büchs, W., 2000. Der Einfluß der Fruchtfolgegestaltung auf die Schlupfabundanzen von Fliegen (Diptera, Brachycera) im Ackerbau. agrarspectrum, Bd. 31, Entwicklung nachhaltiger Landnutzungssysteme in Agrarlandschaften, pp. 197–203.

Rudloff, B., Geier, U., Meudt, M., Urfei, G., Schick, H.-P., 1999. Entwicklung von Parametern und Kriterien als Grundlage zur Bewertung ökologischer Leistungen und Lasten der Landwirtschaft—Indikatorensysteme. In: Umweltbundesamt (Ed.), Texte 42/99.

Schumacher, W., Schick, H.-P., 1998. Rückgang von Pflanzen der Äcker und Weinberge—Ursachen und Handlungsbedarf. Schriftenreihe f. Vegetationskunde 29, 49–57.

Schweizerischer Bundesrat, 2001. Verordnung über die regionale Förderung der Qualität und der Vernetzung von ökologischen Ausgleichsflächen in der Landwirtschaft, Bern.

SRU (Rat der Sachverständigen für Umweltfragen—The German Council for environmental Advisors), 1985. Umweltprobleme der Landwirtschaft. Sondergutachten, Stuttgart.

UNEP (United Nations Environment Programme), Secretariat of the Convention on Biological Diversity (CBD), 1998. Convention on Biological Diversity.

van Elsen, T., 1994. Die Fluktuation von Ackerwildkrautgesellschaften und ihre Beeinflussung durch Fruchtfolge und Bodenbearbeitungszeitpunkt. Ökologie und Umweltsicherung 9, Kassel.

Walz, R., 1997. Grundlagen für eine nationales Umweltindikatorensystem—Weiterentwicklung von Indikatorensystem für die Umweltberichterstattung. In: Umweltbundesamt (Ed.), Texte 37/97, Berlin.

ELSEVIER

Available online at www.sciencedirect.com

Agriculture, Ecosystems and Environment 98 (2003) 435–441

Agriculture Ecosystems & Environment

www.elsevier.com/locate/agee

Method for assessing the proportion of ecologically, culturally and provincially significant areas (OELF) in agrarian spaces used as a criterion for environmental friendly agriculture

Dieter Roth*, Maik Schwabe

Thueringer Landesanstalt fuer Landwirtschaft, Naumburger Strasse 98, D-07743 Jena, Germany

Abstract

Environmental compatibility in agriculture depends to a great extent on the proportion of ecologically, culturally and provincially valuable areas (OELF) in the form of structural elements and virtually undisturbed habitats in fields. A method is presented which enables the assessment of the proportion of "OELF", used as a criterion for environmental friendly agriculture. To this end, both comprehensible objectives for the proportion of OELF and methods for giving evidence of the actual proportion of OELF in agrarian spaces are needed.

The *target proportion of OELF* in agrarian spaces for the various natural areas in Thuringia was derived from precise plans concerning the use and care of agrarian spaces, in co-operation with the authorities responsible for agriculture and nature conservation, and made available to the department for agricultural management in the form of maps and tables.

To calculate the *actual proportion of OELF* in the agrarian spaces of a farm or community as an actual value, for which as a rule there is no complete or exact spatial data available, practical methods are presented. The criterion 'proportion of OELF in agrarian spaces' is seen as fulfilled if the target value of the farm or community can be shown as having been achieved.

Keywords: Environmental friendly agriculture; Structural elements; Habitats; Assessment method

1. Introduction

Environmental compatibility in agriculture is not determined solely by agricultural and agronomic measures such as the amount of fertiliser, nutrient balance or use of plant protection products, but depends to a great extent on the proportion of ecologically, culturally and provincially valuable areas (OELF) in the form of structural elements in typical landscapes, and virtually undisturbed habitats in fields. To be able to use the proportion of "OELF" as a criterion for the environmental compatibility of agriculture, a practical and simple method has been developed and tested for use by farmers. In the following, this method is described and discussed.

2. Presentation of the assessment method

2.1. OELF and their ecological functions in agrarian spaces as a basis for assessment

OELF encompass both agricultural elements, which are not used regularly, such as woody plants in fields, hedges or succession areas, as well as elements which require either regular use or care, such as semidry

* Corresponding author. Tel.: +49-3641-683452; fax: +49-3641-683239.
E-mail address: m.schwabe@jena.tll.de (D. Roth).

0167-8809/$ – see front matter © 2003 Elsevier Science B.V. All rights reserved.
doi:10.1016/S0167-8809(03)00102-6

Table 1
Summary of the ecologically, culturally and provincially significant areas (OELF) in agrarian spaces

Not or hardly of use for agriculture	Subject to extensive use and/or care
Hedges, riparian woody plants, rows of trees etc.	Moist and wet meadows, water catchment fields
Extensive field shrubs and bushes (<3 ha)	Dry and semidry lawn
Succession areas	Acidic dry lawn, e.g. mat-grass lawn
Reed bank, large and small sedge fens	Fertile valley and mountain meadows
Largely natural running and standing water bodies (<3 ha)	Streuobst crops
Dry stone walls, open gravel-pits and others	Extensive fields, field verges
	Heathland, permanent fallow land
	Drainage and irrigation ditches
	Borders and boundaries

lawn, moist and wet meadows or borders dominated by grasses or herbs (Table 1).

Many of the OELF mentioned above were used in historical land use forms for economic purposes. Many of them originate from this time and have been maintained and cared for over centuries. In the meantime, almost all of the OELF (see Table 1) have either lost their use completely or are not profitable to cultivate. Thus, the profits of using extensive grassland are usually way below the cultivation costs. ‘Streuobst’ (scattered fruit trees in one orchard where various types and varieties of fruit are grown together) has not been able to compete with intensive fruit growing for a long time and in addition is difficult to harvest; hedges are not normally needed by farmers any more, they even represent a hindrance as far as cultivation is concerned (Roth and Berger, 1999). Due to economic reasons these agricultural elements do not have the right to exist as a rule. Their ecological, cultural and provincial functions have however remained (Table 2), as a great deal of literature, alone in the German language, will confirm (see e.g. Blab et al., 1995; Jedicke, 1994; Kaule, 1991; Konold et al., 2002; Hoffmann et al., 2001; Plachter, 1991).

A case study from an area of arable farming in the ‘Thuringia Valley’ demonstrates the high significance of OELF, e.g. for local flora. Although the OELF represent only 5% of the study area, 86% of all blossoming plants were found here (Roth and Schwabe, 1998). The decisive consequence of the change in values

Table 2
Significant ecological, cultural and provincial functions of structural elements and virtually undisturbed habitats in agrarian spaces

Landscape water balance and prevention of water pollution	Moist grassland, headwaters and other humid habitats Small bodies of water Riparian edges of running and standing water
Soil protection	Boundaries and borders at right angles to slopes Permanent grassland in more extreme slope locations Woody plants with a wind breaking function
Species protection	Woody habitats (hedges, rows of trees, riparian shrubs, extensive field shrubs) Extensive grassland habitats (dry lawn, moist meadows, dry pastures, Streuobst orchards and others) Path, field and wood margins Extensively used field verges Running and standing water bodies and their embankment areas Succession areas, fallow ground, heath Other small structures, e.g. walls of collected stones
Variety in the landscape and typical characteristics, recreation value	All the elements of a landscape stated

presented—from originally economically used areas and structures to solely or predominantly ecologically, culturally and provincially significant areas—is that maintaining, and even more so restoring these areas no longer takes place automatically during normal agricultural cultivation, but requires specific action. The necessity to make farms realise the ecological significance of OELF and the responsibility of the farmer for their constant maintenance is a result of this. An important step is to develop and to render assessable the proportion of OELF in agrarian spaces as a criterion for environmental friendly agriculture.

2.2. Calculation of landscape-typical target values for the proportion of OELF in agriculturally used areas

The basic principal of the method is that a target value for the proportion of OELF in agrarian spaces is compared to the current proportion of OELF as an actual value. Agriculturally used landscapes means open fields with the respective intermediate landscape structures. The size of an agrarian space as a reference area for the proportion of OELF is calculated by the defined area minus wooded areas (>3 ha), built up areas (including gardens, sports fields, cemeteries and so on), public roads, running water and standing water (>3 ha). The target proportion of OELF in agrarian spaces as a target value depends greatly on the natural conditions, above all on the topography, soil quality and humidity regime of the location, as well as—connected with this—the historical development of land use. The different and often conflicting demands of agriculture and nature conservation hardly allow an exact scientific justification of how high the proportion of the various landscape-typical structures and virtually undisturbed habitats in a certain area should be. However, in order to arrive at quantifiable target figures and thus at an assessment method, the target proportion and the type and distribution of OELF was calculated first of all for the various natural regions in Thuringia, with the help of precise plans on the use and care of agrarian spaces (ANP). These 'ANP' represent a practicable concept agreed on together by farms, nature conservation representatives and community, for the future use and landscaping of agrarian spaces, with the target of securing both agricultural, and ecological–cultural–provincial functions permanently (Roth, 1996; Roth and Schwabe, 1999).

In every ANP, both present and also future OELF targets are designated spatially, and also presented in tables and on maps. The following data and studies in particular are used as a basis for targeting the future OELF content of a particular space:

- A detailed analysis and assessment of the subjects of protection, soil and water, including water quality, present habitats worthy of protection and connective habitats as a whole as an existential basis for local flora and fauna, as well as analysis of the landscape. The analyses include a presentation of the conflicts between agricultural use of the soil and individual protection targets.
- Previous structure of the agrarian space with structural elements and virtually undisturbed habitats, in particular running and standing water before field clearing or melioration, woody structures and (extensive) grassland areas (through the evaluation of historical field maps and land surveys).
- Present landscape concepts and environmental authority plans.
- Data from farms concerning present and future soil use and patterns of production.

Under current production conditions in agriculture, all this will result in a practicable concept for securing and extending the proportion of ecologically and culturally provincially valuable structural elements and habitats, as well as expanding their integration and networking. Practicable means that both the ecological and agricultural functions of the respective agrarian space can be guaranteed to last. In Fig. 1, an example of a concept for the future use and landscaping of agrarian spaces is given by the community of Plothen in the Schiefergebirge (slate mountains) of East Thuringia.

Up to now, plans for the use and care of agrarian spaces (ANP) have been drawn up for approximately 24,000 ha agrarian area in various natural regions in Thuringia. Table 3 contains an extract concerning the increase in the proportion of OELF as proposed, and which has actually been put into practice up to now.

Because the development of the ANP and the calculation of target values for the proportion of OELF in agrarian spaces involves a great deal of work (meaning

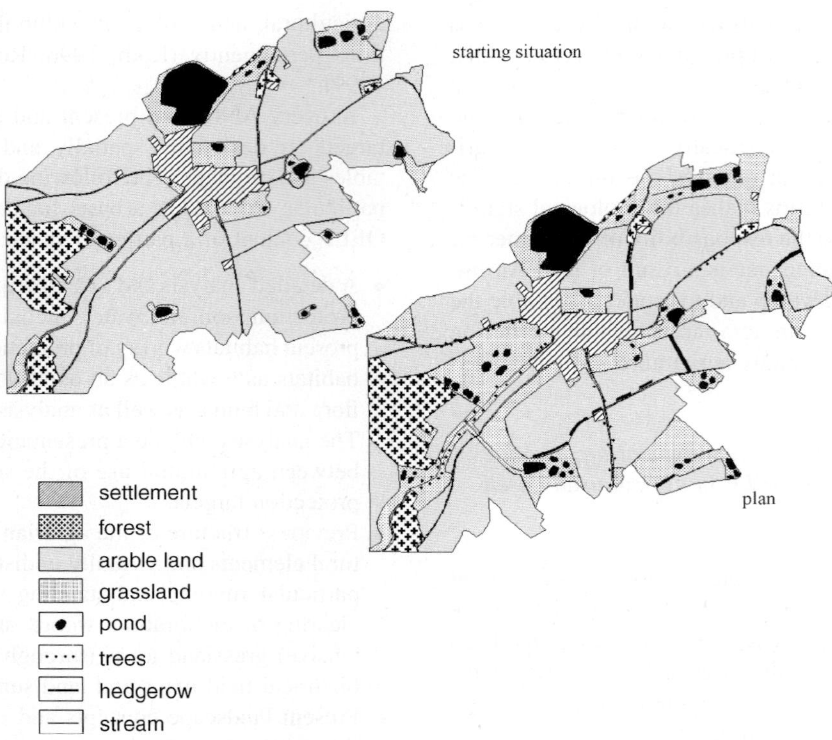

Fig. 1. Plan for the use and care of agrarian spaces (ANP) (section of the ANP plan for the community of Plothen; size: 550 ha).

that such values are only available to a certain extent), the target values calculated for the 24,000 ha up to now have been generalised and transferred to the respective natural regions (Table 4). A differentiation is made between a longer term and a short-term target proportion of OELF, which appears feasible with the help of current promotional possibilities. Experience up to now has shown that the latter forms about 70% of the perspective target proportion of OELF.

The guide values in Table 4 were processed, with reference to the community structure in Thuringia, by way of tables, and cartographically, so that they can be used by farms and communities in Thuringia with no access to ANP or similar precise areal plans.

Table 3
Actual and target values for the proportion of OELF in agrarian spaces (%) in different natural regions of Thuringia (examples, taken from plans concerning the use and care of agrarian spaces)

OELF	Hilly arable land in Thuringia		Slate mountains in East Thuringia		Hilly variegated sandstone area in Thuringia		Meininger lime plateaus with steep valleys	
	Actual	Target	Actual	Target	Actual	Target	Actual	Target
Hedges and rows of trees	0.7	1.3	0.6	1	2	3.1	2.1	2.7
Field shrubs	0.8	1.6	0.9	1.1	3.6	3.6	1.2	1.2
Streuobst crops	0.4	0.4	0.1	0.2	2.1	2.1	1	1.3
Extensive grassland	1.8	5.1	4.3	8.8	0.7	0.7	11.5	16
Borders and boundaries	0.3	1	2	2.5	1.7	1.8	0.5	0.7
Water bodies	0.7	0.7	1.5	1.7	0.4	0.4	0.9	0.9
Total	4.7	10.1	9.4	15.3	10.5	11.7	17.2	22.8

Table 4
Guide values for the target proportion of ecologically, culturally and provincially significant areas (OELF) in Thuringia

Proportion of OELF in % of agrarian space present			Natural regional unit of Thuringia
Present	Targeted		
	Short-term	Long-term	
3–6	6	9	Hilly arable land in Thuringia, in Altenburger Loessgebiet
6–10	9	13	Plateaus of Saale-Sandsteinplatte and Ilm-Saale-Platte, Orlasenke, slate mountains in East Thuringia, eastern part of the North Thuringia variegated sandstone mountains, Waltershaeuser foothills, plateaus of the Meininger lime areas
8–12	11	16	Plothener lakes, Paulinzellaer variegated sandstone mountains, western part of the North and South Thuringia variegated sandstone mountains, Hainich-Duen-Hainleite, Saale-, Werra- and Unstrut-floodplains
12–17	15	22	High-altitude Thuringia forest and Hohe Thueringer slate mountains, steep valley margin areas of Saale-Sandsteinplatte, Hohe Rhoen, Vorderrhoen, steep valley areas of Ilm-Saale-Platte and the Meininger lime plateaus, Werrabergland-Hoerselberge
17–22	18	25	Strong relief areas of the Thuringia forest, upper Thuringia slate mountains and the Harz

Corresponding guide values are also now available for the Federal states of Bavaria, Saxony-Anhalt and Saxony (Diemann and Arndt, 2000; Roth et al., 2001; Unger, 1999).

2.3. Determination of the present proportion of OELF in agrarian spaces

The real proportion of OELF in a farm or in agrarian spaces in the community is compared to the respective target value. Evidence of actual OELF in an agrarian space is only complete if precise areal planning for the agrarian space, comprising all present structural elements and habitats in the sense of the ANP described in Section 2.1, has been carried out. This is however only the case in isolated incidences. Nevertheless, in order to guarantee a reliable and at the same time rational calculation of the present proportion of OELF, several individual steps are necessary.

First of all, the OELF stated in the farm documentation on use of the area are to be used. The list includes rough and dry grassland, mountain pastures rich in species, alpine pastures, saline pastures, moist grazing grassland including areas for meadow-breeding birds, extensive grassland with cutting time restrictions, stands of Streuobst, long-term unproductive fields due to ecological reasons (at least a 10-year fallow period), borders, buffer zones, field verges, areas with woody plants, small bodies of water and other small structures.

To be able to decide whether permanent grassland can be classified as OELF, the simple method by Briemle and Oppermann (1999) can be applied. In Thuringia, grassland areas in the cultural landscape programme/part C with additional nature conservation restrictions, as well as those shown in the contracted nature conservation were assessed as OELF. On the basis of this data, a farm-specific assessment of the proportion of OELF can be made.

Inter-farm evidence of available OELF can also be provided on a community scale. This is above all recommendable for farms which have less than 400–500 ha farming area, which does not represent the natural region satisfactorily, or for farms where a large part of OELF is on community land and can therefore not be assigned to the individual farm (e.g. arable woody plants or small bodies of water). In such cases, the proportion of OELF is taken from the whole agrarian area of the community for each farm operating in this community, regardless of the proportion actually on the individual farm areas. To record OELF which are not documented sufficiently at the farm or in the community, the evaluation of ortho-photos from the land surveying office or interpretative data from aerial photographs for habitat-mapping is recommended. It is also possible to record terrestrial data using satellite-supported surveying methods (GPS; Roth et al., 2001).

As soon as the short-term target value (see Table 4) is reached, the criterion for the proportion of OELF in agrarian spaces is initially fulfilled. Otherwise, every farm and the community are requested to increase the proportion of OELF by creating new landscape-typical structures and habitats leading towards the target value.

3. Examples

In order to demonstrate application of the method for assessing the proportion of OELF in agrarian spaces as a criterion for environmental friendly agriculture, two different examples are described and evaluated in the following.

3.1. Example 1

Cash crop farm A in the hilly arable land of Thuringia cultivates an agrarian area of 860 ha within the community of X. According to Table 4, in farm A at least 6% or 51.6 ha of the agrarian area must be OELF in order to fulfil the criterion 'proportion of OELF in agrarian spaces'. Evidence exists of the following OELF in the farm:

Semidry lawn (on level Keuper slopes with a thin clay layer)[a] (ha)	6.2 (0.72%)
A 3 km long brook with riparian woody plants and narrow brook-meadow[a] (ha)	9.1 (1.07%)
Two groups of field shrubs and woodland[a] (ha)	4.3 (0.50%)
A 14 km tree-bush hedges with an average breadth of 8 m[b] (ha)	11.6 (1.35%)
A 11 km rows of trees lining paths, on average 5 m wide borders[b] (ha)	5.5 (0.64%)
A 6 km borders and boundaries with no woody plants, average width of 1.5 m[b] (ha)	0.9 (0.10%)
Total (ha)	37.7 (4.38% of the farm area)

[a] According to documentation on farm area.
[b] According to own surveying.

With 4.38% proportion of OELF, the farm does not fulfil the criterion mentioned above. It can now check whether the prescribed 6% proportion of OELF can be achieved by including OELF on community areas. The reference size is however the whole agrarian space of the community of X, which is 892 ha. This includes the 860 ha from farm A, a 14 ha arable area cultivated by an inhabitant as a side job, a 4.5 ha Streuobst orchard belonging to the community and also 2.7 ha field shrubs.

The following OELF are therefore present in the 892 ha agrarian space of the community:

OELF in farm A (ha)	37.7 (4.23%)
Community Streuobst orchard (ha)	4.5 (0.50%)
Field shrubs belonging to the community (ha)	2.7 (0.30%)
Total (ha)	44.9 (5.03% of the community area)

This means that the target of 6% is not fulfilled by the community agrarian space either. Both the farm and the community are therefore requested to increase their proportion of OELF at short notice, i.e. in the next 5–10 years.

3.2. Example 2

Grassland farm B with suckler cow husbandry cultivates a 142 ha high-altitude agrarian area in the Thuringia forest of community Y. According to Table 4 it needs at least 15% or 21.3 ha OELF in order that the structure of its habitat be classified as environmental friendly. The farm states the following OELF:

Mat-grass lawn (ha)	3.8 (2.68%)
Water catchment fields (ha)	2.1 (1.48%)
Golden oat-grass meadows (fertile mountain meadows) (ha)	11.8 (8.31%)
Dwarf shrub heath (ha)	0.4 (0.28%)
Brook (1.8 km) lined with tall-forb fields (ha)	7.2 (5.07%)
Four groups of woody plants (ha)	3.8 (2.67%)
Two hedges totalling 1.8 km in length and 10 m in breadth (ha)	1.8 (1.27%)
Total (ha)	25 (17.6%)

This means that farm B fulfils the criterion 'proportion of OELF in agrarian spaces'. If a further 6.3 ha of commercial grassland is extended and plant communities which are valuable in the sense of nature conservation are gradually cultivated on this area, the longer term target proportion of OELF in agrarian spaces (22%) could possibly even be achieved in the near future (see Table 4).

4. Conclusions

Extensively used grassland, copses, hedges and other landscape elements in agrarian landscapes did loose their former economic functions to a large extent but their ecological functions and their role as part of the cultural heritage gained in importance also due to the intensification of agriculture. They became ecological and culturally important areas (OELF) which should be maintained or redeveloped in the agrarian landscape by a certain percentage. Thus, the change in functions and meaning of the landscape elements results in a change in responsibility for these OELF-areas. For fulfilling mainly a task related to the society's interest by protecting ground and surface water, biotopes, species and landscapes and not fulfilling an agricultural task, the requirements for maintenance and development has not been met by agriculture but by the society itself. An important step in that direction is to put a value on and to respect OELF-areas as criterion for environmentally friendly agriculture. By using the method presented here a practicable and comprehensible approach is demonstrated. Expenditures for maintenance and development of OELF going beyond the farmer's proceeds have to be covered by the society. Existing agri-environment programmes are to be further developed and specifically used in that way.

References

Blab, J., Klein, M., Sysmak, A., 1995. Biodiversitaet und ihre Bedeutung in der Naturschutzarbeit. Natur. und Landschaft. 70 (H.1), 11–18.

Briemle, G., Oppermann, R., 1999. Artenreiches Gruenland. Anleitung zur Einstufung von Flaechen fuer die Foerderung im MEKA. II. Faltblatt des Ministeriums Laendlicher Raum Baden-Wuerttemberg (Hrsg.), Stuttgart.

Diemann, R., Arndt, O., 2000. Regionale Bodennutzungstypen und Richtwerte fuer den Biotopverbund im Agrarraum des Landes Sachsen-Anhalt. Hercynia N.F. 33, 43–61.

Hoffmann, J., Kretschmer, H., Pfeffer, H., 2001. Effects of patterning on biodiversity in northeast German agro-landscapes. In: Tenhunen, J.D., Lenz, R., Hantschel, R. (Eds.), Ecosystem Approaches to Landscape Management in Central Europe. Ecological Studies, vol. 147. Springer, Berlin, pp. 325–340.

Jedicke, E., 1994. Biotopverbund. Ulmer, Stuttgart.

Kaule, G., 1991. Arten- und Biotopschutz. Ulmer, Stuttgart.

Konold, W., Boecker, R., Hampicke, U. (Eds.), 2002. Handbuch Naturschutz und Landschaftspflege. Ecomed Publishers, Landsberg/Lech.

Plachter, H., 1991. Naturschutz. G.-Fischer, Stuttgart.

Roth, D., 1996. Agrarraumnutzungs- und -pflegeplaene- ein Instrument zur Landschaftsplan-Umsetzung. Natur. und Landschaft. 18, 237–242.

Roth, D., Berger, W., 1999. Kosten der Landschaftspflege im Agrarraum. In: Konold, W., Boecker, R., Hampicke, U. (Eds.), Handbuch Naturschutz und Landschaftspflege, VIII-6. Ecomed Publishers, Landsberg/Lech, S. 1–18, 2002.

Roth, D., Schwabe, M., 1998. Erfordernisse zum Erhalt und zur Erweiterung von Strukturelementen im Agrarraum als Lebensraeume fuer die heimische Flora. In: Thuer (Ed.), Ministerium fuer Landwirtschaft, Naturschutz und Umwelt, Einfluss der Grossflaechen-Landwirtschaft auf die Flora, pp. 60–71.

Roth, D., Schwabe, M., 1999. Umsetzung von Landschaftspflegemassnahmen im Agrarraum. Natur. und Landschaft. 31 (H.12), 376–381.

Roth, D., Schwabe, M., Unger, H.J., Diemann, R., Pleiner, I., 2001. Standpunkt des VDLUFA 'Der Anteil an oekologisch und landeskulturell bedeutsamen Flaechen ("OELF") im Agrarraum als Kriterium einer umweltvertraeglichen nachhaltigen Landbewirtschaftung'. VdLUFA, Darmstadt.

Unger, H.J., 1999. Flaechenausstattung mit Strukturelementen—Statusquo und Ziel. Z. f. Kulturtechnik und Landentwicklung 40, 113–116.

ELSEVIER

Available online at www.sciencedirect.com

Agriculture, Ecosystems and Environment 98 (2003) 443–451

Agriculture Ecosystems & Environment

www.elsevier.com/locate/agee

Experiences with the application, recordation and valuation of agri-environmental indicators in agricultural practice

Michael Menge*

Saxonian Landesanstalt fuer Landwirtschaft, Department for Soil and Plant Production, Gustav-Kuehn-Str. 8, 04159 Leipzig, Germany

Abstract

The Free State of Saxony has accompanied the programme "Umweltgerechte Landwirtschaft in Sachsen" ("Environmentally sound agriculture in Saxony") and scientifically evaluated it, taking into consideration its effectiveness on market-relief, environmental-relief and income trends.

Along with a-biotic indicators, biotic indicators have been recorded and evaluated as well. The recordation was carried through by the means of plot record lists which were kept in the farms. The valuation of the indicators' effects is taking place using time series and the presentation of trends and variations of each year. This approach is suitable to determine the effects of measures which are aimed at environmental-relief. Furthermore, information about the extent of influence a management system has on the agri-environmental measure can be obtained.

Keywords: Agri-environmental indicators; Biodiversity; Bare fallow land; Valuation

1. Introduction

For the implementation of the accompanying measures of the 1992 EU-CAP (Common Agricultural Policy) reform, the Free State of Saxony has developed and throughout the years regularly updated the programme "Umweltgerechte Landwirtschaft in Sachsen" (UL) (Environmentally sound agriculture in Saxony). The high claim of the programme and the extraordinary readiness of farmers to participate lead to the consideration to have the programme scientifically accompanied by the Saxon Regional Authority for Agriculture since 1994. The aim behind is to evaluate the effectiveness of the programme within the individual incentive measures in terms of market-relief, environmental-relief and development of income. The valuation of the incentive measures was based on abiotic and biotic criteria (agri-environmental indicators).

2. Material and methods

The programme "Environmentally sound agriculture in Saxony" supports the following measures in the field of arable farming:

1. Basic support (BS)

   These are compulsory measures which have to be performed on the farms' entire area of arable land. These are:
   - non-transformation of meadowland into arable land,
   - observance of a compatible well-balanced, at least three-field crop rotation system,

* Tel.: +49-341-9174-120; fax: +49-341-9174-111.
E-mail address: michael.menge@leipzig.lfl.smul.sachsen.de (M. Menge).

0167-8809/$ – see front matter © 2003 Published by Elsevier Science B.V.
doi:10.1016/S0167-8809(03)00103-8

- cultivation of compatible varieties according to variety tests on state level,
- observance of a defined livestock not exceeding 2 LU (livestock unit) per hectare farmland,
- introduction and maintenance of fertilisation based on the consultation programme BEFU (fertilisation input based on the specific needs of a plant),
- continuation of plant protecting measures with decision-guidance/decision-making aids (consultation models).

2. BS and additional support I (BS + AS I)
   20% reduction of N-fertilisation on the entire agricultural area. Additionally, the farmer dispenses with the application of growth regulators.
3. BS and additional support II (BS + AS II)
   Plot-related, the following measures can be chosen: intercropping, undersown crops, mulching.
4. Ecological farming
   In the whole farm, agriculture must be organised according to requirements of the EEC-guideline 2092/91.

A combination of the second and third incentive measure is possible. In the range of grassland/meadowland, measures aiming at a more extensive production (reduced use of production means, late crop use, etc.).

In 72 reference farms with altogether 2900 plots of arable land and a total area of 45,377 ha which are evenly spread in Saxony and which sufficiently represent the different cultivation regions, plot-related production data reports will be evaluated. Additionally, microeconomic data is being collected. Per each incentive measure, 12 farms will be compared with 12 non-participating farms—which makes overall 72 farms.

The following agro-environmental indicators are being recorded:

1. Abiotic factors
   - nitrogen, phosphorous and potassium balance per plot and farm (kg/ha),
   - NO_3-content in autumn on permanent testing plots (kg/ha),
   - quantity of herbicides (per plot and farm) (kg/ha),
   - amount of mechanical treatment measures per farm (per plot),
   - share of conserving soil cultivation measures per farm (in % of the area per year),
   - output date of liquid manure in spring/autumn (date).
2. Biotic indicators
   - diversity of crops per farm (number of cultivated crops per year),
   - variety resistance and diversity per farm (resistance index),
   - legume-share per farm (% and year),
   - wheat-share per farm (% and year),
   - duration of the bare fallow period (date),
   - plot size per farm (ha).
3. Other indicators
   - crop yield per farm (dt/ha),
   - contribution margin per farm (DM/ha),
   - subsidies per farm (DM/farm).

The data analysis will be made using methods of descriptive statistics. All indicators will be subject to the following proceedings:

1st step. First, the plot-related technical data about production measures will be recorded for each farm. Afterwards, the data of each indicator will be geometrically standardised to obtain an area-related mean. All further calculations will be based on this ‘farm value’.

2nd step. Annually, the farm values for each indicator examined in an incentive measure will be arithmetically standardised and their mean variation calculated.

3rd step. The annual mean values of the farms examined and their dispersion in each incentive measure will be illustrated in a diagram or a table in order to show the development trend in the respective incentive measure.

Below, a more detailed description of the recordation of the biotic indicators per farm will be given:

1. Crop diversity
   The number of arable crops cultivated per farm will be detected from the cultivation ratio.
2. Variety resistance and diversity per farm
   The resistance values of different pathogens (mildew, leaf septoria, spike fusariose, etc.)

determined in variety tests on state level will be abstracted to a mean resistance index of one grade. The values are 1 = not diseased, 4–5 = moderately diseased up to 8 = heavily diseased.

Afterwards, the varieties cultivated of each sort of arable crop will be recorded from the plot cards, assigned to the index value of each variety, and presented as overall mean per incentive measure and year. The minimum and maximum number of the cultivated crops is being presented and put in relation to the index value.

3. Legume-share per farm, cereal-share per farm

   The share of legume and cereals in the arable land area of a farm is being derived from the cultivation ratio.

4. Duration of the bare fallow period

   For every plot the length of time between harvest and re-cultivation plus 14 days (day of cultivation until the germination of the seeds) will be recorded.

5. Plot size

   A parcel of land is the smallest unit of a farm. It can consist of one or more plots. Normally, a plot forms a unit which is bordered by paths, hedges, rivers, streams or something comparable. The surfaces of the plot will be recorded and averaged per farm.

3. Results

The crop diversity in a farm can be described by the number of different arable crops which are cultivated. Fig. 1 shows that the participants in the programme cultivate on average 8–10 different crop varieties. This number varies throughout the years and among the farms between 6 and 14 varieties.

In general, the non-participants cultivate less varieties; about 7–8 with a smaller variation. This is due to the mean cereal share in the arable land area (Fig. 2).

On average, in ecological farming only 6–7 different crop varieties are being cultivated; the variation among the farms measures between 4 and 8 varieties. This can be explained by the facts that on one hand, these farms show a higher share in perennial crops and on

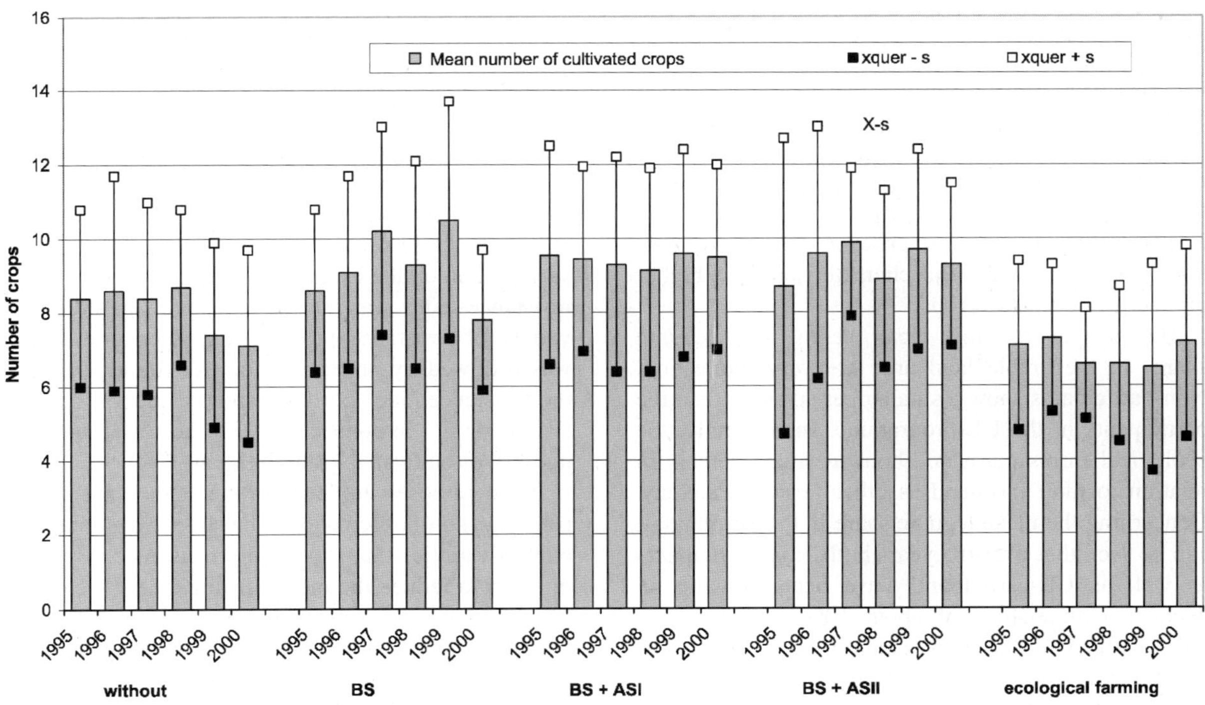

Fig. 1. Mean number of cultivated crops (1995–2000). Basic support (BS) and additional support (AS).

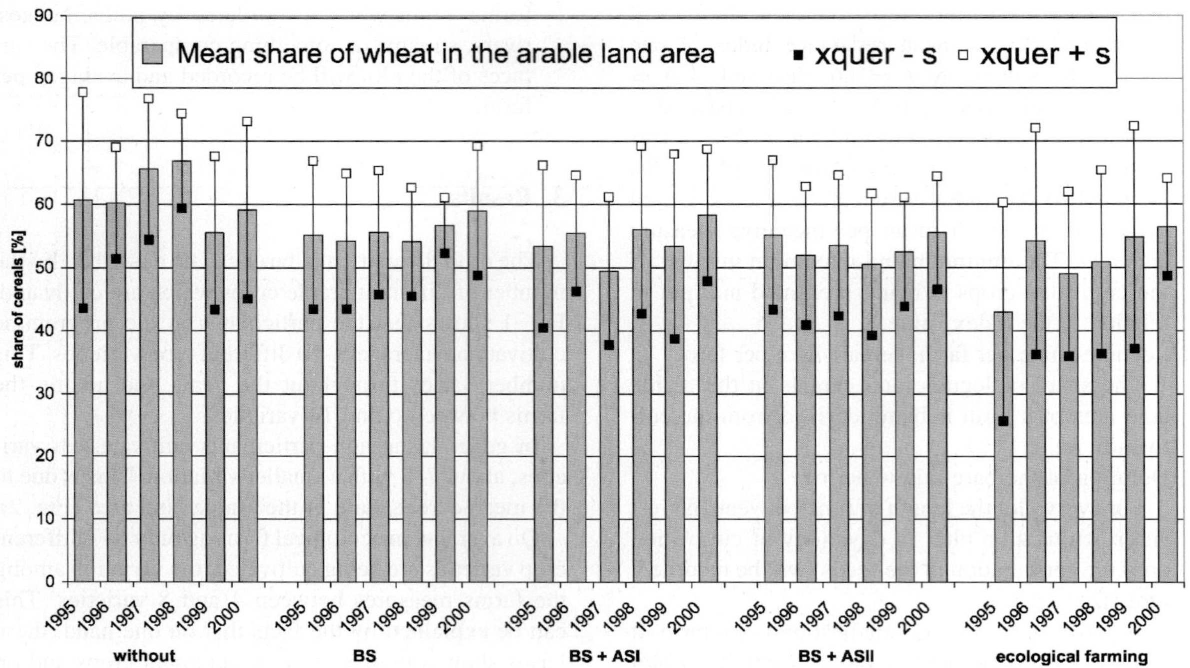

Fig. 2. Mean share of wheat in the arable land area (1995–2000). Basic support (BS) and additional support (AS).

the other hand, some crops like maize, rape and sugar beets are not cultivated in ecological farming. The high number of cultivated crops in Saxony results from the regulation that participants in the UL-programme have to cultivate at least three different crop varieties. Due to a higher share in pasture growing, the mixed farming systems are in general characterised by a higher number of cultivated crops.

The results for the mean cereal and legume shares in the arable land area are very interesting. Non-participants show a smaller cereal share than the participants in the UL-programme. Interestingly, the ecological farms cultivate about as much cereals in relation to their farmland as other farms in Saxony. Concerning the mean legume share in the farmland, it can be seen that over the years both, non-participants as well as UL-participants have been growing an average legume-share of about 10% (Fig. 3). This share varies with the market conditions with a downward trend. The ecological farms cultivate a share of 33–45% of legume, a share which also underlies considerable fluctuations. The reason for this is that market-oriented arable farms only grow grain legume, and mixed farming systems cultivate a high share of lucerne and clover–grass-mixtures for pasture production and in ecological farming systems additionally for nitrogen extraction.

In order to illustrate the genetic diversity within one variety (here at the example of winter wheat), a grade resistance-index was created which was based on the variety of resistance properties of each grade. This was followed by an examination of the elevation of the resistance index per farm and per supporting level (Foerderstufe) in connection with the variation of the number of cultivated grades (Fig. 4).

It can be observed that over the years both, participants and non-participants of the UL-programme, have chosen varieties that show more resistance properties in order to reduce the input of fungicides. It can also be seen that non-participants cultivate a significantly broader spectrum of varieties as the participants of the programme. Participants were not allowed to grow certain grades which have a high demand in plant protective agents. Ecological farming shows about the same

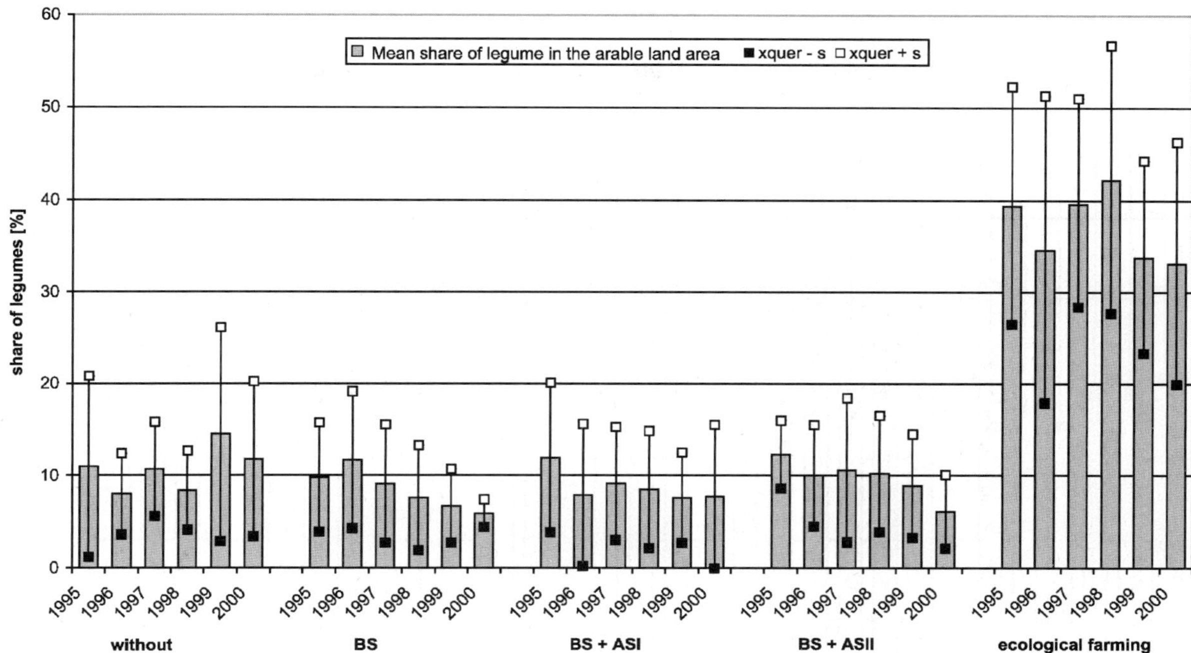

Fig. 3. Mean share of legume in the arable land area (1995–2000). Basic support (BS) and additional support (AS).

resistance index as the rest of the participants but with a significantly lower grade variation. This originates from the fact that in ecological farming, the advisory institutions promote the cultivation of quality wheat in order to obtain higher prices. All in all, in can be seen that not only one but a variety of grades are being cultivated. This contributes vitally to the genetic variety within the variety of winter wheat.

With regard to biodiversity, the duration of the bare fallow period is indirectly of great importance. The duration of the bare fallow period has a bearing on nitrate elutriation, soil erosion, the number of cultivated crops as well as the all-year soil cover with green plants (Table 1).

Compared with the non-participants, the participants in the BS and those in the BS + AS I were able to reduce the duration of the bare fallow period by 12–19%. The BS + AS II could reduce it by 25–40%. This is due to a higher share of intercrops and mulching systems which were used. In ecological

Table 1
Mean duration of the bare fallow period (days) within the different supporting levels[a]

Supporting level	Days[b]					
	1994/1995	1995/1996	1996/1997	1997/1998	1998/1999	1994–1999
Without	52	65	47	53	26	49
BS	45	38	48	40	43	43
BS + AS I	30	52	48	28	40	40
BS + AS II	37	44	36	40	28	37
EF	67	45	44	43	47	49

[a] BS: basic support; AS: additional support; EF: ecological farming.
[b] Period between ploughing and re-cultivation, at the longest until spring of the following year.

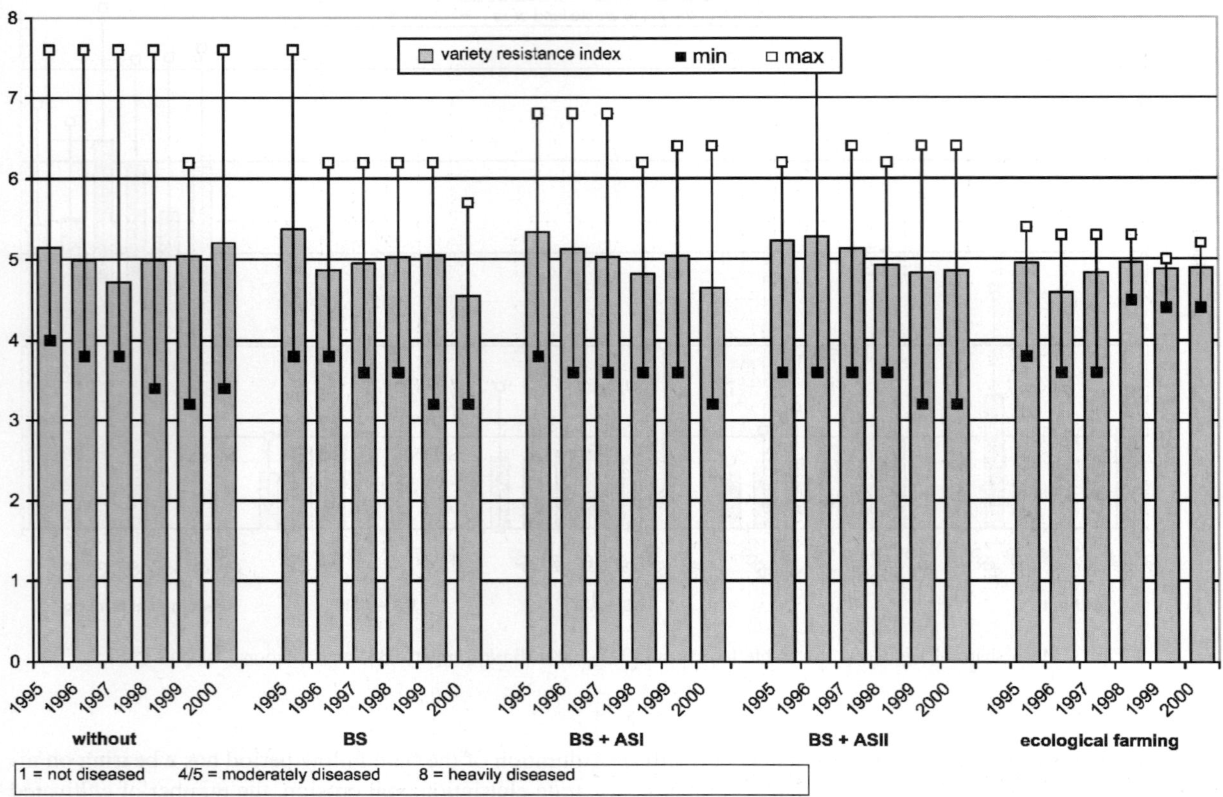

Fig. 4. Variety resistance index of winter wheat depending on the supporting level (1995–2000).

farming, the duration of the bare fallow period averages those of the non-participants.

Between the years, considerable fluctuations can be noticed. However, between the years of 1994 and 1999 the duration of the bare fallow period shows a slight decrease. This shows that supporting level "soil conserving measures" (BS + AS II), mulching, intercropping, undersown crops have a positive effect on both, the farm and the environment.

In Saxony, the average plot size is about 17 ha (Fig. 5). With 48%, plot sizes between 5 and 20 ha are the most common. About 85% of all plots have sizes between 1 and 35 ha. Throughout the years, the plot sizes stay approximately the same but vary between the different supporting levels. Landscape-elements between the plots and within the plots have not yet been recorded. However, they might take an estimate share of 6–10% in the arable land area.

In the range of grassland, the number of plants varieties were permanently recorded on plots with and without KULAP-support between 1995 and 1998 (Fig. 6). It can be noticed, that those not participating in the KULAP show an average of 17 varieties only.

Participants in the KULAP—thus those who practice an extensive farming system with support, are able to show between 19 and 21 varieties on average, the share of herbs being larger than the one of weeds, which itself is on the whole higher than at the non-participants.

Interestingly, it could also be seen that from the 160 plots in Saxony which are submitted to permanent observation on those who are supported by KULAP 15 species which are on the premonition list, three endangered species and one highly endangered species could be verified (Table 2).

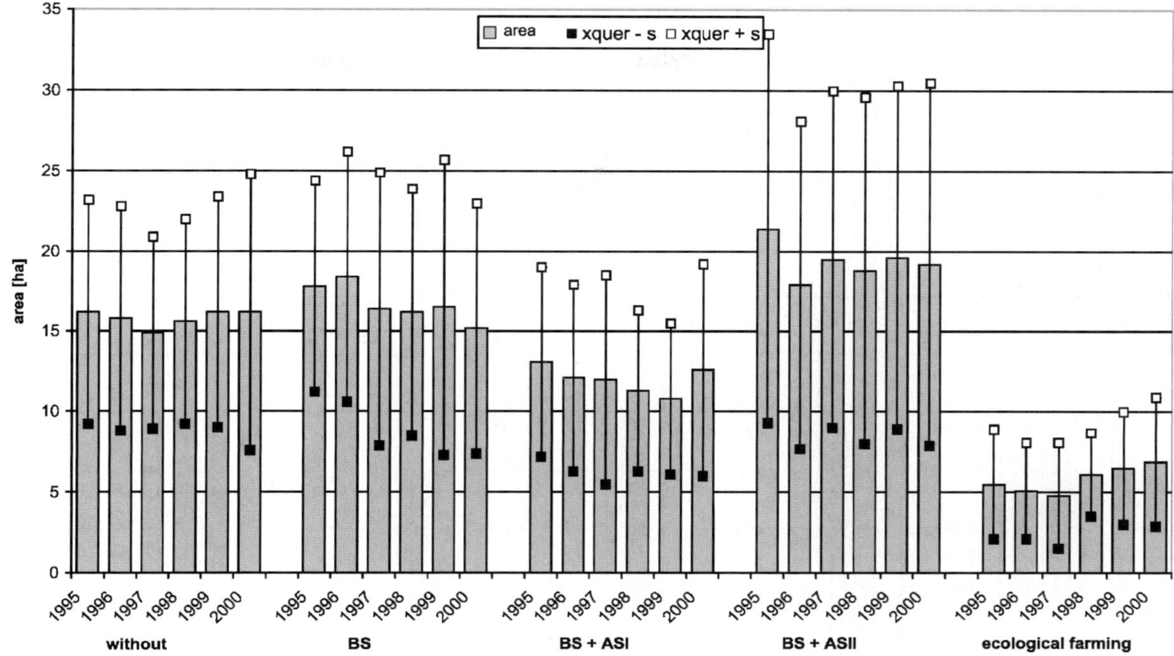

Fig. 5. Average plot size (1995–2000).

Table 2
Species of the Saxon Red list and of its affix (premonition list, list of plants "endangerment suspected")

Number of permanent testing plots with	Total	Thereof on plots supported by KULAP I
Species, threatened by extinction (1)	1	0
Highly threatened species (2)	12	1
Threatened species (3)	16	3
Premonition list (V)	41	15
Endangerment suspected (G)	5	3

4. Discussion

Biodiversity is being characterised by the variety of floristic and faunistic species, the genetic diversity within one species, as well as the variety of biotopes and biocoenosises. A complete illustration of this indicator is extremely complicated and elaborate. An indicator which describes the impact of biodiversity does not yet exist. The recordation of wild living plant and animal species only is very elaborate and costly. A declaration about the minimum degree of diversity in cultural landscape which should be achieved does not yet exist.

In general, studies on this subject only consider one species or genus, mostly disregarding the interaction between species or individuals.

In Saxony, the attempt was made to venture a first step towards the recordation of biodiversity in the fields of agriculture. The recordation of the variety of species, the genetic diversity and the variety of biotopes on agricultural land in the cultural landscape is considerably easier and more cost-effective as on the whole of the biotic environment. In the KUL-system (Eckert et al., 1998) for instance, the share of ecologically–culturally important areas are being evaluated. The acquisition of this data is in fact quite elaborate, but it must only be carried through once and must in the following only be updated or supplemented.

In Saxony, the criteria (indicators) crop diversity and variety resistance and diversity, share of legume and wheat, duration of the bare fallow period, and plot sizes can be quite easily recorded with the aid of plot

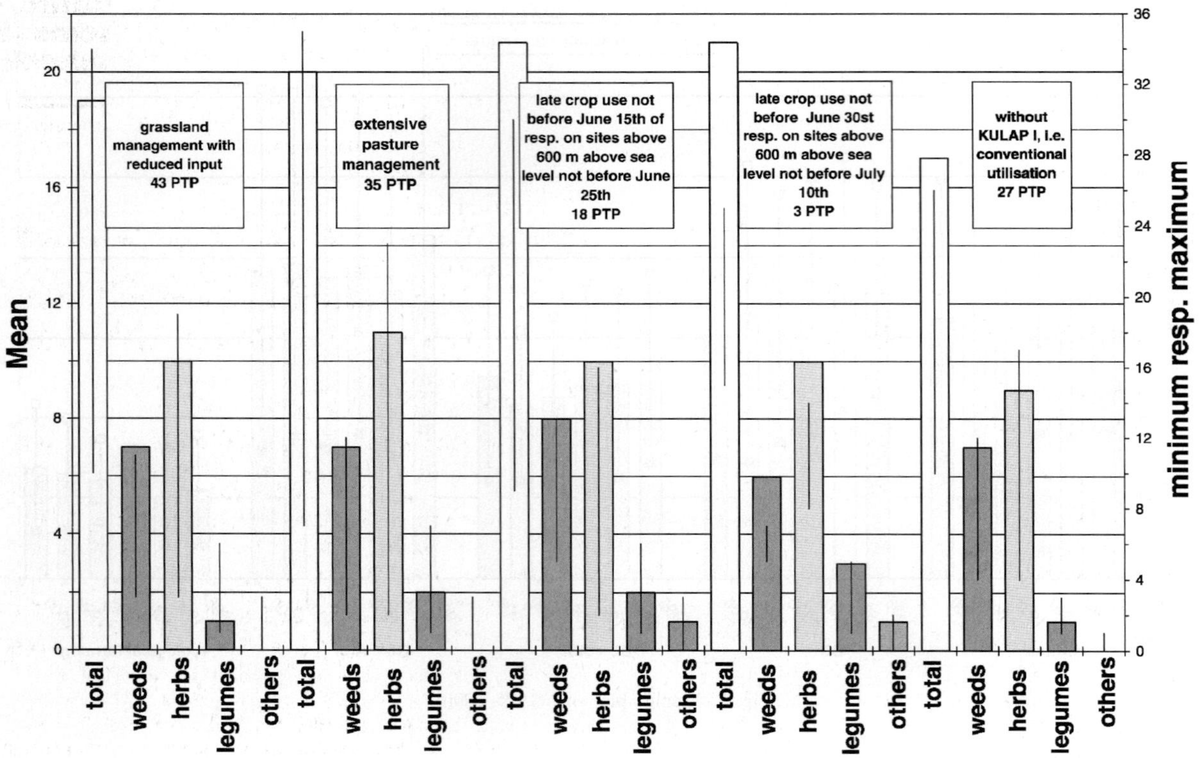

Fig. 6. Number of grades (mean as pillar, minimum and maximum as vertical lines within the pillars) on permanent testing plots (PTPs) with KULAP I-support as well as without KULAP I, i.e. conventional utilisation, in Saxon farms (recordation 1995–1998).

record lists. There is no need for additional efforts. This data will be registered in a data storage, tested for plausibility by experts, and then be available for evaluation.

From literature, a series of valuation approaches are known which act on the assumption that certain marginal and reference/target values may not be exceeded by the farms (Geier, 1999; Dimkic and Schumacher, 1999; Eckert et al., 1998).

A first complex approach is given by the survey "Abschaetzung der Auswirkungen transgener Sorten auf Umweltqualitaetsziele" ("Impact-assessment of transgenetic grades on environmental quality objectives", Werner et al., 1999). In this survey, biotope values for indicators were determined and cultivation measures biotically weighted.

In the Saxon programme "Umweltgerechte Landwirtschaft in Sachsen" ("Environmentally sound agriculture in Saxony", Menge et al., 2001), agro-environmental indicators and their impacts are not being valuated on the basis of marginal, reference or target values. In general, these values are subjectively predetermined, often insufficiently justified and not suitable to achieve the objective. In transient areas and sites, as well as in less favoured regions they lead to substantial problems in case of slightest climatic or local variances. On the contrary, in the UL-programme, mean farm values and their variation are being specified. This makes it possible to determine a present value which furthermore describes how the management of the farms differs during the years (variation of the single values).

If these mean farm values of one incentive measure are illustrated within a fixed timeframe, it is possible to show a trend of the effects the incentive measure has taken. The crucial point is that an environmental-relief has taken place (positive effects for the environment could be achieved). The extent of environmental-relief is subject to regional and local peculiarities and cannot be described by a uniform marginal value. If an environmental-relief cannot be detected, the incentive measure does not take effect and should be modulated. This message would then have to be transferred to agricultural policy.

5. Conclusion

For the implementation of the accompanying measures of the 1992 EU-CAP (Common Agricultural Policy) reform, the Free State of Saxony has developed and throughout the years regularly updated the programme "Umweltgerechte Landwirtschaft in Sachsen" (UL) (Environmentally sound agriculture in Saxony).

Agro-environmental indicators and their impacts are not being valuated on the basis of marginal, reference or target values. In general, these values are subjectively predetermined, often insufficiently justified and not suitable to achieve the objective. In transient areas and sites, as well as in less favoured regions they lead to substantial problems in case of slightest climatic or local variances. On the contrary, in the UL-programme, mean farm values and their variation are being specified. This makes it possible to determine a present value which furthermore describes how the management of the farms differs during the years (variation of the single values).

If these mean farm values of one incentive measure are illustrated within a fixed timeframe, it is possible to show a trend of the effects the incentive measure has taken. The crucial point is that an environmental-relief has taken place (positive effects for the environment could be achieved). The extent of environmental-relief is subject to regional and local peculiarities and cannot be described by a uniform marginal value. If an environmental-relief cannot be detected, the incentive measure does not take effect and should be modulated.

References

Dimkic, C., Schumacher, W.,1999. Umweltbewertungsverfahren fuer die Landwirtschaft. Europaeische Union GD XI.

Eckert, H., Breitschuh, G., Sauerbeck, D., 1998. Kriterien umweltvertraeglicher Landbewirtschaftung—ein Verfahren zur oekologischen Bewertung von Landwirtschaftschaftsbetrieben. Agribiol. Res. 52 (1), 57–76.

Geier, U., 1999. Entwicklung von Parametern und Kriterien als Grundlage zur Bewertung oekologischer Leistungen und Lasten der Landwirtschaft—Indikationssysteme. Forschungsbericht Umweltbundesamt 42/99.

Menge, M., et al., 2001. Ergebnisse und Erfahrungen zum Programm "Umweltgerechte Landwirtschaft in Sachsen", Schriftenreihe der Saechsischen Landesanstalt fuer Landwirtschaft, 1 (5).

Werner, A., Berger, G., Stachow, U., Glemnitz, M., 1999. Abschaetzungen der Auswirkungen transgener Sorten auf Umweltqualitaetsziele. Forschungsbericht ZALF e. V. Muencheberg.

ELSEVIER

Available online at www.sciencedirect.com

Agriculture, Ecosystems and Environment 98 (2003) 453–461

Agriculture Ecosystems & Environment

www.elsevier.com/locate/agee

Field related organisms as possible indicators for evaluation of land use intensity

W. Heyer*, K.-J. Hülsbergen, Ch. Wittmann, S. Papaja, O. Christen

Faculty of Agriculture, Institute of Agronomy and Crop Science, Martin-Luther-University Halle-Wittenberg, Ludwig-Wucherer-Str. 2, D-06108 Halle/Saale, Germany

Abstract

Most farm activities might cause some hazard for the environment. In this context, the effects on the biotic environment are most complex and require a system approach. In this study, we present results on the relation between different farm activities and some important biotic indicators. Farm activities were monitored to determine the socio-economic and abiotic environmental status of an organic farm system, using a farm modelling system based on the REPRO methodology. The project also included field studies on the same site in order to monitor some important biotic indicators. By doing that, the interrelation between abiotic environmental indicators and selected field related organisms (carabids, weeds, earthworm, microbial soil activity) were demonstrated. A complex assessment of the organic farm compared with a conventional farm illustrates the relation between the different indicator groups. In accordance with the OECD-indicator framework these findings are necessary to establish and implement an indirect way to assess the biotic environmental situation on farm level.

Keywords: Evaluation; Farming systems; Biotic indicators; Carabids; Weeds; Earthworm; Microbial soil activity

1. Introduction

The interest to assess the relationship between agricultural practices and the biotic and abiotic environment has increased in recent years. Special impetus came from the climate protection campaign and the reorientation of EU agricultural policy towards more sustainability in agriculture. The implementation of environment-related targets bases on agri-environmental schemes was another major reason for the growing interest in this area of research. The main objective of those programs was to reduce environmental pollution caused by agricultural activities, as well as the preservation of biodiversity on field, farm or landscape level by means of reduction of inputs, less intensive production systems or organic farming (EC, 1992).

A number of indicator sets have been proposed to measure the sustainability of farm operations with focus on the abiotic environment (soil and water quality, greenhouse gas emission etc.) including the socio-economic situation (farm income, gross margins, etc.) on the farm level (Koepf et al., 1989; Faeth, 1993; Jaeger, 1995; Girardin et al., 1996; Müller and Dittrich, 1996; Fox, 1991; Liebig and Doran, 1999; Abraham, 2000). A very comprehensive computer-based approach has been developed by Hülsbergen et al. (2000). Additionally a vast scientific literature is dealing with the biotic environment, however, most described indicators have been developed and used for special proposes in landscape ecology or specific scientific questions (Gruschwitz, 1981; Arnd et al., 1987; Nähring, 1990; Steinborn

* Corresponding author. Tel.: +49-345-55-22-632.
E-mail address: heyer@landw.uni-halle.de (W. Heyer).

0167-8809/$ – see front matter © 2003 Elsevier B.V. All rights reserved.
doi:10.1016/S0167-8809(03)00104-X

and Heydemann, 1990; Luka, 1996; Oñate et al., 2000).

From a practical point of view, those indicators have several limitations:

- Data on farm level is scarce.
- Direct measurements (population densities) are extremely expensive and time consuming.
- Measurements might provide some information about the actual biotic status, but do not allow conclusions about future developments.
- Most measurements do not consider the farm level and therefore separate the results from the socio-economic consequences.
- Cause-effect relations between agricultural production systems and the environment are not determined. As a consequence, farmers or administrators do not have enough information to alter the management system corresponding to environmental needs.

In this paper, we propose a system with pressure indicators to assess the biotic environment on farm level, based on the OECD-indicator framework (pressure-(driving force) state-response model). Most data is readily available on the farm level and allows assessment without direct measurements of the single species or species groups. In order to demonstrate the feasibility of this approach, we present data from some long-term field surveys and accompanied field experiments with special emphasis on the interaction between the abiotic and biotic environmental sector. Carabid beetle (Carabidae) earthworms (Lumbricidae), weed population and microbial soil activity were chosen as organisms and indicators, which show a strong relation to various field cultivation activities. On the other hand, they are regarded as important bioindicators itself providing information about the stability and auto-regulatory processes of agri-ecosystems (Finck, 1952; Chambers et al., 1983; Bryan and Wratten, 1984; Schinner, 1986; Chiverton, 1987; Hance, 1987; Edwards and Bohlen, 1995).

2. Methods

Data were collected during 5-year field studies within the area of a 500 ha farm, situated near the city of Halle (Saale), Saxony-Anhalt (Germany). This area is typical for the loess sites of the Central German Dry Region with a continental climate. During the 5-year period the farm was converted from a conventional intensive to an organic farming system.

Each farm activity was monitored to determine the socio-economic and abiotic environmental status of the farm, using the REPRO-based methodology (p.e. N-balance, humus supply and reproduction) (Hülsbergen et al., 2001). Data on the relevant species were measured at different intervals including the whole field area. Further details on sampling dates and methods are given in Table 1.

Besides the above-mentioned field surveys, experimental plots were established near to the farm area in order to directly measure the impact of plant cover and type of cultivated field plants on the carabid community. These plots allowed special data analysis (Dammer and Heyer, 1997) to estimate the

Table 1
General view on used methods

Organism	Method	Interval	Fields/repetition	Total sample number
Field surveys				
Earthworm	Analyses of soil samples, 0–50 cm depth	Yearly or 6-month interval (six fields), 5 years	23 fields, $10 \times 1/10\,m^2$	>1450
Weeds	Monitoring of weed species and degree of weed coverage	2 times/year, 5 years	$4 \times 100\,m^2$ plots on 21 fields, additional observations	>840
Microbial activity	Soil samples 0–30 cm depth in the spring, determination of enzyme activities	Yearly, 5 years	24 fields, four replications	>480
Experimental plots				
Carabids	Pitfall trapping	9–11 catches/year over 3 years	60 traps each catching period (7 days)	>1600

environmental and anthropogenic impact on beetle population.

The combination of survey data and data from experimental plots gave us a sufficient data basis to describe and evaluate relationships between agricultural activities and the selected organisms.

Detailed information concerning the background of methodology and data analysis are given by Heyer (2000), Hülsbergen and Diepenbrock (2000), Papaja and Hülsbergen (2000), Tischer (2000) and Wittmann and Hintzsche (2000).

3. Results

3.1. Dependencies between farm operations and selected biotic protectables

3.1.1. Plant coverage and ground beetle activity

The first example demonstrates how the time between sowing and crop harvest or forage cuts affects the occurrence of carabids (Fig. 1). When the ground is kept clear (only short-time growth of a weed cover) a total number of 33 carabid species were registered using pitfall traps.

When the plant cover (winter barley, winter rape, field forage) was allowed to grow without interference, even 56–61 species were recorded within 1 year. A similar result was obtained for the relationship between species and individuals. This is noteworthy since the regulatory benefit of this epigeous arthropod group in the agro-ecosystem ensues mainly via the abundance (biomass) of the beetles. Also from this point of view, extended ground coverage is advantageous for the beetle population. The time period is determined by the indicator "time of ground coverage" as showing by REPRO.

3.1.2. Crop number, crop diversity and ground beetle population

Number and growing period of cultivated crops can be easily surveyed on the farm and entered into REPRO producing the indicator "crop diversity". It provides information about the structure of a farm and at the same time points out to developments in the biotic sector of the environment. Fig. 2 shows this relationship by demonstrating the relation between the number of cultivated crops and the number of

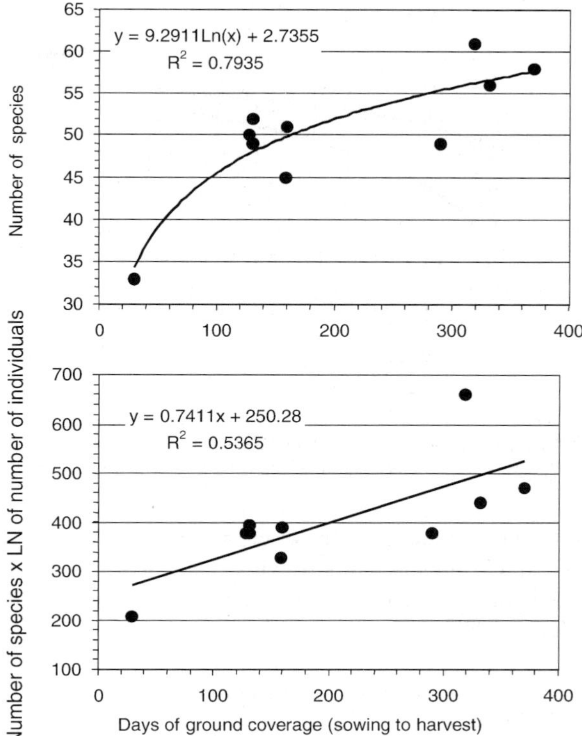

Fig. 1. Occurrence of carabids (number of species or number of species and individuals) in dependence of the period (days) of ground coverage on experimental areas (modified and supplemented according to Baumann, 1994).

carabid species identified in the fields. It is obvious that the species diversity depends largely on the crop species and the number of cultivated crops. In the cited example, 13 (worst case) or 44 (best case) carabid species were recorded in one single crop within 1 year. When several crops were grown, the difference between favourable and unfavourable crop species (or combinations) decreased.

Regarding the evaluation of the findings, it can be underlined that there is no single "bad" or "good" crop, because the growth period involves major changes and effects the different species (Heyer, 2000), thus the niche conditions for the various beetle species within a single crop changes continuously. When seven crop species were grown, the number of captured carabids could no longer be statistically distinguished (method after Poole, 1974) from the number of carabid species in 10 different crop stands. Further surveys revealed

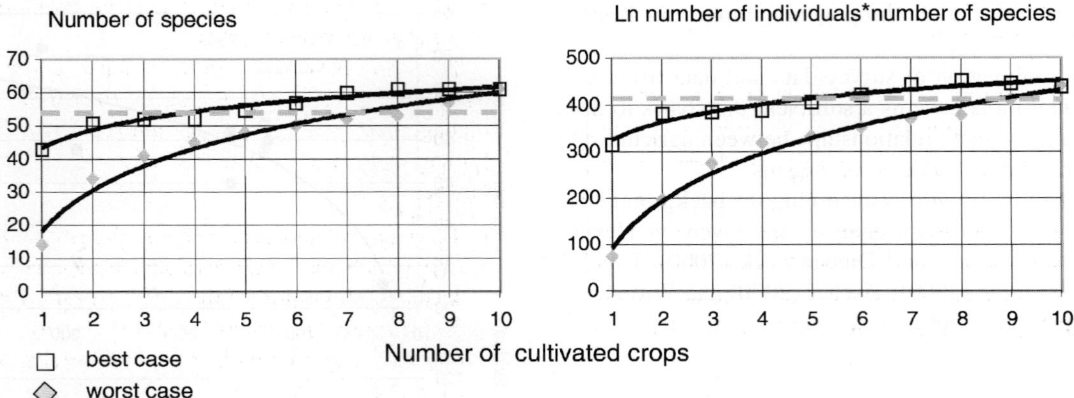

Fig. 2. Dependence of the carabid population (number of species and number of individuals × number of species) on the number of cultivated crops. Points separated by spotted line are significantly different.

that the influence of these crops on the variability of carabid population is roughly equivalent to that of the various annual weather conditions (Dammer and Heyer, 1997).

3.1.3. Resource input, land use and earthworm population

The next example focuses on the reduction of inputs during the shift to organic management and the consequences for the earthworm population after previously differentiated land use (Fig. 3).

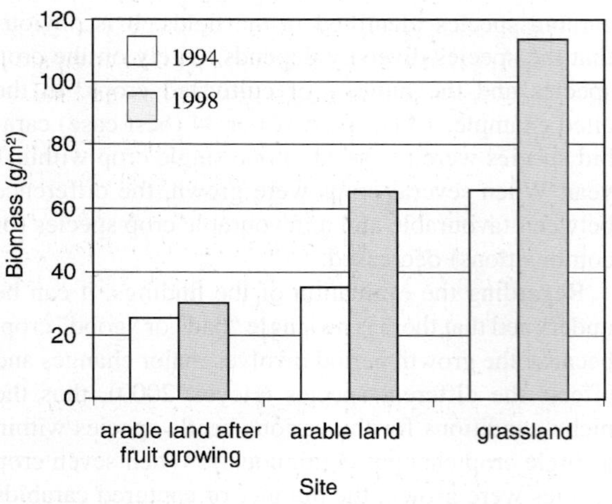

Fig. 3. Abundance of the earthworm population on sites under differing agricultural use and after the shift to organic management (modified according to Papaja and Hülsbergen, 2000).

Earthworms find much better living conditions on grassland than on arable sites (Fig. 3). Old orchards are poor in earthworm populations because of former high pesticide application rates. Despite the marked differences in the basic population density, positive trends of the earthworm numbers due to organic management have been observed on all fields. Although results from one case-study are given here, we can generally conclude that a diverse use of farmland (e.g. ratio between arable land and natural grassland) or a switch to organic farming decreases the pressure of agricultural activities on this indicator group. The respective agronomic and cultural parameters can be successfully analysed by REPRO.

3.1.4. Mineral fertiliser input, N-balance and weed occurrence

Nitrogen is an essential operating resource in agriculture. Yields on the one hand and environmental impacts of a farm on the other depend decisively on the management of the N-cycle. The REPRO model permits to calculate the N-supply by organic fertiliser, crop residues and considers immission from the atmosphere.

Over time, the modified N-supply affected the weed flora, which is given in Fig. 4. The figure shows changes in species number and species composition related to the N-supply to the soil. Declining N-supply slightly increased the species diversity. The number of weed species increased a little, with the percentage of N-seeking species in the total plant cover declining.

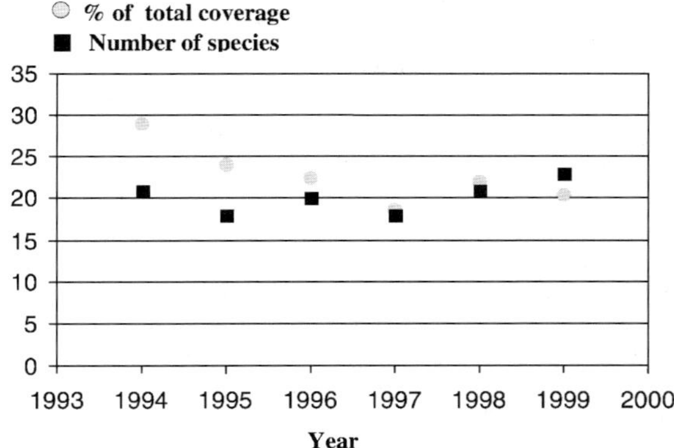

Fig. 4. Development of species diversity and share of nitrophyts (figure $N > 7$) vis-à-vis the total weed coverage in an organically managed farm (modified according to Wittmann and Hintzsche, 2000).

This underlines that the weed flora and their composition response rather sensitively to the indicator "total nitrogen input". There is a marked tendency between nitrophyt share and N-cycle in a system, which will be useful for the description of biotic effects (Fig. 4).

3.1.5. *Humus supply, humus reproduction and soil organisms*

Further examples for relationships between parameters of the abiotic and biotic environment focus on the soil as natural habitat. Table 2 shows the interaction between the microbial biomass (C_{mik}) and its activity (determined on the basis of enzyme activity) and the earthworm colonisation in relation to the humus content of the soil (C_t). Between the mentioned parameters positive correlations have been measured, which are important for the assessment of changes in the biotic sphere of the soil. This implies that agricultural activities increasing an accumulation of organic matter in the soil present simultaneously a stimulation for soil organisms.

3.1.6. *Indicator use in context of the evaluation of farm management systems*

The examples given in this paper underline the usefulness of assessing environmental effects of farm management practices in an indirect way. The following evaluation, however, faces also a number of obstacles:

- A system-related assessment requires data aggregation, which itself requires a standardisation and hierarchical integration of the different indicators. This is especially important for the evaluation of biotic data in connection with abiotic and socio-economic indicators.
- When working towards agricultural policy aims, it is less important to harmonise the impact of a special production system with one single target species but to optimise the entire system towards the sustainability of agricultural production. The selection of biotic indicators is to be made with orientation to this aim.
- Antagonistic effects between the different species are very important. Positive effects for one species frequently cause restrictions for other species. Therefore, production systems are to be evaluated by use of different indicator species. The only exceptions are systems, which were solely designed for the protection of a specific species or habitat.
- There is a need of debate about the evaluation of biotic facts. This concerns target values or the definition of thresholds for the population size of indicator species and also the choice of suitable reference situations. It is not useful to compare the biodiversity of natural areas with farmed land.

The need of research and discussion for solving the mentioned problem areas is enormous and can be realised only step by step in pragmatic approaches. First attempts to include those different aspects have been made by the REPRO designers. Fig. 5 shows

Table 2
Correlation examples (correlation coefficient $r, n = 64, \alpha = 0.05$) between the abiotic and biotic environment sector (modified according to Hülsbergen and Diepenbrock, 2001)

	Microbial biomass and activity			Earthworm abundance and biomass		
	C_{mik}	β-Glucosidase	Catalase	Abundance	Biomass total	Biomass adults
C_t	0.66	0.54	0.83	0.37	0.37	0.33

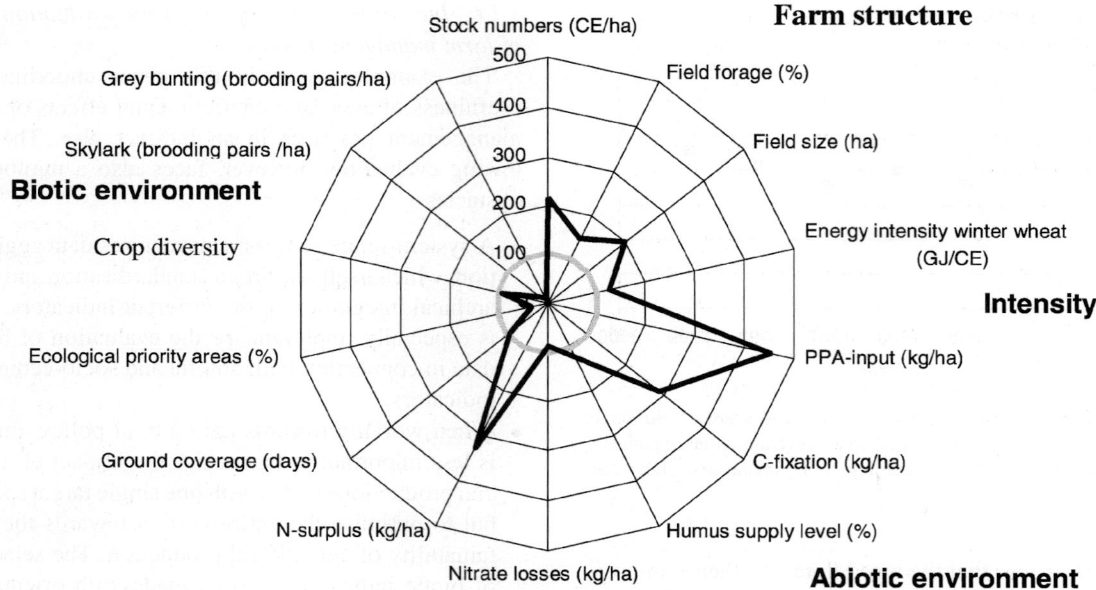

Fig. 5. Relative ecological benefits of a conventional farm (black line) compared to an organic farm (grey line), CE: cattle equivalent.

the integration of biotic environmental indicators into the general description of the ecological effects of agriculture using an example of a conventionally managed farm in the German Central Dry Region. As a reference, an organic farm was chosen whose biotic benefits have been documented in long-term concomitant studies. Furthermore, in this example monitoring data about the Skylark (*Alauda arvensis*) and Grey Bunting (*Milaria calandra*) is used.

A juxtaposition of the two farming systems reveals system-related deviations in the application of plant protection agents, which might involve a potential risk for biotic goods. Several indicators range around the value of reference parameters (crop diversity, nitrate losses, ground coverage); others highlight potentially environmental benefits (priority areas for ecological land use, number of brooding pairs). In comparison, more favourable values are reached for the humus supply and the CO_2-fixing capability.

4. Discussion

The Common Agricultural Policy of the European Union (CAP) has set ecological targets for the agriculture sector. These targets can be reached via agri-environmental schemes based on the EC Regulations 2078/92 and 1257/99 (the latter is in force as from the year 2000). At the same time, the research sector is urged to elaborate the methodical tools for checking the expected ecological benefits. Up to now, a large proportion of the discussion focuses on the choice of suitable indicators (Wood, 1998; Jaeger, 1995) and agreement has been achieved about the indicator lists for describing the economic effects of policy decisions and agri-environmental schemes. Similar findings have been recorded for the abiotic environmental benefits (Rennings and Wiggering, 1997). No consensus, however, exists about the relevant indicators for the biotic environment. One reason are fuzzy ideas about the necessary indicators (e.g. biodiversity) (Wood, 1998), and an indiscriminate transfer of indicators to agriculture, which had initially been established for nature conservation areas, (e.g. Red Data Lists, various environmental schemes (Milon and Shogren, 1995)), but also the high cost and time input required for direct indication (Oñate et al., 2000).

Those considerations make it necessary to evaluate the cause-effect relations between agriculture and environment and to use them in form of appropriate indicators for describing biotic effects. This principle, which was stimulated by the pressure (driving force) state response model of the OECD (Münchhausen and Nieberg, 1997; OECD, 1998), has been strictly implemented in the REPRO model. When assessing biological facts in relation to agricultural activities, the latter and their impact on the environment are principally handled in a complex way. This approach, however, implies also restrictions, mainly in cases where exact population data of the selected indicator species are required. If, on the other hand, such data is available, the REPRO model is able to process information of direct monitoring and consider it for the evaluation.

An unsolved problem is the assessment of ecological risks. For certain sectors guide and threshold values are available (e.g. 50 mg/l water in case of nitrate). These values are consensus for policy makers and actually imply legal power.

Applicable knowledge on biotic goods is also available for plant protection agents (Gutsche and Roßberg, 1997). In most cases, however, such values do not exist, particularly since scientifically based studies are largely missing, or because the biotic sector is assessed by region-specific characteristics. Therefore, comparisons between extended regions or with other sites are normally not helpful. For that reason, more realistic are long-term monitoring data with consideration of the initial situation for reference or the comparison among region-specific management systems, which, if required, can be tailored for concrete biotic situations. Examples of species-related demands to agriculture have been described by Wieland (1997) for the Great Bustard (*Otis tarda*), by Saacke and Fuchs (2001) for the skylark (*A. arvensis*) and by Stubbe et al. (1998) and Weinhold (1998) for the common hamster (*Cricetus cricetus*). This approach of screening farming systems for species-related key facts (disturbance frequency, intermediate cuts, plough depth) is especially interesting when effect analyses for defined target species are to be made. They may be accompanied by farm level evaluations (e.g. share of ecological compensation areas), random field checkups and ratings of the intensity of a given management system.

5. Conclusions

This above-mentioned user-oriented point approach led to the following conclusions for intensified investigations of biotic indicators for environmental situations:

- Proposals are made for various indicator species. In general a single indicator approach is not sufficient to describe the environmental pressure of farming systems. On the other hand, a lag of knowledge in the use of "indicator sets" or in the development of entire "evaluation tools" is observed, considering the system character of agricultural production. The integration of biotic findings to REPRO is to make a contribution to the above-mentioned target.
- The different objectives require different biotic indicators. It is necessary to select indicators, closely related to agricultural activities. Such indicators must provide information on the sustainability of agricultural activities by use of an aggregated indicator "Pressure of the farming system on the biotic environment". For nature conservation, direct indication via target or reference species is recommended.
- The hierarchical order of various farm operations for the populations of potential indicator species in real farm situations is often unclear. Additionally, the population changes from year to year due to the weather conditions. This normally restricts conclusions to specific anthropogenic impact conditions. It is necessary to develop methodical concepts for the rendering of such problems on the basis of the combination of field studies on farmland and exact field plot surveys.

References

Abraham, J., 2000. Auswirkungen von Standortvariabilität auf den Stickstoffhaushalt ackerbaulich genutzter Böden unter Berücksichtigung der Betriebsstruktur, der standortspezifischen Bewirtschaftung und der Witterungsbedingungen. Diss. Inst. Acker- und Pflanzenbau der Martin-Luther-Universität Halle-Wittenberg, 122 pp.

Arnd, U., Nobel, W., Schweizer, B., 1987. Bioindikatoren—Möglichkeiten, Grenzen und neue Erkenntnisse. Verlag Eugen Ulmer, Stuttgart.

Baumann, Ph., 1994. Analyse des Einflusses von Umweltfaktoren auf die Biotopbindung von Carabiden auf Agrarflächen. Martin-Luther-Universität Halle-Wittenberg, Dipl. Arbeit, 90 pp.

Bryan, K.M., Wratten, S.D., 1984. The response of polyphagous predators to prey spatial heterogenity: aggregation by carabid and staphylinid beetles to their cereal aphid prey. Ecol. Entomol. 9, 251–259.

Chambers, R.J., Sunderland, K.D., Wyatt, I.J., Vickermann, G.P., 1983. The effects of predator exclusion and caging on cereal aphids in winter wheat. J. Appl. Ecol. 20, 209–224.

Chiverton, P.A., 1987. Predation of Rhopalosiphum padi (Homoptera: Aphididae) by poyphagous predatory arthropods during the aphids pre-peak period in spring barley. Ann. Appl. Biol. 111, 257–269.

Dammer, K.H., Heyer, W., 1997. Quantifying the influence of the cultivated plant species on the occurrence of carabid beetles within certain species using contingency table analysis. Environ. Ecol. Stat. 4, 321–336.

EC, 1992. Verordnung (EG) Nr. 2078/92. Amtsblatt der Europäischen Gemeinschaften.

Edwards, C.A., Bohlen, P.J., 1995. Biology and Ecology of Earthworms. Chapman & Hall, London.

Faeth, P., 1993. Evaluating agricultural policy and the sustainability of production systems: an economic framework. J. Soil Water Conserv. (March–April), 94–99.

Finck, A., 1952. Ökologische und Bodenkundliche Studien über die Leistungen der Regenwürmer für die Bodenfruchtbarkeit. Z. Pflanzenernährung U. Bodenk. 58, 120–145.

Fox, G., 1991. Agriculture and the environment: economic dimensions of sustainable agriculture. Can. J. Agric. Econ. Workshopproceedings, "Sustainable Agriculture: Economic Perspectives an Challenges", Winnipeg Manitoba, 647–653.

Girardin, Ph., Bockstaller, C., van der Wert, H., 1996. Evaluation of the sustainability of a farm by means of indicators. In: Behl, R.K., Gupta, A.P., Khurana, A.L., Singh, A. (Eds.), Resource Management in Fragile Environments. CCS HAU, Hisar and MMB, New Delhi, 1996, pp. 280–296.

Gruschwitz, M., 1981. Die Bedeutung der Populationsstruktur von Carabidenfaunen für Bioindikation und Standortdiagnose (Coleoptera, Carabidae). Mitt. Dtsch. Ges. Allg. Angew. Entomol. 3, 126–129.

Gutsche, V., Roßberg, D., 1997. SYNOPS 1: a model to assess and to compare the environmental risk potential of active ingredients in plant protection products. Agric. Ecosyst. Environ. 64, 181–188.

Hance, T., 1987. Predation impact of carabids at different population densities on *Aphis fabae* development in sugar beet. Pedobiologia 30, 251–262.

Heyer, W., 2000. Auswirkungen der Fruchtartenvielfalt auf die Selbstregulation von Agrarökosystemen, dargestellt am Beispiel epigäischer Räubergruppen. In: Hälsbergen, K.-J., Diepenbrock, W. (Eds.), Die Entwicklung von Flora, Fauna und Boden nach Umstellung auf ökologischen Landbau, Halle, 2000, pp. 123–134.

Hülsbergen, K.-J., Diepenbrock, W., 2000. Die Untersuchung von Umwelteffekten des ökologischen Landbaus—Problemstellung und Forschungskonzept. In: Hülsbergen, K.-J., Diepenbrock, W. (Eds.), Die Entwicklung von Flora, Fauna und Boden nach Umstellung auf ökologischen Landbau, Halle, 2000, pp. 15–40.

Hülsbergen, K.-J., Diepenbrock, W., 2001. Die Nachhaltigkeit von Düngungssystemen—dargestellt am Seehausener Düngungskombinationsversuch. Arch. Acker- Pfl. Boden. 46, 215–238.

Hülsbergen, K.-J., Diepenbrock, W., Rost, D., 2000. Analyse und Bewertung von Umweltwirkungen im Landwirtschaftsbetrieb—Das Hallesche Konzept. In: Die Agrarwissenschaften im Übergang zum 21. Jahrhundert. Wissenschaftliche Beiträge 8. Hochschultagung, Martin-Luther-Universität, Halle-Wittenberg, 2000, pp. 75–87.

Hülsbergen, K.-J., Feil, B., Biermann, S., Rathke, G.-W., Kalk, W.-D., Diepenbrock, W., 2001. A method of energy balancing in crop production and its application in a long-term fertilizer trial. Agric. Ecosyst. Environ. 86, 303–321.

Jaeger, W.K., 1995. Is sustainability optimal? Examining the differences between economists and environmentalists. Ecol. Econ. 15, 43–57.

Koepf, H., Kaffka, S., Sattler, F., 1989. Nährstoffbilanz und Energiebedarf im landwirtschaftlichen Betriebsorganismus. Verlag Freies Geistesleben, Suttgart.

Liebig, M.A., Doran, W., 1999. Impact of organic production practices and soil quality indicators. L. Environ. Qual. 28, 1601–1609.

Luka, H., 1996. Laufkäfer: Nützlinge und Bodenindikatoren in der Landwirtschaft. Agrarforschung 3, 33–36.

Milon, J.W., Shogren, J.F., 1995. Integrating Economic and Ecological Indicators: Practical Methods for Environmental Policy Analysis. 2. Economic Policy Congress. Praeger, New York, 215 pp.

Müller, D., Dittrich, G., 1996. Ökonomische Bewertung agrarökologischer Maßnahmen, dargestellt am Agrarökologischen Landschaftskonzept Köllitsch, 1996.

Münchhausen, E.V., Nieberg, H., 1997. Agrar-Umweltindikatoren: Grundlagen, Verwendungsmöglichkeiten und Ergebnisse einer Expertenbefragung. In: DBU (Eds.), Umweltverträgliche Pflanzenproduktion—Indikatoren, Bilanzierungsansätze und ihre Einbindung in Ökobilanzen. Zeller Verlag Osnabrück, pp. 13–29.

Nähring, D., 1990. Charakterisierung und Bewertung von Hecken mit Hilfe der Spinnenfauna. Zoolog. Beiträge, N.F. 33, 253–263.

OECD (Organisation for Economic Co-operation and Development), 1998. Report of the OECD Workshop on Agri-environmental Indicators, New York, September 1998. OECD, Paris, COM/AGR/CA/ENVEPOC (98), 136 pp.

Oñate, J.J., Andersen, E., Peco, B., Primdahl, J., 2000. Agri-environmental schemes and the European agricultural landscapes: the role of indicators as valuing tools for evaluation. Landscape Ecol. 15, 271–280.

Papaja, S., Hülsbergen, K.-J., 2000. Die Entwicklung von Regenwurmpopulationen unter dem Einfluß der Bewirtschaftungsumstellung. In: Hülsbergen, K.-J., Diepenbrock, W. (Eds.), Die Entwicklung von Flora, Fauna und Boden nach Umstellung auf ökologischen Landbau, Halle, 2000, pp. 108–122.

Poole, R.W., 1974. An introduction to Quantitative Ecology. McGraw-Hill, New York.

Rennings, K., Wiggering, H., 1997. Steps towards indicators of sustainable development: linking economic and ecological concepts. Ecol. Econ. 20, 25–36.

Saacke, B., Fuchs, S., 2001. Naturschutzorientierte Nutzungsregime im ökologischen Feldfutterbau, Teil a: Naturschutzfachliche Anforderungen aus Sicht der Feldlerche *Alauda arvensis*. Beiträge zur 6. Wissenschaftstagung zum Ökologischen Landbau, 6.–8. März 2001 Freisingen-Weihenstephan, pp. 147–150.

Schinner, F., 1986. Die Bedeutung der Mikroorganismen und Enzyme im Boden. Veröff. Landw.-Chem. Bundesanstalt Linz/Donau 18, 15–39.

Steinborn, H.-A., Heydemann, B., 1990. Indikatoren und Kriterien zur Beurteilung der ökologischen Qualität von Agrarflächen am Beispiel der *Carabidae* (Laufkäfer). Schr. Reihe f. Landschaftspflege u. Naturschutz 32, 165–174.

Stubbe, M., Seluga, K., Weidling, A., 1998. Bestandessituation und Ökologie des Feldhamsters *Cricetus cricetus* (L., 1758). In: Stubbe, M., Stubbe, A. (Eds.), Ökologie und Schutz des Feldhamsters. Martin-Luther-Universität Halle-Wittenberg, Wissenschaftliche Beiträge, pp. 137–182.

Tischer, S., 2000. Veränderung der mikrobiologischen Aktivität nach Bewirtschaftungswechsel. In: Hülsbergen, K.J., Diepenbrock, W. (Eds.), Die Entwicklung von Flora, Fauna und Boden nach Umstellung auf ökologischen Landbau, Halle, 2000, pp. 101–107.

Weinhold, U., 1998. Bau- und Individuendichte des Feldhamsters (*Cricetus cricetus* L. 1758) auf intensiv genutzten landwirtschaftlichen Flächen in Nordbaden. In: Stubbe, M., Stubbe, A. (Eds.), Ökologie und Schutz des Feldhamsters. Martin-Luther-Universität Halle-Wittenberg, Wissenschaftliche Beiträge, pp. 277–288.

Wieland, R., 1997. Habitatmodellierung biotischer Komponenten. Arch. Nat.- Lands. 35, 227–237.

Wittmann, C., Hintzsche, E., 2000. Die Entwicklung der Segetalflora nach Umstellung auf ökologischen Landbau. In: Hülsbergen, K.-J., Diepenbrock, W. (Eds.), Die Entwicklung von Flora, Fauna und Boden nach Umstellung auf ökologischen Landbau, Halle, 2000, pp. 239–258.

Wood, D., 1998. Ecological principles in agricultural policy: but which principles? Food Policy 23, 371–381.

ELSEVIER

Available online at www.sciencedirect.com

Agriculture, Ecosystems and Environment 98 (2003) 463–475

Agriculture Ecosystems & Environment

www.elsevier.com/locate/agee

Nature balance scheme for farms—evaluation of the ecological situation

Rainer Oppermann*

Institut für Landschaftsökologie und Naturschutz (ILN) Singen, c/o NABU, Tübinger Str. 15, D-70178 Stuttgart, Germany

Abstract

Farming needs more and more support in society to get further subsidies. This will only be possible when the ecological benefit can be proved. Up to now a method to measure biodiversity and ecological benefit of farms is missing.

In the state of Baden-Württemberg (Germany), farmers and agri-environment experts have worked together to devise schemes to record the ecological richness and benefits of farms, especially in the field of biodiversity and cultural landscape issues. The scheme uses 47 different indicators divided into four sectors, biodiversity-structural richness, biodiversity-species richness, farm management, and field management (farming methods). Farmers can perform the assessment themselves, on average sized farms in Southwest Germany one working day may be sufficient. After assessing the ecological situation of the farm an evaluation can easily be made in form of a nature balance scheme. The presented nature balance scheme is based on a 100-point target system revealing ecological benefits as well as shortcomings of the farms.

The paper presents results after applying the method to 16 farms in different natural environments and especially the case study Gengenbach (10 farms). Four out of 10 farms achieved the target of 100 points and thus it could be stated that they manage a sustainable agriculture in the sense of biodiversity and environmental sound farming. The results are encouraging because they show that the method works well and farmers become aware of the ecological situation of their farms.

Keywords: Biodiversity; Indicators; Agri-environmental measures; Agriculture; Nature balance

1. Introduction

Farming has long contributed to the biodiversity of rural areas. Until now, high biodiversity and habitats of high ecological value have been a by-product of some types of farming, but with changing policies and attitudes they are becoming more and more important, and for some farms may switch to become the main product (e.g. in nature conservation areas, extensive pastures in the highlands). However, many farmers are not aware of the concrete biodiversity on their farms and neither are the consumers, who, for example, buy food from direct marketing farms.

In discussions, the most common-named measures for nature conservation on farms are "planting a hedge" or "conversion to organic agriculture". But, within the big variety of farm types, landuse types and nature regions, there is a wide range of different possibilities to ensure or to improve naturally sound agricultural management and high biodiversity (Bosshard et al., 2002). Which measures are the most suitable to develop a farm into a sustainable place, which not only guarantees the sustainable protection of the abiotic recourses, but also guarantees biodiversity and nature preservation, is often unknown.

* Tel.: +49-711-966720; fax: +49-7731-996218.
E-mail address: oppermann@iln-singen.de (R. Oppermann).

0167-8809/$ – see front matter © 2003 Elsevier Science B.V. All rights reserved.
doi:10.1016/S0167-8809(03)00105-1

So the idea evolved to develop a nature balance scheme with which it should be possible to get an overview of the nature situation of farms via field surveys. The nature balance scheme is an evaluation scheme that allows a detailed view on the ecological situation but it is not a full scientific analysis. In teamwork of a group of farmers and nature conservation specialists all possible elements of environmentally friendly elements of farming were listed and sorted into different sectors. This list formed the framework for the evaluation of the farms. The objective of this approach is to enable the assessment of the ecological situation for the farmers themselves thus raising consciousness for ecological values and for ecologically sound practices. This may also serve for improving the direct marketing of their products with additional information for the clients.

2. Materials and methods

2.1. Assessing ecological information of the farms

Agriculturally and environmentally educated people normally quickly have an impression and a rather sure appraisal of the ecological situation by going around on the farm and the cultivated areas. The principle of assessing the ecological situation is to record this information which can be observed on the farm in an objective data sheet. The ecological situation and the nature balance scheme are divided into four main sectors (Fig. 1). There are two sectors concerning the field of biodiversity, one is the structural richness and the other is the species richness based on indicator plant species on the farm. Two further sectors concern the management of the farm and the farming methods (field management). These four sectors ensure that both the actual biodiversity and the efforts in environmental friendly practices are considered.

The richness in structures is recorded by adding the acreage of all areas of landscape elements (hedges, stone rows, ditches, . . .) and of extensively used areas (extensively used fruit orchards, meadows, field margins without use of pesticides, etc.). The landscape elements are listed completely in Table 3 (as Tables 3 and 4 mainly contain results, they are arranged behind Tables 1 and 2).

Table 1
Evaluation scheme for the nature balance

Sector of nature balance	Target		Maximum points
	% of area	No. of points	
Biodiversity-landscape structure			
Landscape elements	5	10	15
Extensively cultivated areas	30	15	25
Biodiversity-species richness	30	30	50
Farm management	–	20	25
Field management	100	25	25
Total nature balance		100	140

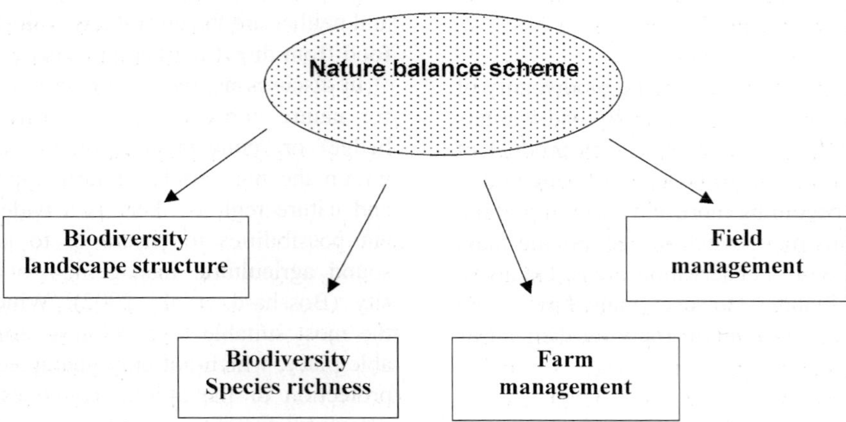

Fig. 1. Main sectors of the nature balance scheme, which are measured and evaluated.

Table 2
Agricultural data of the investigated farms ($n = 10$ farms), Gengenbach, 2000

Characterisation	Total	Average	Size of individual farms
Agricultural used area	143.7 ha	14.4 ha	1.9–32.0
Meadows and pastures	104.1 ha	10.4 ha	0.6–19.9
Arable land	31.4 ha	3.1 ha	0–16.0
Viticulture	4.4 ha	0.4 ha	0–2.2
Fruit plantations	3.8 ha	0.4 ha	0–1.4
Otherwise used areas			
Horticultural used area	7.4 ha	0.7 ha	0–2.5
Forest	190 ha	19 ha	0–79
Livestock	63 milking cows		0–32 milking cows
	24 suckling cows		0–12 suckling cows
	17 cattle		0–5 cattle
	2 horses		0–2 horses
	28 pigs		0–28 pigs
	25 deer		0–25 deer
	15 sheep		0–10 sheep
	150 hens		0–150 hens
Altitude	170–420 m above sea level		

The species richness is recorded with a list of indicator plant species. This list comprises, for meadows and pastures, 28 flower species (or group of species, e.g. *Campanula* sp.) of all nature regions of Baden-Württemberg (Table 4). The selection of the indicator species and the assessment method are described in detail in Briemle (2000) and is based on proposals of Bronner et al. (1997). The flowers can all easily be recognised by farmers. Since the year 2000, the presence of at least 4 of these 28 indicator species is one program position for additional agri-environmental payments for farmers in MEKA II (MLR, 1999a,b). In an analogous way, a list of flower species was developed for assessing arable land and for vineyards comprising 12 indicator species (Bronner et al., 1997). These lists have also been used in the case study presented in this paper (Table 5).

Good farm management practices to be recorded were, for example, balance field margin stripes without use of fertilisers along all water courses and ditches and caring for the biodiversity on the farmyard itself by having a big farm tree and nestling possibilities for swallows, etc. The list of criteria for good farm management practices comprises 20 criteria (Table 3) with each contributing to maximal 2 points to the balance scheme.

The overall field management practice was recorded by considering the percentage of land cultivated without synthetical fertilizer, pesticides and rotation mowing machines as well as the area cultivated with organic farming.

Altogether there are 47 criteria of ecological data (Table 3). The information of the ecological situation of the farm was listed on one data sheet called "Ökologischer Betriebsspiegel" (Ecological information of the farm). The criteria and the Ökologischer Betriebsspiegel are described in detail in Oppermann (2001), the focus of this paper is the presentation of the farm level approach for the evaluation of the ecological situation.

2.2. Nature balance scheme

The Ökologischer Betriebsspiegel forms the database for the evaluation of the nature balance scheme. Each of the four sectors contributes to the evaluation, which is measured on a scale of 100 points. The 100-point scale is a normative approach that has been chosen in order to give a clear orientation for the practice. This evaluation scheme has been worked out under the principle that it should be possible for each farmer to achieve the target of 100 points, when he

Table 3
Ecological data of the investigated farms ($n = 10$ farms), Gengenbach, 2000

Acreages and criteria of the nature balance scheme	All 10 farms		Individual farms		Target (%)
	Acreage (ha)	Part[a] (%)	aaa[b] (ha)	Part of aaa (%)	
Acreage agricultural area of all 10 farms (aaa)	143.7	100	1.9–32	100[a]	
Biodiversity-landscape structure					
Landscape elements	10.0	6.9	0.36–1.46	4.1–19.0	5
Hedges, shrub	1.1	0.7	0–0.33	0–2.1	
Groups of trees and shrubs	0.7	0.5	0–0.20	0–4.2	
Tree-lines (e.g. along roads), forest margins	0.9	0.6	0–0.29	0–2.2	
Dry stone rows	0	–	0	–	
Field banks	2.6	1.8	0.11–0.45	0–4.4	
Unasphalted field roads	2.9	2.0	0.17–0.50	0–11.6	
High herb and reed ranges	0.3	0.2	0–0.12	0–0.8	
Ditches/ditch banks	1.4	1.0	0–0.47	0–2.4	
Ponds	0.1	0.1	0–0.04	0–1.1	
Other elements (e.g. solitaire trees, field sheds)	0.1	0.1	0–0.03	0–0.2	
Extensively used areas	38.5	27	0.04–12.86	1–68	30
Margin stripes arable land	0.5	2	0–0.49	0–8	
Cultivation of seldom/local crop varieties	0	–	0	–	
Extensive crop cultivation	0	–	0	–	
Rotational fallow land with natural greening	0	–	0	–	
Steep viticulture extensive cultivation	3.0	68	0.3–1.20	30–100	
High stem orchards	19.3	13	0–5.25	0–36	
Grassland margins mown max. once per year	0.4	0	0–0.21	0–3	
Extensive grassland use (mown max. twice per year or yield <50 dt/ha or at least four indicator species)	34.6	33	0–9.00	0–76	
Biodiversity-species richness	21.6	15.0	0–6.98	0–29	30
Grassland with at least four indicator species	20.7	19.9	0–6.42	0–46	
Grassland with at least six indicator species	0.3	0.3	0–0.30	0–1	
Grassland with at least four indicator species and at least 100 individuals per area	0.7	0.7	0–0.50	0–5	
Arable land/vineyards with at least two indicator species	0.9	2.6	0–0.56	0–9	
Arable land/vineyards with at least four indicator species	0		0		
Arable land/vineyards with at least six indicator species	0		0		
Field management					
Area cultivated without N-fertilizer	86.0	60	0–28.0	0–100	100
Area cultivated without pesticides and rotation mowing machines	51.7	36	0–14.0	0–82	100
Area cultivated organically (EU-standard)	47.9	33	0–28.0	0–100	100
Farm management[c]	199.6 (292)[d]	20 p. (%)	12–25 p. (points)	12–25 p. (%)	20 p.
Grain growing without CCC	10 (10)	100	2	100	
Crop rotation with at least four rotational elements	4 (10)	40	0–2	0–100	
Moor and floodland adequate grassland use	0 (0)	–	–	–	
Protection of water courses	14.3 (16)	89	1–2	50–100	
Erosion protection in arable land	2 (2)	100	2	100	
Environmental sound ditch maintenance	6 (8)	75	1–2	50–100	
Protection of field road margins	11 (16)	69	0.5–2	25–100	
Harvesting hay	10.9 (18)	68	0.7–2	35–100	
Working with solid manure	11.4 (20)	71	0–2	0–100	
Animal sound stock farming	16 (20)	80	1–2	50–100	

Table 3 (*Continued*)

Acreages and criteria of the nature balance scheme	All 10 farms		Individual farms		Target (%)
	Acreage (ha)	Part[a] (%)	aaa[b] (ha)	Part of aaa (%)	
Preserving seldom local animal breeds	9 (18)	50	0–2	0–100	
Nestling of swallows	14 (20)	70	0–2	0–100	
Nestling of owls or bats	15 (20)	75	0–2	0–100	
Presence of seldom plant or animal species	8 (16)	50	0–2	0–100	
Realization of landscape management	4 (6)	67	1–2	50–100	
Unasphalted farmyard ranges	6 (18)	33	0–2	0–100	
Big farm tree on the farmyard	17 (20)	85	1–2	50–100	
Typical farmyard gardens or stonewalls	12 (20)	60	0–2	0–100	
Region-typical farm house	16 (20)	80	0–2	0–100	
Other elements	13 (20)	65	0–2	0–100	

The left column is the complete list of the assessment and evaluation criteria, a detailed description of these criteria is given in Oppermann (2001).

[a] Reference base for the percent calculation is the relevant acreage: landscape elements: total acreage agricultural area (aaa) of the farms, margin stripes arable land: acreage arable land, steep viticulture extensive cultivation: acreage viticulture, high stem orchards: total aaa (note: being cover trees on other crops they are not added in the sum of extensively used areas), grassland margins and extensive grassland use: grassland acreage).

[b] aaa = acreage agricultural used area of a farm.

[c] The list of criteria for good farm management practices comprises 20 criteria with each contributing to maximal 2 points to the balance scheme.

[d] The sums of points of all 10 farms (in parentheses maximum number of possible points being relevant for the farms according to the natural situation).

really applies naturally sound farming practices on his farm. In Table 1, the evaluation targets are indicated. The base for setting the targets was the objective to have clear figures for a landscape use that may enable the survival of, e.g. populations of meadow and field birds (e.g. whinchat *Saxicola rubetra* and skylark *Alauda arvensis*). These targets refer to experiences and know-how of the involved nature conservation specialists. Comprehensive and detailed studies on the requirements of populations on the landscape and region scale are widely missing. The principle of the point scoring is that both biodiversity and management elements should contribute to the scheme, with a slight dominance of the result-orientated biodiversity sectors (target 55 points, maximum 90 points) versus the measure-orientated management sectors (target 45 points, maximum 50 points).

With exceptional ecological benefits, even more than 100 points and up to a limit of 140 points can be achieved. Missing points in one sector can thus be compensated up to the indicated maximum level by additional points in another sector of the balance. For example, a lack in the species richness target (15 instead of 30 target points) can be partly compensated by extensively used cultivated areas (target 15 points, maximum 25 points).

In 1999, a first field survey was carried out on six farms in different nature regions of Baden-Württemberg (Southwest state of Germany). The results were rather encouraging and have been presented in Oppermann et al. (2000). For the case study presented in this paper, the method used in the 1999 study has been slightly modified by integrating the actual species richness key of the above mentioned MEKA II (MLR, 1999b) and by adapting the evaluation scheme also to small types of farms.

2.3. Case study Gengenbach

In this paper, the results of the case study Gengenbach are presented. The task of the case study was to check the method in only one parish and in an area where very small farms and very differentiated farming practices are carried out.

Gengenbach is situated in the very Southwest of Germany at the edge of the Black Forest (Fig. 2). Ten farms participated, being situated on an altitude from 170 m above sea level (Kinzig valley) to 420 m (slopes

Fig. 2. Geographical location of the commune Gengenbach at the edge of the Black Forest and the Kinzig valley. A part of the rich structured landscape and some typical Black Forest farmhouses can be seen.

of the Middle Black Forest). The farms have been chosen by a local farmers initiative (local agenda 21). Although not being selected randomly the 10 farms represent the typical situation of Gengenbach (e.g. many part-time farms). The data have been collected by field work on all cultivated areas of the farms (concerning landscape structures, species richness, realisation of field margin stripes, etc.) and by interviewing the farmers (concerning the size of the cultivated areas of the farms, the number of livestock, the input of fertilisers, etc.). To map landscape elements and areas with presence of the indicator species, aerial photographs in the scale 1:10000 were used during the field work.

3. Results

The farmers in Gengenbach cultivate areas from 2 to 32 ha (medium 12 ha). Two of the farms are full-time farms, eight farms are part-time farms (a typical situation in some parts of Southwest Germany). The farms have an enormous range of agricultural uses including arable land, meadows and pastures, vineyards, orchards and horticultural crops such as strawberries. Also the 10 farms keep quite a big variety of livestock. Many farmers have some forest areas as well, one farmer even the most of his ground (79 ha). For details of the "technical data" see Table 2. The products are partly sold directly in the local market and to private customers (some of the farms sell bread, wine and other products directly, some also offer apartments for holidays on their farm).

Looking at the ecological data, Table 3 presents the most important figures. There are 4–19% of the agricultural used areas of the farms landscape elements such as hedges, groups of trees and shrubs, field banks, open (not asphalted) field paths and roads, ditches, etc. The most common landscape elements in Gengenbach are unasphalted field paths/roads with 2.0% and field banks with 1.8%—both due to small fields and the relief situation at the edge of the Black Forest. Hedges and groups of trees and shrubs have an account for another 1.3%, ditches and ditch banks 1%. All landscape elements cover 10 ha ground or 6.9% of the agricultural used area.

The extensively used areas have an area of averagely 27% with a very wide variation between 1 and 68%. Some farms have a high percentage (up to 76%) of extensively used grassland (e.g. pastures for suckling cows), some others produce intensively on their farms and only have marginal areas left in extensive use (e.g. margin stripes). In the arable land there are only few extensively used areas, e.g. margin stripes in organic farms (up to 8%).

The areas with a high species richness of indicator species comprise on an average 15% with a variation of 0–29% of the farming area. Looking closer at the biodiversity of the meadows and pastures, 19 indicator species occur in total from the list of 28 species and species groups (Table 4). On the individual farms, 5–14 species occur. The most common indicator species are *Trifolium pratense* and *Chrysanthemum leucanthemum* (96 and 73%, respectively of the areas with a minimum number of four indicator species in the grassland). The most typical species for these nature regions that occur on the farms are *Hieracium pilosella* (11%), *Potentilla erecta* (9%), *Polygala* sp. (8%), *Polygonum bistorta* (6%) and *Thymus pulegioides* (4%). While the grassland has an average of nearly 20% of species rich areas, the arable land and the viticultures together have only 2.6% areas rich in species. Only a few fields were identified as species rich and here indicator species such as *Papaver rhoeas* or *Centaurea cyanea* occur (Table 5).

In the sector farm management it turned out that most farms apply environmentally sound practices: 89% care for the water courses, 57% have solid manure (43% have slurry), and on 61% of the meadows is harvested hay (39% silage). On an average nearly 20 points are achieved and thus the target. With exception of one farm (12 points), all farms reach more than 17 points. Very typical for the Black Forest region is up to now the regional building style of farm houses (80%), often also huge farmyard trees are present (85%). Some of the farmers keep typical local animal breeds such as cows of the type "Hinterwälder" or "Vorderwälder" (50%), which are adapted to the climate and slopes of the Black Forest and are known as endangered breed.

Table 4
Indicator species for agri-environment payments MEKA II in Baden-Württemberg (MLR, 1999b), here with the occurrence of the species that have been found in the investigated farms in all meadows and pastures with the presence of at least four indicator species ($n = 38$ meadows and pastures)

Indicator species	Occurrence (%)
Trifolium pratense	96
Chrysanthemum leucanthemum	73
Centaurea sp.	56
Leontodon sp./*Hypochoeris* rad.	54
Lychnis floscuculi	30
Sanguisorba officinalis	22
Campanula sp.	21
Cardamine pratensis	16
Crepis biennis/*C. mollis*	11
Hieracium pilosella	11
Knautia arvensis	11
Potentilla erecta	9
Polygala sp.	8
Polygonum bistorta	6
Thymus pulegioides	4
Geranium sp.	2
Rhinanthus sp.	2
Silene dioica	2
Tragopogon sp.	1
Cirsium oleraceum	0
Euphrasia rostkoviana	0
Genista sagittalis	0
Geum rivale	0
Meum athamanthicum	0
Phyteuma sp.	0
Salvia pratensis	0
Sanguisorba officinalis	0
Trollius europaeus	0

In species groups, such as *Campanula* sp. all species are valid, but only counted as one indicator species.

In the field management there is a relatively high percentage of areas which are not treated with N-fertilisers and fluid manure (60%) most of the grassland being fertilised only by solid manure or P- and K-fertilisers. Regarding the pesticide situation, the grassland is completely free of application of pesticides but in most meadows rotation mowers are used, causing high damage to animals in the meadows (for example, amphibians more than double loss rate, refer to Claßen et al., 1996; Oppermann et al., 1997). The acreage of organic cultivated land is about 33% (two farms).

Table 5
Indicator species for arable land and viticulture (refer to Bronner et al., 1997; Oppermann, 2001), here with the occurrence of the species that have been found in the investigated farms in all arable land and viticulture with the presence of at least two indicator species ($n = 2$ fields of arable land)

Indicator species	Occurrence (%)
Arable land	
Papaver sp.	100
Anagallis foemina	50
Centaurea cyanus	50
Adonis sp.	0
Anchusa arvensis	0
Delphinium consolida	0
Kickxia sp.	0
Legousia sp.	0
Melampyrum arvense	0
Muscari neglectum	0
Ranunculus arvensis	0
Stachys sp.	0
Viticulture	
Allium sp.	0
Aristolochia clematitis	0
Calendula arvensis	0
Erodium cicutarium	0
Euphorbia helioscopia	0
Gagea sp.	0
Geranium sp.	0
Heliotropum annuum	0
Muscari sp.	0
Ornithogalum sp.	0
Tulipa sylvestris	0
Valerianella carinata/V. locusta	0

In species groups such as Allium sp. all species are valid, but only counted as one indicator species.

In Fig. 3, the results of a 10-farm evaluation in Gengenbach in spring 2000 are demonstrated. It is obvious that some of the targets are achieved by nearly all farms (landscape elements, extensively used areas, farm management). Four out of 10 farms have a good ecological performance even in all 4 sectors (landscape structure, species richness, field management and farm management). They achieve 100 points or even more (up to 110 points). There are two main reasons: one reason is the geographical situation of Gengenbach at the edge of the Black forest to the Kinzig valley with small land plots and many steep slopes which cannot be cultivated intensively. The other reason may be that the farms that have participated voluntarily in the nature balance are mainly part-time farms and not dependent on the agricultural income (but one full-time farm achieved the best result with 110 points).

The sectors species richness and field management are partly not well developed in the ecological sense. Some farms show a very low number of points in these sectors, but others show very good values. So, four farms achieve 24–29 points and thus nearly the target in species richness, two other farms have a good position (10–12 points) to achieve the target with some further efforts (long-term efforts), such as enlarging the species diversity by reducing fertilizer input on sloppy meadows or reducing the pesticide input on the arable land and four farms have an unsatisfactory situation (0–7 points). In the sector field management, three farms have very low values due to intensive cultivation on most of the agricultural used areas. Important may be to point out that in this sector only the acreage of no use of pesticides and chemical N-fertilizer is checked, not the amount of pesticide and fertilizer input (all participating farmers said that they use as less pesticides as necessary).

Analysing the sector showing the best correlation to the total number of points in the nature balance, the sector species richness is correlated highly significantly with the total number of points ($r = 0.92$, $P < 0.001$, Fig. 4). In comparison, all other sectors showed lower correlation, the best being the extensively used areas ($r = 0.84$) and field management ($r = 0.73$).

4. Discussion

Two aspects shall be discussed: the first aspect is the position of the method "nature balance scheme" in the context of other balance schemes; the second aspect are methodical issues.

4.1. Nature balance scheme in comparison to other approaches

The agri-environmental sciences have developed very quickly in the last decade. There are a lot of results for individual sectors, e.g. for the interaction of fertilisers and environment (e.g. Gäth, 1997; Isermann and Isermann, 1997), for pesticides (e.g. Raskin, 1994), for watercourses (e.g. Frede and Dabbbert, 1998) and for landscape issues (e.g. Jenny et al.,

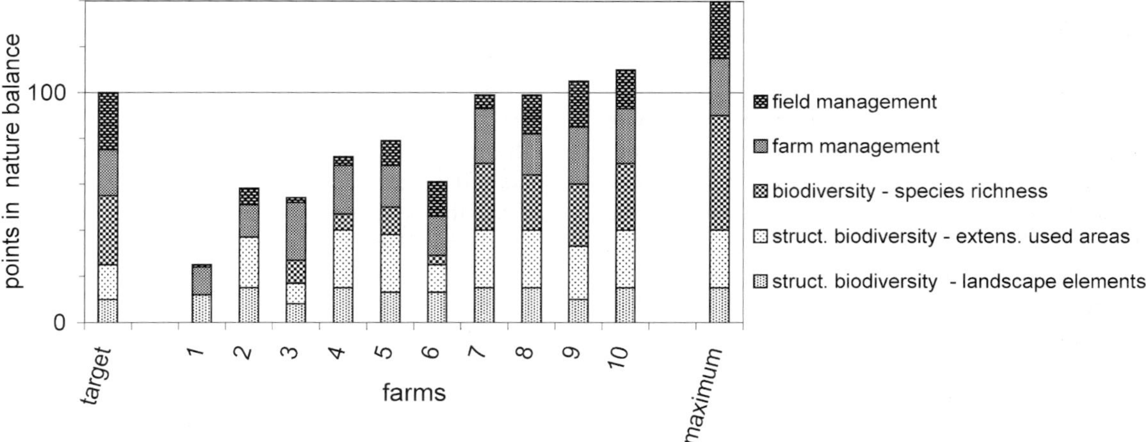

Fig. 3. Results of a 10-farm evaluation in Gengenbach in spring 2000. The evaluation scheme (target points and maximum points) refers to Table 1.

2000; Laußmann and Plachter, 1998). However, an evaluation of the environmental situation needs an approach integrating the different sectors of the environmental situation. There are only a few approaches trying to integrate several sectors of the environmental situation. The best known approach in Germany is the approach KUL (Eckert and Breitschuh, 1997; Eckert, 1998), in Austria, the ecopoint-system of Niederösterreich (NÖA, 1996; Mayrhofer, 1997) and in France, the method of Solagro (1999). All these approaches need highly qualified staff and much time to do the evaluation and normally they cannot be done by the farmers themselves. These approaches consider the biotic resources only marginally in spite of the fact that biodiversity has become substantial in political discussions since the conference of Rio in 1992 (Engelhardt and Weinzierl, 1993). For example, the approach KUL uses 22 criteria for the evaluation of the environmental situation and only two of these indicators deal with biotic resources (crop plant diversity indicator and ecological-cultural-priority areas, refer to Eckert and Breitschuh, 1997; Eckert, 1998). The most developed approach integrating the biotic resources has been done by Wetterich and Haas (1999): they have integrated the number of species in meadows and pastures of Allgäu as one of the eight factors. Beside these general approaches there are a few other approaches taking in focus, especially the biodiversity, such as the method "Frieben" and the "Biodiversity yardstick"; a detailed comparison of different approaches considering biodiversity is given by Braband et al. (2003).

Efforts to achieve more ecological informations, are also made by the EU and the OECD. In

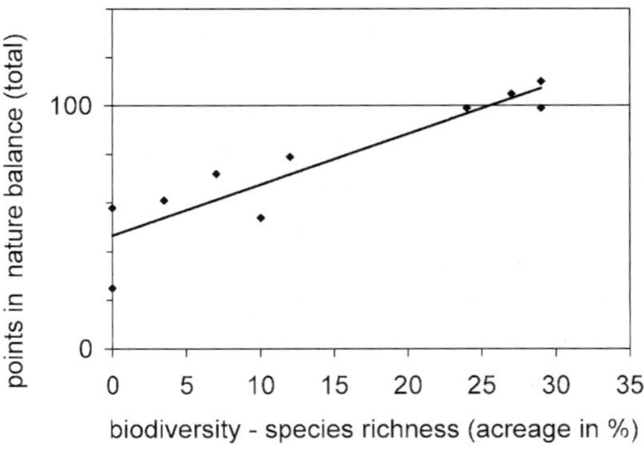

Fig. 4. The result of the nature balance scheme (total number of points) correlates best with biodiversity-species richness (acreage of areas rich in indicator species). Correlation $r =^{***} 0.917$, $P < 0.001$.

the Pressure/Driving Forces–State–Response-model (PSR-model), indicators on the state of the environment are needed to allow responses in policy (Diepenbrock et al., 1997; Von Münchhausen and Nieberg, 1997). The OECD tries to develop indicators that allow comparisons between different countries concerning the environmental situation and enable links to policy instruments (OECD, 1999a,b). Nevertheless, also in the specific reports for landscape, habitats and wildlife concrete indicators, that can be assessed on farm level are missing (OECD, 2001). For example, the species diversity shall be indicated by monitoring the trends in population distributions and number of wild species related to agriculture. This approach is absolutely necessary but there is no link to monitoring possibilities on farm level. The EU-Commission has set up a catalogue of evaluation questions for the mid-term review of the agri-environment schemes throughout all countries of the EU (EU, 2000). It contains 71 indicators for the different criteria and environmental resources. These indicators are kept quite general, especially concerning biodiversity and landscape structure in order to give the possibility for specific adaptations in all EU-countries. Results can be expected in the mid-term review 2004. Considering existing data sources, it is quite clear that there is much effort needed on the regional level in order to get information for these evaluation questions. But as the agri-environment schemes intend payments to farmers the effort of the measures should also be measured on farm level.

Pushed by discussions on OECD and EU-level there are several approaches to measure the impact of agriculture on the environment by indicators in different European countries (e.g. DEFRA, 2001; ECNC, 2001; IFEN, 1999). The most appropriate indicator work on environmental indicators in the European countries is done in the UK. On national level there are 147 indicators for sustainable development and even 1 of the indicators related to biodiversity in agriculture (Populations of wild birds) is one of the 15 key indicators for the UK (DEFRA, 2001). But, also on the regional and local level indicators have been developed (HCC, 2001). For local evaluations it is proposed to assess the "Net change in natural/seminatural habitats" and the "Changes in population of selected characteristic species". These two indicators are similar to the indicators "extensively used areas" and "species richness" in the nature balance scheme that have turned out to be of good indication value for the overall nature situation of the farms.

However, for use in different EU-countries the indicators have to be checked for suitability. For example, the indicators used in the nature balance scheme for assessing the landscape structure may also be used in other parts of Central Europe or even whole Europe and may be completed by some other elements. Measuring the biodiversity–species richness, in any case a regional adaption is necessary, such as developing lists of regional indicator species.

4.2. Methodical aspects

The nature balance scheme approach takes the biodiversity of both the structural biodiversity including spatial, linear and punctual elements and the species richness in the focus of the balance. As this approach is relatively new the question reveals whether the results are reproducible and if assessments carried out by the farmers are reliable. First of all, nearly all the indicators are clearly defined, such as the area of landscape elements, the biodiversity measured by indicator species and the area utilised without input of fertilisers. Following the opinion of the working team that developed the nature balance scheme, it is much easier to assess, for example, the biodiversity of a grassland by having a look at the presence of indicator species than by building a nutrient balance. The experiences of Baden-Württemberg show that in the first years of the new MEKA-program, over 6000 farmers participated with about 7% of the total grassland in the program position "grassland rich in species" (MLR, personal communication, 2002; Oppermann and Briemle, 2002), in which they had to confirm on land plot level the species richness. Secondly, there are experiences throughout many European countries with agri-environment programs: a control of all the measures indicated by the farmers is foreseen and practised at least with sample controls by regional, national and EU-administrations and there are sanctions against those giving wrong information. In the same way a nature balance scheme could be checked on reliability. Nevertheless, the nature balance scheme is up to now a method for awareness raising in the field of environmentally sound agricultural practice and is not linked to any payments or other subsidies.

An important point in discussion is the point scoring in the nature balance scheme, concerning both the targets as well as the criteria. Setting the targets is of normative character but it is quite clear that this should be based on a scientific background as far as possible. For example, the decline of farmland birds and its causes are well known (e.g. Gates and Donald, 2000; Donald et al., 2001), but unfortunately up to now there is a lack in detailed studies on the intensity level of agriculture that allows the existence of stable populations of typical farmland animals, such as the birds whinchat (*S. rubetra*) and skylark (*A. arvensis*). Experiences of whinchat studies, for example, estimate a percentage of 30–50% of extensively used grassland for populations on landscape level (Rebstock, personal communication, 1999; Oppermann, 1993), but further investigations are necessary. The criteria especially in the sector farm management cannot be discussed in detail. For example, the criteria "animal sound stock farming" is subject of several investigations and handbooks, whereas in the nature balance scheme it is measured only with a relatively simple checking and evaluation with 0–2 points. It is obvious that with this approach only a quick view on the different criteria is possible. But with choosing this rapid approach it is possible to get an overview of all sectors in a farm and this on a level that can be carried out and understood by the farmers themselves. The farmers were proud that the analysis showed positive things that they did not realise up to now, for example, the unasphalted field path/roads and the unmown edges of some meadows which they considered negative and that could be shown positive as a refuge, for example, for the hares.

In most balance schemes for farms, nutrient balances play an important role (Eckert and Breitschuh, 1997; Solagro, 1999; Wetterich and Haas, 1999). Getting the nutrient situation into an environmental sound state is a very important goal. Nutrient balances are not integrated in this nature balance scheme due to two reasons. The first reason is that the balance scheme has been worked out under the main rule that there should not be any data that have to be processed in a more or less complicated way (for many farmers nutrient balances are complicated). The whole nature balance should consist of elements which can easily be seen on the farm. The other reason is that there are many small farms in Southwest Germany which do not have any bookkeeping system so that they are not obliged to build in the nutrient balances. The nature balance should be a suitable instrument for all types of farms and also for part-time farms. For the scientific approach it would be very interesting to compare the results of the nature balance scheme method with results of other balance schemes including nutrient balances. Therefore, a larger number of farms in different situations should be investigated.

As the case study has shown, there is a high correlation between the species richness and the general evaluation result, the species richness contributing to only 30% of the total number of target points in the evaluation scheme. The questions reveals whether the indicator species richness is adequate for all nature regions and if the sector biodiversity-species richness would be sufficient in itself to indicate the complete environmental situation of the farms.

In detailed studies of Oppermann et al. (1987) and Waldhardt et al. (2000), the authors could show that the richness of indicator species and groups is highly and significantly correlated with the total plant species richness as well as with some faunistic parameters. Beside assessing some plant species, it would be desirable to investigate as well the fauna in a systematic approach but this takes plenty of time so it is difficult to practice. In this concern, in a comprehensive study in Switzerland for the evaluation of biodiversity in agricultural landscapes there could be found satisfying results that the total number of plant species gives good information on the overall biodiversity of plant and animal species (Duelli, 1997; Duelli and Obrist, 1998).

The indicator plant species selected are species that represent all nature regions of the country Baden-Württemberg (Briemle, 2000). They have been chosen carefully by a team of grassland experts of Baden-Württemberg. Measures applied by the farmer in order to achieve more species richness on meadows, e.g. by reducing the application of fertilisers and the mowing frequency will not lead immediately to more species richness. Often it will take quite a few years until the development of species richness takes place under certain agricultural management. This fact has been considered by taking into account the species richness only up to 30% of the target. The point scoring has a normative character and building the nature balance scheme is a mix of measure-orientated and result-orientated balance sectors.

5. Outlook

Most of the agri-environment programs and assessments still operate only with action-oriented measures and indicators. However, in different scientific and political publications, result-oriented measures and assessments are demanded (SRU, 1996; Bronner et al., 1997; OECD, 1999b). The subsidies based on species richness in grassland MEKA II in Germany (MLR, 1999a) and in the ÖQV in Switzerland (BLW, 2001) demonstrate that biodiversity indicators are suitable for additional payments for the farms.

Reviewing the results of the application of the nature balance scheme on 10 farms in Gengenbach, there are 3 interesting experiences. Firstly, the nature balance scheme method could be applied to all types of farms, from part-time to full-time farms and from specialised to not specialised farms. Secondly, it was rather surprising that a relatively large number of farms (4 of 10) achieved very good results, with the target of about 100 or even more points. It was surprising because the nature balance sets a clear goal for a really good level of environmental sound farming. The third remarkable result was that the farmers reactions were quite positive after initial scepsis. They were astonished on some of the positive results, since from common discussions in the media they are deeply uncertain about farming practices. They were proud that the analysis showed positive things that they did not realise up to now. Emphasising the positive elements the farmers could accept negative points. Showing both the positive and the improvable elements on a farm, one can grow the consciousness for environment.

In this sense, it is to hope that the nature balance scheme can be developed further and applied to many farms, so that it may contribute to an improved understanding of the ecological situation on farm level.

Acknowledgements

The author would like to thank Iris Bohnet (London/Burladingen) for the preliminary check of the English.

References

BLW, Bundesamt für Landwirtschaft, 2001. Öko-Qualitäts-Verordnung vom 01.05.2001. Bern.

Bosshard, A., Oppermann, R., Reisner, Y., 2002. Vielfalt in die Landschaftsaufwertung!—Eine Checkliste für Naturschutz und Landwirtschaft—Naturschutz und Landschaftsplanung 34, 300–308.

Braband, D., Geier, U., Köpke, U., 2003. Bio-resource evaluation within agri-environmental assessment tools in different European countries. Agric. Ecosyst. Environ. 98, 423–434.

Briemle, G., 2000. Ansprache und Förderung von Extensivgrünland. Naturschutz und Landschaftsplanung 32, 171–175.

Bronner, G., Oppermann, R., Rösler, S., 1997. Umweltleistungen als Grundlage der landwirtschaftlichen Förderung. Naturschutz und Landschaftsplanung 29 (12), 357–365.

Claßen, A., Hirler, A., Oppermann, R., 1996. Auswirkungen unterschiedlicher Mähgeräte auf die Wiesenfauna in Nordost-Polen. Naturschutz und Landschaftsplanung 28 (5), 139–144.

Department for Environment, Food and Rural Affairs (DEFRA), 2001. Indicators of Sustainable Development. London, United Kingdom, available at http://www.sustainable-development.gov.uk/.

Diepenbrock, W., Kaltschmitt, M., Nieberg, H., Reinhardt, G. (Eds.), 1997. Umweltverträgliche Pflanzenproduktion: Indikatoren, Bilanzierungsansätze und ihre Einbindung in Ökobilanzen. Initiativen zum Umweltschutz 5, Deutsche Bundesstiftung Umwelt, Osnabrück.

Donald, P.F., Green, R.E., Heath, M.F., 2001. Agricultural intensification and the collapse of Europe's farmland bird populations. Proc. R. Soc. Lond. B 268, 25–29.

Duelli, P., 1997. Biodiversity evaluation in agricultural landscapes: an approach at two different scales. Agric. Ecosyst. Environ. 62, 81–91.

Duelli, P., Obrist, M.K., 1998. In search of the best correlates for local organismal biodiversity in cultivated areas. Biodiv. Conserv. 7, 297–309.

Eckert, H., Breitschuh, G., 1997. In: Diepenbrock, W., Kaltschmitt, M., Nieberg, H., Reinhardt, G. (Eds.), Kritische Umweltbelastungen Landwirtschaft (KUL): Ein Verfahren zur Erfassung und Bewertung landwirtschaftlicher Umweltwirkungen. pp. 185–196.

Eckert, H., et al., 1998. Kriterien umweltverträglicher Landwirtschaft. Standpunkt des VDLUFA, Darmstadt (Selbstverlag), 6 pp.

Engelhardt, W., Weinzierl, M. (Eds.), 1993. Der Erdgipfel-Perspektiven für die Zeit nach Rio. Bonn.

European Centre for Nature Conservation (ECNC), 2001. In: Wascher, D.M. (Ed.), Environmental Indicators for Sustainable Agriculture (ELISA—Final Project Report). ECNC, Tilburg, The Netherlands. Available at http://www.ecnc.nl/.

Europäische Kommission (EU), 2000. Gemeinsame Bewertungsfragen mit Kriterien und Indikatoren. Brüssel, Dokument VI/12004/00 ENDG.

Frede, H.G., Dabbbert, S. (Eds.), 1998. Handbuch zum Gewässerschutz in der Landwirtschaft. Ecomed-Verlag, Landsberg, 451 pp.

Gäth, S., 1997. In: Diepenbrock, W., et al. (Eds.), Methoden der Nährstoffbilanzierung und ihre Anwendung als Agrar-Umweltindikator. pp. 115–126.

Gates, S., Donald, P.F., 2000. Local extinction of British farmland birds and the prediction of further loss. J. Appl. Ecol. 37, 806–820.

Hampshire County Council, 2001. Investigating Appropriate Indicators for Local Biodiversity. Whinchester, Hampshire, United Kingdom. Available at: http://www.hants.gov.uk/tc/biodiversityindicators/.

Institut francais de l'environnement (IFEN), 1999. Agriculture et environnement: les indicateurs. Summary available at http://www.ifen.fr/.

Isermann, K., Isermann, R., 1997. In: Diepenbrock, W., et al. (Eds.), Tolerierbare Nährstoffsalden der Landwirtschaft ausgerichtet an den kritischen Eintragsraten und -konzentrationen der naturnahen Ökosysteme. pp. 127–158.

Jenny, M., Weibel, U., Lugrin, B., Josephy, B., Regamey, J.-L., Zbinden, N., 2000. Förderung von typischen Brutvogelarten der offenen Feldflur durch den ökologischen Ausgleich in intensive genutzten Ackerbaugebieten des Klettgaus SH und der Champagne genevoise GE. Bundesamt für Umwelt, Wald und Landschaft (BUWAL) Bern. 133 pp.

Laußmann, H., Plachter, H., 1998. Der Einfluß der Umstrukturierung eines Landwirtschaftsbetriebes auf die Vogelfauna: Ein Fallbeispiel aus Süddeutschland. Vogelwelt 119, 7–19.

Mayrhofer, P., 1997. In: Diepenbrock, W., et al. (Eds.), Das Ökopunktemodell Niederösterreich—Aufbau und Umsetzung in der Agrarumweltpolitik. pp. 197–208.

MLR (Ministerium Ländlicher Raum, Baden-Württemberg), 1999a. MEKA II—1.Stufe, Informationen zum Programm. Informationsblatt Ministerium Ländlicher Raum, Stuttgart, Drucknummer MLR-60-99, 2 pp.

MLR (Ministerium Ländlicher Raum, Baden-Württemberg), 1999b. Artenreiches Grünland—Anleitung zur Einstufung von Flächen für die Förderung im MEKA II. Informationsblatt Ministerium Ländlicher Raum, Stuttgart, Drucknummer MLR-59-99, 2 pp.

NÖA (Niederösterreichische Agrarbezirksbehörde, Niederösterreichischer Landschaftsfonds) (Eds.), 1996. Regionalprogramm Ökopunkte Niederösterreich. Informationsheft und Erläuterungen, Wien.

OECD, 1999a. Environmental indicators for agriculture. In: Proceedings of The York Workshop on Issues and Design, vol. 2, Paris, 213 pp.

OECD, 1999b. Landwirtschaft und Umwelt- Problematik und strategische Ansätze. Paris, 40 pp.

OECD, 2001. Environmental Indicators for Agriculture: Methods and Results, vol. 3. Paris, 409 pp.

Oppermann, R., 1993. Nahrungspotentiale einer Landschaft für Wiesenbrüter und Konsequenzen für die Grünland-Extensivierung. Verh. Ges. Ökologie 22, 221–227.

Oppermann, R., 2001. Naturschutz mit der Landwirtschaft—Ökologischer Betriebsspiegel und Naturbilanz: Wie naturfreundlich ist mein Betrieb? Stuttgart/Singen, 56 pp.

Oppermann, R., Briemle, G., 2002. Blumenwiesen in der landwirtschaftlichen Förderung. Naturschutz und Landschaftsplanung 34, 203–209.

Oppermann, R., Pfadenhauer, J., Reichholf, J., 1987. Beziehungen zwischen Vegetation und Fauna in Feuchtwiesen. Veröff. Naturschutz Landschaftspflege Bad.-Württ. 62, 347–379.

Oppermann, R., Liczner, Y., Claßen, A., 1997. Auswirkungen von Landmaschinen auf Amphibien und Handlungsempfehlungen für Naturschutz und Landwirtschaft. Institut für Landschaftsökologie und Naturschutz (ILN) -Werkstattreihe, Heft 4. Singen, 119 pp.

Oppermann, R., Nürnberger, M., Kunz, S., 2000. Kriterien zur Messung ökologischer Leistungen in der Landwirtschaft. In: Bericht über die Internationale Tagung "Beiträge zur Entwicklung ökologisch und ökonomisch nachhaltiger Landnutzungssysteme in Agrarlandschaften", Schriftenreihe des Dachverband Agrarwissenschaftliche Forschung (DAF). Band 31, 31–43.

Raskin, R., 1994. Die Wirkung pflanzenschutzmittelfreier Ackerrandstreifen auf die Entomofauna von Wintergetreidefeldern und angrenzenden Saumbiotopen. Diss. Univ. Aachen, 145 pp.

Solagro, 1999. Umweltbewertungsverfahren für die Landwirtschaft. Veröffentlichung im Auftrag der EU, DG XI, ins Deutsche übersetzt von C. Dimkic. Toulouse, 189 pp.

SRU Sachverständigenrat für Umweltfragen, 1996. Sondergutachten Konzepte einer dauerhaft-umweltgerechten Nutzung ländlicher Räume. Drucksache 13/4109, Deutscher Bundestag, Bonn.

Von Münchhausen, H., Nieberg, H., 1997. In: Diepenbrock, W., et al. (Eds.), Agrar-Umweltindikatoren: Grundlagen, Verwendungsmöglichkeiten und Ergebnisse einer Expertenbefragung. pp. 13–30.

Waldhardt, R., Simmering, D., Otte, A., 2000. Standortspezifische Surrogate und Korrelate der α-Artendichten in der Grünland-Vegetation einer peripheren Kulturlandschaft Hessens. Berichte der ANL 24, 79–86.

Wetterich, F., Haas, G., 1999. Ökobilanz Allgäuer Grünlandbetriebe. 1.Aufl., Berlin, 87 pp.

PART V
ECONOMY

Chapter 7 Economy
E. Osinski

Synoptic Introduction

Available online at www.sciencedirect.com

Agriculture, Ecosystems and Environment 98 (2003) 477–482

Agriculture Ecosystems & Environment

www.elsevier.com/locate/agee

Economic perspectives of using indicators

Elisabeth Osinski*, Jochen Kantelhardt, Alois Heissenhuber

Agricultural Economics, Center of Life and Food Sciences, Technical University of Munich, Alte Akademie 14, 85350 Freising, Germany

Abstract

This article deals with the interrelation of ecological and economic aspects through the use of indicators. Two aspects are painted out: the first being the measurement of the economic value of ecological services, and the second being the integration of economic aspects into comprehensive indicator systems.

The measurement of the economic value provides a means of determining consumer-orientated values of landscapes and bio-diversity issues. This measurement, however, is viewed critically by some, especially in nature conservancy circles. For them the intrinsic values of a landscape, which cannot be measured in economic terms, play the central role in the assessment of landscapes [Oekonomische Bewertungsgrundlagen und die Grenzen einer 'Monetarisierung' der Natur. In: Theobald, W. (Ed.), Integrative Umweltbewertung. Springer, Berlin, Heidelberg, pp.95–117]. On the other hand, Braeuers [Money as an indicator: to make use of economic evaluation for biodiversity conservation. In: Buechs, W. (Ed.), Biotic Indicators for Biodiversity and Sustainable Agriculture. Agric. Ecosys. Environ.], whose article presents an overview of current economic assessment methods, provides an alternate point of view. Braeuer [Money as an indicator: to make use of economic evaluation for biodiversity conservation. In: Buechs, W. (Ed.), Biotic Indicators for Biodiversity and Sustainable Agriculture. Agric. Ecosyst. Environ.] argues that through the inclusion of monetary values into the assessment process the whole issue can remain on a factual level. The article by Gerowitt et al. [Rewards for ecological goods—requirements and perspectives for agricultural land use. In: Buechs, W. (Ed.), Biotic Indicators for Biodiversity and Sustainable Agriculture. Agric. Ecosyst. Environ.] even goes further and suggests that a market oriented system be implemented for environmental goods.

In contrast to the subjective measurement of economic value, land use models are based on "objective" expertise. Aim of the articles in this sector is to analyse the enhancement of scientific models by economic aspects. These models would then be used to verify the feasibility of ecological motivated land use claims. The results would form the basis for political decision making by a society. The model of Meyer-Aurich and Zander [Consideration of biotic environmental quality targets in agricultural land use—a case study from the biosphere reserve Schorfheide-Chorin. In: Buechs, W. (Ed.), Biotic Indicators for Biodiversity and Sustainable Agriculture. Agric. Ecosyst. Environ.] reviews land use decision making on the operational level, while the model of Herrmann et al. [Threshold values for nature protection areas as indicators for bio-diversity—a regional evaluation of economic and ecological consequences. In: Buechs, W. (Ed.), Biotic Indicators for Biodiversity and Sustainable Agriculture. Agric. Ecosyst. Environ.] reviews decision making at a regional level. The concluding article by Kantelhardt et al. [Is there a reliable correlation between hedgerow density and agricultural site conditions? In: Buechs, W. (Ed.), Biotic Indicators for Biodiversity and Sustainable Agriculture. Agric. Ecosyst. Environ.] attempts to correlate selected different economic and ecological indicators. The key issue here being how farmers can be compensated for the conservation and preservation of landscapes.

Keywords: Economic value; Ecological value; Economic modelling; Landscape model; Ecological indicator; Economic indicator

* Corresponding author. Tel.: +49-8161-71-4461; fax: +49-8161-71-4426.
E-mail address: osinski@wzw.tum.de (E. Osinski).

0167-8809/$ – see front matter © 2003 Elsevier Science B.V. All rights reserved.
doi:10.1016/S0167-8809(03)00106-3

1. Introduction

Central topic in the following articles is the development of indicators that would allow an nature conservancy assessment of land use practices and landscapes. Many of the generally discussed indicators, such as diversity, naturalness and rareness of habitats and species, stem heavily from ecology. Also included are the biotope areas and their endangerment by people (Usher, 1994). Ecology investigates the relationship between life forms among themselves and with their inanimate surroundings (Haeckel, 1866). Accordingly, that is why ecology provides an important scientific basis for land use assessment.

Eco-centric procedure is at the foreground of an ecological oriented landscape assessment. In the case of biotopes and species, this would mean that they would be protected for their own sake. However, this stands in contrast with the fact that human use of natural landscapes often clashes with conservation of these same landscapes. Therefore, the development of sustainable land use cannot rely only on ecological insight, but must also incorporate anthropogenic objectives and the resulting economic indicators.

The following discussed articles can be divided into two overall categories. The first discusses, which methods can be used to determine land use values. Additionally, the methods of making landscapes and their elements marketable will be analysed. The second category concentrates on the means of integrating ecological and economic criteria into a common indicator system, and the ways of depicting the respective interactions. The key issue, however, is the development of indicator systems. Only through their use is it possible to portrait the complex relationships and system properties occurring at the landscape level. Indicators also allow the deduction of necessary assessment approaches needed for the development of sustainable land use.

2. Economic assessment of landscapes and their elements

The economic assessment of landscapes is defined through the assumptions of the neo-classical economy. Individuals as consumers influence the allocation of resources through their purchase decisions (see Wronka, 1998). Furthermore, economic theory assumes that through these means a societal optimum can be achieved. This of course depends on supply and demand and the assumption that all individuals handle rationally. For landscapes this can be described by the following example: a tourist that favours a certain natural scenery and is willing to pay for its development is in fact part of the development process. The economic value of a landscape is thus not determined by scientists or other experts but by the respective users.

The anthropocentric point of view determines the worth of landscapes through use and non-use values (see Turner et al., 1999; Wronka, 1998). The use value can either be a direct or indirect utility value. A direct utility value, for example, would be the profit made from agricultural or forestal use of a landscape. An indirect value would be where no financial gain is made, like the recreational value of a leisure walk through a forest.

Non-use values include the future utilisation of a landscape, referred to as the options, bequest, and existence values. The options value refers to the value for the current generation, while the bequest value deals with the use for future generations. The existence value is defined as the value assigned to in this case landscapes by individuals who will never make use of them. Here the economic approach meets the eco-centric approach, which emphasises the intrinsic value of landscapes. The significance of the intrinsic value currently has can be seen in the fact that the UN cites it in the preamble of its Convention on Biological Diversity (CBD, 1992).

Economic assessments, however, tend to value the actual use higher than that of future generations. This stands in contrast to nature conservancy assessments which aim to preserve future utilisation of landscapes. Of importance in this case are the non-use values of the landscapes such as option-use, bequest and intrinsic values (see Plachter, 1992).

However, the actual value of a landscape is difficult to determine at best. Landscapes and their elements are considered public goods, for which no competition exists and from whose usage nobody can be excluded. Henceforth landscapes are not marketable, a natural scenery is available for the consumption by any person (Kahn, 1995). This stands in contrast to private goods, such as consumer goods, which are marketable and therefore have a certain value. There are, however, certain methods for determining the value of public goods. Braeuer (2003) categorises these methods into direct and indirect assessment methods. The indirect methods try to estimate the value individuals place on landscapes by analysing their actions on the free market. For example, the travel costs a tourist incurs for a visit to a national park could be used as an indicator for the value of this park. Similarly, the value of an avalanche safe slope could be estimated by the cost of avalanche prevention measures should the slope be shaped differently. This is referred to as an avoidance cost approach.

The direct assessment method determines the economic worth of a landscape by surveying all the concerned as to how much they would be willing to pay for its usage. The most significant direct assessment method is the so called "contingent valuation method" (CVM, see Turner et al., 1999). In contrast to the direct methods it covers with the existence value at least a part of the intrinsic value of landscapes. The CVM requires an accurate hypothetical market for the respective environmental goods. The value of environmental goods is ascertained by increasing the cost of a good until the surveyed persons has reached their limit. The CVM was developed as a response to the environmental damage caused by the Exxon Valdez oil spill. For the first time, compensation payments would not only include use values, like decreased tourism income or cleaning costs, but also the existence value of the landscape itself (Braeuer, 2003).

Similarly, the article by Gerowitt et al. (2003) addresses the problem of landscape based services being considered public goods and thus not being compensated. To address this deficit, government compensates farmers through an agricultural environmental program. The amount of compensation is currently determined by the expenditures needed by the farmer in providing these environmental goods (COM, 1998). The article, however, suggests that the amount of compensation should be the actual value of the environmental good. To calculate this value, one has to first determine the current demand for this good. Farmers would be compensated for their results and not for their actions (see Ahrens, 1992). However, this approach can only be used for goods where the specific producer is identifiable. This is actually the case in the area of bio-diversity, especially concerning the variety of plants on agricultural plots. The whole system would have to be regionally organised in order to account for regional differences.

The above mentioned approach has been partly implemented in the State of Baden-Wuerttemberg. There farmers receive compensation from an agricultural environmental program based on the number of grassland species found on their fields (Oppermann, 2003). It remains to be seen if the amount of compensation is similar to the payment willingness of in this case the government.

3. Consideration of economic aspects in land use modelling

The articles by Braeuer (2003) and Gerowitt et al. (2003) assigned monetary values to environmental goods. The following articles try to combine economic and ecological factors into models for simulating land use. These models would show the consequences of agricultural actions on environmental goods or vice versa the consequences of ecological restrictions on agriculture.

Land-use models can also be understood as interacting indicator systems. They portrait the most important relationships and explain complex systems such as a cultural heritage landscape. Experts select the respective indicators based on scientific knowledge. Aim of these systems is to objectively present the pros and cons of different actions. Indicator systems provide a society with a basis for decision making, but do not actual value different actions. This is done by the society in balancing out the advantages and disadvantages. This stands in contrast to the consumer-orientated economic approach, where the actual assessment is already included (compare Pannell and Schilizzi, 1999).

Land-use models work either at the operational level (see Meyer-Aurich and Zander, 2003) or at the regional level (see Herrmann et al., 2003). The models are usually programmed linearly or quadratically, allowing for an optimum solution to be determined under a given framework for a planning problem. For this, all problem relevant information has to be structured into the framework, that would show all possible combinations with regard to restrictions and decision criteria (Pannell, 1997). The result function is generally the optimisation of profit or minimisation of costs for the landscape user. Often implemented environmental variables replace the restrictions.

Besides, economic control factors (Meyer-Aurich and Zander, 2003) also include biotic factors in their calculations. These factors, that influence type and arrangement of land use forms, are indices, which portrait the "cropping practices" with regard to biotic preservation objectives. These objectives include the "protection of the amphibious population in agricultural areas" as well as the "protection of partridge populations in agricultural areas". For this purpose experts determined the most influential factors for the population of these species. An endangerment factor, composed of endangerment by cultivation and by usage, results for both groups. These factor is then consolidated into an index that is used as an indicator for dangers posed to the species by land use. The accomplished evaluations are then balanced out after being conveyed on to the agricultural areas. However, this balancing is only a static contemplation. If these indices are included in an operational model, then it is possible to undertake assessments concerning the implementation of biotic preservation objectives on farm level. This allows a separate determination of the ramifications on the contribution margin of the farm for the preservation objectives of the partridge and amphibian populations. The model also allows the determination of the interactions to be implemented in the same operation on farm.

A similar approach on a regional level is introduced by Herrmann et al. (2003). The focal point of this approach is not the demand for living space for species, but the implementation of political reference values on the amount of natural green areas. These values, depending on the region, vary from 5 to 30%. An intensively cultivated agricultural area, for example, has a reference value of 10%. With the help of regulations, potential areas for conservation are portrayed. A geographic information system (GIS), linked to the model system (compromising of an economic regional, erosion and biotope development model), is used for the localisation of these areas. This allows a representation of operational, nature and resource conservancy consequences as a result of area transformation. A reduction in operational areas would lead to change in the contribution margin. This would lead to a change in the cultivation spectrum, that would in turn lead to relative increase in erosion fostering cultures in the region. This would then result in an increase of nutrient load on nutrient-sensitive biotopes such as hedges or balks.

Both of the contributions by Meyer-Aurich and Zander (2003) and Herrmann et al. (2003) show a potential combination of ecological relevant factors with economic models. This would allow a cost assessment of ecological dictated measures. The approach by Kantelhardt et al. (2003) goes even further in trying to provide a correlation between ecological and economic factors. This would greatly simplify changes in land use with respect to ecological objectives. The approach also considers a much larger regional level than Herrmann et al. (2003). It was looked for a statistically secure correlation between the existing hedges in landscapes and agricultural suitability on the level of a German federal state. Because of agricultural heterogeneity it is not derivable on the large-scale, but the correlation can be used to verify individual natural areas. The correlation showed that there cannot be a general economic assessment for the preservation of hedges through the furtherance of general measures. Hedges occur with varying densities on sites with different land use. The reasons for their preservation include not only land-use factors but also socio-economic and cultural historic factors. These have to be included in the guidelines for the combination of economic and ecological aspects.

4. Conclusion and perspectives

The following articles show that the contributions of economy to sustainable land use can be divided into assessment of landscapes and their elements and enhancement of land use models. It is also shown that the applicability of the contributions to the two categories differentiate greatly. Thus the consequences for future perspectives have to be made accordingly.

The articles by Braeuer (2003) and Gerowitt et al. (2003), dealing with measurement of value and compensation of environmental services, dealt mostly on a methodical level. Future research should therefore concentrate on the actual implementation of the introduced methods. There already exists substantial experience in the field of economic assessment; Carson et al. (1997) lists around 2000 papers on the subject. It must also be mentioned that the "contingent valuation method" is very work intensive. The actual results and findings for a region are generally not transferable to other regions, even with the Benefit Transfer method (Wronka and Thiele, 2002). The suggested compensation of environmental goods approach by Gerowitt et al. (2003) is hardly in use in Europe, whereas in the United States there already exists a proven approach in the Conservation Reserve Program (CRP). Mello et al. (2002) show that this approach can also be implemented in Germany.

The articles dealing with the second category are on a practical level and show that the development of economic ecological indicator systems is already very advanced. Meyer-Aurich and Zander (2003) show on an operational level how cultivation measures can be ecologically assessed and optimised on an economic and species preservation point of view. Herrmann et al. (2003) demonstrate the consequences of implementation of an conservation objective (10% natural area), by showing operational consequences. Through the referencing the actual area, they also provide suggestions for sustainable regional land use planning. These include integrating restrictions, such as ecological indicators, into operational models to test for reactions. Additionally, the effects on the area by the changing erosion factor is also presented. Finally, Kantelhardt et al. (2003) show that through statistical analysis of economic and ecological indicators means for improving agricultural environmental programs are provided.

Research in this category should concentrate on obtaining more end result orientated data as well as on the transferability of this data and of methods. Currently, indicators have to rely on massive scientific data surveys, which hinders there use in everyday life. Kantelhardt et al. (2003) also show the existing data sets are insufficient to show relationships between economic and ecological insights. More research has to be done on the development of indicators by use of different data bases on different spatial resolutions. Because of limited public resources it must first be proven which data set is of importance for the preservation of landscapes and bio-diversity. On the other hand more emphasis has to be put on the statistical and geometric implications of deriving indicators. Studies relying on the same data base can result in different indicators on different spatial levels.

Summing up, it can be shown that economics play an important role in the development of sustainable land use. Economics, within the framework of a land use model, can show the cost of ecological-orientated measures as well as the optimal use of limited resources. Additionally, economics provide an accurate picture of the amount a society is willing to pay for landscape preservation. Furthermore, economic assessment can also show the benefits for society of ecological measures. Thus economics does not represent a contradiction for assessment on nature conservancy, but in fact provides important indicators for sustainable land use. One can only hope that, economic indicators will play an even greater role in the conservation of nature.

References

Ahrens, H., 1992. Gesellschaftliche Aspekte der Honorierung von Umweltleistungen in der Landwirtschaft. In: Bayerisches Staatsministerium fuer Landesentwicklung und Umweltfragen (Ed.), Untersuchung zur Definition und Quantifizierung von landschaftspflegerischen Leistungen der Landwirtschaft nach oekologischen und oekonomischen Kriterien, Materialien, vol. 84, pp. 117–150.

Braeuer, 2003. Money as an indicator: to make use of economic evaluation for biodiversity conservation. In: Buechs, W. (Ed.), Biotic Indicators for Biodiversity and Sustainable Agriculture. Agric. Ecosyst. Environ. 98, 483–491.

Carson, R.T., Haneman, W.M., Kopp, R.J., Krosnick, J.A., Mitchell, R.C., Presser, S., Ruud, P.A., Smith, V.K., Conaway, M., Martin, K., 1997. Temporal reliability of estimates from contingent valuation. Land Econ. 73 (2), 151–163.

CBD (Convention on biological diversity), 1992. http://www.biodiv.org/doc/legal/cbd-en.pdf.

COM (Commission of the European Communities), 1998. State of Application of Regulation (EEC) No. 2078/92: Evaluation of Agri-Environment Programmes. Commission Working Document, VI/7655/98.

Gerowitt, B., Isselstein, J., Marggraf, R., 2003. Rewards for ecological goods—requirements and perspectives for agricultural land use. In: Buechs, W. (Ed.), Biotic Indicators for Biodiversity and Sustainable Agriculture. Agric. Ecosyst. Environ. 98, 541–547.

Haeckel, E., 1866. Generelle Morphologie der Organismen. 2 Bde. Berlin.

Hampicke, U., 1998. Oekonomische Bewertungsgrundlagen und die Grenzen einer 'Monetarisierung' der Natur. In: Theobald, W. (Ed.), Integrative Umweltbewertung. Springer, Berlin, Heidelberg, pp. 95–117.

Herrmann, St., Dabbert, S., Schwarz-von-Raumer, H.-G., 2003. Threshold values for nature protection areas as indicators for bio-diversity—a regional evaluation of economic and ecological consequences. In: Buechs, W. (Ed.), Biotic Indicators for Biodiversity and Sustainable Agriculture. Agric. Ecosyst. Environ. 98, 493–506.

Kahn, J.R., 1995. The economic approach to environmental and natural resources. Dryden Press, Fort Worth.

Kantelhardt, J., Osinski, E., Heissenhuber A., 2003. Is there a reliable correlation between hedgerow density and agricultural site conditions? In: Buechs, W. (Ed.), Biotic Indicators for Biodiversity and Sustainable Agriculture. Agric. Ecosyst. Environ. 98, 517–527.

Mello, I., Heissenhuber, A., Kantelhardt, J., 2002. Das conservation reserve program der USA—eine moeglichkeit zur effizienten entlohnung von umweltleistungen der landwirtschaft? Berichte ueber Landwirtschaft 80 (1), 85–93.

Meyer-Aurich, A., Zander, P., 2003. Consideration of biotic environmental quality targets in agricultural land use—a case study from the biosphere reserve Schorfheide-Chorin. In: Buechs, W. (Ed.), Biotic Indicators for Biodiversity and Sustainable Agriculture. Agric. Ecosyst. Environ. 98, 529–539.

Oppermann, R., 2003. Nature balance scheme for farms—evaluation of the ecological situation. In: Buechs, W. (Ed.), Biotic Indicators for Biodiversity and Sustainable Agriculture. Agric. Ecosyst. Environ. 98, 463–475.

Pannell, D.J., 1997. Introduction to Practical Linear Programming. Wiley, New York.

Pannell, D.J., Schilizzi, S., 1999. Sustainable agriculture: a matter of ecology, equity, economic efficiency or expedience. J. Sustain. Agric. 13 (4), 57–66.

Plachter, H., 1992. Grundzuege der naturschutzfachlichen Bewertung, Sonderdruck aus: Veroeffentlichungen f. Naturschutz u. Landschaftspflege in Baden-Wuerttemberg, Bd. 67, Karlsruhe.

Turner, R.K., Pearce, D., Bateman, I., 1999. Environmental economics: an elementary introduction. Harvester Wheatsheaf, New York.

Usher, M.B., 1994. Erfassen und Bewerten von Lebensraeumen: Merkmale, Kriterien, Werte. In: Usher, M.B., Erz, W. (Eds.), Erfassen und Bewerten im Naturschutz. Quelle und Meyer, Heidelberg, pp. 17–47.

Wronka, T.C., 1998. Was ist der Preis fuer Umwelt? Moeglichkeiten und Grenzen des kontingenten Bewertungsansatzes. Agribusiness Forschung No. 6. Inst. fuer Agribusiness, Leipzig, 85 p.

Wronka, T.C., Thiele, H., 2002. Transfer von Umweltgueterbewertungen: Moeglichkeiten, Grenzen und empirische Evidenz. Schriften der Gesellschaft fuer Wirtschafts- und Sozialwissenschaften des Landbaues e.V. 37, 287–293.

PART V
ECONOMY

Chapter 7 Economy
E. Osinski

Original Papers

ELSEVIER

Available online at www.sciencedirect.com

Agriculture, Ecosystems and Environment 98 (2003) 483–491

Agriculture Ecosystems & Environment

www.elsevier.com/locate/agee

Money as an indicator: to make use of economic evaluation for biodiversity conservation

Ingo Bräuer*

Institut für Agrarökonomie, Universität Göttingen, Platz der Göttinger Sieben 5, 37073 Göttingen, Germany

Abstract

Environmental economics has developed methods to use a common indicator (money) for environmental policy decisions. This indicator allows cost-benefit analyses (CBAs) for an objective and realistic evaluation of the economic consequences of different development options. The latter is necessary to avoid decisions in disfavour of nature. This paper gives a broad overview about the advantages and disadvantages of this kind of policy evaluation for non-economists.

Economic valuation of natural resources has not been developed for its own sake. It was elicited by political authorities seeking methods to incorporate non-marketable goods in CBA, as well as by the American administration of justice. In Europe claims for environmental damages are less common, and economic evaluation should take place at the level of projects. Most biodiversity conservation projects are suitable for economic valuation. For political reasons they are designed in a way which suits the requirements for an economic evaluation. Economic evaluation cannot only act as a decision aid. In addition, the estimation of benefits of conservation programs, which otherwise might have been neglected, helps to promote acceptance for these programs. Furthermore economic evaluation is necessary for the development of any kind of green accounting or in order to calculate the incremental costs for biodiversity conservation enshrined in the Convention of Biological Diversity. CBAs based on environmental valuation methods are subjected to controversial discussions. Especially methodological (biases in the estimation of values) and ethical (lexicographic preferences) aspects are criticised. By using CBA these problems should always kept in mind. As a response to these critics recommendations for the application of the contingent valuation method (CVM) have been worked out (see NOAA-Panel). To deal with risk and uncertainty as well as with the irreversibly of biodiversity loss, concepts such as save minimum standard (SMS) and the 'precautionary principle' have been developed. Furthermore, there are legal limits for CBA dealing with natural resources. But without economic valuation of the environment, policy decisions are supported which contradict economic rationality, which has been shown in various studies.

Keywords: Cost-benefit analysis; Environmental valuation methods; Contingent valuation; Replacement cost method

1. Introduction

This paper discusses the usefulness of money as an indicator for rational decision making on biodiversity conservation. It highlights the importance of economic valuation and indicates ways for their execution. Therefore the concept of the total economic value (TEV) is presented and two methods to reveal different welfare components of biodiversity are described. Apart from the theoretical background, the requirements for applying them to the evaluation of a conservation program are worked out.

"Valuation can be seen as a method of determining the relative importance of environmental consequences of economic activities. It helps political

* Tel.: +49-551-39-4805; fax: +49-551-39-4812.
E-mail address: ibraeuer@uni-uaao.gwdg.de (I. Bräuer).

0167-8809/$ – see front matter © 2003 Elsevier Science B.V. All rights reserved.
doi:10.1016/S0167-8809(03)00107-5

authorities to make informed decisions about biodiversity conservation" (UNEP, 1995; p. 827). In other words money acts as an indicator to transform environmental problems into policy.

Generally speaking indicators should simplify and allow a ranking of policy options. Most scientific indicators (e.g. for water quality) are applicable exclusively within their scientific context. However, conservation politics needs an indicator which allows cross-sectoral decisions, such as for example nature conservation and land use. The absence of this indicator has lead to the discrepancy that nature plays in decisions only a minor role though everybody assesses nature conservation as being very important. Money could be an indicator with the potential to link these two fields in order to avoid further misallocation. For the consideration of different options of action, scientific knowledge is not enough to make rational decisions as the WBGU (1999) points out. Therefore a dimension of values and norms is necessary. The development of this has been the speciality of economics. Economic values are derived from individual preferences and are based on an anthropocentric ethic. Purpose is an efficient resource allocation. The most common method is the cost-benefit analyses (CBAs).

There are two main areas of application for environmental evaluation. A CBA ex ante acts as decision aid for policy-makers, while an ex post evaluation controls the efficiency of environmental policy.

Schulz and Schulz (1991) have worked out five main reasons why monetary values for environmental goods should be estimated: (1) Comprehensibility: money demonstrates the public the social importance of nature conservation programs in an intelligible and easily understandable way. (2) De-emotionalisation: CBA demands a critical incorporation of all benefits and costs. This systematic and complete listing helps to get a better overview over the range of consequences of projects and promotes a de-emotionalisation in the discussion about its usefulness. (3) Dosificability: only the knowledge of the real costs, which means the costs of the project minus the avoided damage, offers the possibility to make efficient decisions concerning the nature and the amount of environmental protection. (4) Internalisation of external effects needs information about the potential damage to be internalised. (5) Every kind of green accounting needs an economic evaluation of environmental goods. The first two points are particularly relevant for conservation programs, as they help to improve their acceptance. The latter shows the policy relevance of economic evaluation. Its results can be important contributions for a modification and improvement of environmental policy.

2. Political demand of economic evaluation

An important impulse for the development of economic evaluation of natural resources came from politics and public authorities. This call for economic evaluation as a decision aid for environmental policy is not a recent trend. It started in US 60 years ago, when the National Resources Board decided in its Flood Control Act of 1936 that intangible follow-up consequences of their projects have to be considered as well (Marggraf and Stratmann, 2001). Intangible consequences are defined as consequences which cannot be monetarised because they have no effect on marketable goods. In 1950 a report published by the Federal Inter-Agency River Basin Committee and known as the "Green Book" received widespread attention (Hanemann, 1992). The Green Book is a guide for economic evaluation of the effects of river basin projects. It recommends either the use of market prices or, if not possible, the use of alternative methods like accounting the expenditures of a recreationist or his willingness to pay for a further use of the recreation facilities. The conceptual framework for the second kind of evaluation has been developed by Ciriacy-Wantrup (1947). His idea of an estimation of social benefits by questioning the public (referendum) (see paragraph contingent valuation method (CVM)) which was first put into practise by Davis (1963). The main interest lay here on values generated by recreation. While the roots of environmental valuation have been in the water authorities, interest on the method expanded from water recreation into other public goods such as wildlife, air quality or human health. For the following decades the centre of development of environmental evaluation were in the USA. The main reason was the American legislation. In 1981 the National Environmental Policy Act (NEPA) was modified by the Presidential Executive Order (EO) 12 291 which claimed the use of CBA for new regulations. In addition the Comprehensive Environmental Response,

Compensation and Liability Act (1980) brought environmental damage assessment to court (Hanley and Spash, 1993). Guidelines of the US Department of Interior (DOI, 1986) suggested the restoration or opportunity costs as a measure for compensation. A new dimension for this kind of research was the recognition of the importance of non-use values. They had to be considered in damage assessments according to a verdict from 1989 (State of Ohio).

The Exxon Valdez oil spill in 1989 was the first possibility for a large litigation. For the compensation assessment a vast research project was financed dealing with the CVM, whose funds provided by both the governmental as well as from Exxon. As a consequence of this dispute the NOAA-Panel has developed guidelines for the use of CVM in natural resource damage assessments.

In Europe environmental valuation is much less common. There is a smaller number of studies as well as less political influence. In contrast to the USA where policy evaluations are commonplace the EU has just started in the early 1990s to use formal appraisals to asses costs and benefits of EU Directives (Pearce, 1998). One reason is the different legislation which does neither offer the possibility of integrating non-use values into damage assessments nor the requirement of a CBA for new regulations (see EO, 12 291). A second reason is a problem of missing political will (low acceptance of environmental valuation techniques). Though much academic research has been done in the last 20 years (Navrud, 1999 counted 457 studies in the period between 1992 and 1999), hardly any results of environmental economic analyses have found their way into politics (Hackel and Pruckner, 2000). But there is hope for a change. The EU is on the way to take the results of environmental valuation more into consideration. Article 130r in the Maastricht Treaty requires action to take the potential costs and benefits of directives into account (Hackel and Pruckner, 2000). In addition the 1992 Fifth Environmental Action Programme emphasised the role of environmental valuation for a sustainable development and demands a "... development of meaningful cost/benefit analysis methodologies and guidelines in respect to policy measures and actions which impinge on the environment and the natural stock" (European Commission (1992) cited in Pearce, 1998).

3. Economic values as indicators

The term 'value' causes much confusion because it is used in different ways by ecologists and economists. The former are considering values as ethical measures, while the latter consider them as equivalence measures. The second involves a reflection of social costs (Tuner and Pearce, 1993). If we are looking for a decision aid for biodiversity conservation (What should be conserved first and how much?) we have to concentrate on the second concept. A value measured in monetary terms will show the importance of a project in a given society and the society's ability to pay for it. It can be seen as the willingness to commit resources to biodiversity conservation. Prices are in this case indicators for the importance of conservation programs. This information is essential for policy evaluation.

Dealing with biodiversity or other natural resources, economists cannot rely on market prices, though. The products and resources involved are public goods and therefore non-exclusive and non-rival in consumption. In this case market prices are unreliable indicators of social costs, because they may not capture all effects of biodiversity use. These so called external effects give evidence of market failure. Environmental economists have developed methods to reveal values in the presence of market failure which incorporate external effects.

3.1. Values of biodiversity: an economic taxonomy

Even though benefits of biodiversity conservation have not played a role in traditional CBA, there is a huge range of benefits which will be described in this paragraph, using the concept of the TEV (UNEP, 1995) (Fig. 1).

The TEV distinguishes between use and non-use values to show that there is an additional value of biodiversity apart from direct or indirect use. While the use values correspond to the traditional economic concept of benefits, the non-use components are driven by ethical considerations (UNEP, 1995).

Use values are placed on natural goods that are either consumed directly (harvesting, recreation) or used indirectly (ecosystem services). The indirect use value can be derived from ecological functions, like flood control or erosion (Barbier, 1994).

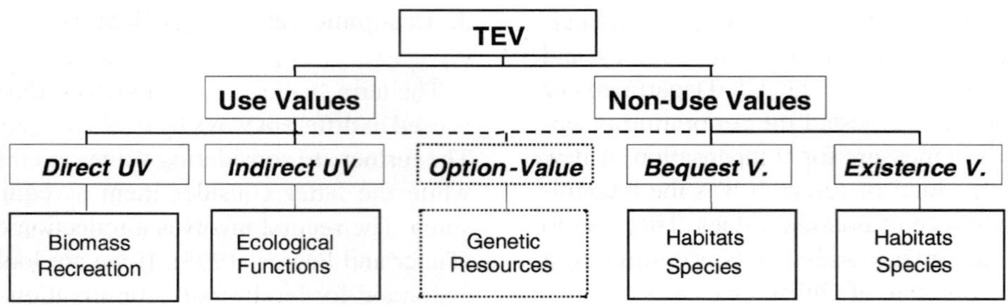

Fig. 1. The total economic value (TEV).

Non-use or passive values refer to the existence of a value even though individuals do not intend to use the resource but feel a 'loss' if it would disappear (UNEP, 1995). They are determined by: (a) the value of keeping a resource intact for one's descendants (bequest value); (b) the fact that people derive benefits from the mere knowledge about the existence and assured survival of a resource like habitats or species (existence value) (Krutilla, 1967).

Between the two groups one can find the option value, the value of the future information or the resource itself will offer. This value will become a use value in the future, e.g. new sorts or drugs derived form genetic information of wild species. For biodiversity conservation the option value is probably the most quoted value in the public discussion. Often biodiversity loss is associated with the loss of genetic information which would be used for breeding or in medical research (Marggraf and Stratmann, 2001).

The decision on which TEV components to take into account depends on what kind of natural resources will be evaluated. For biotic components (chemical/physical parameters) like water or air quality, direct or indirect use values are of central interest. Non-use values may be neglected. But in the case of nature conservation where species or habitats are involved non-use values are known to make up an important part of the economic value (Brouwer and Slangen, 1998). In this case use and non-use components have to be taken into consideration too.

3.2. Evaluation methods

There are two ways economists can choose to estimate the value of non-marketed resources. Either they use surrogate markets to expose the value people implicitly put on the resource or they use simulated markets to force the people to state their preferences explicitly. For each possibility one method will be described.

3.2.1. Indirect methods: the replacement cost method

Surrogate markets can be used for the evaluation of indirect use values. Indirect use values correspond to the ecologist's concept of ecological functions or environmental services (Pearce, 1993). These ecological functions like watershed protection can be valued by the replacement cost method. This technique looks at the costs of replacing a damaged asset to its original state, e.g. a water quality standard (Pearce and Moran, 1994). Therefore market prices are used. The estimated costs are then the lower bound of the true value of the environmental service.

3.2.2. Direct methods: the contingent valuation

There are two reasons why direct methods have to complement indirect ones. First of all if no surrogate can be found for the natural resource to be valuated or secondly if non-use values have to be taken into consideration. As mentioned above non-use values should not be neglected in case of evaluation of biodiversity protection (Brouwer and Slangen, 1998). The CVM is the only method which may elicit these non-use values. Consequently it should play an important role in every evaluation of biodiversity conservation programs. As described above public goods like endangered species have no prices because markets do not exist. The CVM tries to solve this problem by creating a hypothetical market. In the hypothetical market, consumers can express their preferences in monetary terms (Mitchell and Carson, 1989). This means, they reveal their WTP for a given public good, e.g. protecting a specific species in a given location. In this case, the hypothetical market consists of a description of the species and its habitat as well as the program and its consequences (Loomis and White, 1996). The potential consequences of paying or not paying have to be explained. In addition participants have to be informed about the form and frequency of payment (Mitchell and Carson, 1989). For example, for a local conservation program the payment of nature tax has been shown to be the most accepted form (Rommel, 1998; Enneking, 1999). The concept of nature tax is a copy of the health resort tax, which means every visitor of the region has to pay a certain amount of money per day.

The use of CVM is always accompanied with doubts about its viability because various potential biases (strategic, hypothetical, and information bias, as well as embedding) have been worked out (Hausman, 1993). To test the response validity the questionnaire has to contain questions about the social background (recent studies also use environmental attitudes, see Bräuer, 2001) and household income. A carefully designed and worded questionnaire is essential to avoid potential bias. There is a list of recommendations by the NOAA-Panel to minimise these problems (Arrow et al., 1993).

4. How to use economic evaluation?

Which requirements does a biodiversity conservation program have to meet for an economic evaluation? The requirements can be divided into political and methodological aspects.

From a political point of view there is a certain need for priority setting due to the time and money consuming way of environmental valuation:

1. Projects instead of species: to be used as a decision aid for environmental policy economic evaluation is useful only at the level of political decisions. Consequently projects rather than species themselves should be evaluated. In the starting time of CVM species were in the research focus. This makes only sense in countries with claims for compensation like the USA (Enneking, 1999).
2. Concentration on controversial conservation programs: economic evaluation can promote acceptance of conservation programs when their benefits are worked out. Therefore those projects should be evaluated, which are in the public discussion, considered to be expensive due to restrictions to industry or their potential damage level.

   For the realisation of an economic evaluation some methodological conditions are important depending on which evaluation methods will be used:
3. For indirect methods, scientific knowledge must be available about the consequences of the conservation program.
4. If the CVM will be used, the evaluated program should be clear in its aims and consequences and the valued good should be well known and of interest, otherwise people are not willing to take part in the CV.

Most conservation programs fit the mentioned points. This is not a matter of chance. To meet political requirements the programs are designed in a way that they are suitable for a economic evaluation as well. The conservation of single species has changed to more complex tasks like the conservation of biotopes and ecosystems. Of course there is no lack in controversially discussed conservation programs. Acceptance is often low especially by the local people.

For a better marketing most of these programs use flag-ship species or are restricted to a local nature good like a lake, mountain or special landscape (e.g. heather in Lower Saxony or the biosphere reserve Rhön in Hessia). These restrictions on space or species and habitats respectively make the programs easier to understand for the public which is essential for the use of the CVM. While qualitative information about the consequences of programs, which is essential for the CVM is usually available, quantitative information for the use of indirect evaluation methods are rather rare. This is a problem of missing interdisciplinarity in research. While ecologists are interested in what happens, i.e. to demonstrate the links between the different components of the ecosystem, economists need quantitative information for their calculations.

Ecologist often criticise the CVM due to its requirements mentioned in point 4. They ask what will happen to all the less popular but ecologically important species, if they are "un-saleable"? This is not a problem as far as the following reasons are concerned. As mentioned above the aim of every evaluation should be the evaluation of conservation programs rather than single species. In addition there are hardly any programs for small unpopular but ecologically important species. To protect these kinds of species nature conservation policy uses flag-ship species which are also umbrella species. So they provide the necessary acceptance for the general public and guarantee ecosystem conservation.

5. Discussion

Environmental valuation emerged out of CBA, and became a tool for governmental decision making in USA in the beginning of the 20th century (Navrud, 1992). Due to this policy demand for environmental evaluation, different evaluation methods have been developed. Most biodiversity conservation programs should be suitable for economic evaluation. For political reasons they are organised in a way to fulfil the necessary requirements for the different environmental valuation methods. As shown above the CVM can be used in a more general way than any other method. In addition it is the only method to estimate non-use values. The use of RCM, dose-response method, etc. can be less easily applied because a sound database must exist and for many ecosystem services or pollution incidents, the dose-response function is unknown or extremely difficult to work out. But indirect methods have a higher acceptance in the political discussion. The reason for this is that expert estimates or values calculated on the basis of physical parameters have a better confidence than data generated by interviews of the public. Therefore the choice of the right method is highly dependent on the political circumstances in which the results will be used.

Money will and should never replace the different scientific indicators discussed in this volume, because it breaks complex environmental connections down to a single value. It does not offer any physical information about environmental impacts like the other scientific indicators, which is essential for any risk assessment or management plan. But due to its simplicity, money is the only indicator which is compatible (and comparable) with the economic world and offers the possibility to integrate environmental goods to the political decision process. Nevertheless, physical indicators will be an essential prerequisite to CBA (Hanley and Spash, 1993).

In spite of all the pros and cons of the methods of environmental evaluation there are people who reject environmental evaluation in principle on the grounds of lexicographic preferences. (In economic terms, lexicographic preferences mean, that people deny the concept of substitution of nature, because they believe that nature has an intrinsic value, making it more valuable than any other market good.) Of course the value of nature in general is infinite, as humans need functioning ecosystems to survive. These make an evaluation of the world ecosystems, like Costanza et al. (1997) did, very questionable (although it has been quite important for a critical analysis of the value of nature capital). But the antagonists of environmental evaluation have to keep in mind that a lexicographic value for nature would involve that any production process which harms the nature should be forbidden. Consequently a supply of society with market goods would be impossible. On the other hand the renunciation of monetarisation means to attach a zero value to nature, which in turn leads to a misallocation of investment against nature as observed in the past.

The aim of economic evaluation, as described in this article is to investigate the usefulness and efficiency of environmental policy. The yardstick in this case is the maximisation of welfare according to the Haldor–Hicks-criterion. Efficiency from the economic point of view means, that public money is spend according to the preferences of the public (Marggraf and Streb, 1997). Therefore only marginal chances in the supply of natural resources have to be valued and not the natural resources on their own like Costanza et al. (1997) or others did (Marggraf and Birner, 1998). Consequently the results of a CBA helps only to decide whether a project which will change the supply of a natural resource (e.g. an endangered species) is efficient from an economic point of view or not. It says nothing about the value of the endangered species.

With reference to entire range of economic evaluation of natural resources it has to be kept in mind that the social value of natural resources (ecosystem or species) exceeds its marked value for three reasons (UNEP, 1995): (1) Due to a lack of scientific knowledge they will always remain some undiscovered and unvalued ecosystem functions. (2) A functioning ecosystem is more than the sum of its components. (3) Every ecosystem has a 'primary value' because of a range of use and non-use values which are derived from a healthy ecosystem. This primary value is not measurable in conventional economic terms (by consumer preferences).

Another critical point in CBA is incomplete information about the consequences and respectively costs of biodiversity loss due to ecosystem complexity. Considering risk and uncertainty economists have developed the 'precautionary principle' (Haigh, 1993) and the save minimum standard (SMS) approach (Bishop, 1978). The precautionary principle is based on the assumption that the costs of biodiversity loss are uncertain but potentially high and irreversible. This means activities which may particularly harm the environment have to be refrained from, unless the uncertainty issue is resolved. The SMS argues in the same direction. It involves the determination of the minimum desirable level of environmental goods, like wildlife populations or habitat area to maintain ecosystem functions under changing environmental conditions. Decisions based on CBA are only allowed within these environmental standards. These limits are upheld unless the social opportunity costs of doing so are 'unacceptably high'. The definition of unacceptably high has to be decided by policy-makers or through public forums (UNEP, 1995). In addition, sensitivity analysis should be carried out just like for every CBA. This involves to check how much results change with changing assumptions, as well as to determine parameters to which the results are most sensitive. Of particular interest is the point at which the result changes sign, i.e. from a positive to a negative economic value of the project, or vice versa (Schönbäck et al.,1997).

Finally, legislation is fixing the general framework of CBA decisions (e.g. BNatSchG §20ff stipulates the protection of endangered species and habitats), so the extinction of a species cannot be decided, in spite of the 'net benefits' of a particular activity.

Apart from these technical problems in estimating values for non-marketed goods for every CBA, involving environmental goods or not, the following points have to be taken into consideration. The outcome of any CBA is dependent on the distribution of income. Only if this distribution is assumed to be an acceptable basis for decision making, the outcome of a CBA can be used as an acceptable guide for policy (Hanley and Spash, 1993). Decisions according to the results of CBA may result in adverse distributional consequences, because the Kaldor–Hicks-criterion only says that a decision is desirable, if the winners could compensate the losers. But no actual compensation is required to take place.

In contrast to the general opinion, economic evaluation of biodiversity does not mean the sell-out of it. Krutilla (1967) illustrated that not the evaluation of nature itself but incomplete evaluation of nature is responsible for decisions in disfavour of nature conservation. Economic evaluation can act as an indicator to prove whether environmental policy fits the needs of public preferences. Investigations of agricultural wildlife management in The Netherlands or the 'Streuobstanbau' (a traditional fruit production system) in Germany could show that less public money is spend than the public wishes (Zander, 2000; Brouwer and Slangen, 1998). In addition a number of studies have shown that social costs of biodiversity loss are higher than its benefits (Marggraf and Birner, 1998). The DOE comes to the same conclusion when they recapitulate: "Yet environmental costs and benefits have not always been well integrated into government policy assessments, and sometimes they have been forgotten entirely. Proper consideration of these effects will improve the quality of policy making" (DOE, 1991; p. 1).

6. Conclusions

Sound political decisions have to take environmental consequences into account. Economic evaluation provides a useful methodology to integrate environmental goods in the political decision making process. Money is the link that makes environmental goods comparable with the economic world. As an indicator money offers two main advantages: on the one hand it allows the use of environmental sensitive CBAs as a tool for decision making. On the other hand it can promote the acceptance of biodiversity conservation programs, by showing their benefits in an easily understandable way to the public. Most biodiversity conservation programs are suitable for economic evaluation purposes. For political reasons they are designed in a way that fulfils the necessary requirements for the different environmental valuation methods.

But money will never replace traditional physical indicators. It can only help to examine whether environmental policy meets the preferences of the public—which is on its own a very important task—but fails to offer any information about environmental impacts or quality, which is essential for any kind of risk assessment or management plan.

Acknowledgements

Ingo Bräuer is sponsored by the scholarship programme of the German Federal Environmental Foundation (DBU).

References

Arrow, K.J., et al., 1993. Report of the NOAA Panel on Contingent Valuation. 58 Fed. Reg. 4601. Government Printing Office, Washington, DC, January 15.

Barbier, E.B., 1994. Valuing environmental functions: tropical wetlands. Land Econ. 70 (2), 155–173.

Bishop, R.C., 1978. Endangered species and uncertainty: the economics of a safe minimum standard. Am. J. Agric. Econ. 60, 10–18.

Bräuer, I., 2001. Volkswirtschaftlicher Nutzen der Biberwiedereinbürgerung in Hessen: Ein Fallbeispiel zur Anwendung der Kontingenten Bewertungsmethode, Jahrbuch Ökologische Ökonomik: Ökonomische Naturbewertung, Band 2. Metropolis Verlag, Marburg.

Brouwer, R., Slangen, L.H.G., 1998. Contingent valuation of public benefits of agricultural wildlife management: the case of Dutch peat meadow land. Eur. Rev. Agric. Econ. 25, 53–72.

Ciriacy-Wantrup, S.V., 1947. Capital returns from soil conservation practices. J. Farm Econ. 29, 1181–1196.

Costanza, R., d'Arge, R., deGroot, R., Farber, S., Grasso, M., Hannon, B., Limburg, K., Naeem, S., O'Neill, R.V., Paruelo, J., Raskin, R.G., Sutton, P., van den Belt, M., 1997. The value of the world's ecosystem services and natural capital. Nature 387, 253–260.

Davis, R., 1963. The Value of Outdoor Recreation: An Economic Study of the Maine Woods, Diss Harvard University.

Department of the Environment (DOE), 1991. Policy Appraisal and the Environment. HMSO, London.

Enneking, U., 1999. Ökonomische Verfahren im Naturschutz—Der Einsatz der Kontingenten Bewertung im Entscheidungsprozeß. Peter Lang, Frankfurt.

European Commission, 1992. The Fifth Environmental Action Programme: Towards Sustainability. European Commission, Brussels.

Hackel, F., Pruckner, G.J., 2000. Braucht die deutsche umweltpolitik einen Exxon Valdez Tankerunfall? Perspektiven der Wirtschaftspolitik 1 (1), 93–114.

Haigh, N., 1993. The Precautionary Principle in British Environmental Policy. Institute for European Environmental Policy, London.

Hanemann, W.M., 1992. Preface. In: Navrud, S. (Ed.), Pricing the European Environment. Scandinavian University Press, Oslo.

Hanley, N.D., Spash, 1993. Cost-Benefit Analysis and the Environment, Edward Elgar, Hants.

Hausman, J., 1993. Contingent Valuation—A Critical Assessment. North-Holland, Amsterdam.

Krutilla, J.V., 1967. Conservation reconsidered. Am. Econ. Rev. 57, 777–786.

Loomis, J.B., White, D.S., 1996. Economic benefits of rare and endangered species: summary and meta-analysis. Ecol. Econ. 18 (3), 197–206.

Marggraf, R., Birner, R., 1998. The Conservation of Biological Diversity from an Economic Point of View. Diskussionsbeitrag 9801, Institute of Agricultural Economics, University of Göttingen.

Marggraf, R., Streb, S., 1997. Ökonomische Bewertung der natürlichen Umwelt: Theorie, politische Bedeutung, ethische Diskussion. Spektrum, Heidelberg, Berlin.

Marggraf, R., Stratmann, U., 2001. Ökonomische Aspekte der Biodiversitätsbewertung. In: Janich, P., et al. (Hrsg.), Biodiversität—wissenschaftliche Grundlagen und gesellschaftliche Relevanz. Springer, Berlin.

Mitchell, R.C., Carson, R.T., 1989. Using Surveys to Value Public Goods: The Contingent Valuation Method. Resources for the Future, Washington, DC.

Navrud, S. (Ed.), 1992. Pricing the European Environment. Scandinavian University Press, Oslo.

Navrud, S., 1999. Report to EC-DGXI: Pilot Project to Assess Environmental Valuation Reference Inventory (EVRI) and to Expand its Coverage to the EU. Part II. List of European Valuation Studies, 1999.

Pearce, D.W., 1993. Cost-Benefit Analysis, 2nd ed., Macmillan, London.

Pearce, D.W., 1998. Environmental appraisal and environmental policy in the European union. Environ. Res. Econ. 11, 489–501.

Pearce, D., Moran, D., 1994. The Economic Value of Biodiversity. Earthscan, London, UK.

Rommel, K., 1998. Methodik umweltökonomischer Bewertungsverfahren: Kosten und Nutzen der Erhaltung des Biosphärenreservates Schorfheide-Chorin. Transfer Verlag, Regensburg.

Schönbäck, W., Kosz, M., Madreiter, T., 1997. Nationalpark Donau-Auen: Kosten–Nutzen-Analyse. Springer, Wien/New York.

Schulz, W., Schulz, E., 1991. Zur umweltpolitischen Relevanz von Nutzen–Kosten-Analysen in der Bundesrepublik Deutschland. Zeitschrift für Umweltpolitik und Umweltrecht 3, 299–337.

Tuner, R.K., Pearce, D.W., 1993. Sustainable economic development: economic and ethical principles. In: Barbier, E.B. (Ed.), Economics and Ecology. Chapman & Hall, London, pp. 177–194.

United Nations Environment Programme (UNEP), 1995. Global Biodiversity Assessment. Cambridge University Press, Cambridge.

WBGU—Wissenschaftlicher Beirat der Bundesregierung Globale Umweltveränderungen, 1999. Welt im Wandel: Erhaltung und nachhaltige Nutzung der Biosphäre. Jahresgutachten 1999. Springer, Berlin.

Zander, K., 2000. Zahlungsbereitschaft der Bevölkerung für den Erhalt des Streuobstbaus. BDGL-Schriftenreihe 18, 22.

ELSEVIER

Available online at www.sciencedirect.com

Agriculture, Ecosystems and Environment 98 (2003) 493–506

Agriculture Ecosystems & Environment

www.elsevier.com/locate/agee

Threshold values for nature protection areas as indicators for bio-diversity—a regional evaluation of economic and ecological consequences

Sylvia Herrmann[a,*], Stefan Dabbert[a], Hans-Georg Schwarz-von Raumer[b]

[a] *Institute of Agricultural Economics, University of Hohenheim, Schloss Osthof, Stuttgart 70593, Germany*

[b] *Institute of Landscape Planning and Ecology, University of Stuttgart, Breitscheidstrasse 2, Stuttgart D-70174, Germany*

Abstract

Threshold values for nature protection areas are used as indicators for their nature protection value. Several values related to a minimum share of the area in a region or a country have been proposed as basic conditions for a sound development of natural areas. Although the height of these thresholds is still a matter of extended scientific discussion (they range from 5 to 30% of the agricultural area), they have been embraced by political parties and legislation concerning nature protection. This paper highlights the consequences of the practical application of such thresholds. The example of an intensively used agricultural region, the Kraichgau in Baden-Wuerttemberg, is chosen. To improve the actual situation concerning bio-diversity, a selection of sites with a good potential for the development of oligothropic dry grassland was proposed. To evaluate the consequences of this proposal a GIS based modelling tool was used. By the connection of biological, economic and a-biotic models it was possible to describe the reactions of all these factors simultaneously in form of a scenario. The scenario describes an increase of dry grassland area of approximately 10%. Economically, the reduction of acreage leads to a change of crop ratio and a decrease of gross margin. Due to their profitability, row crops remain with their original amount within the rotation of crops and lead, therefore, to intensification within the region. The accompanied increase of erosion rates results in a higher nutrient impact to adjacent bio-topes (hedges, ridges), thus leading to their further deterioration. Consequently, the gain of bio-diversity in some parts of the region is coupled with a loss of bio-diversity in others. Analysing these results, the usefulness of using an integrated economic–ecological modelling system for the evaluation is shown. Later, discussion of the consequences for the economic as well as for the biotic situation provides an idea of the limitations of the thresholds as indicators with respect to their practical performance. Recommendations for the necessary differentiations pertaining to regional context are given to improve the validity of this indicator type.

Keywords: Thresholds; Nature protection area; Bio-tope development; Agricultural area; Modelling system; Scenario; Regional level

1. Introduction

The extent of areas designated for nature conservation purposes is often used as an indicator of the nature protection value of a landscape. Several thresholds have been proposed as the minimum proportion of a region required for the sound development of natural areas. Starting in the 1970s (Haber, 1972; Heydemann, 1981), a number of authors presented different thresholds, with a peak occurring in the 1980s (e.g. Heydemann, 1981; Kaule, 1988; for further references see Horlitz, 1994) culminating with those most

* Corresponding author. Tel.: +49-711-459-2543; fax: +49-711-459-3555.
E-mail address: sh@uni-hohenheim.de (S. Herrmann).

0167-8809/$ – see front matter © 2003 Elsevier Science B.V. All rights reserved.
doi:10.1016/S0167-8809(03)00108-7

recently published by the Scientific Advisory Board of the Federal Government for Global Change (WBGU, 2000) and the amendment of the Law of Nature Protection in Germany. Although the height of these thresholds is still a matter of extended scientific discussion (they range from 5 to 30% of the agricultural area), they have been embraced by political parties and legislation concerning nature protection. The main reasons for this acceptance seem to be the high identification values of these environmental quality targets, their simple control, the opportunity they offer for comparison between different regions through these values and, finally, they are tangible. Nevertheless, several problems might originate from such very general thresholds. First of all, no strict scientific reasons exist for one of the available thresholds. Additionally, no analysis of the effects of these thresholds with regard to economic or social factors for specific regions will be undertaken before presentation. Finally, ecological conflicts could be caused by such very general thresholds, not differentiating the bio-tope types, their relationships and the necessary measures to be performed (Hampicke et al., 1991). These problems could be found especially on the regional level, whereas general space related thresholds have no meaning on local level and are sufficiently explicit for the state level.

This paper aims to show how an impact analysis of such thresholds could be undertaken through use of a regional modelling approach which includes a description of ecological as well as economic consequences. The usefulness of general thresholds should be learned from it. Later, the paper discusses whether the results could provide information concerning the parameters for additional explanations for the practical application of a threshold.

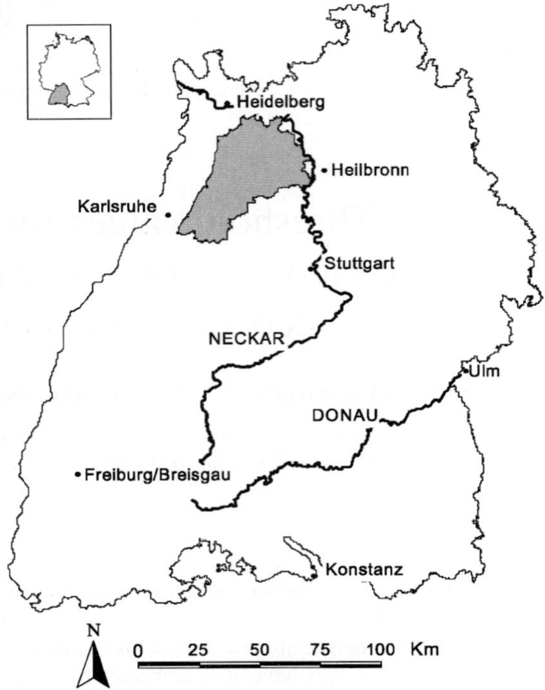

Fig. 1. Location of the research area within Baden-Wuerttemberg.

2. Description of the research area "Kraichgau"

The Kraichgau extends over an area of approximately 45 km × 40 km in the north-western part of Baden-Wuerttemberg (see Fig. 1). As a result of a combination of relatively high average annual temperature (9 °C), an average rainfall of 700 mm per year and extremely fertile soils (Loess), this area has had a long tradition of agricultural use. In most communities of the Kraichgau region, land consolidation and modern agricultural techniques have led to a change of the former landscape structure (e.g. terracing). Despite the hilly topography, most areas are now characterised by large arable fields, with significant volumes of row crops and only a few remaining landscape elements such as hedges and ridges. Under these conditions, run-off of water causing erosion and consequent flooding are increasingly problematic. Nutrient inputs by erosion to the remaining bio-topes cause degradation of the formerly oligotrophic plant communities.

3. Methodology

3.1. Description of the landscape model Kraichgau

The economic–ecological model system Kraichgau consists of six modules. The modules cover the following topics which are explained in detail shortly: 'Nitrogen', 'Erosion', 'Economic Regional Model',

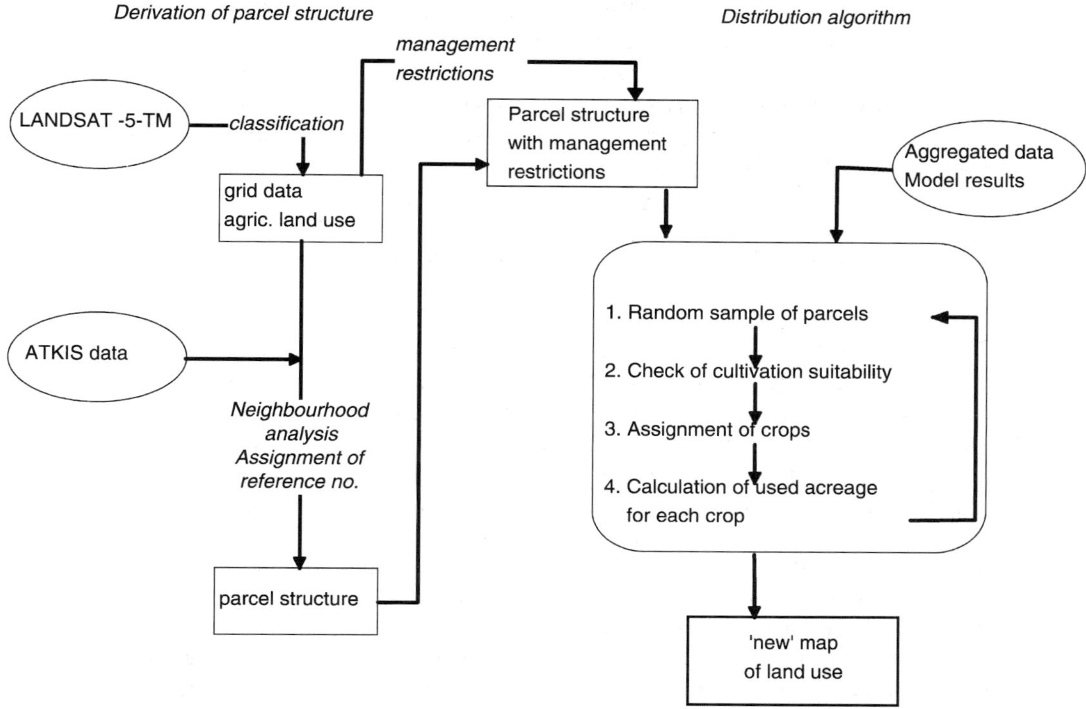

Fig. 2. Data transformation procedure.

'Nitrate', 'Nutrient Input' and 'Spatial Relation'. These modules are brought together under a common user interface and work at a regional scale. Additionally, an economic model reflecting local level values serves as a control unit for the regional level (Dabbert et al., 1999).

The Nitrogen module calculates the nitrate concentration in the seepage water from the nitrate balance as well as the seepage water volume. The Erosion module uses the Universal Soil Loss Equation (USLE) in order to calculate erosion. In the Nitrate module, the amount of organic fertiliser is calculated for the different crops based on a formula that resulted from surveys in the test areas.

The module Spatial Relation consists of a procedure for the modelling of agricultural land use (Moevius, 1999). A parcel structure is derived from a satellite image classification. Following a defined set of rules, the data for land use which is only available from the economic regional modelling on an aggregated level, is transformed to the artificial spatial unit of a 50 m × 50 m grid, which has been agreed to be the common spatial unit used in the project (cf. "proxel", Mauser et al., 2001). Fig. 2 describes the procedure of the data transformation.

The Economic Regional Model used the method of positive mathematical programming. It can be characterised as a municipality-differentiated, comparative-static regional optimisation model (ROMEO, Umstaetter, 1999). It represents the income and the production structure of agriculture at the level of community groups (municipalities) (Umstaetter, 1999).

The Nutrient Input module represents the biotic part of the model complex. Small linear bio-tope structures such as hedges and step ridges that are located horizontal to the slope, are typical for the Kraichgau region. The quantification of nutrient inputs originating from agricultural areas into these structures is the central task of this module. The basis of the input calculations were the erosion rates from the Erosion module. They have been assigned to the single bio-topes within the geographical information system. In a second step, the transport rates of nutrients have been correlated to

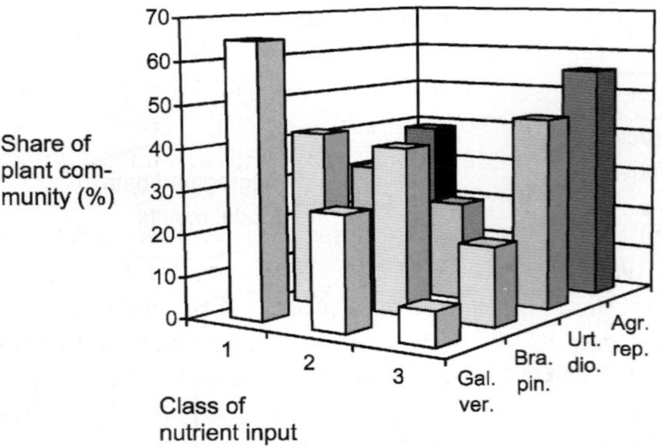

Fig. 3. Relationship between nutrient input and existence of different plant communities (Herrmann, 1995). (class 1 = 50 kg N/(ha per year), class 2 = 50–150 kg N/(ha per year), class $3 \geq 150$ kg N/(ha per year); Gal ver: Galium verum, Urt dio: Urtica dioica, Bra pin: Brachipodium pinnatum, Agr rep: Agropyron repens).

the existing natural vegetation within these bio-topes (Herrmann, 1999a). Fig. 3 shows the relationship between high amounts of nutrient input and the existence of eutropic plant communities (like stinging nettles, Urtica dioica) and low amounts with oligotrophic plants (like yellow or ladies bedstraw, Galium verum), respectively.

All regional modules were fully connected to each other by a GIS environment. The GIS systems ARC VIEW and ARC INFO were used respectively for data storage, visualisation of results, and to fulfil interface tasks and the translation of data between the ecological and economic parts of the model.

The local economic farm model, which was related to three selected test areas, had a specific task. Interviews with farmers provided important data for the landscape model which would have been impossible to obtain through any other means. In addition, the local module allows a result evaluation of the landscape model in terms of practical application by transferring the results to the local level (Koll, 1999). Although not fully connected to the other model parts via the GIS environment, the calculations of the local model are based upon the spatially explicit information stored within the GIS for the three test communities. Therefore, the results of the scenario calculations on regional level could be transferred to the local level also.

Fig. 4 shows the structure of the Kraichgau landscape model. Three disciplines, landscape ecology,

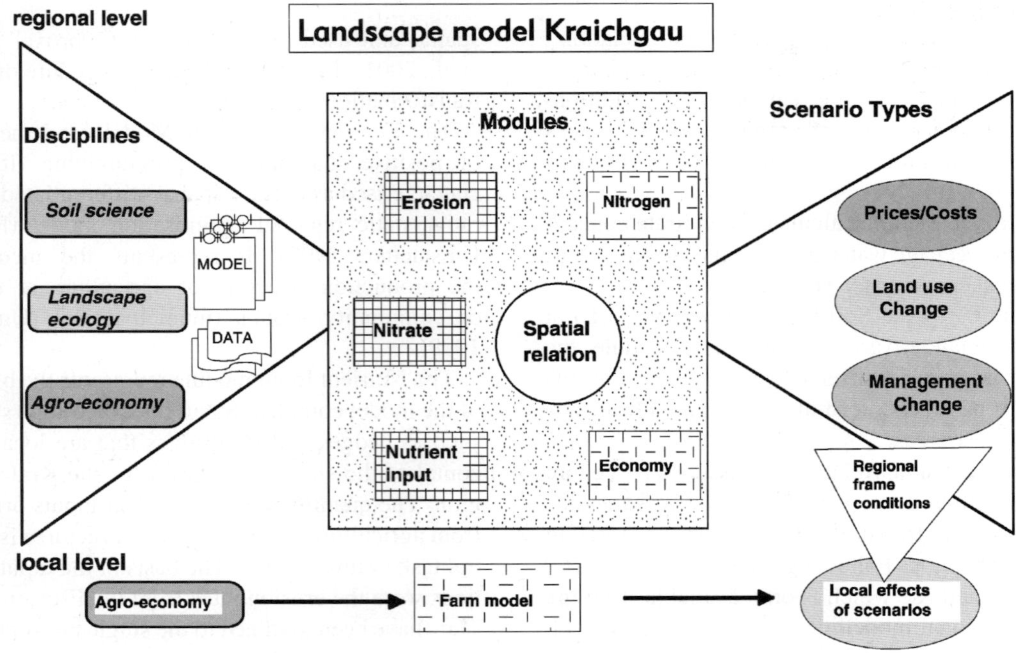

Fig. 4. Structure of the landscape model Kraichgau (Herrmann, 2001).

soil science and agricultural economy, prepared their data sets and their specific models on the regional level. After connection of these modules within a GIS framework the calculation of different types of scenarios was possible. The scenario results formed the framework for the model runs of the local economic model (for further information see Dabbert et al., 1999). By transforming aggregated data (results of the economic model) in a spatially explicit form (as input for the natural science models), the direct connection of all involved disciplines was possible. Consequently, the analysis of effects in different directions via scenario calculations could be performed. This was a basic condition for the evaluation of the following example of thresholds as indicators.

3.2. Scenario description

The scenario belongs to the type "change of actual land use". The intention of the present version was to derive the economic (regional and local) as well as the ecological consequences of the introduction of new nature development areas. Due to the intensive agricultural land use, the state of bio-diversity in the Kraichgau has to be improved urgently. Therefore, the area is well suited for testing such a threshold scenario. The research area, as a Loess region, has a high transport potential for soil and nutrients. To guarantee the development of valuable natural sites, it was necessary to choose non-affected areas. Therefore, areas with the lowest input potential, the hilltops

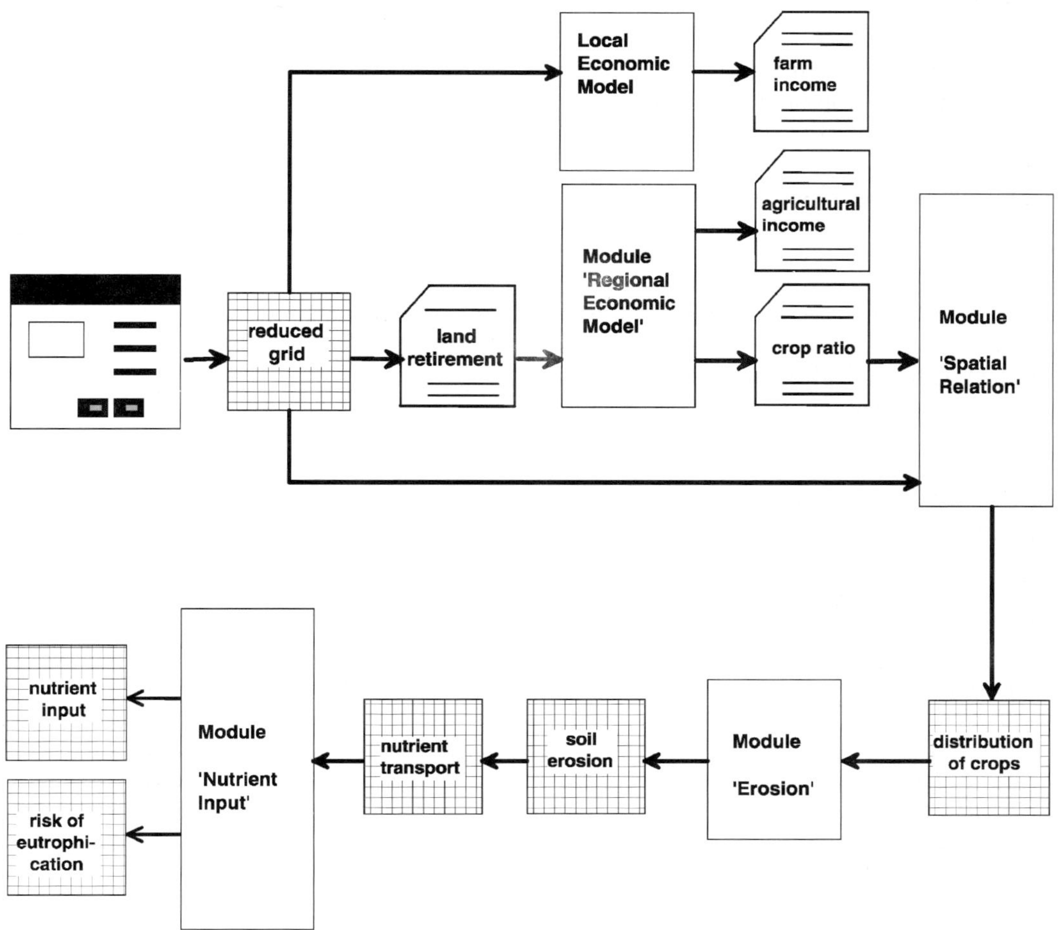

Fig. 5. Calculation scheme of scenario "Introduction of new nature development areas" (Herrmann et al., 1999).

with shallow soils were chosen. Several investigations (Kleyer, 1991; Herrmann, 1995) have shown that after strict reduction of nutrient input within a manageable period of time (5–10 years) site conditions for the reappearance of oligotrophic plants (e.g. *Mesobromion*) could be provided on these formerly arable fields. To optimise the development conditions, only south facing sites had been chosen to guarantee dryness and low nitrogen mineralisation in case nutrient input from the air occurs. The existing neighbouring ridges and sunken roads serve as a seed donor for the dry grassland species.

Fig. 5 describes the procedure of scenario calculation. Steered by a user interface (Schwarz-von Raumer, 1999), the selection of suitable nature development areas (according to the criteria) takes place within the GIS. The amount of reduced arable land for each municipality is transferred to the regional economic model. After a new run the results in terms of farmer's income and crop ratio for the remaining arable land are presented. The municipality related crop ratio is distributed to spatially explicit "fields" as a basis for the further calculations of erosion, nitrate concentration, nutrient input in bio-topes and eutrophication risk for bio-topes. All results can be visualised in tables and maps. Since the local economic model is related to specific fields in the test communities, the proposed nature development sites can be overlaid with the field structure directly. Subsequently, land retirement and resulting decrease of income can be calculated for single farms. Due to the main focus being nutrient transport by erosion, the results of the Nitrogen and Nitrate modules are not discussed further.

4. Results

4.1. Number and size of proposed nature development sites

According to the selection criteria, 10.2% of the agricultural acreage is now dedicated for natural development. It may be concluded then, that the first promising step for the development of a nature protection strategy for this intensively used agricultural area has been achieved. With the value of 10% we reached also the new threshold for desirable bio-tope area which has been established under the updated Federal Law of Nature Protection Legislation in Germany.

The selection in the scenario calculation resulted in 2617 suitable sites for new nature development. The designated areas (see Fig. 6) are spread more or less regularly over the whole region outside of forest or settlement areas, except some focal points with larger occurrences.

The distribution gives the impression of a kind of bio-tope network (see Blab, 1992; Jedicke, 1994), which is created by the computer selection. Nevertheless, nothing could be said about the quality of the space between the bio-topes, which is a fact that could determine the species exchange and survival (see meta population concept, Hanski, 1999). Most of the sites have an area of between 0.25 and 5 ha (see Fig. 7), with a peak in the first size category (0.25 ha). Only single sites reach 20 ha or more (maximum 60 ha). In comparison to the minimal area (Heydemann, 1981) for dry grassland bio-topes (10 ha; DRL, 1983), only a few seem to reach this optimum criteria. Considering the fact that even smaller bio-topes can be suitable as habitats if they are connected to others, the distances between the different sites have to be additionally evaluated (a critical distance of 1–3 km for butterflies and grasshoppers is mentioned by Riess, 1986; Smolis and Gerken, 1987).

4.2. Economic consequences on regional and local level

4.2.1. Regional level

4.2.1.1. Land retirement. The introduction of the nature development sites (see Section 4.1) resulted in a land retirement for the farmers. Since the proportion of land which fulfils the selection criteria is different in each municipality, the share of land to be retired differs considerably between them. Land retirement ranges from 9 to 822 ha per municipality (Fig. 8), with an average of 230 ha. This corresponds with an proportional reduction of the arable land of 0.7 and 12.3%, respectively.

4.2.1.2. Changes in crop ratio. In consequence, the acreage of most crops decreases, too (see Table 1). In general, marginal crops such as oats and rye were

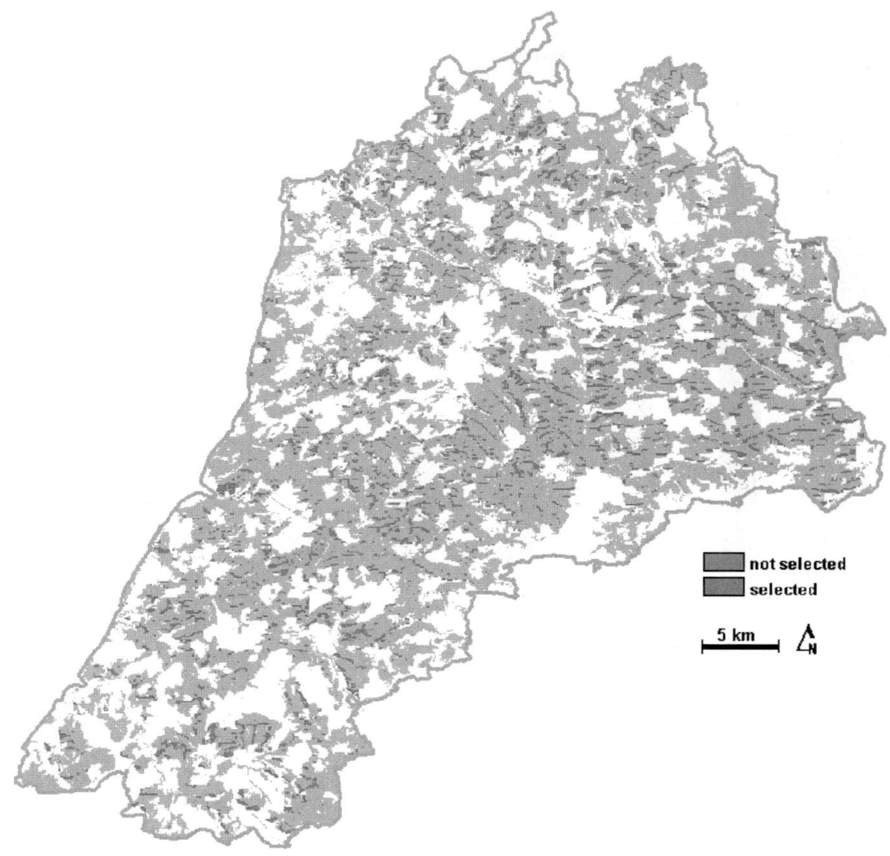

Fig. 6. Distribution of nature development sites.

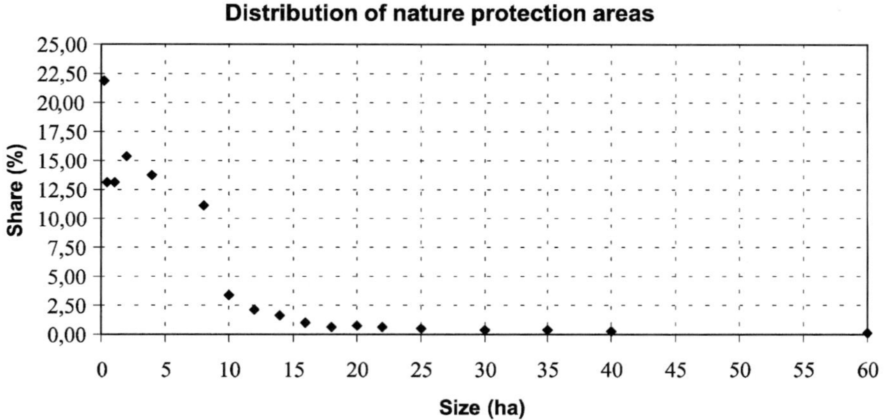

Fig. 7. Size and number of nature development sites.

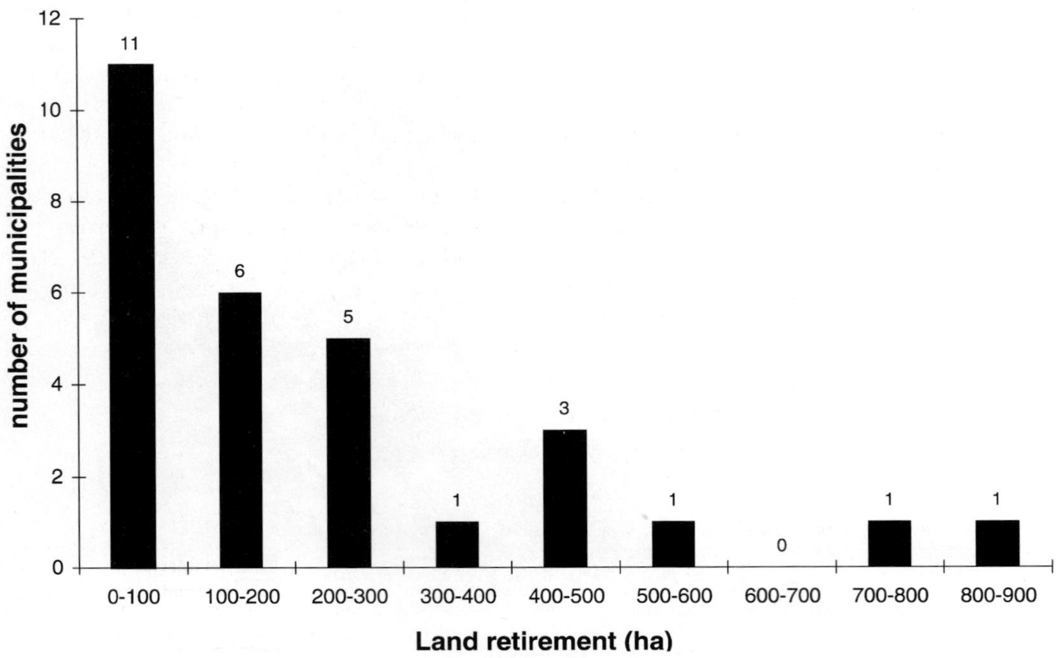

Fig. 8. Land retirement in the scenario "Introduction of new nature development areas".

reduced by more than 16%, whilst cereals like barley and winter wheat were reduced by between 10 and 12%. The highest decrease could be found for the fodder production for animals. The acreage of root crops as sugar beet and potatoes remained unchanged as they are the most profitable crops and their limitation is based on quota rather than on market aspects or aspects of crop rotation. The difference of maize production is also lower (8%). As a consequence, the share of row crops in relation to the available area is increasing (see Table 1), which leads to an intensification of production for the given area. Nevertheless, an

Table 1
Changes in acreage for specific crops, scenario "Introduction of nature development areas"

Crop type	Acreage start situation (ha)	Acreage Scenario (ha)	Decrease (%)	Share of crop rotation start situation (%)	Share of crop rotation scenario (%)
Winter Wheat	21560	19223	−10.8	32.69	32.40
Summer Barley	7540	6644	−11.9	11.43	11.20
Winter Barley	4973	4351	−12.5	7.54	7.33
Maize	4274	3903	−8.7	6.48	6.58
Oats	1913	1597	−16.5	2.90	2.69
Rye	1873	1559	16.8	2.84	2.63
Sugar beet	7101	7101	0	10.77	11.97
Potato	799	799	0	1.21	1.35
Sunflowers	1390	1233	−11.3	2.11	2.08
Rape seed	1280	1120	−12.5	1.94	1.89
Silage Maize	4041	3705	−8.3	6.13	6.24
Clover	1073	871	−18.8	1.63	1.47
Set aside	8144	7231	−11.2	12.35	12.19

Calculation: Thomas Winter, 2001 (oral communication).

increase of pesticide and fertiliser application is not likely to occur as all the areas are already used with the optimum intensity.

4.2.1.3. Changes in gross margin. The total gross margin in the Kraichgau region declined from 158.6 to 151.8 M DM related to the original area. This equates to a reduction of 4.2%. The decrease varies significantly between the municipalities. A majority of municipalities has a decline between 3 and 5% relating the new gross margin to the original acreage. Several have only slight decreases (1–2%), while others suffer from a reduction of more than 5% (see Table 2). This results in financial losses between 600 and 90,000 DM per district if no compensation for the lost acreage is assumed (e.g. remuneration for specific management measures to care for the new bio-topes). These losses have to be balanced by governmental subsidies. Without the concentration on more intensive crops for the remaining acreage (see Section 4.2.1.2 and Table 2), the loss would have been even greater. Despite the strong reduction of fodder production (see Section 4.2.1.2), the effects of the scenario for the animal husbandry have been more slightly (see Table 3). Only suckler cows show a stronger reduction, which does not influence agricultural income very much due to the low distribution in region.

Table 2
Changes in gross margin and acreage on district level, scenario "Introduction of nature development areas"

District	Start situation		Scenario		
	Acreage (ha)	Gross margin, (DM/ha)	Acreage (ha)	Gross margin, scenario conditions (reduced area) (DM/ha)	Gross margin, related to start situation (DM/ha)
1	6179	2500.40	5635	2622.89	2391.97
2	945	1216.93	875	1268.57	1174.60
3	955	2188.48	871	2319.17	2115.18
4	4129	1755.87	3718	1850.46	1666.26
5	3206	1409.86	2917	1494.69	1359.95
6	2476	1716.48	2214	1829.27	1635.70
7	6596	2828.99	5896	3017.30	2697.09
8	1290	2248.06	1281	2224.82	2209.30
9	3529	2275.43	3095	2474.96	2170.59
10	2657	1810.31	2367	1913.81	1704.93
11	2353	1478.96	2239	1509.60	1436.46
12	2042	2820.76	1897	2941.49	2732.62
13	769	1989.60	738	2032.52	1950.59
14	4070	1687.96	3656	1794.31	1611.79
15	1034	2485.49	941	2624.87	2388.78
16	1057	1504.26	1003	1525.42	1447.49
17	2005	1541.15	1961	1540.03	1506.23
18	2287	2050.72	2150	2083.72	1958.90
19	2323	1575.55	2064	1686.05	1498.06
20	1736	1895.16	1599	1982.49	1826.04
21	2196	1580.15	2122	1597.55	1543.72
22	859	1559.95	788	1649.75	1513.39
23	2845	2797.89	2651	2912.11	2713.53
24	7733	2349.67	6911	2497.47	2231.99
25	1949	1410.98	1710	1485.38	1303.23
26	1223	1929.68	1138	2003.51	1864.27
27	1118	1493.74	1080	1527.78	1475.85
28	5136	2260.51	4739	2344.38	2163.16
29	1992	1822.29	1805	1916.90	1736.95
Total	76689	1937	700061	2023	1863

Calculation: Thomas Winter, 2001 (oral communication).

Table 3
Changes in animal husbandry, scenario "Introduction of nature development areas"

Animal husbandry system	Start situation (number of animals)	Scenario (number of animals)	Change (%)
Milk cow	12342	11932	−3.3
Fattening cattle	10822	10253	−5.3
Suckler cow	1547	1268	−18
Fattening pig	44313	44251	−0.1
Breeding pig	9174	9168	−0.1

Calculation: Thomas Winter, 2001 (oral communication).

4.2.2. Local level

On the local level, the land retirement of the single farms could differ significantly. The share ranges between 4.7 and 34.3 ha (average 19.7 ha), which results in a proportional reduction of between 9.7 and 40.6% of the arable land. These values make clear that the focus on single farms results in more drastic economic consequences than on regional level. On the local scale, no balance of very high and very low reduction rates could be performed. The direct relation to the area results in the designation of specific sites which belong to real farmers. These farmers as individuals have to stand the consequences. A loss of 40% of arable land will probably lead to the loss of the farm.

4.3. Ecological consequences in relation to erosion risk and bio-tope quality

4.3.1. Erosion risk

Due to their relative increase, erosion stimulating row crops such as sugar beet and maize have to be distributed on a smaller total acreage. Therefore they have to be planted on slopes previously not used for such crops. Consequently, in comparison to the initial situation the erosion risk in these areas increases. This refers to approximately 40% of the area.

4.3.2. Bio-tope quality

As in some areas the rate of erosion changed (see Section 4.3.1), a new calculation for the nutrient impact for the existing bio-tope structures on the slopes (hedges, ridges) was carried out. The results are surprising. Nearly half of the existing bio-topes will be affected by a higher nutrient input due to the change of crop ratio:

- 57.5% increase of nutrient input;
- 42.5% reduction of nutrient input.

Of the increased 57.5%, more than the half of these bio-topes remain in the highest eutrophication class (>150 kg N/(ha per year); cf. Fig. 3) but their nutrient input still increased. Additional 5% came to this class. The remaining 40% staid in class 1 (18.5%) or 2 (12.5%) showing increased amounts of nutrients or changed from class 1 to 2 (10%).

In the decreasing fraction only 15% appear in a lower input class. Most changes consist of a slight reduction of nutrient input (5–25%) of the already very high class 3 level, not leading to a change of class.

This means that more than half of the oligotrophic plant communities (class 1) would have to stand a higher input of nutrients which could worsen their site conditions. Additionally, the dominating nutrient exploitative stands will be stabilised by further nutrient input. Consequently, the increase of nutrient input in actually oligotrophic stands would constitute a further deterioration of the ecological situation in the region and strengthen the tendency of reducing diversity. This is accentuated as the nutrient depletion of these sites is much more difficult due to their topographic position (erosion barriers on slopes) and their specific soil and vegetation structure. Finally it must be stated, that the proposed measure improves the situation of certain bio-tope types (dry grassland) in the Kraichgau region but results in damage of others (ridge vegetation).

5. Discussion

5.1. Pros and cons of ecological thresholds as indicators for bio-diversity and sustainable land use development

To give a scientific justification for a certain threshold for nature protection areas is very difficult or even not possible (e.g. Kaule, 1988; Wiegleb, 1997; Von Haaren, 1988). Nature protection can only be achieved where based on political consensus or normative commitment. The 10% value of bio-tope area is "generally accepted" (cf. Haber et al., 1993) and this is one of the main reasons to use it.

In order to prepare such a consensus, it is necessary to define the functions of the thresholds for the scales to which they should be related (Horlitz, 1994; Hampicke et al., 1991).

Countrywide, the function of thresholds is to define rough regional ecological guidelines or to get an overview about the potential costs for the implementation of nature protection guidelines for the whole country. On the local level, general thresholds make no sense because the concrete selection of sites takes place.

The most interesting level for the application of such thresholds is the regional level. Here a tightrope walk between too-general recommendations for bio-diversity and a too-tight corset for the development of the communities or the farmers has to be achieved. The latter is politically not practicable because the jurisdiction for the implementation is in the hands of the communities or land owners.

On the regional level, thresholds

- have to be related to different bio-tope types. Otherwise, all possible existing bio-topes could be accumulated to reach the threshold without any relevance to a biotic concept;
- have to be related to the specific conditions of the different regions. In cultural landscapes the relevant factors to influence the bio-diversity are not only the site conditions but even more important the land users. Thresholds, therefore, should also be related to land use. Cultural landscapes like most of those found in middle Europe cannot be treated as natural landscapes. There existence in most cases is due to the activities of farmers and they, therefore, need this management to survive. On the other hand, the quality of the bio-topes is also dependant on the management intensity of the land use (e.g. eutrophication of adjacent bio-topes). A balance must be found between the necessary management intensity for the survival of the farm(s) and the necessary frame conditions for the survival of valuable natural bio-topes. This balance is different in specific regions and has to be included in the discussion and determination of thresholds;
- have to be related to land users, especially in cultural landscapes. Reaction of land users must be included as the land belongs to them. Therefore, compensation has to be financed. This means, the magnitude of costs has to be included (Hampicke et al., 1991). These costs are dependant on the necessary measures (Horlitz, 1994). Generalised thresholds are critical in such cases since it makes a significant monetary difference if a farmer has to care for grassland or for orchards, if he has to manage the nature protection area, or if the total loss of fields has to be compensated.

In general, an effect analysis of planned thresholds has to be done before implementation and—to fulfil the demand of sustainability—ecological as well as economic and social factors have to be included. Tools are necessary to carry out this analysis, especially to test potential measures. The application of the landscape model Kraichgau has shown how the first steps might be undertaken.

5.2. *General conditions and scope of the landscape model Kraichgau*

The presented example describes the possible consequences of a threshold application for a real landscape. The Kraichgau is a cultural landscape and the land users are a very important factor for the development of the landscape features. This caused astonishing model results and unexpected relationships between ecology and economy which only the modelling system revealed. Nevertheless, there are still some gaps and restrictions in the model system.

5.2.1. *Scope of the model system*

Even though the presented model system produces plausible results, it must be stated that this prototype works only within certain frame conditions. The different components and the system configuration are adjusted to specific questions. Existing or easily accessible data have been used, only. A medium time scale (5 years) and a common spatial unit (50 m grid) have been fixed. Land use has been the integrating parameter for the ecological and economic parts. Social aspects had not been included, yet. The time scale is static, even though it is possible to simulate quasi-dynamics by running the scenarios several times with changing parameters. The landscape model refers to agricultural areas and does not include forest and settlements in the modelling process.

5.2.2. *Limits of the ecological parts*

Erosion modelling had been done using the USLE which results in all the well-known problems with this empirical model (cf. Bork and Schroeder, 1996).

The biotic modelling is based on simplified assumptions. The procedure of site selection should be extended by including more information in the selection process (e.g. proximity to existing bio-topes, minimum size requirements, information about separation effect of surrounding agricultural land use, exclusion of highly valued farm land). A mixture of different bio-topes should be planned and the existing bio-topes have to be considered. The modelling of the bio-tope quality should be complemented by a nutrient transport model and a development model for wild plants (e.g. Kleyer, 1998). Another consideration is the connection with the development potential for site adapted fauna (e.g. Wieland and Voss, 2001).

5.2.3. Limits of the economic parts

Comparing the results of the economic models, it is obvious that the land retirement shows a stronger effect on farmer's income at the local level than at the regional one. The maximum reduction of arable land on regional level is only a quarter of the local one. The average reaches one half. Several explanations are possible. In general, the regional model focuses on the average situation in a chosen spatial unit while the local model refers to a single farm. The farm model is based on interviews with farmers and can, therefore, relate to very accurate data. Due to the bigger research unit, the regional model has to refer to statistic data which are in an aggregated form. Additionally, the lower values of the regional model could be produced by the specific PMP version of the model concept. Roehm (2001) points out that some model assumptions might not fully hold for the Kraichgau region. Therefore, the updated version of the regional economic model might produce other results and thus come closer to the local one. Finally, the farm model is directly related to the real sites and no balance within a certain region (and among different farmers) is possible, thus leading to the drastic reduction for individual farms. These results imply a differentiated consideration of local and regional economic aspects respectively due to the different information depth. This is especially true, if the costs for nature protection measures have to be calculated (Hampicke et al., 1991).

5.3. Advantages of the proposed technique/tool

Despite the need for further specific improvements, the modelling system as a whole worked in an acceptable way. The general expectations to describe the main possible effects of a threshold for nature protection areas and to clarify the existing relationships within a cultural landscape, was fulfilled. Additionally, the landscape model offers a tool to negotiate through several options virtually. Through this, an estimation of potential development lines becomes possible. This coincides with the wide spread development of using modelling systems to analyse the potential effects of planning concepts or policy targets "virtually" (e.g. O'Callaghan, 1995; Ravetz, 2000; Tenhunen and Lenz, 2001). Scenario technique is mainly used to describe the possible changes, often coupled with a GIS to visualise the space related results.

For the Kraichgau example, the model shows clearly the effect of a purely sectoral point of view. Only focussing on the nature protection value of an indicator, the consequences can be negative for bio-diversity, too.

The Kraichgau model describes the relationships between ecological demands and economic reactions. Farmers try to maximise their income and, therefore, might react differently to expectations (i.e. keep the most profitable crops and do not reduce all equally according to the land retirement). The landscape model Kraichgau is able to show that an ecological promotion on the one side could cause ecological damage on the other due to the reaction of the land users. It is necessary to describe these relationships for the acceptance of nature protection demands and norms (Von Haaren, 1999). Due to globalisation, normative power is diminishing and win–win situations including economic aspects are needed more than ever (Von Haaren, 1999). This direct connection of economic and ecological factors had been achieved with the parameter "land use", which allows all disciplines to use their specific models and is, therefore, the common "currency" of the model system.

The model is still somewhat mechanistic and rough but, even so, shows unexpected relationships between land use changes and economic reactions so revealing complex economic–ecological relations in a landscape. This simplicity is an advantage as such complex relations should not be modelled totally in all facets (cf. Bugmann, 1997; Hauhs and Lange, 1996;

Wenkel and Schultz, 1999), even if it would be technically possible. The interpretation of the results would not be valuable. Simple models are useful due to their comprehensibility and their greater simplicity of error detection (Wenkel and Schultz, 1999). User oriented systems like agricultural ecosystems are highly suitable for such a modelling strategy. Due to a long experience, they are predictable in their reaction and there already exist evaluated rule frameworks (Hauhs and Lange, 1996). Additionally, such simple models have several parts which are transferable to other landscapes and problems. This enables people to use it for their own purposes (Herrmann, 1999b).

6. Outlook

More tests with these new techniques should be made to familiarise decision makers with them and to highlight the necessary improvements. Later, the connection to a model which describes the demand of bio-diversity in a landscape (e.g. Altmoos, 1999) should be considered.

The results of the model show the consequences of a spatial reference of generally formulated guidelines (i.e. thresholds). But they also should include the behaviour of people as several measures in agricultural landscapes depend on farmer's work. The use of traditional orchards, for example, which is often demanded for nature protection purposes, is only manageable if the farmers concerned find it valuable to care for cultural heritage sites even if, for them, it is economically not very promising. Social factors, therefore, should be included in the modelling process, too.

Acknowledgements

The authors would like to thank Tim Dockerty for checking the English and two unknown reviewers for their valuable comments. Thanks to Thomas Winter for updating the gross margin calculations. The project was carried out with the financial support from the Volkswagen-Stiftung, Germany.

References

Altmoos, M., 1999. Netzwerke von Vorrangflaechen. Nat. Landschaft 31 (12), 357–367.

Blab, J., 1992. Grundlagen des Biotopschutzes fuer Tiere. Kilda, Greven.

Bork, H.R., Schroeder, A., 1996. Quantifizierung des Bodenabtrags anhand von Modellen. In: Fraenzle, O., Mueller, F., Schroeder, W., (Eds.), Handbuch der Umweltwissenschaften. Ecomed, Kap. IV-3.5.

Bugmann, H., 1997. Scaling issues in forest succession modelling. http://www.gcrio.esto.or.jp/ASPEN/science/EOC97/eoc97session1/Bugmann.htm.

Dabbert, S., Herrmann, S., Kaule, G., Sommer, M., 1999. Landschaftsmodellierung fuer die Umweltplanung. Springer, Berlin.

Deutscher Rat fuer Landespflege, 1983. Ein integriertes Schutzgebietssystem zur Sicherung von Natur und Landschaft. Schr.-R. DRL 41, pp. 5–14.

Haber, W., 1972. Grundzuege einer oekologischen Theorie der Landnutzung. Innere Kolonisation 21, 294–298.

Haber, W., Duhme, F., Pauleit, S., Schild, J., Stary, R., 1993. Quantifizierung raumspezifischer Entwicklungsziele des Naturschutzes, dargestellt am Beispiel des Kartenblattes 7435 Pfaffenhofen. Beitraege ARL 125, Hannover.

Hampicke, U., Horlitz, T., Kiemstedt, H., Tampe, K., Walters, M., Timp, D., 1991. Kosten und Wertschaetzung des Arten- und Biotopschutzes. Berichte des Umweltbundesamtes 3.

Hanski, I., 1999. Metapopulation Ecology. Oxford University Press, Oxford.

Hauhs, M., Lange, H., 1996. Perspektiven fuer eine (Meta-)Theorie terrestrischer Oekosysteme. In: Mathes, K., Breckling, B., Ekschmitt, K. (Eds.), Systemtheorie in der Oekologie. Ecomed Landsberg, pp. 95–105.

Herrmann, S., 1995. Quantifizierung von Naehrstoffeintraegen in Kleinstrukturen einer Loess-Agrarlandschaft. Lang, Frankfurt.

Herrmann, S., 1999a. Vegetationsentwicklung unter Nutzungseinfluss. In: Dabbert, S., Herrmann, S., Kaule, G., Sommer, M. (Eds.), Landschaftsmodellierung fuer die Umweltplanung. Springer, Berlin, pp. 105–111.

Herrmann, S., 1999b. Uebertragbarkeit des Ansatzes auf andere Landschaften. In: Dabbert, S., Herrmann, S., Kaule, G., Sommer, M. (Eds.), Landschaftsmodellierung fuer die Umweltplanung. Springer, Berlin, p. 175.

Herrmann, S., 2001. Entscheidungsunterstuetzung in der Landnutzungsplanung mittels GIS-gestuetzter Modellierung—Massstabsbezug, Realitaetsnaehe und Praxisrelevanz. Der Andere. Verlag, Osnabrueck.

Herrmann, S., Umstaetter, J., Koll, H., 1999. Aenderung der Flaechennutzung: Ausweisung von Biotopentwicklungsflaechen. In: Dabbert, S., Herrmann, S., Kaule, G., Sommer, M. (Eds.), Landschaftsmodellierung fuer die Umweltplanung. Springer, Berlin, pp. 156–166.

Heydemann, B., 1981. Zur Frage der Flaechengroesse von Biotopbestaenden fuer den Arten- und Biotopschutz. Jahrbuch. f. Naturschutz Landschaftspflege 31, 21–51.

Horlitz, T., 1994. Flaechenansprueche des Arten- und Biotopschutzes. IHW, Eching.

Jedicke, E., 1994. Biotopverbund. Ulmer, Stuttgart.

Kaule, G., 1988. Abgrenzung und Bewertung von Biotopschutzgebieten. In: ARL (Ed.), Sektorale Entwicklung und raeumliche Planung, pp. 51–78.

Kleyer, M., 1991. Die Vegetation linienfoermiger Kleinstrukturen in Beziehung zur landwirtschaftlichen Produktionsintensitaet. Eine Untersuchung aus dem Kraichgau, einer Loess-Huegellandschaft in Suedwestdeutschland. Diss. Bot. 169, Cramer, Vaduz.

Kleyer, M., 1998. Individuenbasierte Modellierung von Sukzessionen pflanzlicher Wuchstypen bei unterschiedlichen Stoerungsintensitaeten und Ressourcenangeboten. Verh. Ges. f. Oekol. 28, 175–182.

Koll, H., 1999. Testmodul Betriebliche Modellierung. In: Dabbert, S., Herrmann, S., Kaule, G., Sommer, M. (Eds.), Landschaftsmodellierung fuer die Umweltplanung. Springer, Berlin, pp. 88–95.

Mauser, W., Tenhunen, J.D., Schneider, K., Ludwig, R., Stolz, R., Geyer, R., Falge, E., 2001. Remote sensing, GIS and modelling: assessing spatially distributed water, carbon, and nutrient balances in the Ammer river catchment in Southern Bavaria. In: Tenhunen, J.D., et al. (Eds.), Ecosystem Approaches to Landscape Management in Central Europe. Ecological Studies, vol. 147. Springer, Berlin, pp. 583–619.

Moevius, R., 1999. Modul zur Uebertragung aggregierter Daten in raeumlich konkrete Daten. In: Dabbert, S., Herrmann, S., Kaule, G., Sommer, M. (Eds.), Landschaftsmodellierung fuer die Umweltplanung. Springer, Berlin, pp. 112–125.

O'Callaghan, J.R., 1995. NELUP. An introduction. Journal of Environmental Planning and Management 38 (1), 5–20.

Ravetz, J., 2000. City Region 2020. Integrated Planning for a Sustainable Environment. Earthscan, London.

Riess, W., 1986. Konzepte zum Biotopverbund im Arten- und Biotopschutzprogramm Bayern. Laufener Seminarbeitr 10, 102–115.

Roehm, O., 2001. Analyse der Produktions- und Einkommenseffekte von Agrarumweltprogrammen unter Verwendung einer weiterentwickelten Form der Positiven Quadratischen Programmierung. Shaker, Aachen.

Schwarz-von Raumer, H.G., 1999. Integration oekonomischer und oekologischer Module in einem GIS-gestuetzten Landschaftsmodell. In: Dabbert, S., Herrmann, S., Kaule, G., Sommer, M. (Eds.), Landschaftsmodellierung fuer die Umweltplanung. Springer, Berlin, pp. 126–132.

Smolis, M., Gerken, B., 1987. Zur Frage der Populationsgroesse und der intrapopularen Mobilitaet von tagfliegenden Schmetterlingen, untersucht am Bsp. der Zygaenidenarten (Lepidoptera: Zygainidae) eines Halbtrockenrasens. Decheniana 140, 102–117.

Tenhunen, J.D., Lenz, R., Hantschel, R. (Eds.), 2001. Ecosystem Approaches to Landscape Management in Central Europe. Ecological Studies, vol 147. Springer, Berlin.

Umstaetter, J., 1999. Calibrating Regional Production Models Using Positive Mathematical Programming—An Agro-Environmental Policy Analysis in Southwest Germany. Shaker, Aachen.

Von Haaren, C., 1988. Beitrag zu einer normativen Grundlage fuer praktische Zielentscheidungen im Arten- und Biotopschutz. Landschaft Stadt 20 (3), 97–106.

Von Haaren, C., 1999. Landschaftsplanung im Spannungsfeld neuer Anforderungen. Jahrbuch Naturschutz Landschaftspflege 51, 227–250.

Wenkel, K.O., Schultz, A., 1999. Vom Punkt zur Flaeche—das Skalierungs-bzw. Regionalisierungsproblem aus der Sicht der Landschaftsmodellierung. In: Steinhardt, U., Volk, M. (Eds.), Regionalisierung in der Landschaftsoekologie. Teubner, Stuttgart, pp. 19–42.

Wiegleb, G., 1997. Leibildmethode und naturschutzfachliche Bewertung. Nat. Landschaft 73 (1), 9–25.

Wieland, R., Voss, M., 2001. Land use change and habitat quality in Northeast German agro-landscapes. In: Tenhunen, J.D., et al. (Eds.), Ecosystem Approaches to Landscape Management in Central Europe, pp. 341–346.

Wissenschaftlicher Beirat der Bundesregierung Globale Umweltveraenderungen (WBGU), 2000. Dramatischer Verlust biologischer Vielfalt gefaehrdet Chancen zukuenftiger Generationen. Presseerklaerung vom 13.9.2000.

ELSEVIER

Available online at www.sciencedirect.com

Agriculture, Ecosystems and Environment 98 (2003) 507–516

Agriculture Ecosystems & Environment

www.elsevier.com/locate/agee

Comparative assessment of agri-environment programmes in federal states of Germany

Rainer Marggraf*

Institut für Agrarökonomie, Platz der Göttinger Sieben 5, 37073 Göttingen, Germany

Abstract

According to Article 16 of Regulation (EC) No. 746/96 all member states of the EC are required to evaluate their agri-environment programmes to ascertain the environmental and socio-economic impacts. Analysing these impacts is a frontier topic for scientists with much research in progress. This article contributes to this research agenda in developing and applying a scheme for the evaluation of agri-environment measures from an ecological and economic point of view. The agri-environment programmes of all German federal states are systematically compared with regard to their impacts on biotic resources. Expert opinions obtained from a Delphi study serve as indicators for the ecological consequences of the programmes. Two measures for the ecological cost effectiveness of the agri-environment programmes are derived and applied. For the economic evaluation the cost effectiveness ratio and the benefit cost ratio of the programmes are calculated. Thus, this article presents an example for an integrated ecological–economical assessment of policy measures. It is shown that with respect to the protection of biotic resources by agri-environment programmes, no contradiction between ecology and economics exists in Germany. Improved biotic resource conservation is also economically advantageous.

Keywords: Agri-environment programmes; Ecological effectiveness; Cost effectiveness; Cost–benefit analysis

1. Introduction

As in many countries, the condition of the environment in rural areas in Germany is unsatisfactory. In 1985 the independent advisory board of the German Ministry of the Environment presented a detailed report on the environmental effects of the present state of agriculture in Germany (SRU, 1985). The most important problems were identified (e.g., the decline in species and biotope variety, the uniformity of the landscape, the pollution of the groundwater) suggesting that agriculture puts too much pressure on the environment. However, it was also determined that agriculture can have positive effects on the environment, for example, by contributing to half-natural biotopes which advance species sustainment. The government has reacted to these problems by implementing agricultural environmental policy measures. The German agri-environment policy makes use of legal and economic instruments. This article concentrates on the instrument of agri-environment programmes which tries to influence farmers' behaviour using economic incentives and on the impacts of these programmes on agriculturally managed biotic resources. Agri-environment programmes are an important component of the German nature conservation policy and amount to nearly 40% of the German government's expenditure for nature conservation (Stratmann and Marggraf, 2001). The intention of this article is to examine and compare the agri-environment programmes of all German federal states by applying ecological

* Tel.: +49-551-394829; fax: +49-551-394812.
E-mail address: rmarggr@gwdg.de (R. Marggraf).

0167-8809/$ – see front matter
doi:10.1016/S0167-8809(03)00109-9

and economical criteria. The true dimension of the environmental impact due to agricultural land use on Germany's biotic resources is not to be determined, but rather the change in these impacts that occurs based on the agri-environment programmes. Therefore, indicators are developed which can be applied for an integrated ecological–economical assessment of policy measures. The assumptions which must be made for the assessment are explained. The next section provides an overview of the agri-environment programme's structure. A scheme for assessing the consequences of the various implemented measures is presented in Section 3. Expert opinions obtained from a Delphi study serve as indicators for the impact on biotic resources. In Section 4, the effectiveness of the German agri-environment programmes on biotic resource conservation and their cost effectiveness are assayed. In Section 5 the economic efficiency of biotic resource conservation through agri-environment programmes is examined. The last section (Section 6) summarises the most important results.

2. Agri-environment programmes in Germany

German farmers were first offered participation in agri-environment programmes in 1980 (Höll and von Meyer, 1996). These programmes were intended to induce farmers to use natural resources more carefully than was legally prescribed. At first, programmes were directed at particular aims and measures. Examples are the Bavarian programme for the protection of meadow birds, and the programmes in North Rhine-Westphalia and Lower Saxony for the reduction of semi-liquid manure distribution (all established in 1984). Until 1992 the German federal states' programmes and their supplemental regional and local programmes were organised and financed largely independent of EC agricultural programmes. However, in 1985 the EC offered member states support for regional nature conservation programmes and extensification measures 'in ecologically particularly sensitive areas' by granting subsidies on a national level (Article 19 of Regulation 797/85, the 'Efficiency Regulation'; Baldoch and Lowe, 1996). Since 1987, the EC co-finances programmes under certain conditions. As a result the number of agri-environment programmes increased to more than 50 by the end of the 1980s. During the CAP reform of 1992 member states' agri-environment programmes gained new status as part of the supporting measures. First, member states were encouraged to establish agri-environment programmes by Council Regulation (1992). Second, the EC programme financing, typically 50% (and in some less favoured areas up to 75%), is no longer administered by the Guidance Fund, but by the Guarantee Fund of the EAGGF. Financing by the Guarantee Fund is supposed to emphasise the permanence and validity of European agri-environment policy.

EC regulations do not exactly define which production methods are extensive and environmentally favourable and thus deserving of promotion. This definition is largely left to the discretion of the member states. The guidelines for the regulations only state that member states have to define the terms 'extensification' or 'extensive production' when considering positive effects on the environment and natural habitats.

Regulation 2078/92 states eight issues that can be supported by the EC:

- substantial reduction in the use of fertilisers and/ or plant protection products as well as introduction or maintenance of organic farming methods;
- change of farmland to extensive grassland and keep or introduce more extensive forms of crops including forage production;
- reduction of sheep and cattle per forage area;
- use of farming practices compatible with the requirements of environmental protection as well as rearing animals of local, endangered breeds or cultivating threatened agricultural plant varieties;
- ensure the upkeep of abandoned farm land or woodlands;
- set aside farm land for at least 20 years for nature protection and the establishment of biotope reserves;
- management of land for public access and leisure activities;
- improvement of the training of farmers with regard to production methods compatible with the environment.

The regulation demands that farmers voluntarily participate in environmental programmes. However, they must commit to the programme for 5 years. Their obligation must exceed the mere compliance with the principles of 'good agricultural practice' and they must guarantee a considerable reduction of yield-increasing means of production.

As mentioned, the EU contributes 50%, in some cases even 75%, to the financing of the programmes. The premium is determined by a comparison of contribution margins with and without programme participation. Farmers' loss of income (opportunity costs) can be compensated by incentive components of a maximum of 20% of the opportunity costs. Not only the introduction of environmentally favourable measures but also their continuation is remunerable. In this case opportunity costs are determined by incomes from competitive production including all other income components. From the EC's perspective there is no (better) alternative to the agri-environment programmes. Therefore, in 2000 they were continued with minor modifications and a new regulation was added (regulation on promotion of rural areas; Council Regulation, 1999).

More than 25 different agri-environment programmes exist at the federal states' level in Germany. The national programme promotes mainly extensification measures for arable land and grassland as well as organic production methods. It only refers to the first three of the eight issues in EC Regulation 2078/92. The remaining financial costs of these measures after deducting the EC contribution are co-financed by the federal government at 60%. The federal states can support further measures in region-specific programmes. The agri-environment programmes of the

Table 1
The impact of agri-environment measures on biotic resources[a]

Measures		Impact on biotic resources (−5/+5)			
		Region 1	Region 2	Region 3	Average
1	Integrated farming	0.6	0.7	0.3	0.5
2	Environmental spreading of liquid manure on arable land	0.5	0.9	0.9	0.8
3	Integrated horticulture, permanent cops	1.0	0.8	0.9	0.9
4	No use of growth regulating product in arable farming	0.9	0.8	0.9	0.9
5	Mulch seeding	1.0	0.9	0.8	0.9
6	Under sowing	0.9	1.1	0.9	1.0
7	Environmental spread of liquid manure on grassland	1.3	1.4	1.3	1.3
8	Minimum four crop rotation	1.7	1.6	1.6	1.6
9	Extension of row distance	1.7	1.9	2.1	1.9
10	No use of mineral fertiliser in arable farming	1.9	2.1	2.0	2.0
11	Organic horticulture, permanent crops	2.2	2.2	2.4	2.3
12	Livestock density < 1.4 LU/ha forage area	2.1	2.5	2.7	2.4
13	Organic grassland	2.5	2.5	2.6	2.5
14	Reduced mineral fertilising on grassland	2.2	2.7	3.0	2.6
15	Late cut on grassland	2.6	2.6	2.7	2.6
16	Organic arable farming	2.7	3.0	2.7	2.8
17	Maintenance of abandoned land	2.3	2.7	3.6	2.9
18	Extensive use of grassland on single plots	3.2	3.2	3.0	3.1
19	Multiannual field margins	3.0	3.4	3.0	3.1
20	No use of herbicides in arable farming	3.4	3.6	3.5	3.5
21	No use of mineral fertiliser and plant protection on grassland	3.2	3.6	3.6	3.5
22	Conversion of arable land to extensive grassland	3.5	3.8	3.4	3.6
23	No fertilising and plant protection in environmentally sensitive areas	3.6	3.8	3.7	3.7
24	No use of mineral fertiliser and plant protection in arable farming	3.6	4.0	3.9	3.8
25	Multiannual strips along bank and shores	3.8	3.8	3.7	3.8
26	20-Year set aside for biotopes	4.1	3.8	3.5	3.8
27	Single plots under environmental contracts	4.1	3.9	3.9	3.9
28	Protective planting of trees and hedges	4.3	4.2	3.7	4.0
29	Extensive orchards	4.2	4.4	4.3	4.3

[a] Source: Wilhelm (1999b), own calculations.

federal states are very different. Some programmes have a simple structure containing only a few components while others are quite comprehensive. The agri-environment measures which are most often a part of the agri-environment programmes are listed in the first column of Table 1.

3. The impact of agri-environment measures on biotic resources

In order to evaluate the various German agri-environment programmes from the biotic resource conservation stand point, one must refer to information on their environmental impacts. This information can be obtained through estimates made by experts, analysis of the literature or by conducting research projects. Studies on the environment effects of agri-environmental programmes have taken place only for individual states (Wilhelm, 1999a). Obtaining useful results would have taken too long had we conducted our own research. Thus, the task to perform an ecological assessment was done with the help of the first method. The information about the environmental effects of the various agri-environment programmes was derived from experts' opinions in a study prepared in 1998 at the Department of Agricultural Economics of the University of Göttingen with the financial help of the Federal Ministry of Agriculture (Wilhelm, 1999b). Expert knowledge was collected and condensed with the help of the Delphi method, so that it could be used as an indicator for the environmental impacts.

The Delphi method originated in American National Defence Research. It was implemented for the first time in a study called the 'Project Delphi'. The study was to determine Soviet targets in the USA and the atomic bombs necessary for reaching them. A specially structured written method for questioning experts (later called the Delphi method) was developed to solve this problem. In a Delphi study, experts first give their individual opinions. Then all participants are provided with the anonymous results, and once again asked for their opinion. Any number of repetitions can be conducted. The last round of questioning is the relevant one. The Delphi method is being implemented more and more frequently in cases where the data are uncertain or not available. Environmental assessments are such cases (Richey et al., 1985). For example, the Delphi method was already applied in the assessment of elk habitat quality (Schuster, 1985) and for the evaluation of various policy measures for the conservation of wetlands (Bardecki, 1984).

The way the Delphi method operates has not been studied satisfactorily up to now. In particular, no cognitive psychological theory supporting the effectiveness of this method exists. The discussion on standards for conducting a Delphi study have, therefore, not been concluded (Murry and Hammons, 1995). When applying the Delphi method, it must be clear that one is using a method whose possibilities and limitations have not been sufficiently examined. Correspondingly, the results of a Delphi study must be carefully interpreted. Referring to the evaluation of athletic performances one can say: a Delphi study based evaluation is not an exact measurement such as the time in a 100 m dash, the distance in the shot-put, or the weight in weight lifting. It is similar to how figure skating, diving or dancing performances are judged. Thus, the results of a Delphi study certainly are not invariant, but they are substantiated and comprehensible.

For the ecological assessment of the German agri-environment measures 32 experts in agriculture and landscape ecology were questioned. Scientists were chosen who fulfilled the following three criteria: their research concentrates on the ecological effects of intensive agricultural practices, they have published their research results in scientific journals requiring a process of peer review, and they have been presenters at significant conferences. Pre-tests with three of the experts were conducted to develop a questionnaire. Twenty-nine agri-environment measures were to be evaluated based on the reference standard of good agricultural practice. Not only were the effects on the biotic resources asked, but also the effects on abiotic and aesthetic (landscape, cultural aspects) resources. The experts were free to assess agri-environment measures by differentiating between three regions: regions with potential yields above average, low quality agricultural regions, and low mountain ranges and hillsides. This differentiation was the result of the pre-test. The scale for evaluation was also developed based on the pre-test. It is comprised of 11 degrees and ranges from +5 to −5. The assessment +5 indicated that the measure leads to a significant environmental improvement. A score of −5 was given if a measure leads to a significant additional burden for environmental resources. A measure resulting in the same results as good agricultural practice was given a 0. The results of the Delphi study on the impacts on biotic resources are summarised in Table 1.

Integrated farming (measure 1) was defined as follows: maximum of 2.5 LU/ha livestock density; no conversion of grassland; plot related records; soil tests; balance of major nutrients; plant protection pursuant to the threshold principle; obligations concerning rotation and seed varieties. The regional impact on biotic resources of the measures analysed varies from 0.3 (integrated farming in region 3) to 4.4 (extensive orchards in region 2). Due to the lack of space, the scores cannot be discussed in detail. I would merely like to point out the value the experts placed on the 'level-effect' of the agri-environment measures. Important is not only the intended final outcome of the measures, but also from which initial situation the outcome has developed. Thus, for example, the experts award a higher score to converting available land to extensive grassland (measure 22) than the continuation of organic grassland (measure 13).

If one interprets the evaluation scale as an absolute scale, then one can combine the regional impacts and calculate an average ecological value (last column in Table 1). Using this average value as a criterion, one can say that biotic resources are best conserved by extensive orchard preservation, installation and maintenance of protective planting of trees and hedges, and nature protection.

The validity of the questionnaire's results were examined by considering the correlation coefficients and the standard deviation. Most of the correlation coefficients—KENDALL's τ_b—(Bortz et al., 1990) were greater than 0.5 (** $P < 0.01$) and the majority of the standard deviations were under 1.4. According to the literature on the evaluation of Delphi results, these figures confirm the validity of the questionnaire's results (Wilhelm, 1999b).

4. How effective are biotic resources protected by the agri-environment programmes?

The Delphi study results summarised in Table 1 makes it possible to evaluate the ecological effectiveness and the cost effectiveness of the German agri-environment programmes. First, I examine the ecological effectiveness of the agri-environment programmes, i.e., I compare how well they protect the biotic resources. The calculation for ecological effectiveness is based on the following formula.

Ecological effectiveness of the agri-environment programme of the federal state

$$f = \frac{\sum_{r=1}^{3}\sum_{m=1}^{29} a_{mr}^{f} v_{mr}}{100\,\text{ha}} = \frac{\text{EP}}{100\,\text{ha}}$$

where v_{mr} is the ecological value (=the impact on biotic resources as indicated in Table 1) of the measure m ($m = 1, \ldots, 29$) in region r ($r = 1, \ldots, 3$). This value is multiplied by the area upon which the measure is used in region r of the federal state f (a_{mr}^{f}). Then all products are summed up. The results of this calculation are labelled ecological points (EP), which are achieved by the agri-environmental programme of the federal state *f*. Due to the discrepancy in size of the federal states, a comparison of absolute values would not be very helpful. The prerequisite for a meaningful comparison of the effectiveness is a common value of reference. Therefore, the ecological effectiveness of the federal states' agri-environment programme are each calculated for an area of 100 ha. In Table 2 the federal states are ordered according to their EP/100 ha of the total agricultural area. All figures in Table 2 refer to 1999.

The second column of Table 2 shows that the programme in Baden Württemberg is ecologically most effective when protection of the biotic resources is considered in the context of the total agricultural area. Hamburg follows reaching 75% of the ecological effectiveness of Baden Württemberg. The least amount of EP/100 ha agricultural area are achieved in Schleswig-Holstein and Lower Saxony.

The need for the federal states to save money and the fact that promoting agricultural activities with greater ecological effectiveness costs more than conventional agricultural policy have led to two types of agri-environment programmes in Germany as well as in the EU: the shallow-and-wide-approach and the narrow-and-deep approach. The shallow-and-wide-approach strives for a high rate of participation amongst farmers. High degrees of coverage and adoption of schemes are seen as more important than specific ecological outcomes. The narrow-and-deep-

Table 2
Ecological effectiveness and cost effectiveness of the German agri-environment programmes in 1999[a]

		EP/100 ha agricultural area	Application area in % of agricultural area	EP/100 ha application area	Budgetary costs (rounded) in € million	Budgetary costs/EP (rounded)
1	Baden Württemberg	24	53.1	45	79.2	225
2	Hamburg	18	26.9	68	1.12	276
3	Saarland	18	39.7	45	2.4	174
4	Brandenburg	15	27.7	54	45.56	230
5	Thuringia	13	25	53	34.82	332
6	Saxony	10	65	20	54.71	460
7	Bavaria	8	63.3	13	277.94	1023
8	Mecklenburg-Western Pommerania	7	10	70	22.96	240.31
9	Hesse	6	16.2	38	20.35	383
10	North Rhine-Westphalia	6	6.2	96	12.47	138
11	Bremen	5	10.9	46	0.31	404
12	Rhineland-Palatinate	5	12.1	41	21.06	593
13	Saxony-Anhalt	3	10	33	23.93	511
14	Berlin	2	4.1	55	0.05	327
15	Schleswig-Holstein	1	2.1	70	3.48	215
16	Lower Saxony	0.5	1.2	41	11.71	869

[a] Source: Lotz (2001), own calculations.

approach concentrates on agricultural activities with major ecological effects. It is oriented to the solution of discrete and specific problems of environmental conservation in the farmers' countryside and lacks the ambition to involve as many of the country's farmers as possible. The third column in Table 2 shows the total agricultural area in percent upon which the agri-environment programmes are applied in the federal states in 1999, thus illustrating which federal states follow the shallow-and-wide approach and which follow the narrow-and-deep-approach.

The agri-environment programme of Baden Württemberg is a good example for the shallow-and-wide-type. In Baden Württemberg about 53% of agricultural land is subsidised. Extensification of arable land has priority claiming about 70% of funds, while extensification measures on grassland are supported by about 20% of funds. Nature protection and landscape management receive some 10%. The narrow-and-deep-approach is followed, for example, in North Rhine-Westphalia. Here only about 6% of agricultural land is subsidised. The emphasis lies on measures concerning management of nature and landscape which receive one-third of the available funds. Organic farming is also comparatively generously subsidised (30% of funds).

Since the share of the total agricultural area applying the programmes are different from state to state, it is also interesting to compare the ecological effectiveness with respect to the application area. The fourth column of Table 2 shows that when we take the agricultural area to which the programmes are actually applied to as a basis, the rank order of the federal states changes. Now, North Rhine-Westphalia has the top place not Baden Württemberg. Mecklenburg-Western Pommerania holds second place, and Bavaria is last. So with reference to the ecological effectiveness, we have two 'winners'. With respect to the total agricultural area, the biotic resources are protected most effectively in Baden Württemberg. If one considers only the application area, then North Rhine-Westphalia's agri-environment programme is the most ecologically effective.

From 1993 to 1999 a total of approximately € 2.66 billion were spent in Germany for implementing agri-environment programmes. The EU's contribution for this timeframe was approximately € 1.53 billion (almost 55%). The governmental expenditures within the framework of the agri-environment programmes are very different in the German federal states (see fifth column of Table 2). This is also true when one considers the expenditures per hectare. Thus, for example, less than € 3/ha of agriculturally used land were spent in 1999 in Schleswig-Holstein, but in Bavaria it was over € 84. The average amounts per hectare application area are also extremely different. In 1999 the amount ranged from € 79 (in Saarland) to € 366 (in Lower Saxony).

Which federal state spent its money most efficiently; where are the budgetary costs per EP the lowest? The cost effectiveness analysis which sets the ecological effectiveness in relation to the budgetary costs yields answers to these questions.

The cost effectiveness ratio (=budgetary costs/EP) of the agri-environment programmes are presented in the last column of Table 2. As one can see, this ratio is very different from state to state. The most cost effective agri-environment programme can be found in North Rhine-Westphalia. One EP costs approximately € 138 in governmental expenditure. The Saarland follows with € 36 more per EP. Saxony-Anhalt, Rhineland-Palatinate and Lower Saxony can be found in the lower places. Here more than € 500 in government expenditures are necessary to achieve one EP. Bavaria is in last place with budgetary costs of € 1023/EP.

To sum up, the evaluation of the ecological effectiveness and the cost effectiveness of the agri-environment programmes partially lead to the same result. North Rhine-Westphalia, the federal state where the biotic resources are most effectively protected with respect to the application area, also spent government funds for the protection of biological resources most effectively. Baden Württemberg, the federal state with the most effective agri-environment programme with respect to total agricultural area gains in the cost–effectiveness-analysis only placed fourth with € 225/EP.

5. The economic efficiency of biotic resource conservation by agri-environment programmes

From the economic point of view economic efficiency of a policy measure is more important than its cost effectiveness. Therefore, this section concentrates on the evaluation of agri-environment programmes with respect to their economic efficiency. Both German and European agriculture policy are aimed at greater usage of agri-environment programmes. Correspondingly, an ex-ante-analysis will be applied to determine whether expansion of the agri-environment programmes leads to an improvement in efficiency. Greater application of North Rhine-Westphalia's agri-environment programme would be recommended from the ecological perspective. This agri-environment programme is the ecologically most effective based on the area where it has been applied. Would an implementation of this programme on 90% of Germany's agricultural area be an efficiency improvement? This question will be answered below.

Greater application of agri-environment programmes leads to better protection of agriculturally managed biotic resources, but has a negative influence on commercial interests in the agricultural sector, such as job security, less expensive production costs, etc. In Fig. 1 both dimensions of this conflict are found on the axes of the coordinate system, whereby the various commercial interests are simplified by depicting the volume of (produced and consumed) commercial agricultural goods. The decision of the members of society for a certain level of commercial goods (x) simultaneously determines the extent to which biotic resource conservation (R) is possible. The K-curve models all commercial goods-biotic resource combinations that can be maximally achieved. In economic terms K describes the efficient solutions of the conflict "conservation of agriculturally managed biotic resources vs. commercial agricultural goods". Combinations below K are inefficient since society literally "gives" something away. This is so because it is possible either to increase the consumption of commercial goods at a constant level of conservation or realise a better conservation of biotic resources at a constant level of consumption of commercial goods. In Fig. 1 the current situation (A) is described as inefficient. This inefficiency is based on the fact that numerous improvement possibilities of the current agricultural, business, finance or nature conservancy policies exist which would lead to better biotic resource conservation without affecting the supply of commercial goods (Hampicke, 1991; Latacz-Lohmann, 2000).

The condition of nature and the landscape in rural areas, the current situation (A), is primarily determined by how and to what degree land and other natural resources are used. The factors affecting use are determined outside the agri-environment policy

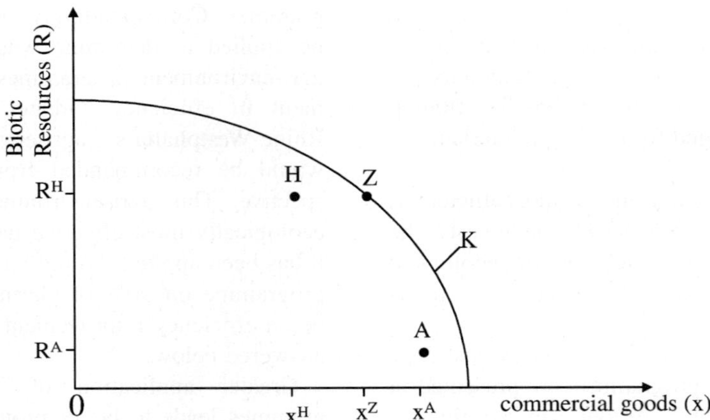

Fig. 1. Conflict of goals: protection of biotic resources vs. commercial interests.

realm to a great extent. This is true not only for the profit-oriented decisions of the farmers, but also for the planning and implementation of public investment and infrastructure projects in rural areas. So an efficient combination of commercial goods and biotic resource protection can only be attained if changes are made in many policy areas. If one concentrates only on agri-environment policy, then one renounces a part of the possible welfare gains. Thus, an increase in agri-environment programmes does not lead to a point on the *K*-curve. This policy measure leads only to another inefficient situation. In Fig. 1 this new situation is depicted as point H. Now, protection of biotic resources has improved (from R^A to R^H) compared to the initial situation. The improvement in resource conservation, however, is reached with a greater decrease in commercial goods ($x^A - x^H$) than necessary. Protection at the level R^H could also be attained by selecting point Z on the *K*-curve, i.e., together with a supply of x^Z (with $x^Z > x^H$).

Is it worthwhile to go from A to H, i.e., does the increase in agri-environment programmes lead to an efficiency improvement? One way of approaching this question is to start with a social welfare function and compare welfare levels before and after the evaluation of the agri-environment programmes. However, from Arrow's theorem we know that an ethically acceptable social welfare function does not exist (Arrow, 1951). The various utility functions (U_i) of the *n* members of the society that result from the respective interests in biotic resource conservation (R) and the commercial goods (x_i)

$$U_i(R, x_i), \quad i = 1, \dots, n \tag{1}$$

cannot be aggregated to a function for the whole society. This fact is taken into account when answering the above question with a cost–benefit analysis. In such an analysis the relevant criterion for decision making—the potential PARETO improvement—does not assume social preferences (Gans and Marggraf, 1997). In a cost–benefit analysis the effects of an action of all of society's members are examined. Those positively sensed consequences lead to economic benefits, and negatively sensed consequences to economic costs. The evaluation of an action occurs after comparing economic benefits and economic costs. Both sides of the coin are considered for the evaluation. That which was achieved and its cost (opportunity costs). For a cost–benefit analysis of agri-environment programmes it is useful to consider the indirect utility function (V_i) which is equivalent to the direct utility function (1):

$$V_i(R, T_i, p, I_i), \quad i = 1, \dots, n \tag{2}$$

Following the indirect utility function (2) the level of an individual's utility depends on how well he is supplied with the environmental good "biotic resources" (R), how much (direct and indirect) taxes (T_i) he must pay, how high the commercial goods' price (p) are and how high his (net) income (I_i) is.

How are these arguments of the indirect utility function affected by agri-environment programmes? These programmes improve the supply of environmental goods ($\Delta R > 0$). If they are financed through additional taxes (and not by higher national debt or a reduction in national expenditures in other areas), then tax payments increase ($\Delta T_i > 0$). Farmers participating in the agri-environment programmes produce less agricultural products. Thus export subsidies can be saved which would reduce the taxes for the individual households ($\Delta T_i < 0$). The price support system of the Common Agricultural Policy is not influenced by the agri-environment programmes. An induced price increase ($\Delta p > 0$) can only occur on agricultural markets without regulated prices. The income of those farmers participating in the programmes increase ($\Delta I_i > 0$), since participation is voluntary and only occurs when the premium is at least high enough to compensate income loss. The restrictions placed on agricultural production and the decrease of agricultural inputs can influence profits and incomes in agribusiness negatively ($\Delta I_i < 0$).

Better conservation of biotic resources ($\Delta R > 0$) leads to economic benefits. The economic benefits of an environmental improvement reflect society's willingness to pay for these improvements. From the results of already existing analyses of willingness to pay for the maintenance of agricultural landscapes, regional management of landscapes and general preservation of nature and species, we know that the average willingness to pay off private households in Germany totals at least € 61 per year (Hampicke, 1999). Thus, this sum corresponds to the minimum economic benefits of $\Delta R > 0$ per household and year.

Higher taxes ($\Delta T_i > 0$) have a negative effect on an individual's benefit level, and thus, lead to economic costs. The economic costs of taxes is due to their so-called excess burden that results from the induced change in the price ratios of commercial goods. Empirical studies have shown that the excess burden cost of taxes lies between 7 and 28% of the tax sum, with 15% being the probable average value (Musgrave and Musgrave, 1980). The economic costs of agri-environment programmes should not be underestimated. I therefore assume a rate of 30% in the following and neglect the decrease in tax burden due to lower export subsidies and the fact that the German agri-environment programmes are being financed by all tax payers in the EU member states.

In 1999, public expenditures for the agri-environment programme in North Rhine-Westphalia amounted to some € 12.42 million. In the same year government expenditures for agri-environment programmes in Germany totalled nearly € 1.02 billion. If all federal states would substitute their agri-environment programmes by the one of North Rhine-Westphalia, and if this programme was applied to 90% of the agricultural area of each state, public expenditures would increase by about € 25.56 billion (Stratmann and Marggraf, 2001). Nation-wide, an ecological and cost effective protection of biotic resources on 90% of the agricultural area, would thus amount to economic costs resulting from a higher tax burden of less than € 7.67 billion per year. For each of Germany's 37 million private households, economic costs of $\Delta T_i > 0$ would be less than € 21 per year. The economic value of price changes corresponds to the compensating variation of these changes. Agricultural products comprise only a small portion of the consumer's basket of goods. Thus, the economic costs of a price increase on agricultural goods are not very high. A 10% price increase on all agricultural goods leads, for example, to an annual economic cost per German household of € 6 (Münch, 1999). An unchanged price support system of the EU is assumed for the cost–benefit analysis of the agri-environment programmes. Thus, changes in price can only occur in a few agricultural products. So we can be sure that the annual economic costs of the price increase is less than € 6 per household.

The economic value of an income change corresponds to the income change. In benefit–cost analysis individual income changes can be evaluated as an aggregate. The income change of the farmers, workers and employers in agribusiness are opposing. When considered all together, they will cancel out ($\sum_i \Delta I_i = 0$) so that the change in income distribution leads to neither economic benefits nor to economic costs.

Thus, we have the following results. If we consider all economic benefit and cost components together, then an annual net-benefit of at least € $(61-21-6) = 34$ per household is gained. All the benefit and cost components exist every year, so discounting would have an influence on the sum, but would not turn the net benefit into a net loss.

6. Conclusion

In Germany more than 25 agri-environment programmes are implemented. All of these programmes have been compared ecologically and economically. The ecological evaluation was based on expert opinions that were gathered in a Delphi study. Two ecological effectiveness ratios were selected as indicators for the ecological evaluation. The indicators relate the ecological values of the agri-environment programmes to the application area or to the total agricultural area. If one considers the ecological effectiveness together with its financial costs (=government expenditures), then the cost effectiveness of the individual programmes can be determined. The following is the result for the German agri-environment programmes. If one considers the ecological effectiveness per hectare agricultural area of the different programmes, the programme of Baden Württemberg is the best. North Rhine-Westphalia ranks on top if the ecological effectiveness per hectare application area is evaluated. The agri-environment programme of North Rhine-Westphalia takes the first place, too, if cost effectiveness is the relevant criterion.

The economic efficiency of greater usage of agri-environment programmes was examined in the last section. It was shown that a better protection of biotic resources is connected with an economic net benefit in Germany. From society's perspective, even the sub-optimal method of desiring to attain greater conservation of biotic resources by strengthened agri-environment programmes alone is better than the status quo. Improved biotic resource conservation is ecologically and economically advantageous.

References

Arrow, K.J., 1951. Social Choice and Individual Welfare. Wiley, New York.

Baldoch, D., Lowe, D., 1996. The development of European agro-environment policy. In: Whitby, M. (Ed.), The European Environment and CAP Reform Policies and Prospects for Conservation. CAB International, Wallingford, pp. 8–25.

Bardecki, M.-J., 1984. Wetland conservation policies in southern Ontario—a Delphi approach. Geographical Monographs 16. York University.

Bortz, J., Lienert, G.A., Boehnke, K., 1990. Verteilungsfreie Methoden in der Biostatistik. Springer, Berlin.

Council Regulation, 1992. Council Regulation (EEC) 2078/92 of June 30, 1992 on agricultural production methods compatible with the requirements of the protection of the environment and the maintenance of the countryside. Off. J. Eur. Communities L215, 85–90.

Council Regulation, 1999. Council Regulation (EC) 1257/1999 of May 17, 1999 on support for rural development from the European Agricultural Guidance and Guarantee Fund (EAGGF) and amending and repealing certain regulations (June 26, 1999). Off. J. L160, 80–101.

Gans, O., Marggraf, R., 1997. Kosten-Nutzen-Analyse und ökonomische Politikbewertung. 1. Wohlfahrtsmessung und betriebswirtschaftliche Investitionskriterien. Springer, Berlin.

Hampicke, U., 1991. Naturschutz-Ökonomie. Ulmer, Stuttgart.

Hampicke, U., 1999. Conservation in Germany's agrarian countryside and the world economy. In: Dragun, A.K., Tisdell, C. (Eds.), Sustainable Agriculture and Environment. Globalisation and the Impact of Trade Liberalisation. Edward Elgar, Cheltenham, pp. 135–152.

Höll, H., von Meyer, H., 1996. Germany. In: Whitby, M. (Ed.), The European Environment and CAP Reform Policies and Prospects for Conservation. CAB International, Wallingford, pp. 70–85.

Latacz-Lohmann, U., 2000. Beyond the green box: the economics of agri-environment policy and free trade. Agrarwirtschaft 49, 342–348.

Lotz, J., 2001. Agrarumweltmaßnahmen 2001—Forschung von Buntbrachen durch den Bund. Informationen für die Agrarberatung, pp. 56–61.

Münch, W., 1999. Effects of CEC-EU accession on agricultural markets: a partial equilibrium approach. Dissertation. University of Göttingen.

Murry, J.W., Hammons, J.O., 1995. Delphi: a versatile methodology for conducting qualitative research. Rev. Higher Educ. 18, 424–436.

Musgrave, R.A., Musgrave, P.B., 1980. Public Finance in Theory and Practice. McGraw-Hill, New York.

Richey, J.S., Man, B.W., Horth, R.R., 1985. The Delphi technique in environmental assessment. Technol. Forecast. Soc. Change 23, 89–94.

Schuster, G., 1985. The Delphi-method—application to Elk habitat quality. Research Paper INT-353. United States Department of Agriculture.

Sachverständigenrat für Umweltfragen (SRU), 1985. Umweltprobleme der Landwirtschaft. W. Kohlhammer, Stuttgart, Mainz.

Stratmann, U., Marggraf, R., 2001. Ausgaben des staatlichen Naturschutzes in Deutschland: Analyse und Bewertung. Jahrbuch für Naturschutz und Landschaftspflege 53, 195–208.

Wilhelm, J., 1999a. Umweltwirkungen von Förderungsmaßnahmen gemäß VO(EWG)2078/92. Schriftenreihe des BMELF, Reihe A: Angewandte Wissenschaft 480, Landwirtschaftsverlag Münster-Hiltrup.

Wilhelm, J., 1999b. Ökologische und ökonomische Bewertung von Agrarumweltprogrammen. Delphi-Studie, Kosten-Wirksamkeits-Analyse und Nutzen-Kosten-Betrachtung. Peter Lang, Frankfurt am Main.

ELSEVIER

Available online at www.sciencedirect.com

Agriculture, Ecosystems and Environment 98 (2003) 517–527

Agriculture Ecosystems & Environment

www.elsevier.com/locate/agee

Is there a reliable correlation between hedgerow density and agricultural site conditions?

Jochen Kantelhardt*, Elisabeth Osinski, Alois Heissenhuber

Chair of Agricultural Economics, Center of Life and Food Sciences, Technical University of Munich, Alte Akademie 14, 85350 Freising, Germany

Abstract

The aim of this study was to determine whether the number of landscape elements and site conditions are reliably associated. As example, we considered the distribution of hedgerows on sites of varying agricultural quality. Experiments were carried out on two spatial levels: the first as the "natural unit", characterised as a region having nearly homogeneous site conditions; the second as the entire area of Baden-Wuerttemberg, comprising various natural units.

The association between ecological and economic information was analysed statistically. We observed that a considerable correlation between "agricultural site quality" and "hedgerow density" (−0.63) existed only on the spatial level of the natural unit, but here exclusively in the case of the less-favoured natural unit. In contrast, no such correlation was found on the spatial level of a federal state (Baden-Wuerttemberg). Although there is still a need for improving indicator deriving processes these results indicate that agri-environmental programmes concerning biotope protection would be best implemented on an administrative level, whose size and landscape homogeneity is comparable to that of the natural units.

Keywords: Agricultural site quality; Agri-environmental indicators; Hedgerow density; Subsidiary principle

1. Introduction

The present cultural-heritage landscape is mainly shaped by agriculture, whereby most of the land in middle Europe is under agricultural use and therefore strongly influenced by farming practices. Only a small percentage of the middle European land remains unused and has, consequently, a semi-natural character; this includes, e.g. landscape elements such as hedgerows and field edges. However, such elements also depend on agricultural land use (i.e. if land adjacent hedgerows remains unused, the hedgerows would develop into a forest). Therefore, it can be said, on the whole, that agriculture plays a multifunctional role: it produces not only food, but also maintains landscapes and biotopes, and consequently contributes to biodiversity (Heissenhuber and Lippert, 2000; OECD, 2001).

The role of agriculture in maintaining landscape elements is certainly not similar for every region. It depends on the respective site conditions, so that the general improvement of site conditions leads to more intensive land use (Fig. 1, Henrichsmeyer, 1977, p. 173). Consequently, a general decrease of landscape elements in favoured regions will result. In addition, nutrients and pesticides of intensive agriculture threaten still-existing landscape elements, whereas in marginal (or less-favoured) regions, the

* Corresponding author. Tel.: +49-8161-71-4046; fax: +49-8161-71-4426.
E-mail address: kantelhardt@wzw.tum.de (J. Kantelhardt).

0167-8809/$ – see front matter © 2003 Elsevier Science B.V. All rights reserved.
doi:10.1016/S0167-8809(03)00110-5

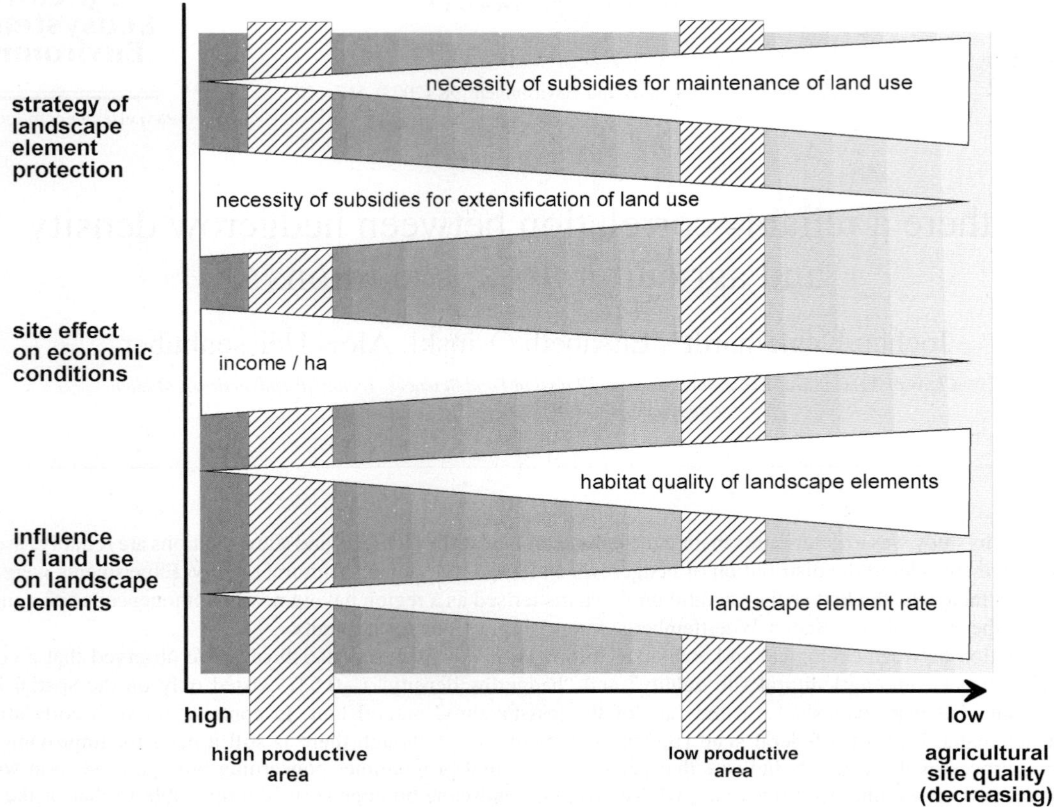

Fig. 1. Economic aspect of biotope protection dependent upon agricultural site conditions.

density of landscape elements will be higher. Landscape elements are also threatened in the latter, since agricultural profits are generally low and land use may be given up at all.

Approaches to protect landscape elements must be differentiated in respect to different existing threats. Thus, securing land use becomes important for regions used extensively, whereas for intensively used regions, it is necessary to reduce management intensity.

The intensification of land use taking place in the last decades has caused a reduction of landscape elements, considered from an agricultural viewpoint, mainly in the more favourable regions (SRU, 1985).

The goal of our study was therefore to determine the statistical association between landscape elements and site conditions, based on empirical data. The distribution of hedgerows in Baden-Wuerttemberg is used as example (cf. Osinski, 2003). Hedgerows are known to decisively shape landscapes in southern Germany. Likewise, there are regions completely lacking hedgerows, irrespective of prevailing site conditions. In such regions, the presence of hedgerows was never important (i.e. avoiding erosion, collecting and removal of stones from arable land); this applies, e.g. to permanent grassland regions.

Hedgerows are thus of high value for the agro-ecosystem. This applies not only to isolated hedgerows, which are important for different plant (see Schulze and Reif, 1984) and insect communities (Stechmann, 1984), but in particular to hedgerow complexes and networks (see Baudry et al., 2000), including hedgerows and adjacent grassland or farmland. Such complexes are also important habitats for wild birds (Heusinger, 1984) or other species types (see references in Baudry et al., 2000).

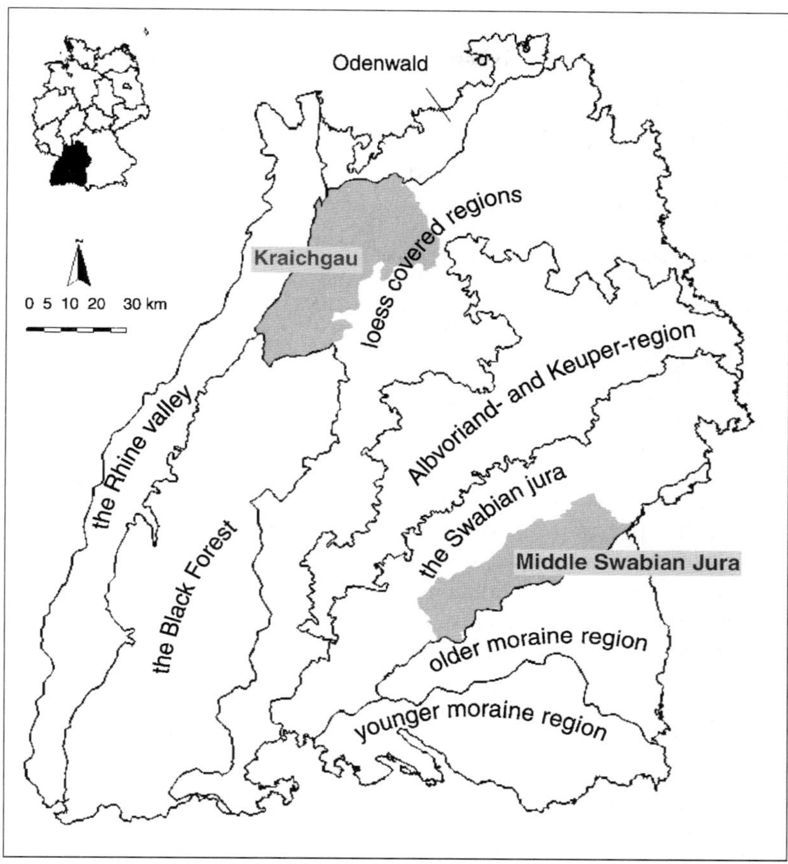

Fig. 2. Landscapes and selected natural units in Baden-Wuerttemberg.

2. Materials and methods

2.1. The study region Baden-Wuerttemberg and two examples of natural units

The study region Baden-Wuerttemberg is one of the German federal states situated in the southwest of Germany. It is approximately 35,000 km^2 in size (Fig. 2) and is subdivided into 1111 communities. Regarded from an agricultural viewpoint, Baden-Wuerttemberg shows a wide variance in site and production conditions. It can be divided into several, clearly distinguishable main landscapes (Fig. 2, cf. Osinski, 2003) and further subdivided into various "natural units", characterized by a nearly homogenous land use. Natural units better suited for agricultural purposes are nowadays dominated by intensively used fields (e.g. Rhine valley, Kraichgau). The less-favourable natural units, situated predominantly in mountainous areas (e.g. the Black Forest, the Swabian Jura), are used as grassland interspersed with arable land.

In addition to examining the federal state Baden-Wuerttemberg, this study also focuses on two natural units, "Kraichgau" and the "Swabian Jura". The natural unit Kraichgau is a highly productive area in the northwest of Baden-Wuerttemberg, approximately 1600 km^2 in size and comprising 83 communities.[1] The agricultural land in Kraichgau is composed primarily of loess soil having a high "production

[1] Because natural and administrative conditions do not coincide, some of the communities not belong completely to the natural unit.

Table 1
Database used for deriving indicators in Baden-Wuerttemberg

Derived GIS map	Resolution	Database	Reference
Area of hedgerow–biotope complexes	1:25,000	Biotope mapping (Biotopkartierung) Baden-Wuerttemberg	Landesanstalt fuer Umweltschutz 80er Jahre (Hoell and Breunig, 1995)
Type of land use (agriculture, forest, etc.)	30 m × 30 m	Satellite image classification	Institut fuer Photogrammetrie und Fernerkundung, 1993[b]
Production potential[a] community size	Community	Statistical database of Baden-Wuerttemberg	Statistisches Landesamt Baden-Wuerttemberg, 1991[c]
Average field size	1 km × 1 km	Aerial photo classification 1993	Fichtner et al., 1994

[a] Production potential: average value per community (German: Ertragsmesszahl, EMZ).
[b] Institut fuer Photogrammetrie und Fernerkundung, Universitaet Karlsruhe, Karlsruhe: Erstellung einer Landnutzungskarte des Landes Baden-Wuerttemberg. Abschlussbericht im Auftrag des MLR Baden-Wuerttemberg, unpublished.
[c] Statistisches Landesamt Baden-Wuerttemberg, 1991. Struktur- und Regionaldatenbank.

potential". In order to avoid erosion, the land was shaped into terraces. Thus, the appearance of shrubs and bushes on the edges of the terraces occurred (compare to Kleyer, 1991). Due to land consolidation measurements, these landscape elements were drastically reduced and are now restricted to some parts of Kraichgau.

The second natural unit, the middle part of the Swabian Jura (Fig. 2), is a less-favoured region located in the central-southern aspect of Baden-Wuerttemberg; its area is approximately 1100 km^2 and comprises 43 communities.[1] The Swabian Jura is characterised by soils originating from limestone, locally shallow and dry. In the past, stones which surfaced during ploughing were collected and relocated to the field margins, whereupon hedgerows developed. Although land consolidation has removed some hedgerows, many still remain.

2.2. Methodology for deriving indicators

Generally, indicators should quantify information and simplify complex phenomena (see Brouwer and Crabtree, 1999). Further, indicators should always be based on data readily available or that which can be easily obtained.

In order to derive indicators that describe site conditions, as well as hedgerow density, it is also necessary to consider the corresponding spatial distribution. Therefore, a geographical information system (GIS) is applied. A GIS is a tool to produce maps for planning purposes, documentation and decision finding (Bill, 1996). Resulting maps and corresponding database are shown in Table 1.

Maps were derived as follows:

- The area comprising the hedgerow complexes was derived from the biotope mapping data of Baden-Wuerttemberg, whereby the complexes consist of hedgerows and the agricultural area in between. A maximum inter-hedgerow-distance of 300 m was defined (Osinski, 2003).
- The predominant land use within the hedgerow complexes was analysed based on a satellite image classification (detailed description of methodology refers to Osinski, 2003).
- The site-specific production potential is based on the "Ertragsmesszahl" (EMZ). This index is used to classify the productivity of the German agricultural surface. The data is derived from a land-wide soil evaluation conducted in the year 1934 and corrected by some site-specific factors. The highest index value is 100, reflecting the best soil in Germany (Bauer, 1993). For purposes of statistical analysis in this study, the average yield index was calculated per community.
- The average field size was based on an aerial photo classification (Fichtner et al., 1994). For this analysis the entire area of Baden-Wuerttemberg was subdivided in 1 km^2 grids and, for each grid, the average field width was estimated (<50, 50–100, 100–200 and >200 m). In a second step the field width was transformed to field sizes, whereby the corresponding field size classes were assumed to be <1, 1–2, 2–5 and >10 ha.

Table 2
Selected ecological and economic indicators on community level in Baden-Wuerttemberg

Indicator	Description (or parameterisation) of indicators
Hedgerow density	Area of hedgerow complexes in the (community (ha)/community size (ha)) × 100
Production potential	Average production potential per community (calculated on basis of the entire agricultural land in the community)
Field size	Average field size per community (calculated on basis of the agricultural land belonging to biotope complexes)
Agricultural site quality	Indicator combining production potential and field size per community (indicator derived with the help of a regression equation)

The indicators actually used in the study were derived by overlaying the different GIS maps (Table 2) and determining a common spatial resolution. Since the municipality was the lowest common data resolution, all other databases having different resolution such as square kilometre (average field size) and irregular single polygons of various sizes (hedgerow complexes) were adapted to that resolution.

Four indicators were used in the present study (Table 2) and defined as follows:

- The hedgerow density (by area percentage of hedgerow complexes per community) indicates the ecological quality of the landscapes under consideration.
- The agricultural site conditions are described using two indicators: field size and production potential, both being of economic relevance for land use. Thus, it is apparent that field size influences machine costs, and that production potential determines profit level (compare to Wechselberger, 2000; Burgmaier et al., 1997). Attention should be directed to the fact that the indicator field size exclusively refers to particular field sizes included in the hedgerow complexes.
- The fourth indicator, agricultural site quality, which was subsequently developed in the study, combines ecological and economic information.

2.3. Testing the correlation between hedgerow density and site conditions

The hypothesis that site conditions and hedgerow density are associated was tested with a bivariate correlation analysis using the Pearson's correlation coefficient. The test is run on the spatial level of Baden-Wuerttemberg, as well as on the level of the natural units (Table 3).

In the following step, an indicator combining economic and ecological information was developed and which was applied in a multiple linear regression analysis considering site conditions and hedgerow density. Field size and production potential were defined as independent variables. The dependent variable was hedgerow density. If the resulting R^2 approached 1.00, then the site conditions could reliably describe the hedgerow density, and the regression equation could be used in the study region as an indicator for hedgerow density. The indicator was designated agricultural site quality.

Furthermore, the extent to which the attributes of the natural units contributed in explaining the distribution of hedgerow density in Baden-Wuerttemberg was also studied. Here, the natural unit itself was introduced as independent variable (in addition to field size and production potential) into the regression analysis by integrating the natural units into the regression equation as "dummy" variables, with the possible values 0 and 1.

Table 3
Statistical approaches for testing indicators

Spatial level	Correlation analysis	Regression analysis
Bundesland (Baden-Wuerttemberg)		
Dependent variable	Hedgerow density	Hedgerow density
Independent variables	Field size Production potential	Field size Production potential Swabian Jura Kraichgau
Natural Units (Swabian Jura, Kraichgau)		
Dependent variable	Hedgerow density	Hedgerow density
Independent variables	Field size Production potential	Field size Production potential

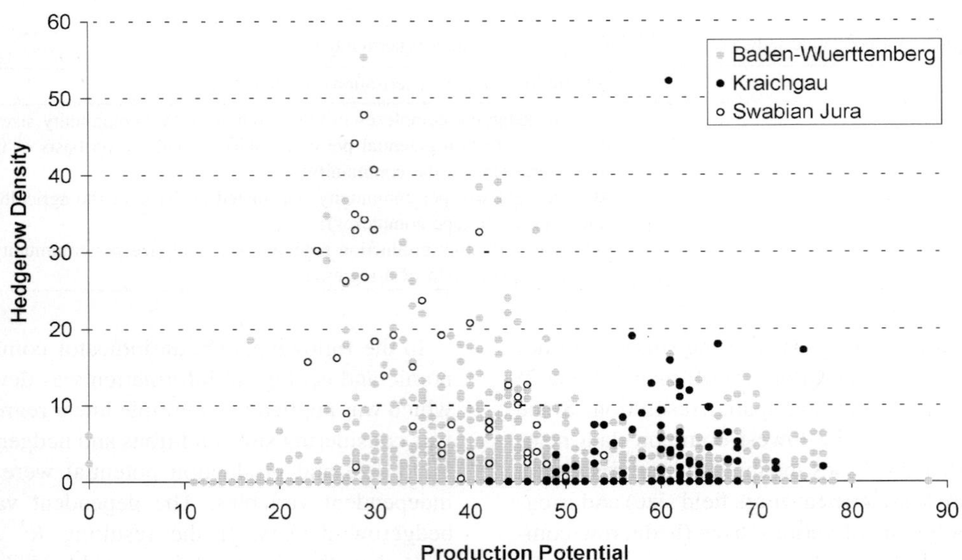

Fig. 3. Scatter data plot of hedgerow density (per community) and average production potential (per community).

3. Results

Fig. 3 is a scatter data plot of hedgerow density (on the basis of hedgerow complexes) and production potential. No clear association could be found between these two factors if hedgerow density was <10%. This finding could be explained by the observation that in many communities of Baden-Wuerttemberg, hedgerows are not important due to the lack of functional significance (i.e. avoidance of erosion). These additional data points (Fig. 3) interfered with the correlation analysis. If hedgerow density exceeded 10%, a correlation appeared as within this area most of the communities are characterised by a low production potential.

The communities belonging to the two natural units chosen as study region (the Swabian Jura and Kraichgau) are also shown here. As expected, the mean hedgerow density in the Swabian Jura communities was higher than those in Kraichgau.

Considering these data (Fig. 3) it may be assumed that a correlation between hedgerow density and production potential in the Swabian Jura exists at best, and which is absent in Baden-Wuerttemberg and Kraichgau; this assumption was corroborated by statistical analysis (Table 4), namely, that hedgerow density is associated with production potential with −0.63 (see Table 5). In contrast to that on the level of Baden-Wuerttemberg, the correlation between production potential and hedgerow density is weaker (−0.11). In Kraichgau the correlation is not significant at a 95% significance level.

The correlation between field size and hedgerow density was with maximally −0.39 in the case of the Swabian Jura generally weaker than that between production potential and hedgerow density, or showed an inverse relationship: in both Kraichgau (0.31) and Baden-Wuerttemberg (0.21) with increasing field size an increasing hedgerow density occurred.

Table 4
Correlation (significance) between hedgerow density and production potential or field size

Region	Correlation between hedgerow density and	
	Production potential[a]	Field size
Baden-Wuerttemberg	−0.11 (0.00)	0.21 (0.00)
Swabian Jura	−0.63 (0.00)	−0.39 (0.01)
Kraichgau	0.20 (0.08)	0.31 (0.01)

[a] Based on Ertragsmesszahl, EMZ (production potential, see Section 2).

Table 5
Regression equation for selected regions to explain hedgerow density

Region	Independent variables	Coefficients	R^2	Significance
Baden-Wuerttemberg (Case I)[a]		4.76[c]	0.06	0.000
	Field size	0.82		
	Production potential	−0.06		
Baden-Wuerttemberg (Case II)[b]		3.90[c]	0.18	0.000
	Swabian Jura	11.82		
	Field size	0.69		
	Production potential	−0.05		
	Kraichgau	1.67		
Middle Swabian Jura[a]		60.74[c]	0.51	0.000
	Production potential	−0.96		
	Field size	−3.71	0.09	0.005
Kraichgau[a]		1.77[c]		
	Field size	0.97		

Dependent variable (variable to explain): hedgerow density.
[a] Possible independent variables (variables to enter): field size, production potential.
[b] Additional possible independent variables (variables to enter): natural unit Middle Swabian Jura, natural unit Kraichgau.
[c] Constant.

The overall of regarding the association between hedgerow density and production potential are shown in Fig. 4. Hedgerows in the Swabian Jura are situated in places where the agricultural site conditions are less favoured, especially in the western part of the region (Fig. 4, upper panel). As expected, hedgerow density in Kraichgau is lower than that in the Swabian Jura, but no clear correlation to site conditions was present. Other factors not considered here might play important roles.

By applying regression equations, we have confirmed the estimation of hedgerow density by factors of agricultural site conditions (Table 5). In the Baden-Wuerttemberg (Case I in Table 5) and Kraichgau regions, the site conditions indicators, field size and production potential, could explain only 6 and 9% of the hedgerow density, respectively (compare R^2, Table 5). In contrast, over 50% of the hedgerow density was explained in the Middle Swabian Jura.

As already indicated by the correlation analyses, the regions Baden-Wuerttemberg, Middle Swabian Jura and Kraichgau differed in respect to the coefficients of the regression equation. In the Middle Swabian Jura, there was a clear negative correlation between hedgerow density and agricultural site conditions. In contrast, within the more productive Kraichgau region, the hedgerow density increased with increasing field size and not decreased as it would have been perhaps expected. The independent variable, production potential, was excluded from the regression analysis due to lack of significance. Further, in the entire area of Baden-Wuerttemberg, field size and hedgerow density were inversely correlated.

It can be stated that agricultural site qualities as an indicator for hedgerow density could only be confirmed in the natural unit Middle Swabian Jura. This finding supports the assumption that natural units themselves or their attributes play a significant role in explaining hedgerow density. We could confirm this assumption by introducing the natural units themselves into the regression analysis (see Baden-Wuerttemberg, Case II in Table 5); both the Swabian Jura and Kraichgau regions become part of the solution. However, the Swabian Jura appears to be a more important factor in explaining hedgerow densities. Taken together, the solution (R^2) increases from 6 to 18% by considering both natural units.

Lastly, we studied the contribution of an additional variable, "product of field size and production potential", in explaining hedgerow densities. According to the assumption of a mutually increasing effect of field size and production potential on hedgerow density, this last variable was included in

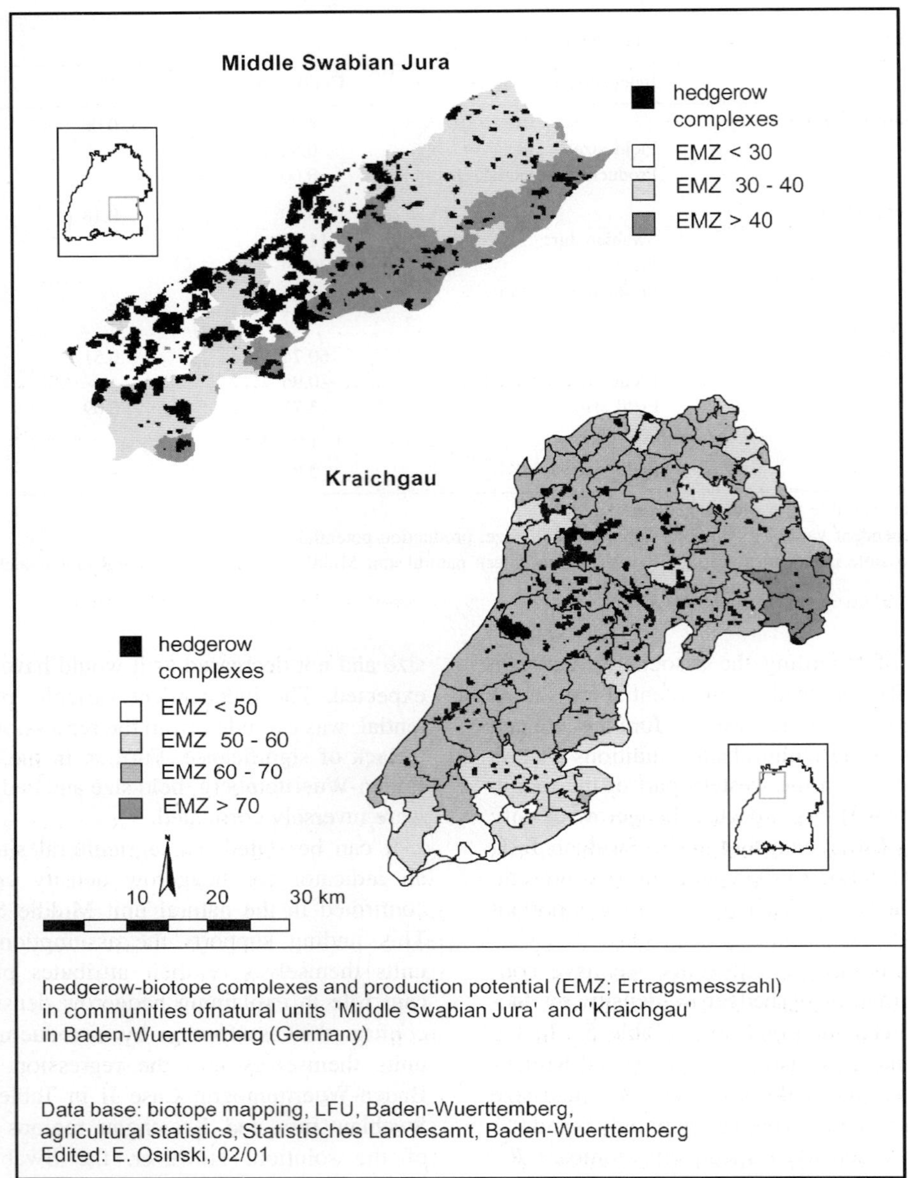

Fig. 4. Distribution of hedgerow biotopes in the natural units "Middle Swabian Jura" and "Kraichgau" (Baden-Wuerttemberg, Germany).

the regression equation. However, no effects on the R^2 were observed.

The result of the regression is shown in Fig. 5. As example, the region showing the highest correlation between agricultural site quality and hedgerow density—the Middle Swabian Jura—was chosen. It becomes clear that in this natural unit, the actual hedgerow density could be thoroughly explained by both the field size and production potential. Hence, the following truth arises: the higher the agricultural site quality within a given community, the lower the hedgerow density observed therein.

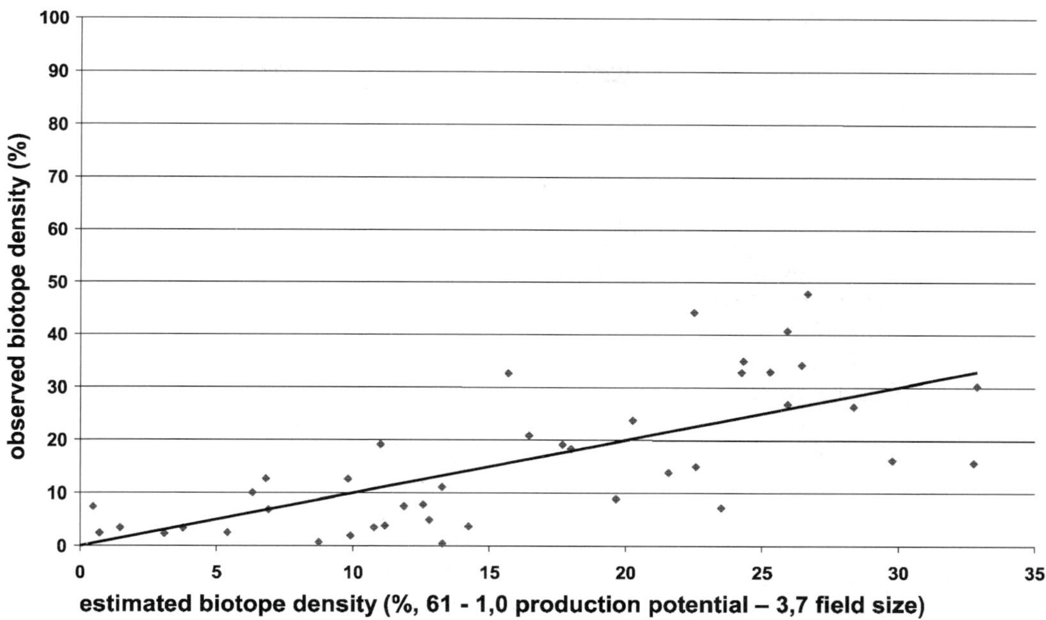

Fig. 5. Relation of hedgerow density and "agricultural site quality" in the natural unit "Middle Swabian Jura".

4. Discussion

The aim of the present study was to determine a reliable association between agricultural site conditions and density of hedgerow complexes, on the spatial level of a German federal state and two of its natural units. However, this could not be reached on all spatial levels using the indicators field size and production potential. A clear correlation could be only found in the less-favoured natural unit Middle Swabian Jura. These results agree with those of Reif et al. (1982), who demonstrated that the highest rates of hedgerow densities occurred in Muschelkalk and Jurassic limestone landscapes in Bavaria, which are comparable to the Swabian Jura.

On the level of the federal state of Baden-Wuerttemberg, as well as within the favoured natural unit Kraichgau, only a weak correlation existed. Hedgerow density cannot be thus generally associated with agricultural site conditions. Nonetheless, as the positive contribution of the natural units themselves to the regression equation show, other factors appeared to be highly correlated with the natural units themselves and seem to be important. One such possible factor could be the relief, as reported for the Muschelkalk and Keuper sandstone regions by Reif et al. (1982). However, historical aspects may have also a high impact.

Our study further shows that in especially intensively used regions, only few hedges still exist. This might be an effect of the land consolidation process that begun in the 1950s (Reif et al., 1982), where nearly half of the hedgerows were removed from agricultural landscapes. Remaining landscapes containing hedgerow biotopes are dissimilar in site qualities. Consequently, efforts to protect remaining hedgerow biotopes by implementing a single programme over the entire area of Baden-Wuerttemberg are likely to fail.

In the case of the Middle Swabian Jura, for example, the attainable agricultural income is low. Hence, land use is in danger of being abandoned. European policies should therefore aim to protect existing hedgerows, maintain agricultural land use and keep it extensive. Since there is a high and significant correlation between hedgerow density and site conditions, the indicator "agricultural site quality" could be applied in this natural unit as a basis for an agri-environmental programme. Subsidies could be thus granted according to agricultural site conditions, with the goal of maintaining land use and consequently preserving hedgerows. This principle is practised in the "eco-point programme", an agri-environmental programme in lower Austria (see Niederoesterreichische Agrarberichtsbehoerde, 1999). In contrast, no correlation between site quality conditions and hedgerow density was found in Kraichgau and, consequently, subsidies cannot be granted for agricultural site conditions. In this natural unit, agri-environmental programmes should first aim to establishing additional hedgerows to avoid erosion.

This study shows that the conservation of hedgerows cannot be solved on the level of the federal state Baden-Wuerttemberg. Furthermore, agri-environmental programmes require a framework on the federal state level, in order to secure targets such as bird protection and which can be only achieved on a large-scale area. These programmes must be nonetheless formulated and refined to meet farmers' needs at a lower level (see the concept of differentiated agri-environmental policy, Heissenhuber, 1995). It would be thus necessary to transfer political and financial competence to an appropriate administrative level, corresponding to the principle of subsidiary (compare to Ewers and Henrichsmeyer, 2000): each political task should be carried out at a level yielding optimum results.

In general, in future studies the database should be improved. For example, the estimation of hedgerow density in this study is based on the area of entire hedgerow complexes including also farm land and small forests. It is to assume that the purely hedgerow area might correspond better with agricultural indicators. Furthermore, the data used are based on different spatial reference units, because it was necessary to adapt the data to the lowest resolution. In order to achieve a higher congruence of the different information sources, a more precise database is needed. In general, the influence of the data type on the result—that is, especially the indicator—has still to be checked thoroughly.

Taken together, we can state that the approach presented by this study is suitable in combining economic and ecological criteria, with regard to a more efficient landscape conservation. This approach corresponds to the request of the OECD (2000) to estimate costs incurred by farmers in landscape improvement and maintenance. Our study further confirms the importance of natural units as a basis for studying landscape protection targets. Altogether, efforts to expand the approach presented here by integrating new indicators, such as topography and historical aspects, merit future attention with a focus on data quality.

Acknowledgements

We hereby thank Prof. G. Kaule, of the Institute for Landscape Planning and Ecology, who gave permission to use the database of land use, field sizes and biotopes and Dr. Ch. Lippert, of the Chair of Agricultural Economics, who gave valuable advises while writing this article.

References

Baudry, J., Bunce, R.G.H., Burel, F., 2000. Hedgerows: an international perspective on their origin, function and management. J. Environ. Manage. 60, 7–22.

Bauer, D., 1993. Landwirtschaftliche Betriebslehre—Produktionsfaktoren. In: BLV Verlagsgesellschaft (Ed.), Die Landwirtschaft. Muenchen, pp. 465–525.

Bill, R., 1996. Grundlagen der Geo-Informationssysteme. Bd. 2: Analysen, Anwendungen und neue Entwicklungen. Wichmann, Heidelberg.

Brouwer, F., Crabtree, B., 1999. Introduction. In: Environmental Indicators and Agricultural Policy. CAB International, pp. 1–11.

Burgmaier, K., Gerner-Haug, I., Wieland, H.-P., 1997. Arbeits- und betriebswirtschaftliche Auswirkungen der Biotopvernetzung in einer Ackerlandschaft—exemplarische Untersuchung in einem 135 ha-Betrieb im Kraichgau. Landinfo 4/97.

Ewers, H.-J., Henrichsmeyer, W. (Eds.), 2000. Agrarumweltpolitik nach dem Subsidiaritaetsprinzip. Schriften zur Agrarforschung und Agrarpolitik, Analytika, vol. 1. Berlin.

Fichtner, K., Osinski, E., Kick, U., 1994. Luftbild-Kartierung von Bewirtschaftungsstruktur und Gehöelzausstattung in den ländlichen Gebieten Baden-Württembergs. Materialien des Instituts für Landschaftsplanung und Ökologie der Universität Stuttgart (unpublished).

Heissenhuber, A., 1995. Betriebswirtschaftliche Aspekte der Honorierung von Umweltleistungen in der Landwirtschaft. In: Dachverband Agrarforschung (Ed.), Schriftenreihe agrarspectrum, vol. 24. pp. 123–138.

Heissenhuber, A., Lippert, C., 2000. Multifunktionalitaet und Wettbewerbsverzerrungen. Agrarwirtschaft 49 (7), 249–252.

Henrichsmeyer, W., 1977. Agrarwirtschaft – Raeumliche Verteilung. In: Albers, W. et al. (Eds.), Handwoerterbuch der Wirtschaftswissenschaften, vol. 1. Stuttgart, pp. 169–185.

Heusinger, G., 1984. Untersuchungen zum Brutvogelbestand verschiedener Heckengebiete. In: Akademie fuer Naturschutz und Landschaftspflege (Eds.), Die tieroekologische Bedeutung und Bewertung von Hecken, vol. 3, No. 2. pp. 99–123.

Hoell, N., Breunig, T., 1995. Biotopkartierung Baden-Wuerttemberg. Ergebnisse der landesweiten Erhebung 1981–1989, Beih. Veroeff. Naturschutz Landschaftspflege Baden-Wuerttemberg 81. Karlsruhe.

Kleyer, M., 1991. Die Vegetation linienfoermiger Kleinstrukturen in Beziehung zur landwirtschaftlichen Produktionsintensitaet. Diss. Botanicae 169, J. Cramer, Berlin, Stuttgart.

Niederoesterreichische Agrarberichtsbehoerde, 1999. Das Regionalprogramm Oekopunkte. In: GD XI (Eds.), Umweltbewertungsverfahren fuer die Landwirtschaft. pp. 59–72.

OECD, 2000. Environmental Indicators for Agriculture. Methods and Results. Vol. 3, OECD, Paris.

OECD, 2001. Multifunctionality—Towards an Analytical Frame work. OECD Publications, Paris.

Osinski, E., 2003. Operationalisation of a landscape oriented indicator. In: Buechs, W. (Ed.), Biotic indicators for biodiversity and sustainable agriculture. Agric. Ecosyst. Environ. 98, 371–386.

Reif, A., Schulze, E.-D., Zahner, K., 1982. Der Einfluss des geologischen Untergrundes, der Hangneigung, der Feldgroesse und der Flurbereinigung auf die Heckendichte in Oberfranken. In: Akademie fuer Naturschutz und Landschaftspflege (Eds.), Berichte, vol. 6. pp. 231–253.

Schulze, E.-D., Reif, A., 1984. Die Bewertung der nordbayerischen Hecken aus botanischer Sicht. In: Akademie fuer Naturschutz und Landschaftspflege (Eds.), Die pflanzenoekologische Bedeutung und Bewertung von Hecken, vol. 3, No. 1. pp. 141–145.

SRU (Rat der Sachverstaendigen fuer Umweltfragen), 1985. Umweltprobleme der Landwirtschaft. Kohlhammer, Stuttgart.

Stechmann, D., 1984. Ergebnisse des Klopfproben-Programms. In: Akademie fuer Naturschutz und Landschaftspflege (Eds.), Die tieroekologische Bedeutung und Bewertung von Hecken, vol. 3, No. 2. pp. 99–123.

Wechselberger P., 2000. Oekonomische und oekologische Beurteilung unterschiedlicher landwirtschaftlicher Bewirtschaftungsmassnahmen und -systeme anhand ausgewaehlter Kriterien. FAM-Bericht 43, Shaker, Aachen.

Available online at www.sciencedirect.com

Agriculture, Ecosystems and Environment 98 (2003) 529–539

Agriculture Ecosystems & Environment

www.elsevier.com/locate/agee

Consideration of biotic nature conservation targets in agricultural land use—a case study from the Biosphere Reserve Schorfheide-Chorin

Andreas Meyer-Aurich [a,*], Peter Zander [b], Mathias Hermann [c]

[a] *Chair of Agricultural Economics, Center of Life and Food Sciences, Technical University of Munich, Alte Akademie 14, 85350 Freising, Germany*

[b] *Department of Socioeconomics, Centre for Agricultural Landscape and Land Use Research (ZALF), Eberswalder Straße 84, 15374 Müncheberg, Germany*

[c] *OEKO-LOG.COM, Hof 30, D-16247 Parlow, Germany*

Abstract

The practices of land use should take into account the demands of nature conservation since several typical species of the agricultural landscape are close to extinction. But there is a lack of indicators which quantify the compatibleness of agricultural land use practices with nature conservation targets as a prerequisite for the economic–ecologic optimization of land use. This paper presents the development and application of indices for cropping practices which assign the contribution of the cropping practices for two biotic nature conservation targets: "preservation of amphibian populations in the agricultural landscape" and "preservation of partridge populations (*Perdix perdix*) in the agricultural landscape". The development of the indices is based on expert knowledge and shows the conformity of cropping practices with the requirements of specific animal species living in the agricultural landscape. The application of the indices on the evaluation of land use of 20 farms within the Biosphere Reserve Schorfheide-Chorin shows that conformity of cropping practices varies substantially between farms. Hence, there seems to be a potential to consider the requirements of the considered species in land use. Model calculations illustrate the economic effects of the integration of nature conservation targets in land use.

Keywords: Environmental impact; Species conservation; Integrated ecological–economic modeling

1. Introduction

The biotic environmental quality of agricultural landscapes is strongly influenced by the prevailing land use systems, because agricultural activities determine the habitats of species living in a specific landscape. With the development of new technologies, the perfection of farm machinery operations and the increased use of chemical inputs in agriculture, as a side effect habitats for wild species have been changed or destroyed and conditions for reproduction have been worsening (Flade et al., 2003; Potts, 1986; Beebee, 1996). Facing these undesired developments, political action is needed to take control over human activities which threaten wild species (Swanson, 1997). Therefore, biotic environmental indicators are needed to support appropriate policy recommendations. However, the derivation of agri-environmental indicators for biodiversity, wildlife habitat and landscape has proved to be a difficult challenge, because of the

* Corresponding author. Tel.: +49-8161-713878; fax: +49-8161-714426.
E-mail address: meyer-aurich@wzw.tum.de (A. Meyer-Aurich).

0167-8809/$ – see front matter © 2003 Elsevier Science B.V. All rights reserved.
doi:10.1016/S0167-8809(03)00111-7

complexity and lack of understanding of many of the ecosystem processes involved (OECD, 1999). Plachter and Korbun (2003) state an increasing "rift between scientific knowledge and application" in nature conservation. The discussion of naturalness, biodiversity, stability and uniqueness lead Plachter and Korbun (2003) to the conclusion, that "site specific targets have to be defined" and "transparent steps and criteria for specification and subsequent ranking of regional targets are needed for the site specific application of the conceptional targets".

To illustrate the steps that have to be undertaken in order to come to considerate agricultural cropping practices we present in the following some results of the "Schorfheide-Project". Within this project, a regional framework of nature conservation targets has been established for the agricultural landscape of the Biosphere Reserve "Schorfheide-Chorin" in Brandenburg, Germany (Plachter and Korbun, 2003; Herrmann, 2003). Agricultural and biological scientists together identified and developed indicators to estimate the degree of conformity of cropping practices with the requirements of endangered species.

This paper presents the development and application of indices for cropping practices. These assign the contribution of the cropping practices for two nature conservation targets: "preservation of amphibian populations[1] in the agricultural landscape" and "preservation of partridge populations in the agricultural landscape". The considered species use the agricultural landscape as habitat and therefore are disposed to agricultural inputs and machinery which can have a lethal effect on them. Recent studies give evidence that agricultural practices may play a crucial role in preserving the considered species (Oldham et al., 1997; Schneeweiß and Schneeweiß, 1997; Herrmann and Mueller-Stieß, unpublished; Pegel, 1987). Furthermore, a significant decline in the populations worldwide as well as in the study region indicate the importance of targeting special efforts to preserve these species (Potts, 1986; Beebee, 1996; Flade et al., 2003).

[1] Considered species were: common toad (*Bufo bufo*), fire-bellied toad (*Bombina bombina*), spadefood toad (*Peolobates fuscus*), moor frog (*Rana arvalis*), edible frog (*Rana esculenta*), smooth newt (*Triturus vulgaris*), crested newt (*Triturus cristatus*).

2. Methods

The study region comprises the agriculturally cultivated area within the Biosphere Reserve "Schorfheide-Chorin" in Brandenburg, Germany. Field specific surveys of the cropping practices including timing and applied inputs were conducted on the major farms within the Biosphere reserve from 1994 to 1997. Land use of the surveyed farms covers 88% of the acreage of agricultural land use in the biosphere reserve and represents the dominating land use of the bigger farms in that area.

Stress indices for partridge and amphibians were derived from agricultural activities which were assumed to determine habitat quality in the agricultural fields, based on the knowledge from experts working on the biology of partridge and amphibians in the study region (Herrmann and Dassow, 2003; Herrmann and Fuchs, 2003; Schneeweiß, 2003). The resulting indices represent partial goal achievements of the cropping practices for the proposed nature conservation targets "preservation of amphibian populations in the agricultural landscape" and "preservation of partridge populations in the agricultural landscape". The index values range from 0 to 1, where "0" indicates no conformity and "1" full conformity with the nature conservation target. The methodology of this approach is described in more detail in Meyer-Aurich (2001).

2.1. Development of a stress indicator for partridge

Partridge populations in the agricultural landscape are affected, first, by direct impacts of hazardous machinery in cropping practices, second, by the quality of the habitat of the production fields and third, by the composition of these habitat elements in the agricultural landscape (Potts, 1986). Four partial match grades were calculated here which determine the goal achievements of the cropping practices:

- specific disturbance impact and frequency of harmful machinery use;
- number of herbicide applications;
- number of insecticide applications;
- amount of applied nitrogen.

Especially during the breeding season partridge is endangered by agricultural machinery operations. The

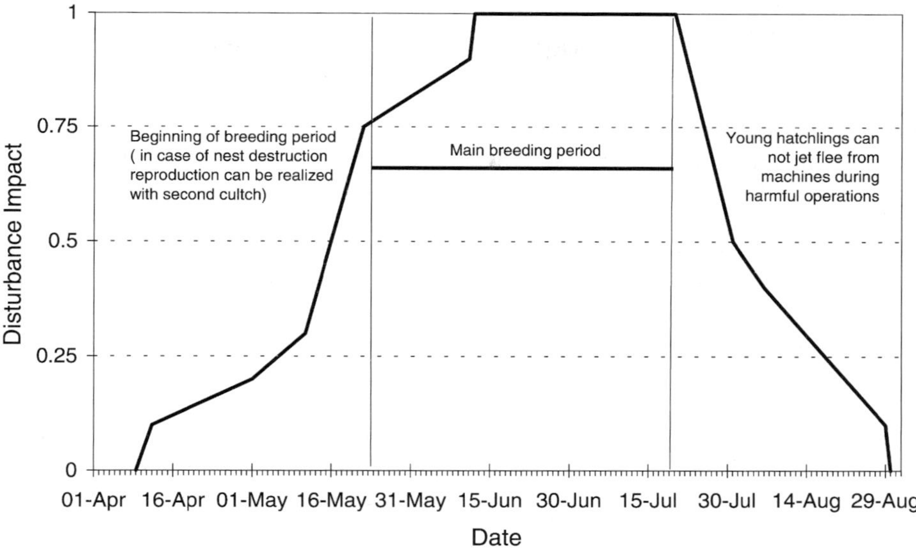

Fig. 1. Assumed disturbance impact of hazardous machinery for partridge in production fields according to timing.

birds often breed within the agricultural production fields or set asides and hens get killed or clutch gets destroyed when machines with rotational implements are used on the fields (Potts, 1986). Therefore, the impact of agricultural machinery which has the potential to destroy the clutch was evaluated according to the timing of the activity.

Fig. 1 shows the relative importance of timing of harmful activities for partridge during the breeding season, as derived from expert knowledge. According

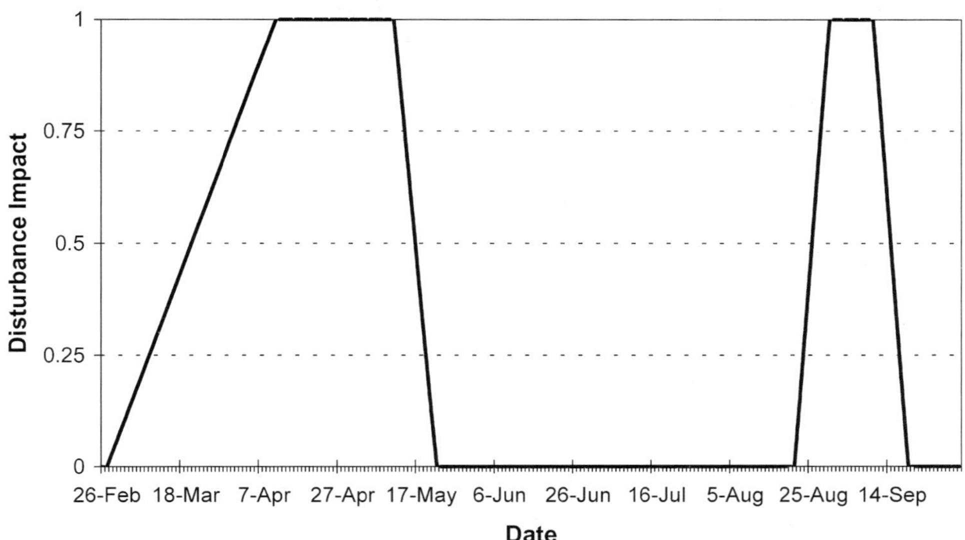

Fig. 2. Assumed disturbance impact of hazardous agricultural practices for amphibians in production fields according to timing.

to Fig. 1, the potential disturbance impact (DI) was assigned to each agricultural activity which has the potential to destroy the clutch or kill the hen. For instance, if a harmful operation has been done at the beginning of the breeding season and the nest is destroyed, there is still time for a second clutch. In this time, the disturbance impact factor ranges between 0 and 0.75, according to the timing of the operation. The sum of the disturbance impacts of all operations of a specific cropping practice represents the disturbance frequency of the cropping practice. According to the disturbance frequencies partial match grades were assigned to the cropping practices (see Eq. (1)):

$$\text{MGPd} = \left\{ \begin{array}{lll} 1 & \text{if} & \sum \text{DI(CP)} = 0 \\ 1 - \sum \text{DI(CP)} & \text{if} & 0 < \text{DI} < 1 \\ 0 & \text{if} & \sum \text{DI(CP)} \geq 1 \end{array} \right\} \quad (1)$$

where MGPd is the partial match grade due to disturbance frequency, DI the disturbance impact, and CP is the cropping practice.

Besides disturbance impacts, habitat quality of partridge depends on the supply of herbs for the adults and insects for the young birds (Potts, 1986). Furthermore, the spacing between the cultivated plants determines the mobility of partridge in the field, and thus habitat quality. Therefore, herbicide and insecticide application rate was considered to indicate food supply and fertilization to indicate the spacing within the field. Partial match grades for herbicides (MGPh) and insecticides (MGPi) were calculated according to the application frequencies and were 0, 0.5 and 1 for frequencies of 2, 1 and 0 applications per year, respectively. Partial match grades for nitrogen fertilization were assumed to be zero for application rates bigger than 200 kg N/ha, 1 for application rates lower than 50 kg N/ha. Match grades for application rates in

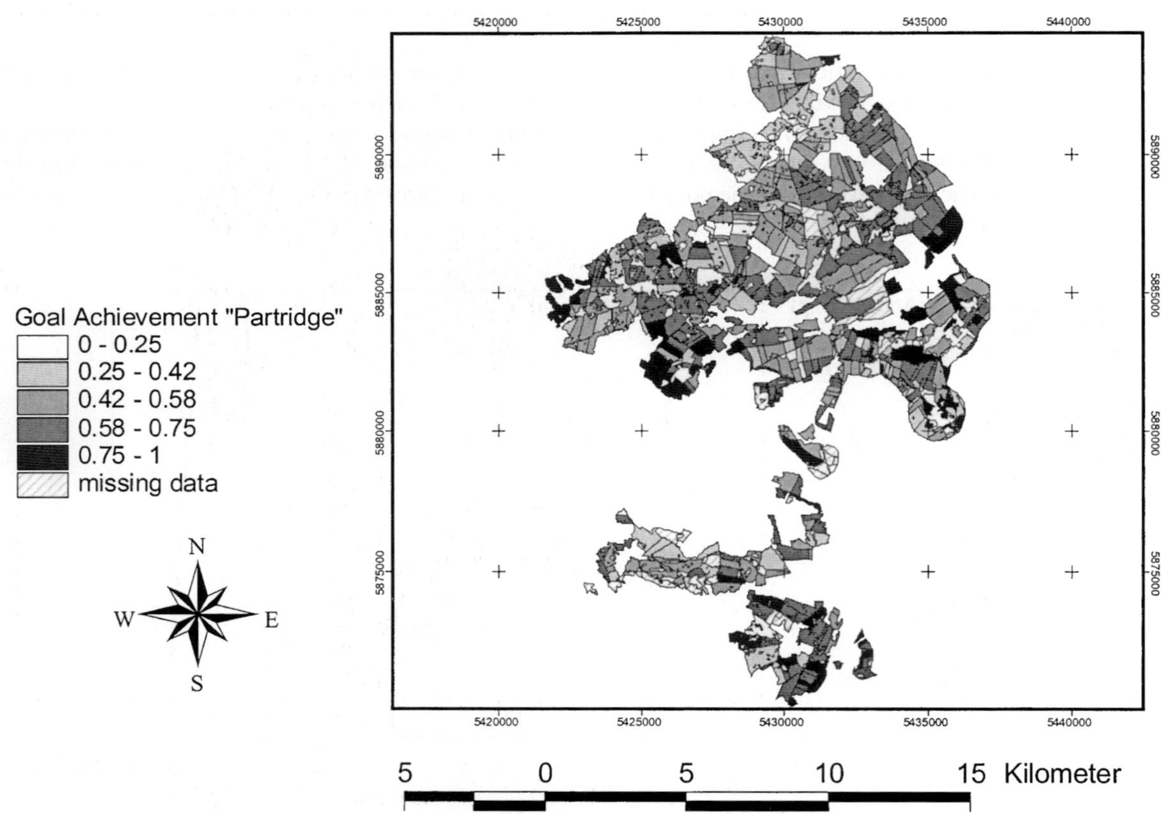

Fig. 3. Goal achievement (partridge conservation) of cropping practices in the study area (1995–1997).

between 50 and 200 kg N/ha were interpolated Eq. (2):

$$\text{MGPn} = \left\{ \begin{array}{lll} 1 & \text{if} & \text{NA} \leq 50 \\ \frac{1}{300}(400 - 2 \times \text{NA}) & \text{if} & 50 < \text{NA} < 200 \\ 0 & \text{if} & \text{NA} \geq 200 \end{array} \right\} \quad (2)$$

where MGPn is the partial match grade due to N-fertilizer applications, and NA is the amount of applied N-fertilizers over one production year (kg/ha).

The partial match grades were weighted according to the supposed importance for partridge (Eq. (3)). It is assumed, that disturbance impacts (MGPd) have the same importance as all intensity match grades together (MGPh, MGPi, MGPn):

$$\text{GA}_\text{P} = 0.5 \times \text{MGPd} + 0.16 \times \text{MGPh} + 0.16 \times \text{MGPi} + 0.16 \times \text{MGPn} \quad (3)$$

where GA_P is the goal achievement for partridge conservation; MGPd, MGPh, MGPi, MGPn are partial match grade due disturbance frequency, herbicide applications, insecticide applications and N-fertilizer applications, respectively.

2.2. Development of a stress indicator for amphibians

While amphibians reside on agricultural fields, they are exposed to operating technical machinery and to agrochemicals. Amphibians most likely confront these dangers while migrating from their winter habitats to their breeding sites, or from the summer to the winter habitat, respectively (Schneeweiß and Schneeweiß, 1997). Fig. 2 shows the disturbance impact due to the expected probability of migration activities throughout the year. It indicates the relative importance of the timing of harmful activities.

Three partial match grades were calculated to indicate the conformity of the cropping practice with the

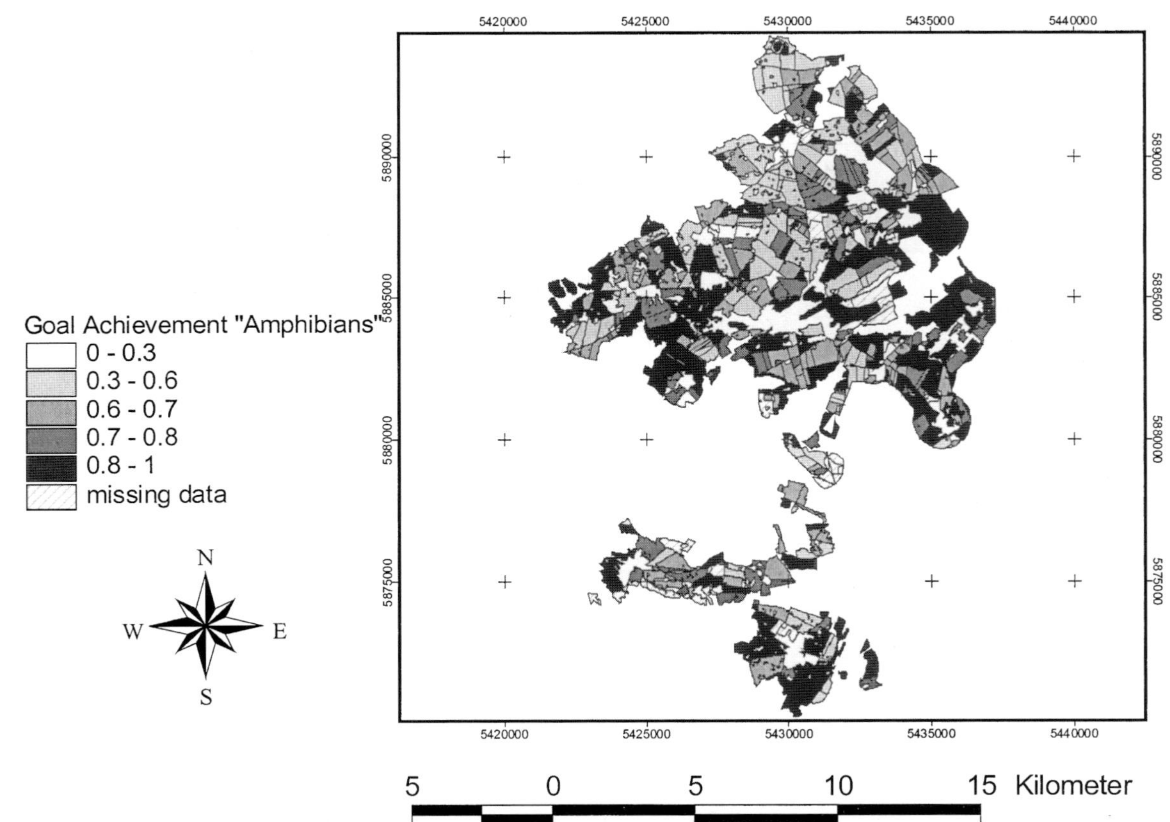

Fig. 4. Goal achievement (amphibian conservation) of cropping practices in the study area (1995–1997).

needs of amphibians (Eq. (4)). The weighting factors were calculated following the assumption that disturbance impact of machines, fertilizers and pesticides weight 100, 90 and 30%, respectively, according to expert knowledge within the research project:

$$\mathrm{MGA} = 0.45\left\{\begin{array}{lll} 1 & \text{if} & \mathrm{DFM}=0 \\ 1-\frac{1}{3}\mathrm{DFM} & \text{if} & 0<\mathrm{DFM}<1 \\ 0 & \text{if} & \mathrm{DFM}\geq 3 \end{array}\right\} +0.41\left\{\begin{array}{lll} 1 & \text{if} & \mathrm{DFF}=0 \\ 1-\frac{1}{3}\mathrm{DFF} & \text{if} & 0<\mathrm{DFF}<1 \\ 0 & \text{if} & \mathrm{DFF}\geq 3 \end{array}\right\} +0.14\left\{\begin{array}{lll} 1 & \text{if} & \mathrm{DFP}=0 \\ 1-\frac{1}{3}\mathrm{DFP} & \text{if} & 0<\mathrm{DFP}<1 \\ 0 & \text{if} & \mathrm{DFP}\geq 3 \end{array}\right\} \quad (4)$$

where MGA is the match grade amphibians, DFM the sum of disturbance impact of harmful implements (plough and intensive tillage implements) for amphibians according to Fig. 2, DFF the sum of disturbance impact of harmful fertilizer applications (any caustic fertilizer), and DFP is the sum of disturbance impact of harmful pesticide applications (any fish-toxic substances).

3. Results

With the help of the indicators land use in the study area was evaluated for the conformity with the considered nature conservation targets. The application of the indicators on agricultural land use in the study area from 1995 to 1997 shows a great spatial heterogeneity of partial goal achievement for both nature conservation targets (see Figs. 3 and 4). Specific areas can be detected in which cropping practices meet with the nature conservation targets and others in which they don't. The spatial patterns of goal achievements are given with the patterns of farms and their fields and the site quality, which determines the suitability of the land for certain crops.

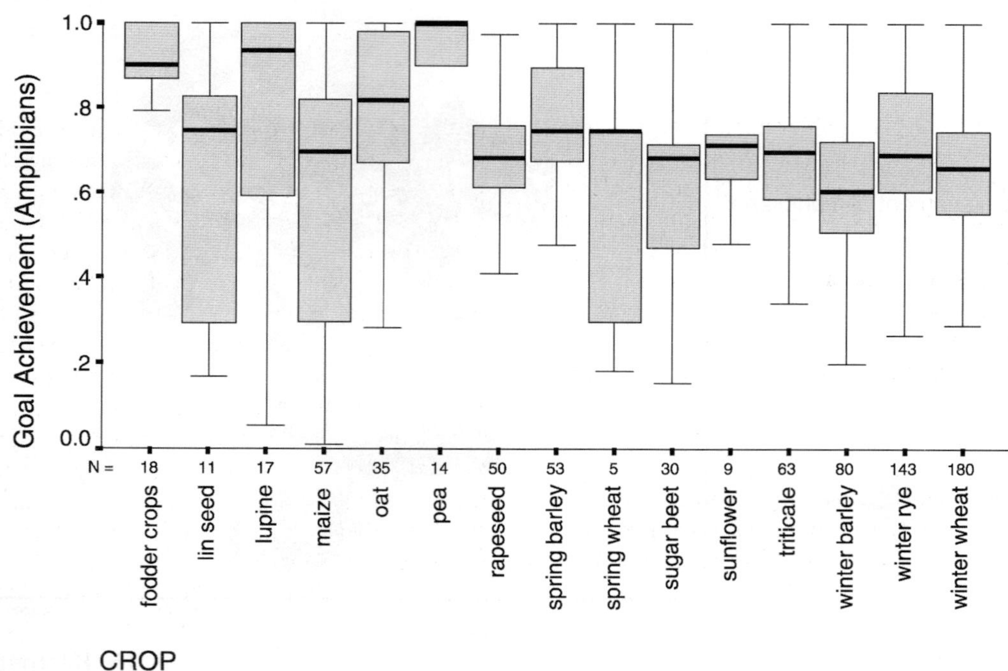

Fig. 5. Goal achievements (preservation of amphibians) of agricultural practices of 19 farms in the study area (by crops; median, 25 and 75% quantiles).

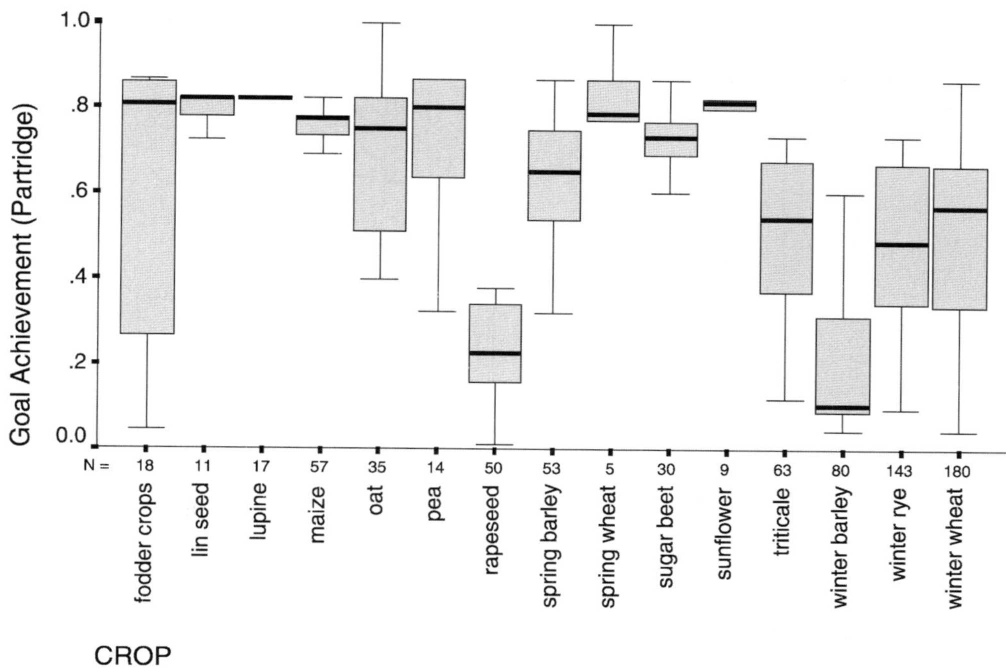

Fig. 6. Goal achievements (preservation of partridge) of agricultural practices of 19 farms in the study area (by crops; median, 25 and 75% quantiles).

The analysis of goal achievements of the cropping practices show that goal achievements vary considerably between crops (Figs. 5 and 6). However, especially for amphibians a large range of goal achievements is evident for most of the crops, which indicates that goal achievements are not only determined by the crop, but also by the farmer and the design of the cropping practices. Nevertheless, it can be seen that some crops like sugar beet, winter barley, winter wheat show smaller goal achievements than others like fodder crops, oat and peas.

Goal achievements for partridge show a clearer differentiation by crops. The most important parameter which determines partridge goal achievement is the disturbance frequency during the breeding season; in particular, the early harvest timing of rapeseed and winter barley had the biggest impact on low match grades.

Figs. 7 and 8 show the variation of goal achievements of all cropping practices by the individual farms. The practices of most farms provide a big range of goal achievements which indicates that each farm operates with cropping practices with high and low goal achievements. However, the medians of goal achievements differ considerably, which indicates, that some structural patterns of the farms have a big impact on goal achievement. It can be concluded that the farm specific patterns of crops cultivated have a big impact of the mean goal achievement of the farm. Figs. 9 and 10 show the area weighted mean of cropping practices by farming systems. The organic and the "sheep" farming system show higher values than the area weighted means of the other farming systems for both nature conservation targets. However, as these farming systems were observed only once, no general conclusion can be drawn from this. The mean goal achievements of the two farming systems with higher numbers of repetitions show no significant differences. These results suggest that the individual design of each farm determines goal achievement in a greater manner than structural patterns do.

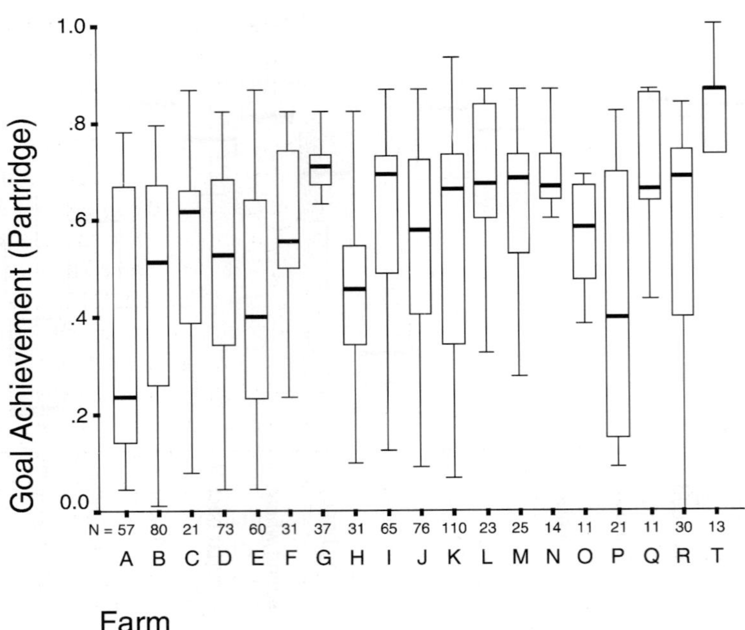

Fig. 7. Goal achievements (partridge conservation) of agricultural practices in the study area (by farms; median, 25 and 75% quantiles, letters indicate individual farms).

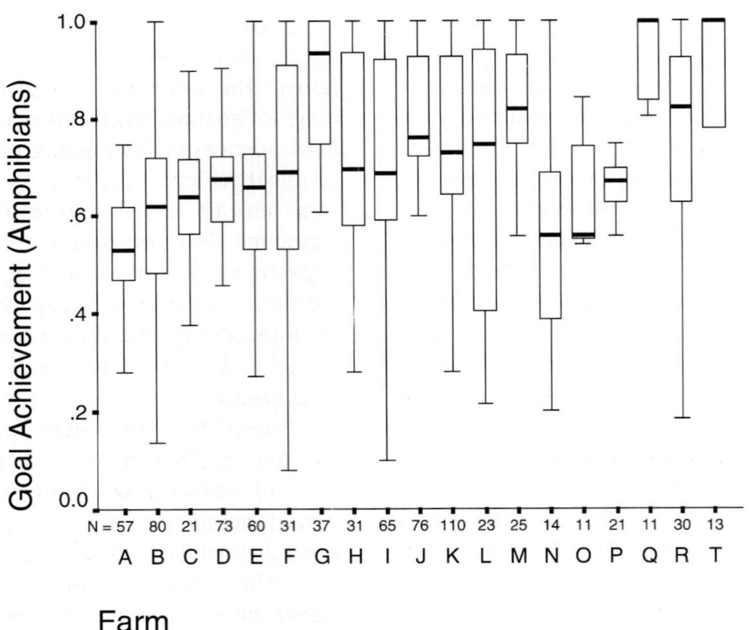

Fig. 8. Goal achievements (amphibian conservation) of agricultural practices in the study area (by farms; median, 25 and 75% quantiles, letters indicate individual farms).

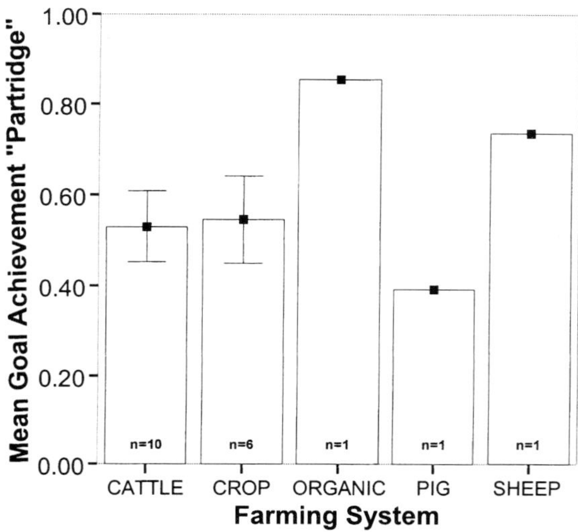

Fig. 9. Mean goal achievements of cropping practices for partridge conservation by farming systems (CATTLE: mixed farming system with cattle husbandry, CROP: crop dominated farming system, ORGANIC: mixed farming system following the principles of organic farming, PIG: mixed farming system with pig husbandry, SHEEP: mixed farming system with sheep husbandry; bars indicate standard deviation).

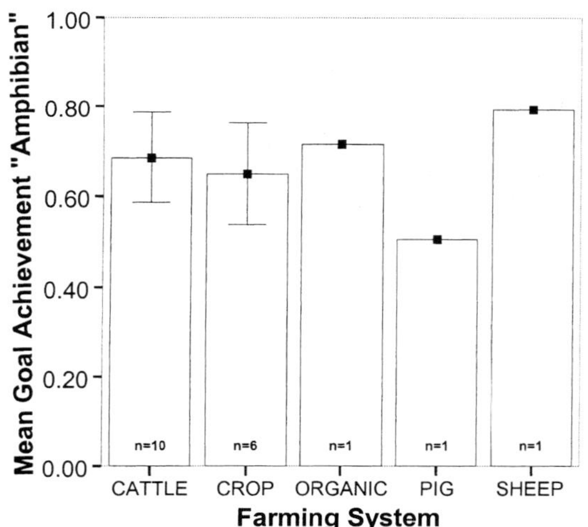

Fig. 10. Mean goal achievements of cropping practices for amphibian conservation by farming systems (abbreviations see Fig. 7).

4. Discussion and conclusions

The presented approach shows the feasibility of an evaluation procedure based on expert judgment. The indices can be interpreted as driving-force indicators following the OECD-classification (OECD, 1997). The application of driving-force indicators on biodiversity in agricultural systems is not without difficulties as the link between agricultural practices and ecological processes within agro-ecosystems especially for the biotic environment is very difficult to model (Tucker, 1999). Nevertheless, if the "right" driving forces are included in the indicators they should be helpful to compare different land use options and optimize land use.

The validation of this approach was restricted to expert judgment and experiences in the study area. Further validation could be achieved by a more thorough study of the interdependencies of agriculture and animal species in agriculturally used landscapes. Unfortunately, only few studies give evidence about the impact of specific agricultural practices on the considered species. A particular difficulty is the limited knowledge of the adaptability of animal species in agricultural landscapes. It can be supposed, that animal species in agricultural landscapes generally are adapted to the impacts of land use. For example the high reproduction and rapid development of the chicks is a pre-adaptation on farming practice. However, as it is evident for many animal species in the agricultural landscape, changes in land use over-stress the adaptability and may result in a decline of species. It can be expected that as agriculture uses the resources more efficiently the chances for species to survive in the agricultural landscape are declining. The difficult question is, to which degree agriculture should provide habitat functions for animals in the agricultural landscape. The presented indices provide a tool with which the increased impact on species in the agricultural landscape can be modeled. Further research has to focus on the quantitative and qualitative demands of the species.

In order to analyze the possibilities of a farm to integrate nature conservation targets into the farm system, a farm model was built, in which the evaluation framework was integrated (Zander et al., 1999). The farm model, which was designed in MODAM (Zander and Kächele, 1999), provides a set of cropping practices with specific goal achievements, which can be considered according to the farm specific requirements.

With the help of a linear programming tool, the optimal allocation of cropping practices can be simulated.

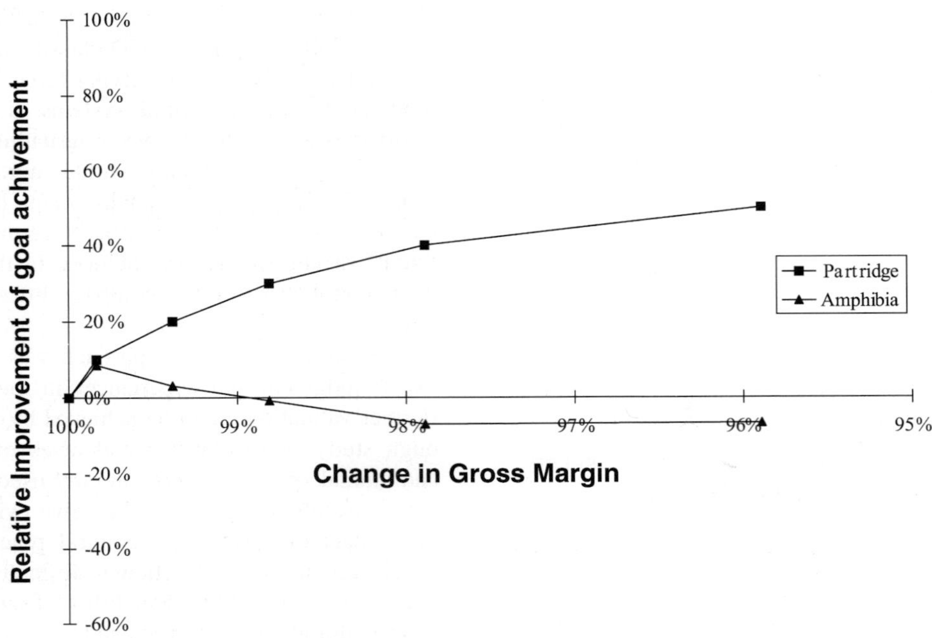

Fig. 11. Trade-off between improvement of goal achievement for preservation of partridge in agricultural landscapes and gross margin for a simulated farm and impact on goal achievement for preservation of amphibians in agricultural landscapes.

The goal function comprises the maximization of the total gross margin over the farm. Introducing stepwise restrictions for environmental goals into the farm model leads to trade-off functions of the total gross margin and the nature conservation target. Fig. 11 shows the loss of gross margin (in %) which has to be taken into account, if the farm model is forced to achieve higher preservation levels of partridge in 10% steps of improvement. It can also be seen that an improvement for partridge has a negative effect on amphibians, even though the effect is quite small. Nevertheless, the main nature conservation targets deduced by the demands of the species were congruent within the Schorfheide-Chorin Project (Flade et al., 2003).

Marginal losses in gross margin combined with high levels of ecological targets could be achieved, because of the heterogeneity of the landscape of the farm model on one hand and the "non-symmetrical" requirements of the target organisms on the other hand. As every field was introduced individually into the linear programming model, the tool was capable to use the comparative advantages of different fields and to reallocate cropping practices under consideration of the ecological targets.

Conclusions from these ecologically optimistic results have to be taken with care as not all costs are included to attain an ecologically optimized allocation of cropping practices over the farm area. The management would have to collect and apply a considerable amount of ecological information. Above, monitoring and control is needed to justify compensation payments for the costs, consisting of higher management efforts, changed measures, input levels and yields.

The evaluation and optimization approach shows, that if specific information on the impact of agricultural management practices is given, it is technically quite easy to implement this into agricultural land use up to a certain point. However, the flexibility to integrate nature conservation targets into the farm is highly depending on the specific farm situation. The impact of structural farm specific factors which determine this flexibility should be analyzed more thoroughly to deduce appropriate policy strategies to integrate nature conservation targets into agriculture.

Acknowledgements

Funding for this research was provided by the German Federal Foundation for the Environment (Deutsche Bundesstiftung Umwelt). The authors thank Andrea Knierim for helpful comments on the manuscript.

References

Beebee, T.J.C., 1996. Ecology and Conservation of Amphibians. Conservation Biology Series. Chapman & Hall, London.

Flade, M., Plachter, H., Schmidt, R., Werner, A., 2003. Nature Conservation in Agricultural Ecosystems. Quelle & Meyer, Wiebelsheim (in press).

Herrmann, M., 2003. Implementation of the results of the Schorfheide-Chorin Project. In: Flade, M., Plachter, H., Schmidt, R., Werner, A. (Eds.), Nature Conservation in Agricultural Ecosystems. Quelle & Meyer, Wiebelsheim (in press).

Herrmann, M., Dassow, A., 2003. Quail (*Coturnix coturnix*). In: Flade, M., Plachter, H., Schmidt, R., Werner, A. (Eds.), Nature Conservation in Agricultural Ecosystems. Quelle & Meyer, Wiebelsheim (in press).

Herrmann, M., Fuchs, S., 2003. Grey partridge (*Perdix perdix*). In: Flade, M., Plachter, H., Schmidt, R., Werner, A. (Eds.), Nature Conservation in Agricultural Ecosystems. Quelle & Meyer, Wiebelsheim (in press).

Meyer-Aurich, A., 2001. Entwicklung von umwelt- und naturschutzgerechten Verfahren der ackerbaulichen Landnutzung für das Biosphärenreservat Schorfheide-Chorin. Verlag Agrarökologie 41, Hannover/Bern.

Organisation for Economic Co-operation and Development (OECD), 1997. Environmental Indicators for Agriculture. OECD, Paris.

Organisation for Economic Co-operation and Development (OECD), 1999. Environmental Indicators for Agriculture, vol. 2. Issues and Design "The York Workshop". OECD, Paris.

Oldham, R.S., Latham, D.M., Hilton-Brown, D., Towns, M., Cooke, A.S., 1997. The effect of ammonium nitrate fertilizer on frog (*Rana temporia*) survival. Agric. Ecosyst. Environ. 61, 69–74.

Pegel, M., 1987. Das Rebhuhn im Beziehungsgefüge seiner Um- und Mitweltfaktoren. Ferdinand Enke Verlag Stuttgart, 198 pp.

Plachter, H., Korbun, T., 2003. A methodological primer for the determination of nature conservation targets in agricultural landscapes. In: Flade, M., Plachter, H., Schmidt, R., Werner, A. (Eds.), Nature Conservation in Agricultural Ecosystems. Quelle & Meyer, Wiebelsheim (in press).

Potts, G.R., 1986. The Partridge: Pesticides, Predation and Conservation. W. Collins Sons & Co. Ltd., London.

Schneeweiß, U., 2003. Fire-bellied toad (*Bombina bombina*). In: Flade, M., Plachter, H., Schmidt, R., Werner, A. (Eds.), Nature Conservation in Agricultural Ecosystems. Quelle & Meyer, Wiebelsheim (in press).

Schneeweiß, N., Schneeweiß, U., 1997. Amphibienverluste infolge mineralischer Düngung auf Agrarflächen. Salamandra 33, 1–8.

Swanson, T., 1997. Global Action for Biodiversity. An International Framework for Implementing the Convention on Biological Diversity. Earthscan Publications, London.

Tucker, G., 1999. Measuring the impacts of agriculture on biodiversity. In: Brouwer, F., Crabtree, B. (Eds.), Environmental Indicators and Agricultural Policy. CABI Publishing, New York, pp. 89–103.

Zander, P., Kächele, H., 1999. Modeling multiple objectives of land use for sustainable development. Agric. Syst. 59, 311–325.

Zander, P., Kächele, H., Meyer-Aurich, A., 1999. Development and application of a multiobjective decision support tool for agroecosystems management (MODAM). Quarterly Bulletin of the International Association of Agricultural Information Specialists XLIV (1–2), 66–72.

ELSEVIER

Available online at www.sciencedirect.com

Agriculture, Ecosystems and Environment 98 (2003) 541–547

Agriculture Ecosystems & Environment

www.elsevier.com/locate/agee

Rewards for ecological goods—requirements and perspectives for agricultural land use

B. Gerowitt[a,*], J. Isselstein[b], R. Marggraf[c]

[a] *Research Centre for Agriculture and the Environment, University of Göttingen, Am Vogelsang 6, D-37075 Göttingen, Germany*
[b] *Department of Agronomy and Plant Breeding, University of Göttingen, von Sieboldstr. 8, D-37075 Göttingen, Germany*
[c] *Department of Agricultural Economics, University of Göttingen, Platz der Göttinger Sieben 5, D-37075 Göttingen, Germany*

Abstract

This paper focuses on the possibilities to reward ecological goods and services of agriculture. Ecological goods in agriculture are achieved by various types of agricultural land use either creating resources or buffering resource consumption. From an economic point of view, positive consequences will result from rewarding the agricultural sector for the production of ecological goods according to market principles. A rewarding system can be developed according to market principles when results instead of actions are rewarded, producer and consumer surpluses are created, demand can be expressed and supply can be provided. Ecological goods are public goods, therefore the expression of demand is up to the community. However, the purchasing of such ecological goods is up to the authority administrating public finances. Establishing regional markets is necessary, since regional differences are evident. Principally, the price of an ecological good should be determined by its value to the public, not by the cost of production.

With respect to the various ecological goods achieved by the agricultural sector, the contribution of agricultural land use to biodiversity, and more specifically to plant biodiversity seems to be the most appropriate for a market-orientated rewarding system. A framework for establishing such a market is outlined in this paper. Deviation from standard economic principles regarding prices for ecological goods are considered to be necessary during an introductory period. Finally based on both, ecological and economical considerations, and regarding the intrinsic and extrinsic characteristics of "plant biodiversity connected with agricultural land use", it seems worth the effort to further examine and develop this concept.

Keywords: Ecological goods and services of agriculture; Market principles; Rewarding concept

1. Introduction

Ecological goods and services produced and undertaken by the agricultural sector result from using the environment in a way which is somehow valued by the society, but not by the market price for the produced food, feed or raw material. These types of goods and services already accompany todays agricultural land use; however, a continuously ongoing discussion indicates, at least in Germany, a general dissatisfaction about their quality and quantity. In addition principal public objectives such as enhancement of biodiversity, environmental protection, leisure and aesthetic perception of the cultural landscape can profit from specific forms of agriculture. Current agricultural environmental policy is trying to influence agricultural land use within this framework.

The major pillar of the Common Agricultural Policy (CAP) at the moment in the EU are direct subsidies

* Corresponding author. Tel.: +49-551-395538.
E-mail address: bgerowi1@gwdg.de (B. Gerowitt).

0167-8809/$ – see front matter © 2003 Elsevier Science B.V. All rights reserved.
doi:10.1016/S0167-8809(03)00112-9

donated to the producers as payments based on the acreage of subsidised crops ("grandes cultures" within the EU). These payments are accompanied by financial transfers with environmental goals—the second pillar. In 1997 direct subsidies for "grand cultures" formed 40.8% of all EAGGF costs (KOM, 1998; Appendix 8), while the financial engagement of the EU in financing the national agri-environmental programmes only covered in 3.7% of this budget (KOM, 1998; Appendix 22c).

Starting with the Agenda 2000, this second pillar has been sequentially strengthened within the European CAP. Additionally, various possibilities for regional adaptation have been offered to realise the desired secondary liability (subsidiarity) of the agricultural policy (KOM, 2000).

Independent of this political process, it is definitely necessary to construct a generally acceptable basis for the transfer of public money into agriculture. The desired ecological goods and services by the agricultural sector, which are not automatically realised by the primary production of food and raw material seem obviously suitable to fulfil this function.

In this paper, the requirements for approaches above and beyond the actual practice in Europe are deduced by applying both ecological considerations and principal economic theories. Based on these prerequisites a perspective is outlined for rewarding suitable ecological goods within a market orientated concept.

2. Ecological goods and services provided by agriculture

Any farming activity is based on the use of abiotic and biotic resources, and thereby, affects the environment. Intensive farming has, in many situations, been shown to exploit and even disturb natural resources and has adverse effects on the environment, i.e. the pollution of ground and surface water, or of the atmosphere, by nitrogenous compounds or pesticides. However, apart from being dependent on natural resources, agriculture also creates resources. Set in a historical perspective agricultural land use has, for example, produced and sustained a major part of the biodiversity found in Central European landscapes. In addition, agricultural land use, rather than forestry or abandoned land facilitates the regeneration of clean ground water resources that are used for society's drinking water supply.

The positive effects of farming become obvious when agricultural land use is no longer maintained. Many attempts connected with conservation contracts for farmers focus on guaranteeing agricultural land use on marginal land in order to maintain biodiversity.

Negative effects of agricultural land use have received considerable attention since the 1970s in agricultural scientific research and its related disciplines, and the reference scientific literature is vast. In Germany extensive reviews on that topic have been repeatedly updated (SRU, 1985, 1994, 2000). According to SRU (1985) the negative effects of farming activities can be clustered as follows (ranked according to importance):

- Loss of species and habitat diversity, removal and disturbance of non-agriculturally used margins within agro-ecosystems, e.g. hedgerows, ditches, grassy margins, etc.
- Pollution of groundwater with nitrates, pesticides and their metabolites leached from farmed land.
- Deterioration of soils and soil function through compaction, erosion and accumulation of pollutants.
- Pollution of surface water with nitrate, phosphorus and pesticides due to farming procedures.
- Pollution of the atmosphere through the gaseous emissions from farming procedures.

Legislation accompanying agricultural land use should prevent any unacceptable environmental effects of farming. However, in specific areas further actions are undertaken to reduce an area-specific unwanted pollution arising from farming activities. For example, in water protection areas in Germany farmers get paid if they employ measures that decrease the nitrate leaching into the groundwater, such as reduced fertiliser use, expanded crop rotation, etc.

Biodiversity plays a major role in both types of the ecological effects caused by agricultural land use, whether synthesising or threatening.

Total biodiversity consists of not only those parts belonging in a wider sense to the cultural landscape but also to the undisturbed natural landscape. The conservation of natural biotopes, i.e. the flora and fauna of areas that have never been managed or used by man, is a widely accepted ecological goal and in Germany, as in many European countries, a certain amount of the tax revenue is spent to maintain such virgin nature. This component of environmental policy and nature conservation is not considered in the present paper.

3. Economic requirements for rewarding the production of ecological goods

Agri-environmental policy can influence agricultural land use directly or indirectly. *Direct control* results through sanctions which have proven to be effective such as restrictions and regulations. *Indirect influence* results through information, appeals or financial incentives.

Methods influencing actions indirectly (e.g. duties and taxes) have more advantages than the methods controlling actions directly (with the help of restrictions and regulations) (OECD, 1994). They are not based on constraints, therefore they guarantee the power of decision to the individuals. The possibility for making individual decisions can be utilised to make nature conservancy measures as inexpensive as possible and to search for new resource-preserving uses of nature. The direct control does not offer any incentive for reducing costs. Restrictions and regulations usually affect the actions only at a specific threshold level. In contrast, the indirect methods influence resource use from the beginning, emphasising that these reasons are scarce and should not be wasted. Therefore, opportunities for implementing indirect methods for influencing behaviour should be preferred for guiding farmers' decisions on environmental effects.

From the economic point of view the most important criterion for assessing a policy is whether it contributes to better satisfying the needs of members of a society. As far as providing society with private goods is concerned, the decentralised, error-absorbing, and receptacle market system has clearly proved superior to central planning. According to the majority of economists, it is also desirable to profit from the advantages of a market economy when supplying the population with agri-environmental goods. Markets for these collective goods are not as easy to establish as markets for private goods; governmental intervention must be stronger. However, market instruments should be given priority. In this context it is recommended to reward ecological goods by farmers as a strategy of agri-environmental programmes (Latacz-Lohmann and van der Hamsvoort, 1998; Heißenhuber and Lippert, 2000). This strategy can only lead to the desired benefits (provision of goods for members of society at lowest cost) when certain requirements are fulfilled and essential constitutive criteria are considered. The success of a rewarding system for ecological goods is determined by the principles of a market system.

Scarcity of goods: Not all goods of individuals that are appreciated by other individuals have to be allocated by the market. The free market allocation mechanism has only to be applied when demand exceeds supply and an expansion of supply is possible and resource binding.

Price as indicator of scarcity: Consumers, by their willingness to pay, signal potential producers as to which goods shall be produced. The price of goods is determined by the costs of production and their valuation by the consumer.

Conforming to the rules: In principle it is clear that suppliers must be compensated for their goods. It is also considered acceptable that demanders must pay money to receive goods.

Transparency of supply and demand relationships: The supplier must be able to calculate his production risk. The demander must be able to recognise which suppliers offer which goods and what the quality is.

The first principle states that ecological goods of agriculture be in short supply. This requirement is undoubtedly fulfilled. Farmers do not 'automatically' use natural resources in such a way that satisfies society; they could use resources, albeit at the expense of the production of agricultural goods, in another way.

The second principle stresses the role of demand in a market economy. It is of particular importance for the establishment of a rewarding system. The demanders of ecological goods from agriculture are interested in the ecological goods provided and not in the activities that led to the provision of them. Therefore, effects on the environment have to be remunerated, not the activities that lead to the desired environmental effects. A rewarding system only has cost-minimising and innovation-stimulating effects when the amount of rewards is determined by the value of ecological activities to society. This requires the integration of (monetary) demand for ecological goods of agriculture in the rewarding system.

Establishment of markets, on which the demand for ecological goods can be directly realised, is restricted as most ecological goods are collective in character. However, the society indirectly pays for ecological goods, whenever arguments like diversity via land-use and the benefit of a cultural landscape are used to justify agricultural subsidies. In the long term it is necessary to determine the population's willingness to pay for ecological goods. For an overview on these methods see Garrod and Willis (1999).

Natural conditions and land use differ widely among regions. The willingness to pay for ecological goods will, therefore, differ between specific regions.

When considering the funding for agri-environmental remuneration, it must be realised that when farmers are the sole recipients, no surplus remains for the demanders creating a suppliers monopoly with discriminatory pricing. Even when the allocation aspects make this irrelevant, distribution aspects argue for establishing producers' and consumers' surpluses on markets for ecological goods. Producer surpluses are essential to induce farmers to treat the land on which they produce agri-environmental goods as a value worth preserving.

According to the principle of conforming to the rules, it must be determined whether all ecological activities in agriculture shall be rewarded by society, or whether farmers should provide certain ecological goods without compensation. Literature frequently suggests to carry out this examination with the help of negative and positive external effects (Heißenhuber, 1995; Hofmann, 1995). It is argued that remunerable ecological activities are positive external effects. Costs connected with a reduction of negative external effects are charged to farmers. From the economic point of view, the concepts of negative and positive external effects are unsuitable for defining remunerable ecological activities. Economically, benefits and damages are equal effects, but with reversed premises; a benefit is averted damage, damage is a missed benefit. Thus all aspired-to changes in the agricultural use of the environment can be interpreted as reduction of a negative external effect or an increase in positive external effects. That is, when society wants farmers to reduce ground water pollution, this can be interpreted as a desired reduction of a negative external effect, or as an increase of positive external effects, i.e. as a contribution toward securing the quality of the water supply.

There is no economic criterion for determining exactly which ecological activities of agriculture are to be rewarded. This becomes clear when interpreting the question of rewarding according to the property right theory (Hampicke, 1996). The individual that sells goods for money acts deliberately since he could also retain them. When an ecological activity of agriculture is rewarded, it simulates that farmers have the natural resources needed for the activity at their disposal. When defining remunerable ecological goods the fundamental issue is the allocation of property rights to natural resources. The determination of property rights is based on value judgements and evolve, thus, they may change. There is no social fundamental principle that defines how to proceed; only suggestions can be made about what must be considered. For example, it should be realised that increasing legal uncertainty leads to negative allocation effects. As a consequence, property rights may not be altered due to political opportunity or even arbitrariness. Another important aspect is the distribution effect of the property right system. When farmers are denied property rights to natural resources, they must provide ecological goods without remuneration. If, as a consequence, economic disadvantages increase to such an extent that farming no longer pays, a small segment of the farming population would be charged with very high costs to the advantage of the majority. Even the favoured majority would consider such a situation as unjust and therefore unacceptable. Additionally the decision as to which ecological goods are remunerable and which are not should always be made by binding and understandable regulation with broad public acceptance. Many forms of extensive land use, which provide agri-environmental goods usually cover only a small portion of fixed costs (Hampicke, 1999). Without a reward, farmers would not permanently provide these activities.

Implications of the last principle, the transparency of supply and demand, are discussed in Section 4.

4. Which ecological goods of agriculture should be rewarded?

As stated from an economic point, the ecological goods of agriculture can be rewarded by honouring the results or by honouring the actions. Action-oriented rewards, dominating in existing programs, credit either actions taken or omitted, for example, elimination of pesticide use or cutting grasslands only after a fixed date. This manner of rewarding ecological goods requires a catalogue of instructions, when the instructions are followed, a pay-off is guaranteed regardless of the results. This system neither stimulates to provide ecological goods by current production methods as cost-effectively as possible, nor does it encourage innovations for production methods focussing ecological goods of agricultural land use.

In the result-oriented approach a result is rewarded. Consequently this result can be called a produced ecological good. In order to initiate market allocation mechanisms ecological goods of agriculture are needed, which can be described as independent goods. With respect to the economic principles, the result of the goods should be scarce, not to achieve with an agricultural practice conforming to the rules and have characteristics which guarantee as much transparency for supply and demand as possible. In order to transfer a market-concept the defined goods must further have some intrinsic characteristics: the producer of the good must be clearly identifiable and the good must be measurable. Production risks are inherent, however, they should be calculable to some degree by the supplier. Thus, for the transparency in the supply and demand, the profiles of the ecological goods to be honoured are crucial.

With respect to the ecological goods of agriculture, various contributions are scarce, however their scarcity is differently ranked by experts (e.g. SRU, 1985) or in interrogations (e.g. Müller and Schmitz, 1999).

Environmental goods like clean water and air provide no transparency for a market-orientated system of rewarding ecological goods, since their status is difficult to relate to individual actions of farmers. Therefore, attempts to regulate these effects are action-orientated. However, these attempts are essential to reduce environmental threats of agriculture, they are not favourable for any market-orientated system. The desired actions can be used for defining standards, being either obligate for any production (conforming to the rules) or facultative for e.g. marketing labels, focussing on skimming off a specialised demand for the way of production.

Contracts to reduce farming intensities in water protection areas have an area-specific focus and are therefore also unsuitable for a market-oriented concept with individual and unconstrained production decisions.

These consequences from the nature of possible goods are also supported by the rules for agricultural land use in Germany: activities with a more environmental impact, like the use of fertilisers and pesticides are at least touched by various legislations (e.g. German Fertilisation Decree, German Plant Protection Act, German Federal Soil Protection Act), while handling wildlife is far less affected. As long as the concrete area or the landscape element has no special protection status, and agri-chemicals are used due to the registered rules, wildlife components can be treated without restrictions.

Thus, goods belonging to the comprised cluster called “biodiversity” are most suitable for any market orientated concept of rewarding ecological goods of agriculture, since they are scarce and require individual, unconstrained production decisions, which are not identical with production conforming to the rules.

From a practical point of view, the total accomplishment “biodiversity” must be subdivided due to the suitability of the components for a market-orientated remuneration system: mobile animals seems inappropriate, because it is almost impossible to relate their appearance (the result) to individual field scaled farming actions. More immobile parts of the fauna seems also unsuitable, since results, which can be detected and valued only by few experts provide no transparency for any concept relying on supply and demand. Therefore, plants seem to be most suitable for a market approach of ecological goods. Many plants of our cultural landscape are connected with the almost innumerable management options of agricultural land use in productive areas by arable farming, grassland farming or orchards of varying intensities and by “unproductive” structuring elements like tractor paths, ditches, hedgerows, copes and puddles.

5. A proposal for rewarding ecological goods in a market approach

The outlined considerations form the conceptional background, which is essential for any new approach in agri-environmental policies. In order to further proceed with the concept, considerations about an implementation are required. At least four crucial aspects are connected with the outlined ideas.

- Catalogue of ecological goods.
- Demand for ecological goods.
- Supply of ecological goods.
- Adjusting and administrating supply and demand.

Catalogue of ecological goods: In the catalogue ecological goods are described

They can be grouped into four types:

- vegetation accompanying grassland use;
- vegetation accompanying arable use;
- vegetation accompanying other forms of land use (e.g. orchards);
- vegetation of boundaries or patches on the property without direct production (e.g. hedgerows, copes, puddles).

These groups must be further sub-clustered to direct applicable goods (e.g. area units with special grassland or arable vegetation types, units of unproductive area with special vegetation), the sub-cluster are to be defined regionally. The catalogue is public. The ecological goods are described up to a detail that allows to evaluate their existence by producer, people and supervisor.

Demand for ecological goods: The decision as to which ecological goods are scarce and thus should be stimulated by a market economy shall be transferred to a decentralised and region-specific public committee comprised of representatives from various interest groups: representatives of the regional community (being e.g. a county in Germany), of nature conservation groups and of farmers groups. Responsible administrative representatives (again representing "nature conservation" and "agriculture") should belong to the board. They decide about their regional demand on ecological goods in quantity terms (m^2, ha). In order to act rationally, the board needs to have an idea about the regionally possibilities for and costs of producing the various ecological goods. For an examination of perspectives of the supply of ecological goods, real transfer payments within the framework of agricultural policy can be used as society's willingness to pay for ecological goods. Therefore, for a starting and establishing period of a market system, both funds available in the budget and production costs for the ecological goods are determinants for their prices.

Supply of ecological goods: The production of demanded goods is up to the producers, who want to serve the market. They are free in the way to produce it, as long as they act conforming to the rules. Since the concept is completely new for all actors, it is worthwhile to offer information about principal production methods and their on-farm costs. If such a concept is established this will be offered by advising institutions, like they offer advice for producing food or primary raw material today. Dogmatically argued in a market system demand is independent from production costs, however, this seems to strict for establishing a new system. The representatives of the demand will also profit from information about the range of production costs for the various goods. Despite of the value of the information, all actors should be aware, that prices for ecological goods are no compensation payments, and can therefore vary dependent on over- or under-performance of the demand. This risk add to the primary production risk i.e. can the good be produced at all. These risks are generally accepted in a market orientated economy, however, farmers will only go into the production of a good, if the two risks are comprehensible.

Adjusting and administrating supply and demand: The market for public goods needs to be organised and supervised, since taxes are used for demanding them. The regional board comes to decision according to a charter. The implementation of the decisions has to be in the response of an administration, who publishes the catalogue of ecological goods and a range for the final price. Producers will announce their possible supply in advance. They receive a feed back before they start the production. Within a determined period they serve the administration with a valid description of the ecological good they have produced, which then can be supervised by the public authorities in spot-checks. As long as the administration is acting within a fixed budget, prices for the ecological goods are used to stimulate some ecological goods while to limit the production of others.

6. Conclusions

As a result of a vital interdisciplinary dialog, we have in this paper outlined ideas, which demand further discussion and attempts for practical implications.

However, the scents of a basic policy change can be widely caught, serving many demands, e.g. of the public, of an expanding EU and its intrinsic interests and extrinsic international integration.

Therefore, characteristics of the outlined proposal are to find to some extend in already realised attempts and programmes. For example, the "biodiversity yardstick" (Oosterveld and Guldemond, 1999) is proposed for farmers to investigate the status of their farm in terms of being habitat for wildlife. The actual agri-environmental programme of the German state Baden–Württemberg accounts for the occurrence of specific wildflower species in grassland (Briemle, 2000).

A market-conform rewarding of ecological goods produced by agriculture is the application of the market idea to a completely new field. Theoretical consideration cannot foresee implementation problems, for an overview of possible implementation mechanisms see Gatto and Merlo (1999). Therefore, such a system must be carefully applied, initially in large-scale trials to gain experience. The advantages of a functioning, market-conforming rewarding system should be worth such trials.

With support of the Federal German Ministry for Education and Science (BMBF) we have actually started an interdisciplinary project to further evaluate and concretise the perspectives and limitations of the concept.

References

Briemle, G., 2000. Ansprache und Förderung von Extensiv-Grünland. Neue Wege zum Prinzip der Honorierung ökologischer Leistungen der Landwirtschaft in Baden–Württemberg. Naturschutz Landschaftsplanung 32, 171–175.

Garrod, G., Willis, K., 1999. Economic Valuation of the Environment—Methods and Case Studies. Edward Elgar, Cheltenham.

Gatto, P., Merlo, M., 1999. The economic nature of stewardship: complementarity and trade-offs with food and fibre production. In: van Huylenbroeck, G., Witby, M. (Eds.), Countryside Stewardship: Farmers, Policies and Markets. Elsevier, Amsterdam.

Hampicke, U., 1996. Perspektiven umweltökonomischer Instrumente in der Forstwirtschaft insbesondere zur Honorierung ökologischer Leistungen. Metzler-Poeschel, Stuttgart.

Hampicke, U., 1999. Conservation in Germany's Agrarian Countryside and the World Economy. In: Dragun, A.K., Tisdell, C. (Eds.), Sustainable Agriculture and Environment. Globalisation and the Impact of Trade Liberalisation. Edward Elgar, Cheltenham, pp. 135–152.

Heißenhuber, A., 1995. Betriebswirtschaftliche Aspekte der Honorierung von Umweltleistungen der Landwirtschaft. Agrarspectrum 24, 123–141.

Heißenhuber, A., Lippert, C., 2000. Multifunktionalität und Wettbewerbsverzerrungen. Agrarwirtschaft 49, 249–252.

Hofmann, H., 1995. Umweltleistungen der Landwirtschaft—Konzepte zur Honorierung. Teubner Verlagsgesellschaft, Leipzig, Stuttgart.

KOM (Commission of the European Union), 1998. 27, Finanzbericht der Europäischen Kommission über die Abteilung "Garantie" des EAGFL.

KOM (Commission of the European Union), 2000. Indikatoren für die Integration von Umweltbelangen in die Gemeinsame Agrarpolitik. Mitteilungen der Kommission an den Rat und das Europäische Parlament.

Latacz-Lohmann, U., van der Hamsvoort, C.P.C.M., 1998. Auctions as a means of creating a market for public goods from agriculture. J. Agric. Econ. 49, 334–345.

Müller, M., Schmitz, P.M., 1999. Der Preis für die Umwelt: Präferenzen und Zahlungsbereitschaft für ausgewählte Landschaftsfunktionen auf der Grundlage der Conjoint-Analyse. Z. f. Kulturtechnik und Landentwicklung 40, 213–219.

OECD (Organisation for Economic Co-Operation and Development), Manageing the Envrionment. The Role of Economic Incentives, Paris.

Oosterveld E., Guldemond, A., 1999. Measuring Biodiversity Yardstick—a Three Years Trial on Farms. http://www.clm.nl/index_uk2.html.

SRU (Rat der Sachverständigen für Umweltfragen), 1985. Umweltprobleme der Landwirtschaft. Sondergutachten, Kohlhammer, Stuttgart.

SRU (Rat der Sachverständigen für Umweltfragen), 1994. Für eine dauerhaft umweltgerechte Entwicklung, Umweltgutachten. Metzler-Poeschel, Stuttgart.

SRU (Rat der Sachverständigen für Umweltfragen), 2000. Schritte ins nächste Jahrtausend, Umweltgutachten. Metzler-Poeschel, Stuttgart.

Available online at www.sciencedirect.com

Agriculture, Ecosystems and Environment 98 (2003) 549–550

Agriculture Ecosystems & Environment

www.elsevier.com/locate/agee

Keyword Index, Volume 98